Hybrid Perovskite Solar Cells

Hybrid Perovskite Solar Cells

Characteristics and Operation

Edited by

Hiroyuki Fujiwara

WILEY-VCH

Editor

Prof. Hiroyuki Fujiwara
Department of Electrical, Electronic and Computer Engineering
Gifu University
1-1 Yanagido
501-1193 Gifu
Japan

Cover: Yoshitsune Kato and Hiroyuki Fujiwara

Library of Congress Card No.:
applied for

British Library Cataloguing-in-Publication Data
A catalogue record for this book is available from the British Library.

Bibliographic information published by the Deutsche Nationalbibliothek
The Deutsche Nationalbibliothek lists this publication in the Deutsche Nationalbibliografie; detailed bibliographic data are available on the Internet at <http://dnb.d-nb.de>.

Print ISBN: 978-3-527-34729-2
ePDF ISBN: 978-3-527-82584-4
ePub ISBN: 978-3-527-82586-8
oBook ISBN: 978-3-527-82585-1

Typesetting Straive, Chennai, India

Printed on acid-free paper

C100747_240921

Contents

Preface

Global climate change is triggering numerous natural disasters, causing heat waves, super storms, floods, and wildfires all over the world. Due to enormous CO_2 emissions caused by human activity, the global temperature rises and the resulting changes to the climate are predicted to become worse in the near future. This is a serious threat for all species living on the earth which needs to be solved quickly through the development of clean-energy technology. Although the historic Paris Agreement was signed in 2016, temperatures have already risen beyond 1 °C, compared with a preindustrial level, and the threshold of 1.5 °C, targeted in the Paris Agreement, could be reached as early as 2030. Accordingly, there is a vital need to push forward the introduction of renewable energy as soon as possible.

Currently, solar electricity is generated primarily by large-area Si solar panels. Over the last 50 years, the fabrication cost of Si solar cells has been reduced drastically and, in some countries, the power-generation cost of Si photovoltaics becomes lower than those of coal and nuclear power plants. Nevertheless, to meet the ambitious target of the Paris Agreement, CO_2 emissions should ultimately be suppressed to zero by 2050 and, therefore, continued reductions in solar-panel cost are indispensable.

Hybrid perovskite solar cells, developed first by Prof. Miyasaka's group in 2009, could play a key role in realizing such a clean energy world in the future. In fact, the development of hybrid perovskite solar cells has revolutionized the prospects of next-generation photovoltaics and is now entirely reshaping the course of solar cell research. "Hybrid perovskite" specifically means that this material consists of an exceptional combination of organic and inorganic components in the form of perovskite crystals. This unique material has allowed us to develop high efficiency devices at a pace never before seen in solar cell history; only 10 years have been needed for the fabrication of a hybrid perovskite solar cell with a high efficiency of 25.2%, even though a half century has been required for Si solar cells to reach a 26.7% efficiency. Importantly, the device structures of hybrid perovskites are quite simple and high-quality hybrid perovskite absorbers can be formed at low temperatures (~100 °C), making the perovskite devices extremely attractive. Now, hybrid perovskites have become the most exciting field of solar cell research, with the largest number of submitted papers among all solar-cell research fields.

High conversion efficiencies realized in hybrid perovskite solar cells need to be understood as a consequence of many favorable characteristics of hybrid perovskites as solar cell materials. In particular, hybrid perovskites satisfy all critical requirements necessary as semiconducting light absorbers. Moreover, by the vast amount of research efforts, better control of material properties and superior device designing have been established, enabling the fabrication of over 20% efficiency devices relatively easily.

This book provides comprehensive descriptions of the fundamental properties of unique hybrid perovskites as well as the principles and applications of various hybrid perovskite photovoltaics. Both basic and advanced concepts of hybrid perovskite devices are covered thoroughly in this book. To serve readers at different levels of expertise, this book consists of two major parts: characteristics of hybrid perovskites (Part I) and hybrid perovskite devices (Part II). In addition to these two parts, an introduction to the whole book (Chapter 1) and an overview of hybrid perovskite solar cells (Chapter 2, by Prof. Miyasaka) are provided.

Importantly, intensive research efforts that have taken place over the last decade have led to remarkable breakthroughs in the understanding and interpretation of hybrid perovskite properties. In Part I of this book (Chapters 3–10), a wide variety of fundamental hybrid perovskite characteristics, including optical, carrier transport, photoluminescence, ferroelectric, and grain boundary properties, are explained. Furthermore, the chapters in Part I attempt to bring some consensus on several controversial issues about band gap structures (direct/indirect), absolute values of absorption coefficients, mechanism of carrier mobility limit and nature of grain boundary.

The Part II of this book (Chapters 11–18) focuses on important aspects of hybrid perovskite solar cells. In Part II, in addition to the overall descriptions for Pb-based hybrid perovskite devices, more specific topics of mixed-cation perovskites and Sn-based perovskites are discussed. The serious problems of hybrid perovskite solar cells (device instability and hysteresis characteristics) are further covered in detail. Within the perovskite research field, new research areas of tandem solar cell applications are growing rapidly. In particular, a tandem device with a perovskite top cell and a crystalline Si bottom cell has attracted significant attention as this device opens a new path for further improving current Si solar-cell technology. Part II of this book will attempt to answer one simple but important question: "why are hybrid perovskite solar cells so special among all solar cell devices?"

For the publication of this book, I would like to thank all chapter authors who put great efforts into preparing manuscripts in the middle of coronavirus pandemic that occurred during 2020. I am especially grateful to Dr. Martin Preuss (Wiley-VCH) who has encouraged me to edit this book. I would like to thank Drs. Taisuke Matsui and Akira Terakawa (Panasonic) for helpful discussion. I wish to express my sincere gratitude to my students at Gifu University, Mitsutoshi Nishiwaki, Yukinori Nishigaki, Yoshitsune Kato, Masayuki Kozawa, Tatsuya Narikuri, Tomoya Kobayashi, Ryo Ishihara, Yosuke Kinden, Sara Maeda, Yusuke Tani, and Ryoki

Toda who helped me greatly to edit this book. I sincerely hope that this book will contribute to suppressing CO_2 emissions and bring a clean-energy world in the very near future.

25 September 2020

Hiroyuki Fujiwara
Gifu

About the Editor

Hiroyuki Fujiwara is a Professor in Department of Electrical, Electronic and Computer Engineering, Gifu University. He received his PhD degree from Tokyo Institute of Technology. He was a research associate at The Pennsylvania State University during 1996–1998. In 1998, he joined the Electrotechnical Laboratory, Ministry of International Trade and Industry, Japan. In 2007, he became a team leader at the Research Center for Photovoltaics, National Institute of Advanced Industrial Science and Technology (AIST), in Japan.

1

Introduction

Hiroyuki Fujiwara

Gifu University, Department of Electrical, Electronic and Computer Engineering, 1-1 Yanagido, Gifu 501-1193, Japan

1.1 Hybrid Perovskite Solar Cells

Hybrid perovskites, unique materials composed of moving organic cations and rigid inorganic framework, show remarkable suitability as light absorbers of photovoltaic devices, enabling us to reach a 25% efficiency in a single-cell configuration [1] and a 29% efficiency in a tandem device configuration [1] in a rather short period of time since the first development of a hybrid perovskite solar cell in 2009 [2]. Compared with structurally complex solar cells such as Si [3, 4], GaAs [5], and $CuInGaSe_2$ [3, 6], the architecture of hybrid perovskite solar cells is surprisingly simple (Figure 1.1a); by essentially sandwiching a perovskite absorber layer between the electron transport layer (ETL) and the hole transport layer (HTL), a quite high efficiency exceeding 20% can be realized without including complicated processing steps [7–22]. For hybrid perovskites, no intentional doping is necessary and the interface recombination is suppressed quite well even though the heterogeneous interfaces are formed with ETL and HTL. Hybrid perovskite polycrystalline absorbers can be formed by a solution-based process at a surprisingly low temperature of ~100 °C [10], making the perovskite materials extremely attractive for both large-area and tandem photovoltaic applications.

Hybrid perovskites, including methylammonium lead iodide ($CH_3NH_3PbI_3$; $MAPbI_3$) and formamidinium lead iodide ($HC(NH_2)_2PbI_3$; $FAPbI_3$), are a unique form of ABX_3 perovskites consisting of organic A-site cations ($A = CH_3NH_3^+$, $HC(NH_2)_2^+$), B-site cations ($B = Pb^{2+}$, Sn^{2+}), and X-site halide anions ($X = I^-$, Br^-, Cl^-) (see Figure 1.1b). The word "hybrid" indicates that the crystal is made specifically by the combination of "organic" and "inorganic" components. In a hybrid perovskite crystal, an organic cation is confined in a cage structure formed by a BX_6 octahedron that is interconnected via all corners to form a three-dimensional perovskite network (Figure 1.1a). The semiconducting properties of hybrid perovskites arise from the inorganic BX_3 component (Section 5.4.3), and all critical phenomena of the light absorption, carrier generation, and carrier transports occur

Hybrid Perovskite Solar Cells: Characteristics and Operation,
First Edition. Edited by Hiroyuki Fujiwara.

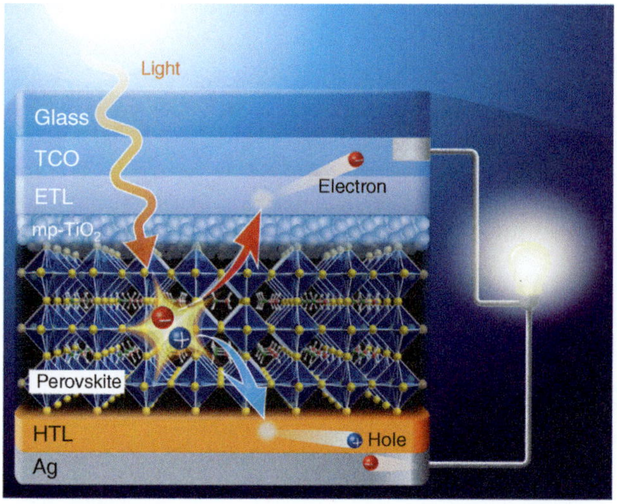

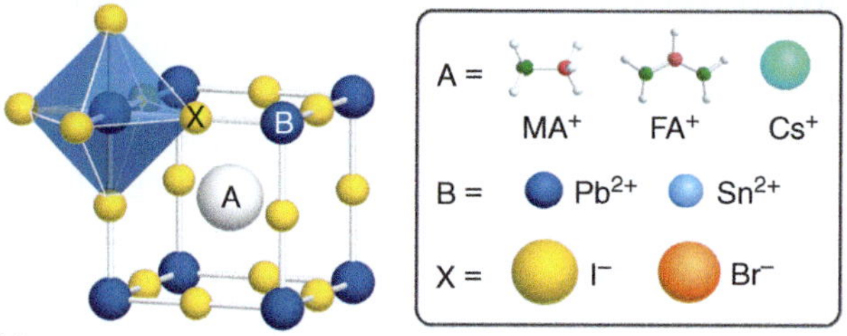

Figure 1.1 (a) Operation of a hybrid perovskite solar cell consisting of a glass/transparent conductive oxide (TCO)/electron transport layer (ETL)/mesoporous TiO_2 (mp-TiO_2)/ perovskite layer ($CH_3NH_3PbI_3$)/hole transport layer (HTL)/metal electrode (Ag) structure, and (b) structure of ABX_3 perovskite crystals, whereas A is an organic (or inorganic) cation (A = $CH_3NH_3^+$ (MA^+), $HC(NH_2)_2^+$ (FA^+) or Cs^+), B is a divalent cation (B = Pb^{2+} or Sn^{2+}), and X is a halide anion (X = I^-, Br^-, or Cl^-). In (b), the structure of a BX_6 octahedron is indicated by blue planes. The ionic radii are summarized in Table 3.3.

within the inorganic framework, while the organic cation plays an important role in determining the fundamental stability of perovskite crystals (Section 3.3).

A standard hybrid perovskite device has a structure composed of glass/ transparent conductive oxide (TCO)/TiO_2 (ETL)/mesoporous TiO_2 (mp-TiO_2)/ perovskite(~500 nm)/HTL/metal (Figure 1.1a). The mp-TiO_2 is the anatase phase of TiO_2 with high porosity (~60%) [23, 24], which is helpful for improving the efficiency [7, 25]. As ETL and HTL of the cells, a variety of inorganic and organic layers have been applied (see Figure 11.2). The operation of the perovskite device is straightforward; namely, the photo-electrons and holes created by light absorption are collected in the ETL and HTL, respectively, and the electrons flow through the outer circuit and recombine with holes at the HTL/metal interface.

Figure 1.2a shows the scanning electron microscope (SEM) image of a state-of-the-art hybrid perovskite solar cell with an efficiency of ~21% [19]. The absorber

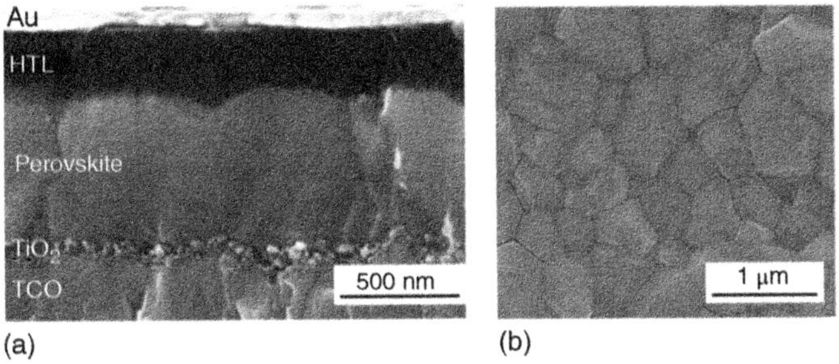

Figure 1.2 (a) Cross-sectional image of a hybrid perovskite solar cell with 20.8% efficiency and (b) surface image of the perovskite absorber layer. The hybrid perovskite layer is a triple-cation-mixed perovskite with $Cs_{0.05}FA_{0.79}MA_{0.16}Pb(I_{0.83}Br_{0.17})_3$. The solar cell in (a) has a standard device structure of TCO(SnO_2:F)/TiO_2(ETL)/mp-TiO_2/perovskite/spiro-OMeTAD(HTL)/Au. *Source*: Singh and Miyasaka [19].

of this cell is a triple-cation-mixed perovskite of $CsFAMAPb(I,Br)_3$, in which three mixed cations of A = Cs^+, FA^+, and MA^+ are incorporated into the $Pb(I,Br)_3$ scaffold (Chapter 3). The structure of this device is exactly the same as that of Figure 1.1a. Even though the grain size of the polycrystalline perovskite is relatively small (~1 μm in Figure 1.2b), the grain boundary does not deteriorate cell characteristics significantly (Chapter 10 and Section 11.5.3) and the high conversion efficiency can be realized.

Historically, in the first hybrid perovskite cell developed by Miyasaka's group in 2009 [2], $MAPbI_3$ nanocrystals were employed as "sensitizers," which behaved only as the photocarrier generator without the active role of carrier transport (Chapter 2). In 2012, the conversion efficiency of the sensitizer concept cells improved markedly by the adoption of a thicker $MAPbI_3$ and a more sophisticate structure of TiO_2/$MAPbI_3$/spiro-OMeTAD, reaching ~10% efficiencies [26, 27]. In 2013, Snaith's group succeeded in developing a hybrid perovskite cell, which utilizes a thick perovskite absorber (~300 nm) for both carrier generation and transport, demonstrating a remarkably higher efficiency of 15% [28].

Figure 1.3 summarizes the development of hybrid perovskite solar cells during 2013–2020 [1, 29–42]. The conversion efficiency of the world-record cell in 2020 is 25.2% with the short-circuit current density (J_{sc}) of 25.1 mA/cm^2, the open-circuit voltage (V_{oc}) of 1.18 V, and fill factor (FF) of 84.8% [1]. A high J_{sc} of ~25 mA/cm^2 has been obtained since the early research stage, and the drastic improvement of the cell efficiency originates mostly from the increase in V_{oc} and FF. One of the main advantages of hybrid perovskite solar cells is a very small J_{sc} loss (Section 11.4.2), and the current perovskite solar cell is limited by V_{oc} and FF (see Figure 12.13) [5]. The carrier collection in the perovskite devices is already 100% (Figure 11.11), and now the experimental V_{oc} (1.18 V) and FF (84.8%) become closer to the theoretical limits of V_{oc} = 1.28 V and FF = 90%, derived for $FAPbI_3$ (Table 12.1). The conversion efficiency of tandem devices is also improving in an unprecedented pace (Figure 17.10), and the mass production of perovskite/Si tandem devices is already in progress [43].

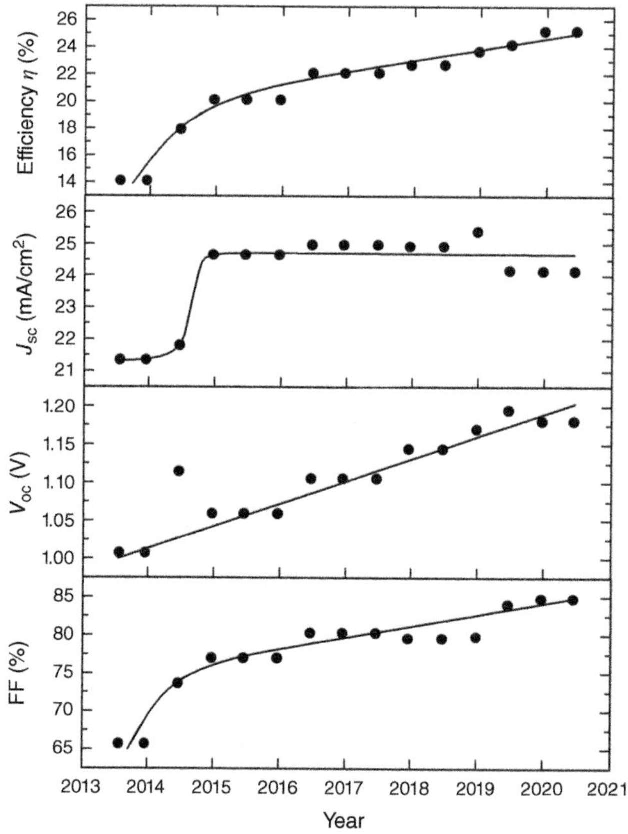

Figure 1.3 Variation of solar cell characteristics reported for record-efficiency hybrid perovskite devices during 2013–2020 [1, 29–42]. The solar cell performances described in "solar cell efficiency tables" are summarized. *Source*: Green et al. [1, 29–42].

1.2 Unique Natures of Hybrid Perovskites

Hybrid perovskites show physical properties that are surprisingly suitable for solar cells. Unlike conventional inorganic semiconductors, organic–inorganic hybrid perovskites exhibit strong dynamical motion at room temperature (Figure 1.4). In $MAPbI_3$, for example, MA^+ organic cations confined in the PbI_3 cage move (reorient) rapidly with a time scale of 5 ps (Figure 3.16). However, the movement of MA^+ is not completely random but is restricted by the hydrogen bond formed with the PbI_3^- inorganic scaffold (Figures 3.15 and 5.5). Moreover, hybrid perovskites are essentially "soft materials" and the PbI_6 octahedra show a large distortion [44, 45], as depicted in Figure 1.4. Furthermore, under the presence of the electric field, halide ions move easily through the three-dimensional inorganic network [46], causing a well-known hysteresis in current–voltage characteristics (Chapter 16). Accordingly, the dynamical structure of hybrid perovskites is quite complex, making the theoretical interpretation more challenging (Chapter 5).

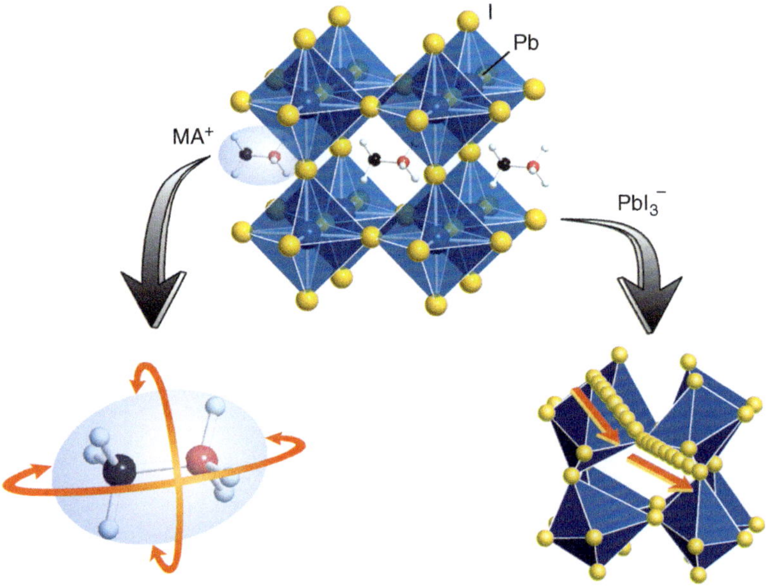

Figure 1.4 Structure of a $MAPbI_3$ perovskite crystal. The rapid reorientation of organic cations (MA^+) within the PbI_3^- inorganic cage and large distortion of the PbI_6 octahedra are illustrated. The movement of I^- under the presence of the electric field is further shown schematically by the orange arrows.

This section introduces the notable characteristics of hybrid perovskites (Section 1.2.1). In particular, the fundamental physical properties of hybrid perovskites are briefly reviewed based on the most studied $MAPbI_3$ perovskites (Section 1.2.2). We further discuss why hybrid perovskite materials are so preferable for photovoltaic devices and why high efficiency can be achieved relatively easily (Section 1.2.3).

1.2.1 Notable Characteristics of Hybrid Perovskites

Solar cells are essentially optical devices, and the strong light absorption within semiconducting absorbers is imperative. Figure 1.5 summarizes the absorption coefficient (α) spectra of representative solar cell absorbers, including $MAPbI_3$ [47], $CsFAPb(I,Br)_3$, and an Sn-based perovskite of $FAMA(Pb,Sn)I_3$. The optical data are taken from Figure 4.1 (see Appendix A for numerical values) and Ref. [48]. All the α spectra in Figure 1.5 are extracted using a reliable spectroscopic ellipsometry technique (Section 4.4). Fortunately, all hybrid perovskite materials show high α of $10^4\ cm^{-1}$ (Figure 4.1), which is comparable to other solar cell absorbers of GaAs, $CuInGaSe_2$ (CIGSe), CdTe, and Cu_2ZnSnS_4 (CZTS). Systematic theoretical and experimental analyses show clearly that hybrid perovskites are direct-transition semiconductors without any indirect nature (Section 5.4.2) and high α near the band gap (E_g) is the direct consequence of the direct optical transition (Figure 5.8). The constant α of $2 \times 10^4\ cm^{-1}$, observed near E_g of hybrid perovskites, leads

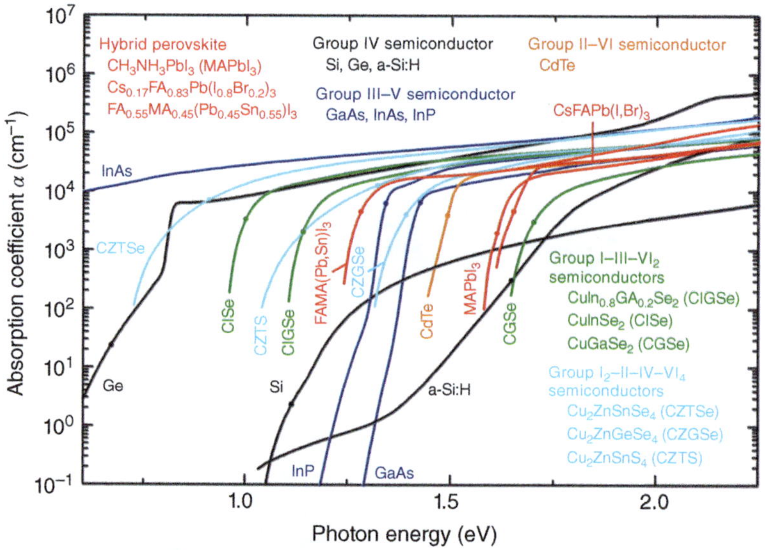

Figure 1.5 Absorption coefficient (α) spectra of representative solar cell absorbers, including $MAPbI_3$, $Cs_{0.17}FA_{0.83}Pb(I_{0.8}Br_{0.2})_3$, and $FA_{0.55}MA_{0.45}(Pb_{0.45}Sn_{0.55})I_3$. The closed circles in the figure show the band gap (E_g) energy of each absorber. All the optical results are extracted from reliable spectroscopic ellipsometry measurements. The hybrid perovskite materials show high α of 10^4 cm^{-1}, which is comparable to other solar cell absorbers of GaAs, CIGSe, and CdTe. The exact numerical values of the perovskites are summarized in Appendix A. The result of $MAPbI_3$ is taken from Ref. [47], whereas the data of $CsFAPb(I,Br)_3$ and $FAMA(Pb,Sn)I_3$ are consistent with those shown in Figure 4.1. The α spectra of all the other inorganic semiconductors are adopted from Ref. [48]. *Source:* Adopted from Fujiwara et al. [48].

to a light penetration depth of $d_p = 1/\alpha = 500$ nm, which is consistent with the experimental perovskite absorber thicknesses. In contrast, in the indirect-transition Si, α is quite low near E_g (~100 cm^{-1}), requiring a thick Si wafer of ~100 μm.

The closed circles in Figure 1.5 show the E_g positions of each absorber. The E_g of $MAPbI_3$ is 1.61 eV [47, 49–51], while this value varies slightly depending on the analysis and characterization methods (Table 4.2). The E_g values of all the different halide perovskites are summarized in Table 4.3. A low-E_g perovskite material (E_g ~ 1.2 eV) can be formed by mixing Sn with Pb, while wider gap perovskites are generated by incorporating lighter halide atoms (Figure 4.9b). The A-site cation species also modifies E_g slightly in the case of $APbI_3$ (see Figure 4.9a). Shockley–Queisser theory (Section 12.2) indicates the favorable E_g range of solar cells to be 1.0–1.6 eV (Figure 12.1), which nicely overlaps the E_g range of hybrid perovskites.

In the operation of solar cell devices, one of the most critical parameters is the photocarrier diffusion length (L_D), which is quantitatively expressed as a product of carrier lifetime (τ) and mobility (μ) (Section 6.1). The larger μ and τ in solar cell active layers directly lead to improvement in the photovoltaic performance. Rather interestingly, μ of hybrid perovskites is not so large; in fact, μ inside the polycrystalline grains of $MAPbI_3$ shows a moderate value of 30 cm²/(V s) (Figure 6.4). This μ is far smaller when compared with that of Si (1500 cm²/(V s), see Figure 6.7). The small μ observed in hybrid perovskites can essentially be attributed to a very large carrier–phonon interaction, as discussed in Section 6.3.4.

What makes hybrid perovskites so attractive for photovoltaic devices is a quite long τ. Figure 1.6 compares the time decays of normalized photoluminescence (PL) intensity (I_{PL}), reported for the polycrystalline thin films of $MAPbI_3$ [52], CIGSe [53], and CZTS [54]. The $MAPbI_3$ result has been obtained for a $MAPbI_3$ thin layer with a passivation layer of tri-n-octylphosphine oxide (TOPO). The I_{PL} of Figure 1.6 shows clearly that, in $MAPbI_3$, the photo-generated carriers have a quite long τ due to a very low carrier recombination. If single exponential decay is assumed for I_{PL}, the position of $I_{PL} = 1/e \sim 37\%$ directly provides τ (i.e. $I_{PL} = \exp(-t/\tau)$). In this case, τ values of $MAPbI_3$, CIGSe and CZTS are estimated to be 8 μs and 180 ns and 2 ns, respectively. Although the absolute τ is quite sensitive to the surface/interface and passivation conditions (Figure 6.5b) [52, 53, 55–58], τ of the hybrid perovskite is far longer than the other polycrystalline films. One remarkable advantage of hybrid perovskites is the absence of deep-level defects as revealed by density functional theory (DFT) calculations (Section 5.6) and a suppressed defect formation within the perovskite bulk layers extends τ substantially. In contrast, in CZTS solar cells, the short τ of the absorber layer makes the photocarrier collection difficult [59], resulting in a low efficiency of 11% [1]. For hybrid perovskites, notable long τ and moderate μ lead to a very large L_D of 50–175 μm in single crystals and 4–23 μm in thin films

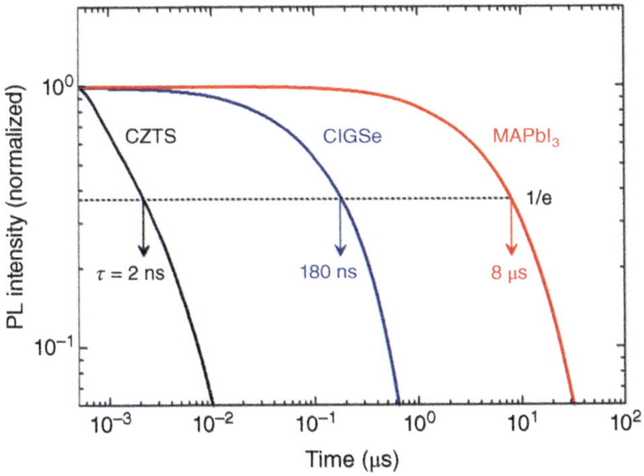

Figure 1.6 Time decays of normalized PL intensities reported for the polycrystalline thin films of $MAPbI_3$ [52], CIGSe [53], and CZTS [54]. The surface of the $MAPbI_3$ layer is passivated by a Lewis-base ligand, tri-n-octylphosphine oxide (TOPO). The time at which the normalized PL intensity becomes $1/e \sim 37\%$ gives the PL lifetime under the assumption of the single exponential decay (i.e. $I_{PL} = \exp(-t/\tau)$). The carrier lifetime τ obtained from this procedure is indicated for each material. Among the materials, $MAPbI_3$ shows an extraordinarily long τ of 8 μs. In the figure, the measurement noises are removed for clarity. *Source*: de Quilettes et al. [52]; Metzger et al. [53]; Phuong et al. [54].

(see Table 6.2). Recall that the perovskite absorber thickness is ~500 nm and L_D is sufficiently larger than the absorber thickness. Thus, in short, quite high efficiency of hybrid perovskite solar cells can be understood by favorable optical properties (strong light absorption and appropriate E_g) as well as the superior carrier transport characteristics represented by long τ (or L_D).

1.2.2 Fundamental Properties of $MAPbI_3$

Table 1.1 shows the room-temperature physical properties of $MAPbI_3$ determined by experimental measurements [47, 49–52, 60–95, 97–99] or theoretical calculations based on DFT [96]. This table summarizes the structural, band, optical, electric, mechanical, and thermal properties of $MAPbI_3$. The corresponding figure and table numbers in this book, from which the numerical values are extracted, are also indicated.

At room temperature, $MAPbI_3$ shows a slightly complicated tetragonal crystal structure (see Figure 3.2). The space group of the tetragonal $MAPbI_3$ has been discussed and confirmed to be *I4cm* [60–62], even though a slightly different assignment exists (i.e. *I4/mcm*) [68, 100]. However, many hybrid perovskites and their alloys exhibit a simple cubic perovskite structure (see Table 3.2). The tolerance factor (see Figure 3.9) is a critical parameter that determines the stability of perovskite crystals and the tolerance factor of $MAPbI_3$ ($t = 0.91$) falls within the ideal range of $t = 0.91–0.98$ (see Figure 3.10).

The $MAPbI_3$ exhibits superior optical properties, such as (i) direct transition, (ii) high α, (iii) sharp absorption edge (i.e. small Urbach energy of $E_U = 15$ meV), and (iv) a low exciton binding energy (Chapter 4). The direct-transition nature of the hybrid perovskite can also be confirmed from a high radiative recombination coefficient ($B = (0.1–2.3)\times10^{-9}$ cm^3/s, see Table 8.1), which is comparable to that of a GaAs direct-transition semiconductor (0.72×10^{-9} cm^3/s). In other words, $MAPbI_3$ is a good luminescence material that can also be applied for light emitting diodes and laser diodes [101] (Chapter 8). Note that good solar cells are essentially good photo emitters (Section 11.5.1), and a high internal PL yield of 91.9% has been reported for $MAPbI_3$ [102]. The sharp band-edge absorption with $E_U = 15$ meV (see Figure 4.17a) is beneficial for lowering V_{oc} loss in the solar cells (Chapter 12) [78]. The exciton binding energy of $MAPbI_3$ (~2 meV) is smaller than the room temperature energy (i.e. $k_BT = 26$ meV) and the exciton formation is negligible in $MAPbI_3$ (Section 4.8). Interestingly, the refractive index n and high-frequency dielectric constant ε_∞ of $MAPbI_3$ are notably lower ($n = \sqrt{\varepsilon_\infty} = 2.23$) than other semiconductors (see Figure 4.3a). The low n can be interpreted by the large optical void created by the presence of the organic cations (Section 4.2). One unique optical characteristic of $MAPbI_3$ is quite strong phonon absorption, which increases the static dielectric constant to a very high value of $\varepsilon_s = 34.9$ (see Figure 4.2).

This strong phonon absorption is a key to understanding the moderate electron and hole mobility of $\mu \sim 100$ $cm^2/(V\,s)$ in $MAPbI_3$ (Section 6.3.4). The intra-grain mobility in polycrystalline $MAPbI_3$ is 30 $cm^2/(V\,s)$, and all the perovskite thin films show similar mobility (Figure 6.10). In the perovskite alloys, however, the

Table 1.1 Physical properties of a $MAPbI_3$ hybrid perovskite at room temperature, determined by experimental measurements and theoretical calculations based on density functional theory (DFT).

Physical property	Value	Figure and Table	References
Structural properties			
Crystal structure	Tetragonal	Table 3.2	[60]
Space group	*I*4*cm*		[60–62]
Lattice parameters	$a = 8.849$ Å ($\alpha = 90°$) $b = 8.849$ Å ($\beta = 90°$) $c = 12.642$ Å ($\gamma = 90°$)	Table 3.2 and Figure 3.3	[60]
Crystal volume	990.0 Å^3		[60]
Density	4.159 g/cm^3		[60]
Tolerance factor	0.91	Figure 3.10	
MA^+ reorientation time	5 ps	Figure 3.16	[63, 64]
Band properties			
Band gap	$E_g = 1.61$ eV	Figure 4.8	[47, 49–51]
E_g variation with temperature T (K)	$E_g = 1.52 + 3.04 \times 10^{-4}\ T$ eV	Figure 4.19c	
Ionization potential	−5.5 eV	Figure 11.4	[65]
Electron affinity	−3.9 eV	Figure 11.4	[65]
Effective mass	$m_e^* = 0.32\ m_0$ (electron) $m_h^* = 0.36\ m_0$ (hole)	Table 6.1	[66]
Optical properties			
Optical transition	Direct transition	Figure 5.8	[67–70]
Absorption coefficient	1.8×10^4 cm^{-1} (at 1.7 eV)	Figure 4.4	[47]
Radiative recombination coefficient	$B = (0.1–2.3) \times 10^{-9}$ cm^3/s	Table 8.1	[71–77]
Urbach energy	$E_U = 15$ meV	Figure 4.17	[78, 79]
Exciton binding energy	~2 meV	Table 4.4	[80, 81]
Refractive index	$n = 2.23$[a)]	Figure 4.2	
High-frequency dielectric constant	$\varepsilon_\infty = 5.0$	Figure 4.2	
Static dielectric constant	$\varepsilon_s = 34.9$	Figure 4.2	
Visible/UV peak	$E_1 = 2.53$ eV, $E_2 = 3.24$ eV	Figure 4.8	[47]
Longitudinal optical (LO) phonon peak	$\omega_{LO1} = 133$ cm^{-1}, $\omega_{LO2} = 40$ cm^{-1}	Figure 4.7	[82]
Transverse optical (TO) phonon peak	$\omega_{TO1} = 32$ cm^{-1}, $\omega_{TO2} = 63$ cm^{-1}	Figure 4.7	[82]

(continued)

Table 1.1 (Continued)

Physical property	Value	Figure and Table	References
Electric properties			
Electron mobility	66 $cm^2/(V\,s)$ (bulk polycrystal)	Table 6.1	[60]
	30 $cm^2/(V\,s)$ (thin film)	Figure 6.4	
Hole mobility	105 $cm^2/(V\,s)$ (single crystal)	Table 6.1	[83]
	30 $cm^2/(V\,s)$ (thin film)	Figure 6.4	
Carrier lifetime	τ = 10–82 μs (single crystal)	Table 6.2	[83–86]
	τ = 8 μs (thin film)	Figure 1.6	[52]
Carrier diffusion length	L_D = 50–175 μm (single crystal)	Table 6.2	[83–85]
	L_D = 4–23 μm (thin film)	Table 6.2	[75, 87, 88]
Carrier density	10^9–10^{10} cm^{-3} (single crystal)	Figure 6.3	[60, 83]
Carrier trap density	~10^{10} cm^{-3} (single crystal)		[83, 86, 89]
Ion density	10^{16}–10^{17} cm^{-3}	Table 16.1	[90–92]
Fröhlich parameter	α = 1.72–2.68	Eq. (6.6)	[82, 93, 94]
Mechanical properties			
Bulk modulus	13.9 GPa	Table 6.1	[95]
Young's modulus	14.3 GPa		[95]
Hardness	0.57 GPa		[95]
Shear modulus	5.4 GPa		[95]
Poisson's ratio	0.33 (DFT)		[96]
Thermal properties			
Debye temperature	131 K (DFT)		[96]
Linear thermal expansion coefficient	43.3×10^{-6} K^{-1}		[97]
Molar heat capacity	C_p = 189.6 J/(K mol)		[98]
Thermal conductivity	0.5 W/(K m)		[99]

The corresponding figure and table numbers in this book, from which the numerical values are extracted, are also indicated.
a) $n = \sqrt{\varepsilon_\infty}$.

tolerance factor affects μ indirectly (Figure 6.11). As shown in Table 1.1, τ of $MAPbI_3$ single crystals is larger than that of the thin films and the resulting L_D also becomes large in single crystals (Figure 6.9). The carrier density of $MAPbI_3$ single crystals is very low (~10^{10} cm^{-3}) [60, 83], while the carrier concentration in the thin films can be controlled widely from p-type to n-type according to the processing condition (i.e. PbI_2/MAI precursor ratio in Figure 6.2) [103]. As mentioned

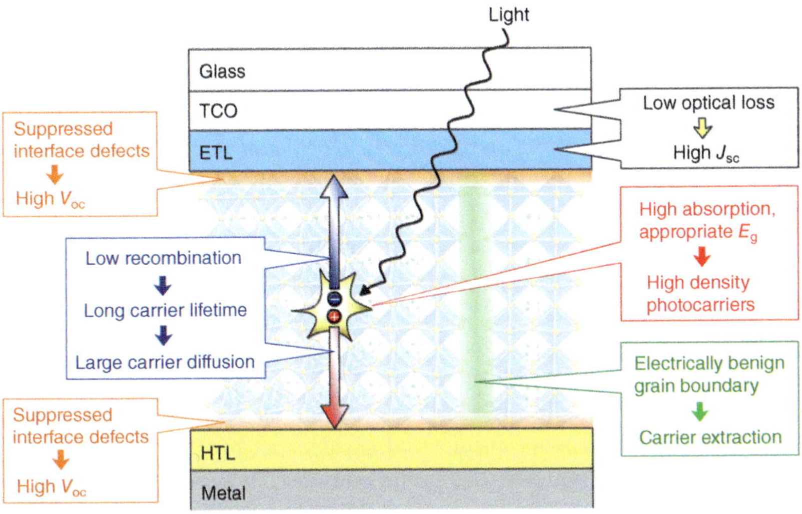

Figure 1.7 Physical features that are quite advantageous for realizing high conversion efficiencies in hybrid perovskite solar cells.

earlier, the formation of bulk defects is suppressed quite well in hybrid perovskites (Section 5.6) and the carrier trap density is very low (~10^{10} cm^{-3} in the single crystal) [83, 86, 89]. There are, however, many mobile ions in $MAPbI_3$ with a concentration of 10^{16}–10^{17} cm^{-3} (Table 16.1).

Hybrid perovskites are soft materials [95] and the bulk modulus of $MAPbI_3$ (13.9 GPa) is far smaller, when compared with conventional inorganic solar cell materials with corresponding values of 40–100 GPa (see Table 6.1). The thermal conductivity of $MAPbI_3$ is also very small [99]. The abovementioned characteristics show that a $MAPbI_3$ hybrid perovskite indeed has a mixed character of both inorganic and organic materials.

1.2.3 Why Hybrid Perovskite Solar Cells Show High Efficiency?

High efficiencies ($\eta > 20\%$) observed widely in hybrid perovskite solar cells can be interpreted as a consequence of several physical processes summarized in Figure 1.7. For the photocarrier generation, (i) appropriate E_g, (ii) strong light absorption (i.e. high α), and (iii) low exciton binding energy of hybrid perovskites allow the generation of high-density carriers within the absorber upon light absorption. Moreover, in standard perovskite devices, the unfavorable light absorption within TiO_2 (ETL) and TCO is suppressed quite well (see Figures 11.3 and 11.11). Sufficient carrier generation within the thin perovskite layer (~500 nm) and low parasitic absorption in the front TCO and ETL have been essential in realizing very high optical gain in the devices (Section 11.4.2).

For the carrier collection, the carrier recombination rate within the perovskite layer is very small due to a low defect (or trap) density, leading to very long τ and L_D. Unlike conventional solar cell semiconductors, in $MAPbI_3$, the electron and hole mobilities are approximately equal (Figure 6.7) and the L_D values for electrons and holes are within the same range. This is a great advantage for solar cells as both electrons and holes need to be collected efficiently in photovoltaic devices. Remarkably, in the perovskite solar cells, the defect formation at the perovskite/ETL and perovskite/HTL interfaces is also suppressed well. The small carrier recombination within the perovskite bulk layer and at the interfaces allows us to reach high V_{oc} (Section 11.5). Note, however, that the performance of the current high-efficiency devices is mainly limited by the interface recombination (see Section 11.5.1).

Importantly, among all polycrystalline-based solar cells, hybrid perovskite devices show the highest performance, even though the grain size of the perovskites is generally small (~1 μm, see Figure 1.2). The long τ and high V_{oc} observed for the small-grain polycrystalline perovskites confirm that the grain boundary is electrically benign in hybrid perovskites. In fact, the efficient extraction of photocarriers in the grain boundary region has been confirmed experimentally (Chapter 9), although L_D is reduced slightly in the grain boundary region [104]. However, FF increases with average perovskite grain size (Figure 11.18), and thus, the larger perovskite grain is still preferable for obtaining higher efficiency.

As mentioned earlier, polycrystalline hybrid perovskite layers are formed at low temperatures (~100 °C) using a spin-coating technique. It is quite surprising that all the favorable physical characteristics shown in Figure 1.7 are obtained for such solution-processed layers.

1.3 Advantages of Hybrid Perovskite Solar Cells

Hybrid perovskite solar cells have strong advantages, in addition to high device performance exceeding 20%. In this section, as distinctive features of the perovskite cells, we focus on the large band gap tunability of hybrid perovskite absorbers (Section 1.3.1), very small V_{oc} losses in the perovskite cells (Section 1.3.2), and a very low efficiency loss at high operating temperatures (Section 1.3.3).

1.3.1 Band Gap Tunability

One important characteristic of hybrid perovskites is the large tunability of E_g. In particular, high-quality perovskite absorbers can be prepared in a wide E_g range by mixing the B-site cation (Pb–Sn) and the X-site anion (Br–I) (Chapter 4), while maintaining the sharp absorption edge (Figures 4.17 and 10.7). Figure 1.8 shows the E_g ranges of Pb- and Sn-based perovskite alloys, together with E_g of non-alloyed perovskites (see Table 4.3 for summary). The conventional E_g range of the Pb perovskites (i.e. $Cs_xFA_yMA_zPb(I,Br)_3$) is 1.55–2.3 eV, whereas the E_g can be lowered down to 1.2 eV in the Sn-based perovskites of $FA_xMA_y(Pb,Sn)I_3$ (Figure 4.10).

The large E_g tunability of hybrid perovskites enables us to develop different single and tandem cell designs. For single cells, perovskite absorbers with E_g = 1.2–1.6 eV are appropriate. Wider gap perovskites of E_g = 1.6–1.7 eV are suitable as the top cells in tandem devices employing Si bottom cells (Figure 12.17 and Chapter 17).

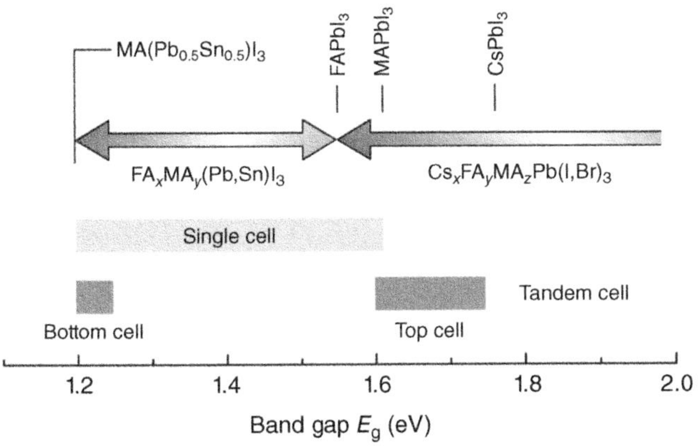

Figure 1.8 Band gap (E_g) ranges of the Pb- and Sn-based perovskite alloys of CsFAMAPb(I,Br)$_3$ and FAMA(Pb,Sn)I$_3$, together with E_g of the non-alloyed perovskites (see also Table 4.1). Note that E_g of the Sn alloys becomes the lowest around Pb/Sn = 1 due to the band gap bowing effect (see Figure 4.10c). The appropriate E_g ranges for the single and tandem cell applications are also indicated. For the perovskite/Si tandem cell, the proper E_g of the perovskite top cell is 1.6–1.7 eV (see Figure 12.18), whereas in perovskite/perovskite tandem cells, the E_g values of the top and bottom cells are 1.7–1.75 and 1.2–1.25 eV, respectively (see Figure 18.6b).

The low-E_g Sn-based perovskites ($E_g \sim 1.25$ eV) further allow the fabrication of perovskite/perovskite tandem cells (i.e. all perovskite tandem cells, Chapter 18).

Figure 1.9 summarizes the current density–voltage (J–V) characteristics reported for the state-of-the-art Sn- and Pb-based perovskite single cells as well as perovskite/Si and perovskite/perovskite tandem cells [1, 12, 21]. The solar cell parameters and absorber E_g values of these devices are also shown in Table 1.2. In the figure, a systematic variation of the J–V curve can be confirmed according to the E_g of the absorber. Remarkably, more than 20% efficiencies have been confirmed for all these cells. In particular, the efficiency of the perovskite/Si tandem device is almost 30%. The realistic efficiency of fully textured perovskite/Si devices is 36% (Table 12.2), and thus, there is a sufficient room for further improvement. Very high V_{oc} values observed already for the tandem cells (~2 V) are particularly promising for the development of higher efficiency devices. One solar cell company forecasted that the levelized cost of electricity (LCOE) of the perovskite/Si device will soon become lower than that of Si devices as early as 2022 [43].

1.3.2 High V_{oc}

One strong advantage of hybrid perovskite solar cells is quite high V_{oc}, which can be explained by the suppression of the defect formation in the bulk layer as well as at the interfaces. Figure 1.10 shows the V_{oc} of the perovskite single cell as a function of the absorber E_g [105], together with the V_{oc} limit derived from Shockley–Queisser theory (see Section 12.2 and Appendix B). For the perovskite cells with $E_g < 1.65$ eV,

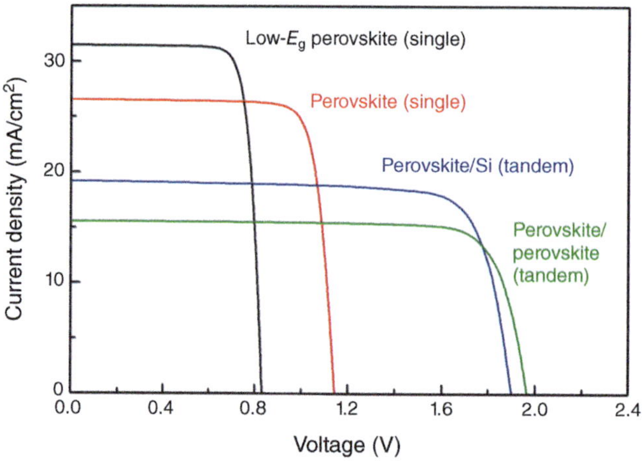

Figure 1.9 Current density–voltage ($J-V$) characteristics reported for the state-of-the-art Pb- and Sn-based perovskite single cells as well as perovskite/Si and perovskite/perovskite tandem devices [1, 12, 21]. The absorber of the low-E_g perovskite (single) is $FA_{0.7}MA_{0.3}(Sn_{0.5}Pb_{0.5})I_3$ [21], whereas the result of the perovskite (single) is obtained with $FAPbI_3$ with a 3.8 mol% of methylenediammonium dichloride [12]. For the perovskite/perovskite tandem cell, the top and bottom cell absorbers are $Cs_{0.05}MA_{0.05}FA_{0.9}Pb(I_{0.95}Br_{0.05})_3$ and $FA_{0.7}MA_{0.3}(Sn_{0.5}Pb_{0.5})I_3$, respectively [21]. The hybrid perovskite solar cells show the hysteresis in the $J-V$ measurements and the $J-V$ characteristics in the figure correspond to those obtained in the reverse scan, which show slightly higher efficiencies, compared with the forward scan. The solar cell characteristics of the corresponding cells are shown in Table 1.2. *Source*: Green et al. [1]; Min et al. [12]; Lin et al. [21].

Table 1.2 Efficiency (η) and characteristics of perovskite-based single and tandem solar cells.

Device	E_g (eV)	η (%)	J_{sc} (mA/cm²)	V_{oc} (V)	FF	References
Sn-based perovskite single[a)]	1.22	21.1[b)]	31.4	0.83	0.808	[21]
Pb-based perovskite single[c)]	1.55[d)]	24.6	26.5	1.14	0.817	[12]
Perovskite/Si tandem	1.68[e)]/1.11	29.1	19.7	1.89	0.778	[1]
Perovskite/perovskite tandem[f)]	1.77/1.22	24.8[b)]	15.6	1.96	0.810	[21]

The band gap (E_g) values of the perovskite absorber layers are also indicated.
a) $FA_{0.7}MA_{0.3}(Sn_{0.5}Pb_{0.5})I_3$.
b) Characteristics obtained from the $J-V$ curve sweep in the reverse direction.
c) $FAPbI_3$ with $MDACl_2$ (3.8 mol%). $MDACl_2$ is methylenediammonium dichloride.
d) E_g of $FAPbI_3$ shown in Table 4.3.
e) E_g of the perovskite shown in Table 17.1.
f) Top cell: $Cs_{0.05}MA_{0.05}FA_{0.9}Pb(I_{0.95}Br_{0.05})_3$, bottom cell: $FA_{0.7}MA_{0.3}(Sn_{0.5}Pb_{0.5})I_3$.

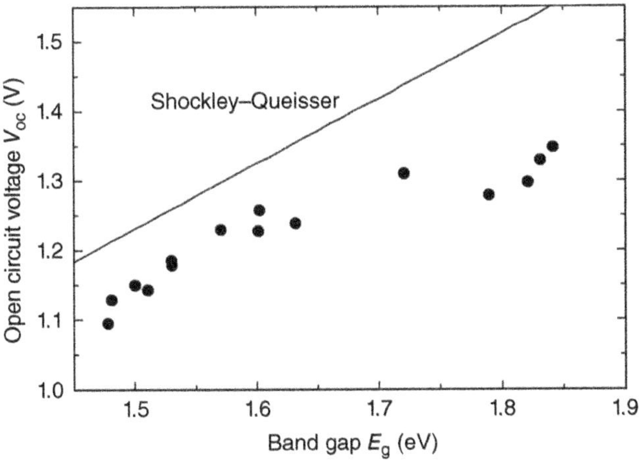

Figure 1.10 V_{oc} of hybrid perovskite single cells as a function of the absorber E_g. More complete data are shown in Figure 17.9. The closed circles show the values of the reported experimental cells, whereas the solid line indicates the V_{oc} limit derived from Shockley–Queisser theory. The numerical values of the theoretical results are shown in Appendix B. *Source*: Adapted from Jošt et al. [105].

the V_{oc} is remarkably high and V_{oc} increases with E_g without significant V_{oc} loss. In particular, a quite high V_{oc} of 1.26 V has been reported for a pure $MAPbI_3$ cell [16]. This V_{oc} is very close to the theoretical limit of 1.32 V (see Table 12.1), with the V_{oc} loss of only 60 mV. In realizing such a high V_{oc} in the perovskite cells, the suppression of the pinhole and the interface defects is critical (see Figure 11.14). Note that the formation of interface defects is particularly sensitive to the selection of ETL and HTL. In other words, the maximum V_{oc} is determined by the total amount of defects formed at the perovskite/ETL and perovskite/HTL interfaces (Figures 11.15 and 11.16), provided that L_D of the bulk layer is sufficiently long. The introduction of sophisticated passivation layers in the perovskite rear interfaces is quite effective in reducing the interface recombination and thus obtaining high V_{oc} (see Figure 11.17).

When the E_g of the perovskite becomes larger (>1.65 eV), however, the V_{oc} tends to become lower than the theoretical value (see also Figures 17.9 and 18.10a). This phenomenon can be interpreted by the strong instability observed in $APb(I,Br)_3$ perovskites with high Br contents; namely, upon light illumination, mixed I–Br perovskites show a phase segregation into the I-rich and Br-rich regions [106, 107], which in turn lowers the effective E_g (see Figure 8.7), leading to a lower V_{oc}. Numerous attempts have been made to suppress the phase separation effects (Section 18.3.3) but a new breakthrough is still necessary to suppress the light-induced segregation observed in wide-gap Br-based perovskites. Recall from Figure 1.8, however, that high-quality perovskites with $E_g < 1.65$ eV fulfill the E_g requirement necessary for top cells in perovskite/Si tandem devices (Figure 12.17), whereas the wide-gap perovskites become particularly important for all perovskite tandem devices (Figures 14.16b and 18.6b).

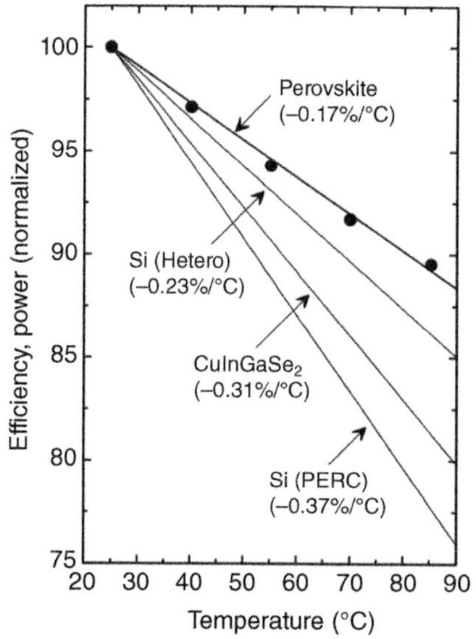

Figure 1.11 Variation of normalized conversion efficiency (or output power) with solar cell (or module) operating temperature for hybrid perovskite [108], conventional Si (i.e. PERC: passivated emitter and rear cell) [109–111], Si heterojunction [112], and $CuInGaSe_2$ [113, 114] solar cells. The temperature coefficients (i.e. the slope of efficiency variation for temperature) obtained for each cell (or module) are indicated in parentheses. The absorber material for the hybrid perovskite cell is $Cs_{0.05}FA_{0.79}MA_{0.16}Pb(I_{0.83}Br_{0.17})_3$, and the closed circles show the measured points for a 18.3% efficient p-i-n device [108].

1.3.3 Low Temperature Coefficient

Photovoltaic devices rarely operate at room temperature and high power output is necessary even under high-temperature operation conditions. Figure 1.11 shows the variation of normalized conversion efficiency (or output power) with solar cell (or module) operating temperature for hybrid perovskite [108], conventional Si (i.e. PERC: passivated emitter and rear cell) [109–111], Si heterojunction [112], and $CuInGaSe_2$ [113, 114] solar cells. Among these photovoltaic devices, the perovskite shows the lowest temperature coefficient (TC) of −0.17%/°C, which is far better than the conventional Si module that shows TC of −0.37%/°C. It is quite remarkable that nearly 90% efficiency is maintained in the perovskite device even at a high operating temperature of 85 °C.

The solar cell efficiency reduction at higher temperature essentially originates from the reduction of V_{oc}. In general, this V_{oc} reduction can be explained by two factors; (i) the decrease of E_g at high temperature and (ii) the increase of blackbody radiation with temperature that in turn increases the J_0 dark current (see Figures 12.5a and 12.6b). Remarkably, unlike conventional tetrahedral-bonding semiconductors (i.e. Si, GaAs, CdTe, and $CuInGaSe_2$), E_g of hybrid perovskites increases with temperature (Figure 4.19), which is quite beneficial to improve TC. The interface properties also affect TC, and indeed, high-efficiency Si heterojunction solar cells exhibit better TC, if compared with the conventional Si PERC cells [112]. Thus, the combination of a hybrid perovskite top cell with a Si heterojunction bottom cell is quite favorable in the view point of the TC.

1.4 Challenges for Hybrid Perovskites

For hybrid perovskite solar cells, critical challenges remain, even though quite high device efficiencies have already been realized. One significant drawback of hybrid perovskites that limit practical applications as photovoltaic materials is strong instability (Chapter 15). Specifically, hybrid perovskites are inherently unstable and exhibit strong phase changes upon light illumination, temperature increase, and exposure to humid air [115–121]. Thus, a major challenge exists for improved stability of the perovskite solar cells. In addition, in a large-area perovskite photovoltaics, the demonstration of high conversion efficiency becomes increasingly difficult, if compared with small-area devices. Furthermore, hybrid perovskites contain a toxic Pb element, which is environmentally detrimental.

In this section, I briefly review weak points of hybrid perovskite solar cells. Section 1.4.1 explains the instability of hybrid perovskite crystals, which is the most serious problem for device applications. In Section 1.4.2, we further see the latest development for the large-area devices. In the final section (Section 1.4.3), we discuss the toxicity of Pb- and Sn-based compounds formed by the decomposition of hybrid perovskites.

1.4.1 Requirement of Improved Stability

Under light illumination, humid-air (H_2O/O_2) exposure and high temperatures, hybrid perovskites transform into PbI_2-dominated phases [115–121]. Figure 1.12 indicates the changes of a thin $MAPbI_3$ polycrystalline layer (50 nm) observed after the temperature increase (i.e. (a) → (c)) [47] and exposure to humid air (i.e. (a) → (f)) [121]. Figure 1.12a shows the surface image of a $MAPbI_3$ layer formed by co-evaporation at room temperature, and this layer consists of very small $MAPbI_3$ grains. Rather remarkably, larger $MAPbI_3$ grains are formed by thermal annealing at a temperature as low as 80 °C (see Figure 1.12b). This temperature is far lower than the thermal annealing (or process) temperatures of other semiconductors including Si, CdTe, and CIGSe. As we have seen in Figure 1.11, a temperature of 80 °C is well within the operating temperature of solar cells, but this temperature corresponds to the annealing temperature for the hybrid perovskite.

When $MAPbI_3$ is maintained at an elevated temperature of 100 °C for 45 minutes, a thermal desorption of volatile MA components occurs, leading to the formation of PbI_2 polycrystals (Figure 1.12c). Note that, in conventional solution-based approaches, the process temperatures of hybrid perovskites are ~100 °C [10] and the strong phase variation observed at ~100 °C is not surprising. Accordingly, easy formation of the hybrid perovskite phase at low temperatures has a natural drawback of thermal instability. Importantly, the thermal stability can be improved if a heavier organic cation (i.e. FA^+) is adopted [122], and indeed, $FAPbI_3$ is stable up to 150 °C (see also Figure 3.14). The $CsFAPb(I,Br)_3$ perovskites, which show better thermal stability than $MAPbI_3$, are frequently employed for solar cell fabrications [123–126], although a triple-cation-mixed perovskite, $CsFAMAPb(I,Br)_3$, is now quite popular as an absorber material [18–21] (Chapter 13).

When hybrid perovskite crystals are exposed to humid air, the perovskite phase is transformed completely to hexagonal-shaped PbI_2/hydrate crystals even at a low relative humidity of 40% (see Figure 1.12d,e). Hybrid perovskites are ionic crystals, and the presence of H_2O leads to the decomposition of the perovskite structures

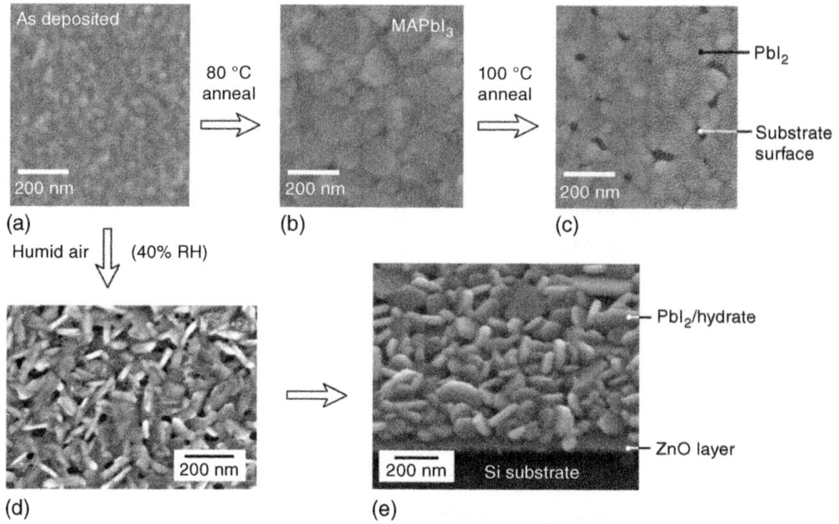

Figure 1.12 Structural changes of a thin $MAPbI_3$ polycrystalline layer (50 nm) observed after the temperature increase (i.e. (a) → (c)) and exposure to humid air (i.e. (a) → (e)). The SEM image of (a) shows the surface morphology of a $MAPbI_3$ layer formed on a ZnO/Si substrate by co-evaporation at room temperature. The surface images of (b) and (c) were obtained after thermal annealing at 80 and 100 °C for 45 minutes, respectively. The thermal annealing at 80 °C leads to the large grain formation, whereas the 100 °C annealing induces the phase transformation to PbI_2. In (d), the phase change of $MAPbI_3$ observed after the exposure to humid air (40% relative humidity, 48 hours) is shown. The image of (e) was obtained by further annealing the structure of (d) at 100 °C, but this thermal annealing does not change the structure and morphology. The exposure to humid air leads to the formation of PbI_2/hydrate crystals. *Source*: (a–c) Shirayama et al. [47]; (d, e) Shirayama et al. [121].

(Section 15.3). For the suppression of the degradation by humid air, the introduction of protective (passivation) layers at the solar cell rear interfaces has been quite effective [127–131].

Hybrid perovskites also show instability for light illumination (Figure 15.9) [115–119], and the strong light-induced phase segregation further occurs for mixed-halide perovskites [106, 107], as mentioned earlier. In the presence of an electric field, the halide ions migrate (see Figure 1.4), modifying the *J–V* characteristics (Chapter 16). The Sn-based perovskites are more unstable, compared with the Pb-based perovskites, due to the rapid oxidation of Sn^{2+} to Sn^{4+} in air [132]. The large instability observed in hybrid perovskites owes to the nature of chemical bonding; namely, for the inorganic framework of PbI_3^-, for example, the overlap of the wave functions between Pb and I atoms is weak, and the electronic charges of Pb and I are isolated (Figure 5.6) due to the antibonding character of the Pb—I bonds (see Figure 5.12). Thus, hybrid perovskites lack the strong bonding character and the Pb—I bonds are easily broken by external factors such as electronic excitation, electric field, temperature, and humidity.

For evaluating the stability of hybrid perovskite solar cells, a single degradation process of either heat, humidity, or light illumination may not be sufficient as the solar cells are operated in more extreme conditions of simultaneous strong light illumination, high temperatures, and high humidity. Accordingly, extensive efforts continue to find the long-term stability of hybrid perovskite solar cells.

1.4.2 Large-Area Solar Cells

Different types of large-area photovoltaic modules (i.e. single and tandem devices) can be formed for hybrid perovskites because of the high E_g tunability. For the perovskite single modules, however, the realization of high module efficiencies becomes more challenging, if compared with wafer-based perovskite/Si tandem devices, as described below.

Figure 1.13 shows the conversion efficiencies of record-efficiency devices (hybrid perovskite, Si, and CIGSe) as a function of the cell (or module) area. The data are taken from Ref. [1]. For hybrid perovskites, a quite high efficiency of $\eta = 25.2\%$ has been obtained for small-area devices (0.09 cm^2). Recently, a very high efficiency of 17.9% has been realized for the large-area module with a size of $30 \times 30\ cm^2$ (aperture area: 802 cm^2, see inset) [133], which was formed by an inkjet printing technology [134]. Over the years, the efficiency of the large-area devices has improved rapidly [35–42], but the efficiency is still lower than the small-area devices. The same trend can also be confirmed for CIGSe modules.

For thin-film large-area modules, it is more difficult to obtain high efficiency due to the essential requirements of (i) very high uniformity for all the constituent layers without the formation of pinhole, (ii) thicker TCO layers, compared with small-area devices, and (iii) increase of dead area (3–10%) due to the formation of

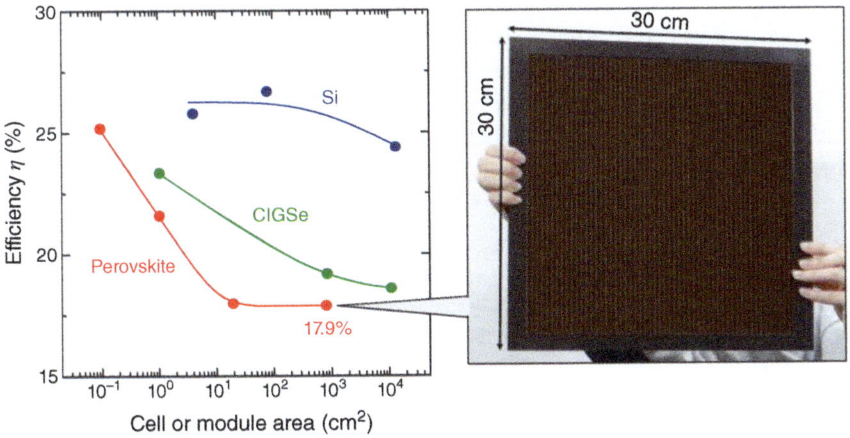

Figure 1.13 Conversion efficiencies of record efficiency devices (hybrid perovskite, Si, and CIGSe) as a function of the cell (or module) area. The data are taken from Ref. [1]. The inset shows the image of a large-area hybrid perovskite module with a size of $30 \times 30\ cm^2$ (aperture area: 802 cm^2), which was fabricated by Panasonic using an inkjet printing technology [133]. A very high efficiency of 17.9% has been reported for the large-area module. *Source*: Photograph of the large-area module is taken from Ref. [133] with permission.

series interconnections [134]. The large-area modules consist of a large number of rectangular-shaped subcells (see Figure 1.13), and the current is flowing by recombination at each subcell. In other words, J_{sc} of the module is determined by the lowest J_{sc} among all the subcells and high-quality control is necessary for all the layers. Moreover, in the large-area module, the carriers need to transport over longer distances, compared with small-area devices, and thicker TCO layers are generally necessary to lower series resistance. Nevertheless, if a thick TCO layer is employed, the J_{sc} loss increases due to the strong light absorption by free carriers in TCO [59]. The free carrier absorption in TCO increases in low energies and its effect becomes significant in the devices with lower E_g [135]. In this point of view, the Pb-based perovskite single module is more advantageous than the Sn-based perovskite single module.

In contrast to the perovskite and CIGSe modules, the efficiency of Si devices remains high in Figure 1.13 even when the module area is increased. This is the result of a simple quality control that can be adopted in wafer-based Si devices with a size of $15 \times 15\,\text{cm}^2$. Specifically, the conversion efficiencies of each Si device can be checked in advance before forming the large module. Thus, high-efficiency modules can be formed by simply selecting high performance cells, while some cells with poor performance are utilized as a lower output module. Consequently, a better quality control can be realized for wafer-based perovskite/Si tandem modules, compared with large-area monolithic single modules. Note, however, that the structure of the tandem cells (see Figure 12.14a, for example) is far more complicated than that of the single cells, and there are technological challenges for the tandem devices.

In the large-area perovskite/perovskite or perovskite/CIGSe modules, the tandem-type subcells are formed, and two current matching conditions for the top and bottom cells as well as each subcell need to be established. The thick TCO layer is also required for this type of module and the high-mobility TCO becomes important to gain sufficient J_{sc} particularly for low-E_g bottom cells [135, 136]. Accordingly, technological hurdles for obtaining better efficiencies are expected to be higher for these modules.

1.4.3 Toxicity of Pb and Sn Compounds

High toxicity of a Pb heavy metal is a serious problem that cannot be avoided in Pb-based hybrid perovskite devices. Although the Pb metal content in a hybrid perovskite solar panel ($\sim 1\,\text{m}^2$) is only a few hundred milligrams, its environmental impact could be very large [137–139]. For CdTe solar modules, the hazardous effect of a Cd heavy metal is also taken seriously. However, CdTe is a thermally and chemically stable compound, while hybrid perovskites are unstable and show a very high solubility for water [137, 138]. For the perovskite modules, worst case scenarios are that (i) the breakage of the solar panels under strong storm and flooding leads to the accumulation of hazardous compounds as a result of the transportation by soil water and (ii) firing of the modules causes the emission of toxic fumes, creating a very damaging environment for living species.

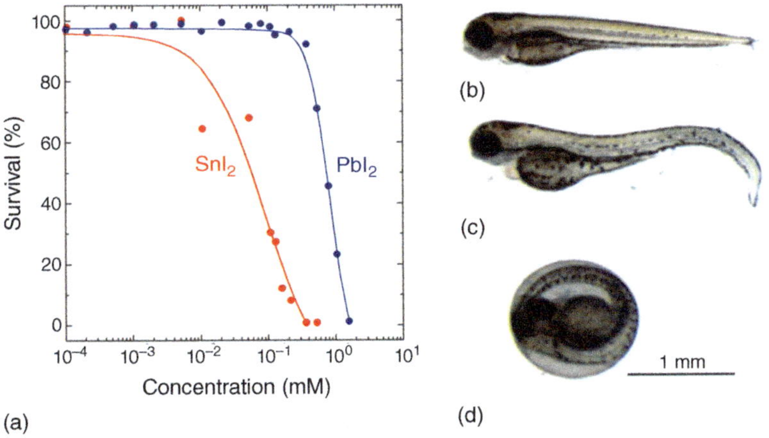

Figure 1.14 (a) Survival rate (day 4) of zebrafishes exposed to aqueous solutions with different concentrations of SnI_2 and PbI_2 toxicant and the photographs of (b) healthy zebrafish, (c) zebrafish exposed to PbI_2 (1×10^{-2} mM), and (d) zebrafish exposed to SnI_2 (1.6×10^{-1} mM). The SnI_2 is more hazardous than PbI_2. *Source*: Babayigit et al. [137].

The toxicity of Pb and Sn is well studied and, when Pb is introduced into a human body, it accumulates in bone with a very long half-life of 20–30 years [138], causing serious health problems. Although Sn-based perovskites are often treated as a more environmentally friendly alternative to Pb-based perovskites, SnI_2 formed by the decomposition of Sn-based perovskites is more lethal than PbI_2 [137, 138]. Figure 1.14 shows the survival rate (day 4) of zebrafishes exposed to aqueous solutions with different concentrations of SnI_2 and PbI_2 toxicant [137]. It can be seen that the survival rate is much lower in the SnI_2 solution than in PbI_2, and both SnI_2 and PbI_2 have serious impacts on the fish growth. The detrimental effect of SnI_2 can be interpreted by the reaction pathways of SnI_2 in water [137]:

$$8SnI_2 + 4H_2O + O_2 \rightarrow 2SnI_4 + 6Sn(OH)I + 2HI \quad (1.1)$$

where SnI_4 further decomposes into

$$SnI_4 + 2H_2O \rightarrow SnO_2 + 4HI \quad (1.2)$$

The harmful influence of SnI_2 largely originates from the formation of hydroiodic acid (HI), which reduces the pH of the solution to a very low level (pH ~ 2 at a 1 mM concentration). In contrast, the reaction of PbI_2 in H_2O is proposed as follows [137]:

$$PbI_2 + H_2O \rightarrow Pb(OH)I + HI \quad (1.3)$$

Thus, the HI formation is more suppressed for PbI_2, with a milder pH of 5 at a 1 mM level. Consequently, environmental health effects are not simply determined by the character of elements but by the nature of chemical compounds.

By considering the above, roof-top application of hybrid perovskite modules is difficult and large-area operations as solar farms are more appropriate. The structural failure of photovoltaic modules could give a significant influence on environment and health; therefore, fail-safe measures for the encapsulation of photovoltaic modules are important. Similar to CdTe solar panels, cost-effective and environmentally friendly 100% recycling programs are also required for hybrid perovskite modules.

1.5 Overview of this Book

This book aims to provide overall understanding for unique hybrid perovskite photovoltaics, covering both the fundamental characteristics of hybrid perovskites (Part I) and the outstanding features of hybrid perovskite photovoltaics (Part II). In this book, after the general introduction (this chapter), the overview of hybrid perovskite solar cells (Chapter 2) is presented. In particular, Chapter 2 describes the short summary for the development and advance of hybrid perovskite solar cells, featuring the key aspects of the perovskite devices.

In Part I of this book (Chapters 3–10), the fundamental properties of hybrid perovskites, which are necessary to interpret the remarkable performance of the perovskite devices, are described. Specifically, in Chapter 3, the crystal structures of various hybrid perovskites and their secondary phase materials are explained. The understanding of the detailed perovskite structures described in Chapter 3 is critical in interpreting the physical properties and stability of hybrid perovskite crystals. Chapter 4 provides descriptions of the optical properties of hybrid perovskites, explaining the strong advantages of hybrid perovskites as light absorbers. Chapter 5 focuses on band and defect properties of hybrid perovskites, determined by DFT. In particular, this chapter emphasizes that hybrid perovskites are direct transition semiconductors and the deep-defect formation is negligible. In Chapter 6, the superior carrier transport properties of hybrid perovskites are described. This chapter provides detailed interpretations for very long carrier diffusion length in hybrid perovskites.

Hybrid perovskite materials show complex ferroelectricity, and their unique ferroelectric characters are described in Chapter 7. The favorable PL properties of hybrid perovskites (Chapter 8) are important characteristics, and the understanding of the PL result is critical for the interpretation of the efficiency loss mechanism in the perovskite solar cells. Chapter 9 describes the physical properties of the perovskite grain boundary. This chapter further reviews the latest developments of the grain-boundary engineering. In Chapter 10, the roles of organic center cations in hybrid perovskite crystals are introduced in the optical point of view.

In Part II of this book (Chapters 11–18), working principles for different types of hybrid perovskite photovoltaics are explained. In this part, the operational principles and important features of hybrid perovskite solar cells are described. Specifically, Chapter 11 shows that hybrid perovskite solar cells have very simple structures and their operation is quite straightforward. The latest approaches for the interface passivation are also introduced in this chapter. In Chapter 12, efficiency limits of single and tandem devices are deduced theoretically, and all the fundamental aspects necessary to understand the theoretical physical limits of the solar cells are explained. With respect to emerging hybrid perovskite devices, mixed-cation perovskites (Chapter 13) and Sn-based perovskites (Chapter 14) are particularly important. In Chapter 13, the authors review the critical variations of hybrid perovskite cell characteristics by A-site cation mixing. The rapid advancement of low-gap Sn-based perovskites and the key factors for the efficiency improvements are explained in detail in Chapter 14.

As mentioned earlier, the stability of hybrid perovskite devices is the most serious problem and this vital issue is treated in Chapter 15. Furthermore, hysteresis that appears in the *J–V* curves of hybrid perovskite cells is a complex phenomenon. The physical origin of this problematic hysteresis character is presented in Chapter 16.

In the final two chapters (Chapters 17 and 18), the basic and latest developments of hybrid perovskite tandem devices are reviewed. In Chapter 17, the tandem devices formed by the combination of hybrid perovskite solar cells with Si and CIGSe devices are introduced. The perovskite/Si tandem technology is developing fast with an unprecedented pace and could become a field of main interests in near future. All perovskite tandem cells with a Pb-based perovskite top cell and an Sn-based perovskite bottom cell also show rapid advancement. In Chapter 18, the fundamental and advanced concepts of all perovskite tandem cells are reviewed. The abovementioned overall descriptions allow us to gain in-depth knowledge for the basic characteristics of hybrid perovskites and to further understand the fundamental and advanced aspects of hybrid perovskite solar cells.

Acknowledgment

The author acknowledges Tatsuya Narikuri of Gifu University for assisting the preparation of this chapter. The author would like to thank the following students of Gifu University for helping the preparation of figures that appear in this chapter: Yoshitsune Kato (Figures 1.1 and 1.7), Yukinori Nishigaki (Figures 1.3, 1.5, and 1.13), Kazuya Shibata (Figures 1.4 and 1.6), Tomoya Kobayashi (Figure 1.9), Ryo Ishihara (Figures 1.10 and 1.14), and Kohei Oiwake (Figure 1.11).

References

1 Green, M.A., Dunlop, E.D., Hohl-Ebinger, J. et al. (2020). *Prog. Photovoltaics Res. Appl.* 28: 629.
2 Kojima, A., Teshima, K., Shirai, Y., and Miyasaka, T. (2009). *J. Am. Chem. Soc.* 131: 6050.
3 Luque, A. and Hegedus, S. (2011). *Handbook of Photovoltaic Science and Engineering*. West Sussex, UK: Wiley.
4 Nakane, A., Fujimoto, S., and Fujiwara, H. (2017). *J. Appl. Phys.* 122: 203101.
5 Kato, Y., Fujimoto, S., Kozawa, M., and Fujiwara, H. (2019). *Phys. Rev. Appl.* 12: 024039.
6 Hara, T., Maekawa, T., Minoura, S. et al. (2014). *Phys. Rev. Appl.* 2: 034012.
7 Correa-Baena, J.-P., Saliba, M., Buonassisi, T. et al. (2017). *Science* 358: 739.

8 Jena, A.K., Kulkarni, A., and Miyasaka, T. (2019). *Chem. Rev.* 119: 3036.
9 Yang, W.S., Park, B.-W., Jung, E.H. et al. (2017). *Science* 356: 1376.
10 Saliba, M., Correa-Baena, J.-P., Wolff, C.M. et al. (2018). *Chem. Mater.* 30: 4193.
11 Shin, S.S., Yeom, E.J., Yang, W.S. et al. (2017). *Science* 356: 167.
12 Min, H., Kim, M., Lee, S.-U. et al. (2019). *Science* 366: 749.
13 Jeon, N.J., Na, H., Jung, E.H. et al. (2018). *Nat. Energy* 3: 682.
14 Chiang, C.-H. and Wu, C.-G. (2018). *ACS Nano* 12: 10355.
15 Saliba, M., Matsui, T., Domanski, K. et al. (2016). *Science* 354: 206.
16 Liu, Z., Krückemeier, L., Krogmeier, B. et al. (2019). *ACS Energy Lett.* 4: 110.
17 Jiang, Q., Chu, Z., Wang, P. et al. (2017). *Adv. Mater.* 29: 1703852.
18 Saliba, M., Matsui, T., Seo, J.-Y. et al. (2016). *Energy Environ. Sci.* 9: 1989.
19 Singh, T. and Miyasaka, T. (2017). *Adv. Energy Mater.* 8: 1700677.
20 Matsui, T., Yamamoto, T., Nishihara, T. et al. (2019). *Adv. Mater.* 31: 1806823.
21 Lin, R., Xiao, K., Qin, Z. et al. (2019). *Nat. Energy* 4: 864.
22 Tong, J., Song, Z., Kim, D.H. et al. (2019). *Science* 364: 475.
23 Docampo, P., Guldin, S., Stefik, M. et al. (2010). *Adv. Funct. Mater.* 20: 1787.
24 Lin, J., Heo, Y.-U., Nattestad, A. et al. (2015). *Sci. Rep.* 4: 5769.
25 Cai, M., Ishida, N., Li, X. et al. (2018). *Joule* 2: 296.
26 Lee, M.M., Teuscher, J., Miyasaka, T. et al. (2012). *Science* 338: 643.
27 Kim, H.-S., Lee, C.-R., Im, J.-H. et al. (2012). *Sci. Rep.* 2: 591.
28 Liu, M., Johnston, M.B., and Snaith, H.J. (2013). *Nature* 501: 395.
29 Green, M.A., Emery, K., Hishikawa, Y. et al. (2013). *Prog. Photovoltaics Res. Appl.* 21: 827.
30 Green, M.A., Emery, K., Hishikawa, Y. et al. (2014). *Prog. Photovoltaics Res. Appl.* 22: 1.
31 Green, M.A., Emery, K., Hishikawa, Y. et al. (2014). *Prog. Photovoltaics Res. Appl.* 22: 701.
32 Green, M.A., Emery, K., Hishikawa, Y. et al. (2015). *Prog. Photovoltaics Res. Appl.* 23: 1.
33 Green, M.A., Emery, K., Hishikawa, Y. et al. (2015). *Prog. Photovoltaics Res. Appl.* 23: 805.
34 Green, M.A., Emery, K., Hishikawa, Y. et al. (2016). *Prog. Photovoltaics Res. Appl.* 24: 3.
35 Green, M.A., Emery, K., Hishikawa, Y. et al. (2016). *Prog. Photovoltaics Res. Appl.* 24: 905.
36 Green, M.A., Emery, K., Hishikawa, Y. et al. (2017). *Prog. Photovoltaics Res. Appl.* 25: 3.
37 Green, M.A., Hishikawa, Y., Warta, W. et al. (2017). *Prog. Photovoltaics Res. Appl.* 25: 668.
38 Green, M.A., Hishikawa, Y., Dunlop, E.D. et al. (2018). *Prog. Photovoltaics Res. Appl.* 26: 3.
39 Green, M.A., Hishikawa, Y., Dunlop, E.D. et al. (2018). *Prog. Photovoltaics Res. Appl.* 26: 427.
40 Green, M.A., Hishikawa, Y., Dunlop, E.D. et al. (2019). *Prog. Photovoltaics Res. Appl.* 27: 3.

41 Green, M.A., Dunlop, E.D., Levi, D.H. et al. (2019). *Prog. Photovoltaics Res. Appl.* 27: 565.
42 Green, M.A., Dunlop, E.D., Hohl-Ebinger, J. et al. (2019). *Prog. Photovoltaics Res. Appl.* 28: 3.
43 Case, C., Beaumont, N., and Kirk, D. (2019). *ACS Energy Lett.* 4: 2760.
44 Bakulin, A.A., Selig, O., Bakker, H.J. et al. (2015). *J. Phys. Chem. Lett.* 6: 3663.
45 Egger, D.A., Rappe, A.M., and Kronik, L. (2016). *Acc. Chem. Res.* 49: 573.
46 Eames, C., Frost, J.M., Barnes, P.R.F. et al. (2015). *Nat. Commun.* 6: 7497.
47 Shirayama, M., Kadowaki, H., Miyadera, T. et al. (2016). *Phys. Rev. Appl.* 5: 014012.
48 Fujiwara, H. and Collins, R.W. (2018). *Spectroscopic Ellipsometry for Photovoltaics: Volume 2: Applications and Optical Data of Solar Cell Materials.* Cham, Switzerland: Springer.
49 Yamada, Y., Nakamura, T., Endo, M. et al. (2014). *Appl. Phys. Express* 7: 032302.
50 Ndione, P.F., Li, Z., and Zhu, K. (2016). *J. Mater. Chem. C* 4: 7775.
51 Zhao, B., Abdi-jalebi, M., Tabachnyk, M. et al. (2017). *Adv. Mater.* 29: 1604744.
52 de Quilettes, D.W., Koch, S., Burke, S. et al. (2016). *ACS Energy Lett.* 1: 438.
53 Metzger, W.K., Repins, I.L., and Contreras, M.A. (2008). *Appl. Phys. Lett.* 93: 022110.
54 Phuong, L.Q., Okano, M., Yamada, Y. et al. (2014). *Appl. Phys. Lett.* 104: 081907.
55 Stranks, S.D., Eperon, G.E., Grancini, G. et al. (2013). *Science* 342: 341.
56 Xing, G., Mathews, N., Sun, S. et al. (2013). *Science* 342: 344.
57 Stolterfoht, M., Wolff, C.M., Márquez, J.A. et al. (2018). *Nat. Energy* 3: 847.
58 Tan, H., Jain, A., Voznyy, O. et al. (2017). *Science* 355: 722.
59 Nakane, A., Tampo, H., Tamakoshi, M. et al. (2016). *J. Appl. Phys.* 120: 064505.
60 Stoumpos, C.C., Malliakas, C.D., and Kanatzidis, M.G. (2013). *Inorg. Chem.* 52: 9019.
61 Dang, Y., Liu, Y., Sun, Y. et al. (2015). *CrystEngComm* 17: 665.
62 Breternitz, J., Lehmann, F., Barnett, S.A. et al. (2020). *Angew. Chem. Int. Ed.* 59: 424.
63 Poglitsch, A. and Weber, D. (1987). *J. Chem. Phys.* 87: 6373.
64 Chen, T., Foley, B.J., Ipek, B. et al. (2015). *Phys. Chem. Chem. Phys.* 17: 31278.
65 Stolterfoht, M., Caprioglio, P., Wolff, C.M. et al. (2019). *Energy Environ. Sci.* 12: 2778.
66 Giorgi, G., Fujisawa, J.-I., Segawa, H., and Yamashita, K. (2013). *J. Phys. Chem. Lett.* 4: 4213.
67 Mosconi, E., Amat, A., Nazeeruddin, M.K. et al. (2013). *J. Phys. Chem. C* 117: 13902.
68 Baikie, T., Fang, Y., Kadro, J.M. et al. (2013). *J. Mater. Chem. A* 1: 5628.
69 Yin, W.-J., Yang, J.-H., Kang, J. et al. (2015). *J. Mater. Chem. A* 3: 8926.
70 Fujiwara, H., Kato, M., Tamakoshi, M. et al. (2018). *Phys. Status Solidi A* 215: 1700730.
71 Yamada, Y., Nakamura, T., Endo, M. et al. (2014). *J. Am. Chem. Soc.* 136: 11610.
72 Yang, Y., Yang, M., Li, Z. et al. (2015). *J. Phys. Chem. Lett.* 6: 4688.
73 Manser, J.S. and Kamat, P.V. (2014). *Nat. Photon.* 8: 737.

74 Yang, Y., Ostrowski, D.P., France, R.M. et al. (2016). *Nat. Photon.* 10: 53.
75 Savenije, T.J., Ponseca, C.S. Jr., Kunneman, L. et al. (2014). *J. Phys. Chem. Lett.* 5: 2189.
76 Wehrenfennig, C., Eperon, G.E., Johnston, M.B. et al. (2014). *Adv. Mater.* 26: 1584.
77 Milot, R.L., Eperon, G.E., Snaith, H.J. et al. (2015). *Adv. Funct. Mater.* 25: 6218.
78 De Wolf, S., Holovsky, J., Moon, S.-J. et al. (2014). *J. Phys. Chem. Lett.* 5: 1035.
79 Nishiwaki, M., Nagaya, K., Kato, M. et al. (2018). *Phys. Rev. Mater.* 2: 085404.
80 Lin, Q., Armin, A., Nagiri, R.C.R. et al. (2015). *Nat. Photonics* 9: 106.
81 Miyata, A., Mitioglu, A., Plochocka, P. et al. (2015). *Nat. Phys.* 11: 582.
82 Sendner, M., Nayak, P.K., Egger, D.A. et al. (2016). *Mater. Horiz.* 3: 613.
83 Dong, Q., Fang, Y., Shao, Y. et al. (2015). *Science* 347: 967.
84 Semonin, O.E., Elbaz, G.A., Straus, D.B. et al. (2016). *J. Phys. Chem. Lett.* 7: 3510.
85 Bi, Y., Hutter, E.M., Fang, Y. et al. (2016). *J. Phys. Chem. Lett.* 7: 923.
86 Shi, D., Adinolfi, V., Comin, R. et al. (2015). *Science* 347: 519.
87 Hutter, E.M., Eperon, G.E., Stranks, S.D., and Savenije, T.J. (2015). *J. Phys. Chem. Lett.* 6: 3082.
88 Chen, Y., Yi, H.T., Wu, X. et al. (2016). *Nat. Commun.* 7: 12253.
89 Saidaminov, M.I., Abdelhady, A.L., Murali, B. et al. (2015). *Nat. Commun.* 6: 7586.
90 Yang, T.-Y., Gregori, G., Pellet, N. et al. (2015). *Angew. Chem.* 127: 8016.
91 Futscher, M.H., Lee, J.M., McGovern, L. et al. (2019). *Mater. Horiz.* 6: 1497.
92 Bertoluzzi, L., Boyd, C.C., Rolston, N. et al. (2020). *Joule* 4: 109.
93 Herz, L.M. (2018). *J. Phys. Chem. Lett.* 9: 6853.
94 Frost, J.M. (2017). *Phys. Rev. B* 96: 195202.
95 Rakita, Y., Cohen, S.R., Kedem, N.K. et al. (2015). *MRS Commun.* 5: 623.
96 Feng, J. (2014). *APL Mater.* 2: 081801.
97 Kawamura, Y., Mashiyama, H., and Hasebe, K. (2002). *J. Phys. Soc. Jpn.* 71: 1694.
98 Onoda-Yamamuro, N., Matsuo, T., and Suga, H. (1990). *J. Phys. Chem. Solids* 51: 1383.
99 Pisoni, A., Jaćimović, J., Barišić, O.S. et al. (2014). *J. Phys. Chem. Lett.* 5: 2488.
100 Weller, M.T., Weber, O.J., Henry, P.F. et al. (2015). *Chem. Commun.* 51: 4180.
101 Xing, G., Mathews, N., Lim, S.S. et al. (2014). *Nat. Mater.* 13: 476.
102 Braly, I.L., de Quilettes, D.W., Pazos-Outón, L.M. et al. (2018). *Nat. Photonics* 12: 355.
103 Wang, Q., Shao, Y., Xie, H. et al. (2014). *Appl. Phys. Lett.* 105: 163508.
104 Hou, Y., Aydin, E., De Bastiani, M. et al. (2020). *Science* 367: 1135.
105 Jošt, M., Kegelmann, L., Korte, L., and Albrecht, S. (2020). *Adv. Energy Mater.* 10: 1904102.
106 Hoke, E.T., Slotcavage, D.J., Dohner, E.R. et al. (2015). *Chem. Sci.* 6: 613.
107 Brennan, M.C., Draguta, S., Kamat, P.V., and Kuno, M. (2018). *ACS Energy Lett.* 3: 204.
108 Jošt, M., Lipovšek, B., Glažar, B. et al. (2020). *Adv. Energy Mater.* 10: 2000454.

109 From product data sheet of "Cheetah HC 72M 390–410 Watt" (Jinko Solar).
110 From product data sheet of "KuMax CS3U-375/380MS" (CanadianSolar).
111 From product data sheet of "ASM-7-PERC-375" (Adani Solar).
112 Mishima, T., Taguchi, M., Sakata, H., and Maruyama, E. (2011). *Sol. Energy Mater. Sol. Cells* 95: 18.
113 From product data sheet of "SF175-S" (Solar Frontier).
114 From product data sheet of "SF170-S" (Solar Frontier).
115 Niu, G., Guo, X., and Wang, L. (2015). *J. Mater. Chem. A* 3: 8970.
116 Berhe, T.A., Su, W.-N., Chen, C.-H. et al. (2016). *Energy Environ. Sci.* 9: 323.
117 Leijtens, T., Eperon, G.E., Noel, N.K. et al. (2015). *Adv. Energy Mater.* 5: 1500963.
118 Wang, D., Wright, M., Elumalai, N.K., and Uddin, A. (2016). *Sol. Energy Mater. Sol. Cells* 147: 255.
119 Wang, R., Mujahid, M., Duan, Y. et al. (2019). *Adv. Funct. Mater.* 29: 1808843.
120 Kim, H.-S., Seo, J.-Y., and Park, N.-G. (2016). *ChemSusChem* 9: 2528.
121 Shirayama, M., Kato, M., Miyadera, T. et al. (2016). *J. Appl. Phys.* 119: 115501.
122 Eperon, G.E., Stranks, S.D., Menelaou, C. et al. (2014). *Energy Environ. Sci.* 7: 982.
123 Lee, J.-W., Kim, D.-H., Kim, H.-S. et al. (2015). *Adv. Energy Mater.* 5: 1501310.
124 Li, Z., Yang, M., Park, J.-S. et al. (2016). *Chem. Mater.* 28: 284.
125 Yi, C., Luo, J., Meloni, S. et al. (2016). *Energy Environ. Sci.* 9: 656.
126 McMeekin, D.P., Sadoughi, G., Rehman, W. et al. (2016). *Science* 351: 151.
127 Chen, P., Bai, Y., Wang, S. et al. (2018). *Adv. Funct. Mater.* 28: 1706923.
128 Cho, K.T., Grancini, G., Lee, Y. et al. (2018). *Energy Environ. Sci.* 11: 952.
129 Jung, E.H., Jeon, N.J., Park, E.Y. et al. (2019). *Nature* 567: 511.
130 Chaudhary, B., Kulkarni, A., Jena, A.K. et al. (2017). *ChemSusChem* 10: 2473.
131 Wang, F., Shimazaki, A., Yang, F. et al. (2017). *J. Phys. Chem. C* 121: 1562.
132 Gu, S., Lin, R., Han, Q. et al. (2020). *Adv. Mater.* 32: 1907392.
133 New Energy and Industrial Technology Development Organization (NEDO) and Panasonic Corporation (2020). Photograph of the large-area module is taken from https://www.nedo.go.jp/english/news/AA5en_100421.html with permission.
134 Li, Z., Klein, T.R., Kim, D.H. et al. (2018). *Nat. Rev. Mater.* 3: 18017.
135 Fuiwara, H. and Collins, R.W. (2018). *Spectroscopic Ellipsometry for Photovoltaics: Volume 1: Fundamental Principles and Solar Cell Characterization*. Cham, Switzerland: Springer.
136 Koida, T. (2017). *Phys. Status Solidi A* 214: 1600464.
137 Babayigit, A., Thanh, D.D., Ethirajan, A. et al. (2016). *Sci. Rep.* 6: 18721.
138 Babayigit, A., Ethirajan, A., Muller, M., and Conings, B. (2016). *Nat. Mater.* 15: 247.
139 Ju, M.-G., Chen, M., Zhou, Y. et al. (2018). *Joule* 2: 1231.

2

Overview of Hybrid Perovskite Solar Cells

Tsutomu Miyasaka and Ajay K. Jena

Toin University of Yokohama, Graduate School of Engineering, 1614 Kurogane-cho, Aoba, Yokohama, Kanagawa 225-8503, Japan

2.1 Introduction

Photovoltaic (PV) devices based on metal halide perovskite semiconductors were first invented by our group in 2006 as an extension of the research on dye-sensitized solar cells, and the first peer review paper was published in 2009 [1]. Similar to the method of fabricating dye-sensitized solar cells, solution coating process is the main technology to prepare photovoltaic active layer (perovskite) and carrier transporting layers on electrode substrates. While having the advantage of a cost-efficient solution processes, perovskite solar cells (PSCs) have evolved as the sole type of organic material-based solar cells that achieve high power conversion efficiency (PCE) (>25%) approaching the highest efficiency Si cells (26%) [2]. Although perovskite materials are generally known as metal oxides or chalcogenides, the composition of perovskites with superior photovoltaic functions is of a family of metal halide-based perovskites, which possess unique chemical and physical properties [3]. Unlike oxide perovskites, halide perovskites are highly ionic crystals characterized as ionic conductors. Here, perovskite crystals composed of lead (Pb) halide frameworks have good optical and physical properties suitable for photovoltaic conversion. Lead halide perovskites and their source materials for crystallization are soluble in polar solvents (alcohol, water, dimethylformamide, etc.). This makes it possible to apply the solution coating as a (printing) process for preparing thin polycrystalline films of perovskites as light absorbers. Owing to the strong light absorption property (high value of absorption coefficient $\sim 10^5\ cm^{-1}$) of halide perovskite semiconductors, photovoltaic cells can be designed with a thin film of light absorber (0.5–1 μm).

The basic structure of lead halide perovskite is $APbX_3$, in which A is a monovalence organic or inorganic cation. The halogen anion (X) is highly electronegative enhancing the ionicity of the perovskite. Its type (I, Br, and Cl) strongly affects the spectral absorption and bandgap of the material. The photovoltaic function of halide perovskites was first discovered with the use of organic cations (A) instead of inorganic alkali cations [1, 3]. Composed of a methylammonium (MA) cation and

Hybrid Perovskite Solar Cells: Characteristics and Operation,
First Edition. Edited by Hiroyuki Fujiwara.

iodine as a halogen, $MAPbI_3$ exhibits excellent photovoltaic properties in the visible light spectrum with a narrow bandgap (approximately 1.55 eV) absorbing up to 800 nm. Various lead halide perovskites have been synthesized by replacing MA and iodine with other organic groups and halides (Br and Cl), respectively. Here, halogen replacement enables continuous tuning of the bandgap due to the ratio of mixed halogens. The bandgap shifts to one of the less energies (optical absorption edge to longer wavelength) in the order of Cl → Br → I. $MAPbI_3$, $MAPbBr_3$, and $MAPbCl_3$ exhibit bandgap energies of 1.55, 2.35, and 3.0 ± 0.1 eV, respectively [4]. In chemical crystallization, ratio of mixed halides, such as the Br–I mixture, can be freely controlled in an attempt to continuously adjust the bandgap. Such synthetic flexibility of compositional design is a rare advantage of halide perovskite semiconductors.

The crystal of $MAPbI_3$ has an isotropic 3-dimensional (3D) structure, which is beneficial for photovoltaic materials that must avoid anisotropic diffusion of photogenerated carriers. Although 3D perovskites are suitable for photovoltaic applications, 1D and 2D structures of perovskites are synthesized by varying the size of the organic cation and their luminescent properties due to photoexcited exciton and quantum confinement effects were studied earlier than those properties with photovoltaic applications. For perovskites of all dimensions, lead halide (PbI_2) that composes its inorganic part is a skeletal structure contributing to the conductivity and photovoltaic properties, while organic cations that modify the crystalline phase and size lattice affect the optical properties of the perovskite. Historically, it is known that PbI_2 itself is a photosensitive halide material that has been used to replace high-cost silver halides in photographic engineering [5]. While PbI_2 has a spectral sensitivity over 500 nm, the organic cation MA drastically enhances the sensitivity over the entire visible light region, up to 800 nm.

PCE of halide perovskite-based achieved an amazing progress over the last 10 years. In 2005, our research group started experiments to crystallize $MAPbI_3$ and $MAPbBr_3$ on a TiO_2 mesoporous layer as the visible light absorber and sensitizer of photoelectrochemical cells. Using $MAPbBr_3$ as a quantum dot-like nanocrystalline absorber, we obtained a PCE of 2.2% in 2006 [6]. $MAPbI_3$ with a narrower bandgap increased the photocurrent density and improved PCE up to 3.8%. This result was reported in the first publication showing a perovskite-based photovoltaic cell in 2009 [1]. In 2012, PCE of the $MAPbI_3$-based photovoltaic cell exceeded 10% when the thickness of the perovskite absorber on the solidification of the cell structure was increased [7]. After that, improvements made to the composition of perovskite and to the quality of charge transport materials resulted in a rapid increase in the PCE of PSCs, reaching the latest record of 25.5% [2], a level almost comparable with the top PCE of crystalline Si solar cells (26.1%). Such rapid progress has not been observed for any other type of photovoltaic material and is unprecedented in the R&D of photovoltaics (PVs). Figure 2.1 shows the progress of PCE in the research of perovskite PV. Here, the advantages gifted to halide perovskites are (i) intrinsically high efficiency in conversion between photonic and electronic quanta and energies, (ii) wavelength (bandgap) tunability based on changing the chemical composition of perovskite by chemical synthesis, (iii) thin (<1 μm) film of perovskite as the active layer, and (iv) low cost of perovskite photovoltaic films (<5 USD/m^2) and

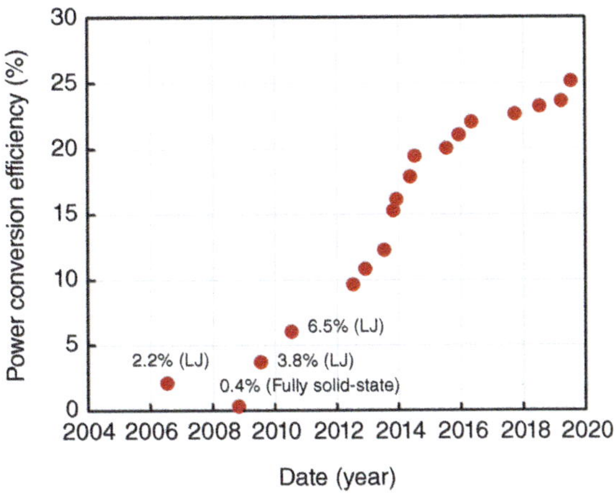

Figure 2.1 Progress of power conversion efficiency (PCE) of halide perovskite-based photovoltaic cells. LJ indicates liquid-junction photoelectrochemical cell, in the early days of research.

manufacturing processes. There is no other photovoltaic device fulfilling all these aspects and enabling high device performance.

The highly interdisciplinary nature of the research enabled such rapid progress of PSCs. While solution-based crystallization of perovskite materials involves a great deal of knowledge and skill in chemistry, optical and electrical characterizations of solid-state crystals (semiconductors) are based on theories of physics. As a result of this interdisciplinary science, many applications beyond photovoltaics have been found for halide perovskites. One such function is as a high-performance light emitter. We observed intense green light emissions from spin-coated thin films of $MAPbBr_3$ under exposure to ambient air at room temperature [8]. Such efficient photoluminescence (PL) of perovskite films is a subject of intense study in fundamental science to assess defect-induced carrier recombination. At the same time, the enhancement of photoluminescence performance leads to the fabrication of wavelength-tunable light-emitting diodes and perovskite-based lasers.

The fact that halide perovskite has excellent properties does not mean that perovskite PV and optoelectronic devices are ready for industrial use. One of the remaining important issues is ensuring high stability and durability of the perovskite and surrounding carrier transporting materials. To overcome this, considerable efforts are being made in both fundamental research and engineering. In fact, this is the main direction of ongoing research on perovskite-based devices. The intrinsic stability of halide perovskites is assumed to be due to the nature of their ionic migration, which is a characteristic of metal halide materials as soft ionic semiconductors. Such ionic conductivity is responsible for the peculiar behavior of carrier transports in the perovskites. For example, ionic migration is assumed to cause hysteretic behavior in current–voltage characteristics of the device. Although ion migration is associated with the stability issue, it plays an important role in screening charges (positive and negative) against Coulombic repulsion to stabilize free carriers for long-range diffusion (>1 μm). Therefore, the question of how to control the behavior of mobile ions by compositional and morphological improvements has become the main focus of study to solve the stability issue.

2.2 Historical Backgrounds of Halide Perovskite Photovoltaics

Research on perovskite materials dates back to 1839 when the Prussian mineralogist and crystallographer, Gustav Rose, discovered the novel material [9, 10]. The perovskite mineral was discovered in the Ural Mountains of Russia by mineral collector August Alexander Kämmerer, who sent the sample to Gustav Rose in Germany for analysis. It was a black mineral composed of calcium titanate ($CaTiO_3$). Kämmerer requested the new mineral be named in honor of the decorated Napoleonic Wars veteran Count Lev A. Perovskiy (1792–1856). Additional information as an interesting story on the discovery of perovskite and collaborations between Russian and German mineralogists has been published as an essay [11]. In 1940s, a high dielectric constant was discovered for an oxide perovskite of $BaTiO_3$. As a ferroelectric ceramic, $BaTiO_3$ has been to date leading the industry of ferroelectrics such as capacitors, superconductors, piezoelectric devices, proton conductors, fuel cells, and memory storage devices [12]. However, oxide-type perovskites have large bandgaps (>2.5 eV) that lack light harvesting ability and exhibit low efficiency in the collection of free charges. Therefore, metal oxide-based perovskites do not exhibit good semiconducting properties suitable for photovoltaic absorbers.

Perovskite generally represents a type of crystal structure with the chemical formula ABX_3, in which A and B are alkali and metal cations, respectively, and X is an anion (Figure 2.2). In an ideal cubic structure, the B cation has a sixfold coordination, surrounded by an octahedron of anions, and the A cation is in 12-fold cuboctahedral configuration. The cubic unit cell of such compounds consists of A cations at the cube corner positions, B sitting at the body-center position, and X anions occupying the face-centered positions. Before halide perovskites attracted attention as photovoltaic semiconductors, perovskites were generally known as inorganic metal oxides (A = alkali cation, X = oxygen).

The class of halide perovskites has halide anions in place of oxide anions (ABX_3; A = monovalent cation, B = divalent metal cation, X = halogen anion) and demonstrates the semiconducting properties and strong visible light absorption desired for photovoltaic applications. Discovery of such halide perovskites dates back to the 1890s. In 1893, Wells et al. conducted a comprehensive study on the synthesis of lead halide crystals from solutions including cesium (Cs)-based $CsPbX_3$ (X = Cl, Br, and I) [13]. In 1957, more than 60 years after the Well's study, Danish researcher

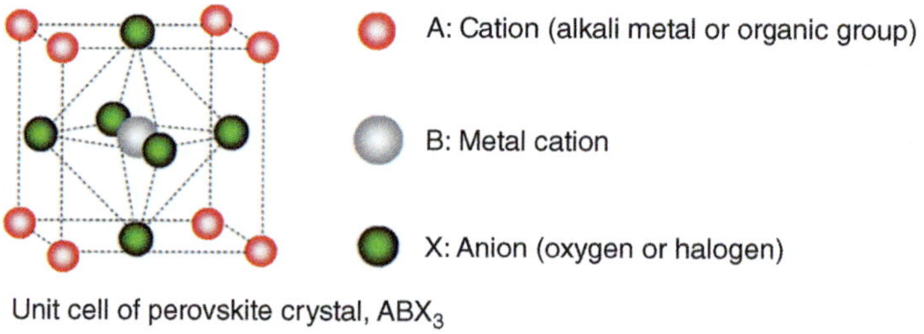

Figure 2.2 General crystal structure of perovskite, ABX_3.

Møller found that $CsPbCl_3$ and $CsPbBr_3$ have a perovskite structure [14, 15], showing a tetragonally distorted structure. Later, Weber found that organic cation ($CH_3NH_3^+$) replaces Cs^+ to form $CH_3NH_3MX_3$ (M = Pb [16] and Sn [17], X = I and Br) and reported the first crystallographic study on organic–inorganic hybrid lead halide perovskites. Toward the end of the last century, the synthesis of a large variety of halide perovskites composed of small and large organic cations was conducted by David Mitzi [18–20]. Mitzi focused his studies on the physical properties of perovskite materials of low-dimensional structures such as two-dimensional (2D) perovskites bearing a large organic group [19]. Based on this study, in the late 1990s, Prof. Kohei Sanui was conducting a project through a Japanese national research program (JST-CREST). This project dealt with self-organized quantum confinement structures using the above perovskites and conducted detailed investigation on the optical properties of 2D [21, 22] and 3D crystals [23, 24]. Although research opened applications of low-dimensional perovskite crystals to non-linear optics and electroluminescence (LED) by utilizing sharp monochromatic optical absorption and luminescence [25, 26], at that time, there was no awareness that these materials could be employed for the utilization of solar energy, which requires a wide spectral absorption to harvest sunlight rather than having a sharp absorption and emission.

During 2005–2006, while we continued putting our efforts toward further development of dye-sensitized solar cells (DSSCs) [27, 28], we explored the use of organic–inorganic lead halide perovskite as an absorber replacing the organic dye in DSSCs. This research began in 2005 as a collaboration between Toin University of Yokohama (TUY) and Tokyo Polytechnic University (TPU) when Akihiro Kojima, then a graduate student at TPU visited Miyasaka's laboratory to examine the ability of $CH_3NH_3PbX_3$ to sensitize TiO_2 electrodes in a photoelectrochemical cell using a Li halide solution as reducing electrolyte. In 2006, we obtained preliminary results which demonstrated the visible light sensitization of TiO_2 with deposited nanocrystals of the perovskite [6]. Assuming that the perovskite would function as a quantum dot-like sensitizer, deposition of the perovskite was done by spin-coating with the precursor solution, in which the loading amount of the perovskite was adjusted so as to obtain the thinnest layer of nanocrystalline perovskite, covering a large surface area of a thick TiO_2 layer ($\sim 10\,\mu m$), similar to the manner of DSSC. This architecture was significantly different from the present PSC that uses a thin TiO_2 film ($<0.1\,\mu m$) as an electron collector (or uses no mesoporous film). After optimizing the cell structure, we fabricated cells that yielded PCEs barely reaching 3.8%. Figure 2.3 shows the performance of perovskite-based photovoltaic cells published in 2009 [1]. Having also realized the necessity for a solid-state cell with a solid-state hole transport material, we attempted to prepare devices by employing a carbon/conductive polymer composite and obtained a low PCE value (<1%). This was the first example of a full-solid-state perovskite photovoltaic cell, which we reported in an international conference (electrochemical society [ECS]) in 2008 [29, 30]. We later realized that such poor performance was apparently due to the significantly low loading amount of perovskite, which was also confirmed by Nam Gyu Park and coworkers who improved our method to obtain a higher PCE by simply increasing the perovskite loading [31].

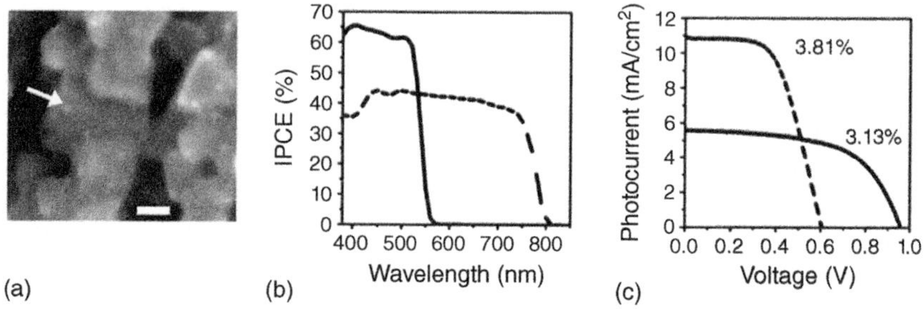

Figure 2.3 (a) SEM image (the scale bar shows 10 nm) of particles (shown by an arrow) of nanocrystalline $CH_3NH_3PbBr_3$ deposited on the TiO_2 surface. (b) IPCE action spectra for photoelectrochemical cells using $CH_3NH_3PbBr_3/TiO_2$ (solid line) and $CH_3NH_3PbI_3/TiO_2$ (dashed line) photoanodes with a liquid electrolyte, 0.4 M LiBr and 0.04 M Br_2 dissolved in acetonitrile for the former, and 0.15 M LiI and 0.075 M I_2 dissolved in methoxyacetonitrile for the latter photoanode. (c) Photocurrent–voltage characteristics for cells using $CH_3NH_3PbBr_3/TiO_2$ (solid line, PCE = 3.13%) and $CH_3NH_3PbI_3/TiO_2$ (dashed line, PCE = 3.81%) under 100 mW/cm^2, air mass (AM) 1.5 irradiation. *Source*: Kojima et al. [1].

In 2011, we started a collaborative work with Henry Snaith (H.S.) of Oxford University for creating solid-state perovskite cells because he had been working on organic hole-transporting materials (HTMs) for the purpose to solidify the structure of DSSC [32]. THM was spiro-OMeTAD (2,2′,7,7′-tetrakis(*N,N*-dimethoxyphenylamine)-9,9′-spirobifluorene), which was spin-coated on the perovskite layer. This method finally led to realize the boost of a PCE of up to 10.9% [7]. A surprising feature of the device was the use of an insulating mesoporous Al_2O_3 as a scaffold for $CH_3NH_3PbI_3$ crystals. A cell using Al_2O_3 exhibited higher voltage and PCE than those using TiO_2 as scaffold. Indeed, it was an indication of the long diffusion length of carriers in perovskite, which was also indicative from our observation of intense green light photoluminescence (PL) from bromide perovskite ($CH_3NH_3PbBr_3$) coated on mesoporous Al_2O_3 [8]. The long carrier diffusion length of the lead halide perovskite, which has been later confirmed by opto-physical analysis [33] and carrier mobility [34] characterizations, has brought radical changes in our understanding of the functioning of PSCs, distinguishing the new technology from any other organic and hybrid material-based solar cells. A considerable number of studies, both in chemistry and physics, have been conducted toward improving the quality of solution-processed perovskite crystalline films and their heterojunction interfaces by reducing the defect density at grain boundaries and interfaces. These monumental efforts have led to the realization of high PCEs of over 25% [35, 36] and a current record PCE of 25.5% [2].

2.3 Semiconductor Properties of Organo Lead Halide Perovskites

The solid-state physics of halide perovskite semiconductors has been investigated in detail regarding their semiconducting properties, carrier transport mechanisms, and theories of defect-related charge recombination. Photovoltaic applications have

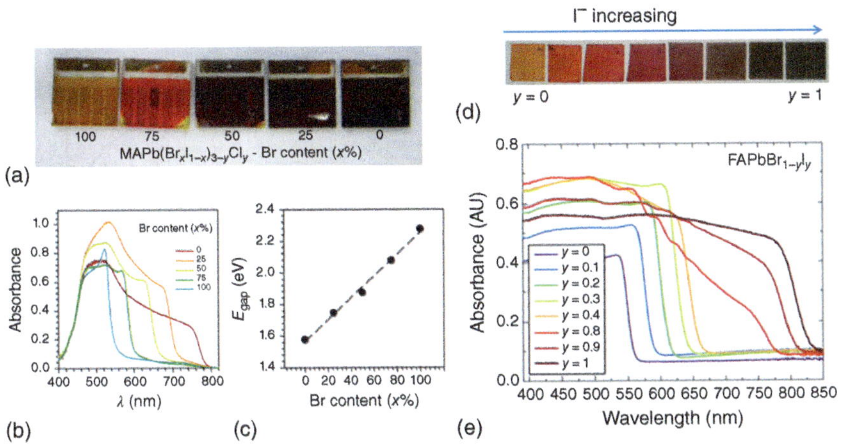

Figure 2.4 (a) Picture of perovskite devices bearing methylammonium lead bromo-iodide perovskite, $CH_3NH_3Pb(Br_xI_{1-x})_{3-y}Cl_y$, ($0 \leq x \leq 1$), films with different Br/I molar ratios on mesoporous TiO_2 substrates. (b) Absorption spectra of the perovskite films and (c) their bandgap energies. *Source*: (a–c) Suarez et al. [37]. (d) Photographs of formamidinium lead bromo-iodide perovskite, $HC(NH_2)_2PbBr_{1-y}I_y$, films with y increasing from 0 to 1 (left to right) and (e) their absorption spectra. *Source*: (d, e) Eperon et al. [38].

been mainly focused on 3D perovskite materials that perform isotropic carrier transport advantageous for power generation. Unlike conventional semiconductors, halide perovskites are ionic crystals and the semiconducting nature of halide perovskites allows the easy tuning of the bandgap and optical absorption by varying the halide ions. Figure 2.4 shows the wavelength tuning for the I and Br mixture of MA lead halide ($CH_3NH_3Pb(Br_xI_{1-x})_{3-y}Cl_y$, $0 \leq x \leq 1$) [37] and formamidinium (FA) lead halide ($HC(NH_2)_2PbBr_{1-y}I_y$, y = 0–1) [38]. The absorption edge wavelength corresponding to the bandgap energy changes proportionally with the content of bromide. When iodide is added to the perovskite structure, $FAPbBr_{1-y}I_y$ (Figure 2.4e) exhibits a continuous red shift from the edge wavelength of ~550 nm for pure bromide ($FAPbBr_3$) to ~830 nm for pure iodide ($FAPbI_3$). A constant emission peak shift similar to that of absorption is also observed in the photoluminescence spectrum. These optical properties corroborate that mixed halide perovskites can form good solid solutions as ionic crystals.

Strong bandgap absorption at the edge wavelength and broad flat absorption covering visible light wavelengths indicates that halide perovskites work as excellent absorbers for sunlight. MA lead halide iodide ($MAPbI_3$), which is a standard perovskite absorber in photovoltaic cells, can be characterized as a rare intrinsic semiconductor [39] exhibiting the excellent mobility of both photogenerated electrons and holes. Density functional theory (DFT) and first principles theory-based calculations have corroborated the superior photophysical properties of $MAPbI_3$, which has long diffusion lengths for free carries. Figure 2.5 shows the lattice structure of MA halide perovskite, $MAPbX_3$ (X = I, Br, and Cl), and the band structures of $MAPbI_3$ based on studies by Kondo et al. [41] and Brivio et al. [40].

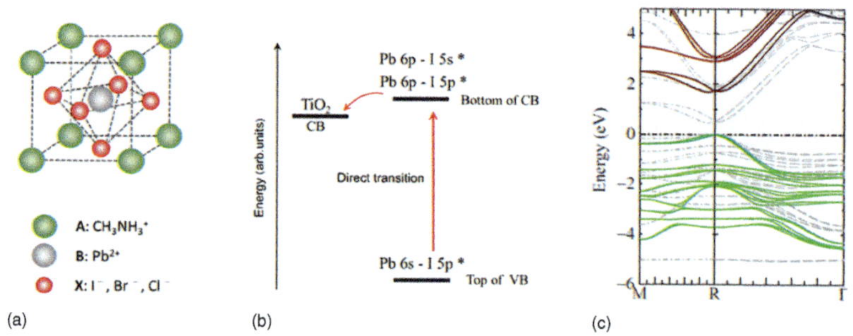

Figure 2.5 (a) Structure of organo-metal halide perovskite crystal. (b) Bandgap structure and energy levels of $CH_3NH_3PbI_3$. (c) Electronic structure of $CH_3NH_3PbI_3$ based on the quasiparticle self-consistent GW approximation. Zero denotes the valence band maximum. Green and red solid lines depict bands of I 5p and Pb 6p, respectively, and light-gray dashed lines show corresponding bands in local-density approximation (LDA). *Source:* Brivio et al. [40].

The valence band (VB) of $MAPbI_3$ consists of approximately 70% of I 5p orbitals and 25% of Pb $6s^2$ orbital (lone pair), whereas the conduction band (CB) consists of a mixture of Pb 6p and other orbitals. Here, VB orbitals have a strong coupling between Pb lone-pair $6s^2$ and I 5p orbitals [42]. This structure is opposite to that of GaAs, in which VB and CB are formed by p and s orbitals, respectively. Here, an important feature of the electronic structure is that both CB and VB consist of antibonding orbitals.

The high symmetry and direct bandgap of $MAPbI_3$ and the p–p electronic transitions from VB to CB, enabled by the Pb s orbital lone pair, contribute to the high optical absorption coefficients of this material ($10^5\,cm^{-1}$) [43]. The unique defect properties of perovskite are attributed to strong Pb s–I p antibonding coupling, weak Pb p–I p coupling, and ionic characteristics [42]. Weak antibonding coupling between Pb p and I p orbitals fixes the conduction band minimum (CBM) close to the Pb p orbital, and strong Pb s–I p antibonding coupling lowers the valence band maximum (VBM) close to the I p orbital. Therefore, for vacancies (defects) formed by the removal of I^-, the defect state lies between the Pb p atomic orbital level and CBM level, and for Pb^{2+} vacancies, the defect state is formed between the I p and VBM levels. Hence, unlike other ionic semiconductors, where localized nonbonding orbitals surrounding the ion vacancies form trap states deep within the bandgap, such defects in perovskites ($MAPbI_3$) generate trap states that either reside within the bands (VB or CB) or exist as shallow traps near the CB and VB. Carriers trapped in shallow defects can be easily detrapped and contribute to current generation. This explains the rare nature of perovskite semiconductors attributed with "defect tolerance". The defect tolerant nature of perovskite is reflected by large carrier diffusion lengths as measured by photoluminescence (PL) lifetime, which are in the range of 1 μm (polycrystalline film) [33] to over 100 μm (single crystal) [34]. Furthermore, as $MAPbI_3$ is an intrinsic semiconductor, the mobility of its carriers is ambipolar, exhibiting a similar effective mass for both electrons and holes (0.23–0.29) [44]. Similar to inorganic photovoltaic semiconductors such as Si and GaAs, photogenerated carriers in $MAPbI_3$ behave as free carriers and these carriers can migrate without recombination for as long as several hundreds of

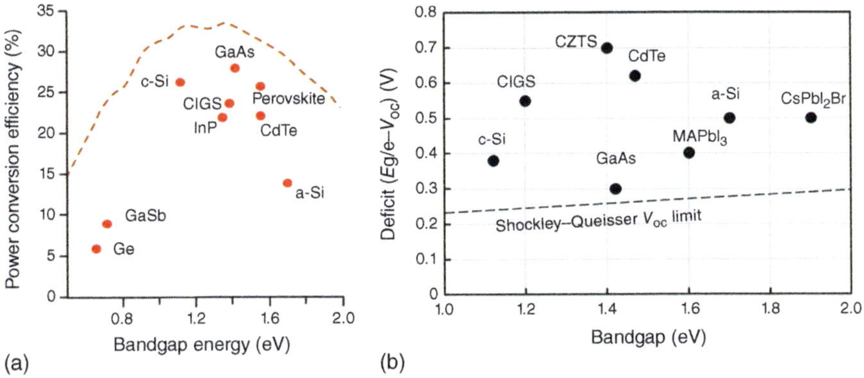

Figure 2.6 Power conversion efficiencies of solar cell as a function of bandgap energy (a) with Shockley–Queisser limit values in dotted line. Average values of V_{OC} deficit (E_g/e–V_{OC}) for different kinds of solar cells (b).

nanoseconds, as shown in PL measurements [33, 45, 46]. It is also considered that the high ionic density in halide perovskites helps to suppress charge recombination by a charge-screening effect against Coulombic interactions.

In summary, the important factors that support superior performance and high efficiency of PSCs are (i) high optical absorption coefficient (10^5 cm^{-1}) that allows the use of thin films (<0.5 μm), (ii) long carrier diffusion length and suppressed recombination (defect tolerance), and (iii) well-balanced charge transfer. In photovoltaic applications, the defect tolerant nature of halide perovskite is especially important and leads to the generation of high voltage. The high PCE of PSCs is a result of high photovoltage (1.1–1.2 V) output rather than the amplitude of the photocurrent [47]. The open-circuit voltage (V_{OC}) of all types of existing solar cells undergoes a large thermal loss from its bandgap energy (eV). Such a voltage loss (V_{OC} deficit) often constitutes more than one-third of the bandgap energy (E_g). Perovskite solar cells of high efficiency usually generate V_{OC} exceeding 1.1 V or in the best cells over 1.2 V (1.26 V obtainable with $MAPbI_3$ [48]) with respect to their E_g of 1.55–1.6 eV. Figure 2.6 shows the average values of the V_{OC} deficit (E_g/e–V_{OC}) for different types of solar cells. The lowest value of the deficit is obtained for a single crystal of GaAs which, as a single-junction solar cell, is known to present the highest PCE value of 27.8%, exceeding the top PCE of crystal silicon (26.1%). With an E_g of 1.42 eV, the GaAs cell is capable of V_{OC} up to 1.12 V. This small deficit, 0.3 eV, is close to the Shockley–Queisser theoretical limit [49, 50]. Compared with GaAs, semiconductors of perovskites and their solar cells are also superior in reducing the V_{OC} deficit to the range of 0.35–0.45 V.

2.4 Working Principle of Perovskite Photovoltaics

Although mesoporous metal oxides such as TiO_2 and Al_2O_3 serve as a scaffold for holding perovskite nanocrystals, the perovskite photovoltaic device can work even without a mesoporous scaffold layer (planar heterojunction with compact

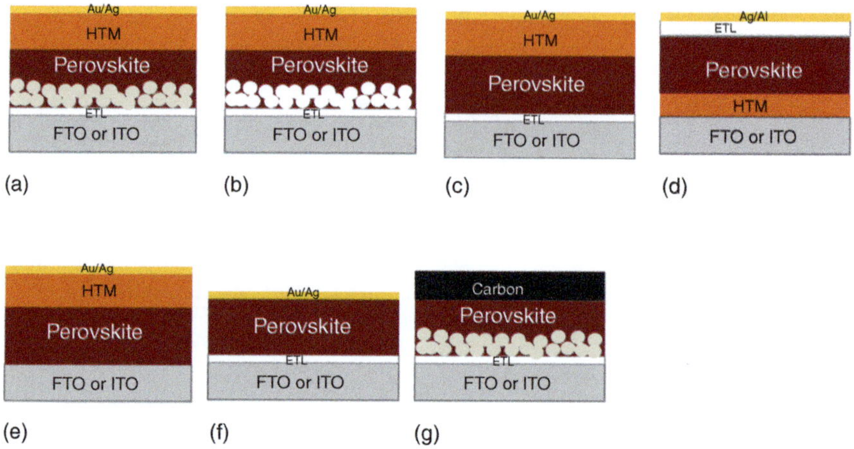

Figure 2.7 Variety of layered architectures for perovskite solar cells. (a) Mesoscopic n–i–p with semiconductor metal oxide. (b) Mesoscopic n–i–p with insulating metal oxide scaffold. (c) Planar n–i–p. (d) Planar n–i–p (inverted). (e) ETL-free. (f) HTM-free. (g) HTM-free carbon electrode. *Source:* Jena et al. [3].

TiO_2 layer for collecting electrons). This makes it clear that the PSCs work more like solid-state p–n junction solar cells in which perovskite works as an intrinsic semiconductor (i) sandwiched between two selective (p and n) contacts. In the n–i–p structure, typically TiO_2 is used as an n-type electron transport material (ETM) and a HTM such as spiro-OMeTAD is used as a p-type contact. Here, a typical device structure using transparent conductive electrodes such as F-doped tin oxide (FTO) or indium tin oxide (ITO) comprises a junction of FTO(ITO)/TiO_2/perovskite/spiro-OMeTAD/metal. In a p-i-n type structure, which is otherwise known as an inverted structure, perovskite is sandwiched by a p-type material (e.g. conductive polymers such as PEDOT-PSS) at the bottom and an n-type layer (e.g. fullerene derivative such as PCBM) at the top, forming a junction of FTO(ITO)/PEDOT-PSS/perovskite/PCBM/metal. Some typical architectures of PSCs are illustrated in Figure 2.7. The perovskite absorbs light to generate electrons and holes, and electrons are selectively collected by the n-type ETM layer, while the holes are collected by the p-type HTM layer. Electrons flow through the external circuit to reach the p-type layer to combine with the holes. Figure 2.8 shows the band diagram of a typical metal oxide ETM-based perovskite device (FTO/TiO_2/perovskite/spiro-OMTAD/Au).

An important factor that leads to the high PCE of PSCs is the open-circuit voltage (V_{OC}). The V_{OC} of some perovskite cells (1.18–1.26 V) [48, 51] approaches the Shockley–Queisser theoretical limit of V_{OC} (~1.32 V) based on the bandgap of the perovskite semiconductors (1.6 eV for $MAPbI_3$) [52, 53]. In photovoltaic cells, V_{OC} is generally determined by the energy gap between the Fermi levels of selective contacts, namely, p-type and n-type conductive layers. For the typical junction structure in Figure 2.8, these contacts are given by spiro-OMeTAD as the HTM and by TiO_2 as ETM, their energy gap corresponds to the difference

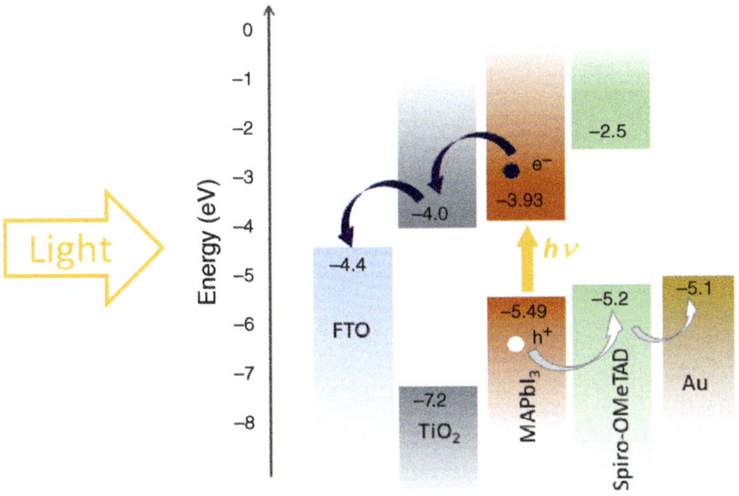

Figure 2.8 Energy diagram of a typical perovskite solar cell using $MAPbI_3$ as the perovskite absorber, TiO_2 as the ETM, and spiro-OMeTAD as the HTM. FTO and Au are front and back contacts, respectively. *Source*: Jena et al. [3].

between the highest occupied molecular orbital (HOMO) level of HTM and the CB level of TiO_2. Although FTO and Au as conductive electrodes have their own work functions, their energy gaps have little influence on V_{OC} because of their low charge selectivity. Furthermore, the CB level of the TiO_2 ETM tends to have less of an effect on controlling the energy gap that determines the maximum value of V_{OC}. This tendency is apparent when TiO_2 is replaced with SnO_2. Despite the low CB level of SnO_2 (approximately 0.25 eV lower than that of TiO_2), comparably high V_{OC} (>1.0 V) can be obtained using both ETMs [54, 55]. This fact may indicate that mesoporous metal oxides (TiO_2 and SnO_2) mainly function as scaffolding for perovskite crystals, and the material that primarily serves as an ETM is a thin compact oxide layer (general thickness <10 nm) that covers the surface of the FTO below the mesoporous layer. However, such a compact layer is considered to be too thin to exhibit the semiconductor property, although it is capable of rectifying electron transfer at the interface by blocking hole injection. As a result, the energy gap that relates to the value of V_{OC} is assumed to be the difference between the CB level of the perovskite absorber and the HOMO level of the HTM.

Theoretical modeling of the electronic structures of $MAPbI_3$ shows that the defect formation influencing carrier recombination is limited to energetically shallow positions near the CB and VB levels rather than deep defects in the bandgap that strongly trap carriers. Material simulation based on first principles theory suggests that such a superior structure is unique to lead iodide-based compositions [42] and is difficult to realize for other metal halide semiconductors. At the time of this review, the only non-lead halide perovskites able to increase PCE beyond 10% are tin (Sn)-based perovskites, which have electronic structures similar to lead halide perovskites and absorb visible light and near-infrared light. Unlike the single crystals of perovskite, solution-processed thin films of perovskites are composed of aggregates of nanocrystalline perovskite particles with dispersed grain size and the large number of defects formed on the grain surface and boundaries substantially influence the photovoltaic properties. Hence, it is essential to minimize the defect density at the grain surface and boundaries by chemically modifying the molecular structure of perovskite at the contact interface. Chemical engineering for this purpose, such as increasing the grain size (decreasing the boundary area), improving the uniformity of film morphology, and passivating the defects using additives, not only enhances the photovoltaic performance but also improves the stability of the perovskite films.

2.5 Compositional Design of the Halide Perovskite Absorbers

Incorporation of different cations in the A site and different halides in the B site influences on both efficiency and stability. A certain mix of cations at the A site and halides at the B site demonstrated superiority over single cation/halide perovskites. The most popular organic cation MA has the issue of thermal stability because of physical evaporation of this small molecule at temperatures above 120° C [56]. In this aspect, the current high-stability development of perovskite PV cells tends to choose MA-free compositions [55] and/or all inorganic compositions [57]. Based on the ionic size and geometrical tolerance factor (τ), which is an empirical index widely used for predicting perovskite crystal structure (see Section 2.6), combinations of different cations such as MA, FA, Cs, Rb, and anions (I, Br, Cl) have been explored. Out of all possible combinations [58], the mixed perovskite that has become most popular in recent years is $(MA/FA/Cs)Pb(I/Br)_3$, which is known as triple cation perovskite. The quadruple cation-based perovskite, which includes Rb (i.e. $(MA/FA/Cs/Rb)\ Pb(I/Br)_3$) as the fourth cation, has also gained interest owing to its high cell efficiency and stability. Here, small inorganic cations such as Cs and Rb are expected to contract the lattice and thus increase the bandgap energy (E_g), while large organic cations such as FA^+ expand the lattice and decrease the E_g. FA^+, having a larger ionic radius ($r = 0.253$ nm) than MA^+ ($r = 0.217$ nm), expands the crystal a small amount, resulting in decreased Pb—I bond distance, which eventually lowers E_g. Pure $FAPbI_3$ shows an E_g of 1.47 eV, while that for pure $MAPbI_3$ is 1.55 eV. With an extended absorption edge (>800 nm), $FAPbI_3$ cells show a larger short-circuit photocurrent (J_{sc}) that leads to a PCE of over 20% [59]. MA and FA affect the intrinsic stability of perovskites. $MAPbI_3$ is fairly stable in its crystal structure temperatures below 120 °C. However, $FAPbI_3$ readily crystallizes into a photo-inactive phase (δ-$FAPbI_3$) at room temperature, while this phase is transformed to the photoactive black phase (α-$FAPbI_3$) at high temperatures (125–165 °C) [60]. Such phase instability often affects the photovoltaic characteristics by reducing the FF of $FAPbI_3$. Despite the phase instability, the thermal stability of $FAPbI_3$ is significantly greater than that of $MAPbI_3$ [56]. $FAPbI_3$, when heated at 150 °C for hours, does not change color, whereas $MAPbI_3$ becomes yellow (due to the formation of PbI_2) at 150 °C (or even lower) for just 30 minutes [38]. Such thermal stability is attributable to the increased tolerance factor and stronger interaction of FA with iodide ions. The partial replacement of MA with FA also significantly improves the thermal stability of $MAPbI_3$. Substitution of a certain amount of MA with Cs also improves efficiency and thermal stability simultaneously. As reported by Niu et al. [61], 9 mol% of Cs in $Cs_xMA_{1-x}PbI_3$ ($x = 0.09$) showed better performance and thermal stability than pristine $MAPbI_3$. Cs inclusion also enables stabilization of $FAPbI_3$. Small and large tolerance factors, respectively, for $CsPbI_3$ ($\tau = 0.85$) and $FAPbI_3$ ($\tau = 0.98$) stabilize the perovskites in orthorhombic and hexagonal structures, respectively, both of which are yellow due to the formation of PbI_2 and not active in photovoltaic function. However, an effective tolerance factor can be tuned to form a cubic structure ($0.9 \leq \tau \geq 1$) by alloying $CsPbI_3$ with $FAPbI_3$ ($FA_xCs_{1-x}PbI_3$) over a range of proportions. Structural/phase stability and the cell performance of $FAPbI_3$ are also improved by mixing 10–20% of Cs. For example, $FAPbI_3$ perovskite thin films, when exposed to a humid environment for 15 days, show continuous degradation, while the addition of 15% of Cs to the film ($FA_{0.85}Cs_{0.15}PbI_3$) sustains performance of the device without reduction of efficiency [62]. These studies led to develop high efficiency and stable cells by using MA-free CsFA cation-based perovskites as hybrid absorbers. Recently, Rb-doped $CsFAPbI_3$ perovskites were used to fabricate highly durable PSCs (PCE > 20.3%) which can function for 1000 hours under continuous sunlight exposure [55].

2.6 Strategy for Stabilizing Perovskite Solar Cells

The performance of the PSCs is reaching the theoretical limit (SQ limit) in terms of the fill factor and output voltage [35, 36]. However, these high PCE record values are not compatible with sufficiently high stability of the device in most cases. This is different from the case of commercialized high-performance devices using inorganic photovoltaic (PV) semiconductors such as Si, GaAs, and CdTe, which can sustain high stability compatible with high efficiency. The instability of halide perovskite materials under high temperature and continuous irradiation is ascribed to the ionic mobility intrinsic to the ionic semiconductors in which highly diffusible halide anions are largely responsible for the instability. Ionic diffusion is also assumed to cause anomalous hysteretic behavior of the photocurrent in I–V performance [63, 64], which can be enhanced in the presence of structural defects or voids at the interface of the perovskite and charge transport layers [65, 66]. Despite these drawbacks, the strong ionic character of the metal halide structure contributes to stabilizing free carriers via the charge screening effect, which is assumed to enable excellent PV performance of the halide perovskite. Therefore, stability and high performance often have a trade-off relationship. To overcome this dilemma, compositional engineering should be directed to not only create the perovskite composition that can reduce the ion diffusion but also to modify the chemical structure of heterojunction interfaces where mobile ions can diffuse beyond the

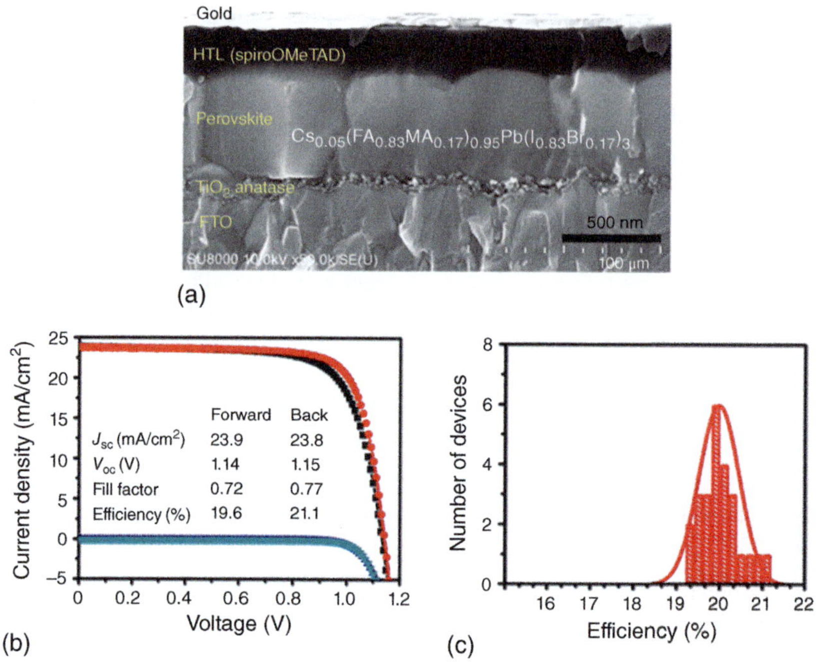

Figure 2.9 Cross-sectional SEM image (a), $J-V$ curves (b), and PCE histogram plot (c) of triple cation perovskite solar cells fabricated in ambient air under controlled humidity (15–25%). *Source*: Based on Singh and Miyasaka [67].

interfaces to interact with the adjacent layers. Simultaneous improvement of efficiency and stability has also been achieved by improving the morphology and physical quality of solution-processed perovskite films. While the use of large grains and high crystallinity of perovskite enable high efficiency, large grains and a dense uniform morphology are advantageous for improving the stability. Figure 2.9 shows an example of a high-efficiency (>21%) triple cation PSC fabricated under ambient conditions in our laboratory [67]. This large-grain perovskite film maintains its high PCE for storage of the cell exposed to ambient air for more than five months.

The triple-cation recipe was then followed by a quadruple-cation mixed perovskite that included Rb as the fourth cation. With the mixed perovskite composition (5% of Rb, bandgap of 1.63 eV), Saliba et al. achieved stabilized efficiencies of up to 21.6% and the polymer-coated cells maintained 95% of their initial performance during 500 hours of operation at the maximum power point at 85 °C [68]. In comparison to Cs/FA/MA triple cation devices, Rb-based quadruple-cation perovskite cells show greater photostability, which can be attributed to reduced recombination and fewer defects [69].

Practical durability issues with perovskite include both intrinsic and extrinsic stability. Here, extrinsic stability is affected by environmental stresses such as heat, light, humidity, and oxygen. The intrinsic structural stability of perovskites can be primarily determined by the Goldschmidt tolerance factor (τ) of the crystal, where $\tau = \frac{rA+rX}{\sqrt{2}\ (rB+rX)}$ and r is the ionic radius. The value of τ varies with the size of the ions

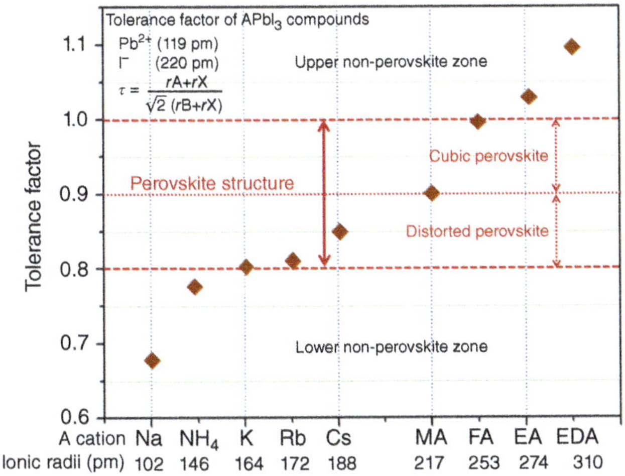

Figure 2.10 Calculated tolerance factors (τ) for different cations (A) in an $APbI_3$ perovskite system. Commonly used cations like Cs, MA, and FA present τ in the range of 0.8–1.0, indicating the cubic perovskite phase structure. Ionic sizes of A cations used in calculation are values referring to XII coordination, not VI coordination. Ethylammonium (EA) and ethylenediamine (EDA) cations are too large, giving rise to τ values >1.0 and thus fall into the "upper forbidden zone" and cannot form perovskite alone. The group I alkali metal cations (Na, K, and Rb) and NH_4 have a τ value <0.8 and fall into the "lower forbidden zone," not forming perovskite. However, all cations of lower or upper "forbidden zones" can be used as additives into those cations in "cubic perovskite formation zone" for phase stabilization. *Source*: Jena et al. [3].

in ABX_3. Since the ionic radii of organic cations (A) cannot be determined accurately, a certain amount of uncertainty lies in the calculated tolerance factors. Figure 2.10 shows the calculated tolerance factors of $APbI_3$ systems where A = Na, K, NH_4, Rb, Cs, MA, FA, EA (ethylamine), and EDA (ethylenediamine). When τ is in the range of $0.8 < \tau < 1$, ideal cubic perovskite structures or distorted perovskite structures with tilted octahedra are favored. Specifically, $0.9 < \tau < 1$ favors a cubic perovskite structure, while for $0.8 < \tau < 0.9$, a distorted perovskite structure is formed. Values of $\tau < 0.8$ and $\tau > 1$ diminish the possibility of the formation of perovskite structures. Therefore, it can be expected that τ close to the middle of the range from 0.8 to 1, away from both the non-perovskite zones, would form a stable perovskite [70]. For $FAPbI_3$, τ is close to 1, which is near the upper boundary of the perovskite structure; therefore, $FAPbI_3$ is prone to the formation of a hexagonal δ-phase, which is photoinactive. An all-inorganic composition, $CsPbI_3$, with $\tau \sim 0.8$, is at the edge of the lower boundary for perovskite structures and normally crystallizes into a δ-phase. $MAPbI_3$ with $\tau \sim 0.9$ is close to the middle of the perovskite zone and forms a black photoactive perovskite phase.

The compositional engineering of perovskite by mixing different cations and anions improves the structural stability of perovskites, essentially by adjusting the value of τ close to the middle of the perovskite zone. The addition of Cs or MA to $FAPbI_3$ moves the τ value down from 1 to stabilize the cubic phase of $FAPbI_3$. The effective tolerance factor (τ_{eff}) of a mixed perovskite, $(MAPbBr_3)_{0.15}(FAPbI_3)_{0.85}$, calculated using the values of r_A(eff) for MA and FA and r_X(eff) for halogen is ~0.98, while that for triple-cation mixed perovskite with 5% Cs (i.e. $Cs_{0.05}(MA_{0.17}FA_{0.83})_{0.95}Pb(I_{0.83}Br_{0.17})_3$) is ~0.97. From such values of τ_{eff}, it can be predicted that more Cs or MA in these mixed perovskite systems should result in even more stable perovskite structures. However, τ_{eff} close to 0.9 is not a sufficient condition for predicting the structural stability of mixed perovskites. The crystal structure difference between $MAPbI_3$ (tetragonal) and mixed perovskites (FA-MA-Cs) (usually cubic) might also explain why mixed perovskites are more stable. As cubic structures with higher symmetry than tetragonal structures are more stable, mixed perovskites with cubic structures show better stability than pure $MAPbI_3$. Moreover, the thermodynamic stability of perovskite structures depends on the halide composition. At high temperatures, $MAPbX_3$ (X = Cl, Br, and I) decomposes into PbX_2(s), MA(g), and HX(g). The energy of formation of these component products of the degradation process confirms that $MAPbI_3$ and $MAPbBr_3$ are thermodynamically more stable than $MAPbCl_3$ [71]. Structural stabilities of mixed-cation perovskites are summarized in review articles [72]. Furthermore, a recent analysis of the multi-cation perovskite decomposition process has confirmed that Cs inclusion increases the thermal stability of $FA_{0.85}MA_{0.15}$ $Pb(I,Br)_3$ perovskite [73].

Ion migration in the halide perovskite crystal also triggers the degradation of materials. It can cause anomalous photocurrent–voltage (*J*–*V*) hysteresis, phase segregation, and deterioration of the long-term stability of PSCs. It is believed that ions in perovskites essentially migrate through the cation and anion defects (i.e. MA^+ and I^- vacancies). Calculation of activation energies of migration for different constituting ions (i.e. I^-, Pb^{2+}, and MA^+ of FA^+) in perovskites suggests that iodide is the most easily migrating ion [74], which is consistent with experimental studies [75]. MA^+ cations also migrate in perovskite crystals [76]. In polycrystalline films, the preferred regions of ion migration are considered to be grain boundaries rather than within the grains. Therefore, it is possible to minimize ion migration by reducing the number of grain boundaries or by increasing the grain size.

In addition to the intrinsic stability of perovskites, the stability of perovskite-based devices is affected by external conditions such as moisture, oxygen (air), heat, light, and external bias, which can degrade the device structure involving the change of surrounding carrier transport materials. Because ionic conductors have some affinity for polar media, the sensitivity of perovskite to moisture (H_2O) has been a matter of concern for its long-term stability. It is known that water molecules easily bind to perovskite by hydrogen bonding to form hydrated compounds, which alter the properties of perovskite locally. The losses caused by these hydrated compounds can be reversed, as shown for $MAPbI_3$ in Eq. (2.1), but further ingression of water can cause irreversible degradation of perovskite to PbI_2 and other components (Eq. (2.2)) [77].

$$\begin{aligned} 4CH_3NH_3PbI_3 + 4H_2O &\leftrightarrow 4[CH_3NH_3PbI_3 \cdot H_2O] \\ &\leftrightarrow (CH_3NH_3)_4PbI_6 \cdot 2H_2O + 3PbI_2 + 2H_2O \end{aligned} \tag{2.1}$$

$$(CH_3NH_3)_4PbI_6 \cdot 2H_2O \rightarrow 4CH_3NH_3I + PbI_2 + 2H_2O \tag{2.2}$$

Therefore, the proper encapsulation of PSCs with films of gas barrier functions is a prerequisite for the prevention of degradation caused by moisture. On the other hand, many efforts have been made to improve the moisture resistance of materials constituting the cell structure, either by introducing hydrophobic layers such as polymer/carbon composites in place of the widely used spiro-OMeTAD [78], or by incorporating non-hygroscopic interlayers between perovskite and HTM [79, 80], or by passivating (modifying) the perovskite surface with small molecules [81]. For instance, tetra-ammonium zinc phthalocyanine has been employed to passivate the grain boundary of $MAPbI_3$ [82], and the stabilized device showed a long lifetime of 1000 hours sustaining 90% of PCE for storage at 85 °C and 45% humidity.

A recent major trend in modifying the surface and grain boundaries of perovskite films is to incorporate a 2-dimensional (2D) perovskite material with a hydrophobic organic group into the general 3-dimensional (3D) perovskites [83]. The method serves two purposes: passivation of the traps and protection against moisture ingression. Large organic cations in 2D perovskites are more hydrophobic than MA and FA and help to increase the moisture resistance. For instance, the incorporation of 0.8 mol% of ethylenediammonium iodide (EDAI) (which forms a 2D perovskite structure $EDAPbI_3$ when mixed with PbI_2) into a 3D $MAPbI_3$ structure improves the PCE by reducing recombination [84] and enables stable operation of the device at 50 °C and 50% relative humidity under illumination. Similarly, phenylethylamine iodide (PEAI) was employed to deposit PEA_2PbI_4 capping layers on top of the 3D perovskite film, which improved the long-term stability of the cell, retaining nearly 90% of the initial PCE (around 18%) after 1000 hours exposure in ambient conditions with a relative humidity of 60 ± 10% [85]. The hydrophobic nature of the trifluoroethylamine (FEA) chain was applied to incorporate 2D perovskites in triple-cation 3D perovskite, $(FA_{0.825}MA_{0.15}Cs_{0.025})Pb(I_{0.85}Br_{0.15})_3$, to reinforce moisture resistance [86]. A cell using FEA-modified perovskites without encapsulation maintained over 90% of the initial PCE (18%) under a humidity of 65% at room temperature for up to approximately 28 days. FEA-based 2D perovskite, $(FEA)_2PbI_4$, was also introduced at the interface of perovskite and HTL to protect $FA_{0.85}MA\ Pb(I_{0.85}Br_{0.15})_3$ from moisture erosion and stabilize the cell under 50 ± 5% humidity for 60 days [87]. Alkyl diammonium-based perovskites can serve as a 2D perovskite interlayer to bridge 3D perovskite layers [88]. Butanediamine iodide ($BEAI_2$) was employed to passivate the grain boundaries of perovskites by vapor phase post-treatment and improved ambient stabilities of the high-performance cell (PCE > 19%) at a high temperature (85 °C) and under continuous illumination [89]. One 2D material, 5-aminovaleric acid iodide (AVAI), has become a popular material for stabilizing $MAPbI_3$ since it has been employed in dopant-free carbon-based PSCs [90]. Three mole percent of AVAI added to $MAPbI_3$ perovskite stabilizes a carbon-based PSC device, and a large module ($10 \times 10\ cm^2$) of the device exhibited stability for more than 10 000 hours (>1 year) at 55 °C under 1-sun illumination [91]. Carbon-based PSCs have relatively high thermal stability owing to the use of dopant-free chemically stable carbon layers. The recent progress of carbonaceous material-based PSCs is summarized in a review [92].

The moisture resistance of PSCs can also be enhanced by modifying the perovskite surface with hydrophobic polymers. For instance, an ultra-thin layer (<10 nm) of polyvinylpyridine (PVP) coated on the perovskite surface ($MAPbI_3$) and in contact with spiro-OMeTAD improves the water resistance of $MAPbI_3$ when dipped in water for up to 100 seconds [79]. A similar result was also reported by Tress and coworkers who found a large effect of suppressing nonradiative recombination using PVP as a thin interlayer and succeeded in elevating V_{OC} to 1.20 V [93]. Furthermore, the continuous chemical functionalization of the grain boundaries of $MAPbI_3$ films by using triblock copolymers that exhibit both hydrophilicity and hydrophobicity has also been found to boost the PCE as well as the stability of $MAPbI_3$ cells [94]. Hence, surface modification using 2D perovskites and hydrophobic polymer materials can dramatically raise the moisture resistance.

Influence of light on the stability (photo-instability) of perovskite is becoming a formidable issue. Almost all chemical processes such as ion migration, defect generation, and phase segregation, which affect the long-term stability of perovskites, are unfortunately found to be influenced by light, which must inevitably shine on the cells for their operation. Photocatalytic degradation by TiO_2 [95], photoinduced ion migration [96], photoinduced trap state generation [97], and photoinduced phase segregation [98] have been found to be the reasons for performance deterioration under constant illumination. UV light-induced photocatalytic degradation by the most widely used ETM, TiO_2, can be prevented by replacement of TiO_2 with an alternative, less photoactive ETL such as SnO_2. However, photoinduced phase segregation or ion redistribution in mixed halide (I and Br) perovskites is a serious issue to be solved. A reversible photoinduced phase separation results in the formation of iodide-rich and bromide-rich regions [99] in mixed-halide perovskites. Such I-rich and Br-rich phase separation leads to the generation of low-bandgap trap states, which causes performance degradation of the solar cells under illumination. It is believed that defects such as halogen vacancies are essentially the main promoters of such phase separation. As ion mobility in these materials is known to be more facile at the grain boundaries [100], reducing the volume of grain boundaries (growing large grains) by compositional tuning with Cs or FA results in improved photostability by slowing down ion migration. Phase segregation can be ignored using pure mono-cation/anion compositions such as $MAPbI_3$, $FAPbI_3$, $CsPbI_3$, etc. In this regard, the use of $MAPbI_3$, which is relatively stable at room temperature and against humidity, is still promising if the device is perfectly protected by encapsulation.

Finally, carrier-transporting materials that contact the perovskite layer also play a substantial role in changing the device stability. The HTM can be very critical for long-term stability because perovskite/HTM junctions are the "beating heart" of the whole device. Spiro-OMeTAD is an excellent HTM in terms of appropriate HOMO levels and is useful as a relatively thick film prepared by simple coating. However, spiro-OMeTAD requires mixing of ionic dopant (lithium bis(trifluoromethanesulfonyl)imide, Li^+TFSI^-, etc.), which promotes the oxidation of this HTM under ambient air to ensure normal hole conductivity. Such dopants, owing to their hygroscopic nature, allow the easy ingress of moisture into the film, which eventually degrades the perovskite film (turning yellow). Furthermore, ionic dopants diffuse to the perovskite layer across the junction interface and trigger deterioration of the device performance. Such diffusion can be accelerated at high temperatures or under ambient humidity. It is known that the spiro-OMeTAD layer undergoes deformation by generating many voids inside the layer upon heating to 100 °C [101, 102]. The mechanism of device degradation includes not only the dopant diffusion of HTM but also ionic migration from perovskites. For example, Carrilo et al. reported that spiro-OMeTAD$^+$ can react with migrating I^- to reduce hole conductivity of the HTM [103]. Zhao et al. found that MA^+ ions from $MAPbI_3$ can diffuse into spiro-OMeTAD and, thereby, cause performance degradation [104]. Therefore, the prevention of ion diffusion between the layers is one of the strategies to improve the thermal stability of PSCs. Apart from instability, another problem of spiro-OMeTAD is high cost of the material. To replace spiro-OMeTAD, a great deal of effort has been made to synthesize various types of organic HTMs for use in PSCs [105]. These HTMs are either small molecules or hole-conducting (p-type) polymers. Conducting polymers are represented by P3HT, PTAA, PEDOT:PSS, etc. PTAA has been employed to record the highest efficiency of PSCs [106]. However, organic HTMs, including these polymers, are intrinsically low in conductivity and require the use of p-dopants and/or additives similar to spiro-OMeTAD. Therefore, their dopants still retain the risk of lowering the device stability.

A solution to avoid the degradation caused by organic HTMs is to develop a dopant-free HTM bearing no ionic moieties in its structure. Inorganic HTMs have been studied to replace organic HTMs. They are p-type oxide or compound semiconductors such as NiO_x, CuO_x, Cu_2O, CuI, Cu(thiourea)I, CuSCN, $CuGaO_2$, $CuCrO_2$, and Cu phthalocyanine. Promisingly, many of them have achieved comparable efficiency and longer life than spiro-OMeTAD. Aside from these inorganic HTMs, the development of dopant-free organic HTMs is becoming a main strategy to stabilize PSCs. Although organic HTMs are thermally less stable than inorganic HTMs, they can have higher chemical and electrochemical stabilities than inorganic materials against oxidation. Among small molecules alternative to spiro-OMeTAD, a truxene core-introduced OMeTAD derivative achieves a high PCE of 18.6% without the addition of dopants and a long lifetime of the device for 33 days of storage [107]. Cao et al. employed a spiro-based dopant-free HTM containing redox-active triphenyl amine units that achieved a high PCE of 20.6% [108]. Donor–π–Acceptor-type (D–π–A) molecules have been developed for new HTMs in which intramolecular charge transfer nature of the D–π–A molecule is assumed to eliminate the need for dopants. Nazeeruddin and coworkers reported that an HTM consisting of a quinolizino acridine donor and terthiophenes and malononitriles as acceptors gave a PCE of 18.9% with superior long-term stability under 1300 hours illumination [109]. Phthalocyanine (Pc) derivatives are known as robust organic pigments and p-type semiconductors. They also work well as dopant-free HTMs with excellent moisture resistance and high thermal stability [110]. For example, a thin film (about 40 nm in thickness) of a butyl derivative of copper Pc (CuPc) was coated on a high efficiency perovskite absorber ($Cs_{0.07}Rb_{0.03}FA_{0.765}MA_{0.135}PbI_{2.55}Br_{0.45}$) by solution process and achieved PCE values exceeding 20%. While the above small-molecule dopant-free HTMs exhibit high efficiency, the use of polymer HTMs is also sought because of their hydrophobic nature to ensure moisture resistance in addition to good thermal stability. Dopant-free polymer HTMs with high solubility would be one of the best choices of HTMs. Unlike poly-triarylamine (PTAA)-type polymers, which work at high efficiency with dopants, these dopant-free polymer HTMs are mostly based on thiophene backbone chains, similar to the structure of P3HT [105]. For example, a dopant-free copolymer comprising ethylhexyloxy phenylene and bithiophene units, which was employed in a PSC device with an inverted structure, showed a PCE of 19.5% (certified) [111]. Park et al. have employed D–A-type copolymers (asy-PBTBDT) comprising benzo-dithiophene and benzothiadiazole (BT) as donor and acceptor units, respectively. Their dopant-free devices yielded PCEs up to 19.8% with a high V_{OC} of 1.14 V [112]. Reflecting on the effect of polymer HTMs as protective layers, most of these polymer HTM-based PSCs have a shelf life of more than 30 days under ambient air. A new dopant-free copolymer was recently developed in our group. Comprising 2,5-pyrroledione and bithiophene units, an alternating copolymer poly(DTSTPD-*r*-BThTPD), has a high hole mobility of 1.5×10^{-3} $cm^2/(V\,s)$ and is thermally stable up to 330 °C [113]. This dopant-free polymer exhibited high V_{OC} performance for all-inorganic PSCs using $CsPbI_2Br$ as an absorber, whose HOMO level is located at approximately −5.58 eV, slightly lower than that of the HTM (Figure 2.11). The dopant-free $CsPbI_2Br$ solar cell yielded a PCE of 15.5% according to its bandgap of 1.92 eV (wider than 1.73 eV of $CsPbI_3$), but achieved a significantly high V_{OC} value of 1.43 V [114]. Combination of these dopant-free and high-efficiency HTMs and 2D-based stabilization of perovskite compositions is expected to realize development of high-performance and robust PSCs toward device implementation.

2.7 All Inorganic and Lead-Free Perovskites

The design of organic-free (all-inorganic) compositions for photovoltaic absorbers can be a way to guarantee the thermal robustness of the device. Cs-based lead perovskites have been extensively studied for this reason. Although Cs perovskites show stability at high temperatures, $CsPbI_3$ undergoes chemical equilibrium changes depending on the temperature. For example, the photovoltaically active black phase of α-$CsPbI_3$ (cubic) can be stabilized at temperatures above 310 °C [115], while the inactive yellow phase δ-$CsPbI_3$ (orthorhombic) tends to dominate the crystal at room temperature. Furthermore, $CsPbI_3$ absorbs moisture due to its hydrophilic nature. Despite this difficulty, the PCE of $CsPbI_3$ PSCs has increased from 2.9% in the very first experiment by Snaith and coworkers [115] in 2015 to 18.4% in a recent work undertaken by Wang et al. [116], who found that the β-phase (tetragonal) of $CsPbI_3$, which is a photoactive black phase (bandgap, 1.68 eV), can be stabilized at room temperature. There are other ways to stabilize the black active phase by introducing 2D-perovskite such as ethylene diamine lead iodide ($EDAPbI_4$) [117]. We found that the inclusion of $Eu^{2+/3+}$ into $CsPbI_3$ can stabilize the black α-$CsPbI_3$ phase at room temperature by reducing the grain size [118]. An

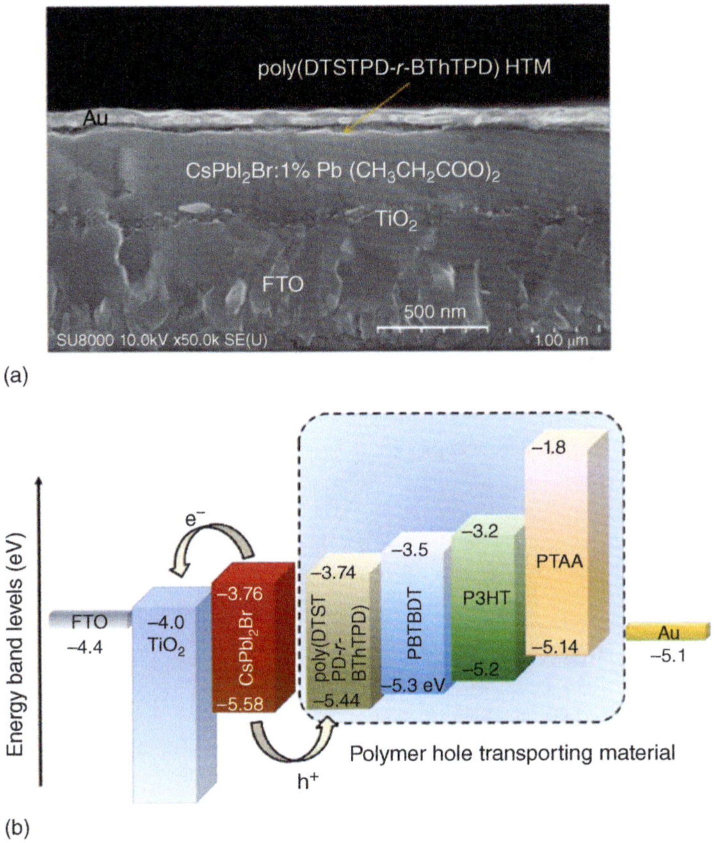

Figure 2.11 Polymer hole-transporting materials (HTMs). A thin film (20–40 nm) of THM coated on the perovskite $CsPbI_2Br$ layer (a) and the relation of polymer HTM energy levels with respect to $CsPbI_2Br$ (b). *Source*: Öz et al. [113].

additional research direction for Cs-based inorganic perovskites is to enhance the photovoltaic functions of Br-mixed $CsPbI_3$, which have wider bandgaps depending on the fraction of Br/I. Br substitution in $CsPbI_3$ lowers the black phase formation temperature from 350 °C to about 250 °C and improves stability against phase transition. Although pure $CsPbBr_3$ having a wide bandgap of ~2.3 eV is not suitable for a sunlight absorber, bromoiodide perovskites $CsPbI_3Br_{3-x}$ with narrower bandgaps show promise for application in the top cell of tandem solar cells or indoor photovoltaic cells. In particular, indoor photovoltaic devices, for use as a power source for IoT goods and equipment, work at high efficiency in harvesting LED illumination (wavelengths, <700 nm) using an absorber (semiconductor) that has a spectral sensitivity only to visible light (bandgap >1.8 eV). The bandgap of $CsPbI_{3-x}Br_x$ can be precisely controlled by the I/Br ratio. $CsPbI_2Br$ with a bandgap of 1.9 eV is an optimal composition for harvesting visible light and is highly stable at room temperature. Using $CsPbI_2Br$, our group obtained a PCE of 15.5% by enhancing the device V_{OC} up to 1.43 V [114], which is the highest V_{OC} so far

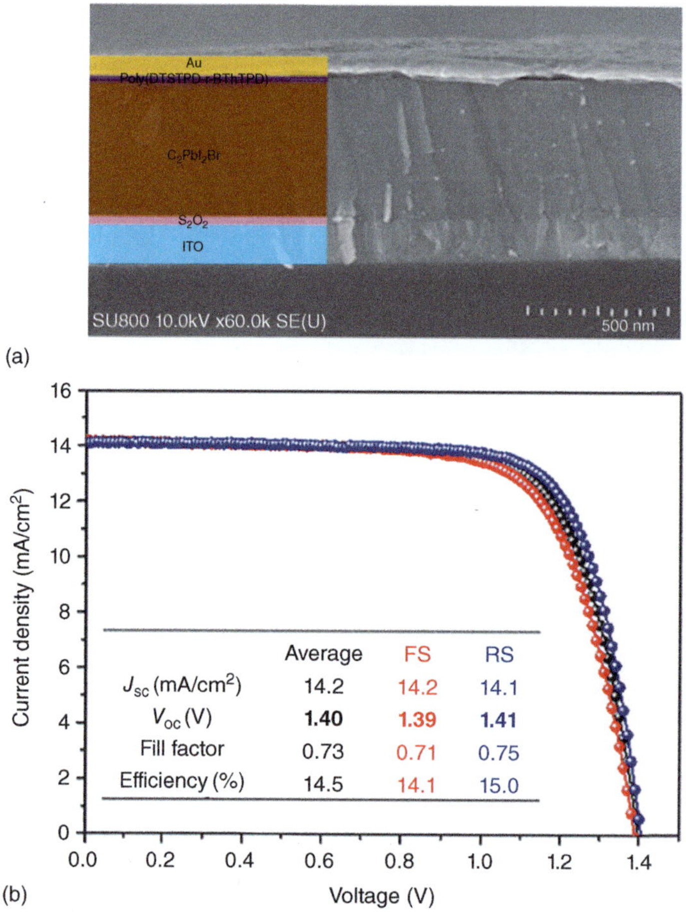

Figure 2.12 Layered structure of a $CsPbI_2Br$-based dopant-free HTM photovoltaic cell comprising ITO/SnO_2-SnO_x/$CsPbI_2Br$/poly(DTSTPD-*r*-BThTPD)/Au, and its *J*–*V* characteristics showing high V_{OC} over 1.4 V. FS and RS indicate forward and reverse scans of voltage, respectively. *Source*: Guo et al. [114].

obtained with visible-light harvesting PSCs. The device was fabricated by using an SnO_2 ETM in which the surface in contact with $CsPbI_2Br$ layer was modified with a thin amorphous SnO_x layer to block hole injection and reduce the interfacial charge recombination. The HTM was a dopant-free polymer DTSTPD-*r*-BThTPD as described above (Figure 2.11). The high V_{OC} obtained with this cell configuration approaches the Shockley–Queisser limit of V_{OC} leaving a voltage loss of <0.5 eV loss. Figure 2.12 shows the SEM image of the device structure and *J*–*V* characteristics; the latter has a small hysteresis between forward and backward scans. The device composed of a fairly thick and flat $CsPbI_2Br$ active layer, and very thin HTM and ETM layers can be characterized as a planar cell.

In the design of all-inorganic perovskites, an important direction toward their commercialization is the preparation of metal halide compounds that do not contain environmentally harmful lead. Lead is one of the 10 most toxic elements for human health as reported by the World Health Organization (WHO). Its toxicity has the potential to cause irreversible health effects affecting the central nervous, hematopoietic, hepatic, and renal systems, producing serious disorders [119]. As lead-free metal halide materials, organic tin halide perovskites such as $MASnI_3$, $MASnI_{3-x}Br_x$, and $FASnI_{3-x}Br_x$ have demonstrated the highest efficiency in the field of lead-free perovskite photovoltaics [120]. The PCE of Sn-based perovskites has now reached 10% by compositional engineering using organic cations. However, Sn-based halide perovskites can function as efficient photovoltaic semiconductors as long as the reduced divalent form of Sn(II) is stabilized in an oxygen-free atmosphere. To overcome this issue, Diau and coworkers showed that guanidinium (GA^+) cation-doped $FASnI_3$ crystal was stabilized in the presence of 1% ethylenediammonium diiodide ($EDAI_2$) as a 2D-forming additive, and the $GA_{0.2}FA_{0.8}SnI_3$-based device reached a PCE of 9.6% [121]. $EDAI_2$ was also applied to passivate $Cs_{0.2}FA_{0.8}SnI_3$ doped with SnF_2 and the device using $Cs_{0.2}FA_{0.8}SnI_3$ achieved a higher PCE (certified) of 10.08% [122]. Along with increasing efficiency, efforts to the enhance oxidation stability of Sn(II) have been made by incorporating reducing agents into the device structure [123]. Hayase et al. studied Sn-based perovskites to demonstrate a method to enhance efficiency and stabilize Sn(II) by compositional engineering. Diaminoethane (DAE) was employed to passivate the defects of EDAI-doped $FASnI_3$ perovskites and DAE-based recombination suppression led to the highest PCE value of 10.18% [124]. Sn-based all-inorganic perovskites have also been studied for photovoltaic applications. They are Cs-based compositions such as $CsSnI_3$ ($E_g \sim 1.3$ eV), $CsSnBr_3$ ($E_g \sim 1.8$ eV), $Cs_2SnI_4Br_2$, or $CsGe_{0.5}Sn_{0.5}I_3$ ($E_g \sim 1.5$ eV). However, their device performance is much lower than that of organic hybrid Sn perovskites [120] and is accompanied by the stability issue of Sn(II). All-inorganic and lead-free perovskites have been studied intensively, focusing on Group 15 metals such as Sb and Bi. In particular, as environmentally benign alternatives to Pb and Sn, Sb and Bi possess similar electronic configurations and comparable ionic radii to Pb, allowing them to incorporate effectively into the perovskite lattice. The crystal structure and optical properties of various ternary bismuth halide perovskite materials have been theoretically and experimentally investigated [125–127]. Although they have high stabilities against heat and moisture, photovoltaic devices using Bi perovskites have lagged behind lead-based or Sn-based perovskite devices in terms of efficiency development because of the large energy loss by carrier recombination, which reduces J_{SC} and V_{OC}. This efficiency loss is primarily related to the poor morphology of the solution-processed polycrystalline film. We found that pinhole-rich poor morphology of $MA_3Bi_2I_9$ results from the fast crystallization of $MA_3Bi_2I_9$ and that retarding the crystallization by adding less volatile solvent (*N*-methyl-2-pyrrolidone) in the precursor improves the compactness of the film and enhances the $MA_3Bi_2I_9$ device performance [128]. However, the quality of Bi-based perovskite films tends to be inferior to those of Pb-based perovskite films, indicating difficulty in controlling the crystallization process based on chemical equilibria. To overcome this, Jain et al. developed a vapor-assisted solution process in which MAI vapor was exposed to the solution-processed BiI_3 film [129]. This method improved the PCE of the $MA_3Bi_2I_9$ device up to 3.17% with V_{OC} = 1.01 V. The performance of non-encapsulated device was stable up to 60 days. All-inorganic analogs of Bi halide perovskites are more important for their robust stability. Silver bismuth halide, for example $AgBi_2I_7$, is one of the potential candidates for all-inorganic lead-free PSCs. However, it is difficult to obtain high-purity single-crystal structures owing to the mixing of different phases (i.e. $AgBi_2I_7$, $AgBiI_4$, $AgBi_2I_5$, or Ag_3BiI_6) formed by the high-temperature annealing process. Addressing this phase impurity issue, we improved the solvent engineering process for obtaining phase-pure $AgBi_2I_7$. As a result, the PCE reached 2.1% [130]. Although this value is one of the highest efficiencies reported for $AgBi_2I_7$, there is a large PCE gap between Bi-based and Pb-based PSCs using all-inorganic perovskite compositions. The reason for this gap is that the use of ionic dopants (LiTFSI, etc.) to HTMs causes the chemical degradation of Ag/Bi perovskites. Therefore, dopant-free HTMs should be selected. Behavior of J–V characteristics indicates the existence of a large charge recombination in quantum conversion, which arises from direct recombination between CB and VB, and from Shockley–Read–Hall (SRH) recombination associated with traps formed within the bandgap, which lead to lower J_{SC} and V_{OC}, respectively [131]. The decrease in V_{OC} is mainly related to an increase in trap density, which has a high value in the grain boundary area rich in defects. Therefore, it is essential to enhance the device performance to improve the quality of the Bi perovskite film by forming a uniform pinhole-free morphology consisting of large grains. This strategy is common for all types of films consisting of lead-free inorganic perovskites which tend to have low quality in terms of uniformity, compactness, and phase purity of crystals. Although PCE of Ag/Bi perovskites are low under sunlight irradiation, we recently found that some Ag/Bi absorbers are capable of high PCE (>10%) under weak indoor light (LED illumination) because of good matching of their bandgap and indoor visible light spectrum. This indicates applications of these lead-free PV devices to IoT devices are highly promising.

Antimony (Sb)-based halide perovskites also exhibit photovoltaic responses with their best efficiencies barely approaching those of Bi-based perovskites. $Cs_3Sb_2I_9$ has a high absorption coefficient with a direct bandgap of 2.05 eV. For examples, mixed cation compositions, $A_3Sb_2I_9$ (A = Cs and MA), are capable of PCE up to 2.04% [132]. Furthermore, the incorporation of Cl^- into $MASb_2I_9$ was found to transform the dimer phase into a 2D phase, exhibiting a PCE of over 2% and stability of the device in dry air for 18 days [133]. Seok and coworkers synthesized a chalcogenide-halide mixed perovskite ($MASbSI_2$) [134], which demonstrated the best PCE of 3.08% with an optical gap of 1.3–1.4 eV. Other interesting lead-free compositions are Ti(IV)-based halide perovskites such as Cs_2TiBr_6 with bandgaps of ~1.38 eV [135]. As a double perovskite, Cs_2TiBr_6 has achieved a PCE of above 3% with a simple device structure (FTO/TiO_2/Cs_2TiBr_6/P3HT/Au). Many other candidates of non-lead perovskites for photovoltaic cells have been summarized in a recent review [136], in addition to the chapters of this book that introduce the progress of new lead-free and/or all-inorganic perovskite materials.

2.8 Development of High-Efficiency Tandem Solar Cells

Wavelength (bandgap) tunability of PSCs enables the design of tandem cells for the purpose to enhance sunlight collection and gain PCE that exceeds the maximum

PCE obtained by any type of single solar cells such as crystalline silicon (Si) cells. To date, significantly high PCEs obtained with triple-junction tandem cells such as InGaP/InGaAs/Ge have been utilized in space satellite missions. In industry, the total cost of fabricating tandem solar cells including materials and process of stacking unit cells becomes more than the cost of any single cell that constitutes the tandem structure. Apart from space applications, cost efficiency (efficiency-to-cost ratio) should be an essential issue in commercializing solar cells for consumers and for society in general. Therefore, it is desirable to boost the PCE by creating tandem cells at no significant additional cost. In this regard, a combination of cost-efficient perovskite and crystalline Si cells or a combination of two perovskite cells is promising candidates for tandem cell fabrication. Technically, the basic rule in the design of a tandem cell is to adjust the photocurrent density of each unit cell constituting the tandem to be equal to that of the other unit cells. Such a current balance can be realized by tuning the bandgap of perovskite absorber (top cell), that is, the composition of halide perovskite. A simple method of making tandem cells is to mechanically superpose two solar cells that are independently fabricated using four terminals for electric combination. For such as the 4-terminal tandem cells, the power generated by the top and bottom cells can be independently processed and combined electrically by the external circuitry so that the highest output is gained. However, the cost of the cell, which is the sum of two cells, becomes less efficient compared to the PCE of the tandem cell. Therefore, two-terminal tandem cells are more desirable in terms of cost efficiency. Recent progress in perovskite-based tandem solar cells is summarized in review articles [137, 138]. Theoretically, by sharing the solar spectrum (photon flux) with a perovskite top cell (1.6 eV bandgap) capable of a V_{OC} of 1.2 V and a crystalline Si bottom cell (1.1 eV bandgap) capable of a 0.8 V as V_{OC}, the PCE of a perovskite/silicon tandem cell can reach close to 30%, depending on the gain in V_{OC} and FF [139, 140]. Figure 2.13 shows an example of the layered structure of a perovskite-Si tandem cell and its solar spectrum absorption and photocurrent density divided by the top and bottom cells [141].

When a perovskite absorber of $Cs_{0.17}FA_{0.83}Pb(Br_{0.17}I_{0.83})_3$ (bandgap of 1.63 eV) was employed with the junction of an Si cell, this 2-terminal tandem cell achieved a PCE of 23.6% with a V_{OC} of 1.65 V [140]. Duong et al. reported a PCE of 26.4% using a 4-terminal tandem cell with a mechanical stack of Rb-containing quadruple cation perovskite (PCE of 17.4% with a bandgap of 1.73 eV) and silicon cells (PCE of 23.9%) [142]. What rendered the PCE higher than the above 2-terminal cell was the higher PCE of the Si cell and the V_{OC} gain, which was 1.81 V (1.12 V for the top, 0.69 V for the bottom). In addition to the challenges of these studies, a venture company, Oxford PV, recently reported a certified PCE of 27.3% with a perovskite and crystalline Si tandem cell (active area was 1 cm^2). The PCE value achieved exceeds the top PCE of a single-junction Si cell (26.1%). Perovskite–perovskite (all-perovskite) type tandem cells have also been studied intensively because of their low-cost advantage. Huang et al. fabricated an all-perovskite tandem cell based on a simplified structure using SnO_{2-x} and fullerene (C60) as carrier transport layers. Achieving PCE of 24.4%, the tandem cells could attain high stability, keeping efficiency almost unchanged for continuous 1-sun illumination for 1000 hours [143].

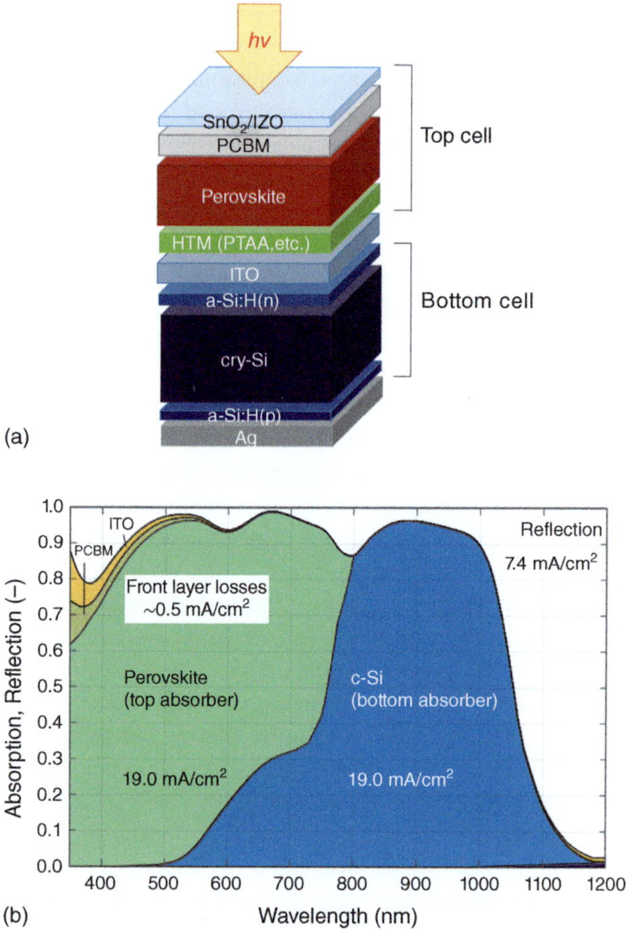

Figure 2.13 Layered structure of the stack of perovskite top cell and Si bottom cell in a 2-terminal tandem solar cell (a), and spectral light absorption performances and corresponding photocurrent densities of top and bottom cells (b). *Source*: Modified from Jäger et al. [141].

2.9 Conclusion and Perspectives

Research of perovskite-based photovoltaics is highly interdisciplinary. Material synthesis is mostly a solution-based chemical process, while the characterization of perovskite semiconductors requires physical methods and device fabrication involves chemistry, physics, and electronics. Progress of the research in terms of efficiency and stability enhancement depends not only on the development of new materials but also efficient collaboration between different disciplines. A high level of defect tolerance endowed to the perovskite materials, as reflected by the long lifetimes of photoexcited free carriers, plays a central role in realizing high efficiency of devices owing to generation high voltage, which becomes especially advantageous when the perovskite device is used as an indoor power source in IoT applications where IoT devices and their secondary batteries (Li-ion batteries, etc.) require higher voltage input under weak light to drive the circuitry. One of the goals of perovskite photovoltaics is to achieve the excellent photophysical properties of GaAs semiconductors, which exhibit an open-circuit voltage and PCE close to the SQ limits with a bandgap of 1.42 eV at the absorption edge of nearby 900 nm, which is the optimum wavelength for a single-junction cell to harvest the power of solar irradiation. Based on the top efficiency of PSCs (25.5%), the efficiency of perovskite modules for practical use is expected to reach more than 22%, which is considered sufficient for commercial devices in terms of cost performance. Besides R&Ds for high stability and durability of the module, ensuring environmental safety of materials is a major issue in industrialization. In the future, lead will be the subject of total replacement with other metals such as Sn, Bi, Sb, Ag, etc. If we look at applications other than photovoltaics, many applications including LED, photodetectors, and lasers are progressing and halide perovskite science and engineering is expected to continue to develop significantly in the future. For comprehensive study on the fundamentals and current advanced applications of perovskite-based photovoltaics and optoelectronics, readers are encouraged to refer to the latest book, entitled *Perovskite Photovoltaics and Optoelectronics – Fundamentals to Advanced Applications* published by Wiley, as a textbook that covers the entire technology of halide perovskites [144].

References

1 Kojima, A., Teshima, K., Shirai, Y., and Miyasaka, T. (2009). Organometal halide perovskites as visible-light sensitizers for photovoltaic cells. *J. Am. Chem. Soc.* 131: 6050–6051.

2 National Renewable Energy Laboratory (2020). Research cell record efficiency chart. Best research-cell efficiencies. https://www.nrel.gov/pv/assets/pdfs/best-research-cell-efficiencies.20190802.pdf.

3 Jena, A.K., Kulkarni, A., and Miyasaka, T. (2019). Halide perovskite photovoltaics: background, status, and future prospects. *Chem. Rev.* 119: 3036–3103.

4 Tao, S.X., Cao, X., and Bobbert, P.A. (2017). Accurate and efficient band gap predictions of metal halide perovskites using DFT-1/2 methods: GW accuracy with DFT expense. *Sci. Rep.* 7: 14386.

5 Tubbs, M.R. and Forty, A.J. (1964). Photographic applications of lead iodide. *Br. J. Appl. Phys.* 15: 1553–1558.

6 Kojima, A., Teshima, K., Shirai, Y., and Miyasaka, T. (2007). Novel photo-electrochemical cell with mesoscopic electrodes sensitized by lead-halide compounds (5). *The Electrochemical Society (ECS), 210th Meeting Abstracts*, MA2007-2, 352.

7 Lee, M.M., Teuscher, J., Miyasaka, T. et al. (2012). Efficient hybrid solar cells based on meso-superstructured organometal halide perovskites. *Science* 338: 643–647.

8 Kojima, A., Ikegami, M., Teshima, K., and Miyasaka, T. (2012). Highly luminescent lead bromide perovskite nanoparticles synthesized with porous alumina media. *Chem. Lett.* 41: 397–399.

9 Rose, G. (1839). Beschreibung einiger neuen Mineralien des Urals. *Annalen der Physik und Chemie* 124: 551–573.

10 Chakhmouradian, A.R. and Woodward, P.M. (2014). Celebrating 175 years of perovskite research: a tribute to Roger H. Mitchell. *Phys. Chem. Miner.* 41: 387–391.

11 Katz, E.A. (2020). Perovskite: name puzzle and German–Russian odyssey of discovery. *Helv. Chim. Acta* 103: e2000061.

12 Acosta, M., Novak, N., Rojas, V. et al. (2017). $BaTiO_3$-based piezoelectrics: fundamentals, current status, and perspectives. *Appl. Phys. Rev.* 4: 041305.

13 Wells, H.L. (1893). Über die cäsium- und kalium-bleihalogenide. *Z. Anorg. Chem.* 3: 195–210.

14 Møller, C.K. (1957). A phase transition in cæsium plumbochloride. *Nature* 180: 981–982.

15 Møller, C.K. (1957). Crystal structure and photoconductivity of cæsium plumbohalides. *Nature* 182: 1436.

16 Weber, D. (1978). $CH_3NH_3PbX_3$, ein Pb(II)-system mit kubischer perowskitstruktur/$CH_3NH_3PbX_3$, a Pb(II)-system with cubic perovskite structure. *Zeitschrift für Naturforschung B* 33: 1443–1445.

17 Weber, D. (1978). $CH_3NH_3SnBr_xI_{3-x}$ (x = 0–3), ein Sn(II)-system mit kubischer perowskitstruktur/$CH_3NH_3SnBr_xI_{3-x}$(x = 0–3), a Sn(II)-system with cubic perovskite structure. *Z. Naturforsch., B* 33: 862–865.

18 Mitzi, D.B. (1999). Synthesis, structure, and properties of organic-inorganic perovskites and related materials. *Prog. Inorg. Chem.* 48: 1–122.

19 Mitzi, D.B. (2000). Organic–inorganic perovskites containing trivalent metal halide layers: the templating influence of the organic cation layer. *Inorg. Chem.* 39: 6107–6113.

20 Mitzi, D.B., Wang, S., Feild, C.A. et al. (1995). Conducting layered organic-inorganic halides containing ⟨110⟩-oriented perovskite sheets. *Science* 267: 1473.

21 Era, M. and Oka, S. (2000). PbBr-based layered perovskite film using the Langmuir–Blodgett technique. *Thin Solid Films* 376: 232–235.

22 Takeoka, Y., Asai, K., Rikukawa, M., and Sanui, K. (2001). Incorporation of conjugated polydiacetylene systems into organic–inorganic quantum-well structures. *Chem. Commun.* (24): 2592–2593.

23 Kondo, T., Iwamoto, S., Hayase, S. et al. (1998). Resonant third-order optical nonlinearity in the layered perovskite-type material $(C_6H_{13}NH_3)_2PbI_4$. *Solid State Commun.* 105: 503–506.

24 Tanaka, K. and Kondo, T. (2003). Bandgap and exciton binding energies in lead-iodide-based natural quantum-well crystals. *Sci. Technol. Adv. Mater.* 4: 599–604.

25 Ema, K., Umeda, K., Toda, M. et al. (2006). Huge exchange energy and fine structure of excitons in an organic-inorganic quantum well material. *Phys. Rev. B* 73: 241310-1-4.

26 Ishi, J., Mizuno, M., Kunugita, H. et al. (1998). Third-order optical nonlinearity due to excitons and biexcitons in a self-organized quantum-well material $(C_6H_{13}NH_3)_2PbI_4$. *J. Nonlinear Opt. Phys. Mater.* 07: 153–159.

27 Miyasaka, T., Ikegami, M., and Kijitori, Y. (2007). Photovoltaic performance of plastic dye-sensitized electrodes prepared by low-temperature binder-free coating of mesoscopic titania. *J. Electrochem. Soc.* 154: A455–A461.

28 Miyasaka, T. (2011). Toward printable sensitized mesoscopic solar cells: light-harvesting management with thin TiO_2 films. *J. Phys. Chem. Lett.* 2: 262–269.

29 Kojima, A., Teshima, K., Shirai, Y., and Miyasaka, T. (2008). Novel photo-electrochemical cell with mesoscopic electrodes sensitized by lead-halide compounds (11). *The Electrochemical Society (ECS), 214th Meeting (PRiME. 2008), Abstracts*, MA2008-2, 27. A.

30 Miyasaka, T. (2015). Perovskite photovoltaics: rare functions of organo lead halide in solar cells and optoelectronic devices. *Chem. Lett.* 44: 720–729.

31 Im, J.H., Lee, C.R., Lee, J.W. et al. (2011). 6.5% efficient perovskite quantum-dot-sensitized solar cell. *Nanoscale* 3: 4088–4093.

32 Snaith, H.J., Petrozza, A., Ito, S. et al. (2009). Charge generation and photovoltaic operation of solid-state dye-sensitized solar cells incorporating a high extinction coefficient indolene-based sensitizer. *Adv. Funct. Mater.* 19: 1810–1818.

33 Stranks, S.D., Eperon, G.E., Grancini, G. et al. (2013). Electron–hole diffusion lengths exceeding 1 micrometer in an organometal trihalide perovskite absorber. *Science* 342: 341–342.

34 Dong, Q., Fang, Y., Shao, Y. et al. (2015). Electron–hole diffusion lengths >175 μm in solution-grown $CH_3NH_3PbI_3$ single crystals. *Science* 347: 967–969.

35 Jeong, M., Choi, I.W., Go, E.M. et al. (2020). Stable perovskite solar cells with efficiency exceeding 24.8% and 0.3-V voltage loss. *Science* 369: 1615–1620.

36 Kim, G., Min, H., Lee, K.S. et al. (2020). Impact of strain relaxation on performance of α-formamidinium lead iodide perovskite solar cells. *Science* 370: 108–112.

37 Suarez, B., Gonzalez-Pedro, V., Ripolles, T.S. et al. (2014). Recombination study of combined halides (Cl, Br, I) perovskite solar cells. *J. Phys. Chem. Lett.* 5: 1628–1635.

38 Eperon, G.E., Stranks, S.D., Menelaou, C. et al. (2014). Formamidinium lead trihalide: a broadly tunable perovskite for efficient planar heterojunction solar cells. *Energy Environ. Sci.* 7: 982–988.

39 Miyano, K., Tripathi, N., Yanagida, M., and Shirai, Y. (2016). Lead halide perovskite photovoltaic as a model p-i-n diode. *Acc. Chem. Res.* 49: 303–310.

40 Brivio, F., Butler, K.T., Walsh, A., and Schilfgaarde, M. (2014). Relativistic quasiparticle self-consistent electronic structure of hybrid halide perovskite photovoltaic absorbers. *Phys. Rev. B* 89: 155204.

41 Tanaka, K., Takahashi, T., Ban, T. et al. (2003). Comparative study on the excitons in lead-halide-based perovskite-type crystals $CH_3NH_3PbBr_3$ $CH_3NH_3PbI_3$. *Solid State Commun.* 127: 619–623.

42 Yin, W.J., Shi, T., and Yan, Y. (2015). Superior photovoltaic properties of lead halide perovskites: insights from first-principles theory. *J. Phys. Chem. C* 119: 5253–5264.

43 Green, M.A., Ho-Baillie, A., and Snaith, H.J. (2014). The emergence of perovskite solar cells. *Nat. Photonics* 8: 506–514.

44 Giorgi, G., Fujisawa, J.I., Segawa, H., and Yamashita, K. (2013). Small photocarrier effective masses featuring ambipolar transport in methylammonium lead iodide perovskite: a density functional analysis. *J. Phys. Chem. Lett.* 4: 4213–4216.

45 De Quilettes, D.W., Vorpahl, S.M., Stranks, S.D. et al. (2015). Impact of microstructure on local carrier lifetime in perovskite solar cells. *Science* 348: 683–686.

46 Stranks, S.D. and Snaith, H.J. (2015). Metal-halide perovskites for photovoltaic and light-emitting devices. *Nat. Nanotechnol.* 10: 391–402.

47 Miyasaka, T. (2018). Lead halide perovskites in thin film photovoltaics: background and perspectives. *Bull. Chem. Soc. Jpn.* 91: 1058–1068.

48 Liu, Z., Krückemeier, L., Krogmeier, B. et al. (2019). Open-circuit voltages exceeding 1.26 V in planar methylammonium lead iodide perovskite solar cells. *ACS Energy Lett.* 4: 110–117.

49 Shockley, W. and Queisser, H.J. (1961). Detailed balance limit of efficiency of p–n junction solar cells. *J. Appl. Phys.* 32: 510–519.

50 Sha, W.E.I., Ren, X., Chen, L., and Choy, W.C.H. (2015). The efficiency limit of $CH_3NH_3PbI_3$ perovskite solar cells. *Appl. Phys. Lett.* 106: 221104.

51 Kim, G.M., Ishii, A., Öz, S., and Miyasaka, T. (2019). MACl-assisted Ge doping of Pb-hybrid perovskite: a universal route to stabilize high performance perovskite solar cells. *Adv. Energy Mater.* 10: 1903299.

52 Tress, W., Marinova, N., Inganäs, O. et al. (2014). Predicting the open-circuit voltage of $CH_3NH3PbI_3$ perovskite solar cells using electroluminescence and photovoltaic quantum efficiency spectra: the role of radiative and non-radiative recombination. *Adv. Energy Mater.* 5: 1400812.

53 Tvingstedt, K., Malinkiewicz, O., Baumann, A. et al. (2014). Radiative efficiency of lead iodide based perovskite solar cells. *Sci. Rep.* 4: 6071.

54 Correa Baena, J.P., Steier, L., Tress, W. et al. (2015). Highly efficient planar perovskite solar cells through band alignment engineering. *Energy Environ. Sci.* 8: 2928–2934.

55 Turren-Cruz, S.H., Hagfeldt, A., and Saliba, M. (2018). Methylammonium-free, high-performance, and stable perovskite solar cells on a planar architecture. *Science* 362: 449–453.

56 Smecca, E., Numata, Y., Deretzis, I. et al. (2016). Stability of solution-processed $MAPbI_3$ and $FAPbI_3$ layers. *Phys. Chem. Chem. Phys.* 18: 13413–13422.

57 Miyasaka, T., Kulkarni, A., Kim, G.M. et al. (2019). Perovskite solar cells: can we go organic-free, lead-free, and dopant-free? *Adv. Energy Mater.* 10: 1902500.

58 Ono, L.K., Juarez-Perez, E.J., and Qi, Y. (2017). Progress on perovskite materials and solar cells with mixed cations and halide anions. *ACS Appl. Mater. Interfaces* 9: 30197–30246.

59 Akin, S., Akman, E., and Sonmezoglu, S. (2020). $FAPbI_3$-based perovskite solar cells employing hexyl-based ionic liquid with an efficiency over 20% and excellent long-term stability. *Adv. Funct. Mater.* 30: 2002964.

60 Jeon, N.J., Noh, J.H., Yang, W.S. et al. (2015). Compositional engineering of perovskite materials for high-performance solar cells. *Nature* 517: 476–480.

61 Niu, G., Li, W., Li, J. et al. (2017). Enhancement of thermal stability for perovskite solar cells through cesium doping. *RSC Adv.* 7: 17473–17479.

62 Li, Z., Yang, M., Park, J.S. et al. (2016). Stabilizing perovskite structures by tuning tolerance factor: formation of formamidinium and cesium lead iodide solid-state alloys. *Chem. Mater.* 28: 284–292.

63 Zhang, Y., Liu, M., Eperon, G.E. et al. (2015). Charge selective contacts, mobile ions and anomalous hysteresis in organic–inorganic perovskite solar cells. *Mater. Horiz.* 2: 315–322.

64 Ravishankar, S., Almora, O., Echeverría-Arrondo, C. et al. (2017). Surface polarization model for the dynamic hysteresis of perovskite solar cells. *J. Phys. Chem. Lett.* 8: 915–921.

65 Weber, S.A.L., Hermes, I.M., Turren-Cruz, S.H. et al. (2018). How the formation of interfacial charge causes hysteresis in perovskite solar cells. *Energy Environ. Sci.* 11: 2404–2413.

66 Almosni, S., Cojocaru, L., Li, D. et al. (2017). Tunneling-assisted trapping as one of the possible mechanisms for the origin of hysteresis in perovskite solar cells. *Energy Technol.* 5: 1767–1774.

67 Singh, T. and Miyasaka, T. (2017). Stabilizing the efficiency beyond 20% with a mixed cation perovskite solar cell fabricated in ambient air under controlled humidity. *Adv. Energy Mater.* 8: 1700677. (1–9).

68 Saliba, M., Matsui, T., Domanski, K. et al. (2016). Incorporation of rubidium cations into perovskite solar cells improves photovoltaic performance. *Science* 354: 206–209.

69 Hu, Y., Hutter Eline, M., Rieder, P. et al. (2018). Understanding the role of cesium and rubidium additives in perovskite solar cells: trap states, charge transport, and recombination. *Adv. Energy Mater.* 8: 1703057. (1–11).

70 Han, G., Hadi, H.D., Bruno, A. et al. (2018). Additive selection strategy for high performance perovskite photovoltaics. *J. Phys. Chem. C* 122: 13884–13893.

71 Brunetti, B., Cavallo, C., Ciccioli, A. et al. (2016). On the thermal and thermodynamic (in)stability of methylammonium lead halide perovskites. *Sci. Rep.* 6: 31896. (1–12).

72 Wang, R., Mujahid, M., Duan, Y. et al. (2019). A review of perovskites solar cell stability. *Adv. Funct. Mater.* 29: 1808843.

73 Shi, L., Bucknall, M.P., Young, T.L. et al. (2020). Gas chromatography–mass spectrometry analyses of encapsulated stable perovskite solar cells. *Science* 368: eaba2412.

74 Eames, C., Frost, J.M., Barnes, P.R.F. et al. (2015). Ionic transport in hybrid lead iodide perovskite solar cells. *Nat. Commun.* 6: 7497. (1–8).

75 Minns, J.L., Zajdel, P., Chernyshov, D. et al. (2017). Structure and interstitial iodide migration in hybrid perovskite methylammonium lead iodide. *Nat. Commun.* 8: 15152. (1–5).

76 Yuan, Y., Chae, J., Shao, Y. et al. (2015). Photovoltaic switching mechanism in lateral structure hybrid perovskite solar cells. *Adv. Energy Mater.* 5: 1500615. (1–7).

77 Leguy, A.M.A., Hu, Y., Campoy-Quiles, M. et al. (2015). Reversible hydration of $CH_3NH_3PbI_3$ in films, single crystals, and solar cells. *Chem. Mater.* 27: 3397–3407.

78 Habisreutinger, S.N., Leijtens, T., Eperon, G.E. et al. (2014). Carbon nanotube/polymer composites as a highly stable hole collection layer in perovskite solar cells. *Nano Lett.* 14: 5561–5568.

79 Chaudhary, B., Kulkarni, A., Jena, A.K. et al. (2017). Poly(4-vinylpyridine)-based interfacial passivation to enhance voltage and moisture stability of lead halide perovskite solar cells. *ChemSusChem* 10: 2473–2479.

80 Hou, Y., Zhou, Z.R., Wen, T. et al. (2019). Enhanced moisture stability of metal halide perovskite solar cells based on sulfur-oleylamine surface modification. *Nanoscale Horiz.* 4: 208–213.

81 Abdelmageed, G., Sully, H.R., Bonabi Naghadeh, S. et al. (2018). Improved stability of organometal halide perovskite films and solar cells toward humidity via surface passivation with oleic acid. *ACS Appl. Energy Mater.* 1: 387–392.

82 Cao, J., Li, C., Lv, X. et al. (2018). Efficient grain boundary suture by low-cost tetra-ammonium zinc phthalocyanine for stable perovskite solar cells with expanded photoresponse. *J. Am. Chem. Soc.* 140: 11577–11580.

83 Grancini, G., Roldán-Carmona, C., Zimmermann, I. et al. (2017). One-year stable perovskite solar cells by 2D/3D interface engineering. *Nat. Commun.* 8: 15684.

84 Lu, J., Jiang, L., Li, W. et al. (2017). Diammonium and monoammonium mixed-organic-cation perovskites for high performance solar cells with improved stability. *Adv. Energy Mater.* 7: 1700444. (1–10).

85 Chen, P., Bai, Y., Wang, S. et al. (2018). In situ growth of 2D perovskite capping layer for stable and efficient perovskite solar cells. *Adv. Funct. Mater.* 28: 1706923.

86 Ye, J., Zheng, H., Zhu, L. et al. (2017). Enhanced moisture stability of perovskite solar cells with mixed-dimensional and mixed-compositional light-absorbing materials. *Solar RRL* 1: 1700125. (1–8).

87 Xu, H., Sun, Y., Zheng, H. et al. (2019). High-performance and moisture-stable perovskite solar cells with a 2D modified layer via introducing a high dipole moment cation. *J. Mater. Chem. C* 7: 15276–15284.

88 Safdari, M., Svensson, P.H., Hoang, M.T. et al. (2016). Layered 2D alkyldiammonium lead iodide perovskites: synthesis, characterization, and use in solar cells. *J. Mater. Chem. A* 4: 15638–15646.

89 Chen, M., Li, P., Liang, C. et al. (2020). Enhanced efficiency and stability of perovskite solar cells by 2D perovskite vapor-assisted interface optimization. *J. Energy Chem.* 45: 103–109.

90 Mei, A., Li, X., Liu, L. et al. (2014). A hole-conductor–free, fully printable mesoscopic perovskite solar cell with high stability. *Science* 345: 295–298.

91 Grancini, G., Roldán-Carmona, C., Zimmermann, I. et al. (2017). One-year stable perovskite solar cells by 2D/3D. *Nat. Commun.* 8: 15684.

92 Fagiolari, L. and Bella, F. (2019). Carbon-based materials for stable, cheaper and large-scale processable perovskite solar cells. *Energy Environ. Sci.* 12: 3437–3472.

93 Yavari, M., Mazloum-Ardakani, M., Gholipour, S. et al. (2018). Reducing surface recombination by a poly(4-vinylpyridine) interlayer in perovskite solar cells with high open-circuit voltage and efficiency. *ACS Omega* 3: 5038–5043.

94 Zong, Y., Zhou, Y., Zhang, Y. et al. (2018). Continuous grain-boundary functionalization for high-efficiency perovskite solar cells with exceptional stability. *Chemistry* 4: 1404–1415.

95 Lee, S.W., Kim, S., Bae, S. et al. (2016). UV degradation and recovery of perovskite solar cells. *Sci. Rep.* 6: 38150. (1-10).

96 Luo, Y., Khoram, P., Brittman, S. et al. (2017). Direct observation of halide migration and its effect on the photoluminescence of methylammonium lead bromide perovskite single crystals. *Adv. Mater.* 29: 1703451. (1–7).

97 Nie, W., Blancon, J.C., Neukirch, A.J. et al. (2016). Light-activated photocurrent degradation and self-healing in perovskite solar cells. *Nat. Commun.* 7: 11574. (1–9).

98 Bischak, C.G., Hetherington, C.L., Wu, H. et al. (2017). Origin of reversible photoinduced phase separation in hybrid perovskites. *Nano Lett.* 17: 1028–1033.

99 Hoke, E.T., Slotcavage, D.J., Dohner, E.R. et al. (2015). Reversible photo-induced trap formation in mixed-halide hybrid perovskites for photovoltaics. *Chem. Sci.* 6: 613–617.

100 Yun Jae, S., Seidel, J., Kim, J. et al. (2016). Critical role of grain boundaries for ion migration in formamidinium and methylammonium lead halide perovskite solar cells. *Adv. Energy Mater.* 6: 1600330. (1–8).

101 Jena, A.K., Numata, Y., Ikegami, M., and Miyasaka, T. (2018). Role of spiro-OMeTAD in performance deterioration of perovskite solar cells at high temperature and reuse of the perovskite films to avoid Pb-waste. *J. Mater. Chem. A* 6: 2219–2230.

102 Jena, A.K., Ikegami, M., and Miyasaka, T. (2017). Severe morphological deformation of spiro-OMeTAD in $(CH_3NH_3)PbI_3$ solar cells at high temperature. *ACS Energy Lett.* 2: 1760–1761.

103 Carrillo, J., Guerrero, A., Rahimnejad, S. et al. (2016). Ionic reactivity at contacts and aging of methylammonium lead triiodide perovskite solar cells. *Adv. Energy Mater.* 6: 1502246. (1–7).

104 Zhao, Y., Zhou, W., Tan, H. et al. (2017). Mobile-ion-induced degradation of organic hole-selective layers in perovskite solar cells. *J. Phys. Chem. C* 121: 14517–14523.

105 Rakstys, K., Igci, C., and Nazeeruddin, M.K. (2019). Efficiency vs. stability: dopant-free hole transporting materials towards stabilized perovskite solar cells. *Chem. Sci.* 10: 6748–6769.

106 Yang, W.S., Park, B.W., Jung, E.H. et al. (2017). Iodide management in formamidinium-lead-halide–based perovskite layers for efficient solar cells. *Science* 356: 1376–1379.

107 Huang, C., Fu, W., Li, C.Z. et al. (2016). Dopant-free hole-transporting material with a C3h symmetrical truxene core for highly efficient perovskite solar cells. *J. Am. Chem. Soc.* 138: 2528–2531.

108 Cao, Y., Li, Y., Morrissey, T. et al. (2019). Dopant-free molecular hole transport material that mediates a 20% power conversion efficiency in a perovskite solar cell. *Energy Environ. Sci.* 12: 3502–3507.

109 Paek, S., Qin, P., Lee, Y. et al. (2017). Dopant-free hole-transporting materials for stable and efficient perovskite solar cells. *Adv. Mater.* 29: 1606555.

110 Feng, Y., Hu, Q., Rezaee, E. et al. (2019). Perovskite solar cells: high performance and stable perovskite solar cells based on dopant-free arylamine-substituted copper(II) phthalocyanine hole-transporting materials. *Adv. Energy Mater.* 9: 1970104.

111 Zhang, L., Liu, C., Zhang, J. et al. (2018). Intensive exposure of functional rings of a polymeric hole-transporting material enables efficient perovskite solar cells. *Adv. Mater.* 30: 1804028.

112 Kim, G.-W., Lee, J., Kang, G. et al. (2018). Donor–acceptor type dopant-free, polymeric hole transport material for planar perovskite solar cells (19.8%). *Adv. Energy Mater.* 8: 1701935.

113 Öz, S., Jena, A., Kulkarni, A. et al. (2020). Lead(II) propionate additive and a dopant-free polymer hole transport material for $CsPbI_2Br$ perovskite solar cells. *ACS Energy Lett.* 5: 1292–1299.

114 Guo, Z., Jena, A.K., Takei, I. et al. (2020). V_{OC} over 1.4 V for amorphous tin oxide-based dopant-free $CsPbI_2Br$ perovskite solar cells. *J. Am. Chem. Soc.* 142: 9725–9734.

115 Eperon, G.E., Paterno, G.M., Sutton, R.J. et al. (2015). Inorganic caesium lead iodide perovskite solar cells. *J. Mater. Chem. A* 3: 19688–19695.

116 Wang, Y., Dar, M.I., Ono, L.K. et al. (2019). Thermodynamically stabilized β-CsPbI3–based perovskite solar cells with efficiencies >18%. *Science* 365: 591–595.

117 Zhang, T., Dar, M.I., Li, G. et al. (2017). Bication lead iodide 2D perovskite component to stabilize inorganic α-$CsPbI_3$ perovskite phase for high-efficiency solar cells. *Sci. Adv.* 3: e1700841. (1–7).

118 Jena, A.K., Kulkarni, A., Sanehira, Y. et al. (2018). Stabilization of α-$CsPbI_3$ in ambient room temperature conditions by incorporating Eu into $CsPbI_3$. *Chem. Mater.* 30: 6668–6674.

119 Babayigit, A., Ethirajan, A., Muller, M., and Conings, B. (2016). Toxicity of organometal halide perovskite solar cells. *Nat. Mater.* 15: 247–251.

120 Ke, W. and Kanatzidis, M.G. (2019). Prospects for low-toxicity lead-free perovskite solar cells. *Nat. Commun.* 10: 965.

121 Jokar, E., Chien, C.H., Tsai, C.M. et al. (2019). Robust tin-based perovskite solar cells with hybrid organic cations to attain efficiency approaching 10%. *Adv. Mater.* 31: 1804835.

122 Liu, X., Wang, Y., Wu, T. et al. (2020). Efficient and stable tin perovskite solar cells enabled by amorphous-polycrystalline structure. *Nat. Commun.* 11: 2678.
123 Tai, Q., Guo, X., Tang, G. et al. (2019). Antioxidant grain passivation for air-stable tin-based perovskite solar cells. *Angew. Chem.* 58: 806–810.
124 Kamarudin, M.A., Hirotani, D., Wang, Z. et al. (2019). Suppression of charge carrier recombination in lead-free tin halide perovskite via Lewis base post-treatment. *J. Phys. Chem. Lett.* 10 (17): 5277–5283.
125 Kawai, T., Ishii, A., Kitamura, T. et al. (1996). Optical absorption in band-edge region of $(CH_3NH_3)_3Bi_2I_9$ single crystals. *J. Phys. Soc. Jpn.* 65: 1464–1468.
126 Arakcheeva, A., Bonin, M., Chapuis, G., and Zaitsev, A.I. (1999). The phases of $Cs_3Bi_2I_9$ between RT and 190 K. *Z. Kristallogr. - Cryst. Mater.* 214: 279–283.
127 Szklarz, P., Pietraszko, A., Jakubas, R. et al. (2008). Structure, phase transitions and molecular dynamics of $[C(NH_2)_3]_3[M_2I_9]$, M = Sb, Bi. *J. Phys. Condens. Matter* 20: 255221.
128 Kulkarni, A., Singh, T., Ikegami, M., and Miyasaka, T. (2017). Photovoltaic enhancement of bismuth halide hybrid perovskite by *N*-methyl pyrrolidone-assisted morphology conversion. *RSC Adv.* 7: 9456–9460.
129 Jain, S.M., Phuyal, D., Davies, M.L. et al. (2018). An effective approach of vapour assisted morphological tailoring for reducing metal defect sites in lead-free, $(CH_3NH_3)_3Bi_2I_9$ bismuth-based perovskite solar cells for improved performance and long-term stability. *Nano Energy* 49: 614–624.
130 Kulkarni, A., Jena, A.K., Ikegami, M., and Miyasaka, T. (2019). Performance enhancement of AgBi2I7 solar cells by modulating a solvent-mediated adduct and tuning remnant BiI_3 in one-step crystallization. *Chem. Commun.* 55: 4031–4034.
131 Kim, H.D., Ohkita, H., Benten, H., and Ito, S. (2016). Photovoltaic performance of perovskite solar cells with different grain sizes. *Adv. Mater.* 28: 917–922.
132 Boopathi, K.M., Karuppuswamy, P., Singh, A. et al. (2017). Solution-processable antimony-based light-absorbing materials beyond lead halide perovskites. *J. Mater. Chem. A* 5: 20843–20850.
133 Jiang, F., Yang, D., Jiang, Y. et al. (2018). Chlorine-incorporation-induced formation of the layered phase for antimony-based lead-free perovskite solar cells. *J. Am. Chem. Soc.* 140: 1019–1027.
134 Nie, R., Mehta, A., Park, B.W. et al. (2018). Mixed sulfur and iodide-based lead-free perovskite solar cells. *J. Am. Chem. Soc.* 140: 872–875.
135 Chen, M., Ju, M.G., Carl, A.D. et al. (2018). Cesium titanium(IV) bromide thin films based stable lead-free perovskite solar cells. *Joule* 2: 558–570.
136 Nie, R., Sumukam, R.R., Reddy, S.H. et al. (2020). Lead-free perovskite solar cells enabled by hetero-valent substitutes. *Energy Environ. Sci.* 13: 2363–2385.
137 Jošt, M., Kegelmann, L., Korte, L., and Albrecht, S. (2020). Monolithic perovskite tandem solar cells: a review of the present status and advanced characterization methods toward 30% efficiency. *Adv. Energy Mater.* 10: 1904102.
138 Zhang, A., Li, Z., Meng, L. et al. (2020). Perovskite-based tandem solar cells: get the most out of the sun. *Adv. Funct. Mater.* 30: 2001904.

139 McMeekin, D.P., Sadoughi, G., Rehman, W. et al. (2016). A mixed-cation lead mixed-halide perovskite absorber for tandem solar cells. *Science* 351: 151–155.

140 Bush, K.A., Palmstrom, A.F., Yu, Z.J. et al. (2017). 23.6%-efficient monolithic perovskite/silicon tandem solar cells with improved stability. *Nat. Energy* 2: 17009.

141 Jäger, K., Kolte, L., Rech, B., and Albrecht, S. (2017). Numerical optical optimization of monolithic planar perovskite-silicon tandem solar cells with regular and inverted device architectures. *Opt. Express* 25: A473–A482.

142 Duong, T., Wu, Y., Shen, H. et al. (2017). Rubidium multication perovskite with optimized bandgap for perovskite-silicon tandem with over 26% efficiency. *Adv. Energy Mater.* 7: 1700228.

143 Yu, Z., Yang, Z., Ni, Z. et al. (2020). Simplified interconnection structure based on C60/SnO_{2-x} for all-perovskite tandem solar cells. *Nat. Energy* 5: 657–665.

144 Miyasaka, T. (2021). *Perovskite Photovoltaics and Optoelectronics – Fundamentals to Advanced Applications*. Wiley.

Part I

Characteristics of Hybrid Perovskites

3

Crystal Structures

Mitsutoshi Nishiwaki, Tatsuya Narikuri, and Hiroyuki Fujiwara

Gifu University, Department of Electrical, Electronic and Computer Engineering, 1-1 Yanagido, Gifu 501-1193, Japan

3.1 What Is Hybrid Perovskite?

Perovskite compounds have a simple basic structure expressed by ABX_3, and a wide variety of oxide, chalcogenide, and halide materials adopt a perovskite-based crystal structure. Figure 3.1 illustrates the structure of a cubic ABX_3 perovskite, where an A-site center cation is confined in a cage structure formed by B-site cations and X-site anions. In particular, the B atom is sixfold coordinated with the X atoms, creating a BX_6 octahedron that is interconnected via all corners to form a three-dimensional perovskite network (Figure 1.1a).

A hybrid perovskite is a unique form of crystal consisting of a monovalent organic A-site cation (A = $CH_3NH_3^+$ and $HC(NH_2)_2^+$), divalent B-site group-IV cations (B = Pb^{2+} and Sn^{2+}), and X-site halide anions (X = I^-, Br^-, and Cl^-). The charge neutrality of hybrid perovskites is expressed as $A^+B^{2+}(X^-)_3$; namely, A^+ is balanced with BX_3^- (or PbI_3^-, for example). In a well-known $BaTiO_3$ perovskite, the charge neutrality is described as $Ba^{2+}Ti^{4+}(O^{2-})_3$, forming a charge state of $A^{2+} - BX_3^{2-}$, which is different from halide perovskites.

When a perovskite crystal has a cubic symmetry, its structure is straightforward. In the perovskite cube in Figure 3.1, there are a 1/8 volume of B atoms at each corner and 1/4 volume of X atoms at each edge center; thus, within a cube, there exist four atoms (i.e. BX_3^-) expressed by $1/8 \times 8 + 1/4 \times 12 = 4$, in addition to the A site contribution. When the A-site cation is methylammonium (MA^+: $CH_3NH_3^+$), there are 12 atoms in the unit cell. In hybrid perovskites, depending on the choices of the A, B, and X atoms, BX_6 octahedra are tilted (distorted) and the B–X–B angle deviates from an ideal 180° (Section 3.2). Rather remarkably, in hybrid perovskites, optoelectronic properties including band gap (Figure 4.9) and mobility (Figure 6.11), as well as the stability of the crystals vary significantly with the selection (or size) of the A-site cation. A unique perovskite structural parameter, Goldschmidt's tolerance factor, is a critical value enabling us to discuss overall perovskite stability (Section 3.3). Hybrid perovskite crystals exhibit strong phase

Hybrid Perovskite Solar Cells: Characteristics and Operation,
First Edition. Edited by Hiroyuki Fujiwara.

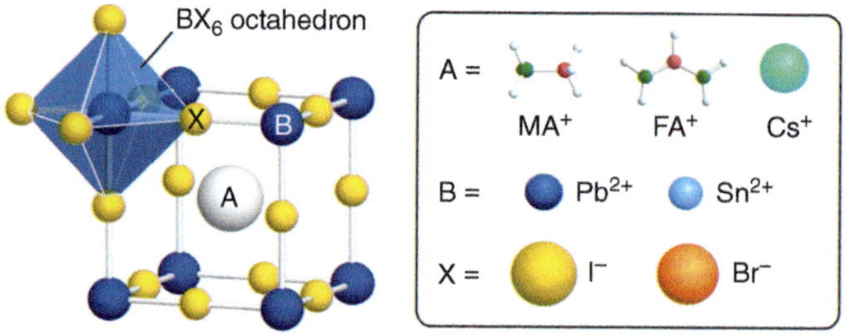

Figure 3.1 Schematic structure of a cubic ABX_3 perovskite. The A, B, and X indicate the A-site cation, B-site cation, and X-site anion, respectively. The blue area represents a BX_6 octahedron of the perovskite structure. The cations and anions employed for hybrid perovskite formation are also summarized. The MA^+ and FA^+ represent $CH_3NH_3^+$ and $HC(NH_2)_2^+$, respectively.

variation with temperature (Section 3.4). Moreover, the organic cations in hybrid perovskites reorient rapidly in a picosecond timescale, while interacting with the BX_3^- scaffold through hydrogen bonding (Section 3.5). An understanding of the detailed perovskite structures described in this chapter is critical to interpreting the physical properties and stability of hybrid perovskite crystals.

3.2 Structures of Hybrid Perovskite Crystals

$MAPbI_3$ hybrid perovskites exhibit three different crystal symmetries (cubic, tetragonal, and orthorhombic) depending on the temperature. In this section, the crystal structures of $MAPbI_3$ and other important hybrid perovskites are introduced. Here, the structures of secondary phase materials, often formed with hybrid perovskites, are also explained.

3.2.1 Crystal Structure of $MAPbI_3$

$MAPbI_3$ crystal phases are categorized into (i) α phase (cubic), (ii) β phase (tetragonal), and γ phase (orthorhombic) [1–5]. Figure 3.2 explains the formation of the β and γ phases from the α phase $MAPbI_3$. The cubic structure in Figure 3.2 is identical to that of Figure 3.1, except that A is MA^+. The tetragonal and orthorhombic phases are the most common non-cubic perovskite structures, which are also characterized by a three-dimensional network of corner-sharing octahedra. In the tetragonal phase, only one tilting angle is different from zero, whereas all the tilting angles are nonzero in the orthorhombic phase. In the β phase, the Pb–I–Pb angle of the *c*-axis is 180°, as in the cubic phase; however, the PbI_6 octahedra are off-centered along the *c*-axis and are rotated around the *c*-axis. Moreover, the corner-sharing octahedral BX_6 units in the β phase have a two-story structure where neighboring octahedra are tilted inward and outward at each level (see also Figure 5.4c).

In the case of $MAPbI_3$, the room-temperature phase is the β phase and the α phase exists only at high temperature ($T > 327$ K). At low temperatures, the orthorhombic

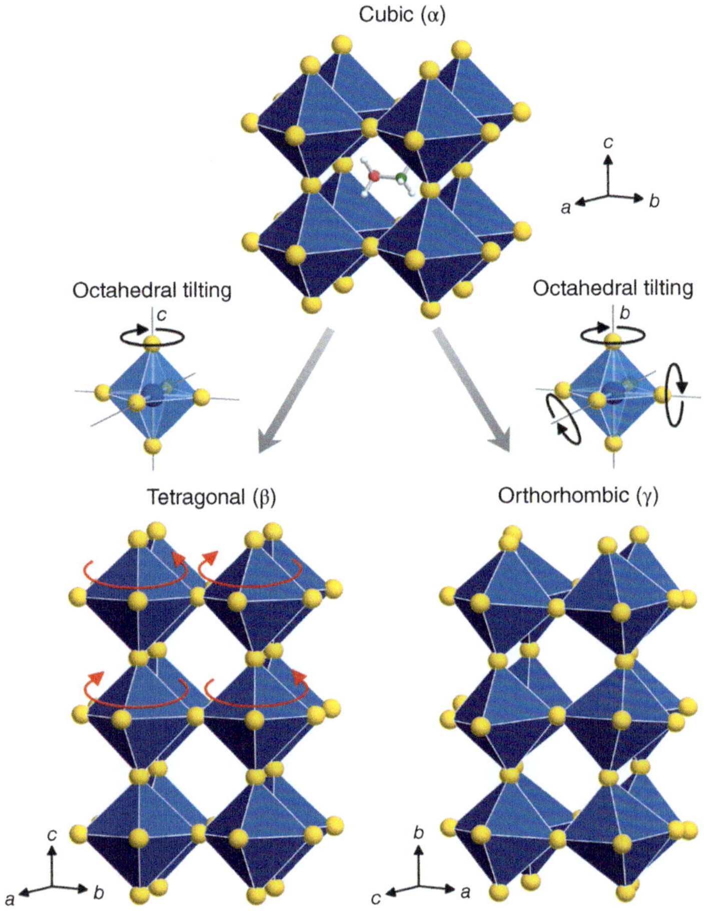

Figure 3.2 Structure of a $MAPbI_3$ cubic (α) phase and formation of the tetragonal (β) and orthorhombic (γ) phases by the octahedral tilting of the α phase crystal. The *a*, *b*, and *c* indicate the axis directions. In the β phase, the PbI_6 octahedra are rotated only around the *c*-axis, while the octahedra are tilted in all directions in the γ phase. Note that the selection of the axis directions is different between the β and γ phases.

phase appears. These phases exhibit a rather sharp transition at the corresponding phase transition temperatures (see Section 3.4).

Figure 3.3 shows the top view of the α, β, and γ phase crystals of $MAPbI_3$. The dotted lines in Figure 3.3 represent the unit cells of each structure, and the crystallographic orientations of each phase are indicated in the inset. The configurations of MA^+ depicted in Figure 3.3 correspond to those obtained from density functional theory (DFT) calculations (see Section 5.3). In the α and β phase crystals, MA^+ orientation is not fixed and its C–N axis is reorienting rapidly. However, these MA^+ cations cannot be rotated freely due to the hydrogen bonding interaction between N—H and I (i.e. N—H···I), whereas, in the γ phase, the dynamic motion of MA^+ is prohibited and MA^+ is fully ordered (Section 3.5). In the Figures 3.1–3.3, the Pb–I–Pb

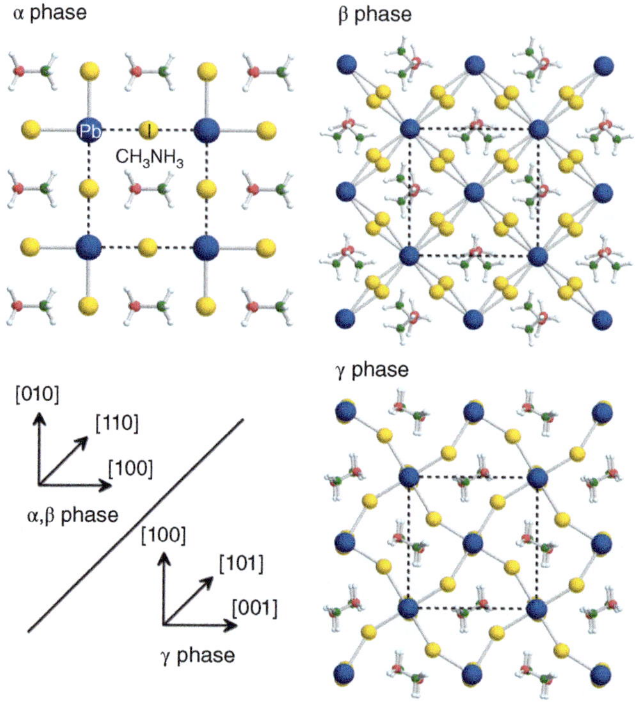

Figure 3.3 Top view of the α, β, and γ phase crystals of $MAPbI_3$. The dotted lines represent the unit cells of each structure, and the crystallographic orientations of each phase are indicated in the inset. The configurations of MA^+ in the figure were obtained from DFT calculations (Figure 5.4), and the orientation of the C–N axis is essentially consistent with the experiment (Section 3.5). Note that the [100] direction in the α phase corresponds to the [110] direction in the β phase, and the unit cell size of the β phase is by a factor of $\sqrt{2}$ larger than that of the α phase.

angle in the α phase is 180°. In the DFT calculation results, the Pb–I–Pb bond angle varies with the size of the A-site cation and the bond angle approaches 180° as the cation size becomes larger (Figure 5.14).

Figure 3.4 shows the X-ray diffraction (XRD) spectra calculated for the different phases of $MAPbI_3$. For the XRD calculations, the crystal structures determined experimentally are used (see Section 3.2.2). The XRD spectrum of the α phase is characterized by a strong (100) peak at $2\theta = 14.0°$, which originates from the perovskite cage structure. In the tetragonal phase, the intense (100) peak in the α phase is split into (002) and (110). As indicated in Figure 3.3, the unit cell size of the β phase is larger than that of the α phase and the [100] direction of the α phase corresponds to the [110] direction in the β phase. For the *c*-axis direction, the α and β phases have the same structure; however, because the β phase has a two-story structure, (001) in the α phase corresponds to (002) in the β phase. The weak (211) diffraction peak at $2\theta = 23.5°$ is a characteristic signature of the β phase formation; this peak is absent in the α phase.

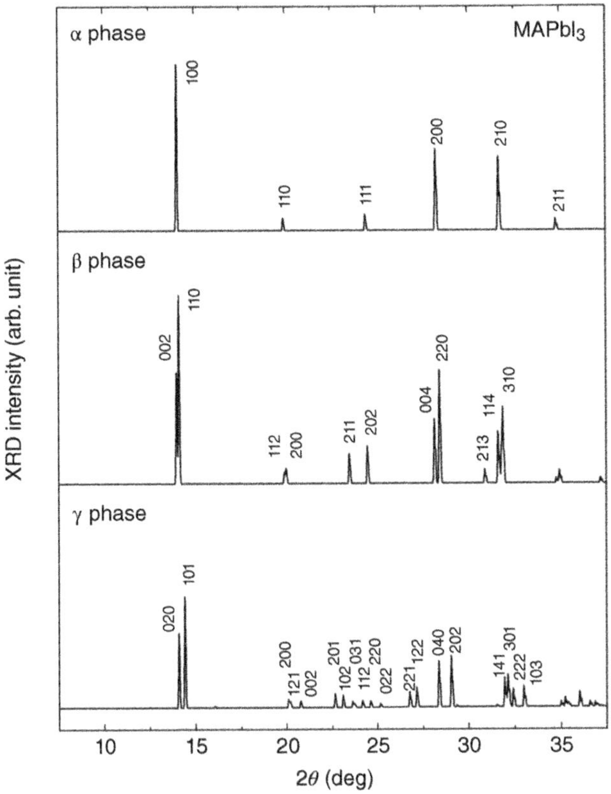

Figure 3.4 XRD spectra calculated for the α, β, and γ phases of $MAPbI_3$. For the XRD calculations, the lattice parameters summarized in Section 3.2.2, determined experimentally, are employed. The crystal plane orientations of major diffraction peaks are also indicated.

In Table 3.1, the XRD 2θ peak positions for the different $MAPbI_3$ phases are summarized. In Table 3.1, the diffraction peak positions of δ-phase formamidinium lead iodide ($HC(NH_2)_2PbI_3$, $FAPbI_3$) and $CsPbI_3$, as well as important secondary phase materials described in Section 3.2.3, are included.

3.2.2 Lattice Parameters of Hybrid Perovskites

Table 3.2 summarizes the structures and lattice parameters of various hybrid perovskite materials reported in Refs. [1, 3, 4, 6–8]. As mentioned above, the crystal structure of $MAPbI_3$ changes with temperature; however, the reported transition temperatures differ slightly. For the β → α transition, temperatures in the range 327–333 K have been reported [1–5], whereas the γ → β transition has been reported to occur in the range 161–165 K [1, 2, 5]. In Table 3.2, the α transition temperature of 327 K and β transition temperature of 165 K reported by Weller et al. [5] are indicated.

Table 3.1 2θ angles of major XRD diffraction peaks observed for $MAPbI_3$ (α, β, and γ) and secondary phases including PbI_2, δ-$CsPbI_3$, δ-$FAPbI_3$, $CH_3NH_3PbI_3 \cdot H_2O$, and $(CH_3NH_3)_4PbI_6 \cdot 2H_2O$.

2θ (degree)	Materials	Orientation
8.10	$CH_3NH_3PbI_3 \cdot H_2O$	0 0 1
8.66	$CH_3NH_3PbI_3 \cdot H_2O$	1 0 0
9.82	δ-$CsPbI_3$	1 0 1
9.95	δ-$CsPbI_3$	0 0 2
10.66	$CH_3NH_3PbI_3 \cdot H_2O$	1 0 −1
11.46	$(CH_3NH_3)_4PbI_6 \cdot 2H_2O$	1 1 −1
11.79	δ-$FAPbI_3$	1 0 0
11.81	$(CH_3NH_3)_4PbI_6 \cdot 2H_2O$	1 0 0
12.08	$(CH_3NH_3)_4PbI_6 \cdot 2H_2O$	1 0 −2
12.60	PbI_2	0 0 1
13.08	δ-$CsPbI_3$	1 0 2
13.97	β-$MAPbI_3$	0 0 2
14.02	α-$MAPbI_3$	1 0 0
14.07	γ-$MAPbI_3$	0 2 0
14.10	β-$MAPbI_3$	1 1 0
14.40	γ-$MAPbI_3$	1 0 1
16.28	δ-$FAPbI_3$	1 0 1
19.88	α-$MAPbI_3$	1 1 0
19.90	β-$MAPbI_3$	1 1 2
20.00	β-$MAPbI_3$	2 0 0
20.08	γ-$MAPbI_3$	2 0 0
20.18	γ-$MAPbI_3$	1 2 1
20.49	δ-$FAPbI_3$	1 1 0
20.75	γ-$MAPbI_3$	0 0 2
20.99	δ-$CsPbI_3$	1 1 1
22.07	PbI_2	1 0 0
22.48	δ-$FAPbI_3$	0 0 2
22.63	γ-$MAPbI_3$	2 0 1
22.70	δ-$CsPbI_3$	2 0 3
23.08	γ-$MAPbI_3$	1 0 2
23.47	β-$MAPbI_3$	2 1 1
23.61	γ-$MAPbI_3$	0 3 1
24.15	γ-$MAPbI_3$	1 1 2
24.40	α-$MAPbI_3$	1 1 1
24.47	β-$MAPbI_3$	2 0 2

(continued)

Table 3.1 (Continued)

2θ (degree)	Materials	Orientation
24.60	γ-$MAPbI_3$	2 2 0
25.07	$(CH_3NH_3)_4PbI_6·2H_2O$	1 3 −1
25.16	γ-$MAPbI_3$	0 2 2
25.37	δ-$CsPbI_3$	1 1 3
25.46	δ-$FAPbI_3$	1 0 2
25.51	PbI_2	0 1 1
25.59	$CH_3NH_3PbI_3·H_2O$	1 1 −2
25.66	PbI_2	1 −1 1
25.72	δ-$CsPbI_3$	2 1 1
26.11	$CH_3NH_3PbI_3·H_2O$	2 0 2
26.28	δ-$FAPbI_3$	2 0 1
26.58	$(CH_3NH_3)_4PbI_6·2H_2O$	1 3 −2
26.75	γ-$MAPbI_3$	2 2 1
27.13	γ-$MAPbI_3$	1 2 2
27.17	δ-$CsPbI_3$	2 1 2
28.15	β-$MAPbI_3$	0 0 4
28.26	α-$MAPbI_3$	2 0 0
28.35	γ-$MAPbI_3$	0 4 0
28.42	β-$MAPbI_3$	2 2 0
28.85	$CH_3NH_3PbI_3·H_2O$	2 1 −2
28.96	$CH_3NH_3PbI_3·H_2O$	3 0 1
29.03	γ-$MAPbI_3$	2 0 2
30.61	δ-$FAPbI_3$	1 1 2
30.89	β-$MAPbI_3$	2 1 3
31.32	δ-$CsPbI_3$	0 1 5
31.54	δ-$FAPbI_3$	2 1 0
31.61	β-$MAPbI_3$	1 1 4
31.67	α-$MAPbI_3$	2 1 0
31.86	β-$MAPbI_3$	3 1 0
31.94	γ-$MAPbI_3$	1 4 1
32.11	γ-$MAPbI_3$	3 0 1
32.40	γ-$MAPbI_3$	2 2 2
32.90	δ-$FAPbI_3$	2 0 2
32.98	γ-$MAPbI_3$	1 0 3
33.56	δ-$FAPbI_3$	2 1 1
33.84	PbI_2	1 0 2
34.79	α-$MAPbI_3$	2 1 1

The details of the secondary phases are described in Section 3.2.3.

Table 3.2 Structures, space groups, and lattice parameters of various hybrid perovskite materials.

Materials	Structure (phase)	Space group	Temperature (K)	Lattice parameter (Å)	References
$MAPbI_3$	Pseudo-cubic (α)	*P4mm*	>327[a)]	$a = 6.3115$ $b = 6.3115$ $c = 6.3161$	[3]
	Tetragonal (β)	*I4cm*[b)]	165–327[a)]	$a = 8.849$ $b = 8.849$ $c = 12.642$	[3]
	Orthorhombic (γ)	*Pnma*	<165[a)]	$a = 8.8362$ $b = 12.5804$ $c = 8.5551$	[4]
$MAPbBr_3$	Cubic (α)	*Pm3m*	>237	$a = 5.901$	[1]
$MAPbCl_3$	Cubic (α)	*Pm3m*	>179	$a = 5.675$	[1]
$FAPbI_3$	Cubic (α)[c)]	*Pm3m*	298	$a = 6.3620$	[6]
	Hexagonal (δ)[d)]	$P6_3mc$	293	$a = 8.6603$ $b = 8.6603$ $c = 7.9022$	[3]
$CsPbI_3$	Orthorhombic (δ)	*Pnma*	293	$a = 10.4342$ $b = 4.7905$ $c = 17.7610$	[3]
$MASnI_3$	Pseudo-cubic (α)	*P4mm*	293	$a = 6.2302$ $b = 6.2302$ $c = 6.2316$	[3]
$MASn_{0.5}Pb_{0.5}I_3$	Pseudo-cubic (α)	*P4mm*	RT	$a = 6.348$ $b = 6.348$ $c = 6.379$	[7]
$FASnI_3$	Orthorhombic (α)	*Amm2*	340	$a = 6.3286$ $b = 8.9554$ $c = 8.9463$	[3]
$CsSnI_3$	Orthorhombic (γ)[e)]	*Pnam*	300	$a = 8.688$ $b = 8.643$ $c = 12.378$	[8]

a) Reported by Weller et al. [5].
b) Different assignments exist for this phase (i.e. *I4/mcm* in Refs. [4, 5]). However, a recent study [9] also confirms that the space group of the β phase is best described by *I4cm*.
c) Metastable structure of $FAPbI_3$. α-$FAPbI_3$ gradually changes into δ-$FAPbI_3$.
d) Stable phase of $FAPbI_3$.
e) Metastable phase.

In the α and β phases of $MAPbI_3$, the internal angles of the unit cells are $\alpha = \beta = \gamma = 90°$. In the high temperature α phase, however, the experimental lattice parameters of a and b are slightly shorter than that of c and the crystal structure becomes pseudo-cubic with the average value of 6.313 Å [3]. The unit cell size of the β phase is larger by a factor of $\sqrt{2}$ than that of the α phase (see Figure 3.3) and, therefore, $a_{\beta\text{-phase}} \sim \sqrt{2}a_{\alpha\text{-phase}}$ in Table 3.2. From the structure of Figure 3.2, we can also notice a relation of $c_{\beta\text{-phase}} \sim 2c_{\alpha\text{-phase}}$. Moreover, in the β phase, $c/2 > a/\sqrt{2}$. Accordingly, the lattice is expanded slightly along the c axis. This is the reason why the (100) XRD peak of the α phase splits into two peaks [i.e. (110) and (002)] in the β phase (see Figure 3.4). For the β phase, different lattice parameters of $a = b = 8.874$ Å and $c = 12.670$ Å have also been reported [4]. For the γ phase, the b axis is traditionally chosen as the z axis (see Figure 3.2), and thus $c_{\beta\text{-phase}} \sim b_{\gamma\text{-phase}}$ in Table 3.2.

Interestingly, $MAPbBr_3$ and $MAPbCl_3$ adopt the cubic α phase at room temperature. $FAPbI_3$ also shows a cubic structure at room temperature; however, this is a metastable phase, and its crystal structure gradually changes into the hexagonal δ phase [6] (see Section 3.2.3). The mixed-cation hybrid perovskites of $CsFAPb(I,Br)_3$, used widely for solar cell fabrication, exhibit the cubic α phase [10, 11]. The α-phase formation is also dominant in $FAMAPb(I,Br)_3$ alloys. The Sn-containing perovskite alloys ($MASn_xPb_{1-x}I_3$)) show the pseudo-cubic structures when $x \geq 0.5$, while the β phase is observed at $x < 0.5$ [7].

3.2.3 Secondary Phase Materials

Hybrid perovskite thin films often contain secondary phases [10–13], such as PbI_2. The suppression of an unfavored impurity phase is critical for the formation of high efficiency devices [10, 12]. Figure 3.5 shows the crystal structures of the secondary phase materials including (a) PbI_2, (b) δ-$CsPbI_3$, and (c) δ-$FAPbI_3$. The dotted lines in Figure 3.5 represent the unit cell structure. The PbI_2 crystal has a two-dimensional plate-like structure, consisting of a PbI_6 edge-sharing octahedron as indicated in Figure 3.5d [14, 15]. In particular, PbI_2 has the continuous stacking of a I–Pb–I unit and the separation of the Pb atom plane along the c axis is 7.0 Å. PbI_2 is the dominant phase formed after the exposure of the hybrid perovskite to humid air, and the hexagonal plate structure is generally confirmed after prolonged H_2O exposure [13, 16, 17].

The $CsPbI_3$ crystal adopts a δ phase shown in Figure 3.5b at room temperature [3]. This crystal has an orthorhombic symmetry consisting of a one-dimensional crystal structure. The cross section of this one-dimensional crystal is formed by double columns of edge-sharing PbI_6 octahedra. The δ-$FAPbI_3$ also has a one-dimensional crystal structure propagating along the [100] direction; however, in this case, the linear crystal structure is made of face-sharing octahedra with a single chain structure shown in Figure 3.5d [3].

Figure 3.6 shows the XRD spectra of PbI_2, δ-$CsPbI_3$, and δ-$FAPbI_3$, calculated from the crystal structures of Figure 3.5. In Figure 3.6, the XRD spectra of the secondary phases are compared with that of $MAPbI_3$ (β phase). The major XRD peak positions are also summarized in Table 3.1. All secondary phases indicate characteristic peaks

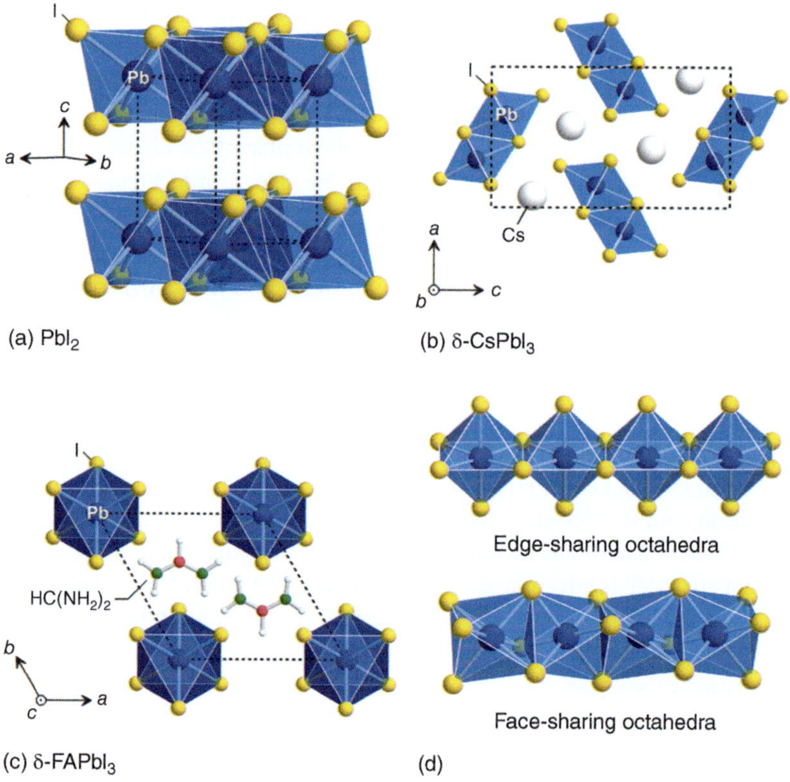

Figure 3.5 Crystal structures of (a) hexagonal PbI_2, (b) orthorhombic δ-$CsPbI_3$, and (c) hexagonal δ-$FAPbI_3$. The dotted lines in the figure represent the unit cell structures. The *a*, *b*, and *c* indicate the axis directions. The PbI_2 and δ-$CsPbI_3$ consist of PbI_6 edge-sharing octahedron shown in (d), while δ-$FAPbI_3$ has a single chain structure made of face-sharing octahedra, also illustrated in (d). The orientation of FA^+ indicated in (c) was determined by DFT.

at low diffraction angles of $2\theta \leq 13.1°$, which are helpful for confirming the presence of impurity phases (Section 3.3.2).

As known well, $MAPbI_3$ shows strong degradation in humid air and the perovskite crystal is converted into a completely different phase by H_2O incorporation (Figure 1.12) [13, 18, 19]. Figure 3.7 shows the crystal structures of two hydrate phases of $CH_3NH_3PbI_3{\cdot}H_2O$ [20] and $(CH_3NH_3)_4PbI_6{\cdot}2H_2O$ [21], determined experimentally. The black lines indicate the unit cell structures. The hydrate phases have monoclinic structures, and the $CH_3NH_3PbI_3{\cdot}H_2O$ phase has a one-dimensional PbI_6 structure, which is the same as δ-$CsPbI_3$ in Figure 3.5b (i.e. double chain of edge-sharing PbI_6 octahedron). In $(CH_3NH_3)_4PbI_6{\cdot}2H_2O$, the PbI_6 octahedron is completely separated, indicating a zero-dimensional structure [22]. The configurations of the MA^+ and H_2O molecules in these hydrate phases can be interpreted by the strong hydrogen bonding interactions generated by H_2O: namely, (i) I atoms (O—H···I) and (ii) MA^+ (N—H···O) [22, 23]. Consequently, the O—H bonds are

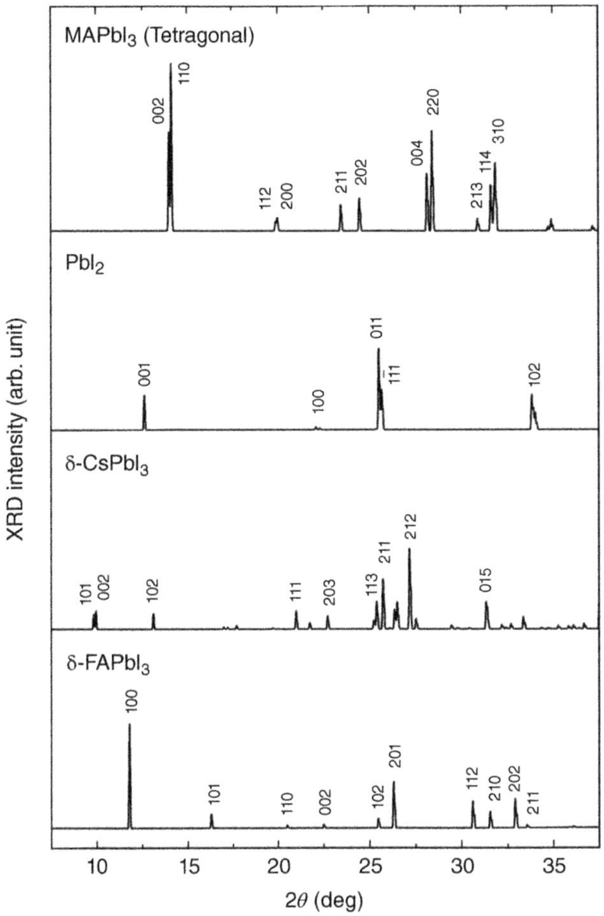

Figure 3.6 XRD spectra of PbI_2, δ-$CsPbI_3$, and δ-$FAPbI_3$ calculated from the crystal structures of Figure 3.5. The major XRD peak positions of this figure are summarized in Table 3.1. The crystal orientations of the major diffraction peaks are also indicated.

oriented in the direction of the I atoms, whereas a N—H· · ·O(H_2)· · ·H—N local structure is generated between MA^+ and H_2O [22].

Figure 3.8 shows the XRD spectra of the hydrate phases calculated from the structures of Figure 3.7. The XRD peak positions of Figure 3.8 are also summarized in Table 3.1. The formation of the hydrate structures can be characterized by low-angle diffraction peaks of (100) at 8.7° in $CH_3NH_3PbI_3{\cdot}H_2O$ and $(11\bar{1})$ at 11.5° in $(CH_3NH_3)_4PbI_6{\cdot}2H_2O$ [16, 18, 19].

3.3 Tolerance Factor

The tolerance factor (*t*), proposed by V. M. Goldschmidt in 1926, is the most important structural parameter for expressing the stability of perovskite compounds

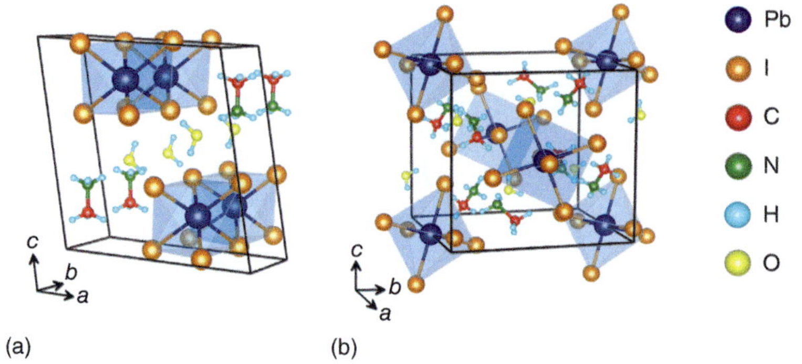

Figure 3.7 Crystal structures of two hydrate phases of $CH_3NH_3PbI_3 \cdot H_2O$ [20] and $(CH_3NH_3)_4PbI_6 \cdot 2H_2O$ [21], determined experimentally. The unit cell structures are denoted by black lines. The arrows represent the directions of the *a*, *b*, and *c* axes of the unit cells. The actual compositions of the unit cells in (a) and (b) are $2(CH_3NH_3PbI_3 \cdot H_2O)$ and $2[(CH_3NH_3)_4PbI_6 \cdot 2H_2O]$, respectively. *Source*: Shirayama et al. [13].

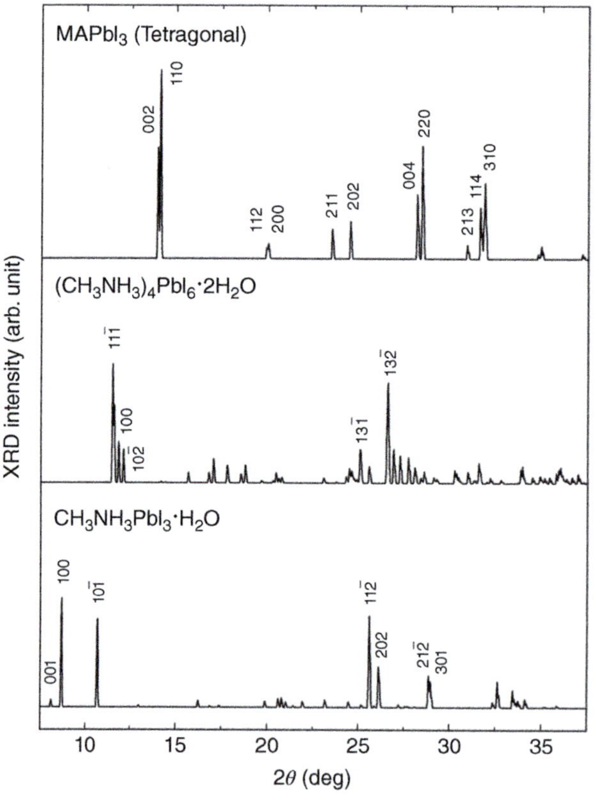

Figure 3.8 XRD spectra of the hydrate phases of $(CH_3NH_3)_4PbI_6 \cdot 2H_2O$ and $CH_3NH_3PbI_3 \cdot H_2O$, calculated from the structures of Figure 3.7. The XRD peak positions of this figure are also summarized in Table 3.1. The crystal orientations of major diffraction peaks are indicated.

and continues to be used as a guiding principle to design hybrid perovskite materials [10, 24–29]. The t is a simple dimensionless value calculated by

$$t = \frac{(r_A + r_X)}{\sqrt{2}(r_B + r_X)} \tag{3.1}$$

where r_A, r_B, and r_X denote the ionic radii of the A-, B-, and X-site atoms. The meaning of Eq. (3.1) is straightforward and can be understood easily from the perovskite crystal structure shown in Figure 3.9. In Figure 3.9, the unit cell was chosen such that the B-site cation was at the center and the perovskites have alternating atomic planes consisting of the A–X and B–X atoms. In the close pack configuration, the diagonal distance from the center of the square to the corner is $r_A + r_X$ for the A–X plane, while we obtain $\sqrt{2}(r_B + r_X)$ in the B–X plane. Accordingly, t simply represents the ratio of the diagonal distances in the alternating A–X and B–X planes. However, it can be observed from Figure 3.9 that r_A needs to be much larger than r_B to form perovskite structures.

To date, almost all known oxide perovskites (i.e. ABO_3) have been confirmed to have t values in the range 0.75–1.00 [30]. However, this is not a sufficient condition for a hybrid perovskite. In this section, the calculation and interpretation of hybrid perovskite t values are discussed (Section 3.3.1) and we further study the t values of mixed-cation perovskites [i.e. $FAMAPb(I,Br)_3$ and $CsFAPb(I,Br)_3$] and their effect on the stability of perovskite crystals (Section 3.3.2).

3.3.1 Tolerance Factor of Hybrid Perovskites

For the calculation of t using Eq. (3.1), the actual r_A, r_B, and r_X values are required. Table 3.3 summarizes the ionic radii that can be applied for the t calculation of hybrid perovskites. In Table 3.3, for the r_A of organic cations, the values calculated by Kieslich et al. [25] are indicated, whereas the r_B and r_X represent those reported by Shannon [31], except for Sn^{2+} [29, 30].

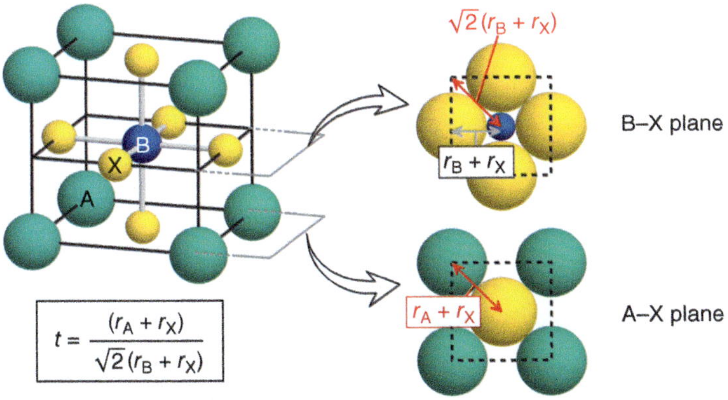

Figure 3.9 Calculation of Goldschmidt's tolerance factor (t) for a perovskite structure. The r_A, r_B, and r_X represent the ionic radii of the A-, B-, and X-site atoms. The t value is calculated as a ratio of the diagonal distances obtained on the A–X and B–X planes.

Table 3.3 Ionic radii (r_A, r_B, and r_X) of monovalent A site cation (A^+), divalent B-site cation (B^{2+}), and monovalent X-site anion (X^-) in ABX_3 halide perovskites.

A^+	r_A (Å)	B^{2+}	r_B (Å)	X^-	r_X (Å)
Ammonium: NH_4^+	1.46	Pb^{2+}	1.19	F^-	1.33
Hydroxylammonium: NH_3OH^+	2.16	Sn^{2+}	1.10	Cl^-	1.81
Methylammonium: $CH_3NH_3^+$	2.17	Ge^{2+}	0.73	Br^-	1.96
Hydrazinium: $NH_3NH_2^+$	2.17	Be^{2+}	0.45	I^-	2.20
Azetidinium: $(CH_2)_3NH_2^+$	2.50	Mg^{2+}	0.72		
Formamidinium: $HC(NH_2)_2^+$	2.53	Ca^{2+}	1.00		
Imidazolium: $C_3N_2H_5^+$	2.58	Sr^{2+}	1.18		
Dimethylammonium: $(CH_3)_2NH_2^+$	2.72	Ba^{2+}	1.35		
Ethylammonium: $(CH_3CH_2)NH_3^+$	2.74	V^{2+}	0.79		
Guanidinium: $(NH_2)_3C^+$	2.78	Mn^{2+}	0.83		
Tetramethylammonium: $(CH_3)_4N^+$	2.92	Fe^{2+}	0.78		
K^+	1.38	Co^{2+}	0.745		
Rb^+	1.52	Ni^{2+}	0.69		
Cs^+	1.67	Pd^{2+}	0.86		
		Pt^{2+}	0.80		
		Cu^{2+}	0.73		
		Zn^{2+}	0.74		

For r_A of organic cations, the values calculated by Kieslich et al. [25] are shown, while the r_A (K^+, Rb^+, and Cs^+), r_B, and r_X values are from the results of Shannon [31], except for Sn^{2+} [29, 30]. The r_A values of K^+, Rb^+, and Cs^+ represent those obtained assuming sixfold coordination [31].
Source: Kieslich et al. [25]; Shannon et al. [31].

Although representative ionic radii are described in Table 3.3, the calculation of ionic radii is in general difficult as an ion is not a rigid sphere and its size depends on the peripheral conditions. For example, r_B changes with the choice of r_X due to the effect of the electronegativity [28], and the coordination number of r_A (i.e. 6-fold or 12-fold coordination) also influences the value [31]. (In Table 3.3, the r_A values of K^+, Rb^+, and Cs^+ represent those of the sixfold coordination reported in Ref. [31].) Moreover, in the case of organic cations, the cation has a nonspherical geometry and hydrogen bonding further varies the bond length, making the determination of r_A difficult. For the r_A of organic cations in Table 3.3, the results have been obtained using a rather simple rigid-sphere model assuming free rotation of the organic cations [25].

Due to the uncertainty of the exact cation and anion sizes, the calculation of an accurate t value is a rather challenging problem. In fact, a variety of ionic radii have been applied for the t estimation of hybrid perovskite [25–29, 32]. Nevertheless, t is a vital parameter, allowing the interpretation of perovskite stability and device performance. In the following, we discuss the t values of hybrid perovskites, calculated according to the representative values indicated in Table 3.3.

For an understanding of structural stability of perovskites, another structural parameter, octahedral factor μ, is often considered. The μ is a simple parameter calculated by the ratio of r_B and r_X [30]:

$$\mu = r_B/r_X \tag{3.2}$$

Figure 3.10 shows the relation of t and μ, calculated for different hybrid perovskites. In Figure 3.10, in addition to (t, μ) of monocation perovskites ($MAPbX_3$, $FAPbI_3$, and $CsPbI_3$), those of mixed-cation perovskites are shown. The t of the mixed-cation perovskites can be calculated by considering the effective radius of r_A (i.e. $r_{A,eff}$) according to

$$r_{A,eff} = xr_{A1} + (1-x)r_{A2} \tag{3.3}$$

where x indicates the molar ratio with r_{A1} and r_{A2} being the radii of two different A-site cations. The effective radius of the X-site anions ($r_{X,eff}$) for $APb(I,Br)_3$ alloys can also be estimated in a similar manner.

In Figure 3.10, the red circles indicate (t, μ) pairs that generate single perovskite phases, whereas the open circles indicate the conditions that lead to the formation of non-perovskite phases (i.e. δ-$CsPbI_3$ and δ-$FAPbI_3$). Moreover, the blue circles represent the formation of the α-phase/δ-phase mixtures, as reported for $Cs_xFA_{1-x}PbI_3$ [10] and $Cs_xFA_{1-x}Pb(I_{0.6}Br_{0.4})_3$ [11]. In general, materials with t = 0.8–1.0 are considered to adopt a perovskite structure [24], while the orthorhombic and hexagonal δ-phases are formed at $t < 0.8$ and $t > 1.0$, respectively. The result of

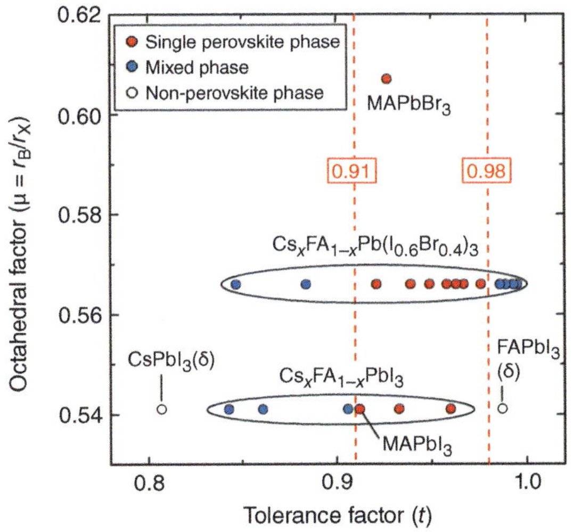

Figure 3.10 Relation of tolerance factor t and octahedral factor μ, calculated for various hybrid perovskites using the numerical values of Table 3.3. The red circles indicate the single perovskite phase, while the open circles show the non-perovskite phases. The blue circles represent the formation of the α-phase/δ-phase mixtures, as reported for $Cs_xFA_{1-x}PbI_3$ [10] and $Cs_xFA_{1-x}Pb(I_{0.6}Br_{0.4})_3$ [11]. This figure indicates that the single perovskite phase is formed in a narrow region $t = 0.91$–0.98 with a center position of $t = 0.945$.

Figure 3.10 essentially agrees with this picture, even though the absolute t values of $CsPbI_3$ (0.807) and $FAPbI_3$ (0.987) are slightly inconsistent. More importantly, the analysis of Figure 3.10 indicates a rather narrow region ($t = 0.91–0.98$) for the formation of the single perovskite phase. Thus, the center position of $t = 0.945$ appears to be the most favorable value for the stabilization of perovskite architecture. In other words, for $Cs_xFA_{1-x}PbI_3$ alloys, the Cs content of $x \sim 0.2$ provides an optimum [10, 33] and a large deviation from the optimum enhances the phase separation into the α and δ phases. The best device performance has also been obtained for hybrid perovskite absorbers in the consistent range $t = 0.94–0.98$ [10]. Conversely, a wide range of μ is possible for the perovskite formation, provided that t is in the range $t = 0.91–0.98$.

3.3.2 Tolerance Factor of Mixed-Cation Perovskites

Mixed-cation hybrid perovskites are now the main materials applied to solar cell fabrication [10, 12, 24, 27, 33–36], and the t value provides a guiding principle for designing proper multi-cation perovskites. Figure 3.11a shows t of $APbX_3$ calculated for different $r_{A,eff}$ and $r_{X,eff}$ and, by following the result of Figure 3.10, the t range 0.91–0.98 is highlighted with the optimum of $t = 0.945$ in red. In Figure 3.11a, the guidelines are indicated for the $APbX_3$ alloys with $A = MA^+$, FA^+, and Cs^+, and $X = I^-$ and Br^-. From similar calculations, the corresponding t contour lines are obtained for $FA_{1-y}MA_yPb(I_{1-x}Br_x)_3$ (Figure 3.11b) and $Cs_yFA_{1-y}Pb(I_{1-x}Br_x)_3$ (Figure 3.11c). Since the r_X values of I^- and Br^- are similar, $r_{X,eff}$ does not vary significantly with the Br content x. In contrast, the r_A of Cs^+ is quite small, compared with FA^+. Accordingly, the t value of the alloys is governed predominantly by $r_{A,eff}$, and the selection of an ideal t can be realized through the mixing of A-site cations. As shown in Figure 3.11a, in $MAPb(I_{1-x}Br_x)_3$, the perovskite structure is maintained in the entire x range, whereas $CsPb(I,Br)_3$ and $FAPb(I,Br)_3$ show unsuitable t, independent of x. For the Cs- and FA-based materials, therefore, the control of $r_{A,eff}$ is critical.

In $FA_{1-y}MA_yPbI_3$, the incorporation of smaller MA^+ moves t toward the favorable value, generating a stable cubic perovskite phase. A small amount of MA^+ ($y > 0.1$) is sufficient to cross the α-phase/δ-phase boundary (see Figure 3.11b). With the inclusion of Br^-, the minimum amount of MA^+ for the α-phase formation increases slightly. Since $FAPbI_3$ exhibits the smallest band gap (E_g) among $APbI_3$ perovskites (see Figure 4.9), $FA_{1-y}MA_yPb(I_{1-x}Br_x)_3$ alloys are generally formed using small x and y to enhance the longer wavelength optical response. Thus, $FAMAPb(I,Br)_3$ absorbers are often made near the border of $t = 0.98$ [36].

In $Cs_yFA_{1-y}Pb(I_{1-x}Br_x)_3$, the incorporation of small Cs^+ has a key role in forming the structurally stable perovskite. By Cs^+ incorporation, the perovskite unit cell shrinks and the (100) XRD peak shifts toward a greater 2θ angle [10, 11]. Since t changes significantly with the Cs content y, t can be tuned toward more structurally stable regions by controlling y. With the optimum Cs^+ content of $y = 0.2$, an ideal t of ~0.95 can be realized (Figure 3.11c). In fact, a number of studies confirm that $y = 0.1–0.3$ in $Cs_yFA_{1-y}Pb(I,Br)_3$-based alloys is essential for generating stable perovskite structures with some great advantages [10–12, 33–35]; i.e. (i) phase

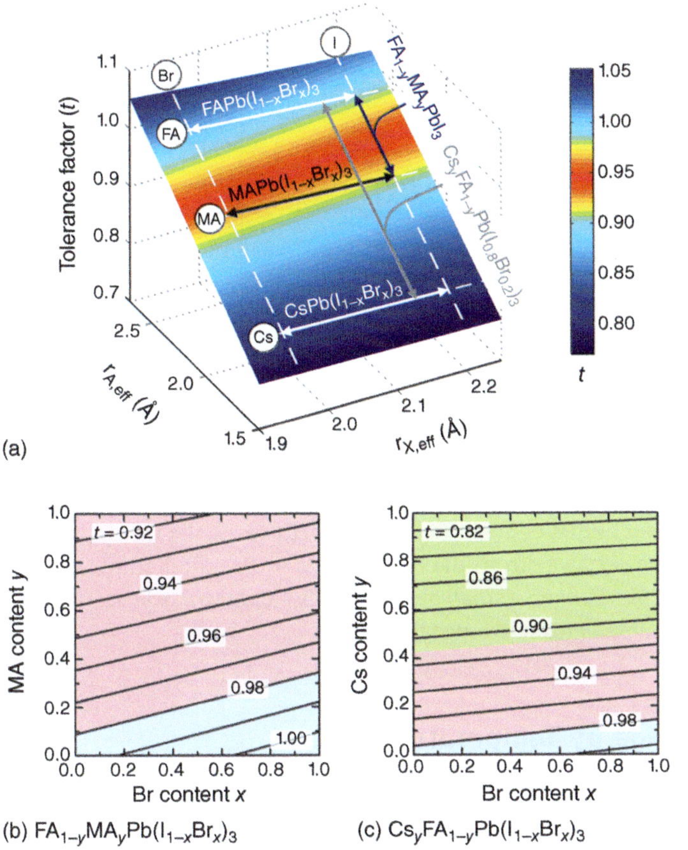

Figure 3.11 (a) Tolerance factor t of $APbX_3$ calculated for different $r_{A,eff}$ and $r_{X,eff}$ and the corresponding t contour lines in (b) $FA_{1-y}MA_yPb(I_{1-x}Br_x)_3$ and (c) $Cs_yFA_{1-y}Pb(I_{1-x}Br_x)_3$. In (a), the favorable t range of 0.91–0.98 is highlighted with the optimum of $t = 0.945$ being in the red color, according to the result of Figure 3.10. For selected $APbX_3$ alloys with $A = MA^+$, FA^+, and Cs^+ and $X = I^-$ and Br^-, the guidelines are shown. In (b) and (c), the calculated t for each composition is shown with the red, blue, and green areas indicating t regions with $t = 0.91$–0.98, $t > 0.98$ and $t < 0.91$, respectively.

impurities can be avoided, (ii) perovskite formation becomes less sensitive to process conditions, and (iii) decomposition into PbI_2 is further suppressed. An x value of ~0.2 is also beneficial for maintaining E_g sufficiently low (see Figure 4.10a). Since $MAPbI_3$ shows lesser thermal stability due to the desorption of small MA^+ at a low temperature of 100 °C [37, 38], research into more thermally stable $CsFAPb(I,Br)_3$ has become popular.

The discussion of t can further be applied for triple-cation-mixed α-phase perovskites, $CsFAMAPb(I,Br)_3$. Figure 3.12 shows two XRD spectra for the Cs content x of 0 and 0.05 in $Cs_x(FA_{0.83}MA_{0.17})_{1-x}Pb(I_{0.83}Br_{0.17})_3$ [12]. When $x = 0$, the alloy is $FA_{0.83}MA_{0.17}Pb(I_{0.83}Br_{0.17})_3$ with $t = 0.98$. As confirmed from Figure 3.11b, the composition of this alloy is in the boundary and the XRD shows the formation of

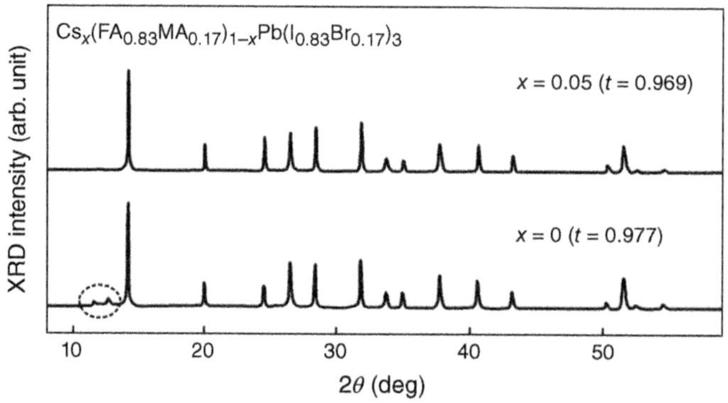

Figure 3.12 XRD spectra for the Cs content x of 0 and 0.05 in α-phase $Cs_x(FA_{0.83}MA_{0.17})_{1-x}Pb(I_{0.83}Br_{0.17})_3$. The corresponding t values are indicated. The XRD peaks in a dotted circle indicate the formation of PbI_2 and δ-$FAPbI_3$. *Source*: Saliba et al. [12].

δ-$FAPbI_3$ and PbI_2 (see a dotted circle in Figure 3.12). However, the Cs^+ incorporation with $x = 0.05$ pushes t away from the boundary ($t = 0.97$) and eliminates the phase impurities. As a result, even a small amount of Cs^+ leads to a drastic improvement in forming a suitable perovskite crystal.

3.4 Phase Change by Temperature

Hybrid perovskite crystals exhibit strong phase variation by temperature [1–5, 10]. The phase change temperature is closely related to material processing temperatures. In this section, the phase transformations of $MAPbI_3$, $FAPbI_3$, $CsFAPbI_3$, and $CsPbI_3$ with temperature are explained.

Figure 3.13 shows the variation of 2θ diffraction patterns with increasing temperature, determined by neutron powder diffraction [5]. In Figure 3.13, a clear γ → β phase transition at 165 K can be seen clearly. The β → α phase transition occurs at 327 K, and the diffraction patterns are simplified notably in the cubic phase. A similar cubic transition temperature of 330 K has also been determined from the temperature-dependent X-ray powder diffraction experiment [4]. In this case, the (211) diffraction peak, which is the strong signature of the tetragonal phase (see Figure 3.4), disappears as the temperature increases.

Figure 3.14 shows the temperature dependences of the XRD for $FAPbI_3$, $Cs_{0.15}FA_{0.85}PbI_3$, and $CsPbI_3$ [10]. In this experiment, the temperature was increased by a heating stage, where the gray areas denoted in Figure 3.14 indicate the range where the temperature was held constant. In the case of $FAPbI_3$, δ-$FAPbI_3$ is formed at room temperature and the sharp δ → α phase transition occurs at 165 °C. At higher temperature (~200 °C), the PbI_2 formation occurs due to the thermal desorption of FA^+.

Remarkably, the δ-to-α transition temperature decreases notably by the incorporation of Cs^+, and the α phase is generated at 125 °C with the Cs content of 15 at.%.

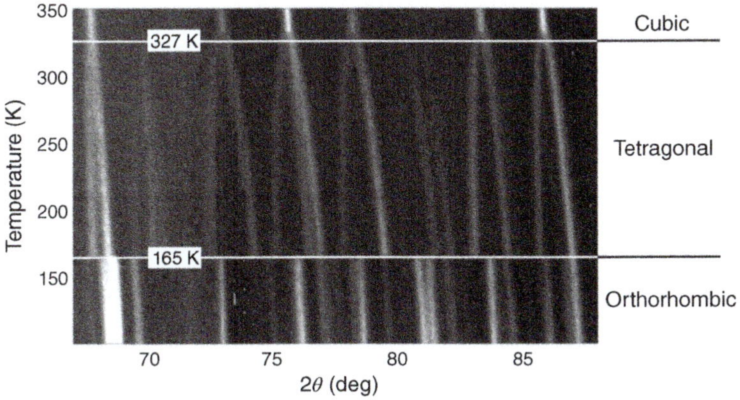

Figure 3.13 Variation of 2θ diffraction patterns with increasing temperature, determined by neutron powder diffraction. In the figure, a clear $\gamma \rightarrow \beta$ phase transition at 165 K and a $\beta \rightarrow \alpha$ phase transition at 327 K can be observed. *Source*: Weller et al. [5].

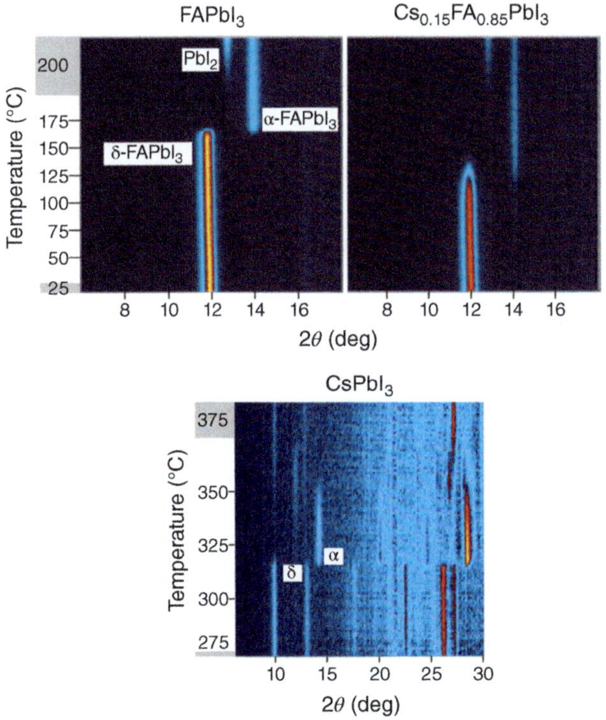

Figure 3.14 Phase changes of $FAPbI_3$, $Cs_{0.15}FA_{0.85}PbI_3$ and $CsPbI_3$ with temperature, determined by XRD. In the experiment, the temperature was increased by a heating stage and the gray areas denoted in the figure indicate the range where the temperature was held constant. *Source*: Li et al. [10].

Moreover, when Cs^+ is introduced, the $\delta \rightarrow \alpha$ transition occurs less sharply and the α phase is formed even below 100 °C at Cs = 30 at.% [10]. In contrast, orthorhombic δ-$CsPbI_3$ is a quite stable phase and the $\delta \rightarrow \alpha$ transition is observed at a very high temperature of 315 °C. It can be observed that, in $CsPbI_3$, the PbI_2 formation is negligible as the desorption of Cs^+ is quite difficult.

3.5 Refined Structures of Hybrid Perovskites

Organic A-site cations incorporated into $APbX_3$ hybrid perovskites show unique dynamical motion within an inorganic cage. In particular, the rotation of organic cations is not completely random and the freedom of the cations is restricted by the formation of strong hydrogen bonding, although such interaction depends strongly on the A-site species and temperature [5, 6]. This section considers more detailed dynamics of organic cations placed within an inorganic scaffold.

3.5.1 Orientation of Center Cations

Based on powder neutron diffraction experiments, Weller et al. determined the precise location of MA^+ and FA^+ within the framework of PbI_3^- at different temperatures [5, 6]. Figure 3.15 shows the refined crystal structures of (a) γ phase (100 K), (b) β phase (180 K), (c) α phase (352 K) of $MAPbI_3$, and (d) α phase of $FAPbI_3$ (298 K) [5, 6]. In the $MAPbI_3$ γ phase, the MA^+ cations are fully ordered. Specifically, the C—N bonds lie normal to the *b*-axis (see Figure 3.3), and the C → N vector has an alternating order in neighboring unit cells along the *b* axis direction. The gray area in Figure 3.15a indicates 90% probability ellipsoid (time-averaged atom positions) for the H atoms of MA^+. Even though the H atoms show the relatively large freedom at both ends of MA^+ at 100 K, the orientation of MA^+ is essentially fixed because of the hydrogen bonding between NH_3 and I (i.e. N—H···I). The strong hydrogen bonding in the γ phase can be evidenced from the short N—H···I distances of 2.6–2.8 Å. In this orthorhombic phase, the tilting of the PbI_6 octahedra results in Pb–I–Pb angles of 161.94° (along the *b* axis) and 150.75° (along the *a* and *c* axes).

Figure 3.15b shows one orientation of MA^+ in the $MAPbI_3$ β phase at 180 K with 50% probability ellipsoids. The β-phase exhibits a smaller octahedral tilting with a Pb–I–Pb angle of 157.92° (along the *a* and *b* axes) and 180° (along *c* axis). In the β phase, the C–N axis of MA^+ is preferentially directed toward the face center position of the distorted cubic structure because of the hydrogen bonding interaction (N—H···I distance: 3.2 Å); however, the cation is disordered around the *c*-axis (see also Figure 3.3). The molecular dynamics (MD) simulations also confirm the preferential orientation of MA^+ even at room temperature [39–43], and the dynamical motion of MA^+ is strongly correlated with the I atoms due to the presence of N—H···I interaction. Indeed, neutron scattering [41] and ultrafast infrared-vibrational spectroscopy [42] experimentally confirmed the strong

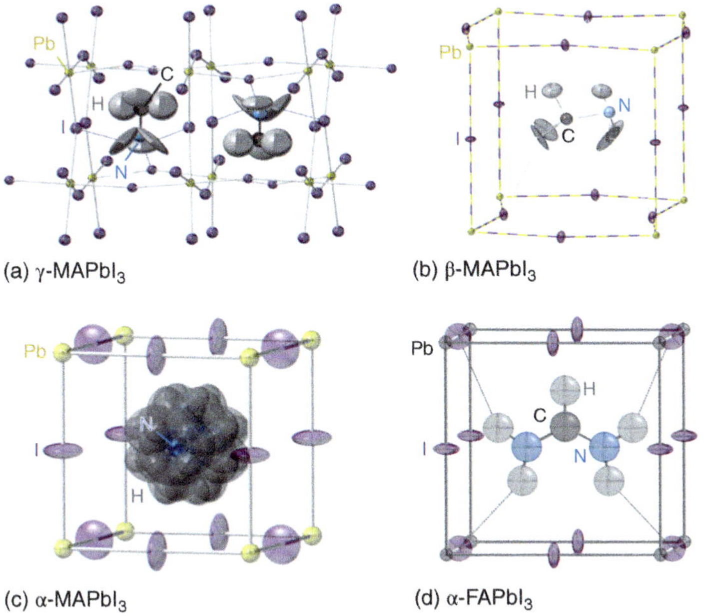

Figure 3.15 Refined crystal structures of (a) γ phase (100 K), (b) β phase (180 K), (c) α phase (352 K) of $MAPbI_3$, and (d) α phase of $FAPbI_3$ (298 K), as reported by Weller et al. [5, 6]. The result in (a) indicates 90% probability ellipsoids for the constituent atoms, while those in (b) and (c) represent 50% probability ellipsoids. In (d), the result of 30% probability is shown. The dotted lines indicate hydrogen bonding interaction generated between N–H and I (N–H· · ·I). *Source*: (a–c) Weller et al. [5]. Licensed under CC BY 3.0; (d) Weller et al. [6].

coupling of MA^+ with PbI_3^-. Moreover, a large-scale MD simulation of $MAPbI_3$ shows that roughly half of MA^+ is participating in the hydrogen bonding [40].

At a high temperature of 352 K, the perovskite crystal becomes cubic with all Pb-I-Pb angles being 180°. The 50% probability ellipsoid of the α phase in Figure 3.15c shows that the Pb atom position is fixed, whereas the I atoms move with greater freedom. Hydrogen bonding interaction becomes weaker at 352 K (N—H· · ·I distance: 3.1–3.5 Å), and MA^+ is orientationally more disordered, rotating rapidly near the unit cell center.

In Figure 3.15d, the 30%-probability ellipsoid representation of α-$FAPbI_3$ (cubic) at room temperature is shown. In this crystal, the atomic configuration of PbI_3^- is almost the same as that of the cubic $MAPbI_3$ (see Figure 3.15c). In α-$FAPbI_3$, however, the N-(C)-N axis of FA^+ is directed toward the center of the cube face and the free motion of FA^+ is hindered effectively by the strong N—H· · ·I bonding that occurs at both ends of FA^+ (see dotted lines in Figure 3.15d). As a result, the N—H· · ·I distance in α-$FAPbI_3$ (2.8–3.0 Å) is considerably shorter, if compared with $MAPbI_3$ at the same temperature. Thus, the hydrogen bonding interaction is significant in α-$FAPbI_3$, and the preferential rotation of FA^+ around the N-(C)-N axis is the dominant motion in $FAPbI_3$ [44, 45].

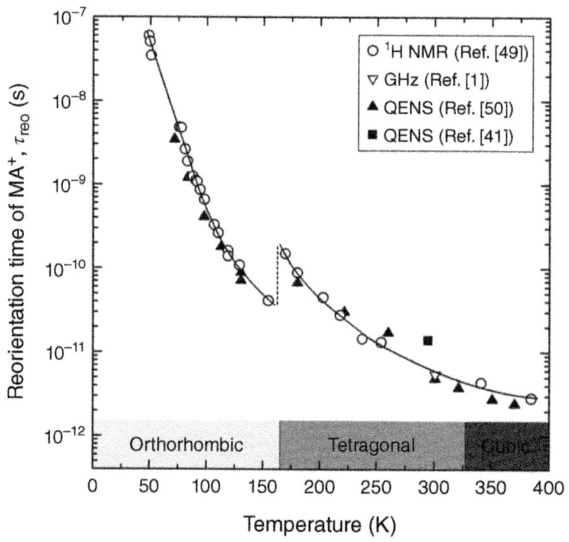

Figure 3.16 Variation of MA^+ reorientation time (τ_{reo}) in $MAPbI_3$ with temperature, as summarized by Fabini et al. [45]. In this figure, τ_{reo} determined by nuclear magnetic resonance (NMR) [49], GHz spectroscopy [1], and quasi-elastic neutron scattering (QENS) [41, 50] is shown. *Source*: Fabini et al. [45].

In the DFT calculations, the hydrogen bonding in $APbI_3$ discussed above is well reproduced [37, 46, 47] (see Figure 5.3).

3.5.2 Relaxation of Center Cations

The organic center cations reorient rapidly within the inorganic scaffold, even though organic cations interact strongly with the PbI_3^- framework through hydrogen bonding. It is now well established that MA^+ and FA^+ in PbI_3^- show similar reorientation with a time scale of 0.5–14 ps at room temperature [1, 39–42, 44, 45, 48–50]. Figure 3.16 shows the variation of MA^+ reorientation time (τ_{reo}) in $MAPbI_3$ with temperature, as summarized by Fabini et al. [45]. In Figure 3.16, τ_{reo} determined by nuclear magnetic resonance (NMR) [49], GHz spectroscopy [1], and quasi-elastic neutron scattering (QENS) [41, 50] is shown. As confirmed from Figure 3.16, all the measurements show consistent results and the relaxation time becomes longer with decreasing temperature as the thermal motion becomes less significant. It can be observed that the $\gamma \rightarrow \beta$ transition varies τ_{reo} significantly, whereas the effect of the $\beta \rightarrow \alpha$ transition on τ_{reo} is negligible. This stems from the fact that the crystal structure of the γ phase is quite different from those of the β and α phases.

Because the free motion of the organic cations is prohibited by the intense hydrogen bonding, τ_{reo} in Figure 3.16 needs to be interpreted as the time interval from one configuration with the hydrogen bonding to another with a similar configuration, yet with a different orientation. In other words, MA^+ jumps between preferential orientations within the PbI_3^- cage [41, 42]. Moreover, MA^+ is wobbling around the crystal axis before the reorientation [42]. Accordingly, the atomic arrangements, represented by Figure 3.15b,d, are important for the interpretation of hybrid perovskite crystals.

Acknowledgment

The authors acknowledge Yuuki Nagasaki for the preparation of Figures 3.12 and 3.14.

References

1 Poglitsch, A. and Weber, D. (1987). *J. Chem. Phys.* 87: 6373.
2 Onoda-Yamamuro, N., Matsuo, T., and Suga, H. (1990). *J. Phys. Chem. Solids* 51: 1383.
3 Stoumpos, C.C., Malliakas, C.D., and Kanatzidis, M.G. (2013). *Inorg. Chem.* 52: 9019.
4 Baikie, T., Fang, Y., Kadro, J.M. et al. (2013). *J. Mater. Chem. A* 1: 5628.
5 Weller, M.T., Weber, O.J., Henry, P.F. et al. (2015). *Chem. Commun.* 51: 4180.
6 Weller, M.T., Weber, O.J., Frost, J.M., and Walsh, A. (2015). *J. Phys. Chem. Lett.* 6: 3209.
7 Hao, F., Stoumpos, C.C., Chang, R.P.H., and Kanatzidis, M.G. (2014). *J. Am. Chem. Soc.* 136: 8094.
8 Yamada, K., Funabiki, S., Horimoto, H. et al. (1991). *Chem. Lett.* 20: 801.
9 Breternitz, J., Lehmann, F., Barnett, S.A. et al. (2020). *Angew. Chem. Int. Ed.* 59: 424.
10 Li, Z., Yang, M., Park, J.-S. et al. (2016). *Chem. Mater.* 28: 284.
11 Rehman, W., McMeekin, D.P., Patel, J.B. et al. (2017). *Energy Environ. Sci.* 10: 361.
12 Saliba, M., Matsui, T., Seo, J.-Y. et al. (2016). *Energy Environ. Sci.* 9: 1989.
13 Shirayama, M., Kato, M., Miyadera, T. et al. (2016). *J. Appl. Phys.* 119: 115501.
14 Doni, E., Grosso, G., Harbeke, G. et al. (1975). *Phys. Status Solidi B* 68: 569.
15 Bordas, J., Robertson, J., and Jakobsson, A. (1978). *J. Phys. C: Solid State Phys.* 11: 2607.
16 Yang, J., Siempelkamp, B.D., Liu, D., and Kelly, T.L. (2015). *ACS Nano* 9: 1955.
17 Philippe, B., Park, B.-W., Lindblad, R. et al. (2015). *Chem. Mater.* 27: 1720.
18 Leguy, A.M.A., Hu, Y., Campoy-Quiles, M. et al. (2015). *Chem. Mater.* 27: 3397.
19 Christians, J.A., Miranda Herrera, P.A., and Kamat, P.V. (2015). *J. Am. Chem. Soc.* 137: 1530.
20 Hao, F., Stoumpos, C.C., Liu, Z. et al. (2014). *J. Am. Chem. Soc.* 136: 16411.
21 Wakamiya, A., Endo, M., Sasamori, T. et al. (2014). *Chem. Lett.* 43: 711.
22 Vincent, B.R., Robertson, K.N., Cameron, T.S., and Knop, O. (1987). *Can. J. Chem.* 65: 1042.
23 Dong, X., Fang, X., Lv, M. et al. (2015). *J. Mater. Chem. A* 3: 5360.
24 Correa-Baena, J.-P., Saliba, M., Buonassisi, T. et al. (2017). *Science* 358: 739.
25 Kieslich, G., Sun, S., and Cheetham, A.K. (2014). *Chem. Sci.* 5: 4712.
26 Chen, Q., De Marco, N., Yang, Y.M. et al. (2015). *Nano Today* 10: 355.
27 Jacobsson, T.J., Correa-Baena, J.-P., Pazoki, M. et al. (2016). *Energy Environ. Sci.* 9: 1706.

28 Travis, W., Glover, E.N.K., Bronstein, H. et al. (2016). *Chem. Sci.* 7: 4548.
29 Hoefler, S.F., Trimmel, G., and Rath, T. (2017). *Monatsh. Chem.* 148: 795.
30 Li, C., Lu, X., Ding, W. et al. (2008). *Acta Cryst.* B64: 702.
31 Shannon, R.D. (1976). *Acta Cryst.* A32: 751.
32 Ferdani, D.W., Pering, S.R., Ghosh, D. et al. (2019). *Energy Environ. Sci.* 12: 2264.
33 Lee, J.-W., Kim, D.-H., Kim, H.-S. et al. (2015). *Adv. Energy Mater.* 5: 1501310.
34 Yi, C., Luo, J., Meloni, S. et al. (2016). *Energy Environ. Sci.* 9: 656.
35 McMeekin, D.P., Sadoughi, G., Rehman, W. et al. (2016). *Science* 351: 151.
36 Jeon, N.J., Noh, J.H., Yang, W.S. et al. (2015). *Nature* 517: 476.
37 Shirayama, M., Kadowaki, H., Miyadera, T. et al. (2016). *Phys. Rev. Appl.* 5: 014012.
38 Conings, B., Drijkoningen, J., Gauquelin, N. et al. (2015). *Adv. Energy Mater.* 5: 1500477.
39 Mattoni, A., Filippetti, A., Saba, M.I., and Delugas, P. (2015). *J. Phys. Chem. C* 119: 17421.
40 Carignano, M.A., Kachmar, A., and Hutter, J. (2015). *J. Phys. Chem. C* 119: 8991.
41 Leguy, A.M.A., Frost, J.M., McMahon, A.P. et al. (2015). *Nat. Commun.* 6: 7124.
42 Bakulin, A.A., Selig, O., Bakker, H.J. et al. (2015). *J. Phys. Chem. Lett.* 6: 3663.
43 Deretzis, I., Mauro, B.N.D., Alberti, A. et al. (2016). *Sci. Rep.* 6: 24443.
44 Carignano, M.A., Saeed, Y., Aravindh, S.A. et al. (2016). *Phys. Chem. Chem. Phys.* 18: 27109.
45 Fabini, D.H., Siaw, T.A., Stoumpos, C.C. et al. (2017). *J. Am. Chem. Soc.* 139: 16875.
46 Fujiwara, H., Kato, M., Tamakoshi, M. et al. (2018). *Phys. Status Solidi. A* 215: 1700730.
47 Kato, M., Fujiseki, T., Miyadera, T. et al. (2017). *J. Appl. Phys.* 121: 115501.
48 Wasylishen, R.E., Knop, O., and Macdonald, J.B. (1985). *Solid. State Commun.* 56: 581.
49 Xu, Q., Eguchi, T., Nakayama, H. et al. (1991). *Z. Naturforsch., A: Phys. Sci.* 46: 240.
50 Chen, T., Foley, B.J., Ipek, B. et al. (2015). *Phys. Chem. Chem. Phys.* 17: 31278.

4

Optical Properties

Hiroyuki Fujiwara[1], Yukinori Nishigaki[1], Akio Matsushita[2], and Taisuke Matsui[2]

[1] *Gifu University, Department of Electrical, Electronic and Computer Engineering, 1-1 Yanagido, Gifu 501-1193, Japan*
[2] *Panasonic Corporation, Technology Innovation Division, 3-1-1 Yagumo-naka-machi, Moriguchi City, Osaka 570-8501, Japan*

4.1 Introduction

Strong light absorption within absorbers is one of the critical conditions for the efficient operation of solar cells. For hybrid perovskite absorbers that are formed typically by a spin coating process, the formation of a thicker layer (>1 μm) is difficult and a high absorption coefficient (α) is particularly important. Fortunately, various hybrid perovskites exhibit high α in the visible/ultraviolet (UV) region as shown in Figure 4.1, which summarizes the α spectra of methylammonium lead iodide ($CH_3NH_3PbI_3$; $MAPbI_3$) [1], formamidinium lead iodide ($HC(NH_2)_2PbI_3$; $FAPbI_3$) [2], $MAPbBr_3$ [3], $CsFAPb(I_{1-x}Br_x)_3$ [4], and $FAMA(Sn,Pb)I_3$ (see also tabulated data in Appendix A). In particular, hybrid perovskites indicate α of ~10^4 cm^{-1} in the band gap (E_g) region, which is comparable to representative solar cell materials including GaAs, $CuInGaSe_2$, and Cu_2ZnSnS_4 (see Figure 1.5).

Hybrid perovskites show unique light absorption properties in a wide energy region extending from 10^{-6} to 6 eV, due to the interband transition in the visible/UV region, chemical bond absorption in the infrared (IR) region, optical phonon absorption in the terahertz (THz) region, and organic cation orientation in a GHz range (Section 4.2). Remarkably, because hybrid perovskites consist of an organic cation and inorganic scaffold, these materials exhibit the optical signature characteristics of both organic and inorganic parts.

To date, numerous results have been reported for the optical properties of hybrid perovskites in the visible/UV region [1–53]. The E_g of hybrid perovskites changes largely by the choice of the A-, B-, and X-site species in ABX_3 perovskites (Section 4.3). Furthermore, for hybrid perovskites, a wide variety of alloys can be formed, allowing for a large variation of E_g by the mixture of different cations and

Hybrid Perovskite Solar Cells: Characteristics and Operation,
First Edition. Edited by Hiroyuki Fujiwara.

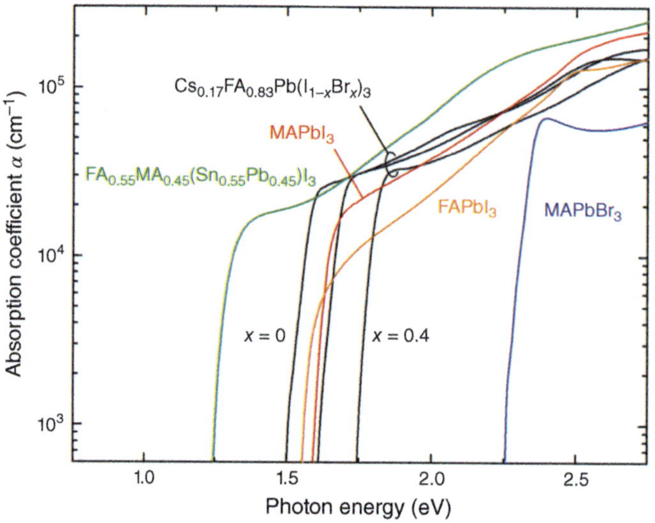

Figure 4.1 α spectra of various hybrid perovskites, determined by spectroscopic ellipsometry for $MAPbI_3$ [1], $FAPbI_3$ [2], $MAPbBr_3$ [3], $CsFAPb(I_{1-x}Br_x)_3$ and $FAMA(Sn,Pb)I_3$. For $CsFAPb(I_{1-x}Br_x)_3$, the optical constants reported in Ref. [4] were adjusted slightly by extra roughness correction [5]. In the figure, three spectra of $CsFAPb(I_{1-x}Br_x)_3$ with x of 0.0, 0.2, and 0.4 are shown. The numerical optical-constant values in the figure are summarized in Appendix A.

anions (Section 4.3). For the absolute α values of the perovskite materials, however, controversial results have been reported [1, 8–17]. The reason for the inconsistent optical data is an important and complex issue (Section 4.4). Nevertheless, it is now established that careful measurement and analysis based on spectroscopic ellipsometry (SE) provides accurate α values [1, 5]. In Figure 4.1, the reliable α spectra determined by SE have been summarized.

In $APbX_3$ perovskites, when the A-site cation species is varied, the magnitude of α in the E_g region changes and the variation of the X-site halide anion shifts the entire spectrum in the energy direction (Section 4.5). On the other hand, hybrid perovskites exhibit a sharp absorption tail (Section 4.6), which is a significant advantage of hybrid perovskite absorbers. For temperature variation, unlike other conventional semiconductor materials, the perovskite compounds show a slight E_g increase with temperature (Section 4.7).

As evidenced from Figure 4.1, $APbI_3$ hybrid perovskites show negligible excitonic band-edge absorption, whereas $MAPbBr_3$ exhibits an absorption peak due to exciton formation. The absence of an exciton peak in $APbI_3$ confirms the efficient generation of electron–hole carriers, and numerous studies have consistently demonstrated the weak exciton character of $MAPbI_3$ perovskites (Section 4.8).

This chapter describes the fundamental and unique optical characteristics of hybrid perovskite materials in detail. We emphasize that, in interpreting device characteristics and performing device simulations, accurate optical spectra are of critical importance.

4.2 Light Absorption in $MAPbI_3$

The interaction of light with materials can be described completely by two important physical quantities: refractive index n and extinction coefficient k [54]. The n is related to the wavelength λ of light (or electromagnetic wave), and λ of light reduces to λ/n in a material with n. In contrast, k is related to the magnitude of an electromagnetic wave and light intensity reduces strongly in a material with high k. The above (n, k) can be considered as a single complex number (complex refractive index) defined by

$$N = n - \mathrm{i}k \tag{4.1}$$

From k, a well-known α is obtained:

$$\alpha = \frac{4\pi k}{\lambda} \tag{4.2}$$

From Maxwell's equations, N is defined by a complex dielectric constant ($\varepsilon = \varepsilon_1 - \mathrm{i}\varepsilon_2$) [54]:

$$\varepsilon \equiv N^2 \tag{4.3}$$

Thus, from Eqs. (4.1) and (4.3), we obtain

$$\varepsilon_1 = n^2 - k^2 \tag{4.4}$$

$$\varepsilon_2 = 2nk \tag{4.5}$$

Although the representation is slightly different, all k, α, and ε_2 show the light absorption characteristics. In particular, $k = \alpha = \varepsilon_2 = 0$ when the light absorption is zero and we further obtain $\varepsilon_1 = n^2$ from Eq. (4.4) in this condition.

Figure 4.2 shows the dielectric function ($\varepsilon = \varepsilon_1 - \mathrm{i}\varepsilon_2$) of $MAPbI_3$, obtained experimentally in a quite wide energy region extending from 10^{-6} to 6 eV. In this figure, ε_1 is shown by a linear scale, whereas the ε_2 values are indicated by a logarithmic scale. The $MAPbI_3$ hybrid perovskite shows unique light absorption characteristics (i.e. $\varepsilon_2 > 0$) in different energy regions, which can be categorized into four distinct regions: (i) semiconductor-type light absorption in the visible/UV region, (ii) chemical bonding absorption in the IR region, (iii) optical phonon absorption in the THz region, and (iv) orientational absorption induced by the organic cation rotation in the GHz region. The experimental spectra of each corresponding region, shown in Figure 4.2, are taken from Ref. [1] (visible/UV), Ref. [55] (IR), Ref. [56] (THz), and Ref. [57] (GHz).

Physically, all the light absorption in Figure 4.2 can be explained by considering the dipole formation. An electromagnetic wave is essentially the wave of an electric field **E**, and when the light enters a material, the electric field separates the charge existing within the material, creating the positive-charge-rich and negative-charge-rich regions (i.e. electric dipole). In the visible/UV region, a nonzero dipole matrix element for the optical transition from the valence to conduction bands (i.e. interband transition) leads to the light absorption and the onset of $\varepsilon_2 > 0$

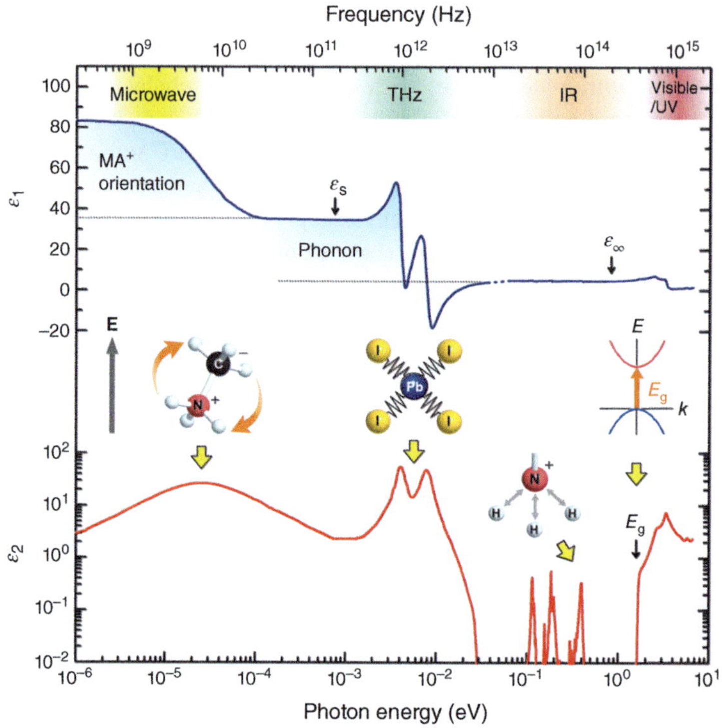

Figure 4.2 Dielectric function ($\varepsilon = \varepsilon_1 - i\varepsilon_2$) of $MAPbI_3$ obtained experimentally in a quite wide energy region extending from 10^{-6} to 6 eV. The ε_2 spectrum shows the light absorption characteristics. In the figure, the experimental spectra reported for the visible/UV [1], IR [55], GHz [56], and THz [57] regions were summarized, whereas the dotted line indicates the unreported region. All the spectra were obtained at room temperature, except for the dielectric function in the GHz region, which was measured at 210 K. The ε_1 is shown by a linear scale, while ε_2 is indicated by a logarithmic scale. The ε_s and ε_∞ represent the static dielectric constant and high-frequency dielectric constant, respectively. *Source*: Shirayama et al. [1]; Glaser et al. [55]; Sendner et al. [56]; Anusca et al. [57].

essentially corresponds to E_g of $MAPbI_3$ (Section 4.2.1). The dipole formation by the vibration of the atoms in the IR region generates sharp absorption features, yet with a weak ε_2 amplitude of ~0.5 (Section 4.2.2).

One distinctive feature of $MAPbI_3$ light absorption is an extraordinary strong optical phonon absorption with a considerably large amplitude of $\varepsilon_2 \sim 50$ (Section 4.2.3). Another unique feature, which is absent in conventional semiconductors, is a very strong light absorption by orientational dipole formation. The MA^+ cation has a permanent dipole of C(δ^-)–N(δ^+), where δ^+ and δ^- represent the positive and negative partial charges, respectively. Accordingly, when light enters (i.e. when the electric field is applied), MA^+ rotates so that the orientation of the C–N dipole is aligned along the oscillation direction of the electric field. This slow collective motion of MA^+, which occurs in the GHz range, leads to rather strong light absorption with an ε_2 peak amplitude of 27. A similar orientational absorption occurs for H_2O molecules with a dipole of H(δ^+)–O(δ^-) [58]. In microwave ovens, the microwave absorption of the H_2O molecules in the GHz range is utilized as the heating system. A large ε_2 absorption by MA^+ orientation is consistent with the fact that MA^+ is preferentially aligned under the presence of the electric field.

The ε_1 spectrum of $MAPbI_3$ in Figure 4.2 has a complex structure. As known well, the ε_1 and ε_2 spectra are not independent and the ε_1 spectrum is calculated from the ε_2 spectrum by applying Kramers–Kronig relations (see Appendix A). The ε_1 optical response is generally characterized by two physical values: namely, the static dielectric constant (ε_s) and high-frequency dielectric constant (ε_∞). These values are obtained from the flat ε_1 regions observed between the IR and visible/UV region (ε_∞) and between the GHz and THz region (ε_s). The absolute ε_1 value essentially shows the total amount of electric dipoles created by the electric field and, as the frequency of the alternating electric field increases, ε_1 gradually decreases. For example, we observe $\varepsilon_1 = 83.6$ at $E = 10^{-6}$ eV. This includes all the dipole-formation effects of the MA^+ orientational, phonon, IR, and interband absorption. However, the motion of MA^+ orientation is essentially slow, and the MA^+ orientational dipoles disappear when the polarity of the electric field switches too fast at a higher frequency of light. At $E = 10^{-3}$ eV, therefore, by the elimination of the orientational dipole, ε_1 decreases to $\varepsilon_s = 34.9$. At higher E (i.e. higher frequency), the disappearance of the phonon absorption further reduces ε_1 to $\varepsilon_\infty = 5.0$ and ε_1 finally becomes one (i.e. $\varepsilon_1 = n = 1$) at a quite high frequency. At very low frequency ($<10^4$ Hz), on the other hand, ε_1 is reported to increase to ~100 due to the ion migration [16, 51].

Figure 4.3 summarizes (a) ε_∞ and (b) ε_s of $MAPbX_3$ ($X = I^-$, Br^-, and Cl^-), which are compared with those reported for tetrahedral-bonding crystals [59]. For ε_∞, all the hybrid perovskites show notably small values. Note that $\varepsilon_\infty = \varepsilon_1 = n^2$ because $k = 0$ in the ε_∞ region. If compared with the values of GaAs ($\varepsilon_\infty = 10.9$, $n = 3.30$),

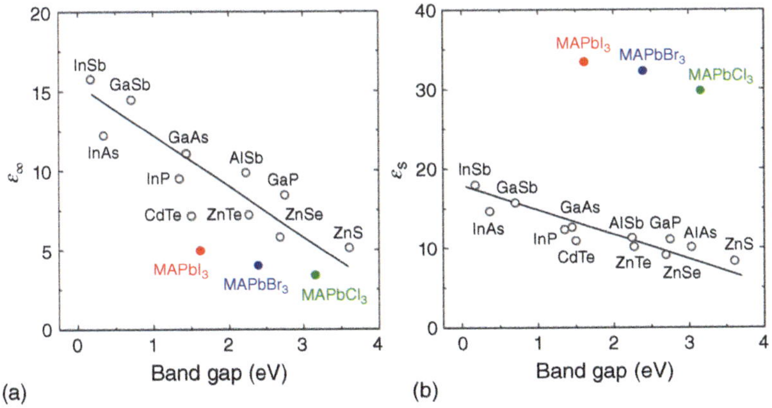

Figure 4.3 (a) High-frequency dielectric constant ε_∞ and (b) static dielectric constant ε_s vs. E_g of $MAPbX_3$ ($X = I^-$, Br^-, and Cl^-) and tetrahedral-bonding crystals. The ε_∞ value of $MAPbI_3$ from Figure 4.2 was used, whereas those of $MAPbBr_3$ and $MAPbCl_3$ were extracted from the dielectric function model parameters in Ref. [7] assuming $E = 0$ eV. The ε_s values of $MAPbX_3$ [56] were determined by the THz measurements (Section 4.2.3). For tetrahedral-bonding semiconductors, the values of Ref. [59] were adopted. *Source*: Fujiwara and Collins [7]; Adachi [59].

the corresponding values of $MAPbI_3$ ($\varepsilon_\infty = 5.0$, $n = 2.23$) are remarkably low. The small ε_∞ (or n_∞) of the hybrid perovskites can be interpreted by the large optical void created by the presence of the organic cations, as the polarizability of organic cations is negligible in the visible region. In other words, the optical density of hybrid perovskites is intrinsically low.

In contrast, ε_s of hybrid perovskites determined by Sendner et al. [56] is surprisingly high, compared with other conventional compound semiconductors. The high ε_s in hybrid perovskites is consistent with quite strong optical phonon absorption in the materials. This extraordinary strong optical photon absorption is a unique feature of hybrid perovskites and is proposed to lower carrier mobility by enhanced phonon scattering (Section 6.3.4).

In the following sections, further details regarding the light absorption in the visible/UV region (Section 4.2.1), in the IR region (Section 4.2.2), and in the THz region (Section 4.2.3) are described.

4.2.1 Visible/UV Region

Figure 4.4 shows the dielectric function ($\varepsilon = \varepsilon_1 - i\varepsilon_2$), ($n$, k) spectra, and α spectrum of $MAPbI_3$ in the visible/UV region [1]. The determination of accurate optical

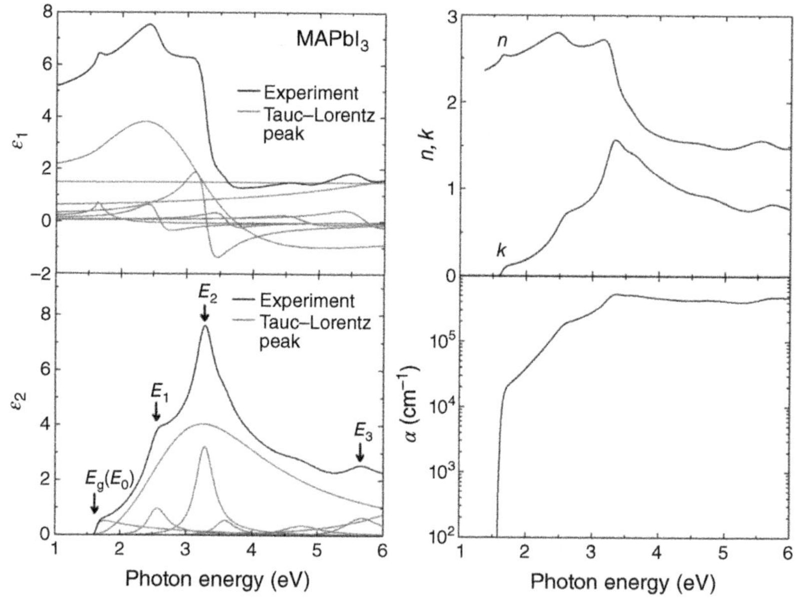

Figure 4.4 Dielectric function ($\varepsilon = \varepsilon_1 - i\varepsilon_2$), ($n$, k) spectra, and α spectrum of $MAPbI_3$ in the visible/UV region, determined by spectroscopic ellipsometry characterization [1]. These spectra are consistent with those of Figures 4.1 and 4.2. In the figure, the dielectric function expressed by the Tauc–Lorentz model is shown (see Appendix A). The gray lines indicate the contribution of each Tauc–Lorentz peak. The peak parameter values and the numerical optical-constant values of $MAPbI_3$ are summarized in Appendix A. *Source*: Modified from Shirayama et al. [1].

data for hybrid perovskites has been challenging because of the (i) extensive roughness, which corresponds to λ of the probe light (~500 nm) [10, 60] and (ii) quite fast degradation that occurs by humid air [61]. To avoid these serious problems, the optical spectra of Figure 4.4 were obtained from an ultra-smooth $MAPbI_3$ layer with a roughness of 5 nm and its measurement was performed in a N_2-globe bag without exposing the sample to air at all [1].

In Figure 4.4, it can be seen that the spectra shapes of ε_1 and ε_2 resemble those of n and k, respectively, and the α spectrum is calculated from k by Eq. (4.2). The onsets of ε_2, k, and α spectra essentially correspond to the E_g of $MAPbI_3$ (1.61 eV). In the high energy region of $MAPbI_3$, there exist clear absorption peaks at 2.53, 3.24, and 5.65 eV, which are labeled as the E_1, E_2, and E_3 optical transitions, respectively. Note that a small peak at ~4.5 eV is caused by the presence of PbI_2 at the surface and interface. This PbI_2 peak becomes stronger with the perovskite degradation that occurs near the surface region [61]. The thin lines for (ε_1, ε_2) represent the results of the parameterization using the Tauc–Lorentz model (Appendix A). In this scheme, the ε_2 spectrum is expressed as the sum of all the Tauc–Lorentz peaks (see Appendix A for more details).

To understand the light absorption phenomena in hybrid perovskites, quite extensive density functional theory (DFT) analyses have been performed [1–3, 17, 38, 42]. Nevertheless, the DFT calculation results vary significantly depending on the calculation (or approximation) method (Section 5.4). It has been found, however, that DFT optical spectra calculated based on a rather simple approximation (generalized gradient approximation within the Perdew–Burke–Ernzerhof, i.e. PBE scheme) demonstrate remarkable agreement with the experimental spectra (Figure 5.8), enabling us to perform detailed optical transition analyses.

Figure 4.5 summarizes the result of the optical transition analysis performed for $MAPbI_3$ by DFT (PBE): (a) band structure and (b) charge density profiles near the conduction band minimum (CBM) and valence band maximum (VBM) [1]. Importantly, the light absorption of $MAPbI_3$ in the visible/UV region can be expressed by semiconductor-type optical transitions (i.e. the electron transition from the valence band to the conduction band). Moreover, the E_0, E_1, and E_2 transition peaks observed experimentally in Figure 4.4 are consistent with the direct optical transitions at high symmetry points at R, M, and X points in the Brillouin zone, respectively (see arrows in Figure 4.5). For the E_2 peak, however, there are three transition components labeled as $E_{2,a\text{–}c}$. The above assignment of the optical transitions was determined by calculating the ε_2 spectral contributions for each pair of the valence and conduction bands [1, 5].

The band structure of Figure 4.5 clearly shows the direct-transition-type band structure; the VBM and CBM are positioned at the same symmetry point of R. When the spin–orbit coupling (SOC) effect is considered in the DFT calculations, however, the conduction band splits into two bands (Rashba splitting) and the band structure becomes slightly indirect (see Figure 5.8). Nevertheless, the DFT optical spectra calculated by incorporating the Rashba splitting effect disagree with the experimental spectra; in particular, (i) the light absorption strength is seriously underestimated and (ii) the absorption tail broadens excessively when the

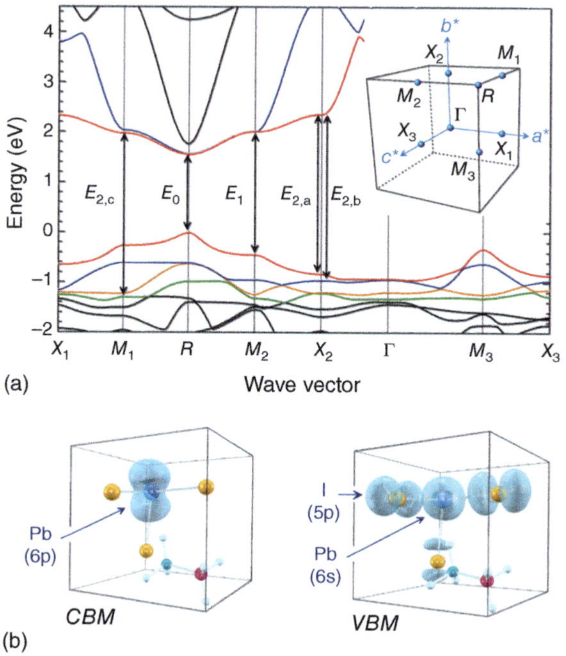

Figure 4.5 Optical transitions in $MAPbI_3$: (a) band structure and (b) charge density profiles near CBM (1.4–1.6 eV) and VBM (−0.2 to 0 eV) of $MAPbI_3$, calculated within PBE. The inset of (a) shows the high symmetry points in the Brillouin zone of the assumed pseudo-cubic structure (Figure 5.4a). The arrows in (a) indicate the direct semiconductor-type optical transitions, corresponding to the E_0–E_2 transitions in Figure 4.4. *Source*: Shirayama et al. [1].

Rashba splitting is considered (Figures 5.8 and 5.9). Because the DFT results change significantly depending on the approximation method, the DFT results must be justified according to the experimental results, particularly for hybrid perovskites that include the dynamic motion of organic cations. Consequently, the direct-gap formation in $MAPbI_3$ is quite appropriate to interpret the experimental results correctly (see Section 5.4.2 for further details).

As shown in Figure 4.5b, the CBM of $MAPbI_3$ consists of Pb 6p orbital, whereas the VBM is composed of Pb 6s and I 5p states (Figure 5.11a). Since the CBM and VBM of $MAPbI_3$ are anti-bonding states (Figure 5.12), the charge profiles of Pb 6s, Pb 6p, and I 5p are separated and localized near the Pb and I atoms. The light absorption in semiconductors can be understood simply from the overlap of the electronic charge in the valence and conduction states [62]. In $MAPbI_3$, the Pb 6s charge in the VBM overlaps with the Pb 6p charge in the CBM and thus the light absorption at the VBM → CBM occurs by Pb 6s → 6p transition [62]. In the energy region slightly above CBM, the dispersive I charge is present [1] and its contribution becomes important. Importantly, there are no electronic charges (or states) of MA^+ near CBM and VBM (Figure 5.11a); thus, the direct light absorption by MA^+ is negligible in the visible/UV region. In other words, the visible/UV light absorption in hybrid perovskites occurs only in the inorganic framework (i.e. PbI_3^- scaffold in $APbI_3$).

4.2.2 IR Region

In hybrid perovskites, weak light absorption occurs in the IR region due to the dipole moments created by the vibration of the organic cations. Figure 4.6 shows the ε_2

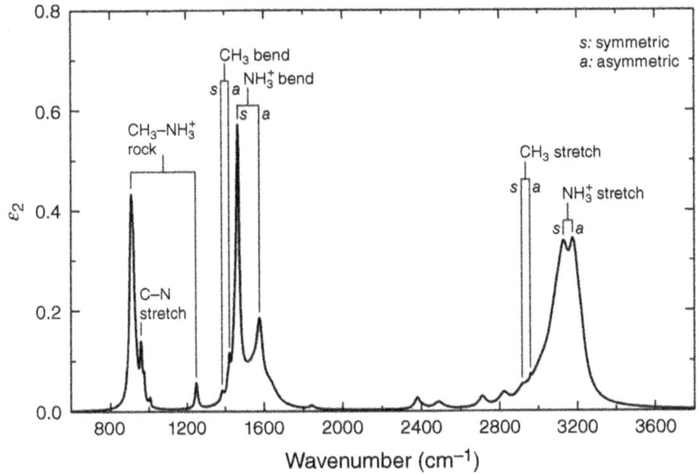

Figure 4.6 ε_2 spectrum of $MAPbI_3$ in the IR region determined from transmission measurements. The result of the peak assignment analysis based on theoretical calculations (second-order Møller–Plesset perturbation theory) is also indicated. There exist symmetric (s) and asymmetric (a) modes for stretching and bending vibrations. *Source*: Glaser et al. [55].

spectrum of $MAPbI_3$ obtained in the IR region [55]. In $MAPbI_3$, the main IR peaks originate from NH_3^+ vibrational modes: NH_3^+ bending (~1550 cm^{-1}) and stretching (~3150 cm^{-1}) modes. As discussed by Glaser et al. [55], the peak position of the N–H stretching mode is sensitive to hydrogen bonding and the shift of the N–H stretching peak toward lower energy indicates the formation of the strong hydrogen bonding in $MAPbI_3$ (see also Figure 3.15).

The ε_2 peaks in the IR region have a Lorentz shape, and the dielectric function can be expressed as a sum of a Lorentz model peak:

$$\varepsilon = \varepsilon_\infty + \sum_j \frac{A_j}{\omega_{0,j}^2 - \omega^2 - \mathrm{i}\omega\gamma_j} \tag{4.6}$$

where A_j, $\omega_{0,j}$, and γ_j represent the amplitude, peak frequency, and broadening parameters of the jth peak, respectively. The peak positions (i.e. $\omega_{0,j}$) obtained from the above analysis are summarized in Table 4.1. All the IR peak positions are essentially independent of the X-site halide anion, although the IR peak shifts slightly toward a higher wavenumber in the order of I → Br → Cl [55].

4.2.3 THz Region

One of the important features of hybrid perovskite optical properties is the quite strong optical phonon absorption by the inorganic framework. Figure 4.7 shows the ε_2 and $\mathrm{Im}(-1/\varepsilon)$ spectra of $MAPbI_3$ [56] and GaAs [59]. The ε_2 peaks show transverse optical (TO) phonon modes, where $\mathrm{Im}(-1/\varepsilon)$ is called the loss function, representing the longitudinal optical (LO) phonon modes. The TO and LO phonon peaks can be described by the Lorentz line shape shown in Eq. (4.6). Based on a DFT study, the

Table 4.1 IR peak positions for major vibrational modes in $MAPbI_3$, reported by Glaser et al. [55].

Peak position (cm^{-1})	Vibrational mode
911	CH_3–NH_3^+ rocking
960	C–N stretching
1250	CH_3–NH_3^+ rocking
1469	NH_3^+ bending (symmetric)
1577	NH_3^+ bending (asymmetric)
3132	NH_3^+ stretch (symmetric)
3179	NH_3^+ stretch (asymmetric)

The result of the peak assignment is also shown.
Source: Modified from Glaser et al. [55].

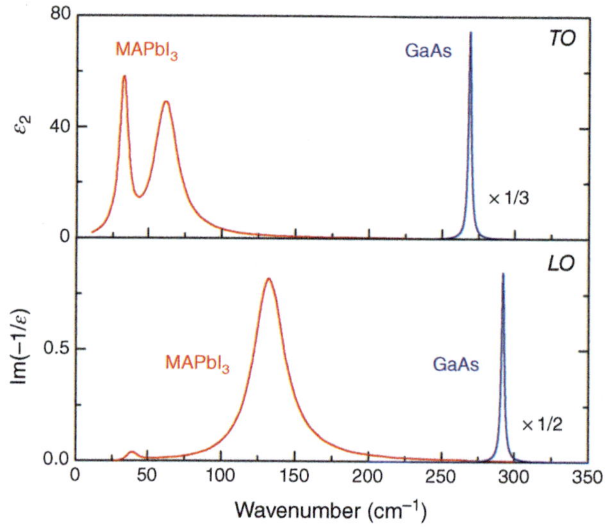

Figure 4.7 ε_2 and Im($-1/\varepsilon$) spectra of $MAPbI_3$ [56] and GaAs [59] at room temperature, which show the transverse optical (TO) phonon and longitudinal optical (LO) phonon modes, respectively. The ε_2 amplitude of GaAs is reduced to a third of the original, and Im($-1/\varepsilon$) amplitude of GaAs is also reduced to half for clarity. *Source*: Sendner et al. [56]; Adachi [59].

strong TO peak at $\omega_{TO} = 63\,cm^{-1}$ has been assigned to a Pb–I stretching-like vibration, whereas the peak of $\omega_{TO} = 32\,cm^{-1}$ is due to a Pb–I–Pb rocking-like vibration [56]. The peak positions of the two LO phonon modes are 40 and 133 cm^{-1}. The slight change in the phonon frequencies has been observed in $MAPbBr_3$ and $MAPbCl_3$ by the effect of ionic masses [56].

In Figure 4.7, the TO and LO absorption peaks of $MAPbI_3$ are significantly broader, compared with GaAs. In fact, the peak half width of $\omega_{LO} = 133\,cm^{-1}$ ($25\,cm^{-1}$) is 10 times greater than that of the GaAs LO peak ($2.4\,cm^{-1}$). The extraordinary phonon peak broadening observed in $MAPbI_3$ has been attributed to the dynamic disorder of PbI_6 octrahedra [56]. In other words, the "soft and flexible" inorganic cage (see Table 6.1) allows unusually strong phonon absorption. The enhanced phonon absorption in $MAPbI_3$ can be evidenced from the considerably larger integrated peak area in $MAPbI_3$, compared with GaAs.

The very large ε_s of $MAPbI_3$, shown in Figures 4.2 and 4.3, can also be understood from the phonon absorption spectra. In general, ε_s is estimated by a simple theory known as the Lyddane–Sachs–Teller relation [59]:

$$\frac{\varepsilon_s}{\varepsilon_\infty} = \prod_j \frac{\omega_{LO,j}^2}{\omega_{TO,j}^2} \tag{4.7}$$

In non-polar materials (Si, for example), $\omega_{LO} = \omega_{TO}$; thus, we obtain $\varepsilon_s = \varepsilon_\infty$ from Eq. (4.7). In $MAPbI_3$, the peak separation between ω_{TO} and ω_{LO} is considerably greater than that in GaAs, leading to a remarkably large ε_s, which has been estimated to be $\varepsilon_s = 33.5$ from Eq. (4.7) [56].

Important points in $MAPbI_3$ phonon absorption include the (i) low energy peak position of ω_{LO} and (ii) strong LO-phonon absorption that leads to $\varepsilon_s \gg \varepsilon_\infty$. Carriers in semiconductors interact with LO phonons (phonon coupling), and the small ω_{LO} and $\varepsilon_s \gg \varepsilon_\infty$ observed in $MAPbI_3$ are indicative of the carriers being scattered by LO-phonon vibration (see Eq. (6.6)). In fact, it has been proposed that the low mobility of ~100 $cm^2/(V\,s)$ observed in $MAPbI_3$ (see Figure 6.4) originates from the strong LO phonon scattering, although different interpretations exist (Section 6.3.4).

4.3 Band Gap of Hybrid Perovskites

The E_g of a light absorber is one of the most critical physical parameters directly influencing the solar cell conversion efficiency, as only limited E_g (E_g = 1.0–1.6 eV) is allowed for efficient devices (Figure 12.1). Nevertheless, the reported E_g values of hybrid perovskites are rather controversial [1, 10, 12, 18–46]. The origin of such controversy can be attributed to the different analysis methods and contributions of inaccurate α spectra from which the E_g values are extracted. In this section, we review the E_g analysis of $MAPbI_3$ (Section 4.3.1) and see the E_g values of different hybrid perovskites (Section 4.3.2) and perovskite alloys (Section 4.3.3).

4.3.1 Band Gap Analysis of $MAPbI_3$

Table 4.2 summarizes various E_g values of $MAPbI_3$ reported using different measurement and analysis techniques [1, 10, 12, 18–27]. The reported E_g depends strongly on the measurement technique, independent of the sample forms (i.e. thin film or single crystal). In high-precision ellipsometry characterization, consistent values of 1.60–1.61 eV have been obtained [1, 8, 10, 12, 18]. The transmission (T) measurements result in slightly reduced E_g values (1.57–1.61 eV) [20–23]. Note that E_g can be underestimated seriously, particularly when the baseline of the α spectrum

Table 4.2 Various E_g values of $MAPbI_3$, reported using different measurement and analysis techniques.

Measurement	Analysis	Sample	E_g (eV)	References
Ellipsometry	Critical point	Thin film	1.61	[1]
Ellipsometry	Critical point	Thin film	1.61	[18]
TA[a)]	—	Thin film	1.61	[19]
Transmission	Tauc plot	Thin film	1.61	[20]
Ellipsometry	DF Modeling[b)]	Single crystal	1.60	[12]
Ellipsometry	DF Modeling[b)]	Thin film	1.60	[10]
Transmission	Tauc plot	Thin film	1.60	[21]
Transmission	Linear fitting	Thin film	1.57	[22]
Transmission	ND[c)]	Thin film	1.57	[23]
EQE	Linear fitting	Thin film[d)]	1.55	[24]
Reflection	Kubelka–Munk	Single crystal	1.52	[25]
Reflection	Kubelka–Munk	Single crystal	1.51	[26]
Reflection	Kubelka–Munk	Thin film	1.50	[27]

The results are shown in order of decreasing E_g.
a) Transient absorption spectroscopy.
b) E_g is determined from modeling (parameterization) of dielectric function (DF).
c) Not described.
d) Reported as the E_g of $MAPbI_2Cl$.

is elevated by roughness scattering (Section 4.4.2). In the reflectance (R) measurements, α spectrum is deduced using a relation known as the Kubelka–Munk function: $\alpha \propto F(R) = (1-R)^2/(2R)$. The E_g values obtained from this scheme are, however, low (1.50–1.52 eV) [25–27].

The E_g of $MAPbI_3$ is often represented by 1.55 eV based on the work of Lee et al. [24]. This E_g is deduced as the onset of the external quantum efficiency (EQE) spectrum at $\lambda \sim 800$ nm. However, the E_g obtained from this indirect approach reflects the absorption onset, which shifts toward a reduced energy by tail absorption. The EQE spectrum varies by external effects including film thickness, light scattering, and back-side reflection, and accurate E_g determination is rather difficult from EQE measurements.

In general, the most reliable E_g analysis is performed by critical point analysis, where E_g is extracted from the theoretical fitting of the second derivative (ε_1, ε_2) spectra [54, 63]:

$$\frac{d^2\varepsilon}{dE^2} = \begin{cases} j(j-1)A\exp(i\phi)(E-E_p+i\Gamma)^{j-1} & (j \neq 0) \\ A\exp(i\phi)(E-E_p+i\Gamma)^{-2} & (j = 0) \end{cases} \tag{4.8}$$

where A, ϕ, E_p, and Γ are the amplitude, phase, position, and width of the peak, respectively. Depending on the band structure, the critical point is classified into

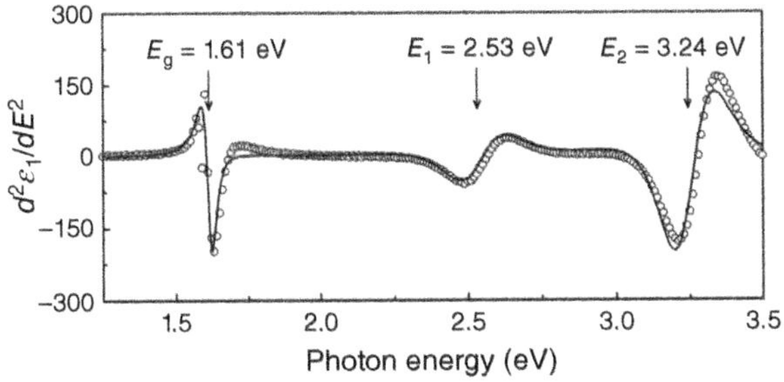

Figure 4.8 Result of the critical point analysis performed for the experimental dielectric function of $MAPbI_3$ shown in Figure 4.4. A similar good fitting is also obtained for $d^2\varepsilon_2/dE^2$. From the analysis, E_g of $MAPbI_3$ is determined to be 1.61 ± 0.01 eV. *Source*: Shirayama et al. [1].

one dimension ($j = -1/2$), two dimensions ($j = 0$), three dimensions ($j = 1/2$), or excitonic ($j = -1$).

Figure 4.8 shows the results of the $MAPbI_3$ critical point analysis performed for the experimental dielectric function of Figure 4.4 [1]. In the second derivative spectra, the spectral feature in only the transition peak (critical point) region is extracted, enabling the reliable determination of the transition energy. From the fitting analysis based on Eq. (4.8), E_g has been determined to be 1.61 ± 0.01 eV. The exact same E_g was reported in another critical point analysis [18]. The transition absorption (TA) spectroscopy also shows the negative TA peak at 1.61 eV due to photobleaching observed for a direct-gap semiconductor [19]. Thus, the appropriate E_g of $MAPbI_3$ is 1.61 eV.

4.3.2 Band Gap of Basic Perovskites

The E_g values of various hybrid perovskites reported in Refs. [1, 2, 4, 18, 20, 32–36] are summarized in Table 4.3. Due to the variation of the measurement and analysis methods, inconsistent E_g values have been reported, particularly for Sn-based materials including $FASnI_3$ [32, 34, 37], $MASnI_3$ [20, 34, 38–42], and $CsSnI_3$ [33, 34, 43–46]. Moreover, strong excitonic peaks are present in the absorption spectra of $APbBr_3$ and $APbCl_3$. In these cases, E_g can no longer be determined from the onset of the absorption; rather it must be deduced by separating the contribution of the excitonic peak from the band-to-band (interband) absorption. Such analyses have been performed for $MAPbBr_3$ [35] and $MAPbCl_3$ [36] in Table 4.3. Note that, if conventional E_g analyses are performed for excitonic absorption spectra, E_g is seriously underestimated.

Figure 4.9 shows the variation of E_g with the (a) A-site cation and (b) X-site anion in hybrid perovskites. In $APbI_3$ materials, E_g increases slightly in the order of FA → MA → Cs. A systematic DFT study reveals that this E_g variation with the A-site cation is caused by the change in the Pb–I–Pb bond angle; the larger A-site cation (Table 3.3) leads to a Pb–I–Pb angle nearer to 180°, which in turn reduces the E_g (Figure 5.14). This phenomenon has been interpreted by the anti-bonding character of the Pb and I orbitals (see Section 5.5.1).

Table 4.3 E_g of various perovskites.

Material	Measurement	Analysis	E_g (eV)	References
$FASnI_3$	Ellipsometry	Critical point	1.40	[32]
$MASnI_3$	Transmission	Tauc plot	1.28	[20]
$CsSnI_3$	Transmission	Steepest gradient	1.36	[33]
$FAPbI_3$	Ellipsometry	Critical point	1.55	[2, 18]
$Cs_{0.17}FA_{0.83}PbI_3$	Ellipsometry	Tauc plot	1.55[a)]	[4]
$MAPbI_3$	Ellipsometry	Critical point	1.61	[1]
$CsPbI_3$	Transmission	Steepest gradient	1.76	[33]
$CsSnBr_3$	Transmission	ND[b)]	1.81	[34]
$MASnBr_3$	Transmission	ND[b)]	2.13	[34]
$FAPbBr_3$	Ellipsometry	Critical point	2.30	[18]
$MAPbBr_3$	Ellipsometry	Elliott's formula	2.39	[35]
$CsPbBr_3$	Transmission	ND[b)]	2.31	[34]
$FASnBr_3$	Transmission	ND[b)]	2.63	[34]
$CsSnCl_3$	Transmission	ND[b)]	2.88	[34]
$CsPbCl_3$	Transmission	ND[b)]	2.99	[34]
$FAPbCl_3$	Transmission	ND[b)]	3.02	[34]
$MAPbCl_3$	Transmission	Exciton model	3.16	[36]
$MASnCl_3$	Transmission	ND[b)]	3.50	[34]
$FASnCl_3$	Transmission	ND[b)]	3.55	[34]

The E_g of the perovskite materials is shown in order of increasing E_g. The absolute E_g is influenced by the measurement and analysis methods (see Table 4.2); the E_g values derived from transmission measurements could be underestimated slightly, compared with ellipsometry results.

a) E_g was determined after performing extra roughness correction [5] for the optical spectrum reported in Ref. [4].

b) Not described.

In contrast, there is no clear trend for E_g when the A-site cation is changed in $ASnI_3$. However, unlike $MASnI_3$, which exhibits the pseudo-cubic structure, $FASnI_3$ and $CsSnI_3$ have orthorhombic structures (see Table 3.2) and their E_g values cannot be compared directly. The DFT study of $CsSnI_3$, for example, shows that the crystal structure changes E_g significantly [64]. Moreover, due to the small ionic radius of Sn^{2+}, the Sn–I–Sn bond angle in $MASnI_3$ is almost 180° [33]. In $ASnI_3$ perovskites, therefore, the E_g variation by the lattice expansion is considered more important [65].

It can be observed from Figure 4.9b that the influence of the X-site anion is considerably greater than those of the A-site and B-site cations; the variation of the halide anion from I^- to Br^- increases E_g drastically by ~0.8 eV. The increase in E_g by the order of I → Br → Cl follows the general trend that a lighter atom leads to E_g widening.

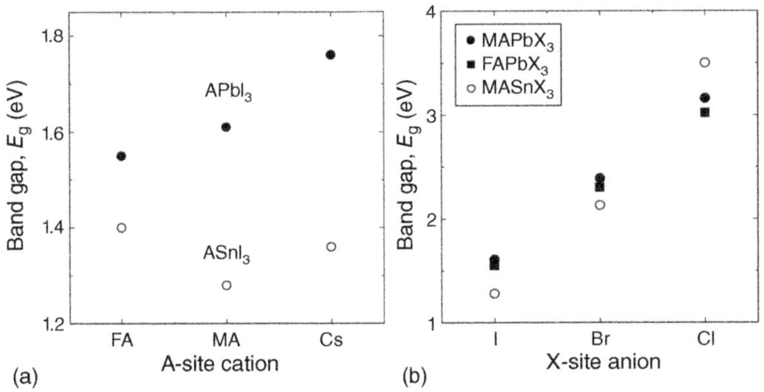

Figure 4.9 Variation of E_g with (a) the A-site cation and (b) the X-site anion in various hybrid perovskites. In (a), E_g variations in $APbI_3$ and $ASnI_3$ with the A-site cation are shown, whereas the E_g variations in $MAPbX_3$, $FAPbX_3$, and $MASnX_3$ with the X-site anion are summarized in (b). The E_g values are adopted from those presented in Table 4.3.

4.3.3 Band Gap Variation in Perovskite Alloys

To control E_g in a wide range, the alloying by I–Br and Pb–Sn mixing has been performed for hybrid perovskites. Figure 4.10a shows the E_g variations with the Br content x in $Cs_{0.17}FA_{0.83}Pb(I_{1-x}Br_x)_3$ and $Cs_{0.05}FA_{0.8}MA_{0.15}Pb(I_{1-x}Br_x)_3$. The E_g values have been extracted from the optical data in Appendix A using the Tauc analysis. Up to $x = 0.4$, the E_g of the perovskites increases linearly with x:

$$E_g(x) = 1.54 + 0.57x\ \text{eV} \tag{4.9}$$

Figure 4.10b shows the E_g variation of $MAPb(I_{1-x}Br_x)_3$ in a wider x range ($x = 0$–1), which indicates a similar E_g increase, yet with a slight E_g bowing [22]:

$$E_g(x) = 1.57 + 0.39x + 0.33x^2\ \text{eV} \tag{4.10}$$

For the E_g determination, a simple analysis has been employed (see $E_g = 1.57$ eV of Ref. [22] in Table 4.2). For $FAPb(I_{1-x}Br_x)_3$ and $CsPb(I_{1-x}Br_x)_3$, similar E_g variations have been confirmed (see Figure 18.7). In the case of $MAPb(Br_{1-x}Cl_x)_3$, a slight E_g bowing expressed by $E_g = 2.415 + 0.53x + 0.30x^2$ has also been reported [36].

Interestingly, for the Sn–Pb mixed alloys (i.e. $A(Sn,Pb)I_3$), a significant E_g bowing has been confirmed. Figure 4.10c summarizes the E_g change for the Sn content x in $MA(Pb_{1-x}Sn_x)I_3$ for which E_g is expressed by the following equation [42]:

$$E_g(x) = (1-x)E_{g,MAPbI_3} + xE_{g,MASnI_3} - 1.06x(1-x)\ \text{eV} \tag{4.11}$$

where $E_{g,MAPbI_3}$ and $E_{g,MASnI_3}$ show the E_g values of $MAPbI_3$ (1.6 eV) and $MASnI_3$ (1.3 eV), respectively. Notably, E_g bowing has also been confirmed when different A-site cations are adopted in $A(Pb,Sn)I_3$ (see Figures 10.8 and 18.12).

The origin of the monotonous E_g increases in $APb(I,Br)_3$, and the strong E_g bowing in $A(Sn,Pb)I_3$ can be explained by the energy alignment of the VBM and CBM [42] (Figure 4.10d). When the X-site anion is changed in $MAPbX_3$, for

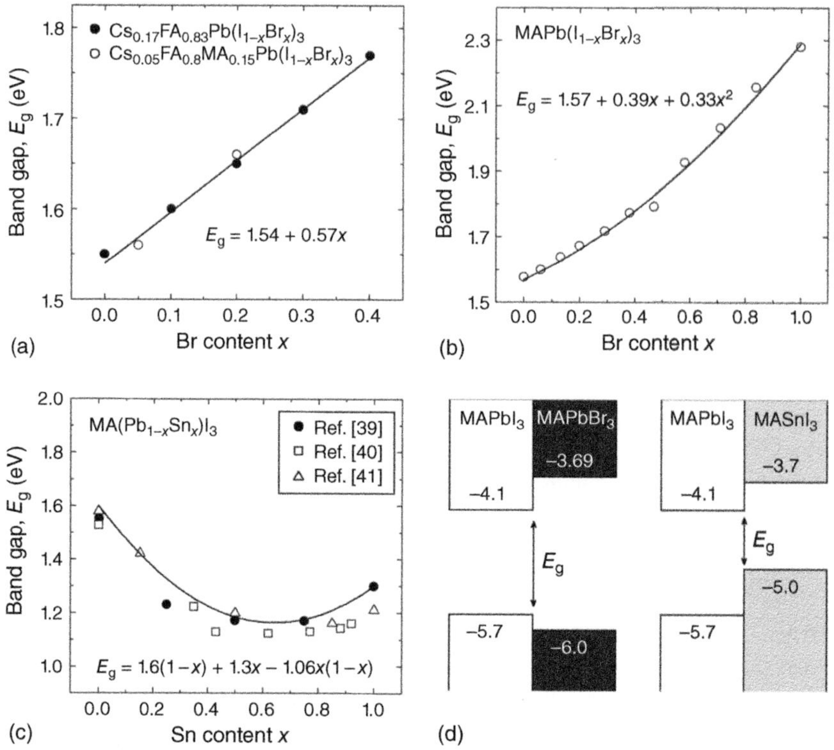

Figure 4.10 Variations of E_g in (a) $CsFAPb(I_{1-x}Br_x)_3$ and $CsFAMAPb(I_{1-x}Br_x)_3$ with x, (b) $MAPb(I_{1-x}Br_x)_3$ with x, (c) $MA(Pb_{1-x}Sn_x)I_3$ with x, and (d) band edge positions of $MAPbI_3/MAPbBr_3$ and $MAPbI_3/MASnI_3$ relative to the vacuum level in eV. The result of $CsFAPb(I_{1-x}Br_x)_3$ in (a) was obtained after performing the extra roughness correction [5] for the optical data of Ref. [4]. *Source*: (b) Noh et al. [22]. (c, d) Goyal et al. [42].

example, the energy position of the CBM moves upward, whereas that of the VBM shifts downward by I → Br exchange. When Br is incorporated into $APbI_3$, therefore, the E_g widening occurs by the energy shift of VBM and CBM. In contrast, the band alignment between $MAPbI_3$ and $MASnI_3$ is staggered (type-II alignment) and the Sn alloying with Pb leads to a smaller gap material, whose optical transition occurs from the Sn-derived VBM state to the Pb-based CBM state. This E_g reduction is helpful for an all-perovskite tandem application, as the E_g reduction in the Sn-based alloy increases the photosensitivity in the long λ region (see Chapter 18).

4.4 True Absorption Coefficient of $MAPbI_3$

Light absorption in the visible/UV region is of critical importance in operating solar cells. Although extensive studies have been made to determine the α spectra of hybrid perovskites in the visible/UV region, the absolute α values obtained using spectroscopic ellipsometry and transmission/reflectance (T/R) measurements are

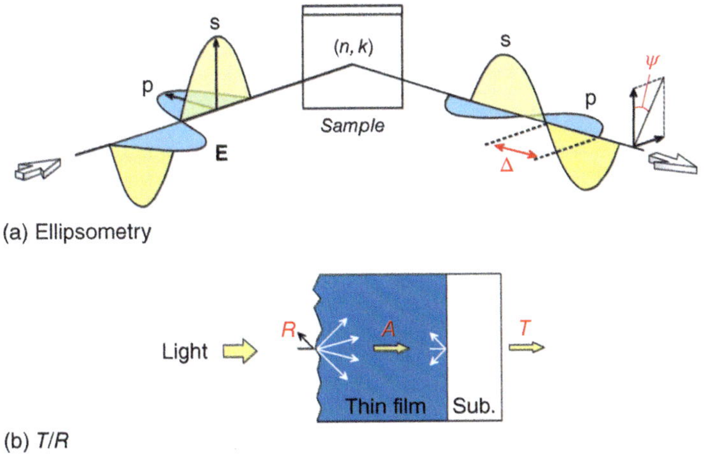

Figure 4.11 Principles of (a) ellipsometry and (b) transmission/reflectance (*T*/*R*) measurements. Ellipsometry measures two angles (ψ, Δ), which express the relative amplitude and phase difference between the p- and s-polarized light (electromagnetic) waves. In (b), the white arrows show non-ideal surface light scattering and the light reflection at the thin-film/substrate interface, which enhance the nominal absorptance (*A*) if simple analyses are performed.

seriously controversial [1, 8–17]. In this section, we review the basic principles of optical measurements (Section 4.4.1) and discuss how we can obtain accurate optical constants (Section 4.4.2).

4.4.1 Principles of Optical Measurements

Figure 4.11 shows the principles of (a) ellipsometry and (b) T/R optical measurements. Ellipsometry is a high-precision optical characterization technique, in which the optical constants are determined by the change in polarized light upon reflection [54]. This technique employs p- and s-polarized light waves, whose oscillation directions of the electric field are parallel and perpendicular to the incident plane, respectively (see Figure 4.11a). Depending on the (n, k) values, the polarized light undergoes different changes upon light reflection and ellipsometry measures two angles (ψ, Δ), which express the relative amplitude and phase difference between the p- and s-polarized waves, respectively. The ψ is essentially related to the n component, whereas the phase variation Δ is generated by the light absorption (i.e. k). In ellipsometry, therefore, two defining values of optical constants can be determined from a single measurement.

Although ellipsometry allows the determination of highly accurate values, its data analysis tends to be complicated. In particular, the exact optical modeling of an entire sample structure as well as all the (n, k) values of the constituent layers is necessary [54]. Moreover, ellipsometry spectra (particularly Δ spectrum) are extremely sensitive to surface/interface structures; improper modeling of these structures generates strong artifacts and, often, the adjustment of roughness thickness on a few-Å scale is required [5].

In T/R measurements, on the other hand, the absolute light intensities (i.e. square of the electric field; $I \propto |\mathbf{E}|^2$) of a thin-film/substrate structure are generally characterized. In this case, the phase information obtained in ellipsometry is lost and, therefore, α is obtained analytically. The simplest approach often adopted to determine α is the application of Beer's law: $I(d) = I_0 \exp(-\alpha d)$, where I_0 and $I(d)$ show the light intensity on the surface ($d = 0$) and at the film thickness of d, respectively. In this case, α is obtained from the measured absorbance (A_b) based on a simple relation:

$$A_b = -\ln[I(d)/I_0] = -\ln[T(d)/T_0] = \alpha d \tag{4.12}$$

Thus, from a known d, α can be determined using A_b. In another approach, the following relation has been adopted [14]:

$$\alpha = \ln(1 + A/T)/d \tag{4.13}$$

where A is the absorptance that can be estimated from $A = 1 - R - T$ (i.e. $T + R + A = 1$). Eq. (4.13) is particularly useful to eliminate the effect of interference.

In the above simple T/R approaches, however, the estimation of the accurate α of hybrid perovskites is difficult because the (i) light scattering that occurs on a film surface and (ii) optical interference caused by the light reflection at a thin-film/substrate interface are completely neglected. In particular, hybrid perovskite layers formed by one-step spin coating generally exhibit very large roughness with a size comparable to the wavelength of incident light ($\lambda \sim 500$ nm) [10, 60]. Accordingly, the strong light scattering occurs in these layers, which in turn enhances the nominal α due to the increased light pass length. Note that, in Eqs. (4.12) and (4.13), the αd component is considered and, when the analytical film thickness d is fixed, the enhanced absorption by light scattering is misinterpreted as high α in the film. The optical interference effect, caused by the finite light reflection on the surface and at the interface, also enhances the effective optical pass length within the layer and, similarly, nominal α increases if a simple layer thickness d is adopted in the analysis.

As a result, the α values derived particularly from simple T/R analyses are generally overestimated. The T/R measurements themselves are straightforward and are useful for obtaining the overall shape of the α spectra and rough α values; however, the extra care is necessary to interpret the absolute α value. On the other hand, ellipsometry measurements are far less sensitive to light scattering as only the specular reflection component is measured [6, 7]. In ellipsometry analysis, the optical interference effect is also modeled explicitly. Nevertheless, rigorous optical modeling of the layer structures is necessary for accurate optical-constant determination by ellipsometry.

4.4.2 Interpretation of α Variation

Figure 4.12a shows the α spectra of $MAPbI_3$ determined by spectroscopic ellipsometry [1, 8–12] and T/R-based techniques [13–16]. The reported $MAPbI_3$ α spectra are quite different, although the spectral shape is rather similar. The highly controversial α results do not originate from measurement techniques; all the ellipsometry

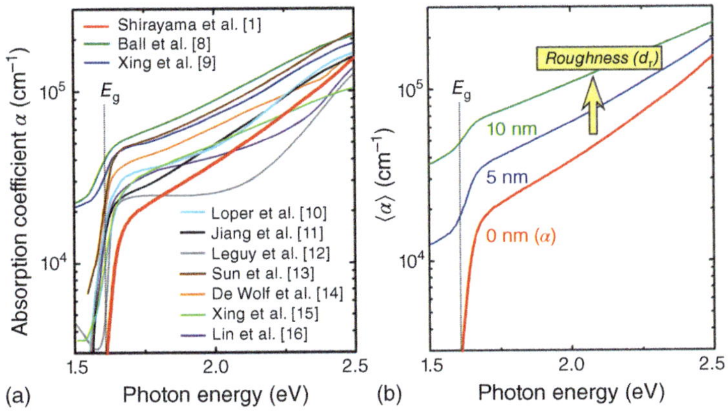

Figure 4.12 (a) α spectra of $MAPbI_3$ determined by spectroscopic ellipsometry [1, 8–12] and T/R-based techniques [13–16] and (b) α simulation result obtained when a hypothetical roughness thickness (d_r) is assumed for Shirayama's result. The dotted line indicates the E_g position of 1.61 eV. In (b), the α spectrum of $d_r = 0$ nm corresponds to the α spectrum of Shirayama et al. and the nominal α (pseudo-absorption coefficient, $\langle\alpha\rangle$) increases largely with increasing d_r, which is modeled by the Bruggeman effective medium theory with a 50 vol% of void in the roughness layer. *Source*: Shirayama et al. [1].

and T/R results are essentially inconsistent. In particular, some results show very high α values even below the E_g of $MAPbI_3$.

The large α variation observed for the ellipsometry results can be explained by the extensive roughness in $MAPbI_3$ layers. Specifically, the α spectrum of Shirayama et al. which is shown in Figure 4.4, was obtained from an ultra-smooth $MAPbI_3$ layer where the effect of the roughness is minimized. Notice that the α values of Shirayama et al. show the smallest values among the results. Figure 4.12b shows the α simulation result obtained when a hypothetical roughness thickness (d_r) is assumed for the Shirayama's result [1]. In the figure, the α spectrum of $d_r = 0$ nm corresponds to the spectrum reported by Shirayama et al. and the nominal α (pseudo-absorption coefficient, $\langle\alpha\rangle$) increases largely with increasing d_r. As explained previously in Figure 4.11a, Δ in ellipsometry can be related to the light absorption. However, the roughness also induces the large phase lag and, in simplified ellipsometry analyses, the Δ change generated by the roughness is misinterpreted as the α increases [5]. As confirmed from Figure 4.12, the ellipsometry results of Refs. [8, 9] are well reproduced by the roughness simulation. In fact, when the roughness correction is performed for the α spectra of Refs. [8, 9], the recalculated α spectra show excellent agreement with Shirayama's result [5]. In some ellipsometry studies, on the other hand, (T, R) spectra are analyzed simultaneously, in addition to (ψ, Δ) ellipsometry spectra [10, 66, 67]. However, when extensive sample roughness modifies the T/R spectra, the validity of such (ψ, Δ, R, T) analyses must be examined carefully.

It is highly likely that the large α confirmed for the T/R results is caused by strong light scattering due to large roughness. In fact, some α spectra obtained from the T/R analyses show large non-zero α values in the region below E_g, which can be understood by the light scattering effect of the roughness. Note that any transmission loss generated by roughness light scattering can be misinterpreted as light absorption in the simple analysis described above. In a well-cited α spectrum reported by De Wolf et al. [14], the α was determined from Eq. (4.11); therefore, the influence of the light scattering and optical interference was not included in this analysis, and the large analysis error is expected. It is often argued that $MAPbI_3$ shows the highest α among solar cell materials based on the result of De Wolf et al. However, when accurate α spectrum is compared, the α of $MAPbI_3$ is comparable to those of GaAs and CdTe (see Figure 1.5). In the past, $CuInSe_2$ (CISe) had been considered to have the highest α ($\sim 10^5$ cm^{-1}) among all the solar cell materials [68]. Nevertheless, this was also artifact caused by the roughness scattering in the T/R measurements [5], and actual α of CISe is comparable to other solar cell materials [69].

To justify the α result of Shirayama et al., complementary analyses were performed. Figure 4.13 shows the experimental α spectrum of Shirayama et al., and additional analysis results were obtained from spectroscopic ellipsometry (SE),

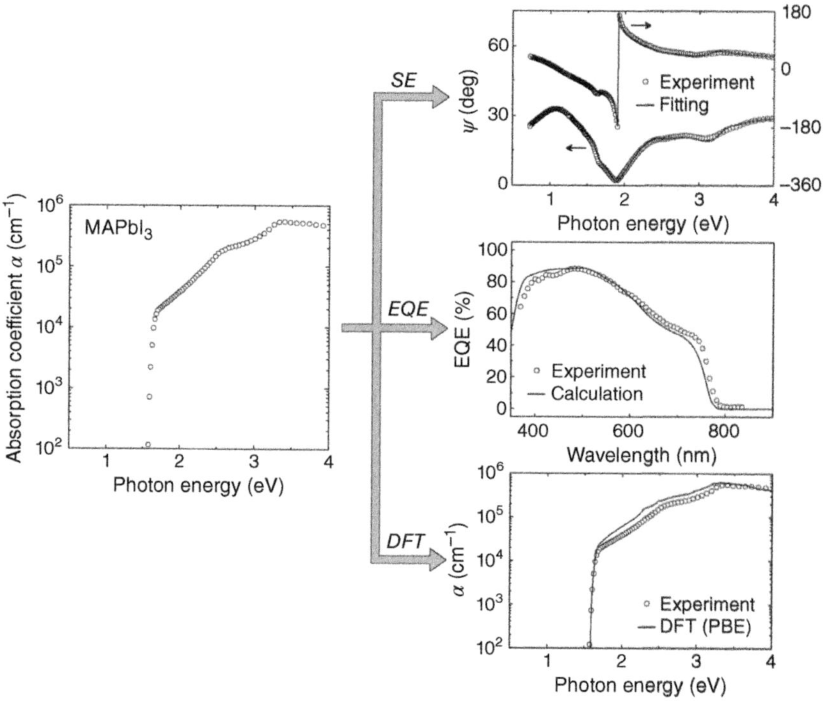

Figure 4.13 Experimental α spectrum of $MAPbI_3$ reported by Shirayama et al. [1]. The α spectrum is consistent with other analysis results obtained for spectroscopic ellipsometry (SE) [1], external quantum efficiency (EQE) [1], and DFT [70] analyses. In the SE analysis, a $MAPbI_3$ layer with different thickness (85 nm) is analyzed. In the EQE analysis, a semi-transparent $MAPbI_3$ solar cell with a structure of MgF_2/glass/SnO_2:F/TiO_2/$MAPbI_3$/spiro-OMeTAD/MoO_x/ZnO:Al/MgF_2 [71] is analyzed by adopting the α spectrum in the figure (see Section 11.4). The calculated J_{sc} (16.5 mA/cm^2) shows excellent agreement with the experimental value (16.7 mA/cm^2). The DFT calculation is performed by PBE using a very high mesh density of $16 \times 16 \times 16$ k mesh [70] (see Section 5.2). *Source*: Nishiwaki et al. [70]; Shirayama et al. [1].

EQE, and DFT analyses. In SE, a multi-sample analysis was performed [1] and the calculated (ψ, Δ) spectra (solid lines) show almost perfect agreement with the experimental spectra (open circles).

In the EQE analysis, a semi-transparent $MAPbI_3$ solar cell reported in Ref. [71] was analyzed using the α spectrum of Shirayama et al. In this solar cell, a rear metal electrode is not present and the light absorption properties can be studied with greater accuracy due to the lack of metal back-side reflection. As shown in the figure, the calculated EQE spectrum (solid line) shows excellent agreement with the measured spectrum (open circles) and the short-circuit current density (J_{sc}) obtained from the analysis agrees quite well with the experimental value [1]. It should be emphasized that, when an α spectrum with greater absolute values is applied for the EQE simulation, J_{sc} is seriously overestimated [1].

In Figure 4.13, the experimental α spectrum of Shirayama et al. is further compared with the DFT α spectrum [70]. For the DFT calculation, the PBE functional was employed (see Section 5.2). The magnitude of α determined theoretically by DFT shows excellent overall agreement with the experimental spectrum; therefore, the experimental light absorption is consistent with the dipole transition expected from rigorous theoretical calculation.

As discussed above, there are inherent difficulties for obtaining reliable optical spectra of hybrid perovskites and the extracted experimental optical spectrum must be justified based on other complementary methods including optical device simulations and DFT calculations. The key points for accurate optical-constant characterization are the (i) preparation of smoother perovskite layers to reduce the roughness effect, (ii) optical measurements without exposing the samples to air to avoid rapid surface degradation, and (iii) application of a high-precision ellipsometry technique with exact optical modeling.

4.5 Universal Rules for Hybrid Perovskite Optical Properties

The optical properties of hybrid perovskite materials vary with the choice of A-site cations and X-site anions. The A-site cation has a notable effect on absolute α values (Section 4.5.1), while the X-site anion shifts the α spectrum in the energy direction (Section 4.5.2). In this section, we discuss the universal rules for the optical property variation with the A-site cation and X-site anion species.

4.5.1 Variation with Center Cation

Figure 4.14 summarizes the α spectra obtained with the variation of the A-site cations. All the α spectra were determined by high-precision ellipsometry characterization. These results show that the α in the band-edge region changes with the choice of organic cation; $FAPbI_3$ shows a low α, whereas the α in $MAPbI_3$ increases. The incorporation of a small amount of Cs (17 at.%) into $FAPbI_3$ increases

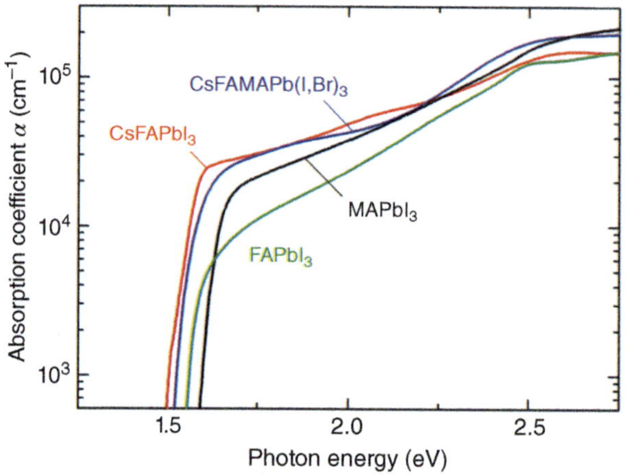

Figure 4.14 α spectra obtained with the variation of the A-site cations. The α spectra of $MAPbI_3$ [1], $FAPbI_3$ [2], $Cs_{0.17}FA_{0.83}PbI_3$, and $Cs_{0.05}FA_{0.8}MA_{0.15}Pb(I_{0.95}Br_{0.05})_3$ are shown. The α spectra are consistent with those shown in Figure 4.1. For $CsFAPbI_3$, the optical constants reported in Ref. [4] were adjusted slightly by extra roughness correction [5].

the band-edge α. High α is also maintained in a $CsFAMAPb(I,Br)_3$ triple-cation perovskite even though the Cs and MA contents in the material are relatively low.

The unexpected large influence of the A-site cation on α has been discussed based on DFT calculations [2]. As mentioned above, in the visible/UV region, there is no direct light absorption by the organic cations and the optical transition is characterized by Pb(6s) + I(5p) → Pb(6p). Nevertheless, the distribution of the I-5p charge densities varies largely depending on the cation species due to the hydrogen bonding interaction (I· · ·H—N). Interestingly, when the hydrogen bonding is formed, the I-5p charge disappears due to the reduction of the N–I distance. This electrostatic interaction (the anti-coupling effect [2]) can be adopted to explain the α variation in Figure 4.14. In particular, the α in the perovskite increases in the order of FA → MA → Cs due to the weakening hydrogen bonding interaction (see Figure 10.11). Nevertheless, what is less apparent is the high α observed for the FA-based mixed-cation perovskites. The molecular dynamics simulations of $CsFAPbI_3$ show that rather strong distortion occurs for the $FAPbI_3$ cubic structure by incorporation of Cs^+ [72]. Thus, a large-scale DFT calculation is necessary to further verify the effect of the mixed A-site cations on α.

4.5.2 Variation with Halide Anion

Figure 4.15 summarizes the variations of hybrid perovskite α spectra with the X-site anion, determined by spectroscopic ellipsometry characterization. All the optical constant data of the figure are summarized in Appendix A. In $MAPbX_3$, the entire spectrum shifts toward higher energy as the X-site halide anion becomes lighter [3]. In the case of $MAPbBr_3$ and $MAPbCl_3$, the presence of the excitonic absorption increases the α in the E_g region. For the $APb(I_{1-x}Br_x)_3$ alloys in Figure 4.15, the entire α spectrum slides toward higher energy without an amplitude change. A similar trend can be confirmed for the $A(Sn,Pb)I_3$ material shown in Figure 4.1. This simple horizontal α-shift can be explained by applying the sum rule [73]:

$$\int E\varepsilon_2(E)\,dE = \text{const.} \tag{4.14}$$

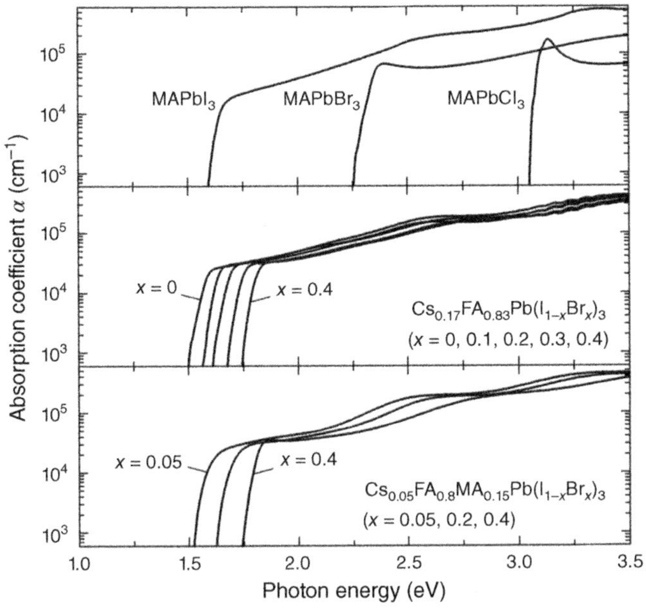

Figure 4.15 Variations of hybrid perovskite α spectra with the X-site anion, determined by spectroscopic ellipsometry characterization. In the figure, the reported α spectra of $MAPbI_3$ [1], $MAPbBr_3$ [3], and $MAPbCl_3$ [3] are shown. For $CsFAPb(I,Br)_3$, the optical constants reported in Ref. [4] were adjusted slightly by extra roughness correction [5]. All the optical constant data in this figure are summarized in Appendix A.

The sum rule requires that, when the ε_2 spectrum is shifted toward a higher energy by ΔE, the ε_2 amplitude is reduced by a factor of $f = E/(E + \Delta E)$ [2, 74].

Figure 4.16a shows an example of the ε_2 shift calculation based on the sum rule. In this figure, the ε_2 spectrum of $MAPbI_3$ is shifted toward higher energy while reducing the ε_2 amplitude using f. From the shifted ε_2 spectra, the corresponding ε_1 spectra are calculated by applying the Kramers–Kronig integration [54, 73]:

$$\varepsilon_1(E) = 1 + \frac{2}{\pi} P \int_0^{\infty} \frac{E' \varepsilon_2(E')}{E'^2 - E^2} dE' \tag{4.15}$$

where P shows the principal value of the integral. From a set of $(\varepsilon_1, \varepsilon_2)$ obtained from the above calculation, $\alpha(E)$ can further be deduced. Figure 4.16b shows the α spectra calculated from the above approach. It can be seen that the resulting α spectrum shifts horizontally in the energy direction without the amplitude change. The above scheme can be applied to calculate the optical constants of an alloy with an arbitrary composition (or ΔE) and is particularly helpful in designing a perovskite top cell in a tandem architecture.

When the dielectric function is modeled by the Tauc–Lorentz model (see Figure 4.4), the alloy dielectric function can be modeled by simply increasing the corresponding peak energy positions toward higher energy. From the Tauc–Lorentz model parameters described in Appendix A, the optical constants

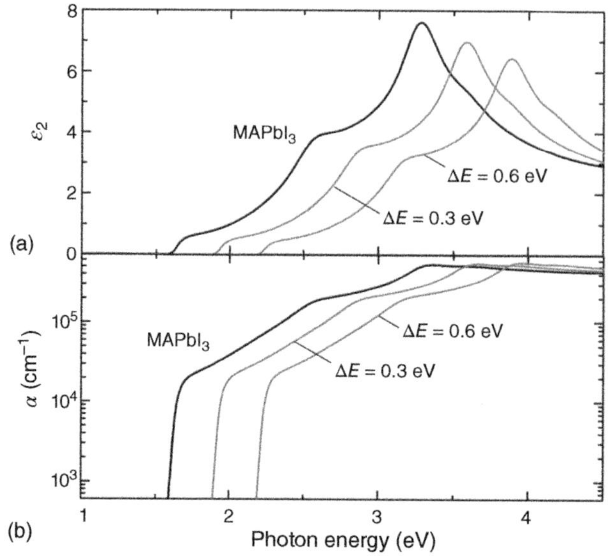

Figure 4.16 Spectral shift calculation of $MAPbI_3$ based on the sum rule: variation of (a) ε_2 and (b) α spectra. In the figure, the ε_2 spectrum of $MAPbI_3$ [1] is shifted toward higher energy by $\Delta E = 0.3$ and 0.6 eV. The horizontal shift of the α spectrum by alloying of the X-site anions can be reproduced by this model.

of various perovskite alloys can be calculated. Such an approach has previously been adopted in the EQE simulation program (e-ARC software) [75], explained in Figure 11.13.

4.6 Subgap Absorption Characteristics

The sharp absorption feature at the E_g edge (subgap region) is critical for solar cell absorbers as strong tail absorption increases the saturation current density (J_0), leading to significant open-circuit voltage (V_{oc}) reduction [76]. The tail absorption in semiconductors can be analyzed quantitatively by the Urbach energy (E_U), obtained by considering an exponential tail state [77]:

$$\alpha = \alpha_0 \exp\left(\frac{E}{E_U}\right) \tag{4.16}$$

where α_0 shows a proportional constant. Thus, E_U can be estimated easily from the relation of $\ln\alpha \propto E/E_U$.

Figure 4.17 shows the (a) result of the Urbach analysis performed for $MAPbI_3$ and (b) variation of E_U with E_g of various hybrid perovskites in Figures 4.1 and 4.15. $MAPbI_3$ exhibits a sharp band-edge characterized by a low E_U of 15 meV [14, 70]. Photothermal deflection spectroscopy (PDS) further shows that the sharp absorption tail is maintained over a wide α range ($\alpha = 10$–10^4 cm^{-1}) [14]. A low E_U is also observed for the mixed-cation perovskite alloys (Figure 4.17b), confirming that the

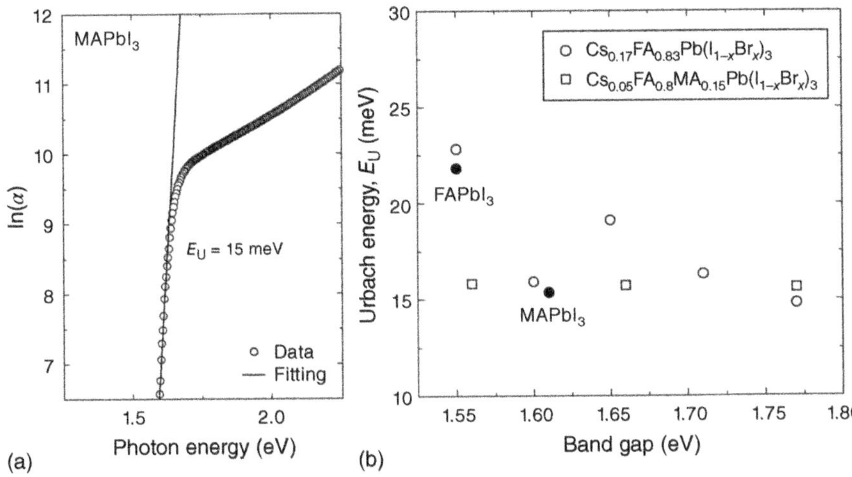

Figure 4.17 (a) Urbach analysis performed for $MAPbI_3$ [70] and (b) variation of E_U with E_g obtained for various hybrid perovskites shown in Figures 4.1 and 4.15. The line in (a) indicates the result of linear fitting based on $\ln\alpha \propto E/E_U$. *Source*: Based on Nishiwaki et al. [70].

A-site cation and X-site anion mixing does not lead to band-edge fluctuation. A similar low E_U has also been observed in Sn-based materials (see Figure 10.7). Note that, in $APbBr_3$ and $APbCl_3$, the E_U analysis becomes difficult as the band-edge excitonic absorption prevents the tail state analysis.

A sharp absorption edge observed in hybrid perovskites is supported by the theoretical calculations. Figure 4.18a compares the experimental and DFT α spectra of $MAPbI_3$ in the band-edge region [70]. Both spectra overlap almost perfectly even though a perfect crystal structure at 0 K is assumed in the DFT calculation. This result indicates that the formation of an unintentional tail state is almost completely suppressed in the experimental perovskite crystal. A DFT study further shows that small E_U in $MAPbI_3$ is an inherent character. Figure 4.18b shows the relation of the experimental E_U and DFT absorption edge energy (E_{DFT}) derived from $\ln(\alpha_{DFT}) \propto E/E_{DFT}$ [70]. This result shows that the E_U and E_{DFT} of $MAPbI_3$ are notably smaller, compared with the other solar cell materials in the figure. The small E_{DFT} of $MAPbI_3$ has been attributed to the sharp density-of-states (DOS) distribution in the band edge region (see Figure 5.11b). In the case of $Cu_2ZnSnSe_4$ (CZTSe) and Cu_2ZnSnS_4 (CZTS), E_U is considerably greater than E_{DFT} due to intense cation mixing in the experimental crystals [70], severely limiting V_{oc} of these solar cells [76].

It should be emphasized that the above DFT results were obtained for the direct-gap band structure shown in Figure 4.5 (i.e. within PBE). When the SOC interaction is considered, E_{DFT} increases significantly to 37 meV due to the conduction band splitting (i.e. Rashba splitting) and the calculated E_{DFT} deviates from the experimental E_U (see Figure 5.9). Accordingly, Rashba splitting is inconsistent with the sharp absorption edge confirmed experimentally.

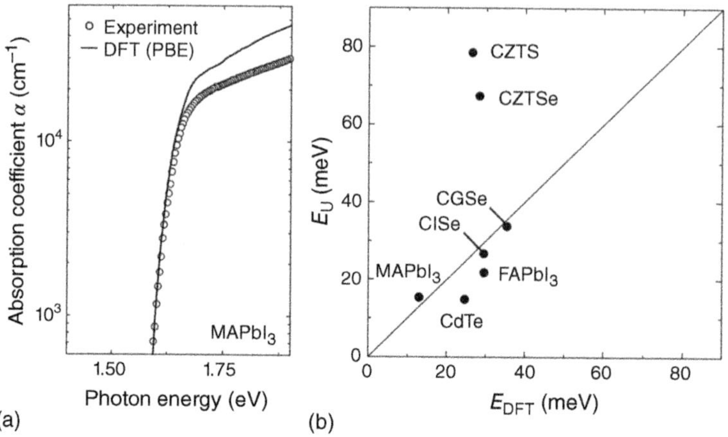

Figure 4.18 (a) Experimental and DFT α spectra of $MAPbI_3$ in the band-edge region and (b) relation of the experimental E_U and DFT absorption edge energy (E_{DFT}). *Source:* Nishiwaki et al. [70].

4.7 Temperature Effect on Absorption Properties

Almost all semiconductors show E_g reduction with a temperature increase [77]. Nevertheless, unlike conventional semiconductors, hybrid perovskites show E_g widening with temperature (T) [28–31]. In this section, we discuss the unique E_g variation of $MAPbI_3$ with T.

Figure 4.19a shows the ε_2 spectral variation with T in $MAPbI_3$ [29]. The ε_2 absorption feature changes dramatically by the γ (orthorhombic) → β (tetragonal) phase change that occurs at $T \sim 160$ K (see also Figure 3.13). In the low-T γ phase, a clear excitonic peak is present, whereas the interband absorption dominates at higher temperatures. In both regions, the band-edge absorption shifts toward higher energy with increasing T, although E_g shows a rapid reduction by the γ → β phase change. A similar spectral variation can be confirmed in Figure 4.19b, where α values in the band-edge region are shown for different T [30]. In the figure, a high-energy spectral shift with increasing T and a sudden drop of E_g at $T \sim 160$ K can be confirmed more clearly. The black and red regions observed at low T are consistent with the excitonic peak.

Figure 4.19c shows the variation of the E_g position with T. In this figure, three results [28–30] are summarized to cover a wider T range and each result is shifted slightly such that the E_g at room temperature shows a consistent value of 1.61 eV. The large E_g reduction by 0.1 eV at 160 K can be attributed to the change in the Pb–I–Pb bond angle by the γ → β transition [30]. Similar to the effect of the A-site cation in $APbI_3$ (Figure 4.9a), E_g decreases when the Pb–I–Pb bond angle becomes collinear. Since the Pb–I–Pb angle of the β phase is nearer 180° (see Figure 3.15), the E_g decreases by the sharp phase transition at $T \sim 160$ K. On the other hand, the β → α transition at 327 K does not influence E_g, as the crystal structures are essentially similar.

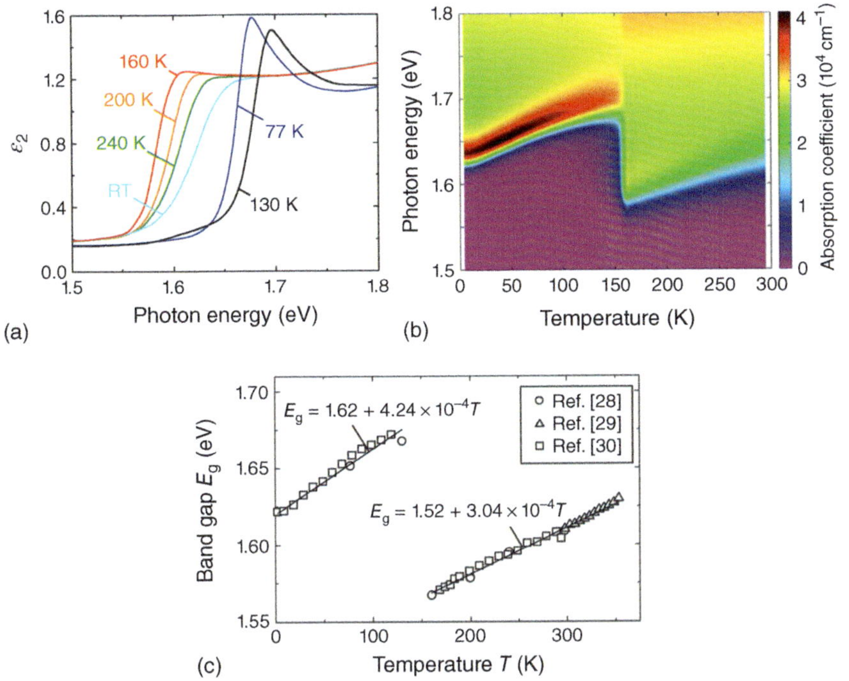

Figure 4.19 (a) ε_2 spectral variation of $MAPbI_3$ with temperature T, (b) change in band-edge α with T, (c) variation of the E_g position with T. In (a), non-zero ε_2 values observed below E_g are most likely due to the contribution of the roughness. In (c), the results of three studies [28–30] are summarized but each result was shifted slightly so that E_g at room temperature shows a consistent value of 1.61 eV. *Source*: (a) Jiang et al. [29]. (b) Davies et al. [30]. Licensed under CC BY 4.0. (c) Foley et al. [28]; Jiang et al. [29]; Davies et al. [30].

The increase of E_g with T, observed in two different regimes in $MAPbI_3$, can be fitted by a linear model of

$$E_g(T) = 1.62 + 4.24 \times 10^{-4}\, T \text{ eV } (T < 160\text{ K}) \tag{4.17}$$

$$E_g(T) = 1.52 + 3.04 \times 10^{-4}\, T \text{ eV } (T > 160\text{ K}) \tag{4.18}$$

The observed increase in E_g with T has been attributed to the lattice expansion and enhanced electron–phonon coupling [31]. The E_g widening at high T has also been confirmed in other hybrid perovskites including $FAPbI_3$, $FAPbBr_3$, and $MAPbBr_3$ [53].

4.8 Excitonic Properties of Hybrid Perovskites

To explain the efficient carrier generation in hybrid perovskites, the excitonic character of $MAPbI_3$ has been investigated and discussed extensively [16, 30, 35, 47–52].

Table 4.4 Exciton binding energy (E_X) in hybrid perovskites at room temperature.

Material	Method	E_X (meV)	References
$MAPbI_3$	Impedance analysis	~2	[16]
	Magneto-optical characterization	~2	[47]
	Absorption analysis	6	[48]
	Absorption (Elliott's theory)	~8	[30]
	Absorption (Elliott's theory)	25	[49]
$MAPbBr_3$	Absorption (Elliott's theory)	41.6	[35]
	Absorption (Elliott's theory)	64	[49]
	Absorption (f-sum rule)	60	[50]
$MAPbCl_3$	Absorption (Elliott's theory)	69	[49]

Analysis methods for each report are also indicated.

The exciton is a hydrogen-like structure generated for a hole in the valence band and an electron in a conduction band, which are attracted to each other through the Coulomb force. In general, the exciton formation in materials is characterized from the exciton binding energy (E_X) derived for an electron–hole Coulomb attraction and the stability of the exciton can be gauged simply from the magnitude of E_X. Table 4.4 summarizes E_X values and their analysis methods reported for $MAPbX_3$ [16, 30, 35, 47–50]. One simple way to determine E_X is to apply the Wannier equation:

$$E_X = \frac{\mu}{m_0}\frac{1}{\varepsilon^2}\frac{m_0 q^4}{2(4\pi\varepsilon_0\hbar^2)} \tag{4.19}$$

where m_0, q, and ε_0 indicate the electron mass, electron charge, and vacuum permittivity, respectively [16]. The two important variables in Eq. (4.19) are the exciton effective mass μ and dielectric constant ε (i.e. $\varepsilon = \varepsilon_1$). By assuming $\mu = 0.6\text{–}0.7\, m_0$ and $\varepsilon = 70$ determined experimentally, a very low E_X of ~2 meV has been derived for $MAPbI_3$ [16]. Since the thermal energy at room temperature (i.e. $k_BT = 26$ meV) is considerably greater than E_X, the exciton formation is negligible in $MAPbI_3$. Consistent low E_X values in the range 2–8 meV have also been reported for $MAPbI_3$ (see Table 4.4), although the discussion of the appropriate E_X values continues [51, 52]. On the other hand, in the DFT calculation within PBE, excitonic interaction (i.e. electron–hole Coulomb attraction) is neglected completely. Thus, the excellent agreement observed between the experimental and DFT spectra in Figure 4.13 confirms that the excitonic character is negligible in $MAPbI_3$.

In $MAPbBr_3$ and $MAPbCl_3$, E_X values become larger than the room-temperature thermal energy. Thus, for these perovskites, the exciton is stable at room temperature, generating the clear excitonic feature in the absorption spectra (Figure 4.15).

References

1 Shirayama, M., Kadowaki, H., Miyadera, T. et al. (2016). *Phys. Rev. Appl.* 5: 014012.
2 Kato, M., Fujiseki, T., Miyadera, T. et al. (2017). *J. Appl. Phys.* 121: 115501.
3 Leguy, A.M.A., Azarhoosh, P., Alonso, M.I. et al. (2016). *Nanoscale* 8: 6317.
4 Tejada, A., Braunger, S., Korte, L. et al. (2018). *J. Appl. Phys.* 123: 175302.
5 Fujiwara, H., Fujimoto, S., Tamakoshi, M. et al. (2017). *Appl. Surf. Sci.* 421: 276.
6 Fujiwara, H. and Collins, R.W. (2018). *Spectroscopic Ellipsometry for Photovoltaics: Fundamental Principles and Solar Cell Characterization*, vol. 1. Cham: Springer Nature.
7 Fujiwara, H. and Collins, R.W. (2018). *Spectroscopic Ellipsometry for Photovoltaics: Applications and Optical Data of Solar Cell Materials*, vol. 2. Cham: Springer Nature.
8 Ball, J.M., Stranks, S.D., Hörantner, M.T. et al. (2015). *Energy Environ. Sci.* 8: 602.
9 Xing, G., Mathews, N., Lim, S.S. et al. (2014). *Nat. Mater.* 13: 476.
10 Löper, P., Stuckelberger, M., Niesen, B. et al. (2015). *J. Phys. Chem. Lett.* 6: 66.
11 Jiang, Y., Green, M.A., Sheng, R., and Ho-Baillie, A. (2015). *Sol. Energy Mater. Sol. Cells* 137: 253.
12 Leguy, A.M.A., Hu, Y., Campoy-Quiles, M. et al. (2015). *Chem. Mater.* 27: 3397.
13 Sun, S., Salim, T., Mathews, N. et al. (2014). *Energy Environ. Sci.* 7: 399.
14 De Wolf, S., Holovsky, J., Moon, S.-J. et al. (2014). *J. Phys. Chem. Lett.* 5: 1035.
15 Xing, G., Mathews, N., Sun, S. et al. (2013). *Science* 342: 344.
16 Lin, Q., Armin, A., Nagiri, R.C.R. et al. (2015). *Nat. Photonics* 9: 106.
17 Fujiwara, H., Kato, M., Tamakoshi, M. et al. (2018). *Phys. Status Solidi A* 215: 1700730.
18 Ndione, P.F., Li, Z., and Zhu, K. (2016). *J. Mater. Chem. C* 4: 7775.
19 Yamada, Y., Nakamura, T., Endo, M. et al. (2014). *Appl. Phys. Express* 7: 032302.
20 Zhao, B., Abdi-jalebi, M., Tabachnyk, M. et al. (2017). *Adv. Mater.* 29: 1604744.
21 Colella, S., Mosconi, E., Fedeli, P. et al. (2013). *Chem. Mater.* 25: 4613.
22 Noh, J.H., Im, S.H., Heo, J.H. et al. (2013). *Nano Lett.* 13: 1764.
23 Eperon, G.E., Stranks, S.D., Menelaou, C. et al. (2014). *Energy Environ. Sci.* 7: 982.
24 Lee, M.M., Teuscher, J., Miyasaka, T. et al. (2012). *Science* 338: 643.
25 Stoumpos, C.C., Malliakas, C.D., and Kanatzidis, M.G. (2013). *Inorg. Chem.* 52: 9019.
26 Baikie, T., Fang, Y., Kadro, J.M. et al. (2013). *J. Mater. Chem. A* 1: 5628.
27 Kim, H.-S., Lee, C.-R., Im, J.-H. et al. (2012). *Sci. Rep.* 2: 591.
28 Foley, B.J., Marlowe, D.L., Sun, K. et al. (2015). *Appl. Phys. Lett.* 106: 243904.
29 Jiang, Y., Soufiani, A.M., Gentle, A. et al. (2016). *Appl. Phys. Lett.* 108: 061905.
30 Davies, C.L., Filip, M.R., Patel, J.B. et al. (2018). *Nat. Commun.* 9: 293.
31 Saidi, W.A., Poncé, S., and Monserrat, B. (2016). *J. Phys. Chem. Lett.* 7: 5247.
32 Ghimire, K., Zhao, D., Yan, Y., and Podraza, N.J. (2017). *AIP Adv.* 7: 075108.

33 Rajagopal, A., Stoddard, R.J., Hillhouse, H.W., and Jen, A.K.-Y. (2019). *J. Mater. Chem. A* 7: 16285.

34 Tao, S., Schmidt, I., Brocks, G. et al. (2019). *Nat. Commun.* 10: 2560.

35 Yang, Y., Yan, Y., Yang, M. et al. (2015). *Nat. Commun.* 6: 7961.

36 Kumawat, N.K., Dey, A., Kumar, A. et al. (2015). *ACS Appl. Mater. Interfaces* 7: 13119.

37 Koh, T.M., Krishnamoorthy, T., Yantara, N. et al. (2015). *J. Mater. Chem. A* 3: 14996.

38 Umari, P., Mosconi, E., and De Angelis, F. (2015). *Sci. Rep.* 4: 4467.

39 Hao, F., Stoumpos, C.C., Chang, R.P.H., and Kanatzidis, M.G. (2014). *J. Am. Chem. Soc.* 136: 8094.

40 Im, J., Stoumpos, C.C., Jin, H. et al. (2015). *J. Phys. Chem. Lett.* 6: 3503.

41 Anaya, M., Correa-Baena, J.P., Lozano, G. et al. (2016). *J. Mater. Chem. A* 4: 11214.

42 Goyal, A., McKechnie, S., Pashov, D. et al. (2018). *Chem. Mater.* 30: 3920.

43 Shum, K., Chen, Z., Qureshi, J. et al. (2010). *Appl. Phys. Lett.* 96: 221903.

44 Yu, C., Chen, Z., Wang, J.J. et al. (2011). *J. Appl. Phys.* 110: 063526.

45 Chen, Z., Yu, C., Shum, K. et al. (2012). *J. Lumin.* 132: 345.

46 Sabba, D., Mulmudi, H.K., Prabhakar, R.R. et al. (2015). *J. Phys. Chem. C* 119: 1763.

47 Miyata, A., Mitioglu, A., Plochocka, P. et al. (2015). *Nat. Phys.* 11: 582.

48 Yamada, Y., Nakamura, T., Endo, M. et al. (2015). *IEEE J. Photovoltaics* 5: 401.

49 Saba, M., Quochi, F., Mura, A., and Bongiovanni, G. (2016). *Acc. Chem. Res.* 49: 166.

50 Sestu, N., Cadelano, M., Sarritzu, V. et al. (2015). *J. Phys. Chem. Lett.* 6: 4566.

51 Herz, L.M. (2018). *J. Phys. Chem. Lett.* 9: 6853.

52 Jiang, Y., Wang, X., and Pan, A. (2019). *Adv. Mater.* 31: 1806671.

53 Wright, A.D., Verdi, C., Milot, R.L. et al. (2016). *Nat. Commun.* 7: 11755.

54 Fujiwara, H. (2007). *Spectroscopic Ellipsometry: Principles and Applications*. West Sussex: Wiley.

55 Glaser, T., Müller, C., Sendner, M. et al. (2015). *J. Phys. Chem. Lett.* 6: 2913.

56 Sendner, M., Nayak, P.K., Egger, D.A. et al. (2016). *Mater. Horiz.* 3: 613.

57 Anusca, I., Balčiūnas, S., Gemeiner, P. et al. (2017). *Adv. Energy Mater.* 7: 1700600.

58 Miyata, K. and Zhu, X.-Y. (2018). *Nat. Mater.* 17: 379.

59 Adachi, S. (1999). *Optical Properties of Crystalline and Amorphous Semiconductors: Materials and Fundamental Principles*. Norwell: Kluwer Academic Publishers.

60 Jung, H.S. and Park, N.-G. (2015). *Small* 11: 10.

61 Shirayama, M., Kato, M., Miyadera, T. et al. (2016). *J. Appl. Phys.* 119: 115501.

62 Kato, M., Nishiwaki, M., and Fujiwara, H. (2020). *Phys. Rev. Mater.* 4: 035402.

63 Lautenschlager, P., Garriga, M., Vina, L., and Cardona, M. (1987). *Phys. Rev. B* 36: 4821.

64 Huang, L.Y. and Lambrecht, W.R.L. (2013). *Phys. Rev. B* 88: 165203.

65 Prasanna, R., Gold-Parker, A., Leijtens, T. et al. (2017). *J. Am. Chem. Soc.* 139: 11117.

66 Guerra, J.A., Tejada, A., Korte, L. et al. (2017). *J. Appl. Phys.* 121: 173104.

67 Manzoor, S., Häusele, J., Bush, K.A. et al. (2018). *Opt. Express* 26: 27441.

68 Jaffe, J.E. and Zunger, A. (1984). *Phys. Rev. B* 29: 1882.

69 Minoura, S., Maekawa, T., Kodera, K. et al. (2015). *J. Appl. Phys.* 117: 195703.

70 Nishiwaki, M., Nagaya, K., Kato, M. et al. (2018). *Phys Rev. Mater.* 2: 085404.

71 Kranz, L., Abate, A., Feurer, T. et al. (2015). *J. Phys. Chem. Lett.* 6: 2676.

72 Ghosh, D., Smith, A.R., Walker, A.B., and Islam, M.S. (2018). *Chem. Mater.* 30: 5194.

73 Marder, M.P. (2010). *Condensed Matter Physics*. Hoboken, NJ: Wiley.

74 Nishiwaki, M. and Fujiwara, H. (2020). *Comput. Mater. Sci.* 172: 109315.

75 Nakane, A., Tampo, H., Tamakoshi, M. et al. (2016). *J. Appl. Phys.* 120: 064505.

76 Kato, Y., Fujimoto, S., Kozawa, M., and Fujiwara, H. (2019). *Phys Rev. Appl.* 12: 024039.

77 Pankove, J.I. (1971). *Optical Processes in Semiconductors*. New York: Dover.

5

Physical Properties Determined by Density Functional Theory

Hiroyuki Fujiwara, Mitsutoshi Nishiwaki, and Yukinori Nishigaki

Gifu University, Department of Electrical, Electronic and Computer Engineering, 1-1 Yanagido, Gifu 501-1193, Japan

5.1 Introduction

Density functional theory (DFT) is a revolutionary method developed in the last century, enabling the interpretation and prediction of material properties through computational calculation without requiring experimental parameters [1–3]. The rapid advancement in computer technology now allows the implementation of satisfactory DFT calculations using inexpensive computer cluster systems, resulting in the vast expansion of theoretical studies. For material science, DFT is an indispensable method and a variety of DFT calculations performed for hybrid perovskites have revealed the unique physical properties of the materials [4–38]. Nevertheless, unlike conventional inorganic materials, hybrid perovskites have organic cations that reorient rapidly within the inorganic scaffold (Figure 3.16), making the DFT calculations more challenging. Moreover, for hybrid perovskites, DFT results vary significantly with calculation (approximation) methods [4–25] and controversial interpretations have been made, particularly for band structures [4, 10, 12, 20–27]. Understanding the physical properties derived from DFT requires an overall knowledge of DFT. One crucial feature of hybrid perovskite compounds is the absence of deep-level defects, which contributes significantly to the enhancement of carrier lifetime and the resulting diffusion length (Section 6.4). Such defect physics in hybrid perovskites can be revealed directly by DFT [15, 19, 28, 29].

This chapter starts with the basic principles of DFT (Section 5.2). The hybrid perovskite crystal structures derived from DFT are described in Section 5.3. We discuss in Section 5.4 the controversial band structures and the most appropriate DFT method that reproduces the overall experimental results. In Section 5.5, the formation of the band gap in hybrid perovskites is further explained. Finally, the unique properties of hybrid perovskites for defect formation are presented in Section 5.6.

Hybrid Perovskite Solar Cells: Characteristics and Operation,
First Edition. Edited by Hiroyuki Fujiwara.

5.2 What Is DFT?

DFT (or first principles) is a quantum mechanical method in which the wave function and its energy state are calculated based on the density of electron clouds [1–3]. This section introduces the fundamental principles of DFT (Section 5.2.1) and explains the assumptions and limitations of various DFT methods (Section 5.2.2).

5.2.1 Basic Principles

For electronic structure calculations, DFT is now the most widely used method. From DFT, characteristic properties of materials can be determined directly from the fundamental equations derived for electrons and nuclei in materials. Figure 5.1 illustrates electrons surrounding a nucleus from the perspective of (a) a many-body problem and (b) a DFT problem. In a rigorous many-body approach, mutual interactions of the electrons and their interaction with the nuclei are calculated explicitly. Unfortunately, this many-body problem is rarely solvable, and the computation time becomes enormous even for a simple material system.

In 1965, Kohn and Sham developed a remarkable scheme [39], now called DFT, which eventually made computational materials science possible. In DFT, the electronic properties of a many-electron system are uniquely determined by the spatially dependent electron density $[\rho(\mathbf{r})]$ and the total energy of the system is calculated as a function of $\rho(\mathbf{r})$ (i.e. $E[\rho(\mathbf{r})]$). A function of another function is called a "functional" and, therefore, a theory of electronic-state calculations based on $E[\rho(\mathbf{r})]$ is referred to as "density functional theory." DFT is a theory for many-body systems, but the electron in DFT is treated as an independent non-interacting particle. In other words, DFT is a one-electron approximation method and its essence is to replace a difficult interacting many-body problem with an independent one-electron problem that can be solved more easily by considering the three-dimensional distribution of $\rho(\mathbf{r})$. Specifically, in DFT, an n-electron system ($n = 5$ in Figure 5.1a) becomes a set of non-interacting one-electron systems, where each electron moves within a total density of $\rho(\mathbf{r})$. This reduces the computation time drastically to the level of practical use.

For the electronic structure calculation, DFT employs the following Kohn–Sham equation [39]:

$$\left[-\frac{1}{2}\nabla^2 + V_{\text{eff}}(\mathbf{r})\right]\psi_i(\mathbf{r}) = E_i\psi_i(\mathbf{r}) \tag{5.1}$$

This equation uses atomic units and, in this case, $-(\hbar^2/2m)\nabla^2 = -(1/2)\nabla^2$. The $\psi_i(\mathbf{r})$ is the time-independent wave function of electron i, whereas E_i indicates the corresponding energy of $\psi_i(\mathbf{r})$. Equation (5.1) is analogues to the famous Schrödinger equation but, in the Kohn–Sham equation, the effective potential V_{eff} is given by

$$V_{\text{eff}} = V_{\text{Nuclei}} + V_{\text{Electron}} + V_{\text{XC}} \tag{5.2}$$

where V_{Nuclei} and V_{Electron} describe the potential created by positive-charged nuclei and negative-charged electrons. The importance of the Kohn–Sham equation lies in the introduction of the exchange-correlation potential (V_{XC}). In fact, all the

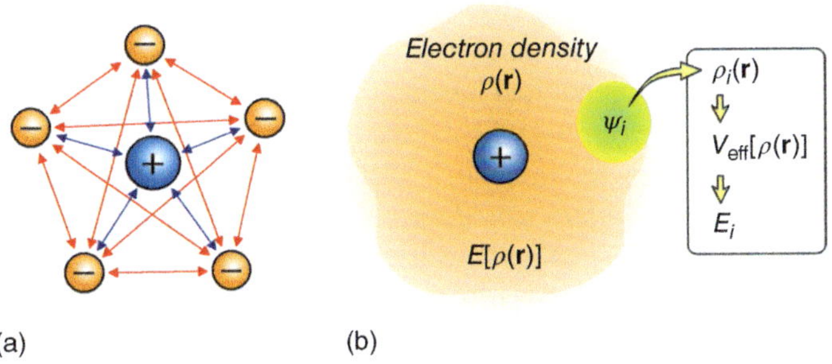

Figure 5.1 Quantum mechanical problems from the perspective of (a) a many-body problem and (b) a DFT problem. In DFT, electronic structures of a many-electron system are determined by a spatially dependent electron density [$\rho(\mathbf{r})$] and the total energy of the system is calculated as the density function of $\rho(\mathbf{r})$ (i.e. $E[\rho(\mathbf{r})]$). The ψ_i in (b) represents the non-interacting wave function of electron i, from which the partial density $\rho_i(\mathbf{r})$ is calculated. The $V_{\text{eff}}[\rho(\mathbf{r})]$ is the effective potential of the Kohn–Sham equation in Eq. (5.1), which further allows the determination of the energy E_i of electron i.

complex many-body effects in Figure 5.1a are incorporated into the V_{XC} term, which can reasonably be approximated as a local or nearly local functional of the electron density. All the potential terms that appear in Eq. (5.2) are functionals of $\rho(\mathbf{r})$: i.e. $V_{\text{eff}}[\rho(\mathbf{r})] = V_{\text{Nuclei}}[\rho(\mathbf{r})] + V_{\text{Electron}}[\rho(\mathbf{r})] + V_{\text{XC}}[\rho(\mathbf{r})]$. Accordingly, Eq. (5.1) becomes solvable if $\rho(\mathbf{r})$ is known. Note that in an independent n one-electron system,

$$\rho(\mathbf{r}) = \sum_{i=1}^{n} \rho_i(\mathbf{r}) = \sum_{i=1}^{n} |\psi_i(\mathbf{r})|^2 \tag{5.3}$$

where $\rho_i(\mathbf{r})$ is the partial electron density, given by $\rho_i(\mathbf{r}) = |\psi_i(\mathbf{r})|^2$. Naturally, we obtain that

$$\int \rho(\mathbf{r})\, d\mathbf{r} = n \tag{5.4}$$

In the DFT approach, from $\psi_i(\mathbf{r})$, we first obtain $\rho_i(\mathbf{r})$. By including other density distributions, we calculate $\rho(\mathbf{r})$, from which $V_{\text{eff}}[\rho(\mathbf{r})]$ is determined. Thus, by assuming $\psi_i(\mathbf{r})$ and $V_{\text{eff}}[\rho(\mathbf{r})]$, Eq. (5.1) can be solved to determine E_i (Figure 5.1b). In DFT, this calculation process is carried out self-consistently (i.e. iteratively) by minimizing the total energy of $E[\rho(\mathbf{r})]$. Since $E = \int \psi^* \hat{H} \psi\, d\mathbf{r}$, $E[\rho(\mathbf{r})]$ is expressed by the following equation [1–3, 39]:

$$\begin{aligned} E[\rho(\mathbf{r})] &= E_{\text{Kin}}[\rho(\mathbf{r})] + E_{\text{Nuclei}}[\rho(\mathbf{r})] + E_{\text{Electron}}[\rho(\mathbf{r})] + E_{\text{XC}}[\rho(\mathbf{r})] \\ &= -\frac{1}{2}\int \psi_i^*(\mathbf{r})\nabla^2\psi_i(\mathbf{r})\, d\mathbf{r} + \int V_{\text{Nuclei}}(\mathbf{r})\rho(\mathbf{r})\, d\mathbf{r} \\ &\quad + \frac{1}{2}\iint \frac{\rho(\mathbf{r})\rho(\mathbf{r}')}{|\mathbf{r}-\mathbf{r}'|}\, d\mathbf{r}d\mathbf{r}' + E_{\text{XC}}[\rho(\mathbf{r})] \end{aligned} \tag{5.5}$$

The first term in the above equation is the kinetic energy, whereas the second and third terms are energies corresponding to the attractive interaction between the electron and nuclei and the repulsive interaction between the electrons, respectively.

The E_{XC} in Eq. (5.5) represents the exchange-correlation energy that originates from the interaction of electron spins. In particular, E_{XC} can be separated into two contributions: (i) exchange energy, generated by electrons with the same spin, and (ii) correlation energy, created by electrons with different spins. The Pauli exclusion principle prohibits two electrons with the same spin from occupying the same orbital, generating less overlap of the electron densities around the electron. Even when two electrons have different spins, these electrons repel each other because of the Coulomb repulsion force, which also results in less overlap in electron density. The $E_{XC}[\rho(\mathbf{r})]$ includes the correction terms for many-body effects, and the quality of a DFT calculation depends critically on how accurately $E_{XC}[\rho(\mathbf{r})]$ is calculated. However, the major challenge of DFT is that the exact form of $E_{XC}[\rho(\mathbf{r})]$ in Eq. (5.5) is unknown except for the case of a free-electron gas, and thus $E_{XC}[\rho(\mathbf{r})]$ needs to be appropriately approximated. Many different methods have been proposed for the calculation of $E_{XC}[\rho(\mathbf{r})]$ (see Section 5.2.2) and physical properties including the change in band gap (E_g) depend strongly on the exchange-correlation energy functional used in the DFT calculation [3, 40]. Accordingly, interpreting DFT results requires knowing which exchange-correlation functional is adopted.

5.2.2 Assumptions and Limitations

In DFT, the positions of the nuclei are "frozen" and the motions of atomic nuclei and electrons are treated separately (Born–Oppenheimer approximation). In other words, the electron–lattice (or phonon) interaction is completely neglected, and thus DFT is an approximation method for the ground-state calculation at 0 K.

A variety of exchange-correlation energy functionals exist for DFT calculations. Furthermore, up to now, more advanced approximation methods have been developed [1, 3]. Figure 5.2 summarizes various theoretical methods, which are shown with a level of approximation (or computational cost). The theoretical calculation methods of hybrid perovskites can be separated into two large groups depending on whether the spin–orbit coupling (SOC) interaction, described below, is considered.

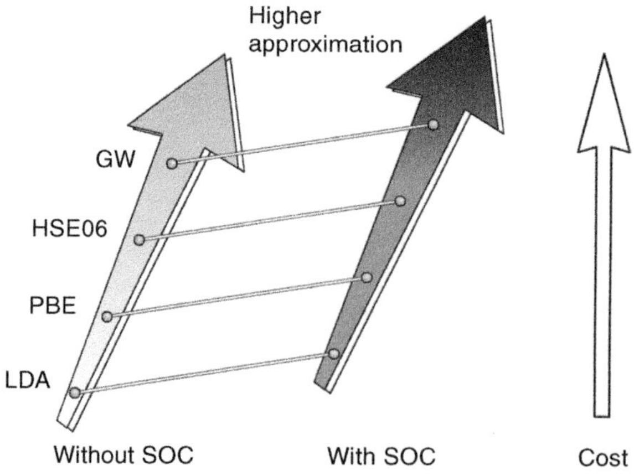

Figure 5.2 Various theoretical calculation methods shown with a level of approximation (or computational cost). In actual computations, the calculation cost of the spin–orbit coupling (SOC) is quite high and the computational time increases significantly.

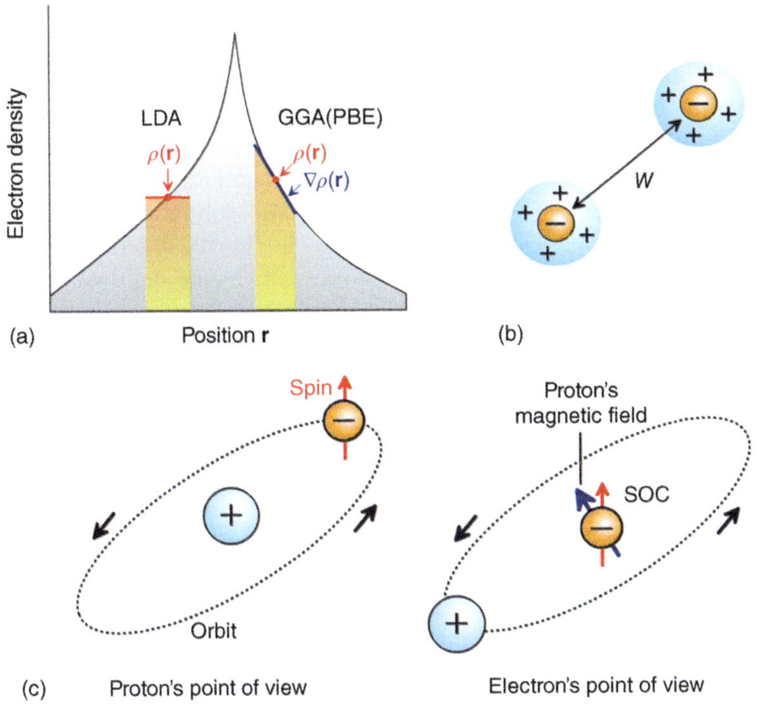

Figure 5.3 (a) Assumptions in local density approximation (LDA) and generalized gradient approximation (GGA), (b) illustration of quasiparticle, formed by the screening of the electron charge by a positive charge contribution, assumed in the GW approximation, and (c) spin–orbit coupling (SOC) interaction generated by proton. In (a), $\nabla\rho$ shows the gradient of the electron density $\rho(\mathbf{r})$ and PBE (Perdew, Burke, and Ernzerhof) is a well-known method of GGA. In (b), "W" denotes the screened Coulomb interaction.

The local density approximation (LDA) is the simplest DFT scheme [1–3]. In this method, $E_{\mathrm{XC}}[\rho(\mathbf{r})]$ is calculated from the local electron density while assuming a uniform electron gas (see Figure 5.3a). However, the assumption used for this method is quite simple and the unrealistic approximation often leads to calculation errors. In the generalized gradient approximation (GGA), in addition to the local electron density, its gradient (i.e. $\nabla\rho$) is also considered to account for the inhomogeneity of the electron density distribution (Figure 5.3a). Among various GGA methods proposed so far [1–3], a scheme developed by Perdew, Burke, and Ernzerhof (PBE) presents a simplified GGA method with no empirical elements [41]. The PBE scheme is now the most popular method of DFT. In the above DFT methods, however, E_g of semiconductor materials is generally seriously underestimated [3, 40, 42].

To reproduce more accurate E_g for materials, Heyd, Scuseria, and Ernzerhof developed a hybrid functional, which is referred to as HSE06 [43, 44]. Here, "06" indicates the year of 2006. In HSE06, the exchange-correlation energy functional is expressed by the PBE functional, but the accurate exchange energy derived from the Hartree–Fock (HF) method is further incorporated as $E_{\mathrm{XC}}^{\mathrm{HSE06}} = E_{\mathrm{X}}^{\mathrm{PBE+HF}} + E_{\mathrm{C}}^{\mathrm{PBE}}$. Here, $E_{\mathrm{X}}^{\mathrm{PBE+HF}}$ gives the exchange energy determined by mixing the PBE and HF results, whereas E_C^{PBE} is the correlation energy obtained from PBE. The contribution of E_X^{PBE+HF} is further divided into short- and long-range components, and the HF contribution is considered only in the short-range term to the amount of 25%. The inclusion of the HF component leads to spatially closer electrons and smaller bond lengths, which in turn increase E_g [40]. Although HSE06 provides accurate calculation results, the computation time is much longer than PBE (×50).

The GW method is a state-of-the-art method that also solves the E_g-underestimation problem [3]. Here, "G" is the single-particle Green's function, while "W" is a screened Coulomb interaction. As illustrated in Figure 5.3b, when an electron is present, it attracts positive charges, which induce the screening of the electron charge. In the GW approximation, this quasiparticle is considered based on the Dyson equation, and thus the GW method is not a DFT method that employs the Kohn–Sham equation shown in Eq. (5.1). In particular, unlike DFT, the GW method contains the features of electron excitation and E_g can be estimated more accurately [42]. Nevertheless, the computation takes even longer than with HSE06.

One important aspect of the DFT calculations of hybrid perovskites is the SOC interaction [4, 13, 17, 20, 26]. In particular, the approximation of DFT improves significantly by incorporating the SOC effect (Figure 5.2). Figure 5.3c illustrates the SOC interaction, as explained in Ref. [45]. Imagine that an electron with a spin orbits around the nucleus of a proton; this is the same situation as the earth orbiting around the sun. When we are on the earth, however, we feel that the sun is rotating around the earth. Similarly, from the electron's point of view, the proton rotates around the electron. The rotation of the ion (proton) generates a magnetic field that interacts with the spin of the electron. This is called the SOC interaction. In heavy atoms, the SOC effect becomes significant because the flow of a heavy ion induces a stronger magnetic field. In Pb-containing hybrid perovskites, the strong SOC effect is expected due to heavy Pb atoms ($Z = 82$), while the influence of SOC becomes weak in the Sn-based perovskites. The band structures of Pb-based hybrid perovskites change drastically when the SOC effect is considered (Section 5.4). Nevertheless, in this chapter, we show that, among all the DFT approximation methods, a simple DFT method of PBE without the SOC provides the best results in representing room-temperature experimental results of hybrid perovskites.

5.3 Crystal Structures Determined by DFT

From DFT, ground-state crystal structures can be determined by minimizing $E[\rho(\mathbf{r})]$ and forces generated between the constituent atoms (i.e. $F = -\partial E/\partial R$). Thus, the stable perovskite structures can be deduced from DFT theoretical calculations. This section shows hybrid perovskite structures obtained from DFT (Section 5.3.1), and we also discuss the orientation of organic-center cations in DFT (Section 5.3.2).

5.3.1 Hybrid Perovskite Structures

Figure 5.4 shows crystal structures of methylammonium lead iodide ($MAPbI_3$; $CH_3NH_3PbI_3$) and formamidinium lead iodide [$FAPbI_3$, $HC(NH_2)_2PbI_3$]

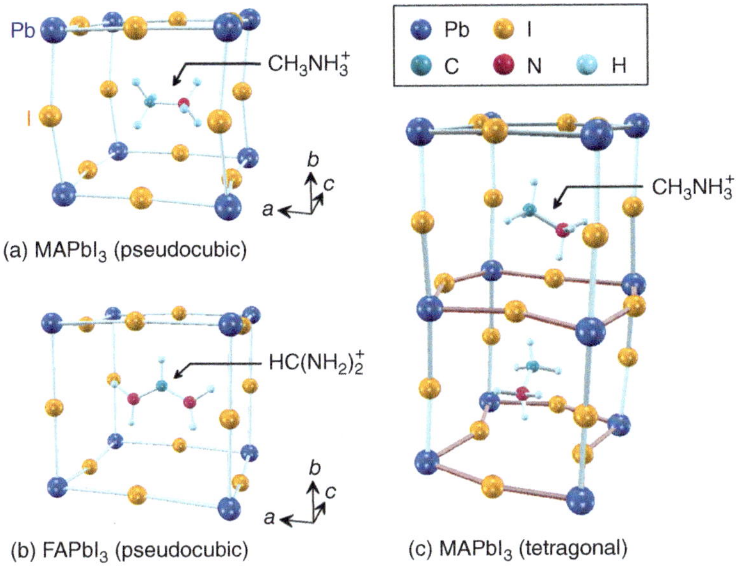

Figure 5.4 Crystal structures of (a) $MAPbI_3$ (pseudocubic) [16], (b) $FAPbI_3$ (pseudocubic) [30], and (c) $MAPbI_3$ (tetragonal), obtained from the structural optimization of PBE without the SOC. The arrows indicate the *a*, *b*, and *c* axes of the pseudocubic unit cells. In (c), the in-plane Pb–I networks are shown by the thin red lines. *Source*: (a) Shirayama et al. [16]. (b) Modified from Kato et al. [30].

determined by DFT: (a) $MAPbI_3$ (pseudocubic) [16], (b) $FAPbI_3$ (pseudocubic) [30], and (c) $MAPbI_3$ (tetragonal). Those DFT structures show the unit-cell structures obtained by using PBE without SOC. The arrows in Figure 5.4 indicate the directions of the *a*, *b*, and *c* axes of the perovskite unit cells (see Figure 3.2). As mentioned above, DFT-optimized structures correspond to those at 0 K. In actual experiments, however, the low-temperature phase of $MAPbI_3$ is the orthorhombic (γ) phase, whereas the tetragonal (β) and cubic (α) phases appear only at higher temperatures (>165 K and >327 K, respectively, see Figure 3.13). To calculate the physical properties of the tetragonal and cubic phases, the initial structures in these DFT calculations are assumed to be tetragonal- and cubic-based crystals. In other words, the optimized DFT structures shown in Figure 5.4 are obtained as the lowest-energy local-minimum structures, rather than the global minimum structure.

The DFT optimization of $MAPbI_3$ and $FAPbI_3$ crystals within cubic symmetry leads to the pseudocubic structures shown in Figure 5.4a,b. The lattice parameters of the pseudocubic $MAPbI_3$ are $a = 6.306$ Å, $b = 6.291$ Å, and $c = 6.310$ Å with $\alpha = \gamma = 90.00°$ and $\beta = 90.35°$ for the unit cell [16], whereas the exchange of MA^+ with FA^+ leads to slightly larger lattice parameters of $a = 6.416$ Å, $b = 6.236$ Å, and $c = 6.353$ Å [30]. The expansion of the unit-cell volume in $FAPbI_3$ originates from the incorporation of the larger center cation (Table 3.3). Notice that the Pb–I–Pb bond angle of the PbI_3^- framework becomes more collinear when the larger organic cation is introduced. The detailed DFT shows that the variation of the Pb–I–Pb bond angle further modifies E_g of $APbI_3$-type perovskites (see Section 5.5). In Table 5.1,

Table 5.1 Lattice parameters of $MAPbI_3$ (α, β, and γ phases) and $FAPbI_3$ (α phase), deduced from crystal structure optimization based on PBE.

Structure (phase)	Lattice parameter (Å)	
	DFT(PBE)	Experiment[a)]
$MAPbI_3$		
Pseudocubic (α)	$a_{ave} = 6.30$[b)]	$a_{ave} = 6.31$[c)]
Tetragonal (β)	$a = 8.80$[d)]	$a = 8.85$
	$b = 8.80$[d)]	$b = 8.85$
	$c = 12.99$[d)]	$c = 12.64$
Orthorhombic (γ)	$a = 8.84$[d)]	$a = 8.84$
	$b = 12.97$[d)]	$b = 12.58$
	$c = 8.77$[d)]	$c = 8.56$
$FAPbI_3$		
Pseudocubic (α)	$a_{ave} = 6.34$[e)]	$a_{ave} = 6.36$

The experimental values taken from Table 3.2 are also shown.
a) See Table 3.2.
b) Average of the *a*, *b*, and *c* parameters reported in Ref. [16].
c) Average of the *a*, *b*, and *c* parameters of Table 3.2.
d) Reference [12].
e) Reference [30].

the lattice parameters of different $MAPbI_3$ crystals and $FAPbI_3$, determined from the PBE optimizations [12, 16, 30], are summarized. It can be seen that the PBE lattice parameters show remarkable overall good agreement with the experimental results, confirming that the simple PBE method is sufficient for the structural determination.

In Figure 5.4, on the other hand, the C–N axis of MA^+ and the N–N axis of FA^+ are parallel to the *a* axis. These configurations reproduce the actual alignments of the organic cations observed experimentally at room temperature (see Figure 3.15). The experimental A-site cation orientation is determined by the hydrogen bonding interaction expressed by N—H· · ·I (Section 3.5.1). Accordingly, DFT successfully reproduces the interaction between the organic cation and inorganic scaffold (i.e. I atoms). However, the position of the A-site cation is "frozen" completely in the case of DFT and the dynamic cation reorientation observed in experiments (Figure 3.16) is not accounted for in DFT.

In the DFT calculation of β-$MAPbI_3$, a variety of cation alignments are possible, although $E[\rho(\mathbf{r})]$ does not change significantly upon a slight change of the cation orientation. The top view of the tetragonal phase in Figure 5.4c is shown in Figure 3.3. Unfortunately, DFT computations become increasingly difficult when the total number of atoms in a unit cell increases. Thus, many DFT

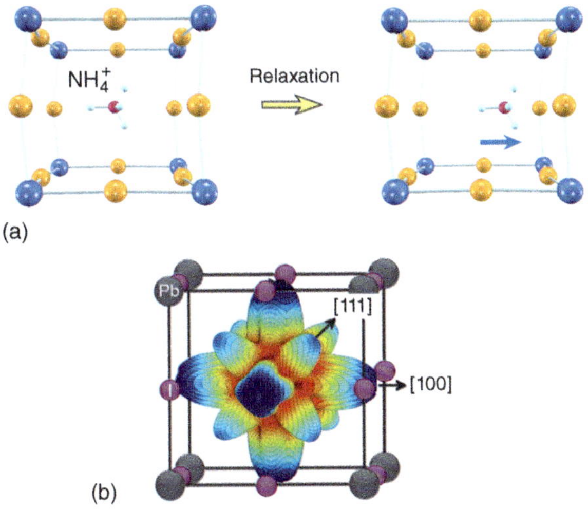

Figure 5.5 (a) Relaxation of NH_4^+ located at the center position of the PbI_3^- scaffold, which induces the shift of NH_4^+ due to the hydrogen bonding interaction N—H···I [16] and (b) orientational energy surface obtained when the orientation the C-N axis of MA^+ is varied within the $MAPbI_3$ cubic structure [34]. In (b), at each orientation, the translational position of the MA^+ cation is also optimized within the cubic unit cell. The energy increases in the order of blue → yellow → red. The total energy decreases when the C-N axis of MA^+ is aligned parallel to the ⟨100⟩ direction at the off-center position and the second-lowest energy is obtained when the axis is oriented in the ⟨111⟩ direction. *Source*: (a) Shirayama et al. [16]. (b) Bechtel et al. [34].

calculations for hybrid perovskites have been performed within cubic symmetry, because the DFT calculation of the tetragonal structure is computationally more demanding.

5.3.2 Organic-Center Cations

In the DFT of hybrid perovskites, the orientation and location of the organic-center cation are determined primarily by the hydrogen bonding interaction [7, 16, 25, 30, 32–34]. For example, if NH_4^+ located at the center position of PbI_3^- framework is relaxed (or structurally optimized), NH_4^+ moves to the off-center position so that the I···H—N interaction is maximized (Figure 5.5a) [16]. In fact, the energy-minimum position of MA^+ has been studied by a systematic PBE calculation [34] and $MAPbI_3$ shows the lowest energy when the C–N axis of MA^+ is aligned parallel to the ⟨100⟩ direction at the off-center position, whereas the second-lowest energy is obtained toward the ⟨111⟩ direction (Figure 5.5b). The large stabilization observed for the ⟨100⟩ direction results from the reduced N—H···I distance allowed for this orientation. Note that, in the case of inorganic $CsPbI_3$, Cs^+ locates at the center position of the PbI_3^- cage, as there is no hydrogen bonding interaction [25].

5.4 Band Structures

Absolute E_g and the type of optical transition (i.e. direct or indirect) are critical properties that are characterized by band structures. Numerous DFT studies have so far been made to determine the band structures of hybrid perovskites [4–27]. Nevertheless, E_g and band structures vary drastically depending on the approximation level of DFT (Section 5.4.1). Due to the substantially different DFT results obtained based on a variety of assumptions, a quite controversial interpretation has been made, particularly for the nature of the band structure at the E_g position (Section 5.4.2). The calculation of the density of states in hybrid perovskites (Section 5.4.3) reveals the unique characters of the valence and conduction bands. The curvature of the bands at the E_g position further determines the effective mass (Section 5.4.4). This section reviews the band structures calculated by various DFT methods and discusses the most appropriate DFT method that can reproduce the E_g and light absorption characteristics of hybrid perovskites. In contrast to a general hypothesis, we demonstrate that a simple PBE method without the SOC provides the best overall agreement with experimental results.

5.4.1 Band Structures of Hybrid Perovskites

Figure 5.6 shows the band structure of $MAPbI_3$ calculated from PBE (without SOC) assuming the pseudocubic structure of Figure 5.4a [16]. Similar band structures have been reported widely [9, 12, 15, 17, 18, 21, 26]. The inset shows high-symmetry points in the corresponding Brillouin zone. Since the assumed crystal is not perfectly cubic, the resulting reciprocal lattices (i.e. a^*, b^*, c^*) are slightly different, which in turn changes the energy positions of M_{1-3} and X_{1-3}. As shown in Figure 5.6, $MAPbI_3$ is a direct-transition semiconductor, with the direct gap being formed at the R point in the Brillouin zone [9, 12, 15, 17, 18, 21, 26]. The E_g (E_0 in Figure 5.6) obtained from the PBE calculation is 1.56 eV, which shows excellent agreement with the experimental value of 1.61 eV (see Table 4.3).

The arrows in the band structure (E_{0-2}) represent major optical transitions that occur in $MAPbI_3$, which are determined by the detailed transition analysis based on PBE [16]. In the experimental absorption spectrum of $MAPbI_3$, three clear optical transitions can be seen (see Figure 4.4) and these features originate from the absorption transitions at the E_{0-2} points denoted by the arrows, although the E_2 transition peak consists of three contributions (i.e. $E_{2,a\text{-}c}$). In the insets of Figure 5.6, the charge density profiles for the band energy regions indicated by the gray color are shown. The valence band minimum (VBM) consists of the Pb 6s and I 5p states, whereas the conduction band minimum (CBM) is dominated by isolated Pb 6p [5, 9, 10, 12, 13]. In the energy position deeper into the conduction band, Pb and I orbitals become quite dispersive.

We now look at how DFT calculation methods affect the band structure. Figure 5.7 shows the variation of the $MAPbI_3$ band structure with the DFT calculation method: (a) PBE, (b) PBE + SOC, and (c) HSE06 + SOC [21]. For the calculations, the orthorhombic crystal is assumed and, in the case of the orthorhombic and

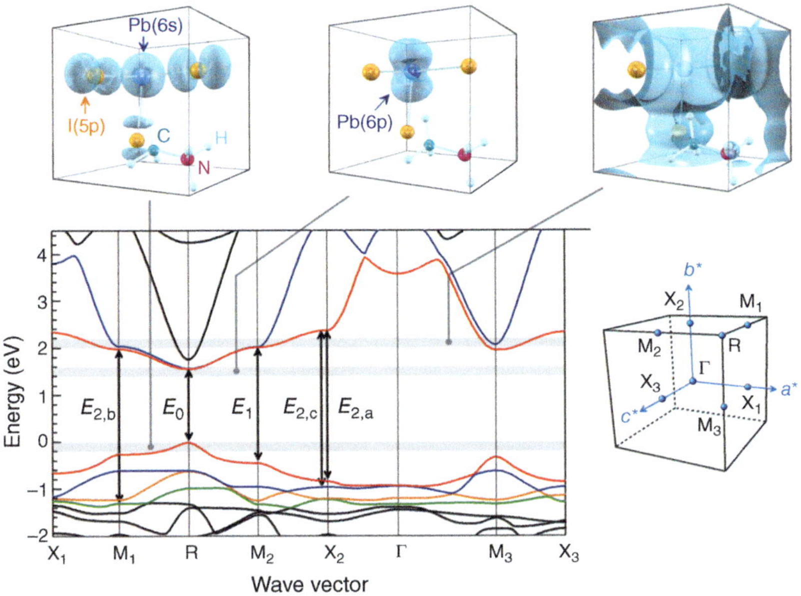

Figure 5.6 Band structure of $MAPbI_3$ calculated by PBE without the SOC assuming the pseudocubic structure of Figure 5.4a. High-symmetry points in the corresponding Brillouin zone are shown. Since the assumed crystal is not perfectly cubic, the reciprocal lattices (a^*, b^*, c^*) differ, resulting in a slight variation in the energy of M_{1-3} and X_{1-3}. In $MAPbI_3$, a direct transition occurs at the R point. The arrows in the band structures (E_{0-2}) represent major optical transitions in $MAPbI_3$ (see also Figure 4.4) and the E_0 transition corresponds to E_g (1.56 eV). The figure shows the charge density profiles for the energy regions of $E = -0.2$ to 0.0 eV (VBM), $E = 1.4–1.6$ eV (CBM), and $E = 2.0–2.2$ eV. The VBM consists of the Pb 6s and I 5p states, whereas the CBM is dominated by isolated Pb 6p. *Source*: Shirayama et al. [16].

tetragonal crystals, the direct transition occurs at the Γ point in the Brillouin zone. The PBE calculation shows $E_g = 1.71$ eV, which is similar to that of the pseudocubic crystal in Figure 5.6. When the SOC is considered, E_g decreases drastically down to $E_g = 0.8$ eV, which is unrealistically low, compared with the experimental E_g. Nevertheless, when HSE06 + SOC is adopted, E_g increases, leading to better agreement with the experimental value.

As described above, a PBE calculation generally leads to a severe underestimation of E_g, which is why HSE06 and GW methods are widely employed to deduce more accurate E_g [40, 42]. In the case of Pb-based hybrid perovskites, however, PBE provides a good estimate of E_g. It is now widely considered that this fortunate E_g agreement in PBE is caused by the cancelation of errors induced by two effects: i.e. (i) the underestimation of E_g caused by the PBE functional and (ii) the overestimation of E_g generated without considering the SOC [10, 20]. Accordingly, when the SOC is incorporated into more advanced calculations such as HSE06 and GW methods, the experimental E_g is reproduced. Note that the strong SOC effect is expected for Pb-based perovskites due to the heavy Pb atoms, while in Sn-based hybrid perovskites the influence of SOC becomes smaller and E_g is underestimated [13, 46].

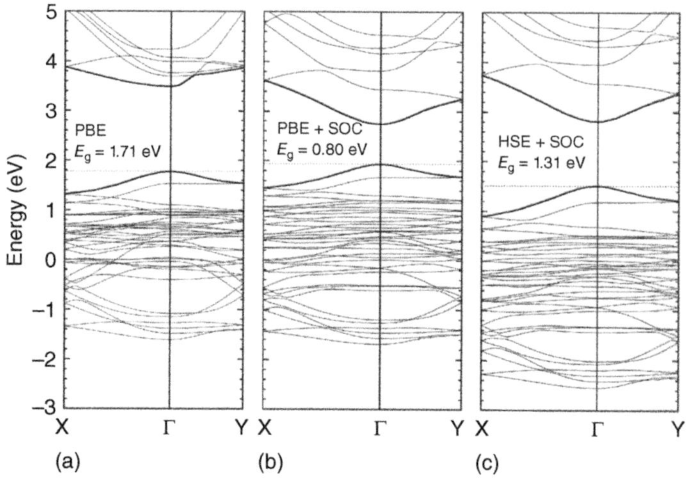

Figure 5.7 Variation of the $MAPbI_3$ band structure with the DFT calculation method: (a) PBE, (b) PBE + SOC, and (c) HSE06 + SOC. The E_g values of the assumed $MAPbI_3$ orthorhombic crystal, obtained from the different methods at the Γ point, are indicated. *Source*: Menéndez-Proupin et al. [21].

The above trend can be confirmed more systematically in Table 5.2, in which the E_g values of $MAPbI_3$ (α, β, and γ phases) obtained from various theoretical calculations [4–22] are summarized. It can be seen that the overall trend of E_g is independent of the crystal structure and the incorporation of SOC induces the large E_g reduction in all the calculation methods, including LDA, PBE, and GW, while experimental values of E_g can be reproduced by LDA, PBE, HSE + SOC, and GW + SOC. Nevertheless, the reported E_g fluctuates even when the same approximation method is adopted. This is most likely due to differences in the optimized crystal structures. In fact, some studies have confirmed that E_g changes by 0.2–0.4 eV depending on the orientation of the organic cation [17, 22, 34]. The effect of MA^+ orientation on E_g becomes larger, particularly when the SOC is included [17]. In theoretical calculations that assume 0 K, the orientation of the organic cation is assumed to be perfectly uniform with its motion completely frozen. Accordingly, there are obvious limitations for the structural modeling in computational calculations and the calculation results need to be verified by experimental results.

5.4.2 Direct–Indirect Issue of Hybrid Perovskite

One of the biggest controversies in the theoretical calculations of hybrid perovskites is the question of direct- or indirect-gap formation [4–27]. The direct-gap formation is quite favorable for solar cell materials, because the direct optical transition generates strong and sharp light absorption, whereas the indirect transition leads to weaker and broader optical transitions [47, 48], which are detrimental as light absorbers. As we have seen in Figure 5.6, $MAPbI_3$ is a direct-transition semiconductor. However, this interpretation varies when the SOC effect is considered [4, 7, 17, 22, 25, 27].

Table 5.2 E_g for three different crystal phases of $MAPbI_3$ calculated by different DFT and GW approximation methods.

Calculation method	E_g (eV)		
	Cubic (α)	Tetragonal (β)	Orthorhombic (γ)
LDA	1.46[a)]	1.51–1.61[b)]	1.58[c)]
LDA + SOC	0.53[a)]		
PBE	1.30–1.64[d)]	1.43–1.85[e)]	1.46–1.82[f)]
PBE + SOC	0.07–0.52[g)]	0.57–0.77[h)]	0.59–0.80[i)]
HSE + SOC	1.14–1.53[j)]	1.60[k)]	1.31–1.90[i)]
GW	2.73[a)]	2.68[l)]	
GW + SOC	1.16–1.67[m)]	1.67[l)]	

"SOC" indicates the calculation that incorporates the spin–orbit coupling interaction (see Figure 5.3c). Variations in E_g reported for each method are shown by the ranges obtained for the references indicated.
a) Reference [4].
b) References [5, 6].
c) Reference [6].
d) References [7–12, 14–18].
e) References [7, 13, 19].
f) References [9, 11, 12, 19–21].
g) References [7, 8, 11, 12, 18].
h) References [7, 8, 11–13].
i) References [11, 12, 21].
j) References [11, 12].
k) Reference [12].
l) References [8, 13].
m) References [4, 8, 22].

Figure 5.8a–c shows the band structures calculated by PBE [16, 25], PBE + SOC, and GW + SOC [22], respectively, together with ε_2 (Figure 5.8d) and absorption coefficient (α) (Figure 5.8e) spectra obtained from each theoretical calculation. All the results are obtained by assuming the pseudocubic structure with the C–N axis of MA^+ directing toward ⟨100⟩. In the PBE result, the direct-gap formation can be seen clearly. In contrast, when PBE + SOC is assumed, the conduction and valence bands split into two bands near the R point. This band splitting induced by the SOC interaction is generally referred to as "Rashba splitting." In the case of PBE + SOC, E_g is seriously underestimated, as mentioned above. When the band structure is calculated by GW + SOC, E_g increases and Rashba splitting appears clearly, particularly in the conduction band, generating a slight indirect-gap formation. Since GW + SOC is the highest-level approach in calculating non-excitonic properties (see Figure 5.2), GW + SOC is generally considered as the most appropriate method.

Nevertheless, only the absolute E_g has yet been considered to justify the theoretical results and a rough agreement in E_g is not sufficient to guarantee that strong Rashba splitting is present in experimental $MAPbI_3$ crystals held at room temperature.

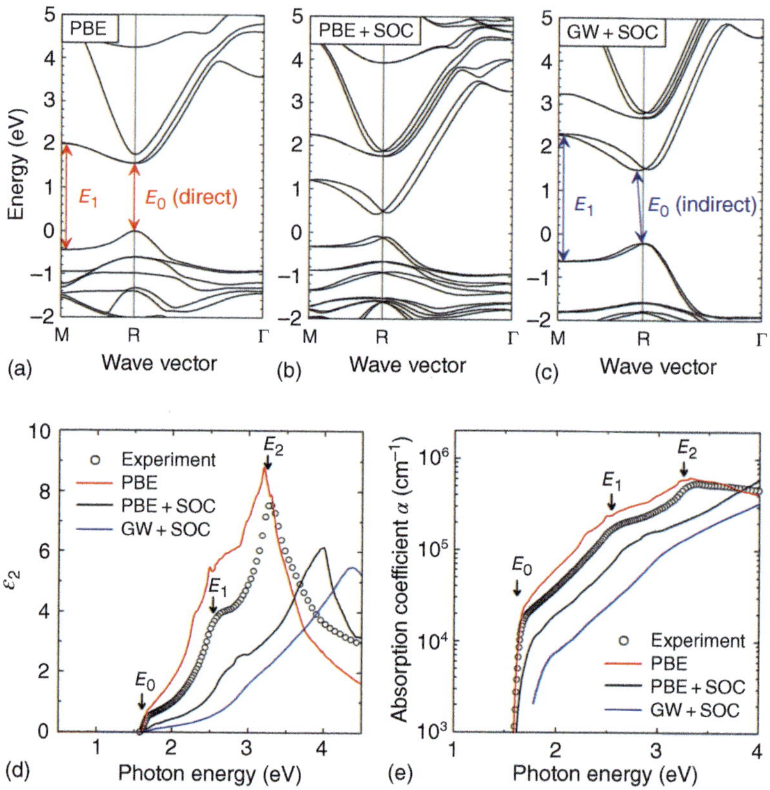

Figure 5.8 Band structures calculated by (a) PBE [16, 25], (b) PBE + SOC, and (c) GW + SOC [22], together with (d) ε_2 and (e) α spectra obtained from each theoretical calculation. The direct transition is obtained in the PBE calculation, while the band structures become indirect when the SOC interaction is considered. The α spectrum calculated from PBE assuming the direct transition shows the best agreement with the experimental α, while the assumption of the indirect transition seriously underestimates α as shown in (e). All results are obtained by assuming the pseudocubic structure with the C–N axis of MA^+ directed along ⟨100⟩. The PBE result of (a) corresponds to that shown in Figure 5.6. In (d) and (e), the solid lines indicate the theoretical results, whereas the open circles show the experimental result shown in Figure 4.4. The E_0 position corresponds to E_g. For the PBE + SOC results in (d) and (e), the optical spectra are shifted toward higher energy to compensate for the underestimated E_g contribution. *Source*: (a) Fujiwara et al. [25]. (c) Adapted from Leguy et al. [22].

The best way to verify valence- and conduction-band structures is to determine the light absorption properties directly from theoretical calculations and to compare the theoretical absorption spectrum with the experimental spectrum. The light absorption characteristics can be expressed by the following equation [49–51]:

$$\varepsilon_2 = A \int f\delta(E_{\mathrm{C},\mathbf{k}} - E_{\mathrm{V},\mathbf{k}} - E)\, d\mathbf{k} \tag{5.6}$$

Here, ε_2 is the imaginary part of the dielectric function, which shows the magnitude of the light absorption (see Section 4.2). The factor A in Eq. (5.6) is a constant. More importantly, f represents the oscillator strength, which shows the strength of the optical transition at each **k** point in the Brillouin zone [49–51]. The f value can be calculated explicitly from the wave functions of the conduction and valence states (i.e. ψ_C and ψ_V, respectively) [50, 51]. The δ function in Eq. (5.6) indicates the transition energy at each **k** point, given by $E = E_{C,\mathbf{k}} - E_{V,\mathbf{k}}$, where $E_{C,\mathbf{k}}$ ($E_{V,\mathbf{k}}$) shows the conduction (valence) band energy at a specific **k** point. Equation (5.6) shows that light absorption (ε_2) is determined by integrating f in the Brillouin zone for an optical transition energy determined by the exact energy positions of the conduction and valence bands [51]. Accordingly, the agreement between theoretical and experimental ε_2 spectra can justify all $E_{C,\mathbf{k}}$ and $E_{V,\mathbf{k}}$ as well as f derived from ψ_C and ψ_V.

Figure 5.8d compares the experimental ε_2 spectrum of $MAPbI_3$ (open circles) with the theoretical results obtained by using PBE [16, 25], PBE + SOC, and GW + SOC [22] (solid lines). For the PBE + SOC result, the ε_2 spectrum is shifted toward higher energy to compensate for the underestimated E_g contribution. It can be seen that PBE provides the best overall agreement among the various calculations; in particular, all the transition energies observed experimentally (E_0 = 1.61 eV, E_1 = 2.53 eV, and E_2 = 3.24 eV in Figure 4.8) are reproduced almost perfectly and the ε_2 amplitude is also reproduced quite well even though a perfect periodic structure with unrealistic uniform MA^+ orientation is assumed in the DFT calculation. In contrast, the agreement with the experimental spectrum deteriorates when the SOC effect is included; namely both E_1 and E_2 transition energies are overestimated notably with PBE + SOC and GW + SOC. This deviation originates from the widening of the conduction band due to SOC, as confirmed in the results of Figures 5.7 and 5.8.

More importantly, the absolute ε_2 near E_g (E_0) is underestimated seriously in the SOC results. This effect can be confirmed more clearly in the corresponding α spectra (Figure 5.8e); namely the near-band-edge α of GW + SOC becomes unrealistically low compared with the experimental result. The drastic α reduction in the GW + SOC result can be interpreted as Rashba splitting in the conduction band, which lowers f significantly. In contrast, the PBE α spectrum reproduces the experimental result surprisingly well, supporting the direct-gap formation at the R point.

More evidence for the direct transition in $MAPbI_3$ can be found in the absorption-edge properties. Figure 5.9 compares the experimental α spectrum in the E_g region with the theoretical spectra (PBE and PBE + SOC) [50]. As discussed in Figure 4.17, experimental hybrid perovskites show very sharp absorption edges with the Urbach energy (E_U) of ~15 meV. The PBE absorption spectrum reproduces the sharp absorption edge with a DFT absorption-edge energy of E_{DFT} = 13 meV [50], derived from $\ln(\alpha_{DFT}) \propto E/E_{DFT}$ (see also Figure 4.18). On the contrary, in PBE + SOC, E_{DFT} increases to 37 meV, as the band splitting by the Rashba effect allows the optical transition in the wider **k** points, broadening the band-edge absorption notably.

Figure 5.10 compares a direct-type band structure with a slightly indirect band structure generated by Rashba splitting. When the direct gap is formed, α increases and the band-edge absorption becomes sharper. Furthermore, the photon emission rate R, which characterizes photoluminescence (PL) characteristics, is proportional

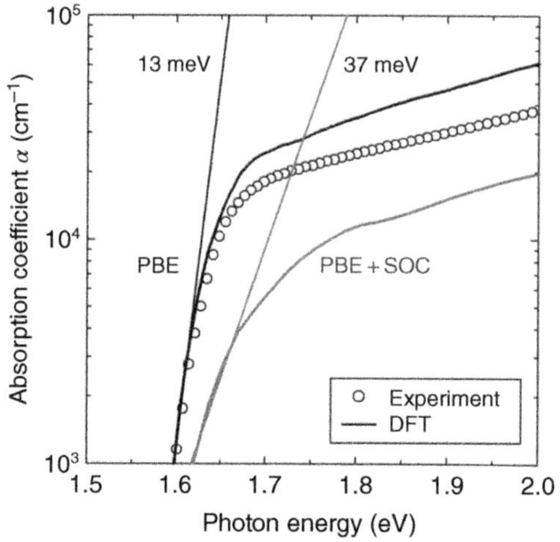

Figure 5.9 α spectra of $MAPbI_3$ obtained from DFT calculations (solid lines) and experiment (open circles). The experimental result is adopted from Ref. [16] (see Figure 4.4). The DFT calculations were implemented by using PBE and PBE + SOC. The α spectrum of PBE + SOC is blueshifted to compensate for the underestimated E_g contribution. The solid lines indicate the fitting results for the DFT α spectra according to $\ln(\alpha_{DFT}) \propto E/E_{DFT}$. The E_{DFT} of PBE (13 meV), derived by assuming a direct gap, is quite close to the experimental value (E_U = 15 meV), whereas E_{DFT} of PBE + SOC (37 meV) is seriously overestimated due to fictitious Rashba band splitting.

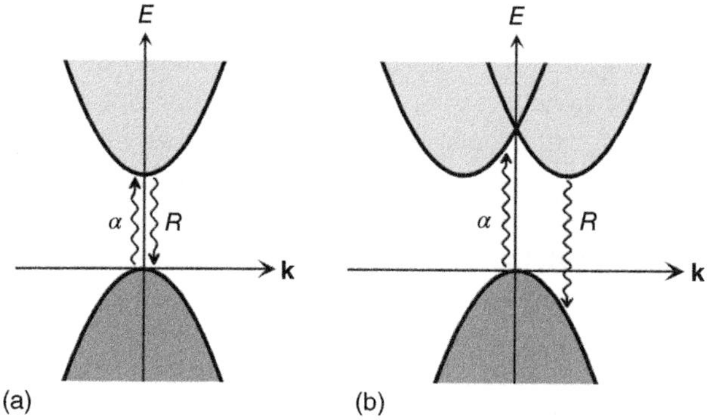

Figure 5.10 Comparison of a (a) direct-type band structure with a (b) slightly indirect band structure generated by Rashba splitting. The α and R are the light absorption coefficient and photon emission rate, respectively.

to α (see Eq. (8.3)). This implies a higher PL efficiency in a more absorbing material. From R, we can further express the radiative recombination coefficient as $B = R/n_i^2$, where n_i is the intrinsic carrier concentration. The B values of $MAPbI_3$, determined experimentally, fall in the range $(0.1–2.3) \times 10^{-9}$ cm^3/s (see Table 8.1), which is comparable to those of direct-gap semiconductors (Section 8.4). Moreover, these experimental values show good agreement with a DFT value of $B = 1.6 \times 10^{-9}$ cm^3/s, which is deduced by assuming direct-gap formation in $MAPbI_3$ [6].

In the indirect-gap material formed by Rashba splitting, a low α with a broader absorption edge is expected. The implication of the indirect-gap formation in the SOC results is a critical issue as several authors attributed quite long carrier lifetime observed in $MAPbI_3$ (Figure 1.6) to a low recombination rate (i.e. B) originating from the indirect-gap nature [7, 17, 27, 52, 53]. However, the low B is consistent with the low α and low PL efficiency, which is in sharp contrast with the experimental results. In fact, the B value deduced from GW + SOC assuming an indirect gap ($B = 4 \times 10^{-13}$ cm^3/s in Ref. [27]) is far lower than the experimental values.

All the experimental evidences of (i) a high α of 10^4 cm^{-1} in the E_g region (Figure 4.4), (ii) a sharp absorption feature characterized by E_U = 15 meV (Figure 4.17), and (iii) quite high PL efficiency with high B (Table 8.1) observed for $MAPbI_3$ strongly support the direct band-gap formation, which is reproduced quite well within the PBE approximation. In other words, Rashba splitting of the conduction bands observed in the SOC results is quite fictitious from the optical point of view. The fact that the simple PBE provides the best result is rather surprising because there is generally an assumption that a higher-level approximation method almost always leads to a better result. Nevertheless, since many fundamental assumptions are made for all theoretical calculations, theoretical results always need to be justified based on experimental results, particularly for hybrid perovskites, which include the dynamical motion of organic cations. The above results confirm that the PBE within the pseudocubic basis provides quite a good approximation for hybrid perovskites. The PBE optical spectra obtained for $MAPbBr_3$ and $FAPbI_3$ also agree well with the experimental results [30], although the excitonic feature observed in $MAPbBr_3$ cannot be reproduced by DFT methods.

5.4.3 Density of States

DFT allows the calculation of the density-of-states (DOS) distribution, and the consistent DOS results have been reported widely for hybrid perovskites [5–10, 12, 13, 15–17]. Figure 5.11a shows the total DOS and partial DOS contributions of the Pb 6s, Pb 6p, and I 5p states in $MAPbI_3$, determined from PBE [16]. This result is consistent with the band structure shown in Figure 5.6, and the position of $E = 0$ eV corresponds to the VBM. The valence band of $MAPbI_3$ mainly consists of Pb 6s and I 5p orbitals, whereas the conduction-band DOS is dominated by Pb 6p. The clear s-orbital shape of Pb 6s and the p-orbital shape of I 5p and Pb 6p can be confirmed in the charge density distributions shown in Figure 5.6. When the DOS obtained from DFT is broadened appropriately, the DOS distribution agrees well with that observed experimentally from photoelectron spectroscopy [54].

The critical conclusion extracted from the DOS calculation is that there are no contributions of C, N, and H atoms (i.e. MA^+) near the VBM and CBM [5, 12, 15]; in other words, the semiconducting properties of $MAPbI_3$ are essentially derived from the inorganic cage, and organic cations are not involved in the formation of valence and conduction bands, even though A-site cations play a decisive role in determining the stability of perovskite crystals (Section 3.3). Consequently, direct light absorption by MA^+ is also negligible in the visible and ultraviolet regions.

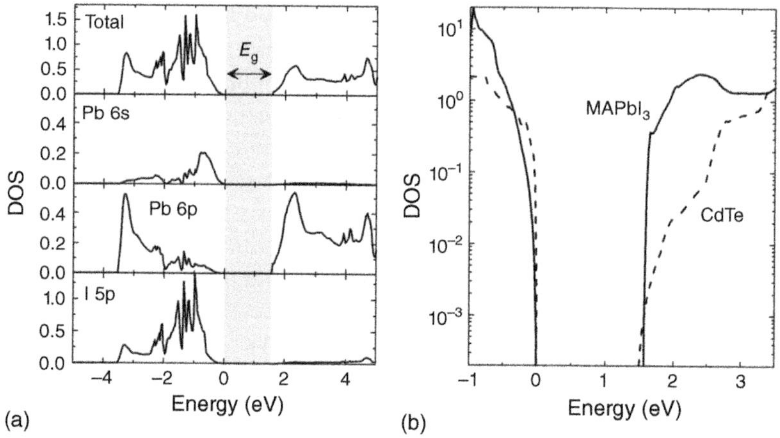

Figure 5.11 (a) Total DOS and partial DOS contributions of the Pb 6s, Pb 6p, and I 5p states in $MAPbI_3$ [16] and (b) total DOS of $MAPbI_3$ and CdTe on a logarithmic scale [50], as determined from PBE. The DFT result is consistent with the band structure given in Figure 5.6. *Source*: Modified from Shirayama et al. [16]; Nishiwaki et al. [50].

In Figure 5.11b, the total DOS of $MAPbI_3$ is shown by a logarithmic scale [50]. It can be seen that, near the VBM and CBM, the DOS of $MAPbI_3$ changes quite sharply, if compared with the DOS of CdTe. The very sharp absorption edge (i.e. low E_U) observed experimentally in $MAPbI_3$ can be understood partly by this quite steep DOS distribution. In fact, among many solar cell materials, $MAPbI_3$ shows the sharpest absorption edge theoretically due to the abrupt DOS feature (see Figure 4.18).

5.4.4 Effective Mass

The effective mass m^* is an important physical parameter that ultimately governs carrier mobility. The effective mass m^* for electrons (m_e^*) and holes (m_h^*) can be calculated directly from the curvature of the band structures near CBM and VBM according to the following equation [18]:

$$m_{e,h}^* = \hbar^2 \left[\frac{\partial^2 E(k)}{\partial k^2} \right]^{-1} \tag{5.7}$$

Table 5.3 summarizes m_e^*/m_0 and m_h^*/m_0 of $MAPbI_3$, calculated for the three crystal phases of cubic (α), tetragonal (β), and orthorhombic (γ) using different approximation methods [13, 18, 21]. Here, m_0 denotes the free-electron mass and m^* of semiconductors generally becomes much lighter than m_0 because the band formation facilitates the movement of electrons and holes. However, m^* changes depending on how the curvature between the high-symmetry points in the Brillouin zone is selected. For example, in the tetragonal $MAPbI_3$, m_e^* obtained along the **k** pass of $\Gamma \rightarrow$ R ($0.28m_0$) is much lower than that of $\Gamma \rightarrow$ X ($1.14m_0$) within PBE [13]. In Table 5.3, m_e^* and m_h^* obtained along the **k** pass of R $\rightarrow \Gamma$ are shown, except for the results of Ref. [21]. To justify the results of the effective-mass calculation, the reduced mass μ_R, defined by $1/\mu_R = 1/m_e^* + 1/m_h^*$, is calculated (see Table 5.3).

Table 5.3 Effective mass of electron (m_e^*) and hole (m_h^*), reported for $MAPbI_3$ based on DFT and GW calculations.

Method	Phase[a)]	Effective mass		Reduced mass (μ_R)	References
		m_e^*/m_0	m_h^*/m_0		
PBE	α[b)]	0.32	0.36	0.17	[18]
PBE	β	0.28	0.26	0.13	[13]
PBE	γ	0.22	0.26	0.12	[21]
PBE + SOC	α[b)]	0.23	0.29	0.13	[18]
PBE + SOC	β	0.14	0.20	0.08	[13]
PBE + SOC	γ	0.17	0.21	0.09	[21]
HSE + SOC	γ	0.18	0.22	0.10	[21]
GW + SOC	β	0.15	0.18	0.08	[13]
Experiment	γ			0.15	[55]
Experiment	γ			0.10	[56]

The values of m^* from Refs. [13, 18] represent those obtained from the E–**k** relation along the **k** pass of $\Gamma \to R$. The reduced mass μ_R is calculated by $1/\mu_R = 1/m_e^* + 1/m_h^*$. The experimental μ_R values indicate those obtained from magneto-optical measurements performed at low temperatures.

a) α: Cubic; β: tetragonal; γ: orthorhombic.
b) Pseudocubic structure.

The theoretical μ_R values show excellent agreement with the experimental values of 0.10–0.15, obtained from the magneto-optical characterization of excitonic γ-phase $MAPbI_3$ [55, 56].

The results of Table 5.3 show clearly that m_e^* and m_h^* do not vary significantly with the crystal structure and calculation method and, more importantly, $m_e^* \sim m_h^* \sim (0.2–0.3)m_0$. The relation of $m_e^* \sim m_h^*$ can be confirmed from similar band curvatures near VBM and CBM (see Figure 5.8). The trend of $m_e^* \sim m_h^*$ is quite unusual and advantageous for carrier transport. In particular, the carrier mobility μ is expressed as $\mu = q\langle\tau\rangle/m^*$, where $\langle\tau\rangle$ is a mean carrier scattering time [47] and μ is inversely proportional to m^*. In the general case of $m_e^* < m_h^*$, observed for tetrahedral-bonding semiconductors (see Table 6.1), the collection of holes becomes more difficult. In contrast, both electrons and holes can be collected efficiently in the hybrid perovskite due to the unique characteristic of $m_e^* \sim m_h^*$ (see Section 6.3.3). The small m_h^* observed for $MAPbI_3$ can be understood from the antibonding character of the VBM, which enhances the dispersion of the VBM (see Section 5.5.1).

5.5 Band Gap

In this section, we discuss the generation of E_g in $MAPbI_3$ as revealed by a PbI_6 cluster calculation (Section 5.5.1). This section also explains E_g variations with

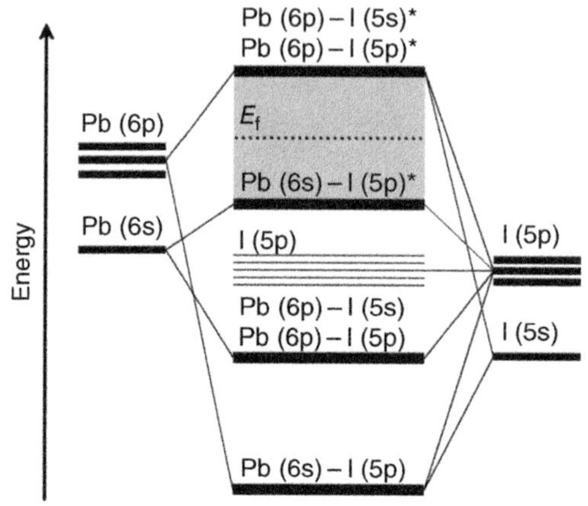

Figure 5.12 Bonding diagram of an isolated $[PbI_6]^{4-}$ cluster. The interaction of the Pb (6s and 6p) and I (5s and 5p) orbitals creates the bands indicated at the center of the figure. The gray area shows the band gap of the assumed $[PbI_6]^{4-}$ cluster and "*" denotes antibonding states. *Source:* Umebayashi et al. [57].

the A-site cation (Section 5.5.2) and the X-site anion (Section 5.5.3) based on DFT calculations.

5.5.1 What Determines Band Gap?

To understand the E_g formation in semiconductors, electronic-state calculations of isolated clusters are important. Figure 5.12 shows the bonding diagram for the isolated $[PbI_6]^{4-}$ cluster [57]. The bars indicated in the left and right sides of the figure represent the orbitals of Pb (6s and 6p) and I (5s and 5p), while the bars at the center indicate the bands formed by the interaction of Pb and I orbitals. The gray area shows the band gap of the assumed $[PbI_6]^{4-}$ cluster and "*" denotes that the specific band is formed by antibonding interaction. It can be seen that the VBM consists of the antibonding state generated by Pb(6s) and I(5p) interactions, whereas the CBM is also the antibonding state induced by Pb(6p)–I(5p) and Pb(6p)–I(5s) interactions. The antibonding nature of the VBM can be confirmed directly from the charge density distribution in Figure 5.6, which shows the localization of the Pb(6s) and I(5p) orbitals with negligible charge overlap between the Pb and I atoms. In the CBM, the contribution of Pb(6p) is dominant (see also Figure 5.11a), but the conduction bands are still made by the antibonding coupling with the I orbital.

One crucial effect that varies the E_g of $APbI_3$ perovskites is Pb–I–Pb bond bending [11, 35, 36]. Figure 5.13 shows the electronic state of the VBM in $MAPbI_3$. When the Pb–I–Pb bond angle is 180°, the repulsive interaction generated by σ-antibonding between the Pb(6s) and I(5p) is maximized. When the bond angle is not collinear, however, the stabilization of the antibonding state occurs, shifting the VBM toward the deeper position in the energy scale, which in turn widens E_g [35]. By bond bending, the CBM state is also lifted due to the antibonding character of the CBM [11, 36]. As a result, the Pb–I–Pb bond angle exerts a significant influence on E_g. When the orthorhombic-tetragonal ($\gamma \rightarrow \beta$) phase transition occurs in $MAPbI_3$, the E_g decreases significantly by 0.1 eV (see Figure 4.19), which can be interpreted by the change in the Pb–I–Pb bond angle.

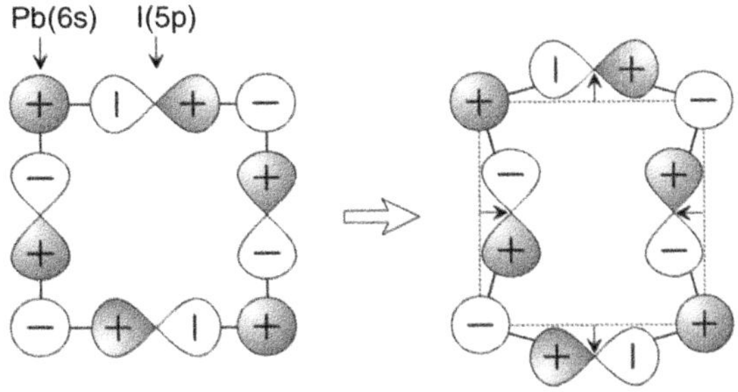

Figure 5.13 Wave functions of Pb(6s) and I(5p) at the VBM in $MAPbI_3$, illustrated based on the result of Ref. [35]. The Pb and I wave functions for the Pb–I–Pb bond angles of 180° and <180° are shown. The Pb(6s)–I(5p) antibonding interaction is stabilized by the Pb–I–Pb bond bending.

5.5.2 Effect of Center Cation

The E_g of $APbI_3$ perovskites changes with the size of the A-site cation (see Figure 4.9), which can be understood as the effect of the Pb–I–Pb angle variation. Figure 5.14 shows the LDA calculation results for $APbI_3$ perovskites with different A-site cations [36]. In Figure 5.14a, the Pb–I–Pb bond angle is shown as a function of the steric size of the A-site cation. As we have seen in Figure 5.4, the bond angle becomes closer to 180° with increasing A-site cation size. In other words, the PbI_6 octahedral tilting decreases as the cation size becomes larger. In Figure 5.14b, E_g of $APbI_3$ is shown for the maximum Pb–I–Pb bond angle. Since larger bond bending leads to an increase in E_g, the perovskite E_g is determined essentially by the maximum bond angle in the unit cell. The $APbI_3$ shows a large reduction of E_g with increasing bond angle, although the data scatter slightly due to the variation in the Pb–I bond length in the DFT-optimized structures [36].

Figure 5.14c summarizes the calculated E_g values of various $APbI_3$ perovskites. In the DFT results, the E_g difference between $FAPbI_3$ and $CsPbI_3$ is 0.16 eV, which reproduces well the experimental value of 0.21 eV (see Table 4.3). Note, however, that a stable perovskite crystal can be formed only when the size of the A-site cation falls in a narrow range (Figure 3.11) and, experimentally, there is a limitation for the choice of the A-site cation. In the case of $ASnI_3$, the Sn–I–Sn bond angle is already 180° due to the small size of Sn atoms [58] and the contribution of Sn–I lengths is more important to understand the E_g variation in $ASnI_3$ [59].

5.5.3 Effect of Halide Anion

The E_g of $APbX_3$ depends strongly on the X-site anion (Figure 4.9b and Table 4.3). Figure 5.15 shows the band alignment of $MAPbX_3$ ($X = I^-$, Br^-, and Cl^-) calculated by a PBE-based method [37]. This band lineup was deduced by using the Pb 1s core level energy as a reference. $MAPbX_3$ shows type-I alignments, in which both conduction and valence bands shift so that E_g simply increases. For the band offsets, slightly different values have also been reported [38]. An increasing E_g in the order of I → Br → Cl is consistent with the general trend that E_g increases for lighter constituent atoms; in particular, the valence- and conduction-band halide orbitals change from 5p (I) to 4p (Br) to 3p (Cl). The type-I band alignment in Figure 5.15 has been confirmed experimentally (see Figure 4.10d).

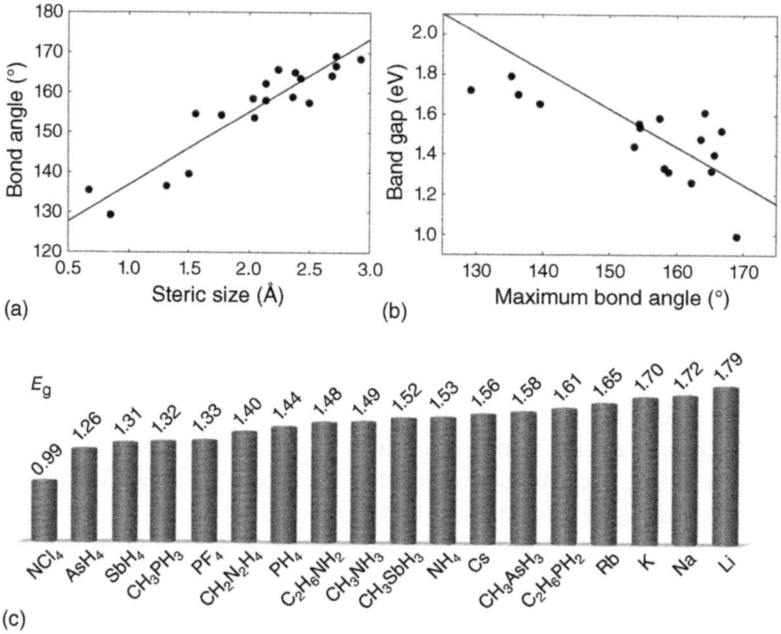

Figure 5.14 LDA calculation results for $APbI_3$ perovskites with different A-site cations: (a) Pb—I—Pb bond angle as a function of the steric size of the A-site cation, (b) E_g variation as a function of maximum Pb—I—Pb bond angle in the perovskite, and (c) E_g calculated for various $APbI_3$ perovskites. The steric size of the A-site cation shown in (a) does not refer to the ionic radius shown in Table 3.3 but is defined as the radius of the sphere that contains 95% of the DFT electron density. *Source*: Filip et al. [36].

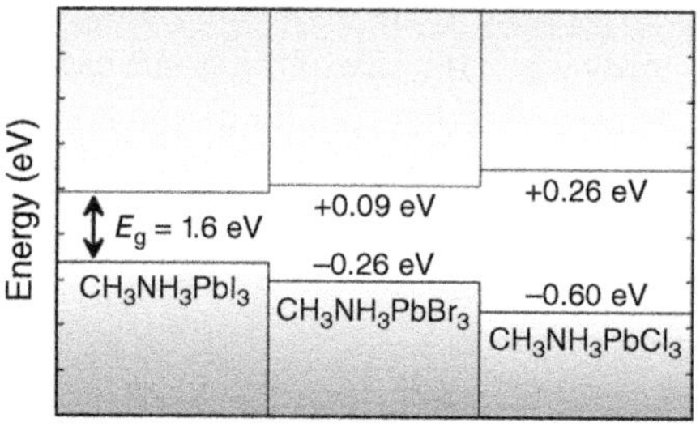

Figure 5.15 Band alignment of $MAPbX_3$ ($X = I^-$, Br^-, and Cl^-) calculated by a PBE-based method. This band lineup has been obtained by using the Pb 1s core level energy as a reference. The $MAPbX_3$ shows the type-I band alignments. *Source*: Butler et al. [37]. Licensed under CC BY 3.0.

5.6 Defect Physics

One of the remarkable features of hybrid perovskites is a quite low density of deep carrier traps in the crystals [60–62]. The very long carrier diffusion length observed in $MAPbI_3$ (Section 6.4) can be well understood by the absence of deep-level defects in the bulk component. Moreover, for hybrid perovskites, the carrier concentration in the layers can be varied by controlling defects that act as shallow donors and acceptors [63]. This section describes the unique defect physics of $MAPbI_3$ revealed by GGA (PBE) calculations [15, 19, 24, 28, 29].

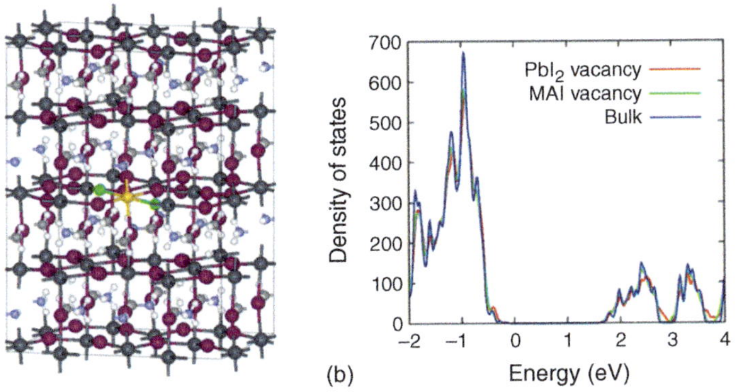

Figure 5.16 Results of PBE calculation for vacancy-type defects in $MAPbI_3$: (a) a tetragonal $MAPbI_3$ structure with a PbI_2-vacancy defect introduced at center position of a supercell and (b) variation in the $MAPbI_3$ DOS due to formation of PbI_2 and MAI vacancy defects. In the DFT calculations, a supercell created by a 2 × 2 × 2 cell of an optimized tetragonal $MAPbI_3$ is relaxed after removing a Pb atom (yellow) and two I atoms (green) indicated in (a). *Source*: Kim et al. [19].

Figure 5.16 shows the DFT calculation results for vacancy-type defects in $MAPbI_3$: (a) a tetragonal $MAPbI_3$ structure with a PbI_2-vacancy (V_{PbI_2}) defect introduced at the center position of the supercell and (b) variation of the $MAPbI_3$ DOS by the formation of V_{PbI_2} and MAI vacancy (V_{MAI}) defects [19]. For the DFT calculation of defects, a supercell approach, in which a defect is introduced into a repetitive unit-cell structure, is generally employed. The supercell structure of Figure 5.16a was created as a 2×2×2 cell of an optimized tetragonal $MAPbI_3$ (a total of 384 atoms), which was relaxed after removing the Pb atom and two I atoms indicated by yellow and green in the figure. Figure 5.16b compares the DOS distribution of the ideal $MAPbI_3$ with those of V_{PbI_2} and V_{MAI}. The V_{MAI} is created by simply removing MA^+ and a neighboring I atom. Quite surprisingly, the formation of V_{PbI_2} and V_{MAI} has little effect on the DOS, revealing the fact that the removal of the PbI_2 and MAI does not create deep-level defects at all. The formation energy of V_{PbI_2} is also quite low (27 meV), suggesting that the vacancy-type defects are easily formed, but the quite good semiconducting properties are maintained due to the absence of the gap state formation.

The formation of elemental defects has also been investigated in detail [15, 24, 28, 29]. Figure 5.17 summarizes the transition energy levels of various defects, including MA, Pb, and I vacancies (i.e. V_{MA}, V_{Pb}, and V_I), MA, Pb, and I interstitial defects (MA_i, Pb_i, I_i), defects created by cation substitutions (MA_{Pb}, Pb_{MA}), and cation–anion substitutions (MA_I, Pb_I, I_{MA}, and I_{Pb}) [28]. Here, MA_{Pb} indicates that MA is introduced into a Pb site, for example. The transition energy levels, indicated by the bars in Figure 5.17, represent the Fermi level at which an acceptor (donor) defect receives (releases) the electron. Specifically, the two numbers denoted at each energy level indicate the defect charge states; the minus (plus) sign shows the acceptor (donor) level. The transition energy level indicated by 1−/2− for V_{Pb}, for example, represents the acceptor level that can accept one electron, which in

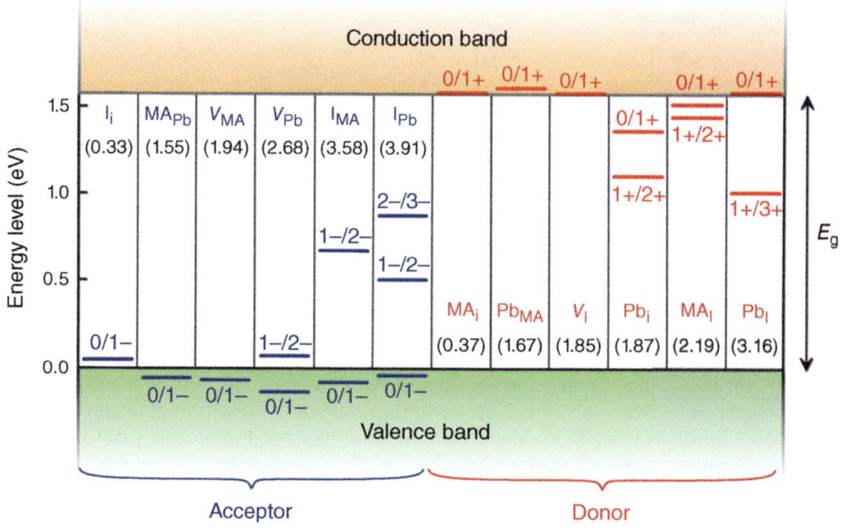

Figure 5.17 Results of PBE calculation for the transition energy levels of various point defects, including MA, Pb, and I vacancies (i.e. V_{MA}, V_{Pb}, and V_I), MA, Pb, and I interstitial defects (MA_i, Pb_i, I_i), defects created by cation substitutions (MA_{Pb}, Pb_{MA}), and cation–anion substitutions (MA_I, Pb_I, I_{MA}, and I_{Pb}). For the calculation, a 3 × 3 × 3 supercell of the cubic perovskite structure with MA^+ lying along the [111] direction is assumed. The transition energy levels indicated by the bars for each defect represent the Fermi level at which an acceptor (donor) defect receives (releases) the electron, and the corresponding charge states of each defect are also indicated. The formation energies of the neutral defects are given in parentheses. *Source*: Yin et al. [28].

turn changes the single-negative charge state (1−) to the double negative charge state (2−). Furthermore, the formation energies of the neutral defects are shown in parentheses.

The DFT calculations of Figure 5.17 show important results: (i) most of the point defects form the shallow acceptor and donor states and (ii) all defects that create deep levels (i.e. I_{MA}, I_{Pb}, and Pb_i) have high formation energies. In $MAPbI_3$, therefore, the generation of deep carrier traps is energetically unfavorable. This, in turn, reduces the non-radiative recombination effectively, leading to the improvement of the carrier diffusion length and the resulting solar cell performance (Section 1.2.3).

The formation of the shallow acceptors and donors by the introduction of the point defects further explains the self-doping effect in hybrid perovskites. The carrier concentration or Fermi level in $MAPbI_3$ can be controlled by changing the PbI_2/MAI precursor ratio [63, 64] and the MAI-rich condition leads to the p-type absorber formation, whereas the PbI_2-rich condition enhances the n-type conductivity (Figure 6.2). Based on the DFT result of Figure 5.17, Wang et al. concluded that the enhanced p-type character observed under MAI-rich growth conditions is due to the formation of V_{Pb} acceptors, whereas the V_I formation realizes the n-type doping in PbI_2-rich films [63]. Although a precise relation between carrier concentration and defect formation has not been established, it is now widely accepted that point defects are the source of electron–hole carriers generated in $MAPbI_3$.

The unique defect properties confirmed for $MAPbI_3$ have been attributed to the strong antibonding character discussed in Figure 5.12 and high ionicity of the perovskite [15, 19]. In particular, the low formation energy of V_{Pb} originates from the Pb(6s)–I(5p) antibonding coupling [15]; since there is no effective overlap of the Pb and I orbitals near the VBM (see Figure 5.6), the energy cost for the removal of the Pb atom is relatively low.

For the absence of deep gap states, the ionic bonding of $MAPbI_3$ is considered to play a significant role [15, 19]. In a covalent Si semiconductor, for example, dangling bond formation generates deep mid-gap states [65], as the isolated Si dangling bond forms a local charge state, which is quite different from the Si bulk state. In the ionic $MAPbI_3$, the Pb and I orbitals form the local charge, partly due to the strong antibonding character and thus the energy state caused by the removal or introduction of a point defect does not differ significantly from that of the $MAPbI_3$ bulk state. The above DFT results show that abundant defects exist in $MAPbI_3$. However, such defects do not act as carrier traps and even allow the control of carrier properties by adjusting the ratio of source materials. Accordingly, $MAPbI_3$ is a representative material that has a so-called "defect tolerance."

Acknowledgment

The authors would like to thank Kazuya Shibata (Figures 5.10, 5.12, and 5.17) and Yuuki Nagasaki (Figures 5.14 and 5.15) of Gifu University for helping the preparation of figures that appear in this chapter.

References

1 Martin, R.M. (2004). *Electronic Structure: Basic Theory and Practical Methods*. Cambridge: Cambridge University Press.
2 Lee, J.G. (2012). *Computational Materials Science: An Introduction*. New York: CRC Press.
3 Bechstedt, F. (2015). *Many-Boday Approach to Electronic Excitations: Concepts and Applications*. Heidelberg: Springer-Verlag.
4 Brivio, F., Butler, K.T., Walsh, A., and van Schilfgaarde, M. (2014). *Phys. Rev. B* 89: 155204.
5 Filippetti, A. and Mattoni, A. (2014). *Phys. Rev. B* 89: 125203.
6 Filippetti, A., Delugas, P., and Mattoni, A. (2014). *J. Phys. Chem. C* 118: 24843.
7 Amat, A., Mosconi, E., Ronca, E. et al. (2014). *Nano Lett.* 14: 3608.
8 Mosconi, E., Umari, P., and De Angelis, F. (2016). *Phys. Chem. Chem. Phys.* 18: 27158.
9 Baikie, T., Fang, Y., Kadro, J.M. et al. (2013). *J. Mater. Chem. A* 1: 5628.

10 Mosconi, E., Amat, A., Nazeeruddin, M.K. et al. (2013). *J. Phys. Chem. C* 117: 13902.
11 Kim, J., Lee, S.-C., Lee, S.-H., and Hong, K.-H. (2015). *J. Phys. Chem. C* 119: 4627.
12 Yin, W.-J., Yang, J.-H., Kang, J. et al. (2015). *J. Mater. Chem. A* 3: 8926.
13 Umari, P., Mosconi, E., and De Angelis, F. (2015). *Sci. Rep.* 4: 4467.
14 Brivio, F., Walker, A.B., and Walsh, A. (2013). *APL Mater.* 1: 042111.
15 Yin, W.-J., Shi, T., and Yan, Y. (2014). *Appl. Phys. Lett.* 104: 063903.
16 Shirayama, M., Kadowaki, H., Miyadera, T. et al. (2016). *Phys. Rev. Appl.* 5: 014012.
17 Motta, C., El-Mellouhi, F., Kais, S. et al. (2015). *Nat. Commun.* 6: 7026.
18 Giorgi, G., Fujisawa, J.-I., Segawa, H., and Yamashita, K. (2013). *J. Phys. Chem. Lett.* 4: 4213.
19 Kim, J., Lee, S.-H., Lee, J.H., and Hong, K.-H. (2014). *J. Phys. Chem. Lett.* 5: 1312.
20 Even, J., Pedesseau, L., Jancu, J.-M., and Katan, C. (2013). *J. Phys. Chem. Lett.* 4: 2999.
21 Menéndez-Proupin, E., Palacios, P., Wahnón, P., and Conesa, J.C. (2014). *Phys. Rev. B* 90: 045207.
22 Leguy, A.M.A., Azarhoosh, P., Alonso, M.I. et al. (2016). *Nanoscale* 8: 6317.
23 Geng, W., Zhang, L., Zhang, Y.-N. et al. (2014). *J. Phys. Chem. C* 118: 19565.
24 Xiao, Z. and Yan, Y. (2017). *Adv. Energy Mater.* 7: 1701136.
25 Fujiwara, H., Kato, M., Tamakoshi, M. et al. (2018). *Phys. Status Solidi A* 215: 1700730.
26 Even, J., Pedesseau, L., and Katan, C. (2014). *J. Phys. Chem. C* 118: 11566.
27 Azarhoosh, P., McKechnie, S., Frost, J.M. et al. (2016). *APL Mater.* 4: 091501.
28 Yin, W.-J., Shi, T., and Yan, Y. (2014). *Adv. Mater.* 26: 4653.
29 Du, M.H. (2014). *J. Mater. Chem. A* 2: 9091.
30 Kato, M., Fujiseki, T., Miyadera, T. et al. (2017). *J. Appl. Phys.* 121: 115501.
31 Shirayama, M., Kato, M., Miyadera, T. et al. (2016). *J. Appl. Phys.* 119: 115501.
32 Lee, J.-H., Bristowe, N.C., Bristowe, P.D., and Cheetham, A.K. (2015). *Chem. Commun.* 51: 6434.
33 Lee, J.H., Lee, J.-H., Kong, E.-H., and Jang, H.M. (2016). *Sci. Rep.* 6: 21687.
34 Bechtel, J.S., Seshadri, R., and Van der Ven, A. (2016). *J. Phys. Chem. C* 120: 12403.
35 Lee, J.-H., Bristowe, N.C., Lee, J.H. et al. (2016). *Chem. Mater.* 28: 4259.
36 Filip, M.R., Eperon, G.E., Snaith, H.J., and Giustino, F. (2014). *Nat. Commun.* 5: 5757.
37 Butler, K.T., Frost, J.M., and Walsh, A. (2015). *Mater. Horiz.* 2: 228.
38 Buin, A., Comin, R., Xu, J. et al. (2015). *Chem. Mater.* 27: 4405.
39 Kohn, W. and Sham, L.J. (1965). *Phys. Rev.* 140: A1133.
40 Marsman, M., Paier, J., Stroppa, A., and Kresse, G. (2008). *J. Phys. Condens. Matter* 20: 064201.
41 Perdew, J.P., Burke, K., and Ernzerhof, M. (1996). *Phys. Rev. Lett.* 77: 3865.
42 Shishkin, M., Marsman, M., and Kresse, G. (2007). *Phys. Rev. Lett.* 99: 246403.
43 Heyd, J. and Scuseria, G.E. (2003). *J. Chem. Phys.* 118: 8207.

44 Heyd, J. and Scuseria, G.E. (2006). *J. Chem. Phys.* 124: 219906.
45 Chapman, M. and Sá de Melo, C. (2011). *Nature* 471: 41.
46 Goyal, A., McKechnie, S., Pashov, D. et al. (2018). *Chem. Mater.* 30: 3920.
47 Fujiwara, H. and Collins, R.W. (2018). *Spectroscopic Ellipsometry for Photovoltaics: Fundamental Principles and Solar Cell Characterization*, vol. 1. Cham: Springer Nature.
48 Fujiwara, H. and Collins, R.W. (2018). *Spectroscopic Ellipsometry for Photovoltaics: Applications and Optical Data of Solar Cell Materials*, vol. 2. Cham: Springer Nature.
49 Yu, P.Y. and Cardona, M. (2010). *Fundamentals of Semiconductors: Physics and Materials Properties*. Heidelberg: Springer-Verlag.
50 Nishiwaki, M., Nagaya, K., Kato, M. et al. (2018). *Phys. Rev. Mater.* 2: 085404.
51 Kato, M., Nishiwaki, M., and Fujiwara, H. (2020). *Phys. Rev. Mater.* 4: 035402.
52 Zheng, F., Tan, L.Z., Liu, S., and Rappe, A.M. (2015). *Nano Lett.* 15: 7794.
53 Hutter, E.M., Gélvez-Rueda, M.C., Osherov, A. et al. (2017). *Nat. Mater.* 16: 115.
54 Tao, S., Schmidt, I., Brocks, G. et al. (2019). *Nat. Commun.* 10: 2560.
55 Tanaka, K., Takahashi, T., Ban, T. et al. (2003). *Solid State Commun.* 127: 619.
56 Galkowski, K., Mitioglu, A., Miyata, A. et al. (2016). *Energy Environ. Sci.* 9: 962.
57 Umebayashi, T., Asai, K., Kondo, T., and Nakao, A. (2003). *Phys. Rev. B* 67: 155405.
58 Rajagopal, A., Stoddard, R.J., Hillhouse, H.W., and Jen, A.K.-Y. (2019). *J. Mater. Chem. A* 7: 16285.
59 Prasanna, R., Gold-Parker, A., Leijtens, T. et al. (2017). *J. Am. Chem. Soc.* 139: 11117.
60 Shi, D., Adinolfi, V., Comin, R. et al. (2016). *Science* 347: 519.
61 Saidaminov, M.I., Abdelhady, A.L., Murali, B. et al. (2015). *Nat. Commun.* 6: 7586.
62 Dong, Q., Fang, Y., Shao, Y. et al. (2015). *Science* 347: 967.
63 Wang, Q., Shao, Y., Xie, H. et al. (2014). *Appl. Phys. Lett.* 105: 163508.
64 Emara, J., Schnier, T., Pourdavoud, N. et al. (2016). *Adv. Mater.* 28: 553.
65 Park, J.S., Kim, S., Xie, Z., and Walsh, A. (2018). *Nat. Rev. Mater.* 3: 194.

6

Carrier Transport Properties

Hiroyuki Fujiwara and Yoshitsune Kato

Gifu University, Department of Electrical, Electronic and Computer Engineering, 1-1 Yanagido, Gifu 501-1193, Japan

6.1 Introduction

Transport of photo-generated carriers is a critical factor in the operation of photovoltaic devices [1–3]. For carrier transport in solar cells, the most important physical parameter is carrier diffusion length defined by

$$L_D = \sqrt{D\tau} \tag{6.1}$$

where D and τ are the diffusion coefficient and carrier lifetime, respectively [1, 2]. The D is a simple linear function of carrier mobility μ, which is expressed by the Einstein relation:

$$D = \left(\frac{k_B T}{q}\right)\mu \tag{6.2}$$

where k_B, T, and q are the Boltzmann constant, temperature, and electron charge, respectively [1, 2]. Note that $k_B T/q = 25.9\,\text{mV}$ at 300 K. Consequently, L_D is determined by the pair (μ, τ), and numerous efforts have been made to determine these characteristic values for various hybrid perovskites, including methylammonium lead iodide ($CH_3NH_3PbI_3$; $MAPbI_3$) [4–35] and formamidinium lead iodide ($HC(NH_2)_2PbI_3$; $FAPbI_3$) [36–39].

Quite fortunately, hybrid perovskites show favorable high (μ, τ) that lead to large L_D (Table 1.1). The L_D of the perovskites is sufficiently longer than an absorber layer thickness of ~500 nm, which is the typical value used for spin-coated hybrid perovskites. One of the remarkable features of hybrid perovskites is their quite high τ that can be confirmed easily based on photoluminescence (PL) characterization (Figure 1.6 and Chapter 8) [40]. It is now accepted that the long τ observed in hybrid perovskites originates from the absence of deep defects [41, 42]. Specifically, density functional theory (DFT) calculations indicate that the generation of deep traps is energetically unfavorable in $MAPbI_3$ (see Figure 5.17). In fact, trap densities reported for $MAPbI_3$ single crystals are surprisingly low at $\sim 10^{10}\,\text{cm}^{-3}$ [10, 23, 24]. Some authors argue that the high τ is caused by the indirect-gap formation (Rashba

Hybrid Perovskite Solar Cells: Characteristics and Operation,
First Edition. Edited by Hiroyuki Fujiwara.

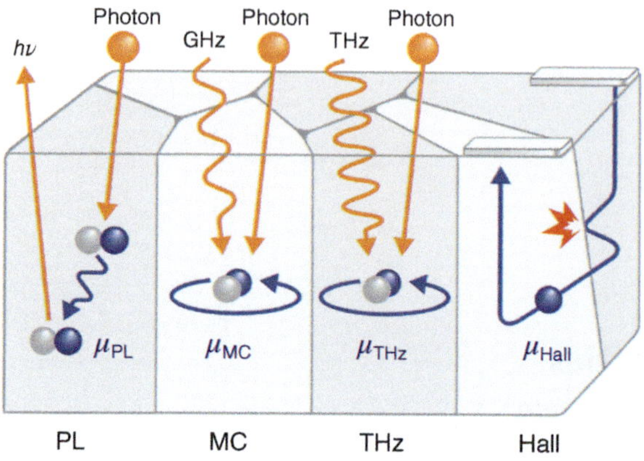

Figure 6.1 Principles of carrier transport characterization by photoluminescence (PL), microwave conductivity (MC), terahertz (THz) spectroscopy, and Hall measurements. The carrier mobility (μ) determined by each method is denoted as μ_{PL}, μ_{MC}, μ_{THz}, and μ_{Hall}, respectively. All the noncontact techniques (PL, MC, and THz) characterize carriers excited by short laser pulses. In actual Hall measurements, a total of four or six electrodes are used, although only two electrodes are shown in the figure for simplicity. Note that, for polycrystalline thin films, μ_{MC} and μ_{THz} characterize the intragrain mobility without the significant influence of grain boundaries, whereas μ_{Hall} includes carrier scattering at the grain boundaries.

splitting in the conduction band) [43–47]; however, this is highly unlikely because the direct-gap formation in the perovskites is evident experimentally (Section 5.4.2).

For the absolute values of (L_D, μ, τ) for hybrid perovskites [4–39], however, there is widespread confusion. The intense controversy over hybrid-perovskite transport properties arises essentially from differences in characterization techniques; namely (μ, τ) depend strongly on how these quantities are measured. Figure 6.1 illustrates the principles of carrier transport characterization by PL, microwave conductivity (MC), terahertz (THz) spectroscopy, and Hall measurements. In the PL, MC, and THz techniques, electron and hole carriers are generated first by short-wavelength laser pulses, following which the time decay of the photoemission (PL) or the photoconductivity (MC and THz) is characterized (i.e. the optical pump-probe method). The PL mobility (μ_{PL}) is determined analytically from the time decay of the PL intensity [5–7]. The origin of the MC mobility (μ_{MC}) and THz mobility (μ_{THz}) is essentially similar, and these are evaluated by using free-carrier absorption phenomena in the GHz and THz regions. Specifically, free carriers (i.e. free electrons in metals, for example) absorb light particularly in the THz and GHz regions as expressed by the Drude model [48–50]. The variation of the electromagnetic-wave transmission (or reflection) in the GHz and THz ranges can be related directly to the conductivity [22, 51, 52], from which μ is extracted with a known carrier concentration estimated from the optical-pump laser pulse [13, 15, 19].

The PL, MC, and THz methods are noncontact methods and μ can be determined relatively easily. However, μ deduced from these methods includes contributions from both electrons and holes; thus, data interpretation is slightly difficult. In contrast, only one carrier type (i.e. electrons or holes) is characterized in the

Hall measurements. Among all the techniques shown in Figure 6.1, the Hall measurement is the most reliable method because the Hall mobility (μ_{Hall}) can be determined accurately based on the Hall effect [53]. This is the exceptional technique that further allows the simultaneous determination of carrier type (i.e. n- or p-type) and carrier concentration.

Even though the Hall measurement is the favored technique, μ_{Hall} is affected severely by grain boundaries [49, 54, 55] because the conductivity is measured in the direction perpendicular to the grain boundaries (see Figure 6.1). In contrast, μ_{MC} and μ_{THz} are negligibly influenced by grain boundaries, as free carriers do not exist in the region of grain boundaries (see Figure 9.1). Accordingly, when grain boundaries make a significant contribution, we find $\mu_{THz} > \mu_{Hall}$; therefore, comparing μ_{THz} (or μ_{MC}) with μ_{Hall} elucidates the effect of grain boundaries on carrier transport.

In this chapter, we review the carrier transport properties of hybrid perovskites, in an effort to gain consensus on the absolute values of (L_D, μ, τ). Section 6.2 explains the carrier properties of hybrid perovskites, using $MAPbI_3$ as an example. We focus on the carrier mobility of $MAPbI_3$ in Section 6.3, which presents controversial physical models to explain what limits the maximum mobility in $MAPbI_3$. Based on the understanding of measurement techniques, we further discuss the appropriate L_D of hybrid perovskites (Section 6.4). Section 6.5 explains unique transport properties of various hybrid perovskites.

6.2 Carrier Properties of Hybrid Perovskites

The semiconducting properties of hybrid perovskites arise essentially from the inorganic scaffolds (i.e. PbI_3^- in $APbI_3$ perovskites), and the unique defect characteristics of hybrid perovskites allow the control of carrier conductivity by the selection of process conditions. This section treats self-doping in hybrid perovskites by the formation of shallow donor and accepter defects (Section 6.2.1) and discusses its effect on carrier mobility (Section 6.2.2).

6.2.1 Self-Doping in Hybrid Perovskites

Hybrid perovskites are rarely doped intentionally because the carrier concentration can be varied widely from p- to n-type by so-called "defect engineering," in which the carrier properties are controlled by introducing proper defects [9, 56, 57]. Figure 6.2 shows (a) the carrier concentration and the corresponding μ_{Hall} and (b) the energy separation between the Fermi level (E_F) and valence-band maximum (i.e. $\Delta E = E_F - E_{VBM}$) in $MAPbI_3$ thin films as a function of the precursor ratio (PbI_2/MAI) [9]. Here, the precursor ratio represents the molar ratio of PbI_2/MAI in the precursor mixture solution, adopted for a one-step spin-coating process. The results shown in Figure 6.2a,b were obtained from six-contact Hall measurements with an electrode spacing of $\lesssim$1 mm and ultraviolet photoelectron spectroscopy (UPS), respectively. In the figure, the carrier type switches from p to n at the

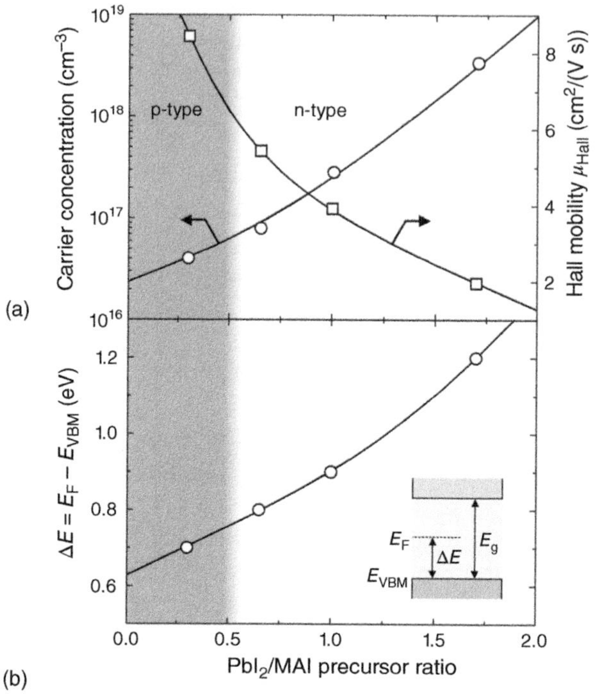

Figure 6.2 (a) Carrier concentration and Hall mobility (μ_{Hall}) and (b) energy separation between the Fermi level (E_F) and valence-band maximum (i.e. $\Delta E = E_F - E_{VBM}$) in $MAPbI_3$ thin films as a function of the precursor ratio (PbI_2/MAI) adopted in the thin film formation. The carrier type changes from p-type to n-type at PbI_2/MAI ~ 0.5. The lines are to guide the eyes. *Source*: Modified from Wang et al. [9].

precursor ratio of ~0.5 and the electron carrier density increases in the PbI_2-rich film. Upon increasing the precursor ratio, the E_F position systematically shifts upward due to the enhanced n-type character [9, 57].

Importantly, p-type $MAPbI_3$ films can be converted to n-type films by thermal annealing at temperatures $T \leq 150\,°C$ [9, 56, 58]. This carrier-type conversion by annealing originates from MAI desorption from the $MAPbI_3$ surface. Accordingly, it is now widely accepted that $MAPbI_3$ has n-type conductivity in MAI-poor (PbI_2-rich) films, whereas p-type conductivity occurs in PbI_2-poor (MAI-rich) layers [9, 56–59]. Based on DFT results obtained for the point-defect formation in $MAPbI_3$ (see Figure 5.17), Wang et al. concluded that the n-type character of MAI-deficient layers originates from shallow donor formation by I vacancies (V_I), whereas p-type hole generation in the PbI_2-deficient layers occurs because of shallow accepter formation due to Pb vacancies (V_{Pb}) [9]. At the present stage, no direct quantitative correlations between defect and carrier densities have been established. Nevertheless, the systematic variation of carrier concentration with the processing condition strongly supports the generation of shallow donor and accepter levels by simple point (or elementary) defects.

The perovskite defects mentioned earlier (V_I and V_{Pb}) are charged defects (see Figure 5.17) and thus free electrons and holes in the perovskites can be scattered by these defects through the Coulomb interaction (i.e. ionized-defect scattering). The strong μ_{Hall} reduction with carrier concentration, shown in Figure 6.2a, can therefore be understood in terms of enhanced carrier scattering by ionized defects.

In Figure 6.2, the carrier properties are shown based on the PbI_2/MAI precursor ratio. However, this precursor ratio is only a relative value and the actual Pb/MA ratio in the perovskite layers varies according to (i) growth conditions (deposition methods, solvent) [9, 57], (ii) thermal annealing process [9, 56, 58], and (iii) substrate (or underlying layers) [57]. Thus, the precursor ratio must be considered as a process parameter, rather than as a physical parameter.

For Pb-based hybrid perovskites, thin layers with low carrier concentrations can be fabricated. Nevertheless, Sn-based hybrid perovskites often exhibit quite high p-type carrier densities in the range 10^{18}–10^{19} cm^{-3} [60–62]. This significant self-doping effect has been attributed to the oxidation of the Sn^{2+} B-site cation to Sn^{4+} [63, 64]. In fact, it has been demonstrated that the incorporation of Sn^{4+} in the form of SnI_4 notably enhances p-type conductivity [65]. Importantly, the stability of the 2+ oxidation state deteriorates as the group-IV B-site cation becomes lighter [63]. Thus, Sn^{2+} in $ASnI_3$ is unstable and is easily oxidized into the more stable Sn^{4+} state. In photovoltaic devices, high carrier concentrations are quite detrimental and, to reduce the p-type carrier density in $ASnI_3$ compounds, the additives of SnF_2 [62] and Sn metal powders [66] were used.

6.2.2 Effect of Carrier Concentration on Mobility

Figure 6.3 shows μ_{Hall} as a function of carrier concentration in $MAPbI_3$ [8–10] and $MASnI_3$ [8, 60, 61]. In the figure, the results obtained from bulk crystals (single crystals and polycrystals) and thin films are shown separately. The $MAPbI_3$ bulk crystals show a very low carrier concentration of $\sim 10^{10}$ cm^{-3} [8, 10], which is significantly less than that in the thin films [9], most likely due to the suppression of the shallow-defect formation at equilibrium conditions. It can be seen that μ_{Hall} of $MAPbI_3$ bulk crystals is $\sim$100 $cm^2/(V\,s)$ and decreases significantly with increasing carrier concentration, independent of carrier type. As discussed in Section 6.3.1, μ_{Hall} of $MAPbI_3$ thin films decreases in the presence of grain boundaries. However, the result of Figure 6.2 shows a clear trend that μ_{Hall} of $MAPbI_3$ is also limited by strong carrier scattering by ionized defects, particularly at high carrier concentrations. Thus, L_D of $MAPbI_3$ decreases notably at high carrier densities, resulting in reduced solar cell performance. Consequently, the carrier concentration in the perovskite device needs to be controlled properly (see also Section 11.2).

As mentioned earlier, $MASnI_3$ shows strong p-type character. Rather surprisingly, μ_{Hall} of $MASnI_3$ bulk crystals is substantially larger than that of $MAPbI_3$ bulk crystals and exhibits >300 $cm^2/(V\,s)$ at the carrier concentration of 10^{15} cm^{-3} [8]. Accordingly, μ increases notably when Pb is replaced with Sn. We attribute the higher μ_{Hall} in $MASnI_3$ to a smaller effective mass in $MASnI_3$ than in $MAPbI_3$ (see Section 6.3.3).

6.3 Carrier Mobility of $MAPbI_3$

The mobility μ is a universal physical parameter that allows for the direct evaluation of carrier transport properties. Nevertheless, the μ reported for $MAPbI_3$ varies in a

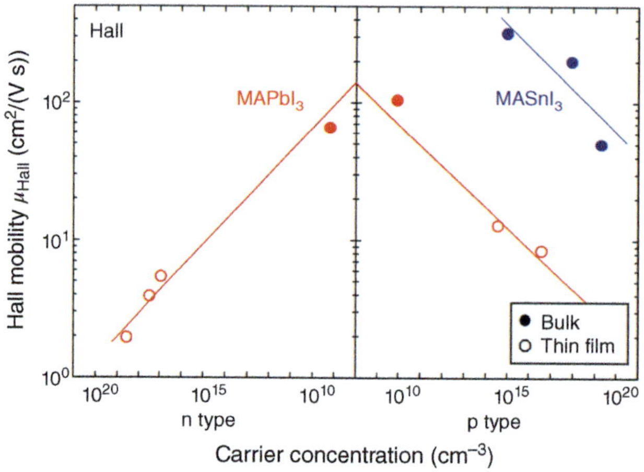

Figure 6.3 Variation of Hall mobility (μ_{Hall}) with carrier concentration in $MAPbI_3$ [8–10] and $MASnI_3$ [8, 60, 61]. The solid symbols show the results for bulk samples (single crystals and polycrystals), whereas the open circles indicate the results of thin-film samples. The lines are to guide the eyes. The μ_{Hall} of the thin films is strongly affected by ionized-defect scattering and grain-boundary scattering.

quite wide range of 0.5–100 $cm^2/(V\,s)$, depending on the sample and characterization method [4–21]. Section 6.3.1 discusses the origin of variation in μ and what is the appropriate μ for $MAPbI_3$. In Section 6.3.2, we see the relationship between the effective mass and μ by making further comparisons with other solar cell materials. The temperature dependence of μ (Section 6.3.3) provides important information on the carrier scattering mechanism in $MAPbI_3$. Based on various results for μ, we discuss what determines the maximum μ of $MAPbI_3$ (Section 6.3.4).

6.3.1 Variation of Mobility with Characterization Method

Figure 6.4 summarizes μ of $MAPbI_3$, as determined by various characterization methods, which are categorized into three groups: (i) PL [5–7], (ii) Hall [8–10], and (iii) MC [11–16] and THz [16–20] measurements. Here, μ_{THz} and μ_{MC} are treated as the same group because these methods characterize μ based on the free carrier absorption that occurs primarily within the grains. In other words, μ_{THz} and μ_{MC} represent the intragrain mobility without the significant influence of grain boundaries. In addition to the aforementioned methods, μ obtained from a time-of-flight (TOF) measurement [21] is also shown in Figure 6.4.

For $MAPbI_3$ bulk crystals (single crystals and polycrystals), consistent μ values of ~100 $cm^2/(V\,s)$ (66–135 $cm^2/(V\,s)$) have been confirmed by Hall [8, 10], MC [15], and TOF [21] measurements, as indicated by the red circles in Figure 6.4. The result of μ_{Hall} in the figure was evaluated from samples with low carrier concentrations of 10^{10} cm^{-3} (see Figure 6.3) and the influence of charged-defect scattering is minimized. Thus, $\mu \sim 100$ $cm^2/(V\,s)$ in Figure 6.4 represents the maximum μ of $MAPbI_3$ at room temperature.

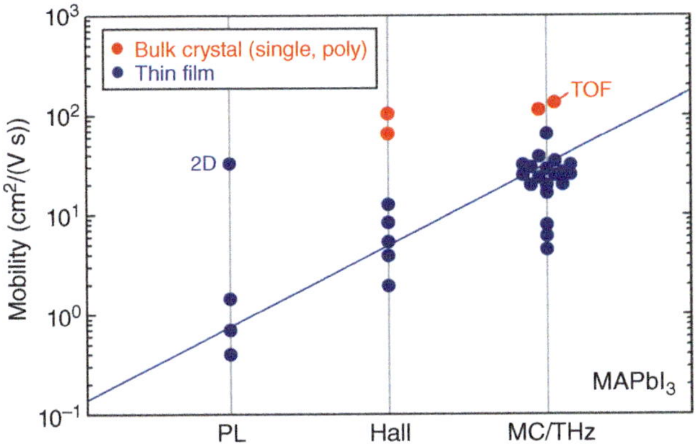

Figure 6.4 Mobility of $MAPbI_3$ determined by various characterization methods, which are categorized into the three groups of (i) PL [5–7], (ii) Hall [8–10], and (iii) MC [11–16] and THz [16–20] measurements. The red circles show the results of the bulk samples (single crystals and polycrystals), whereas the blue circles indicate the results for the thin-film samples. The plot labeled "TOF" represents μ deduced by using the time-of-flight technique [21]. In addition, the PL mobility denoted as "2D" shows that this mobility is determined by the two-dimensional lateral diffusion of carriers within a $MAPbI_3$ grain [7]. The blue line for the thin-film results is to guide the eyes.

In contrast, μ obtained from $MAPbI_3$ thin films (blue circles in Figure 6.4) differs significantly depending on the measurement method; in particular, there exists a clear trend of $\mu_{MC} \sim \mu_{THz} > \mu_{Hall} > \mu_{PL}$. The results of $\mu_{MC} \sim \mu_{THz}$ show clearly that the intragrain mobility of $MAPbI_3$ thin films is $\sim$30 cm^2/(V s). Note, however, that μ_{MC} varies with optical pump intensity (fluence) [11–13, 15, 16], probe microwave frequency [11], and interface structures [11, 14], and thus some ambiguity remains. When bare $MAPbI_3$ samples are probed at the microwave frequency of $\sim$10 GHz, μ_{MC} agrees quite well with μ_{THz}. Quite importantly, in a THz ellipsometry measurement, which characterizes the Drude-type free-carrier absorption directly without the pump light, the same μ_{THz} of $\sim$30 cm^2/(V s) is observed [20]. Accordingly, $\mu \sim 30$ cm^2/(V s) can be taken as the maximum intragrain μ in $MAPbI_3$ thin films. This upper limit observed in the thin films ($\sim$30 cm^2/(V s)) is much less than that confirmed in the bulk crystal ($\sim$100 cm^2/(V s)), indicating that enhanced scattering centers exist within $MAPbI_3$ grains due to imperfect growth of $MAPbI_3$ thin films.

The μ_{Hall} of polycrystalline thin films is significantly affected by grain boundaries (see Figure 6.1), and μ_{Hall} of $\sim$5 cm^2/(V s) observed in $MAPbI_3$ is notably lower than μ_{MC} and μ_{THz}. In hybrid perovskite thin films, grain sizes exceed the film thickness and there are no grain boundaries along the film-thickness direction (see Figure 6.1). To represent the vertical carrier transport in the thin-film devices, therefore, an intragrain μ of 30 cm^2/(V s) is appropriate.

The result $\mu_{PL} \sim 0.5$ cm^2/(V s) in Figure 6.4 was extracted by assuming a one-dimensional diffusion model:

$$\frac{\partial n(x,t)}{\partial t} = D\frac{\partial^2 n(x,t)}{\partial x^2} - k(t)n(x,t) \tag{6.3}$$

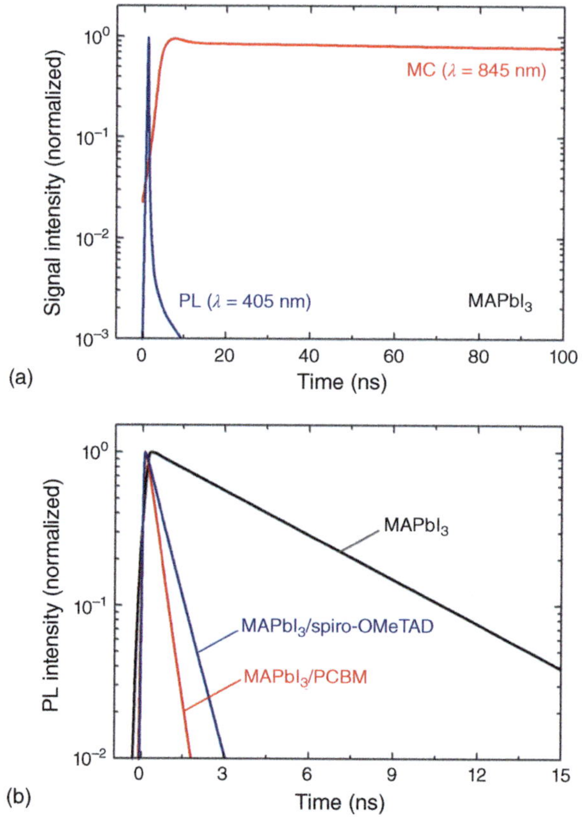

Figure 6.5 (a) Normalized time decays of PL and MC signal intensities for a $MAPbI_3$ single crystal [22] and (b) normalized PL time decays of $MAPbI_3$/quartz, PCBM/$MAPbI_3$/quartz, and spiro-OMeTAD/$MAPbI_3$/quartz samples [6]. In the figure, the smoothed results of Refs. [6, 22] are shown for clarity. In (a), the MC measurement was made by using a pump laser at 845 nm, whereas the PL was excited at 405 nm. Note that the lifetime tends to decrease when using an excitation laser with a shorter wavelength because carriers are generated nearer the surface due to reduced optical penetration depth. *Source*: (a) Bi et al. [22]. (b) Xing et al. [6].

where $n(x, t)$, D, and $k(t)$ are the carrier density distribution, diffusion coefficient, and PL decay rate, respectively [5, 6]. The problem of determining μ by PL is that the carrier lifetime tends to be underestimated seriously in this technique. Figure 6.5 shows (a) the time decays of PL and MC signal intensities for a $MAPbI_3$ single crystal [22] and (b) the PL time decays of $MAPbI_3$/quartz, PCBM/$MAPbI_3$/quartz, and spiro-OMeTAD/$MAPbI_3$/quartz samples [6]. What is striking in Figure 6.5a is that the PL lifetime ($\tau_{PL} < 10$ ns) is drastically shorter than the MC lifetime ($\tau_{MC} = 15\,\mu$s). The τ_{PL} is strongly influenced by interface conditions [5, 6, 67, 68] and, in the example of Figure 6.5b, τ_{PL} decreases significantly due to interface formation with the PCBM and spiro-OMeTAD layers. Importantly, τ_{PL} of the bare $MAPbI_3$ sample in Figure 6.5b is governed primarily by surface recombination and τ_{PL} improves drastically upon passivating the $MAPbI_3$ surface (Figure 1.6) [40]. As a result, if the nominal short PL decay is adopted for Eq. (6.3), D is seriously underestimated, leading to the drastic reduction in μ. When the lateral diffusion of carriers within a grain is analyzed by PL, however, μ_{PL} increases to 33 $cm^2/(V\,s)$ (indicated by "2D" in Figure 6.4) [7], which is consistent with the intragrain μ determined by the MC and THz techniques.

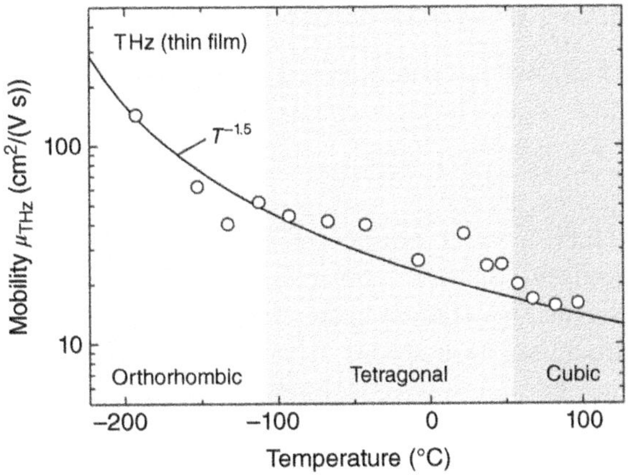

Figure 6.6 Mobility of a $MAPbI_3$ thin film determined by THz measurements (μ_{THz}) as a function of temperature T. The solid line is the fitting result, which shows a clear $T^{-1.5}$ dependence of the experimental data. The phase transition temperatures of 165 K (orthorhombic → tetragonal) and 327 K (tetragonal → cubic) are also indicated (see also Figure 3.13). *Source*: Milot et al. [19]. Licensed under CC BY 4.0.

To investigate μ of $MAPbI_3$, the field effect transistors (FETs) were fabricated [28–32]. Nevertheless, the FET mobilities are surprisingly low (10^{-5}–0.5 $cm^2/(V\,s)$) [28–31] and thus an accurate determination of μ is difficult from FET measurements, which use quite high applied voltages of <60 V.

6.3.2 Temperature Dependence

Figure 6.6 shows the variation of μ_{THz} in a $MAPbI_3$ thin film with a measurement temperature T, reported in Ref. [19]. The μ_{THz} increases strongly with decreasing T and reaches a high value of 100 $cm^2/(V\,s)$. As explained in Figure 3.13, the crystal structure of $MAPbI_3$ shows distinct phase transitions at 165 K (orthorhombic → tetragonal) and 327 K (tetragonal → cubic). However, no significant discontinuity appears for μ_{THz} as a function of T. This is in sharp contrast with the MA^+ reorientation time (Figure 3.16) and the band gap (Figure 4.19), which show irregular discontinuities at the orthorhombic-tetragonal transition temperature. In other words, the organic cation reorientation and the Pb–I–Pb bond angle change have little effects on μ. The solid line in Figure 6.6 shows the result of fitting to the experimental data and a clear $T^{-1.5}$ dependence can be seen. Similar variations of $\mu \propto T^m$ with $m = -1.8 \sim -1.4$ have been confirmed from PL [7], MC [11, 13], and THz [19] analyses. For μ_{THz} of $FAPbI_3$, different μ variations ($m = -1.5 \sim -0.5$) have been reported [38, 39].

As known well [1, 2], when μ is limited by lattice scattering, μ decreases with increasing T because stronger lattice vibrations occur at higher T. In contrast, when μ is determined by ionized-impurity (or defect) scattering, μ increases with T [1, 2], because higher thermal speeds of the carriers at high T reduce the Coulombic interaction between charged scattering center and carrier. The negative exponent observed for μ_{THz} of $MAPbI_3$ ($\mu \propto T^{-1.5}$) clearly shows that carrier scattering is dominated by thermal-induced lattice vibrations (see Section 6.3.4 for more discussion).

6.3.3 Effect of Effective Mass

Physically, μ is determined by two factors: (i) an effective mass m^* and (ii) a mean time interval between carrier scattering ($\langle\tau\rangle$), as follows [2, 48, 49]:

$$\mu = q\frac{\langle\tau\rangle}{m^*} \tag{6.4}$$

If there are a large number of defects or scattering centers, the time interval between scattering shortens, thus reducing μ. However, even when the relaxation time $\langle\tau\rangle$ is same, we observe higher μ for smaller m^*. Thus, semiconductors with low m^* are quite advantageous in terms of carrier transport. Note that m^* is an inherent material value derived from the band structure (Section 5.4.4) and, in semiconductor materials, m^* becomes smaller than the electron mass in vacuum (m_0) due to the band formation (i.e. $m^*/m_0 < 1$).

Table 6.1 summarizes the effective mass of electrons (m_e^*) and holes (m_h^*), the maximum electron mobility (μ_e) and hole mobility (μ_h), and the bulk modulus of representative solar cell materials. The results of $m_e^* = 0.32m_0$ and $m_h^* = 0.36m_0$ for $MAPbI_3$ correspond to the results of DFT in Table 5.3 [70], whereas μ_e and μ_h of $MAPbI_3$ indicate μ_{Hall} in Figure 6.3 [8, 10]. All other values in Table 6.1 are taken from Ref. [69]. The bulk modulus shows how a material compresses under external pressure and represents the hardness (or softness) of materials. It can be seen that $MAPbI_3$ is a very soft material (Table 1.1) [69, 71–73].

Figure 6.7 shows the variation of μ with m_e^*/m_0 and m_h^*/m_0 obtained from Table 6.1. This figure contains a plot for $MASnI_3$ that is based on the results of Refs. [8, 74, 75]. All the tetrahedral-bonding semiconductors show the trend of $m_e^* < m_h^*$ and thus $\mu_e > \mu_h$. In particular, for these materials, due to the large difference between m_e^* and m_h^*, μ_h is significantly less than μ_e. In contrast, $MAPbI_3$ has the unique property of $m_e^* \sim m_h^*$ and thus we find $\mu_e \sim \mu_h$. In solar cells, since both electron and hole carriers need to be collected, well-balanced $\mu_e \sim \mu_h$ in $MAPbI_3$ provides a great advantage for efficient carrier collection. Nevertheless, the maximum μ_e and μ_h of $MAPbI_3$ (~100 $cm^2/(V\,s)$) are rather low. Comparing

Table 6.1 Effective masses of electrons (m_e^*/m_0) and holes (m_h^*/m_0), electron mobility (μ_e) and hole mobility (μ_h), and bulk moduli of representative solar cell materials.

Material	m_e^*/m_0	m_h^*/m_0	μ_e ($cm^2/(V\,s)$)	μ_h ($cm^2/(V\,s)$)	Bulk modulus (GPa)
Si	0.32	0.54	1500	500	98
GaAs	0.06	0.53	8000	400	75
CdTe	0.11	0.72	1100	100	42
$CuInSe_2$	0.16	1	150	20	75
$MAPbI_3$	0.32	0.36	66	105	<20

All the numerical data were taken from Ref. [69], except for the effective mass [70] and mobility [8, 10] of $MAPbI_3$ (see Table 5.3 for m^* of $MAPbI_3$).
Source: Egger et al. [69].

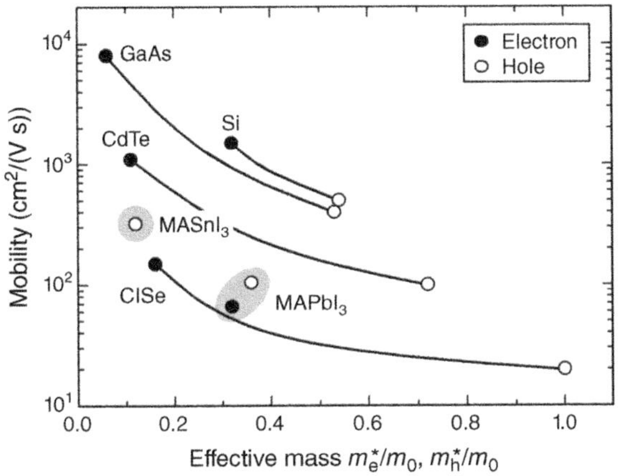

Figure 6.7 Variation of mobility (μ) with m_e^*/m_0 (closed circles) and m_h^*/m_0 (open circles), obtained from Table 6.1. The results for $MASnI_3$ are taken from Refs. [8, 74, 75]. The gray areas highlight the results of $MABI_3$ (B = Pb^{2+}, Sn^{2+}). For inorganic semiconductors, the plots for the electrons and holes are connected by curves for clarity. Note that the result of $MAPbI_3$ is unique in that $m_e^* \sim m_h^*$, which leads to well-balanced electron and hole mobilities of $\mu_e \sim \mu_h$.

μ_e values for Si and $MAPbI_3$ shows that μ_e of $MAPbI_3$ is one order of magnitude smaller, even though both materials have similar $m_e^* \sim 0.3m_0$. The reason for low μ in $MAPbI_3$ is discussed in Section 6.3.4.

We have seen in Figure 6.3 that μ_h of $MASnI_3$ is substantially larger than that of $MAPbI_3$. This result can be interpreted from a lower m_h^* of $MASnI_3$ (see Figure 6.7), which in turn increases μ according to Eq. (6.4). Thus, the effect of m^* is significant in hybrid perovskite materials.

6.3.4 What Determines Maximum Mobility of $MAPbI_3$?

The experimental μ of $MAPbI_3$ described earlier reveals the important facts that (i) the maximum μ of nondoped $MAPbI_3$ single crystals at room temperature is $\sim$100 cm^2/(V s), (ii) the intragrain μ of $MAPbI_3$ thin films at room temperature is $\sim$30 cm^2/(V s), and (iii) the T dependence of μ is $\mu \propto T^{-1.5}$. The μ in semiconductor materials is influenced by a variety of factors and, by incorporating carrier scattering models proposed for $MAPbI_3$ [76–88], the experimental μ of hybrid perovskites can be expressed as

$$\frac{1}{\mu} = \frac{1}{\mu_{\text{ion}}} + \frac{1}{\mu_{\text{LO}}} + \frac{1}{\mu_{\text{AP}}} + \frac{1}{\mu_{\text{grain}}} + \frac{1}{\mu_{\text{TF}}} + \left(\frac{1}{\mu_{\text{polaron}}} + \frac{1}{\mu_{\text{dis}}}\right)_{\text{Model}} \tag{6.5}$$

The above equation is based on Matthiessen's rule [2], which indicates that the observable μ is given by the inverse relation and is limited by the lowest μ factor. The major factors determining μ in semiconductors include ionized-impurity scattering (μ_{ion}) [1, 89], longitudinal optical (LO) phonon scattering (μ_{LO}) [76–80], acoustic phonon scattering (μ_{AP}) [1, 81–83], and grain-boundary scattering (μ_{grain})

[49, 54, 55]. In Eq. (6.5), μ_{TF} is incorporated to express the reduced μ in the perovskite thin films. The two terms denoted as "Model" represent the proposed scattering mechanisms of polaron formation ($\mu_{polaron}$) [84–86] and anharmonic structural disorder (μ_{dis}) [69, 87], as explained below.

When carrier density is low, the value of μ_{ion} becomes quite large [1, 89] and the term $1/\mu_{ion}$ vanishes (i.e. $1/\mu_{ion} \rightarrow 0$). In single crystals of $MAPbI_3$ with low carrier densities, the terms $1/\mu_{ion}$, $1/\mu_{grain}$, and $1/\mu_{TF}$ disappear. Moreover, DFT calculations show that μ_{AP} of $MAPbI_3$ is quite high (>1500 $cm^2/(V\,s)$) [81–83] and $1/\mu_{AP}$ is also eliminated. Consequently, several studies have proposed that the maximum μ is determined by μ_{LO} (i.e. $\mu \sim \mu_{LO}$ in Eq. (6.5)) [76–80]. Note that LO phonons displace the atomic positions in a way that can create local electric fields, which further scatter free carriers through strong Coulomb interactions. This coupling between LO phonons and free carriers is referred to as the Fröhlich interaction. The strength of the electron–phonon coupling can be expressed by a dimensionless Fröhlich parameter α [77, 84]:

$$\alpha = \left(\frac{1}{\varepsilon_\infty} - \frac{1}{\varepsilon_s}\right)\left(\frac{R_y}{\hbar\omega_{LO}} \cdot \frac{m^*}{m_0}\right)^{1/2} \tag{6.6}$$

where ε_∞ and ε_s show high frequency and static dielectric constants (see Figure 4.2). The quantities R_y, $\hbar$, c, and ω_{LO} are the Rydberg constant (R_y = 13.6 eV), Dirac's constant, speed of light, and LO-phonon frequency. The prefactor $(1/\varepsilon_\infty - 1/\varepsilon_s)$ reflects the effect of the ionic screening.

As explained in Chapter 4, $MAPbI_3$ shows unique phonon absorption characteristics, including (i) quite strong phonon absorption, which leads to $\varepsilon_s \gg \varepsilon_\infty$ (Figure 4.2) and (ii) very low peak frequency of ω_{LO} (Figure 4.7). These specific features lead to a quite high coupling constant α, which is estimated to be 1.72–2.68 in $MAPbI_3$ [77, 84, 85]. This value is significantly larger than those of tetrahedral-bonding semiconductors, such as GaAs (α = 0.07) and CdTe (α = 0.31) [90], and is comparable to crystals with high ionicity, including AgCl (α = 1.8) and KI (α = 2.5) [90]. Accordingly, the LO-phonon coupling in the hybrid perovskite is indeed strong and μ_{LO} likely governs the room temperature μ of $MAPbI_3$. Theoretical attempts show that the maximum μ_{LO} ranges from 30 to 200 $cm^2/(V\,s)$ with a T dependence of $T^{-1.37}$ to $T^{-1.5}$ [78–80].

To explain the maximum μ of ~100 $cm^2/(V\,s)$ in $MAPbI_3$, two different models have been proposed further: (i) the formation of polarons [84–86] and (ii) carrier scattering by anharmonic lattice vibration [69, 87]. The physical pictures of these models are shown in Figure 6.8. When a hole conducts inside a polar (ionic) semiconductor, for example, the positive charge of the hole attracts anions, while expelling the cations in the opposite direction. Thus, a conduction hole (or electron) displaces the constituent ions from their equilibrium positions, creating a quasiparticle called a polaron (Figure 6.8a). The formation of a polaron lowers μ through a widespread lattice deformation, which can be interpreted as an increase in m^*. The polaron effective mass (m_p) is derived from the Fröhlich parameter α and m_p increases with α [84, 85]. For $MAPbI_3$, m_p is 28–64% heavier than m^*

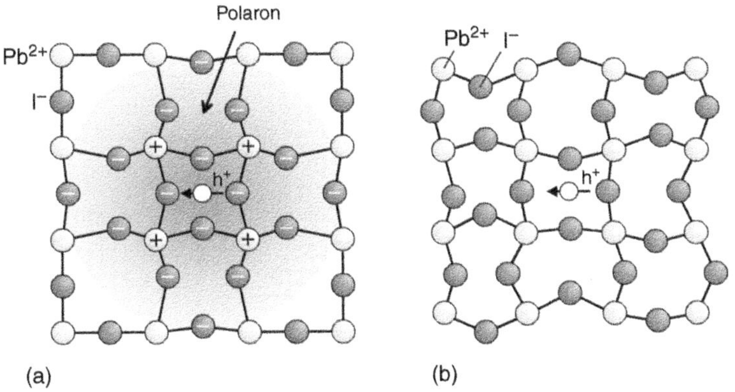

Figure 6.8 Physical pictures of (a) polaron formation and (b) carrier scattering by anharmonic lattice vibrations in $MAPbI_3$. The h^+ in the figure indicates a hole. In (a), the hole carrier conduction inside a polar (ionic) semiconductor displaces cations and anions through the Coulomb interaction, creating a quasiparticle called a polaron. In (b), carriers in $MAPbI_3$ are scattered by anharmonic (disordered) vibrations of the PbI_3^- inorganic cage, as proposed in Refs. [69, 87]. *Source*: Egger et al. [69]; Mayers et al. [87].

[84–86] and the polaron-limited mobility ($\mu_{polaron}$ in Eq. (6.5)) is estimated to be 133–200 cm²/(V s) [84, 85]. The corresponding polaron radii range from 26 to 51 Å [84, 85]. However, although $\mu_{polaron}$ quantitatively agrees with the maximum μ of ~100 cm²/(V s) at room temperature, the theoretical calculation shows a T dependence of $\mu_{polaron} \propto T^{-0.46}$ [85]. Thus, μ variation for T is underestimated with the polaron concept.

If the PbI_3^- cage deforms anharmonically, a local potential fluctuation is created, scattering carriers (see Figure 6.8b). A molecular-dynamics-based calculation shows that the maximum μ_e of $MAPbI_3$, limited by such an anharmonic lattice disorder (μ_{dis} in Eq. (6.5)), is ~135 cm²/(V s) with the T dependence of $\sim T^{-2}$ [87]; thus, this model provides the correct transport parameters for $MAPbI_3$. It has been argued that carrier scattering by dynamic lattice disorder becomes particularly important in compounds with "soft lattices" [69, 87] (see bulk modulus in Table 6.1).

All the above discussions, however, are based on a quite simple assumption that μ is determined totally by a single scattering mechanism. In the LO–phonon scattering model, for example, $\mu(T) = \mu_{LO}(T)$ is assumed in Eq. (6.5). However, this is highly doubtful, because the T dependence of $\mu(T) \propto T^{-1.5}$ is reported for $MAPbI_3$ thin films [7, 11, 13, 18, 19], in which the intragrain μ is lowered by μ_{TF} and μ_{ion}. In particular, it is likely that the intragrain μ of $MAPbI_3$ thin films (~30 cm²/(V s)) is described by $1/\mu = 1/\mu_{ion} + 1/\mu_{TF} + 1/\mu_{LO}$, or other similar models. Accordingly, more quantitative experimental data, including $\mu(T)$ of single crystals, are necessary to conclude the absolute limiting factor of μ in $MAPbI_3$ at room temperature. We mention, however, that the strong LO–phonon carrier scattering is quite likely because the LO–phonon absorption is extraordinarily strong in hybrid perovskites, leading to a large Fröhlich parameter.

Table 6.2 Carrier diffusion length (L_D), lifetime (τ), and mobility (μ), reported for $MAPbI_3$ single crystals [10, 15, 22–24] and thin films [5, 6, 13, 14, 19, 25–27].

Method	L_D (μm)	τ (μs)	μ ($cm^2/(V s)$)	References
Single crystal				
TPV[a)]	175	82	105	[10]
MC	55	10	115	[15]
MC	50	15	64.5	[22]
SPCM[b)]	10–28			[15]
TA[c)]	10	0.57	67.2	[23]
PL	8	9.9	2.5	[24]
Thin film				
Opt-Hall[d)]	23	30	8	[25]
MC	9	5	6.2	[13]
MC	4.1	0.22	30	[14]
THz	2.5	0.07	35	[19]
IMPS[e)]	1.2–1.5			[26]
EBIC[f)]	1			[27]
PL	0.13	0.005	1.45	[6]
PL	0.13	0.01	0.66	[5]

The characterization results obtained from various techniques are summarized. Some of the τ and μ values are calculated by applying the relations of Eqs. (6.1) and (6.2).
a) Transient photovoltaic spectroscopy.
b) Scanning photocurrent microscopy.
c) Transient absorption spectroscopy.
d) Optical Hall measurement.
e) Intensity-modulated photocurrent/photovoltage spectroscopy.
f) Electron-beam-induced current technique.

6.4 Diffusion Length

One of the greatest advantages of hybrid perovskite solar cells is long L_D of carriers [10, 13–15, 19, 22–27]. Nevertheless, absolute L_D values differ significantly depending on the characterization technique. Table 6.2 summarizes L_D in $MAPbI_3$ single crystals and thin films, together with their corresponding μ and τ. Figure 6.9 further summarizes the relation of μ and τ for the values of Table 6.2. For the single crystals, L_D varies in a range of 8–175 μm, while L_D of the thin films is 0.1–23 μm.

In Figure 6.9, we can confirm $\mu \sim 100\,cm^2/(V\,s)$ for $MAPbI_3$ single crystals, even though the PL result shows much lower μ, as discussed earlier. In other words, the large variation in L_D observed for the single crystals originates from the variation in τ. The MC measurements consistently show $\tau \sim 10$ μs [15, 22], whereas the transient photovoltaic (TPV) measurements [10] and transient absorption (TA)

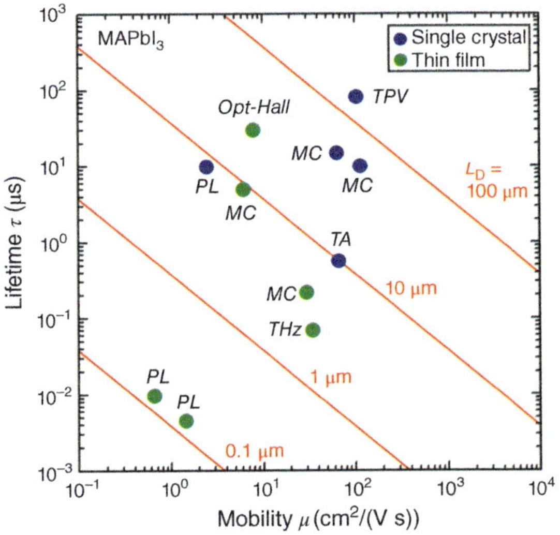

Figure 6.9 Carrier mobility μ and carrier lifetime τ for each characterization of $MAPbI_3$, obtained by using different techniques summarized in Table 6.2. The blue circles show results reported for the single crystals [10, 15, 22–24], whereas the green circles indicate results obtained from the thin-film samples [5, 6, 13, 14, 19, 25]. The red lines indicate the constant-L_D lines obtained for $L_D = 0.1, 1, 10,$ and 100 μm, calculated from Eqs. (6.1, 6.2).

spectroscopy [23] give different τ values. On the other hand, L_D can be determined directly from spatially resolved techniques, including scanning photocurrent microscopy (SPCM) [15] and the electron beam-induced current (EBIC) technique [27]. The SPCM characterization of $MAPbI_3$ single crystals results in L_D of 10–28 μm [15], which is consistent with ~50 μm determined by MC [15, 22]. Accordingly, $L_D \sim 50$ μm with $\mu \sim 100\,cm^2/(V\,s)$ and $\tau \sim 10$ μs can be considered as a representative L_D of $MAPbI_3$ single crystals.

For $MAPbI_3$ thin films, we observe $\mu = 10–30\,cm^2/(V\,s)$, with $\mu_{PL} \sim 1\,cm^2/(V\,s)$, which are consistent with the results of Figure 6.4. However, τ reported for the thin films is quite different ($\tau = 5 \times 10^{-3}$–30 μs in Table 6.2), most likely due to the strong influence of interfaces in the thin-film geometry. The τ_{PL} is particularly sensitive to the interface (see Figure 6.5b) and the PL technique seriously underestimates L_D. The direct characterization of L_D in $MAPbI_3$ thin-film solar cells has been performed by EBIC, resulting in $L_D \sim 1$ μm [27]. The MC results of the thin films provide L_D in a similar range ($L_D \sim 5$ μm) [14]. Accordingly, L_D of 1–5 μm appears to be appropriate for $MAPbI_3$ thin films. Nevertheless, an extremely long τ_{PL} of 8 μs has been reported for a $MAPbI_3$ thin film with optimized surface [40] (see Figure 1.6). If $\tau_{PL} = 8$ μs and $\mu = 30\,cm^2/(V\,s)$ are applied, we obtain $L_D \sim 25$ μm, which is comparable to that of the single crystals. Thus, the L_D of 4–23 μm is shown in Table 1.1.

It should be emphasized that the above L_D sufficiently exceeds the absorber thicknesses of $MAPbI_3$ solar cells (300–500 nm). In this condition, the effect of carrier recombination within the bulk component becomes less significant and the interface recombination becomes more important. In fact, the open-circuit voltage (V_{oc}) of hybrid perovskite solar cells depends critically on the interface structure (see Section 11.5.1). Accordingly, it can be concluded that the carrier transport characteristics of $MAPbI_3$ thin films are quite satisfactory, which is one of the main reasons why solar cells with over 20% efficiencies can be fabricated using $MAPbI_3$ as a light absorber (Chapters 1 and 11).

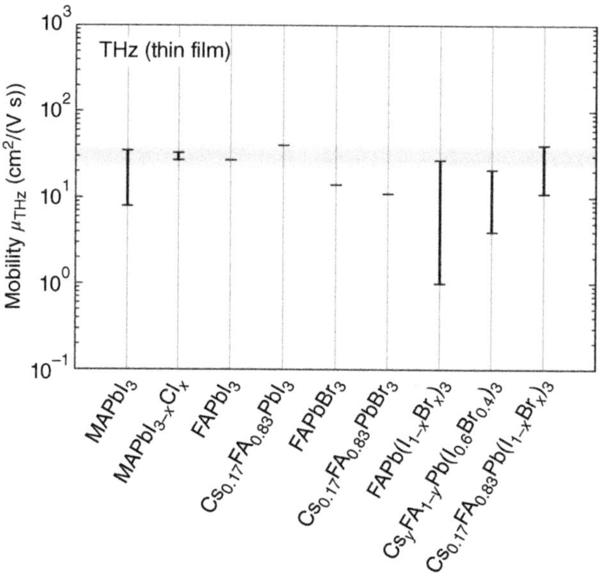

Figure 6.10 Mobility of Pb-based hybrid perovskite thin films, as determined by THz spectroscopy (μ_{THz}). In the figure, the maximum and minimum ranges of μ_{THz} reported for each material [17–19, 37, 91–94] are indicated by the bars. The gray region represents the possible upper limit of $\mu_{THz} \sim 30\,cm^2/(V\,s)$ for hybrid perovskite thin films. *Source*: Wehrenfennig et al. [17]; Karakus et al. [18]; Milot et al. [19]; Rehman et al. [37]; Wehrenfennig et al. [91]; Rehman et al. [92]; McMeekin et al. [93, 94].

6.5 Carrier Transport in Various Hybrid Perovskites

A variety of perovskite compounds and alloys have been used to fabricate hybrid perovskite solar cells. Here, we see the carrier transport properties of representative perovskite materials. Figure 6.10 summarizes μ_{THz} values reported for various Pb-based hybrid perovskite thin films [17–19, 37, 91–94]. In the figure, the maximum and minimum ranges of μ_{THz} reported for each material are indicated by bars. This result shows an important result; the upper limit of μ_{THz} is ~30 cm²/(V s) for all the perovskite thin films. Based on the discussion of Section 6.3.4, the observed limit on μ_{THz} could be expressed as

$$\frac{1}{\mu_{THz}} = \frac{1}{\mu_{ion}} + \frac{1}{\mu_{TF}} + \frac{1}{\mu_{LO}} \tag{6.7}$$

Moreover, μ_{THz} values for $MAPbI_3$ [19], $FAPbI_3$ [37], and $Cs_{0.17}FA_{0.83}PbI_3$ [92] are quite similar, confirming that the effect of the A-site cation is essentially negligible.

Nevertheless, the A-site cation affects μ indirectly through the perovskite crystal stability. Figure 6.11 shows (a) μ_{THz} and the width of the (100) X-ray diffraction (XRD) peak and (b) tolerance factor (t), as a function of the Cs content y in $Cs_yFA_{1-y}Pb(I_{0.6}Br_{0.4})_3$ thin films. The result of Figure 6.11a is taken from Ref. [92], whereas t in Figure 6.11b is calculated by using Eq. (3.1) with the values of Table 3.3.

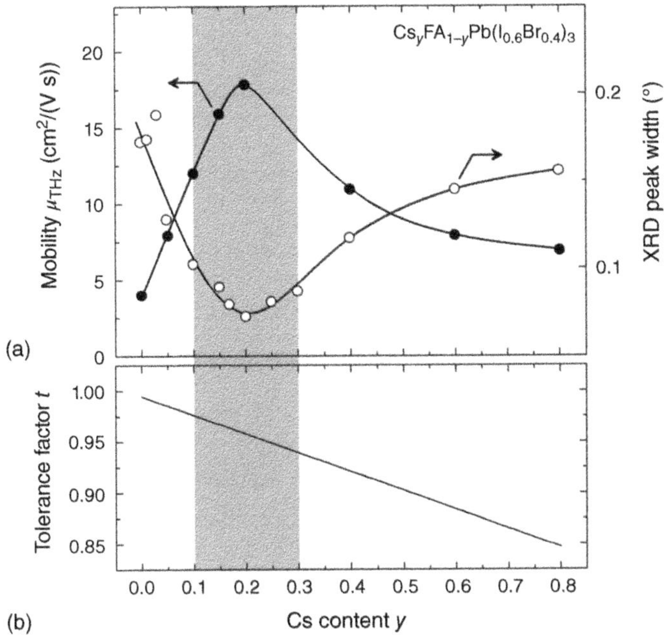

Figure 6.11 (a) μ_{THz} and XRD width of a (100) peak and (b) tolerance factor (t), as functions of the Cs content y in $Cs_yFA_{1-y}Pb(I_{0.6}Br_{0.4})_3$ thin films. The t in (b) is calculated by using Eq. (3.1) with the values given in Table 3.3. The gray region shows the optimum t range of $t = 0.91–0.98$, as established in Figure 3.10. *Source*: (a) Adapted from Rehman et al. [92].

It can be seen from Figure 6.11a that the crystallinity of the $CsFAPb(I,Br)_3$ alloys improves notably upon the slight incorporation of Cs ($y \sim 0.2$) into $FAPb(I,Br)_3$, as evidenced by the sharper XRD peak. More importantly, the crystallinity confirmed by XRD shows an almost perfect correlation with the transport characteristics and μ_{THz} reaches a maximum when the peak width is minimized.

The high μ_{THz} and small peak width at $y \sim 0.2$ can be interpreted by t, which is a global indicator that describes the stability of perovskite crystals (Section 3.3). For Pb-based hybrid perovskites, $t = 0.945$ is an ideal value (Figure 3.10), allowing the formation of the stable perovskite phase. Note that secondary-phase materials, such as PbI_2, δ-$CsPbI_3$, and δ-$FAPbI_3$ (see Figure 3.5), form when t deviates largely from the optimum t range of 0.91–0.98 (Figure 3.10), as indicated in Figure 6.11. For $CsFAPb(I_{0.6}Br_{0.4})_3$, $t \sim 0.95$ is obtained when $y = 0.24$ (Figures 6.11b and 3.11c). Thus, the Cs content of $y \sim 0.2$ not only provides higher crystallinity but also prevents the inclusion of the secondary phase. Moreover, τ_{PL} improves at this Cs composition [92]. As a result, the proper control of the A-site cation species is critical for the improvement of μ_{THz} and L_D. Note that the variation of μ_{THz} in $CsFAPb(I,Br)_3$ thin films can best be represented by μ_{TF} in Eq. (6.7), provided that the crystallinity of the single crystals is much better than that of the thin films.

For the variation of the Br content x in $FAPb(I_{1-x}Br_x)_3$ and $Cs_{0.17}FA_{0.83}Pb(I_{1-x}Br_x)_3$ thin films, a large change in μ_{THz} is also reported (Figure 6.12) [37, 92]. In the case of $FAPb(I_{1-x}Br_x)_3$ thin films, μ_{THz} has a minimum value of ~1 cm²/(V s) at

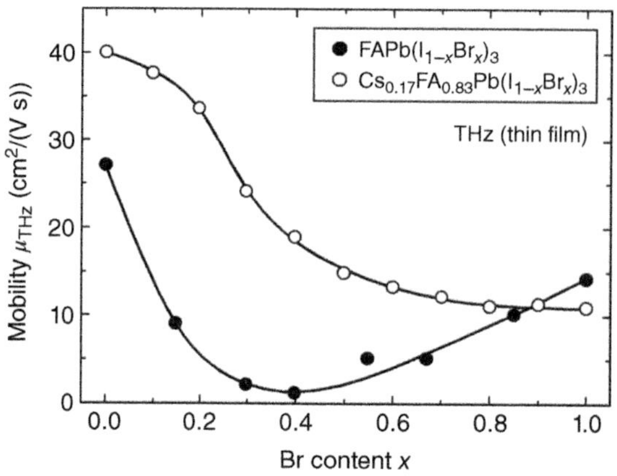

Figure 6.12 μ_{THz} for $FAPb(I_{1-x}Br_x)_3$ and $Cs_{0.17}FA_{0.83}Pb(I_{1-x}Br_x)_3$ thin films as a function of the Br content x, reported in Refs. [37, 92]. The very low μ_{THz} observed at $x = 0.3–0.4$ in $FAPb(I_{1-x}Br_x)_3$ has been attributed to amorphous phase formation. The result of $FAPb(I_{1-x}Br_x)_3$ was adapted from Ref. [37]. The result of $Cs_{0.17}FA_{0.83}Pb(I_{1-x}Br_x)_3$ was adapted from Ref. [92]. *Source*: Rehman et al. [37, 92].

$x = 0.3–0.4$, which is attributed to the formation of an amorphous phase in this region. The μ_{THz} of $Cs_{0.17}FA_{0.83}Pb(I_{1-x}Br_x)_3$ also varies strongly with x. For this alloy, μ_{THz} decreases monotonically from 40 to 11 $cm^2/(V\,s)$ with increasing x. This reduction is suggested to originate from the strong LO–phonon scattering in Br-based perovskite compounds [4]. However, μ of a $MAPbBr_3$ single crystal evaluated from the TOF measurement (115 $cm^2/(V\,s)$ in Ref. [24]) is comparable to μ_{TOF} of a $MAPbI_3$ single crystal (135 $cm^2/(V\,s)$ in Figure 6.4). In addition, the Fröhlich parameter of $MAPbBr_3$ ($\alpha = 1.69$) is almost the same as that of $MAPbI_3$ ($\alpha = 1.72$) [84]. Accordingly, μ reduction by the Br incorporation is not evident in $MAPbX_3$. The Br content has negligible effect on t in $Cs_{0.17}FA_{0.83}Pb(I_{1-x}Br_x)_3$ (see Figure 3.11). Since the absolute μ_{THz} at $x = 1.0$ is rather small (11 $cm^2/(V\,s)$), it is possible that μ_{TF} or μ_{ion} plays a dominant role in the carrier scattering of the $Cs_{0.17}FA_{0.83}PbBr_3$ thin film. For the increase of x in these thin films, the large reduction in L_D from 4.4 μm ($x = 0$) to 0.8 μm ($x = 1$) is also confirmed [92].

References

1 Sze, S.M. (1981). *Physics of Semiconductor Devices*. New York: Wiley.
2 Anderson, B.L. and Anderson, R.L. (2005). *Fundamentals of Semiconductor Devices*. New York: McGraw-Hill.
3 Smets, A., Jager, K., Isabella, O. et al. (2016). *Solar Energy*. Cambridge: UIT Cambridge.
4 Herz, L.M. (2017). *ACS Energy Lett.* 2: 1539.
5 Stranks, S.D., Eperon, G.E., Grancini, G. et al. (2013). *Science* 342: 341.
6 Xing, G., Mathews, N., Sun, S. et al. (2013). *Science* 342: 344.
7 Biewald, A., Giesbrecht, N., Bein, T. et al. (2019). *ACS Appl. Mater. Interfaces* 11: 20838.
8 Stoumpos, C.C., Malliakas, C.D., and Kanatzidis, M.G. (2013). *Inorg. Chem.* 52: 9019.
9 Wang, Q., Shao, Y., Xie, H. et al. (2014). *Appl. Phys. Lett.* 105: 163508.

10 Dong, Q., Fang, Y., Shao, Y. et al. (2015). *Science* 347: 967.
11 Oga, H., Saeki, A., Ogomi, Y. et al. (2014). *J. Am. Chem. Soc.* 136: 13818.
12 Reid, O.G., Yang, M., Kopidakis, N. et al. (2016). *ACS Energy Lett.* 1: 561.
13 Savenije, T.J., Ponseca, C.S. Jr., Kunneman, L. et al. (2014). *J. Phys. Chem. Lett.* 5: 2189.
14 Hutter, E.M., Eperon, G.E., Stranks, S.D., and Savenije, T.J. (2015). *J. Phys. Chem. Lett.* 6: 3082.
15 Semonin, O.E., Elbaz, G.A., Straus, D.B. et al. (2016). *J. Phys. Chem. Lett.* 7: 3510.
16 Ponseca, C.S. Jr., Savenije, T.J., Abdellah, M. et al. (2014). *J. Am. Chem. Soc.* 136: 5189.
17 Wehrenfennig, C., Eperon, G.E., Johnston, M.B. et al. (2014). *Adv. Mater.* 26: 1584.
18 Karakus, M., Jensen, S.A., D'Angelo, F. et al. (2015). *J. Phys. Chem. Lett.* 6: 4991.
19 Milot, R.L., Eperon, G.E., Snaith, H.J. et al. (2015). *Adv. Funct. Mater.* 25: 6218.
20 Uprety, P., Wang, C., Koirala, P. et al. (2018). *IEEE J. Photovoltaics* 8: 1793.
21 Shrestha, S., Matt, G.J., Osvet, A. et al. (2018). *J. Phys. Chem. C* 122: 5935.
22 Bi, Y., Hutter, E.M., Fang, Y. et al. (2016). *J. Phys. Chem. Lett.* 7: 923.
23 Saidaminov, M.I., Abdelhady, A.L., Murali, B. et al. (2015). *Nat. Commun.* 6: 7586.
24 Shi, D., Adinolfi, V., Comin, R. et al. (2015). *Science* 347: 519.
25 Chen, Y., Yi, H.T., Wu, X. et al. (2016). *Nat. Commun.* 7: 12253.
26 Zhao, Y., Nardes, A.M., and Zhu, K. (2014). *J. Phys. Chem. Lett.* 5: 490.
27 Edri, E., Kirmayer, S., Henning, A. et al. (2014). *Nano Lett.* 14: 1000.
28 Chin, X.Y., Cortecchia, D., Yin, J. et al. (2015). *Nat. Commun.* 6: 7383.
29 Li, F., Ma, C., Wang, H. et al. (2015). *Nat. Commun.* 6: 8238.
30 Li, D., Wang, G., Cheng, H.-C. et al. (2016). *Nat. Commun.* 7: 11330.
31 Senanayak, S.P., Yang, B., Thomas, T.H. et al. (2017). *Sci. Adv.* 3: e1601935.
32 Wang, G., Li, D., Cheng, H.-C. et al. (2015). *Sci. Adv.* 1: e1500613.
33 Buin, A., Pietsch, P., Xu, J. et al. (2014). *Nano Lett.* 14: 6281.
34 Johnston, M.B. and Herz, L.M. (2016). *Acc. Chem. Res.* 49: 146.
35 Valverde-Chávez, D.A., Ponseca, C.S. Jr., Stoumpos, C.C. et al. (2015). *Energy Environ. Sci.* 8: 3700.
36 Eperon, G.E., Stranks, S.D., Menelaou, C. et al. (2014). *Energy Environ. Sci.* 7: 982.
37 Rehman, W., Milot, R.L., Eperon, G.E. et al. (2015). *Adv. Mater.* 27: 7938.
38 Gélvez-Rueda, M.C., Renaud, N., and Grozema, F.C. (2017). *J. Phys. Chem. C* 121: 23392.
39 Davies, C.L., Borchert, J., Xia, C.Q. et al. (2018). *J. Phys. Chem. Lett.* 9: 4502.
40 de Quilettes, D.W., Koch, S., Burke, S. et al. (2016). *ACS Energy Lett.* 1: –438.
41 Yin, W.-J., Shi, T., and Yan, Y. (2014). *Appl. Phys. Lett.* 104: 063903.
42 Kim, J., Lee, S.-H., Lee, J.H., and Hong, K.-H. (2014). *J. Phys. Chem. Lett.* 5: 1312.
43 Amat, A., Mosconi, E., Ronca, E. et al. (2014). *Nano Lett.* 14: 3608.
44 Motta, C., El-Mellouhi, F., Kais, S. et al. (2015). *Nat. Commun.* 6: 7026.
45 Zheng, F., Tan, L.Z., Liu, S., and Rappe, A.M. (2015). *Nano Lett.* 15: 7794.
46 Azarhoosh, P., McKechnie, S., Frost, J.M. et al. (2016). *APL Mater.* 4: 091501.

47 Hutter, E.M., Gélvez-Rueda, M.C., Osherov, A. et al. (2017). *Nat. Mater.* 16: 115.
48 Fujiwara, H. (2009). *Spectroscopic Ellipsometry: Principles and Applications*. West Sussex: Wiley.
49 Fujiwara, H. and Collins, R.W. (2018). *Spectroscopic Ellipsometry for Photovoltaics: Fundamental Principles and Solar Cell Characterization*, vol. 1. Cham: Springer.
50 Fujiwara, H. and Collins, R.W. (2018). *Spectroscopic Ellipsometry for Photovoltaics: Applications and Optical Data of Solar Cell Materials*, vol. 2. Cham: Springer.
51 Beard, M.C., Turner, G.M., and Schmuttenmaer, C.A. (2000). *Phys. Rev. B* 62: 15764.
52 Ulbricht, R., Hendry, E., Shan, J. et al. (2011). *Rev. Mod. Phys.* 83: 543.
53 Schroder, D.K. (2006). *Semiconductor Material and Device Characterization*. Hoboken, NJ: Wiley-Interscience.
54 Faÿ, S., Steinhauser, J., Nicolay, S., and Ballif, C. (2010). *Thin Solid Films* 518: 2961.
55 Sago, K., Kuramochi, H., Iigusa, H. et al. (2014). *J. Appl. Phys.* 115: 133505.
56 Cui, P., Wei, D., Ji, J. et al. (2017). *Sol. RRL* 1: 1600027.
57 Emara, J., Schnier, T., Pourdavoud, N. et al. (2016). *Adv. Mater.* 28: 553.
58 Naikaew, A., Prajongtat, P., Lux-Steiner, M.C. et al. (2015). *Appl. Phys. Lett.* 106: 232104.
59 Cai, M., Ishida, N., Li, X. et al. (2018). *Joule* 2: 296.
60 Mitzi, D.B., Feild, C.A., Schlesinger, Z., and Laibowitz, R.B. (1995). *J. Solid State Chem.* 114: 159.
61 Takahashi, Y., Hasegawa, H., Takahashi, Y., and Inabe, T. (2013). *J. Solid State Chem.* 205: 39.
62 Kumar, M.H., Dharani, S., Leong, W.L. et al. (2014). *Adv. Mater.* 26: 7122.
63 Noel, N.K., Stranks, S.D., Abate, A. et al. (2014). *Energy Environ. Sci.* 7: 3061.
64 Hao, F., Stoumpos, C.C., Cao, D.H. et al. (2014). *Nat. Photonics* 8: 489.
65 Takahashi, Y., Obara, R., Lin, Z.-Z. et al. (2011). *Dalton Trans.* 40: 5563.
66 Lin, R., Xiao, K., Qin, Z. et al. (2019). *Nat. Energy* 4: 864.
67 Stolterfoht, M., Wolff, C.M., Márquez, J.A. et al. (2018). *Nat. Energy* 3: 847.
68 Tan, H., Jain, A., Voznyy, O. et al. (2017). *Science* 355: 722.
69 Egger, D.A., Bera, A., Cahen, D. et al. (2018). *Adv. Mater.* 30: 1800691.
70 Giorgi, G., Fujisawa, J.-I., Segawa, H., and Yamashita, K. (2013). *J. Phys. Chem. Lett.* 4: 4213.
71 Rakita, Y., Cohen, S.R., Kedem, N.K. et al. (2015). *MRS Commun.* 5: 623.
72 Sun, S., Fang, Y., Kieslich, G. et al. (2015). *J. Mater. Chem. A* 3: 18450.
73 Feng, J. (2014). *APL Mater.* 2: 081801.
74 Umari, P., Mosconi, E., and Angelis, F.D. (2015). *Sci. Rep.* 4: 4467.
75 Ma, Z.-Q., Pan, H., and Wong, P.K. (2016). *J. Electron. Mater.* 45: 5956.
76 Wright, A.D., Verdi, C., Milot, R.L. et al. (2016). *Nat. Commun.* 7: 11755.
77 Herz, L.M. (2018). *J. Phys. Chem. Lett.* 9: 6853.
78 Filippetti, A., Mattoni, A., Caddeo, C. et al. (2016). *Phys. Chem. Chem. Phys.* 18: 15352.

79 Yu, Z.-G. (2016). *J. Phys. Chem. Lett.* 7: 3078.

80 Poncé, S., Schlipf, M., and Giustino, F. (2019). *ACS Energy Lett.* 4: 456.

81 He, Y. and Galli, G. (2014). *Chem. Mater.* 26: 5394.

82 Wang, Y., Zhang, Y., Zhang, P., and Zhang, W. (2015). *Phys. Chem. Chem. Phys.* 17: 11516.

83 Zhao, T., Shi, W., Xi, J. et al. (2016). *Sci. Rep.* 6: 19968.

84 Sendner, M., Nayak, P.K., Egger, D.A. et al. (2016). *Mater. Horiz.* 3: 613.

85 Frost, J.M. (2017). *Phys. Rev. B* 96: 195202.

86 Schlipf, M., Poncé, S., and Giustino, F. (2018). *Phys. Rev. Lett.* 121: 086402.

87 Mayers, M.Z., Tan, L.Z., Egger, D.A. et al. (2018). *Nano Lett.* 18: 8041.

88 Brenner, T.M., Egger, D.A., Rappe, A.M. et al. (2015). *J. Phys. Chem. Lett.* 6: 4754.

89 Ellmer, K. (2001). *J. Phys. D. Appl. Phys.* 34: 3097.

90 Devreese, J.T. (1996). *Encycl. Appl. Phys.* 14: 383.

91 Wehrenfennig, C., Liu, M., Snaith, H.J. et al. (2014). *Energy Environ. Sci.* 7: 2269.

92 Rehman, W., McMeekin, D.P., Patel, J.B. et al. (2017). *Energy Environ. Sci.* 10: 361.

93 McMeekin, D.P., Sadoughi, G., Rehman, W. et al. (2016). *Science* 351: 151.

94 McMeekin, D.P., Wang, Z., Rehman, W. et al. (2017). *Adv. Mater.* 29: 1607039.

7

Ferroelectric Properties

Tobias Leonhard, Holger Röhm, Alexander D. Schulz, and Alexander Colsmann

Karlsruhe Institute of Technology (KIT), Material Research Center for Energy Systems (MZE), Strasse am Forum 7, 76131 Karlsruhe, Germany

7.1 On the Importance of Ferroelectricity in Hybrid Perovskite Solar Cells

Among the distinct properties of light-harvesting organic metal halide (OMH) perovskites are their low charge carrier recombination at traps states (Shockley–Read–Hall [SRH] recombination) [1] and their excellent charge carrier transport [2–4]. Recent theoretic and experimental findings have pointed out that some OMH perovskites are ferroelectric and that the corresponding microscopic internal fields may partly be the origin of the low charge carrier recombination and hence the excellent device performance [5–9]. Not only does a ferroelectric light-harvesting layer produce solar cells with different charge carrier dynamics. The inherent switchability of ferroelectric materials between at least two states under electrical fields also strongly affects solar cell operation, characterization, and stability [10, 11].

This chapter reviews the characterization techniques toward the investigation of ferroelectricity in (light-harvesting) semiconductors, including common pitfalls, and summarizes different data interpretations. Not least, important implications of ferroelectric light-harvesting semiconductors will be discussed with respect to solar cell performance and characterization.

In the literature, the term "ferroelectric solar cells" has been used in conjunction with solar cells that utilize the bulk-photovoltaic effect: [12] microscopic electric fields in insulating ferroelectric domain assemblies produce higher-than-usual photovoltages (similar to tandem solar cells), but their negligible device currents, to date, render the concept rather inefficient. In contrast, this chapter deliberately focuses on semiconducting thin films, which operate as single light-harvesting layers in solar cells.

Hybrid Perovskite Solar Cells: Characteristics and Operation,
First Edition. Edited by Hiroyuki Fujiwara.

7.2 Ferroelectricity

The (macroscopic) textbook definition of ferroelectricity is the existence of a spontaneous polarization that is switchable between at least two polarization states under an external electric field [13]. Microscopically, ferroelectric materials form domains with spontaneous polarization. These properties of ferroelectric materials find their origin in the crystal structure.

7.2.1 Crystallographic Considerations

The crystal structure and its symmetry determine most of the fundamental properties of a crystalline material such as its conductivity, its band structure, and its ferroic properties, the latter of which includes ferromagneticity, ferroelasticity, and ferroelectricity. Symmetry elements, including rotational symmetries, mirror symmetries, and inversion, define 32 fundamental crystal classes. In general, ferroelectricity requires a unit cell that exhibits a polar axis, which can only occur in non-centrosymmetric crystals. Non-centrosymmetry is synonymous with a lack of inversion symmetry in the unit cell of the crystal. Figure 7.1a illustrates non-centrosymmetry in a hypothetic two-dimensional crystal comprising two different atomic constituents. The polar axes prevent the congruent lattice inversion at the center of the unit cell [14].

Out of the 32 crystal classes, 20 exhibit one or more polar axes in the unit cell, making the crystal *piezoelectric*. By applying an external mechanical force along one of the polar axes as illustrated in Figure 7.1b, atoms are displaced, inducing an electric polarization and hence a charging of the crystal faces perpendicular to the polar axis. Similarly, an externally applied electrical field deforms the crystal along one of the polar axes (*inverse* or *converse piezoelectric effect*). In 10 out of the 20 piezoelectric crystal classes, a unique polar axis exists and intrinsic dipole moments in the

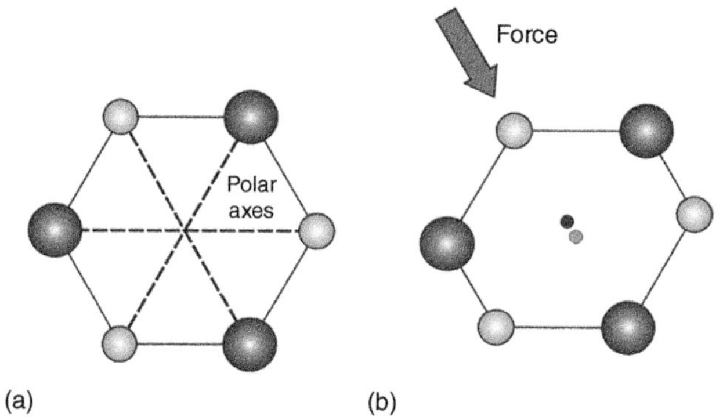

Figure 7.1 Illustration of the piezoelectric effect in a hypothetic two-dimensional, non-centrosymmetric crystal comprising two types of atoms. (a) The configuration of atoms forms three polar axes, which preclude inversion symmetry. (b) Upon application of a mechanical force along one of the polar axes, the displacement of atoms creates a polarization.

unit cell form a spontaneous polarization even in the absence of an external electric field [15]. In these crystal classes the polarization magnitude changes upon application of a temperature gradient, and hence the crystal is considered *pyroelectric*. Changes in the permanent dipole moments become detectable in variations of the surface potential until surface charges have screened and compensated the pyroelectric polarization. Among the pyroelectric materials, some crystals exhibit a permanent spontaneous polarization $\vec{P}_S$ with at least two different equilibrium states. The crystal can switch between those states if a sufficiently large external electric field is applied. The occurrence of different equilibrium states of spontaneous polarization $\vec{P}_S$ characterizes *ferroelectricity*. Thus, every ferroelectric material is also pyroelectric and piezoelectric, but not necessarily vice versa. The spontaneous polarization in ferroelectrics can originate either from ordered intrinsic electric dipole moments or from displaced atoms in the unit cells, thus creating a polar axis [14]. For example, in a tetragonal crystal (axes $c_T > a_T = b_T$; angles between axes: 90°), displaced atoms along the c_T-axis can induce ferroelectricity. Above the *Curie temperature* T_C, which is a material-specific property, the crystal transitions to a non-ferroelectric (*paraelectric*) state with high symmetry. In the example of a tetragonal crystal, a transition from the tetragonal structure to a cubic structure is observed when exceeding T_C. Similarly, upon cooling below T_C, the symmetry is reduced and modifications in the crystal structure and deformations of the unit cells induce mechanical strain in the crystal lattice and generate a polar axis. Strain produces a uniform parallel orientation of electric dipoles within a group of ferroelectric unit cells, which adds up to strong polarization fields in the crystal (in contrast to *antiferroelectrics* where the antiparallel polarization of neighboring unit cells diminishes the macroscopic effect of polarization). To compensate the electric fields from the internal polarization $\vec{P}_S$, screening charges accumulate at the crystal surface and form depolarization fields $\vec{E}_D$ [15]. The presence of at least two distinct polarization states with different orientation fosters the subdivision of ferroelectric materials into energetically more favorable domains. Such domains represent groups of unit cells with uniformly oriented polar axes. The formation of domains minimizes the energy of the system as an interplay of mechanical stress, internal polarization, and screening charges. Domain walls separate domains. Figure 7.2 illustrates different domain configurations for a cubic to tetragonal phase transition. 180°-domains with antiparallel polarization orientation between neighboring domains compensate strong depolarization fields $\vec{E}_D$. Mechanical stress originating from the phase transition or external pressure can render other domain configurations energetically more favorable. 90°-domains, where the polar axes between neighboring domains form an angle of 90°, reduce both mechanical strain and polarization fields. Hence, the observation of 90°-domains often indicates that both mechanical strain and polarization coexist and drive the crystal formation [16, 17]. Details of the interaction of mechanical strain and polarization as well as further background on the mechanisms behind domain formation and domain characteristics are discussed in the literature [18, 19].

Upon phase transition at T_C, crystals can also form states of spontaneous mechanical strain as a mechanical equivalent to ferroelectricity [20]. This *ferroelasticity* requires the existence of two strain states in the material, which are switchable

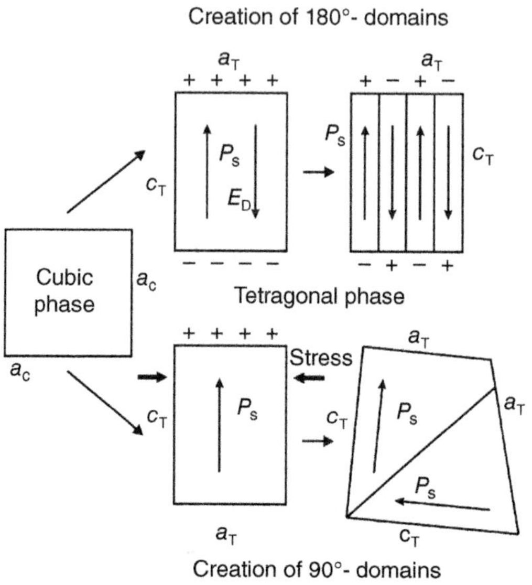

Figure 7.2 Illustration of principal ferroelectric domain formation in a tetragonal crystal. Upon cooling below the Curie temperature T_C, a non-ferroelectric cubic crystal transitions to a tetragonal ferroelectric phase by deformation, whereby two types of ferroelectric domain configurations can form. Top: In order to compensate spontaneous polarization and induced surface charges, alternating domains with antiparallel polarization orientation form, creating 180°-domains. Bottom: Mediated by mechanical stress, the crystal forms 90°-domains, partially compensating stress and polarization. Source: Damjanovic [15]. Reprinted (adapted) with permission from Damjanovic 1998, DOI: 10.1088/0034-4885/61/9/002. © IOP Publishing. Reproduced with permission. All rights reserved.

under sufficiently strong external mechanical stresses [21, 22]. These strain states are equivalent in crystal structure [23] and are induced by small displacements of atoms in the crystal [24]. Internal strain often coexists with spontaneous polarization: common ferroelectric materials exhibit ferroelastic properties, while ferroelasticity can occur without ferroelectricity. Telling apart ferroelastic and ferroelectric crystals only by measurements of their piezoelectric properties can bear some pitfalls as discussed further. Commonly, ferroelectric and purely ferroelastic (non-ferroelectric) materials can be distinguished by poling experiments, since only the internal polarization $\vec{P}_S$ of a ferroelectric material responds to an external electric field. Hence, poling events in which the polarization switches between its equilibrium states following the external electric field are considered a classical hallmark of ferroelectricity [13].

In ferroelectric insulators, poling is investigated by measurements of the characteristic polarization hysteresis as a function of the electric field as illustrated in Figure 7.3. Microscopically, the hysteresis corresponds to a reorganization of the ferroelectric domains with distinct polarization orientations, which are represented by arrows in Figure 7.3. In the unpoled state (A), domains organize to minimize the formation energy by minimization of the net polarization, strain, and screening charges. Upon application of an increasing external electric field, the polar dipoles gradually align, and the domains change their shape and position, until most of the domains exhibit uniform orientation. Macroscopically, the latter shows as a

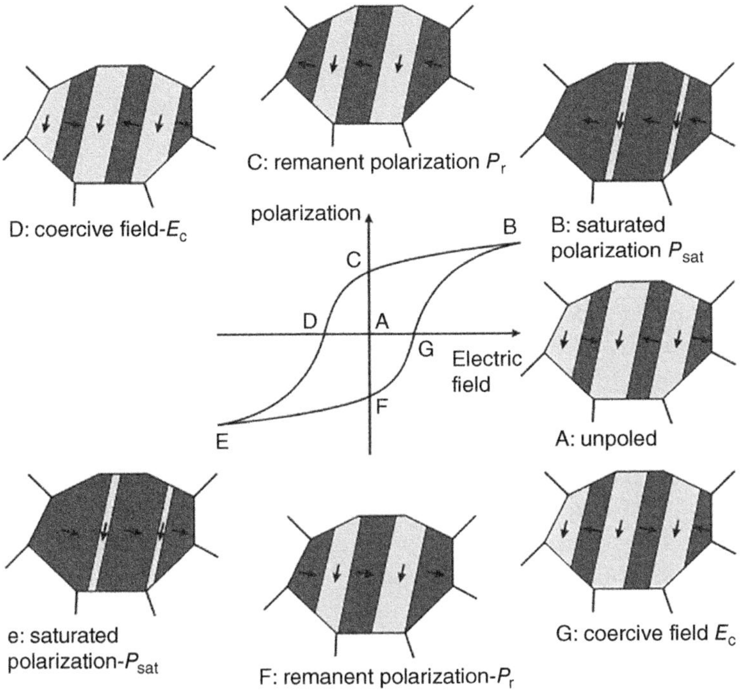

Figure 7.3 Electric poling in a ferroelectric crystal. The switching of the spontaneous polarization can be observed macroscopically in typical hysteresis curves and in modifications of domain patterns on the microscale. The macroscopic effect is connected to characteristic states of the crystal describing distinct polarization orientations, labeled A–G. The polarization aligns toward the external applied electric field, changing domain sizes and moving domain walls.

saturation of the polarization magnitude $|\vec{P}_{sat}|$ for further increasing electric fields (B). During this process of unit cell ordering, the strain in the crystal can increase, enabled by the energy provided by the electric field. Upon a subsequent gradual release of the electric field, the domains return to their original state to minimize strain. However, some preferential polarization orientation persists as *remanent polarization* $\vec{P}_R$, even if the external electric field is reduced to zero (C). In order for zero internal net polarization ($P = 0$), additional energy is required from a coercive field E_C in reverse direction (D). By increasing the reverse electric field, the same process of polarization and domain pattern alignment occurs (E), but in reverse direction, followed by a negative remanent polarization (F) under a decreasing reverse electric field, altogether generating a point-symmetric hysteresis curve. The shape of the hysteresis depends not only on the strength of spontaneous polarization, but also on sample features such as the geometry, crystal defects, mechanical stress, and temperature [15]. Although the investigation of the polarizability of the sample by hysteresis measurements is well established to prove or disprove ferroelectricity, its application to certain sample geometries such as thin films, as commonly used in modern solar cells, can become challenging. Local thickness variations can induce strongly enhanced electric fields, producing local thermal damage before any switching of polarization is accomplished. Switching becomes even more challenging if the material under investigation is a semiconductor that exhibits conductivity of photo-generated charge carriers and mobile ions, both of which can be found in modern OMH perovskites.

7.2.2 Ferroelectricity in Thin Films

Thin films are characterized by high aspect ratios with large lateral dimensions (*x*- and *y*-direction) and a small vertical dimension (*z*-direction, film thickness). This geometry reduces the symmetry of the system and leads to anisotropic stress upon changes in temperature and in particular upon phase transition, driving the formation of domains and their width within crystalline thin films [16, 25].

Thin film properties are strongly affected by the interfaces toward neighboring layers, substrate surfaces, or surrounding atmosphere. Each interface can be considered a defect of the crystal and, therefore, influences the thin film properties [26]. The binding forces between a crystalline thin film and neighboring solid materials lead to *clamping*, i.e. a transfer of stress and restriction of lattice expansion. Additional defects at the thin film surface, at interfaces, or in the bulk can modulate the microstructure of the crystal and thereby influence domain formation as well as shape and position of domain walls (*pinning*) [27].

7.2.3 Crystallography of $MAPbI_3$ Thin Films

A fundamental requirement for the presence of a spontaneous (ferroelectric) polarization is the lack of centrosymmetry [18]. At room temperature, $MAPbI_3$ has a tetragonal crystal structure. At $T_C = 327$ K and beyond, a transition occurs from the tetragonal to the cubic phase [28]. Notably, both phases can appear in the temperature regime in which solar cells usually operate [29]. The centrosymmetric cubic phase above T_C is described consistently [28, 30], and the literature also widely agrees on a tetragonal crystal structure at room temperature (below T_C) [28, 31]. However, while most reports conclude on a non-centrosymmetric nature of $MAPbI_3$ [32–34], some describe a centrosymmetric crystal structure [35].

In general, most ferroelectrics with perovskite structure exhibit high intrinsic dielectric constants, reflecting their intrinsic polarization [36]. The intrinsic polarization typically originates from the distortion of the BX_6 octahedron, i.e. displacements of the B-site cations vs. the X-site anions [37, 38]. In common ferroelectrics such as barium titanate ($BaTiO_3$) or lead zirconate-titanate ($Pb(Zr,Ti)O_3$, PZT), the displacement can be rather strong resulting in a significant polarization [39, 40]. In contrast, $MAPbI_3$ possesses only a small tetragonality with a c/a ratio close to unity ($c/a = 1.009$) [34]. In addition, neighboring PbI_6 octahedra are alternatingly tilted [33, 41], and additional distortion of the crystal lattice from the organic molecule occurs [42]. Simulations have shown that the polar moment in tetragonal $MAPbI_3$ is expected to be rather small [9], limiting its experimental accessibility. Breternitz et al. recently pointed out that the formation of twin domains may be one of the reasons for the contradicting reports [34]. This twinning may lead to seemingly high symmetry on a macroscopic scale, impeding a correct classification to symmetry elements [34]. They proposed a microscopic symmetry breaking in the $MAPbI_3$ unit cell due to the distinct shift of iodine atoms caused by the interaction with the organic molecule methylammonium (MA), which would only be visible to specific measurement techniques. Others reported that the orientation of MA, which comprises an intrinsic dipole moment, might contribute to a polarization of the unit cell [6, 42, 43].

7.3 Probing Ferroelectricity on the Microscale

The presumably small polarization coupled with other special features of hybrid perovskites, such as the intrinsic conductivity for both charge carriers and mobile ions, calls for advanced measurement techniques to probe ferroelectricity and effects thereof on the microscale. Techniques based on *atomic force microscopy (AFM)* offer a number of versatile opportunities to probe the genuine features of ferroelectricity while mostly avoiding sample damage or changes to the sample properties during the measurement process. In particular, *piezoresponse force microscopy (PFM)* is one of the most relevant characterization tools for the investigation of ferroelectricity.

7.3.1 Atomic Force Microscopy

Generally, AFM enables the nondestructive investigation of the sample topography from the nanometer to the micrometer regime by measuring the interacting forces between the tip and the sample [44, 45]. For this purpose, the tip is mounted onto a cantilever, which possesses a certain spring constant and a specific resonance frequency. When the tip approaches the sample surface, the sample produces a force onto the tip and the cantilever deforms. This deflection is monitored with a laser beam that is reflected from the cantilever onto a four-quadrant diode as illustrated in Figure 7.4. In contact mode, the tip scans the sample surface in permanent contact.

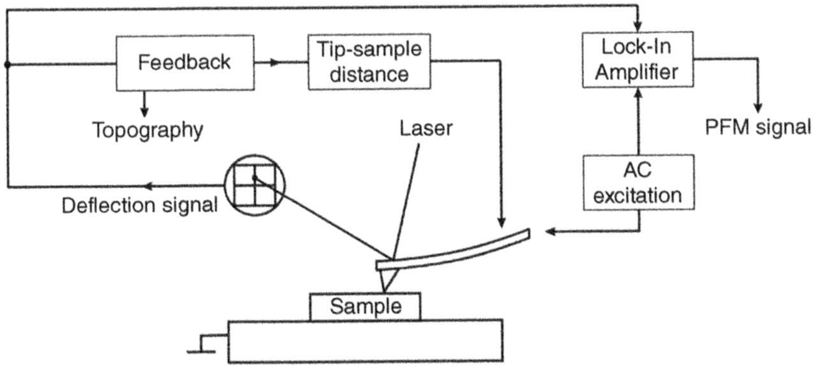

Figure 7.4 Schematic of the implementation of an AFM measurement setup with additional components for PFM. The cantilever deflection is tracked with a laser, which is monitored on a four-quadrant photodetector. Left: The control unit receives the laser signal from the photodetector, which reproduces the interaction of tip and sample, and adjusts the height of the tip for constant force onto the sample in contact mode. The topography signal is then obtained from tracing the tip adjustment. Right: To conduct PFM, an additional AC bias is applied and the piezoelectric response of the sample is measured through the tip deflection. To disentangle piezoresponse and topography, lock-in amplifiers are used.

The left side of the schematic in Figure 7.4 summarizes the processing of the information on the tip movement. In a nutshell, the displacement of the laser beam, which is monitored on the photodiode, is processed by a feedback system. This feedback system then adjusts the tip height through the tip–sample distance control to hold the interacting forces and hence a specific cantilever deflection constant [46]. The feedback information yields a topographic map for each position of the sample surface.

7.3.2 Piezoresponse Force Microscopy

Being part of the AFM family, PFM is a powerful characterization tool to explore the piezoelectric response of samples with high spatial resolution [47, 48]. In particular, it allows the characterization of ferroelectric domains with distinct polarization by their different piezoresponse on the nanometer scale [49]. In contrast to other microscopy techniques, PFM does not require any labor-intensive sample preparation such as polishing and etching [50]. PFM utilizes the inverse piezoelectric effect by application of an external AC-bias V_{AC} between the conductive AFM tip and the sample and by simultaneous measurement of the changes of the sample dimensions. As visualized in Figure 7.5, the time-dependent excitation bias $V_{AC}(t) = V_0 \cdot \sin \omega t$ and hence the electric field $\vec{E}_{AC}(t)$ follow a sine function. The ideal undelayed response of the sample to this bias shows periodic expansions and contractions following the piezoelectric strain [51].

Figure 7.5b depicts the corresponding time-dependent deformation of an individual ferroelectric domain with its polarization $\vec{P}_S$ pointing downward. While the

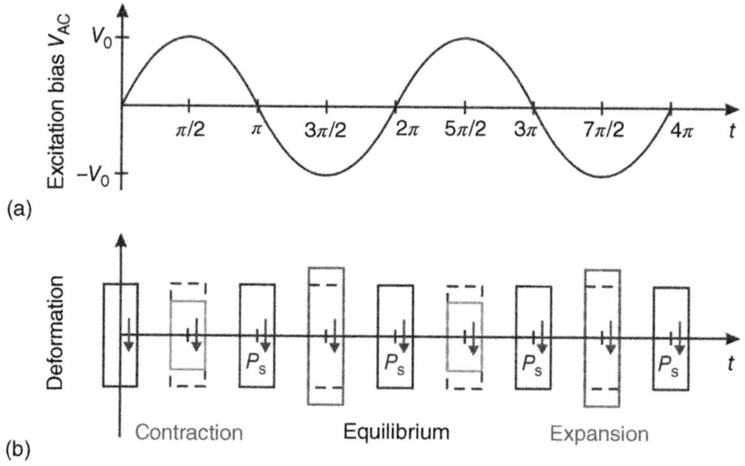

Figure 7.5 The measurement principle of PFM utilizes the response of the sample to the time-dependent AC-bias V_{AC} (inverse piezoelectric effect). (a + b) The sinusoidal V_{AC} causes periodic expansions and contractions of an individual domain with the spontaneous polarization $\vec{P}_S$ pointing downward. Antiparallel $\vec{E}_{AC}$ and $\vec{P}_S$ incur sample contraction, while parallel $\vec{E}_{AC}$ and $\vec{P}_S$ cause sample expansion relative to the equilibrium state, where no electric field is applied.

sample is in equilibrium at $V_{AC} = 0$ (for $\omega t = 0, \pi, 2\pi, \ldots$), the sample expands if $\vec{E}_{AC}(t)$ is parallel to $\vec{P}_S$ and contracts if $\vec{E}_{AC}(t)$ is antiparallel to $\vec{P}_S$. In PFM, the tip acts as both stimulator and detector. The expansions and contractions of the sample under the electric bias become detectable by deflections of the cantilever, which are generally monitored by the reflected laser on the photodiode, and hence superimpose the topography signal [52]. Thus, PFM is susceptible to topographic cross talk, which is why smooth sample surfaces help toward the interpretation of data.

Figure 7.6 depicts four principal orientations of spontaneous sample polarization and illustrates the respective displacement of the AFM tip by the piezomechanical strain induced by the applied electric field $\vec{E}_{AC}$. Figure 7.6a,b illustrate the expansion or contraction of the sample and the corresponding up- and down-deflection of tip and laser, if the polarization (or components thereof) is oriented parallel or

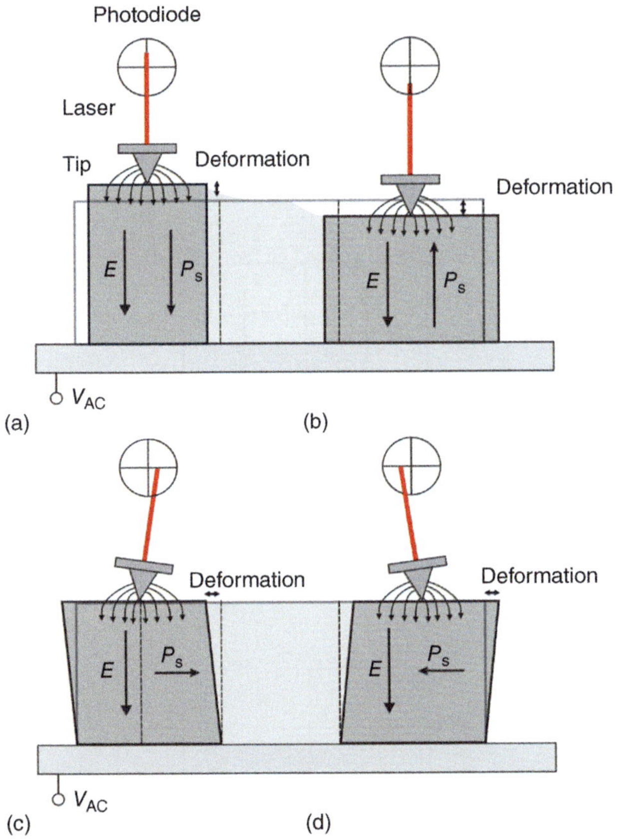

Figure 7.6 Probing the piezoelectric sample response. Depending on the relative orientation of the electric excitation field $\vec{E}_{AC}$ and the sample polarization $\vec{P}_S$, the sample expands or contracts, inducing tip deflections monitored by a laser beam on a photodiode. (a + b) If $\vec{P}_S$ is oriented out-of-plane and (anti-)parallel to $\vec{E}_{AC}$, a vertical piezoresponse force microscopy (VPFM) signal is produced. (c + d) In case of in-plane polarization with an orientation perpendicular to $\vec{E}_{AC}$, shearing of the sample causes a torsion of the cantilever and produces a lateral piezoresponse force microscopy (LPFM) signal.

antiparallel to the vertical external electric field. If the polarization (or components thereof) is oriented perpendicular to the vertical external electric field, as depicted in Figure 7.6c,d, shearing of the material becomes visible in a torsion of the cantilever, which is tracked by a left and a right deflection of the laser. Hence, deflection and torsion of the cantilever allow the differentiation of vertical and horizontal piezoresponses of the sample and thus the differentiation of vertical and horizontal ferroelectric polarization components, commonly referred to as *vertical piezoresponse force microscopy* (*VPFM*) and *lateral piezoresponse force microscopy* (*LPFM*). In order to obtain PFM data and to disentangle the deflection of the cantilever caused by the piezoelectric effect and the topography, lock-in amplifiers are used (Figure 7.4, right side) [51]. The additional lock-in amplifiers transform the piezoelectric signal from a deflection into an output voltage with amplitude A and phase φ [50]. Ideally, the PFM amplitude represents the magnitude of the internal polarization of the sample, whereas the phase shows the orientation of the polarization. This behavior is illustrated in Figure 7.7a on an exemplary ideal ferroelectric sample with vertical domains of alternating (up- and downward) polarization. Upon scanning across domain assemblies with an AFM tip, VPFM produces a constant PFM amplitude

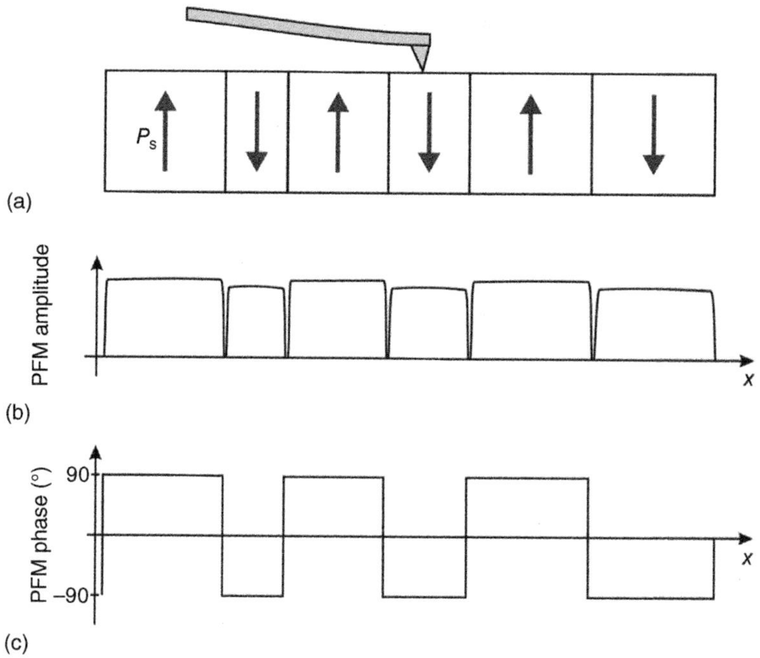

Figure 7.7 (a) Illustration of PFM measurements on a sample with alternating polarization orientation, exemplified on 180° vertical polarization (alternating up- and downward polarization). (b) Applying PFM on the sample, the domains produce a constantly high PFM amplitude, which represents the strength of spontaneous polarization. At domain boundaries where the polarization changes, the amplitude decreases. (c) The corresponding PFM phase allows conclusions on the relative orientation of the spontaneous polarization, here showing a 180° contrast between domains of different polarization.

A (Figure 7.7b), only decreasing at domain walls. The corresponding PFM phase in Figure 7.7c depends on the orientation of $\vec{P}_S$ relative to the electric excitation field and yields a phase contrast of 180° between neighboring domains. Hence, a PFM micrograph represents a map of the piezomechanical sample response and assigns a distinct PFM amplitude and phase to each pixel. The lateral resolution of the PFM image is in the nanometer regime [49, 53] and depends on the tip radius r, which is usually in the range of $r = 20$–30 nm, [50]; however, it is also affected by the (local) elastic, dielectric, and topographic properties of the sample [51, 54].

Sometimes, weak piezoelectric responses may conceal the true nature of the sample under investigation. A weak signal may originate from weak spontaneous polarization in the crystal or from a thin layer comprising only a small number of unit cells in z-direction that contribute to the signal in a thin film. In particular, the conductivity of the sample can limit the applicable AC bias magnitude V_{max} and hence the PFM contrast. Increasing the AC bias beyond V_{max} can induce reorientations of $\vec{P}_S$, i.e. ferroelectric poling, or damage the sample by electric currents [55]. To avert weak PFM signals and to enhance the signal-to-noise ratio, samples are often measured near contact resonance of the tip–sample system (Figure 7.8) [57]. The mechanically oscillating system of tip and sample in contact resonates to a certain frequency f_1 of the excitation bias (Figure 7.8a,b). If the AC frequency is constant during a PFM scan, this PFM technique is referred to as resonance-enhanced *single-frequency piezoresponse force microscopy* (sf-PFM). On the downside, sf-PFM is very susceptible to topographic cross talk, because variations of the topography can change the tip–sample system and its resonance frequency, possibly producing variations in the PFM amplification. Consequently, resonance-enhanced sf-PFM can impede both the quantitative analysis of the PFM amplitude and the assignment of polarization orientations in ferroelectric domains. This is why the measurement parameters must be selected carefully to allow the interpretation of resonance-enhanced sf-PFM data for ferroelectric features, and potential features from topographic cross talk must be considered during PFM data interpretation.

Dual AC resonance tracking piezoresponse force microscopy (DART-PFM) is an advanced PFM mode that enables a consistent resonance amplification with less cross talk from topography [56]. Figure 7.8c illustrates this tracking of the resonance frequency by dual excitation of the cantilever during the measurement. The cantilever is driven by the sum of two AC potentials with frequencies close to the resonance of the tip–sample system. The first frequency f_1 is chosen slightly below, and the second frequency f_2 is chosen slightly above the contact resonance frequency (Figure 7.8d), generating two PFM amplitudes, A_1 and A_2. A feedback system controls both driving frequencies for a constant difference $A_1 - A_2$.

7.3.3 Characterization of $MAPbI_3$ Thin Films with sf-PFM

Although the measurement principle of PFM is rather comprehensible, the interpretation of the measurement data requires a thorough analysis [54]. In contrast to most ferroelectric insulators, $MAPbI_3$ is conductive and comparatively soft. In early works on the probing of the piezoresponse of $MAPbI_3$, the samples showed strong

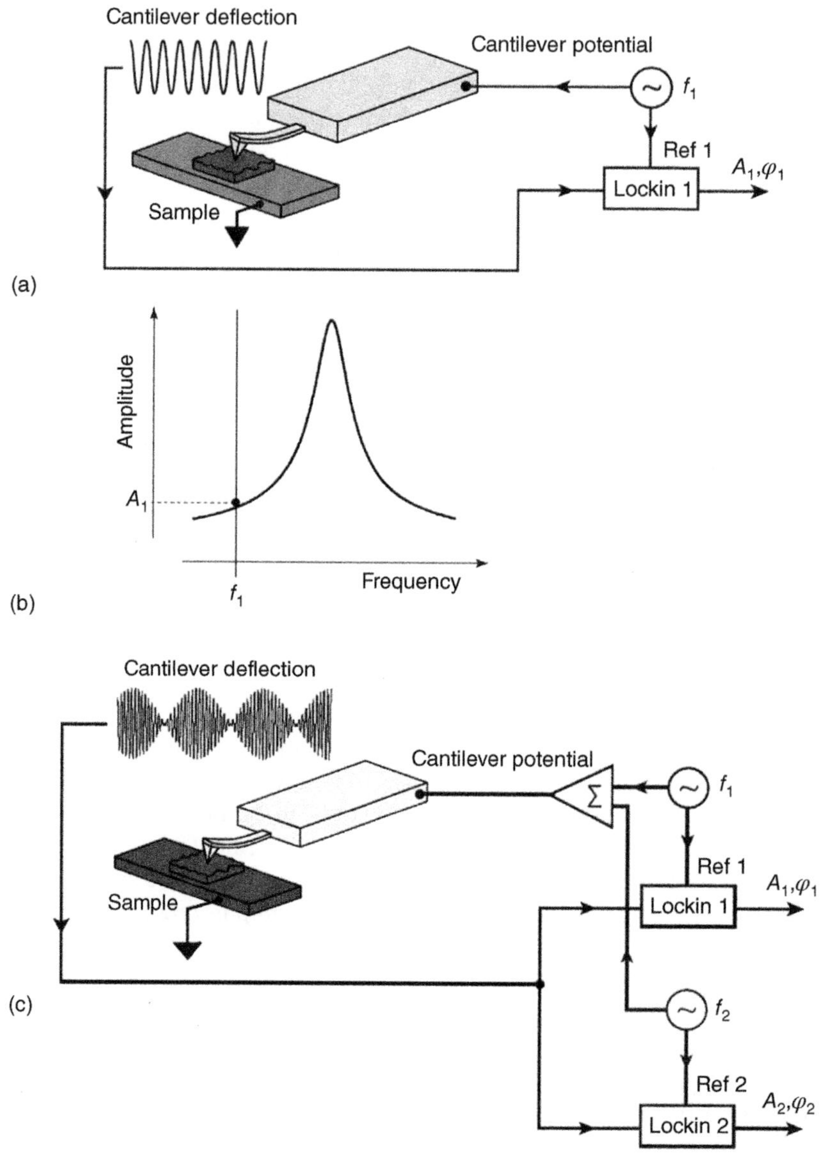

Figure 7.8 Schematic implementation of single-frequency enhanced piezoresponse force microscopy (sf-PFM) and Dual AC resonance tracking piezoresponse force microscopy (DART-PFM). (a, b) In sf-PFM, the frequency f_1 of the AC excitation bias is set close to the contact resonance of the tip–sample system, such that the PFM signal is amplified and the signal-to-noise ratio is enhanced. Yet, topography features may change the resonance conditions, producing artifacts. Source: Ch. 1. Reprinted (adapted) with permission from Oxford Instruments GmbH Asylum Research, Asylum User Guide, Chapter Rev. 621. Copyright (2010) Oxford Instruments GmbH Asylum Research. (c, d) In DART-PFM, the cantilever is excited at the sum of two frequencies f_1 and f_2, which are set below and above the contact resonance. Using a second lock-in amplifier allows tracking of the resonance, warranting a constant amplification and a decoupling of the PFM signal and the topographic cross talk. Source: Rodriguez et al. [56]. Reprinted (adapted) with permission from Rodriguez et al. 2007, DOI: 10.1088/0957-4484/18/47/475504. © IOP Publishing. Reproduced with permission. All rights reserved.

Figure 7.8 (*Continued*)

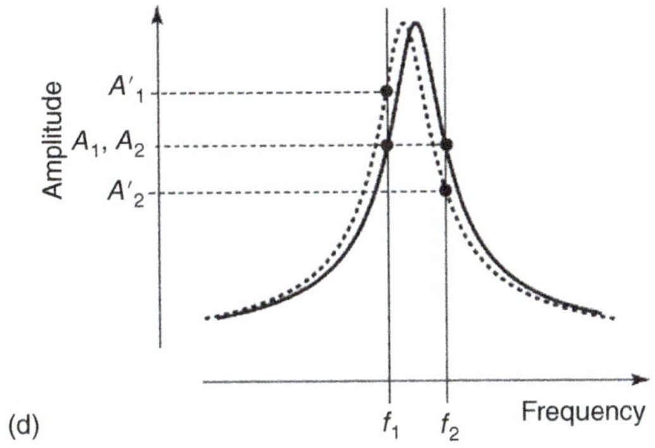

wear due to the PFM measurement, which may have originated from too strong AC biases [58, 59]. While generally an increase of excitation bias enhances the piezoresponse and thereby improves the measurement contrast, excessive electric fields can induce sample damage from electric currents and hence corrupt the measurement. Similarly, the mechanical force exerted by the AFM tip on the soft $MAPbI_3$ crystals can erode their surface. To prevent layer damage, the applied force between the tip and the sample has to be adjusted accordingly. Strelcov et al. examined various loading forces and advised a maximum force of 40 nN that can be applied during PFM measurements without damaging the polycrystalline $MAPbI_3$ thin films [60]. This is why most works on $MAPbI_3$, to date, have used tips with low spring constants of 0.1–3 N/m [61–63]. While soft cantilevers protect the sample from damage, the resonance-enhanced PFM scans are more prone to artifacts that do not stem from the piezoresponse of the sample [64].

Figure 7.9 depicts a typical result of resonance-enhanced sf-PFM on common $MAPbI_3$ thin films (thickness 300 nm). The LPFM images show large $MAPbI_3$ grains comprising domains with alternating piezoresponse and a clear 180° phase contrast between neighboring domains. The domains have a width of 90 nm and

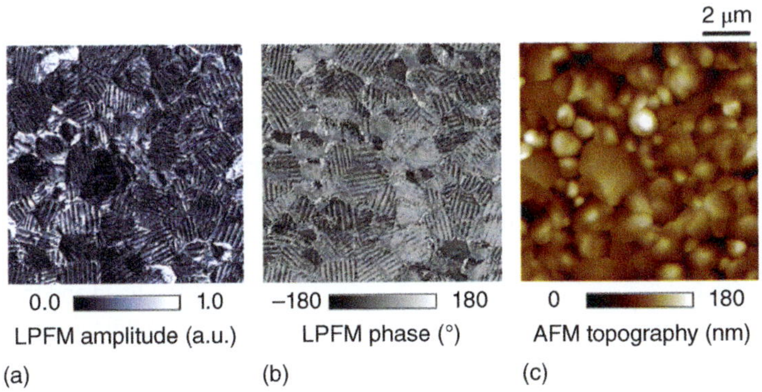

Figure 7.9 (a) The lateral sf-PFM amplitude image (LPFM) of a 300 nm thick $MAPbI_3$ thin film shows characteristic 90 nm wide domains. (b) The corresponding phase image reveals 180° contrast between neighboring domains stemming from alternatingly oriented polarization. (c) Topography of the polycrystalline $MAPbI_3$ thin film for reference.

occasionally continue at angles of 90° within grains. The 180° phase difference between domains is indicative for a change in polarity and hence of ferroelectricity. Yet, sf-PFM measurements can become susceptible for topographic cross talk and artifacts when operating close to the contact resonance of the tip–sample system [57]. According to the model of a driven damped harmonic oscillator, a 180° phase shift occurs at the resonance frequency of the tip–sample system, which coincides with the peak of the amplitude over frequency graph (Figure 7.10). Since unnoticed shifts of the resonance frequency may cause variations in the signal amplification and thus induce false amplitude and phase contrast, it is important to verify the domain contrast obtained from sf-PFM.

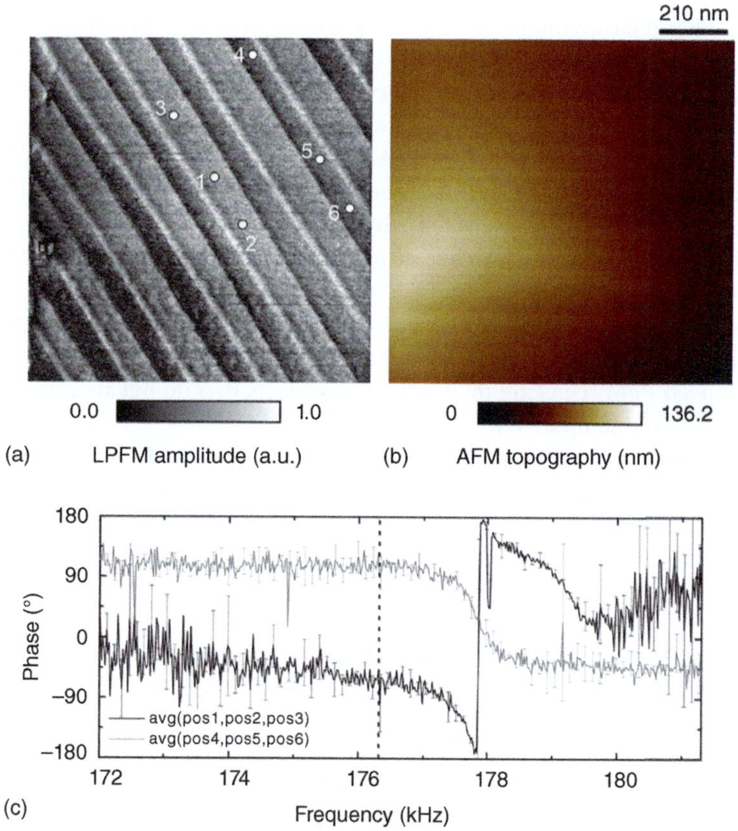

Figure 7.10 (a) Magnified lateral sf-PFM amplitude of a strictly alternating domain configuration (V_{AC} = 1.1 V). (b) Sample topography for reference. (c) Average of multiple phase responses of a high-amplitude domain (positions 1, 2, and 3) as well as the average of multiple phase responses of a low-amplitude domain (positions 4, 5, and 6). The tip–sample resonance frequency remains at 177.8 kHz at all measurement positions, hence excluding changes to the resonance frequency as the origin of the domain contrast. The dashed line indicates the frequency used to measure subfigure (a).

Figure 7.10a depicts a magnified amplitude image of domains recorded in LPFM mode on a $MAPbI_3$ thin film sample. For reference, the corresponding topography is depicted in Figure 7.10b. This sf-PFM measurement can be validated by measuring the average frequency response of different domains in the sample. The black and gray lines in Figure 7.10c are averaged frequency sweeps of three local phase scans close to the (lateral) contact resonance (here: f_{RES} = 177.8 kHz). Both curves differ by about 180° and show a distinct phase shift at the contact resonance, indicating opposite torsion of the cantilever and hence alternating polarization orientation of the domains at positions 1–3 and positions 4–6. Notably, the error bars of each curve indicate only minor deviations within the particular domain, and no significant change of the resonance frequency is observed, verifying that the 180° phase contrast between domains stems from the piezoresponse of the sample.

It is imperative to note that fabrication methods and hence sample properties often differ from lab to lab. Thus, measurement parameters cannot readily be transferred to other samples or measurement setups but must be adjusted appropriately.

In addition to misinterpretations of artifacts stemming from contact-resonance enhancement, the PFM signal can be strongly affected by various physical, electrochemical, electrostatic, and capacitive effects [65, 66]. This has led to controversial discussions on the ferroelectric nature of OMH perovskites [60, 61, 67–72]. Many effects must be considered, including piezoresponse, electric currents mediated by semiconductivity and photoconductivity, electrostatic charging, and shifting of mobile ionic charges. External electric fields that are applied in PFM measurements can drive all of these effects. In particular, ion migration and surface charging were suspected to corrupt the PFM signals, concealing the true nature of $MAPbI_3$ [69, 73].

Several groups implemented advanced PFM techniques in order to investigate the true nature of the domains in $MAPbI_3$, to minimize cross talk, to enhance the measurement resolution, and to disentangle all the effects influencing PFM contrast [68, 74].

For example, by using DART-PFM on $MAPbI_3$ thin films, Vorpahl et al. have obtained similar domain patterns as previously shown with sf-PFM, avoiding shifts of the resonance frequency and hence supporting the claim of alternating ferroelectric polarization [68]. Huang et al. compared vertical sf-PFM with vertical DART-PFM and confirmed alternating domain patterns; however, they concluded that these patterns stem from alternatingly polar and nonpolar domains [74].

Being ferroelectric, the domains disappear when heating $MAPbI_3$ to beyond T_C. Figure 7.11a depicts domain patterns in polycrystalline $MAPbI_3$ thin films that vanished upon sample heating to 65 °C (Figure 7.11b) [68]. This coincides with the phase transition of the crystal from tetragonal to cubic. In Figure 7.11c, the domain pattern reappears upon cooling below T_C. The recurring domain shape may be a result of pinning and clamping effects [75], where defects and impurities act as seeds for domain formation. These effects are also indicative of the interplay between mechanical strain and polarization, which is typical for ferroelectrics.

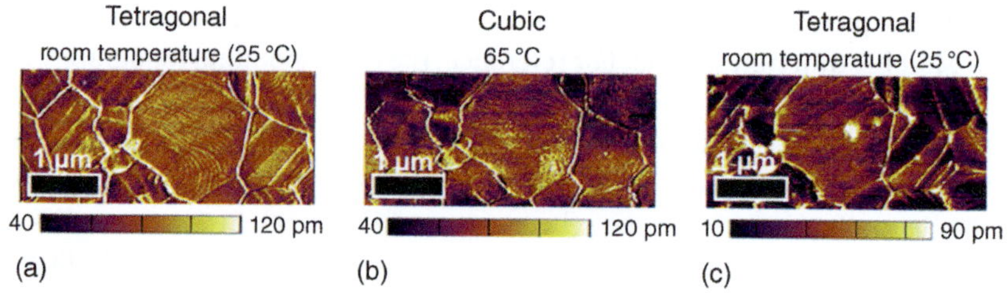

Figure 7.11 Formation and disappearance of domain patterns upon phase transition. (a) DART-PFM amplitude image of alternatingly polarized domain patterns at room temperature. (b) Upon heating above T_C, the domains vanish and the crystal transforms into a cubic structure. (c) After cooling back down to the tetragonal phase, the domains reappear almost identical in size and position, indicating pinning and clamping effects in the polycrystalline thin film. Source: Reprinted (adapted) with permission from Vorpahl et al. [68]. Copyright (2018) American Chemical Society.

7.3.4 Correlative Domain Characterization

The ferroelectric domains and their properties also show in other experiments using complementary characterization techniques, thereby completing the picture of the ferroelectric properties of $MAPbI_3$ and related microstructural features.

7.3.4.1 Transmission Electron Microscopy

Rothmann et al. investigated polycrystalline $MAPbI_3$ thin films of 300 nm thickness by *transmission electron microscopy (TEM)* [76]. The TEM image of $MAPbI_3$ in Figure 7.12 shows the same distinct domain patterns that are also observed in PFM images. In TEM, the electrons travel through the sample, interacting with the crystal lattice. Notably, not all grains exhibit strong domain contrast, but some appear rather blurred, which may stem from a tilt of the domain assembly.

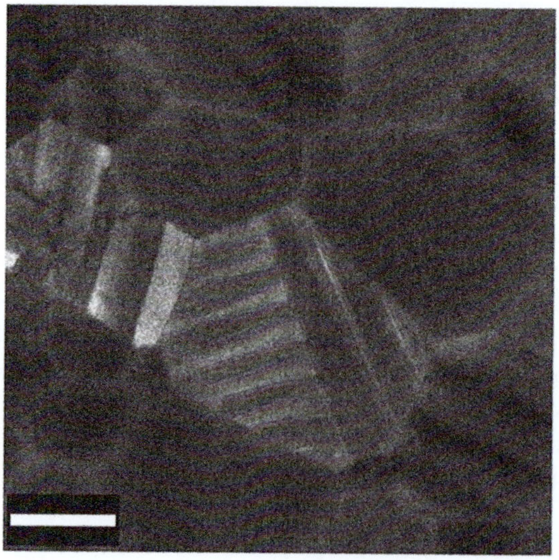

Figure 7.12 The domains of polycrystalline $MAPbI_3$ thin film samples also show in transmission electron microscopy (TEM), confirming their bulk character (scale bar, 500 nm). Source: Reprinted (adapted) with permission from Rothmann et al. [76]. Published under Creative Commons Attribution 4.0 International License (2017).

7.3.4.2 X-ray Diffraction

X-ray diffraction (XRD) is the prevalent technique to determine the crystal phase and orientation of a powder or polycrystalline sample. This measurement uses the Bragg condition for constructive interference. If the difference in path length of X-rays that are diffracted by parallel crystal planes is equal to an integer multiple of the wavelength, constructive interference occurs. Therefore, the characteristic angles at which constructive interference produces diffraction peaks correspond to the spacing of crystal planes. This lattice spacing is specific for the crystal structure and orientation. In the commonly used Bragg–Brentano measurement geometry, the X-ray source and the detector are held at a constant radius from the sample with the beam focused on the detector. Consequently, at the sample position, the beam is defocused. Simultaneous rotation of the source and the detector by θ relative to the sample plane during the measurement produces X-ray diffractograms of intensities vs. 2θ with high angular resolution, but typically results in low spatial resolution on the sample.

Jacobsson et al. put the high angular resolution to use by analyzing lattice parameter changes of $MAPbI_3$ across a range of temperatures, including T_C [77]. Remarkably, they found a large thermal expansion coefficient of $1.57 \times 10^{-4}\ K^{-1}$ and determined the phase transition temperature T_C from the cubic to the tetragonal phase to $54\,°C \pm 1\,°C$.

Another important use of XRD for the investigation of OMH perovskite thin films is the determination of the texture (preferential crystal orientation) of samples. Strong textures often correlate with high-power conversion efficiencies of perovskite solar cells [78, 79]. Figure 7.13 shows an XRD diffractogram of a $MAPbI_3$ thin film with a (110) texture. The (110) plane is parallel to the *c*-axis. Therefore, the *c*-axis of the grains and their polarization is oriented in-plane. Notably, XRD diffraction measurements provide an integrated information of the texture of the sample surface. If local resolution of the crystal orientation at grain size level is required, other diffraction techniques are better suited.

7.3.4.3 Electron Backscatter Diffraction

Electron backscatter diffraction (EBSD) can be used to gather spatially resolved crystal orientations with sub-grain resolution [81]. EBSD utilizes the beam of

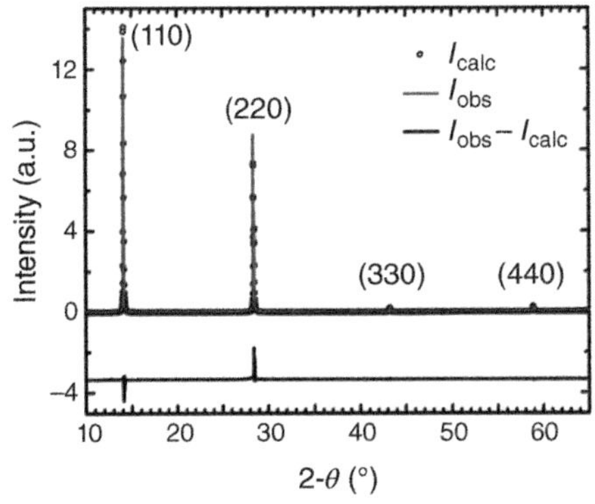

Figure 7.13 XRD diffractogram showing a strong (110) texture of a typical light-harvesting $MAPbI_3$ thin film as often employed in solar cells. Source: Reprinted (adapted) with permission from Leonhard et al., DOI: 10.1002/ente.201800989. Copyright (2019) WILEY-VCH Verlag GmbH & Co. KGaA. Leonhard et al. [80].

a scanning electron microscope to probe the crystal lattice of a sample. The interaction with the crystal lattice generates characteristic diffraction patterns that represent the orientation of the crystal lattice planes [82]. By comparing the diffraction patterns to theoretic models, a distinct crystal orientation is assigned to each sample position. EBSD can therefore produce a map of crystal orientations, which correlates with domain patterns probed by PFM measurements. Figure 7.14

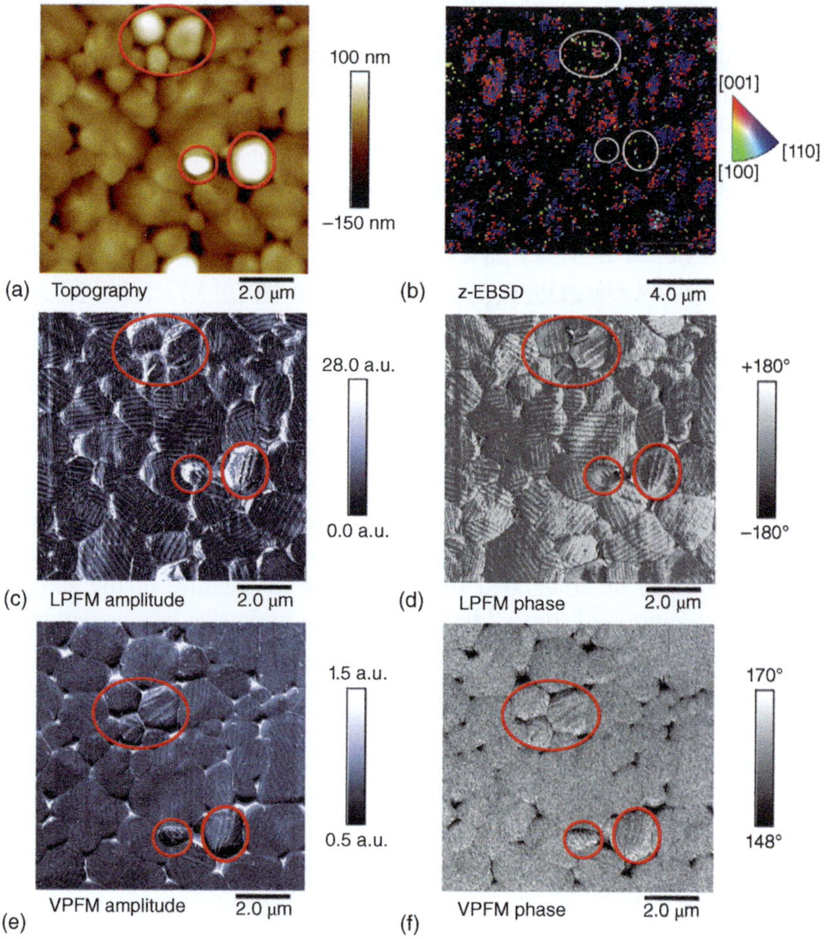

Figure 7.14 Correlative microscopy study combining PFM and EBSD. (a) In the AFM topography image of the polycrystalline $MAPbI_3$ layer, most grains appear with sizes of several micrometers and flat surfaces, while some others protrude from the layer. (b) The EBSD micrograph shows the (110) crystal orientation of flat grain surfaces with high spatial resolution. The protruding grains exhibit a different orientation indicating a tilt of the grains. (c)–(f) LPFM and VPFM amplitude and phase images. Flat grains with (110) orientation do not show any PFM contrast in vertical direction but distinct in-plane polarization. Tilted grains exhibit both in-plane and out-of-plane polarization. Source: Reprinted (adapted) with permission from Leonhard et al. [80]. Copyright (2019) WILEY-VCH Verlag GmbH & Co. KGaA.

shows a typical correlation of EBSD and PFM measurements together with the AFM topography of a polycrystalline $MAPbI_3$ sample [80]. In Figure 7.14b, the vast majority of grains exhibit a (110) crystal orientation (texture), which is in accordance with many reports on XRD measurements of $MAPbI_3$ thin films [83–85]. A (110) crystal orientation is synonymous for the *c*-axis of the tetragonal perovskite unit cell being oriented in-plane. The corresponding PFM images show matching domain patterns: on (110) oriented grains, LPFM images exhibit strong amplitude contrast and distinct phase contrast of 180° (Figure 7.14c,d). VPFM images of the same grains, as exemplified in Figure 7.14e,f, show no or only minor contrast, which may stem from cross talk by cantilever buckling. Very few grains (highlighted) exhibit a tilted orientation relative to the substrate (Figure 7.14b). Concurrently, these grains show a clear domain contrast in VPFM (Figure 7.14e,f). This anisotropy of LPFM and VPFM indicates a predominant in-plane orientation of the polarization of the majority of grains. Together, these observations allow two important conclusions: Firstly, the polarization correlates with the *c*-axis of the tetragonal crystal structure, which is typical for many ferroelectrics and which nicely matches the work by Breternitz et al. on $MAPbI_3$ [34]. Secondly, the few grains that are tilted relative to the substrate plane exhibit both in-plane and out-of-plane polarization components.

7.3.4.4 Kelvin Probe Force Microscopy

If individual grains exhibit alternatingly polarized domains with out-of-plane polarization components, the surface potential must be spatially modulated. Such a surface potential modulation can be monitored with *Kelvin probe force microscopy (KPFM)*. Figure 7.15 depicts a correlation of topography, EBSD, LPFM, VPFM, and KPFM of a representative tilted grain surrounded by grains with (110) orientation. The surrounding grains with (110) orientation consistently show 90-nm-wide and in-plane polarized domains in LPFM. The protruding grain in the center of the image has a different orientation and hence a tilted domain assembly. It shows in both LPFM and VPFM, indicating out-of-plane polarization components, and the domains are seemingly widened, which is a projection effect of the tilted domain assembly. The out-of-plane polarization components locally induce compensating surface charges (screening charges), and hence the contact potential difference (CPD) in the KPFM images is modulated, precisely resembling the domain pattern in the PFM images. These surface charges can be either free charge carriers or ions from the crystal lattice, both of which occur in $MAPbI_3$ [86].

7.3.5 Polarization Orientation

The data discussed earlier produces a consistent picture of domains with almost exclusive in-plane polarization on large, flat grains. Yet, the geometry and symmetry of the ferroelectric domain assemblies can provide further information on the polarization direction. Figure 7.16a depicts a characteristic pattern of alternating domains with an in-plane polarization and a bright contrast in LPFM amplitude.

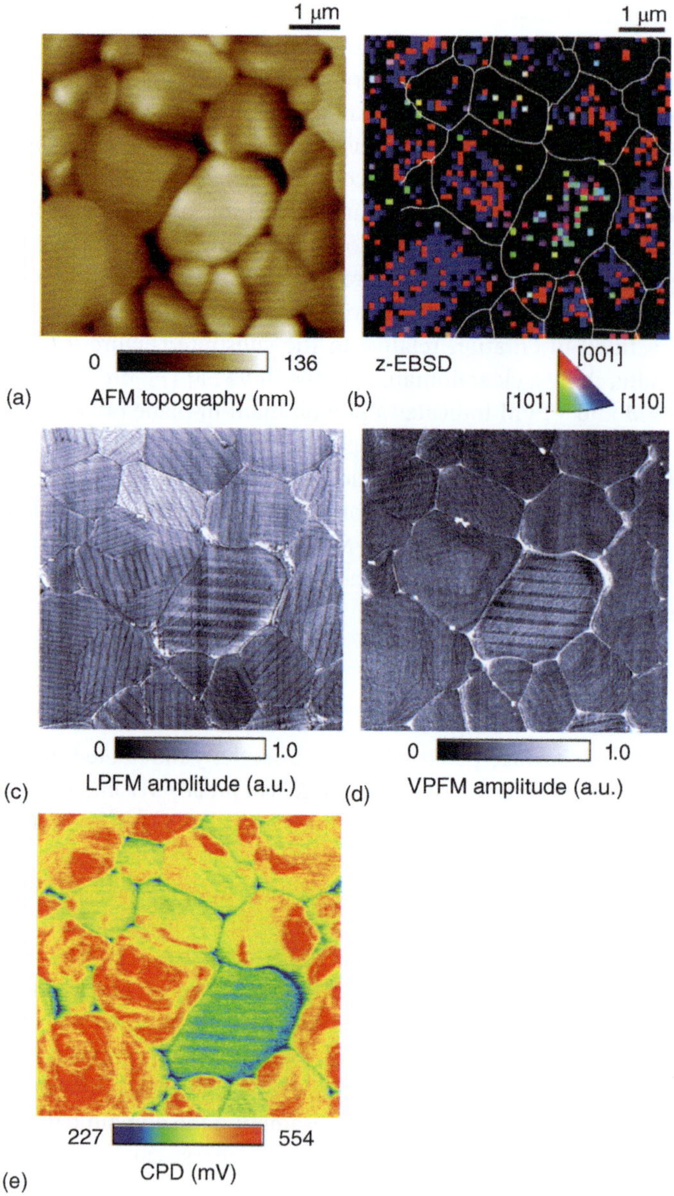

Figure 7.15 Correlative study on the grain orientation in $MAPbI_3$ thin films, the ferroelectric domain polarization, and their influence on surface charges. (a) The central grain protrudes from the surrounding flat grains with diameters of several micrometer. (b) The corresponding EBSD measurement reveals that this grain is tilted while the other grains exhibit a preferential (110) orientation. (c, d) Only the tilted grain exhibits strong contrast in both LPFM and VPFM, indicating in-plane as well as out-of-plane polarization components. At the same time, the domains appear wider due to projection effects on the surface of the tilted grain. (e) The vertical polarization component produces different compensating surface charges, which become visible in the contact potential difference (CPD) in KPFM. Source: Reprinted (adapted) with permission from Leonhard et al. [80]. Copyright (2019) WILEY-VCH Verlag GmbH & Co. KGaA.

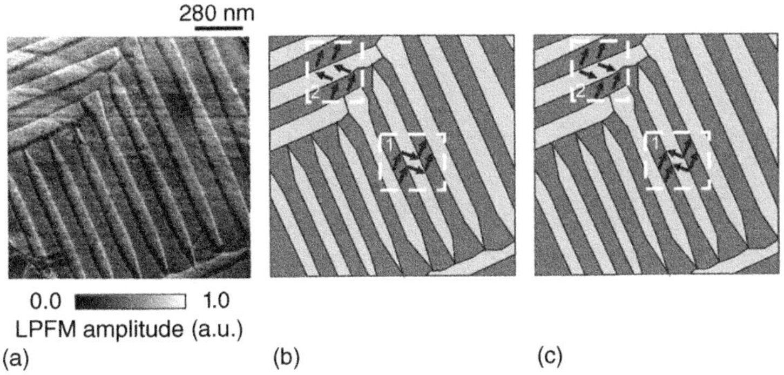

Figure 7.16 (a) LPFM micrograph of domains with alternating polarization on an individual grain of $MAPbI_3$. The in-plane polarized domains allow conclusions on the polarization orientation toward the domain boundaries. Two principal domain configurations are possible: (b) head-to-tail orientation and (c) head-to-head orientation. While head-to-tail orientations of ferroelectric polarizations across domains are generally energetically favorable, the low polarization strength of $MAPbI_3$ may allow the occurrence of both.

Occasionally, the domains continue at an angle of 90°. To be consistent throughout these alternating domains, the polarization must form a 45° angle toward the domain boundaries as illustrated in Figure 7.16b,c, hence forming 90°-domains (see Figure 7.2; not to be confused with the continuation angle of 90° of these domains). These 90°-domains cause the distinct contrast in LPFM amplitude between neighboring domains (180°-domains with antiparallel in-plane polarization would create a uniform PFM amplitude).

To date, it remains subject of further research whether these domain assemblies in $MAPbI_3$ exhibit head-to-tail or head-to-head polarization or if both configurations can coexist. Since both configurations compensate energetically unfavorable charge carrier accumulation and mechanical strain in the polycrystalline layer, they are often referred to as ferroelastic domain configurations in a ferroelectric material [15].

7.3.6 Ferroelastic Effects in $MAPbI_3$ Thin Films

Common ferroelectric materials are also ferroelastic, which is why $MAPbI_3$ is also often described and characterized by its ferroelastic properties. Mechanical stress, e.g. during thin film formation, triggers the formation of 90°-domains. Strelcov et al. pointed out that mechanical switching of domains by the loading forces of the AFM tip is unlikely to occur during scanning in contact mode, but that the tip would rather damage the layer [60].

Using DART-PFM, Strelcov et al. observed changes to the domain patterns in $MAPbI_3$ upon application of external mechanic stress [60] The topography and PFM amplitude images in Figure 7.17 show how the shape and amount of domains change in response to mechanical stress.

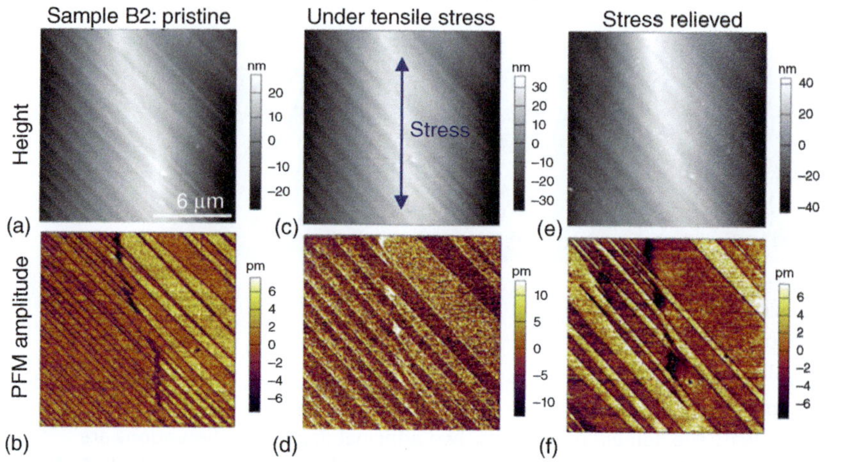

Figure 7.17 Effects of tensile mechanical stress on the domain patterns in a $MAPbI_3$ single crystal. (a, b) Both the topography and the PFM amplitude of a pristine sample exhibit alternating domains with varying position and width. (c, d) Upon tensile stress, the domain pattern changes to wider but fewer domains. (e, f) When the stress is relieved, some domains retain their shape, but also new domains form. Reprinted with permission of AAAS from [Strelcov et al. [60]]. © The Authors, some rights reserved; exclusive license American Association for the Advancement of Science. Distributed under a Creative Commons Attribution NonCommercial License 4.0 (CC BY-NC).

Early works on the ferroic properties of $MAPbI_3$ concluded that $MAPbI_3$ is purely ferroelastic (non-ferroelectric) [60, 67]. Employing PFM, Hermes et al. found domain patterns in $MAPbI_3$; since the authors did not observe clear hysteresis curves switching of the polarization orientation under an externally applied electrical field, they concluded on a ferroelastic nature of the domains [67]. The samples were visible in PFM despite their supposedly non-ferroelectric character, back then, being attributed to ion migration and charging of surfaces [87].

Jariwala et al. recently reported on the relative orientation of grains in a $MAPbI_3$ thin film by using an advanced EBSD detector to obtain high resolution [75]. In their study, they showed that local variations of the crystal orientation within individual grains of a polycrystalline thin film create local heterogeneities in crystal structure and strain. Especially at grain boundaries, the orientation of individual grains relative to their neighbors can change, which results in local strain. Although Jariwala et al. did not correlate their results to ferroelectric domains, these findings hint at certain mechanisms behind the domain pattern formation in $MAPbI_3$. The mechanical strain that occurs during thin film deposition and drying can determine the orientation of grains and the formation of domains. In light of different sample fabrication procedures in different labs, this would explain and consolidate seemingly inconsistent conclusions in the literature whether $MAPbI_3$ is ferroelectric or purely ferroelastic (non-ferroelectric) [69–72, 87]. All data discussed here support a combined ferroelectric–ferroelastic nature of $MAPbI_3$.

Although the discussions whether $MAPbI_3$ is ferroelectric or purely ferroelastic (non-ferroelectric) continue to fill the scientific literature, there is, to date, no holistic theory explaining all the microscopic observations earlier in the absence of ferroelectricity.

7.4 Ferroelectric Poling of $MAPbI_3$

The macroscopic and classical hallmark of ferroelectric materials is their polarizability in an external electric field. While most common ferroelectrics (e.g. $BaTiO_3$) are insulators, the semiconducting nature of $MAPbI_3$ fundamentally changes the methods of characterization and the interpretation of data. The poling of ferroelectric insulators typically requires an external electrical poling field that exceeds the coercive field. Readily applying strong electric fields to the semiconducting $MAPbI_3$, however, would cause significant electric currents and eventually destroy the crystal or thin film. Irrespective of this sample damage, the ionic conductivity and photoconductivity of $MAPbI_3$ would further complicate the characterization.

Ferroelectric poling requires a certain amount of activation energy provided by an electric field in order to switch the unit cells in the crystal lattice from one stable polarization state to the other. Depending on the direction of the electrical field, this transformation may enforce a deformation of the unit cell and hence induce stress and strain in the crystal.

In insulating ferroelectrics, the poling can be tracked by monitoring a current of poling charge carriers, flowing upon ferroelectric switching under a sufficiently strong electric field. In $MAPbI_3$, this ferroelectric switching current is much smaller than the ionic and electronic currents that are driven through the semiconductor by the applied bias. The ionic conductivity increases with temperature and the electronic conductivity increases with generation of free charge carriers under illumination. Unlike the electronic current, the ionic current is limited by the number of free ions or the supply of ions that can be released from the crystal lattice. Therefore, under a DC field, the ionic current must cease at some point while an alternating voltage modulates a continuous ionic current in the sample.

Since both the ionic and electronic currents can damage the $MAPbI_3$ sample during poling, a couple of mitigation strategies exist:

(1) The contribution of ionic conductivity in $MAPbI_3$ can be reduced by reducing the crystal temperature [8].
(2) Photoconductivity can be suspended by keeping the sample in the dark [88].
(3) An alternating electric field, which is applied for a limited duration, can provide the energy that is required for poling, while minimizing the electric currents and hence the thermal damage (*AC poling*) [89].
(4) Ferroelectric poling can be accomplished with electric poling fields of reduced magnitude, if the electric field is continuously applied for a longer time. This process is referred to as *creeping poling* and involves thermally activated shifting of domain walls and domains [90].
(5) In order to maximize the effect of the electric field on the ferroelectric polarization, ideally, the poling field should be aligned parallel or antiparallel to the polarization dipole of the crystal.

Some of these strategies or combinations thereof have been employed for ferroelectric poling of $MAPbI_3$. For example, following strategy 1, Rakita et al. investigated $MAPbI_3$ single crystals at low temperatures (204 K), where the tetragonal ferroelectric phase still prevails, and recorded a hysteresis curve vs. an electric poling field in the range of $\pm 200\,\text{mV}/\mu\text{m}$ [8]. Strategy 3 and a combination of strategies 2, 4, and 5 turned out to be most successful to characterize $MAPbI_3$ at room temperature, which is why they are described in more detail below.

7.4.1 AC Poling of $MAPbI_3$

The most commonly employed technique to probe a sample for ferroelectricity is the application of an AC voltage and the tracking of poling charges through the sample via a reference capacitor.

Common setups for this AC poling of ferroelectrics build on the Tower–Sawyer circuit in Figure 7.18. This circuit comprises a reference capacitance C_{ref} that limits the total amount of charge carriers, which can be driven through the sample during one cycle. By measuring the voltage of this capacitor, the total displacement of charge carriers from both ferroelectric poling and leakage currents can be calculated. In order to ensure that a significant part of the poling voltage V_{pol} is applied to the sample, C_{ref} must be chosen much larger than C_{sample}. Yet, a smaller reference capacitance allows a more accurate measurement of the total charge during one cycle, hence requiring a careful choice of C_{ref} and C_{sample}.

This measurement principle produces best results on insulator samples with a high shunt resistance R_{sample} because the majority of charge carriers flowing onto the reference capacitor must then stem from poling charges. The reliability of the measurement principle on samples with low shunt resistance, however, is limited. A low shunt resistance can originate from defects in the sample such as pinholes that interconnect both electrodes, from conductive impurities within the sample, or from a high conductivity of the sample itself. In $MAPbI_3$ thin films, all three effects can occur, but its relatively high intrinsic conductivity prevails. According to Fan et al. applying the Tower–Sawyer approach to OMH perovskites can produce sample damage and lead to a misinterpretation of data [59].

Figure 7.19a shows the polarization magnitude P of a $MAPbI_3$ thin film sample that was calculated from the charge on the reference capacitor C_{ref}, assuming a high

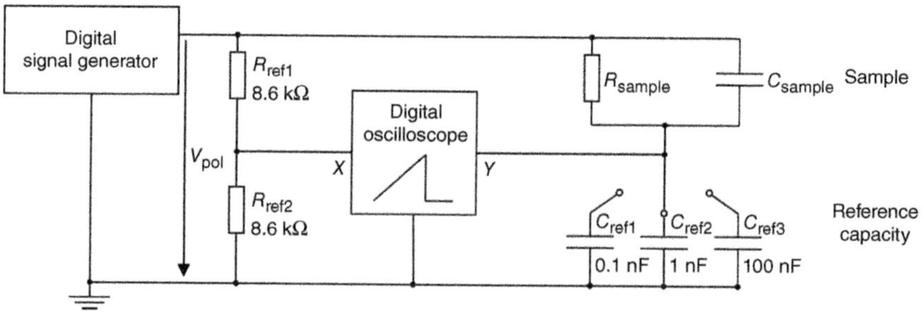

Figure 7.18 Exemplary Tower–Sawyer setup for tracking ferroelectric AC poling. Source: Röhm [91]. Licensed under CC BY SA-4.0.

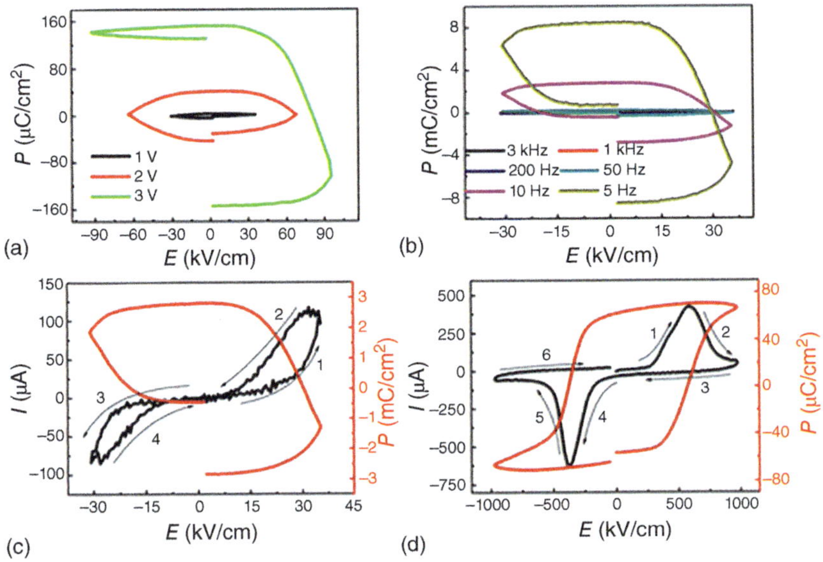

Figure 7.19 Challenges of measuring *P*–*E* hysteresis loops on a semiconducting, ionic compound such as $MAPbI_3$ using a reference capacitor. (a) While the nonlinear increase of charge carriers through the sample creates a hysteresis loop, (b) the frequency dependency reveals a rather slow charging of the reference capacitor. (c) The sample current shows a hysteresis but no saturation, so that an unadulterated poling current can be excluded. (d) For reference, the hysteresis of a ferroelectric material with a low leakage current ($BiFeO_3$) shows an increase of the current upon ferroelectric switching that decreases afterward. This is synonymous with a saturation of the polarization P_S. Source: Reprinted (adapted) with permission from Fan et al., DOI: 10.1021/acs.jpclett.5b00389. Copyright (2015) American Chemical Society. Fan et al. [59].

shunt resistance of the sample. The alternating poling voltage amplitude V_{pol} was varied between 30 and 90 kV/cm², leading to a nonlinear increase of the reference capacitor charging. The strong frequency dependence of this charging process cannot be traced to any poling processes within the sample, which would occur on a much shorter timescale, but rather to a low shunt resistance (Figure 7.19b). This is further exemplified by studying the current *I* through the sample (Figure 7.19c): At an electric field of approximately 20 kV/cm, the current increases exponentially but does not saturate (region 1), and even with decreasing electric field (region 2), the high current is retained, indicating an enhanced sample conductivity and hence a reduced shunt resistance. In contrast, a typical ferroelectric reference (Figure 7.19d) shows a poling current that declines after the switching process, hence producing a characteristic hysteretic ferroelectric polarization curve. This comparison demonstrates the ambiguities in the macroscopic current characteristics under alternating poling voltages, thus not allowing sound conclusions on the ferroelectricity or non-ferroelectricity of the semiconducting $MAPbI_3$.

7.4.2 Creeping Poling and Switching Events on the Microscopic Scale

Due to its semiconducting nature, electric poling experiments on $MAPbI_3$ require careful design. In contrast to classical insulating ferroelectrics, the non-negligible sample conductivity can produce dissipative thermal energy from the flow of charge carriers, eventually destroying the sample. The mitigation strategy of choice is *creeping poling* under small DC voltages over several minutes, which was successfully employed on oxidic perovskites in the past [90]. In creeping poling, the small device currents do not destroy the sample, while the moderate electric field can be used to change the polarization orientation of ferroelectric samples over a longer period of time. By carrying out the experiment in the dark, additional photocurrents can also be avoided for minimal device currents.

For the design of the experiment, further attention must be paid to the polarization direction of $MAPbI_3$ thin films. The (110) texture of most $MAPbI_3$ thin films implies an in-plane orientation of the *c*-axis as well as an in-plane polarization of the tetragonal unit cell. Therefore, lowest poling fields are expected if the field is applied to the sample in-plane. For reasons of symmetry, electric fields perpendicular to the crystal polarization do not lead to directional switching in a crystal comprising tetragonal unit cells.

To account for all these boundary conditions, Röhm et al. designed an elaborated poling setup, which is depicted in Figure 7.20 [88]. This sample architecture is commonly used for thin film field-effect transistors, employing a conductive (n-doped) silicon wafer with an insulating native silicon dioxide (SiO_2) layer and two gold electrodes on top, which are realized in an interdigitated structure. The two gold electrodes allow the lateral application of the poling voltage over a well-defined poling channel. At the same time, the setup enables PFM measurements before and after each poling process using the conductive silicon wafer as counter electrode. The native SiO_2 barrier avoids leakage currents between the two electrodes flowing through the conductive silicon and similarly diminishes parasitic currents through the sample during PFM measurements.

Figure 7.21a,b depict LPFM measurements on pristine $MAPbI_3$ thin films inside the lateral poling channel at two different positions, before and after applying a

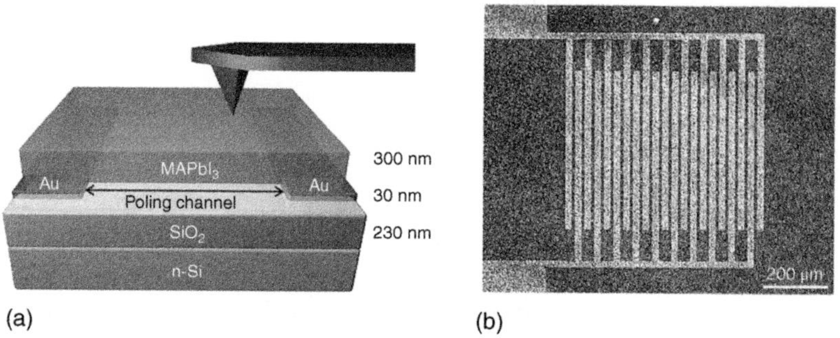

Figure 7.20 (a) Schematic device architecture for the application of lateral poling fields. Two gold electrodes form a poling channel atop an n-doped silicon wafer, on top of which the $MAPbI_3$ layer is applied. A SiO_2 layer isolates the conductive substrate from the electrodes. The conductive substrate can be used as a counter electrode for PFM measurements. (b) Light-microscopy image of the $MAPbI_3$ layer atop the interdigitated electrodes that form 10 µm wide poling channels. Source: Reproduced with permission from Dissertation Röhm [91], (CC BY-SA 4.0).

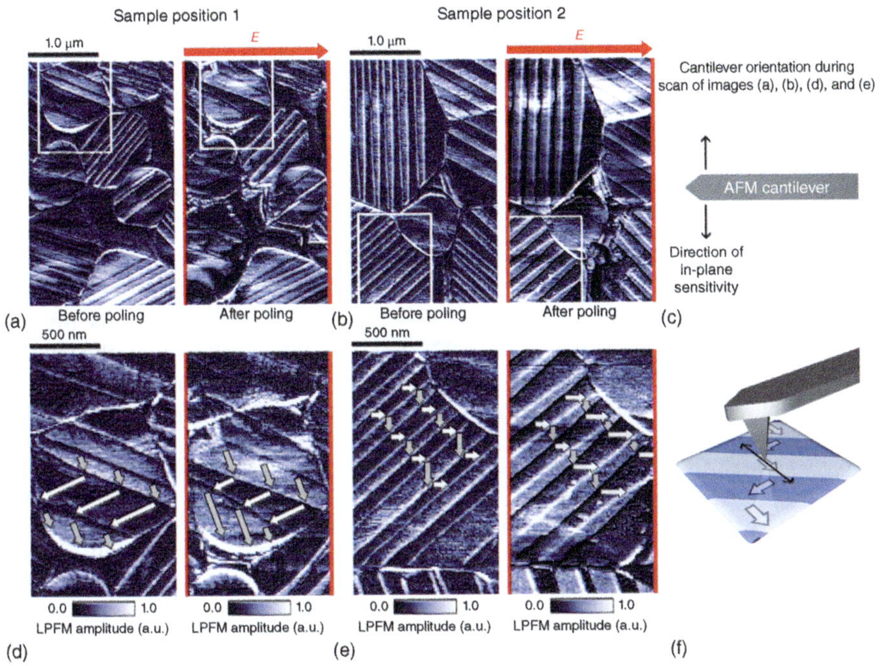

Figure 7.21 (a, b) LPFM measurements at two different positions on a $MAPbI_3$ thin film before (left) and after (right) electric poling by application of a 4.5 V/μm electric field for 11 minutes. (c) The orientation of the AFM cantilever indicates the direction of measurement sensitivity during the scan in subfigures (a) and (b). Areas of bright color represent a significant piezoresponse, where the polarization is perpendicular to the cantilever axis. (d, e) Magnified areas of subfigures (a) and (b), with arrows indicating the polarization configuration. (f) Illustration of the torsion of the AFM cantilever during LPFM measurement. Source: Reproduced with permission from Dissertation Röhm [91], (CC BY-SA 4.0).

lateral DC poling field of 2.0 V/μm for 11 minutes [88]. The red arrow indicates the direction of the electric poling field. Upon application of the poling field, some domains shrunk while others expanded. This effect corresponds to unit cells of the crystal changing their orientation to fit neighboring domains, which is energetically more favorable under the applied bias. Consequently, these energetically more favorable domains grow and domain walls are shifted, which is the hallmark of ferroelectric materials exposed to external poling fields. Similar effects can be observed when reversing the poling direction. Outside the poling channel, above the electrodes, the domain pattern remained unaffected (data not shown here) ruling out any thermal effects to trigger the shaping of domains.

The static lateral electric field is well defined within the poling channel, allowing conclusions on the reorganization of the orientation of the polarization direction within domains.

The widening and narrowing of domains depend on the orientation of the electric field (red arrow). If domains exhibit polarization components parallel to the electric field, the domains persist and expand. If domains exhibit polarization components antiparallel to the electrical field, the domains or parts thereof change their polarization orientation and hence shrink [92]. This process is illustrated in the magnified sample areas in Figure 7.21d,e. Domains with polarization components that are aligned with the electric field grow, while others shrink. From the changes of the domain dimensions, the polarization direction of domains can be derived as indicated by the gray and white arrows. This interpretation is also in accordance with the theoretic background of PFM measurements: high LPFM amplitudes can only be measured if the piezoresponse of the sample is oriented perpendicular to the AFM cantilever, which is the direction of the highest LPFM sensitivity (Figure 7.21c). Figure 7.21f depicts a schematic of the AFM tip during the LPFM measurement, which is only sensitive to a piezoresponse of the sample perpendicular to the cantilever and hence results in high LPFM amplitude (bright) while sample positions with polarization parallel to the cantilever show only low LPFM amplitude (dark).

7.4.3 Macroscopic Effects of Poling

As demonstrated in Section 7.4.2, creeping poling for several minutes with an electric field of 2 V/μm changes the arrangement of ferroelectric domains in a $MAPbI_3$ thin film. As a consequence of the dipoles having a preferential orientation, screening charges must accumulate at the grain boundaries in order to compensate the potential difference, and the corresponding built-in electrical field is expected to show in current–voltage sweeps. The sample setup depicted in Figure 7.22 can also be used to monitor the macroscopic effects of poling. Measuring the current through the poling channel while sweeping the poling field between +2 and −2 V/μm produces the *I–E* characteristic in Figure 7.22a [88]. A strong symmetric hysteresis of the current reveals a change of the sample conductivity at a characteristic electric field of E_{on} = ±1.6 V/μm. The curve is similar to the characteristics of a Schottky diode for both polarities, which could stem from energetic barriers at the grain boundaries or at the semiconductor–metal interface of the poling channel. Since E_{on} is independent of the total width of the channel (reported for 2.5, 5, and 10 μm), the former explanation appears more likely. Such a barrier for charge carriers may stem from ionic or electronic screening charges at the grain boundaries of $MAPbI_3$ grains within the poling channel (black outlines in the bottom illustration). Upon creeping poling under +2 V/μm, the current through the channel first rises to 800 nA within 10 seconds and subsequently drops to a stabilized magnitude of 120 nA within several minutes (data not shown here). Afterward, the *I–E* characteristic in Figure 7.22b changed dramatically: for negative poling fields *E*, the channel blocks charge carriers, while for positive *E*, the rising point of the current decreases to E_{on} = +0.5 V/μm. Similarly, reverse creeping poling under −2 V/μm leads to a stabilized channel current of −120 nA, but unlike poling of the pristine sample, the current starts at low amplitude and slowly rises upon poling (data not shown here). Afterward, the *I–E* curve is inverted compared

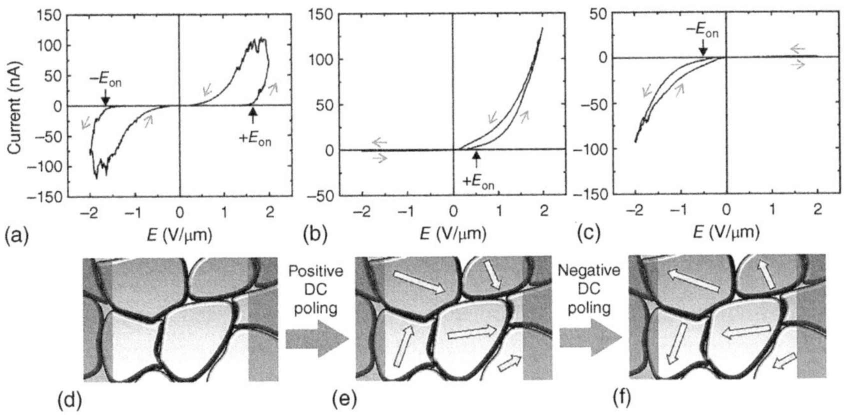

Figure 7.22 *I–E* characteristic of the $MAPbI_3$ thin film inside the poling channel as described in Figures 7.20 and 7.21. (a) Before poling, the Au/$MAPbI_3$/Au structure produces symmetric characteristics in forward and backward direction, comparable to a double Schottky diode. (b) After creeping poling for 10 minutes at 2 V/μm, the characteristic becomes asymmetric, similar to the *I–E* curve of a diode. (c) Similar creeping poling with opposite polarity inverts the characteristic. (d–f) The illustrations underneath each curve show the deductive remanent ferroelectric net polarization of the $MAPbI_3$ grains. Source: Reproduced with permission from Dissertation Röhm 2019, (CC BY-SA 4.0). doi: 10.5445/IR/1000099065 [91].

with forward poling (Figure 7.22c). This observation led to the conclusion that ferroelectric poling leads to an accumulation of screening charges at the grain boundaries and the metal-to-perovskite interface at the electrodes, resulting in an overlap of a remanent polarization and the Schottky diode behavior of the $MAPbI_3$ thin film (Figure 7.22e,f). Hence the sizeable ionic conductivity of OMH perovskites must result in intertwined effects of ferroelectricity, ionic and electronic conductivity.

7.5 Impact of Ferroelectricity on the Performance of Solar Cells

7.5.1 Pitfalls During Sample Measurements

The remarkably low electric field under which ferroelectric poling can occur in $MAPbI_3$ raises questions to what extent creeping poling can occur during common device measurements, which would then affect the electric characteristics and eventually the performance of the solar cell [93]. Nearly all *J–V* sweeps exceed the earlier reported $E_{pol} = 2\,V/\mu m$. In particular, during maximum power point (MPP) tracking or long-term stability measurements, such small electric fields are applied to solar cells for a longer period of time.

Röhm et al. applied (vertical) electric fields of ±3 V/μm to $MAPbI_3$ layers in the dark for 10 minutes, which exhibited mostly large grains with predominant in-plane polarization, sandwiched between the electrodes of solar cells [88]. Before and after the electrical stress, they performed *J–V* and MPP measurements under illumination. As shown in Figure 7.23, the electric bias produced no notable changes to the solar cell performance after poling. This observation fits very well earlier poling experiments where in-plane polarized domains could not be switched by vertical electric fields (see Section 7.4.2 for details). However, the situation may be fundamentally different, if the $MAPbI_3$ layer contains domains with out-of-plane polarization components as it occurs in layers comprising smaller grains. Then polarization effects may well overlay the photovoltaic response of the solar cell, which would require a careful analysis of the measurement data.

Other effects of unwanted ferroelectric poling are expected to occur during lifetime testing. During accelerated aging tests at elevated temperatures (typically 85 °C [29]), $MAPbI_3$ presents itself in its cubic paraelectric form. Any electric monitoring of the solar cell performance during this test may produce misleading results while the microscopic electric fields of the ferroelectric domains are absent or, at least, the data may be difficult to compare with room temperature aging. This effect may become even more drastic, if the solar cell is subdued to temperature cycling, where a frequent crossing of T_C from lower to higher and from higher to lower temperatures occurs together with permanent changes from the ferroelectric to the paraelectric state and back. Notably, the switching from the ferroelectric to the paraelectric state and back may also occur during outdoor operations or under the solar simulator, where $T_C = 55$ °C of $MAPbI_3$ can be exceeded easily.

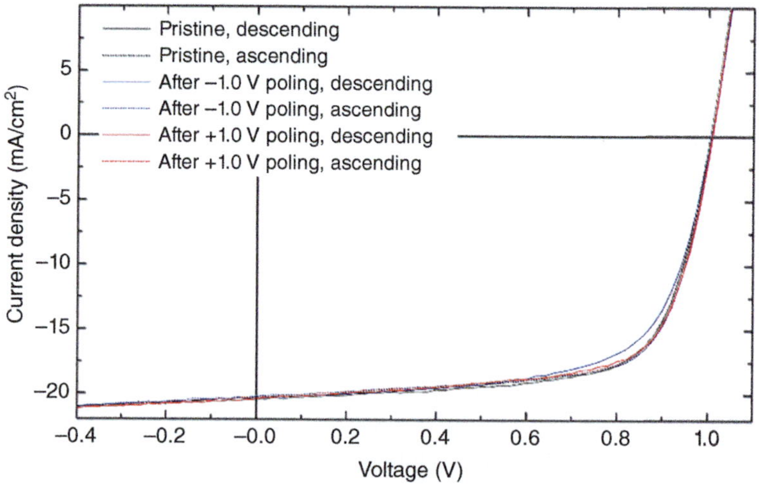

Figure 7.23 *J–V* characteristics of a typical $MAPbI_3$ solar cell with predominant in-plane polarization before and after poling for 10 minutes. Neither positive nor negative vertical electric poling fields have a notable influence on the *J–V* curve. Source: Röhm, H. (2019). Ferroelektrizität in Methylammoniumbleiiodid-Solarzellen. Karlsruhe. https://doi.org/10.5445/IR/1000099065. Licensed under CC BY SA-4.0 [91].

7.5.2 Charge Carrier Dynamics in Solar Cells

One of the most relevant loss mechanisms in hybrid perovskite solar cells is SRH recombination [94]. SRH recombination occurs at intermediate bandgap states, which stem from defects and impurities in the crystal lattice. Depending on the energetics, electrons or holes can occupy these trap states and hence increase the probability of recombination. Controlling SRH recombination is therefore one of the most important handles to enhance the performance of perovskite solar cells.

To date, no clear experimental picture exists about the charge carrier dynamics in $MAPbI_3$ under influence of ferroelectric polarization. Part of the experimental challenge lies in the convolute of many different material properties that can produce similar macroscopic effects. The fill factor, which is the most important solar cell parameter to describe charge carrier recombination, is a complex parameter. It also unfolds series and parallel device resistance, polarization, and other effects of electrical fields as well as the ideality factor – notwithstanding that more than one recombination mechanism occurs in solar cells. In addition, no experimental evidence exists whether the polarization of neighboring domains is oriented head-to-head, producing charged domain walls, or head-to-tail, producing uncharged domain walls. In early device simulations, Sherkar and Koster investigated how a ferroelectric polarization might influence the solar cell operation and if it can explain the outstanding performance of perovskites solar cells [7]. They suggested that charged domain walls may locally collect charge carriers from the bulk, driven by the electric field within the ferroelectric domains, and act as charge carrier "highways" to the electrodes. This would directly translate into low recombination losses and enhanced fill factors. However, also uncharged domain walls with head-to-tail orientation of the polarization of neighboring domains can assist the charge carrier extraction through separation of charge carriers at a potential step along the domain walls [95]. In a different set of simulations based on experimental domain patterns, Rossi et al. pointed out that only an in-plane polarization of domains would translate into improved charge carrier separation and extraction, whereas any out-of-plane polarization, which is aligned with the electric field between the electrodes, would have no effect [9]. By fitting the corresponding experimental *J*–*V* curves of solar cells, taking into account the widely accepted material properties of $MAPbI_3$, they estimated that the polarization strength of $MAPbI_3$ must be on the order of $0.2\,\mu C/cm^2$. This polarization strength is significantly lower than the polarization strength of most of the classical ferroelectric materials, more difficult to detect, and may therefore partly explain the controversy about the ferroelectric nature of $MAPbI_3$.

References

1 Stranks, S.D., Burlakov, V.M., Leijtens, T.J. et al. (2014). *Phys. Rev. Appl.* 2: 034007.
2 Green, M.A., Dunlop, E.D., Hohl-Ebinger, J. et al. (2020). *Prog. Photovolt. Res. Appl.* 28: 3–15.

3 Wehrenfennig, C., Eperon, G.E., Johnston, M.B. et al. (2014). *Adv. Mater.* 26: 1584–1589.
4 Herz, L.M. (2017). *ACS Energy Lett.* 2: 1539–1548.
5 Seidel, J., Fu, D., Yang, S.-Y. et al. (2011). *Phys. Rev. Lett.* 107: 126805.
6 Frost, J.M., Butler, K.T., and Walsh, A. (2014). *APL Mater.* 2: 081506.
7 Sherkar, T.S. and Koster, L.J.A. (2016). *Phys. Chem. Chem. Phys.* 18: 331.
8 Rakita, Y., Bar-Elli, O., Meirzadeh, E. et al. (2017). *Proc. Natl. Acad. Sci. U.S.A.* 114 (28): E5504–E5512.
9 Rossi, D., Pecchia, A., Auf der Maur, M. et al. (2018). *Nano Energy* 48: 20–26.
10 Yang, S.Y., Seidel, J., Byrnes, S.J. et al. (2010). *Nat. Nanotechnol.* 5: 143–147.
11 Nechache, R., Harnagea, C., Li, S. et al. (2015). *Nat. Photonics* 9: 61–67.
12 Fix, T. (2019). Oxide and ferroelectric solar cells. In: *Advanced Micro- and Nanomaterials for Photovoltaics* (eds. D. Ginley and T. Fix), 19–34. Elsevier.
13 Jaffe, B., Cook, W.R. Jr., and Jaffe, H. (1971). *Piezoelectric Ceramics*. London: Academic Press.
14 Gross, R. and Marx, A. (2012). *Festkörperphysik*. München: Oldenbourg.
15 Damjanovic, D. (1998). *Rep. Prog. Phys.* 61: 1267.
16 Arlt, G. (1989). *Ferroelectrics* 91: 3–7.
17 Chen, L., Ouyang, J., Ganpule, C.S. et al. (2004). *Appl. Phys. Lett.* 84: 254.
18 Lines, M.E. and Glass, A.M. (1977). *Principles and Applications of Ferroelectrics and Related Materials*. Oxford: Oxford University Press.
19 Arlt, G. (1990). *J. Mater. Sci.* 25: 2655–2666.
20 Wadhawan, V.K. (1982). *Phase Transit.* 3: 3–103.
21 Aizu, K. (1969). *J. Phys. Soc. Jpn.* 27: 387–396.
22 Salje, E.K.H. (2012). *Annu. Rev. Mater. Res.* 42 (1): 265–283.
23 Barsch, G.R. and Krumhansl, J.A. (1984). *Phys. Rev. Lett.* 53: 1069.
24 Aizu, K. (1970). *J. Phys. Soc. Jpn.* 28: 706–716.
25 Schilling, A., Adams, T.B., Bowman, R.M. et al. (2006). *Phys. Rev. B* 74: 024115.
26 Steinhausen, R., Hauke, T., Seifert, W. et al. (1998). Eleventh IEEE International Symposium on Applications of Ferroelectrics -ISAF 1998. *Proceedings of the Eleventh IEEE International Symposium on Applications of Ferroelectrics (Cat. No.98CH36245)*, pp. 93–96. Clamping of piezoelectric thin films on metallic substrates: influence on the effective piezoelectric modulus d_{33}.
27 Gruverman, A., Auciello, O., Ramesh, R., and Tokumoto, H. (1997). *Nanotechnology* 8: A38–A43.
28 Poglitsch, A. and Weber, D. (1987). *J. Chem. Phys.* 87: 11.
29 Khenkin, M.V., Katz, E.A., Abate, A. et al. (2020). *Nat. Energy* 5: 35–49.
30 Brivio, F., Frost, J.M., Skelton, J.M. et al. (2015). *Phys. Rev. B* 92: 144308.
31 Kojima, A., Teshima, K., Shirai, Y., and Miyasaka, T. (2009). *J. Am. Chem. Soc.* 131: 6050–6051.
32 Stoumpos, C.C., Malliakas, C.D., and Kanatzidis, M.G. (2013). *Inorg. Chem.* 52: 9019–9038.
33 Dang, Y., Liu, Y., Sun, Y. et al. (2015). *CrystEngComm* 17: 665.
34 Breternitz, J., Lehmann, F., Barnett, S.A. et al. (2019). *Angew. Chem. Int. Ed.*: 1433–7851.

35 Baikie, T., Fang, Y., Kadro, J.M. et al. (2013). *J. Mater. Chem. A* 1: 5628–5641.
36 Homes, C.C., Vogt, T., Shapiro, S.M. et al. (2001). *Science* 293 (5530): 673–676.
37 Megaw, H.D. (1952). *Acta Cryst.* 5: 739.
38 Haertling, G.H. (1999). *J. Am. Ceram. Soc.* 82 (4): 797–818.
39 Sharma, H.B., Sarma, H.N.K., and Mansingh, A. (1999). *J. Mater. Sci.* 34: 1385–1390.
40 Frantti, J., Fujioka, Y., Zhang, J. et al. (2014). *Rev. Sci. Instrum.* 85: 083901.
41 Whitfield, P.S., Herron, N., Guise, W.E. et al. (2016). *Sci. Rep.* 6: 35685.
42 Leguy, A.M.A., Frost, J.M., McMahon, A.P. et al. (2015). *Nat. Commun.* 6: 7124.
43 Weller, M.T., Weber, O.J., Henry, P.F. et al. (2015). *Chem. Commun.* 51: 4180.
44 Binnig, G., Quate, C.F., and Gerber, C. (1986). *Phys. Rev. Lett.* 56: 930.
45 Giessibl, F.J. (2003). *Rev. Mod. Phys.* 75: 949.
46 Voigtländer, B. (2015). *Scanning Probe Microscopy*. Berlin, Heidelberg: Springer-Verlag.
47 Güthner, P. and Dransfeld, K. (1992). *Appl. Phys. Lett.* 61: 1137.
48 Gruverman, A., Auciello, O., and Tokumoto, H. (1998). *Annu. Rev. Mater. Sci.* 28: 101–123.
49 Jungk, T., Hoffmann, A., and Soergel, E. (2008). *New J. Phys.* 10: 013019.
50 Soergel, E. (2011). *J. Phys. D: Appl. Phys.* 44: 464003.
51 Gruverman, A., Alexe, M., and Meier, D. (2019). *Nat. Commun.* 10: 1661.
52 Proksch, R. (2015). *J. Appl. Phys.* 118: 072011.
53 Denning, D., Guyonnet, J., and Rodriguez, B.J. (2016). *Int. Mater. Rev.* 61: 1.
54 Collins, L., Liu, Y., Ovchinnikova, O.S., and Proksch, R. (2019). *ACS Nano* 13 (7): 8055–8066.
55 Strelcov, E., Kim, Y., Yang, J.C. et al. (2012). *Appl. Phys. Lett.* 101: 192902.
56 Rodriguez, B.J., Callahan, C., Kalinin, S.V., and Proksch, R. (2007). *Nanotechnology* 18: 475504.
57 Jesse, S., Mirman, B., and Kalinin, S. (2006). *Appl. Phys. Lett.* 89: 022906.
58 Kutes, Y., Ye, L., Zhou, Y. et al. (2014). *J. Phys. Chem. Lett.* 5: 3335–3339.
59 Fan, Z., Xiao, J., Sun, K. et al. (2015). *J. Phys. Chem. Lett.* 6: 1155–1161.
60 Strelcov, E., Dong, Q., Li, T. et al. (2017). *Sci. Adv.* 3: e1602165.
61 Röhm, H., Leonhard, T., Hoffmann, M.J., and Colsmann, A. (2017). *Energy Environ. Sci.* 10: 950.
62 Garten, L.M., Moore, D.T., Nanayakkara, S.U. et al. (2019). *Sci. Adv.* 5 (1): eaas9311.
63 Hermes, I.M., Best, A., Winkelmann, L. et al. (2020). *Energy Environ. Sci.* 13 (11): 4168–4177.
64 Harnagea, C., Pignolet, A., Alexe, M., and Hesse, D. (2004). *Integr. Ferroelectr.* 60: 101–110.
65 Kalinin, S.V. and Bonnell, D.A. (2002). *J. Mater. Res.* 17: 5.
66 Henrichs, L.F., Bennett, J., and Bell, A.J. (2015). *Rev. Sci. Instrum.* 86: 083707.
67 Hermes, I.M., Bretschneider, S.A., Bergmann, V.W. et al. (2016). *J. Phys. Chem. C* 120: 5724–5731.
68 Vorpahl, S.M., Giridharagopal, R., Eperon, G.E. et al. (2018). *ACS Appl. Energy Mater.* 1: 1534–1539.

69 Liu, Y., Collins, L., Proksch, R. et al. (2018). *Nat. Mater.* 17: 1013–1019.
70 Schulz, A.D., Röhm, H., Leonhard, T. et al. (2019). *Nat. Mater.* 18: 1050.
71 Gomez, A., Wang, Q., Goni, A.R. et al. (2019). *Energy Environ. Sci.* 12: 2537–2547.
72 Colsmann, A., Leonhard, T., Schulz, A.D., and Röhm, H. (2020). *Energy Environ. Sci.* 13: 1888–1891.
73 Vasudevan, R.K., Balke, N., Maksymovych, P. et al. (2017). *Appl. Phys. Rev.* 4: 021302.
74 Huang, B., Kong, G., Esfahani, E.N. et al. (2018). *NPJ Quantum Mater.* 3: 30.
75 Jariwala, S., Sun, H., Adhyaksa, G.W.P. et al. (2019). *Joule* 3: 1–13.
76 Rothmann, M.U., Li, W., Zhu, Y. et al. (2017). *Nat. Commun.* 8: 14547.
77 Jacobsson, T.J., Schwan, L.J., Ottosson, M. et al. (2015). *Inorg. Chem.* 54 (22): 10678–10685.
78 Wang, S., Ma, Z., Liu, B. et al. (2018). *Sol. RRL* 2: 1800034.
79 Jang, G., Kwon, H.-C., Ma, S. et al. (2019). *Adv. Energy Mater.* 9: 1901719.
80 Leonhard, T., Schulz, A.D., Röhm, H. et al. (2019). *Energ. Technol.* 7: 1800989.
81 Muscarella, L.A., Hutter, E.M., Sanchez, S. et al. (2019). *J. Phys. Chem. Lett.* 10: 6010–6018.
82 Schwartz, A.J. (2000). *Electron Backscatter Diffraction in Materials Science*. New York: Kluwer Academic.
83 Haruyama, J., Sodeyama, K., Han, L., and Tateyama, Y. (2014). *J. Phys. Chem. Lett.* 5 (16): 2903–2909.
84 Yu, H., Wang, F., Xie, F. et al. (2014). *Adv. Funct. Mater.* 24: 7102–7108.
85 Ma, Y., Hangoma, P.M., Park, W.I. et al. (2019). *Nanoscale* 11: 170.
86 Lin, Q., Armin, A., Nagiri, R.C.R. et al. (2015). *Nat. Photonics* 9: 106–112.
87 Liu, Y., Collins, L., Belianinov, A. et al. (2018). *Appl. Phys. Lett.* 113: 072102.
88 Röhm, H., Leonhard, T., Hoffmann, M.J., and Colsmann, A. (2020). *Adv. Funct. Mater.* 30: 1908657.
89 Qiu, C., Wang, B., Zhang, N. et al. (2020). *Nature* 577: 350–354.
90 Gao, Z., Zhang, H., Liu, Y. et al. (2017). *Curr. Appl. Phys.* 17: 495.
91 Röhm, H. (2019). *Ferroelektrizität in Methylammoniumbleiiodid-Solarzellen*. Karlsruhe: https://doi.org/10.5445/IR/1000099065.
92 Daniels, J.E., Cozzan, C., Ukritnukun, S. et al. (2014). *J. Appl. Phys.* 115: 224104.
93 Colsmann, A. and Röhm, H. (2019). *J. Phys. Energy* 2: 011003.
94 Sherkar, T.S., Momblona, C., Gil-Escrig, L. et al. (2017). *ACS Energy Lett.* 2: 1214–1222.
95 Liu, S., Zheng, F., Koocher, N.Z. et al. (2015). *J. Phys. Chem. Lett.* 6: 693–699.

8

Photoluminescence Properties

Yasuhiro Yamada[1] *and Yoshihiko Kanemitsu*[2]

[1] *Chiba University, Graduate School of Science, Department of Physics, 1-33 Yayoi-cho, Inage, Chiba 263-0022, Japan*
[2] *Kyoto University, Institute for Chemical Research, Gokasho, Uji, Kyoto 611-0011, Japan*

8.1 Introduction

Many studies have been devoted to understanding the high photon-to-electricity conversion efficiency of perovskite solar cells. Since the transiently photoexcited electrons and holes dominate the photoconversion process, photoluminescence (PL) spectroscopy is one of the most suitable methods for evaluating solar cell materials and devices [1–5]. In addition, the energy conversion efficiency in solar cells depends on the external PL quantum efficiency of the semiconductor material, as derived from detailed balance calculation proposed by Shockley and Queisser [6]. Thus, accurate evaluation of both the radiative and nonradiative recombination channels is required. Hybrid perovskites demonstrate efficient band-to-band PL, suggesting large radiative recombination. In addition, time-resolved PL measurements have revealed the extremely long carrier lifetime of halide perovskites [7, 8]. The carrier lifetime recently reached >30 μs even in solution-processed perovskite thin films [9]. Long carrier lifetime means small nonradiative recombination, inducing long carrier diffusion length and small carrier recombination loss, which are the great advantage of perovskite-based devices [10, 11]. The low nonradiative recombination rate in hybrid perovskites has been discussed in terms of defect tolerance and is related to the unique band structure [12–14]. PL spectroscopy is also a powerful tool to investigate the working principles of solar cells. The charge transfer process of photoexcited carriers from the perovskite layer to the electron/hole transport layer in perovskite solar cells can be examined by employing time-resolved PL [15–18]. This chapter describes the fundamental PL properties of hybrid perovskite materials and explains the physics behind high-performance of perovskite solar cells. Examples of materials and device characterization by PL spectroscopy are then introduced.

Hybrid Perovskite Solar Cells: Characteristics and Operation,
First Edition. Edited by Hiroyuki Fujiwara.

8.2 Overview of Luminescent Properties

Luminescence is a phenomenon in which a substance absorbs energy and then emits light. Several types of luminescence are classified based on the excitation source, such as PL (photoexcitation), electroluminescence (EL) (electron-beam excitation), cathodoluminescence (CL) (electric field), mechanoluminescence (mechanical deformation), and chemiluminescence (chemical reaction). Among them, PL has been used as the most convenient evaluation tool to clarify the nature of the excited state of luminescent materials due to the high spatial and temporal controllability of light. For instance, time-resolved measurements are possible by using a pulsed laser as well as spatial-resolved spectroscopy through using a microscope.

As the simplest case describing optical absorption and PL processes, let us consider a two-level system, as shown in Figure 8.1a: optical excitation occurs from the ground state to the excited state because of light absorption, followed by optical transition with light emission (PL) from the excited state to the ground state. Here, the resonance energy of the optical absorption and PL is the same, $\hbar\omega_0$. However, even when there is no fluctuation in the energy of the two levels, the absorption and PL spectra have nonzero linewidth when the lifetime of the excited state is finite. It is known that the optical spectrum has the following Lorentz shape function.

$$f(\omega) = \frac{A}{(\omega - \omega_0)^2 + (\gamma/2)^2} \tag{8.1}$$

The shape function is a Fourier transform of the motion of an electric dipole with damped oscillation ($X(t) = A \exp(-\gamma t/2) \cos(\omega_0 t)$), where the spectral linewidth (γ) is the relaxation rate of the excited state and corresponds to the homogeneous linewidth.

Next, let us consider a case where the energies of optical absorption and PL are different (Figure 8.1b). The energy difference between light absorption and light emission is called the Stokes shift, which is often observed in organic materials because structural changes occur owing to photoexcitation.

In semiconductors, it is necessary to consider an energy band structure composed of closely spaced orbitals that are usually considered as a continuum. Consider the optical transition between the high-energy conduction band and the low-energy valence band separated by the band gap energy E_g, as shown in Figure 8.1c. When the excitation light energy exceeds E_g, electrons and holes are generated in conduction and valence bands, which rapidly fall to the conduction band bottom and valence band top by thermalization, typically within the timescale of picoseconds. The radiative recombination of electrons and holes leads to light emission (PL). Although optical absorption and PL are reverse processes, the optical absorption and PL spectra are largely different because of thermalization. At thermal equilibrium, the principle of detailed balance requires the following relation between emission and absorption spectra, which is known as van Roosbroeck–Shockley relation [19]:

$$R(\nu) = P(\nu)\varrho(\nu)\mathrm{d}\nu \tag{8.2}$$

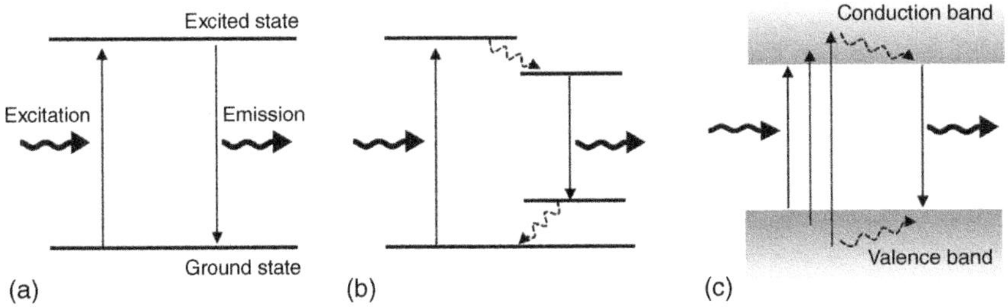

Figure 8.1 Schematic illustration for excitation (optical absorption) and emission (PL) in (a) a two-level system, (b) a four-level system displaying PL Stokes shift, and (c) a semiconductor.

where $R(\nu)$, $P(\nu)$, and $\rho(\nu)$ are the emission rate per unit volume, photon absorption rate, and density of photons with frequency ν, respectively. We can convert the photon absorption rate into the optical absorption coefficient, α, by $P(\nu) = \alpha\, c/n_r$, where c and n_r are the speed of light and refractive index. Further, $\rho(\nu)$ is given by the Planck distribution at temperature T. Then, we obtain the following relation,

$$R(\nu) = \frac{8\pi\nu^2 n_r^2}{c^2[\exp(h\nu/k_B T) - 1]}\alpha(\nu) \tag{8.3}$$

where h and k_B are Planck constant and Boltzmann constant, respectively. This equation shows that the PL efficiency is high in a strongly photo-absorbing material and that the PL spectrum emphasizes low-energy optical transition because of $(\exp\,(h\nu/k_B T)-1)$ term, which is more dominant at low temperatures. Note that this equation applies to only intrinsic semiconductors (not to heavily doped semiconductors).

In actual semiconductors, recombination without light emission, so-called nonradiative recombination, also occurs (see Figure 8.2). Nonradiative recombination usually involves midgap states induced by defects or impurities and consists of multiphonon emission processes in which bandgap energy is transferred to a number of phonons rather than emitted as light. Shockley–Read–Hall model is often used to describe the defect-mediated nonradiative recombination processes [20]. Auger recombination, which involves three carriers (electron–electron–hole or electron–hole–hole) occurring under intense photoexcitation, is also a nonradiative process; electron–hole recombination energy is used to excite another carrier to the higher state [21]. The ratio of radiative and nonradiative recombination directly determines the PL quantum efficiency and also the photoconversion efficiency of solar cells.

8.3 Room-Temperature PL Spectra of a Hybrid Perovskite Thin Film

This section describes the PL spectra of $CH_3NH_3PbX_3$ [CH_3NH_3 = MA; X = I, Br, and Cl] at room temperature. Figure 8.3 shows the room-temperature PL spectra

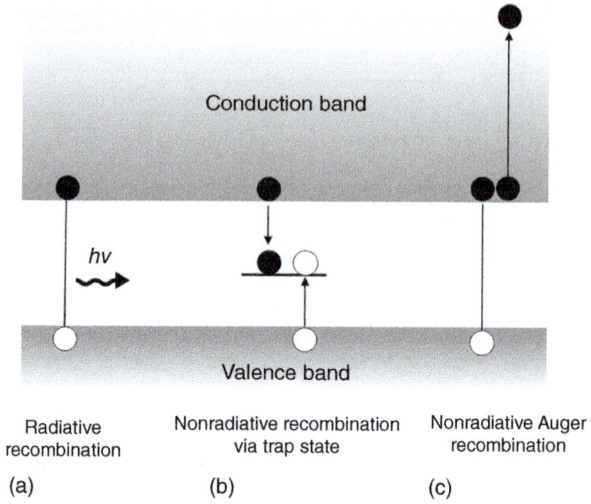

Figure 8.2 Schematic illustration for (a) radiative and (b) nonradiative recombination processes via trap state, and (c) nonradiative Auger recombination involving two electrons and a hole.

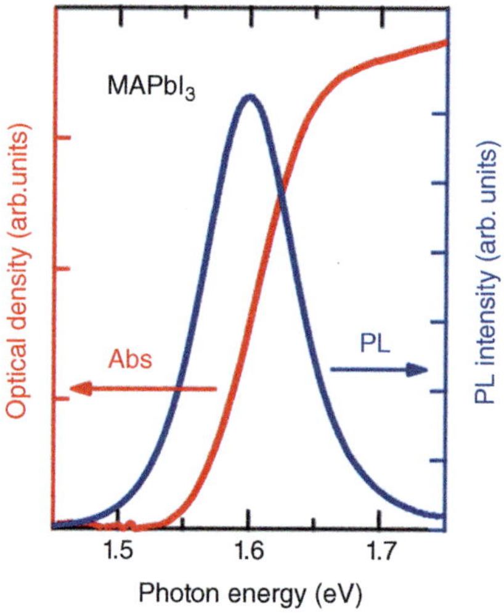

Figure 8.3 Room-temperature PL and optical absorption spectra of a $CH_3NH_3PbI_3$ thin film. Source: Kanemitsu and Handa [22]. Copyright (2018) The Japan Society of Applied Physics.

of a $CH_3NH_3PbI_3$ thin film [22]. The optical absorption spectrum is also plotted in Figure 8.3. The PL spectrum has a single peak at 1.60 eV, which coincides with the onset of optical absorption. Almost no PL Stokes shift suggests the band-to-band PL, which contrasts with organic semiconductors that typically show large Stokes shifts due to the structural changes in photoexcited states. The optical absorption spectrum shows an abrupt onset near the band gap energy (1.61 eV) with a small Urbach energy [23–25]. The absorption spectrum shape near the band edge is well explained by assuming small exciton binding energy (~5 meV) and spectral broadening [26]. The van Roosbroeck–Shockley relation (Eq. (8.2)) holds between the optical absorption and PL spectra, suggesting that the PL origin is also the band-to-band optical transition.

One of the unique features of the PL spectral shape of a hybrid perovskite is its broad linewidth (~100 meV for $MAPbI_3$). In general, the origin of spectral broadening is categorized into homogeneous and inhomogeneous broadening. Inhomogeneous broadening appears when the emission energy has a fluctuation, while homogeneous broadening is natural (lifetime) broadening and appears even when all oscillations have the same eigenfrequency. When the linewidth is inhomogeneous, the PL spectral shape varies with changing excitation photon energy within the PL emission band. As shown in Figure 8.4, the PL spectral shape is broad and independent of the excitation photon energy, meaning that the spectral broadening is homogeneous [27].

Intriguingly, PL can be observed even when the excitation energy is slightly below the band gap energy. In this case, light emission above the excitation energy (anti-Stokes PL) leads to semiconductor-based emission cooling [28, 29]. The extremely strong anti-Stokes PL occurring under low-energy excitation conditions suggests a strong electron–phonon interaction in $CH_3NH_3PbI_3$.

The PL emission energy of hybrid perovskites changes with halogen substitution, as shown in Figure 8.5a (PL peak energy $MAPbI_3$: 1.6 eV; $MAPbBr_3$: 2.3 eV; $MAPbCl_3$: 3.2 eV). All these materials display strong room-temperature PL, as shown in Figure 8.5b. The halogen substitution (I ➔ Br ➔ Cl) shifts the band gap energy to the higher energy side, which covers from the near-UV to the near-infrared spectral region [30]. While the room-temperature PL mechanism is dominated by band-to-band recombination between free electrons and holes for $MAPbI_3$ and $MAPbBr_3$, exciton PL is observed for $MAPbCl_3$ owing to the large exciton binding energy (~40 meV) exceeding thermal energy at room temperature [31].

Mixed halide perovskite $MAPbX_3$ has also been fabricated and characterized [32–34]. Figure 8.6a shows the photographs of $MAPbBr_{3-x}Cl_x$ and $MAPbBr_{3-x}I_x$

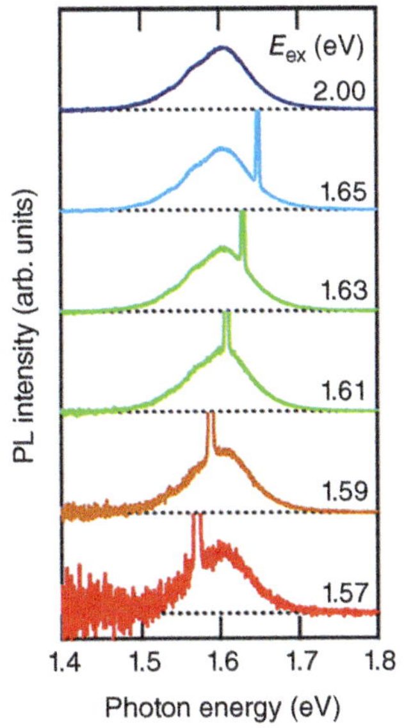

Figure 8.4 PL emission spectra of a $MAPbI_3$ thin film under resonant excitation. The excitation energies are shown in the figure. Source: Reproduced from Yamada et al. [27], with the permission of The Chemical Society of Japan.

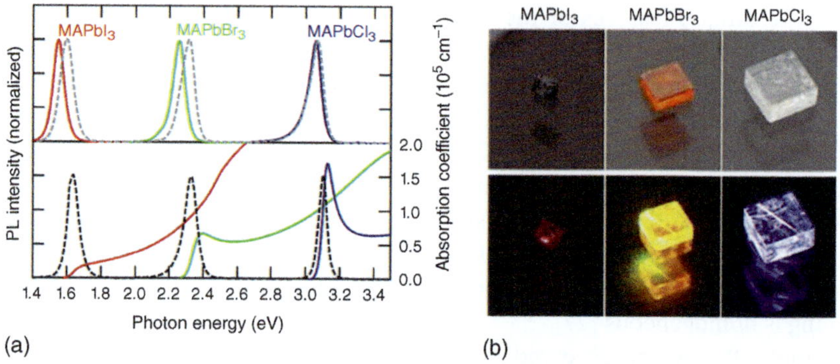

Figure 8.5 (a) PL spectra of $MAPbI_3$, $MAPbBr_3$, and $MAPbCl_3$ thin films (dotted curves) and single crystals (solid curves). Optical absorption spectra are also plotted in the lower panel. The different PL spectral shape between thin films and single crystals is attributable to the photon reabsorption effect. Source: Yamada et al. [30]. (b) Photographs of $MAPbI_3$, $MAPbBr_3$, and $MAPbCl_3$ single crystals. Source: Reproduced from Kanemitsu and Handa [22], with the permission of IOP publishing group.

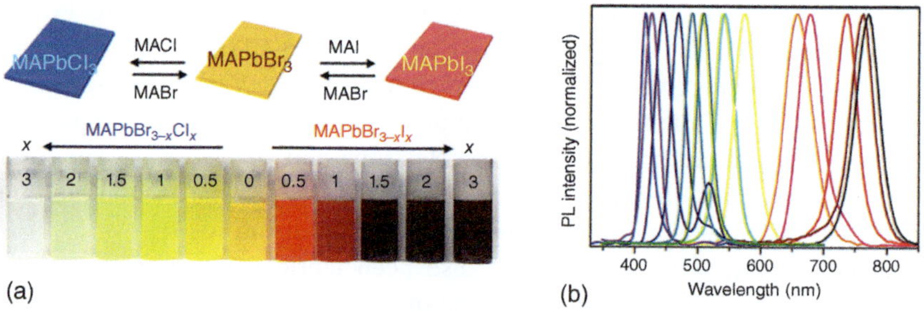

Figure 8.6 (a) Schematic illustration for the reversible anion exchange reaction of $MAPbX_3$ nanocrystals and photographs for $MAPbBr_{3-x}Cl_x$ and $MAPbBr_{3-x}I_x$ colloidal solutions. (b) Steady-state photoluminescence spectra for the same films. Source: Reprinted with permission from Ref. [32]. Copyright 2015, American Chemical Society.

nanocrystals colloidal solutions fabricated by anion exchange reaction [32]. The corresponding PL spectra are also displayed in Figure 8.6b. The PL energy can be changed continuously, covering all the visible range (400–850 nm corresponding to 1.5–3.1 eV). However, it should be noted that phase segregation occurs in mixed halide $MAPbX_3$ due to ion migration under photoexcitation, which results in the appearance of new PL band at the higher energy [35–40]. As reported by E.T. Hoke et al., an additional PL peak forms at ~1.68 eV for thin films of $MAPb(Br_xI_{1-x})_3$ and grows in intensity under continuous illumination (see Figure 8.7a) [35]. As schematically illustrated by Figure 8.7b, bromide and iodide ions are phase-segregated upon illumination into higher-bandgap Br-rich and lower-bandgap I-rich domains.

The substitution of A-site cation into $CH(NH_2)_2$ (= FA) and Cs also slightly changes the PL properties such as emission energy and linewidth; however, the PL mechanism itself is almost unchanged [41–44]. Pb^{2+} can also be replaced with Sn^{2+}, and $MASnI_3$ also shows band-to-band PL peaking at 1.25 eV, which shows redshift compared with $MAPbI_3$ [45–48]. However, tin halide perovskites currently display

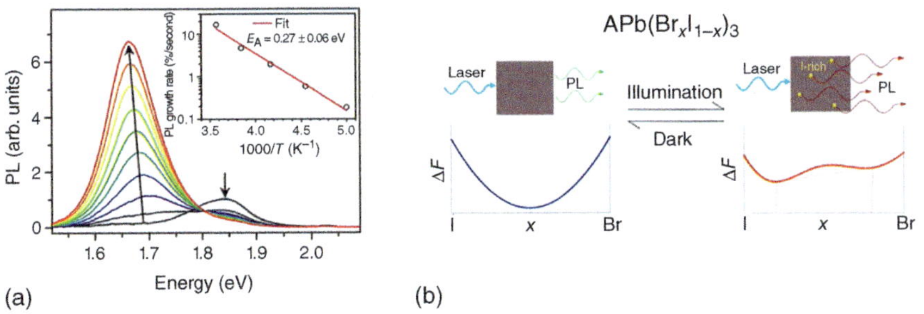

Figure 8.7 (a) PL spectra of a $MAPb(Br_{0.4}I_{0.6})_3$ thin film over 45 seconds in five seconds increments under above-bandgap excitation at 300 K. The inset shows the temperature dependence of initial PL growth rate. Source: Hoke et al. [35]. Licensed under CC BY-3.0. (b) Schematic of the proposed mechanism for photo-induced phase segregation. Source: Reprinted with permission from Ref. [36]. Copyright 2016, American Chemical Society.

Figure 8.8 PL spectra of the $MASnI_3$ thin films prepared with SnF_2 and without SnF_2. The inset shows the PL spectra normalized at the peak intensity. Source: Reprinted with permission from Ref. [49]. Copyright 2017, American Chemical Society.

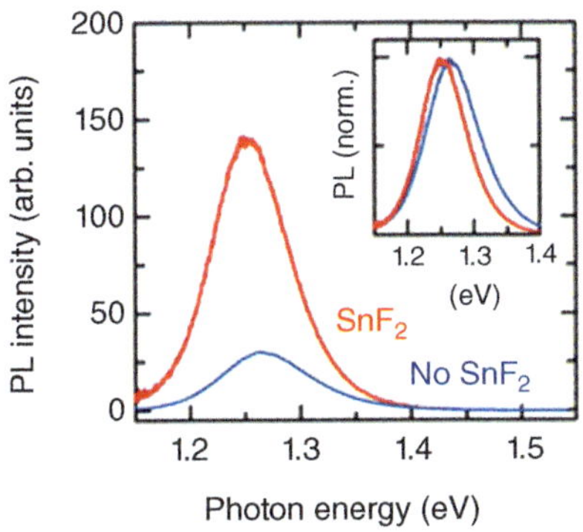

poor PL efficiency compared with lead halide perovskites. The PL intensity from $MASnI_3$ thin film is enhanced by the addition of SnF_2 [49], which significantly reduces the unintentionally doped hole density (Figure 8.8).

8.4 Time-Resolved PL of a Hybrid Perovskite

Since transiently photoexcited electrons and holes drive photoconversion processes, it is necessary to study the dynamical optical responses and reveal the optical properties of hybrid perovskites to improve the performance of perovskite-based devices. Time-resolved PL measurements provide information about electron–hole recombination processes in semiconductor materials. Here we discuss the carrier recombination processes of hybrid perovskites.

Figure 8.9 shows the time-resolved PL spectrum (two-dimensional contour image) of a solution-processed $CH_3NH_3PbI_3$ thin film obtained by a streak camera [50]. As shown in the inset, the PL spectrum shape is independent of time elapsed after pulsed laser excitation, and the PL peak appears at the onset energy of the optical absorption spectrum. The spectrum shape is independent of excitation laser fluence and photon energy.

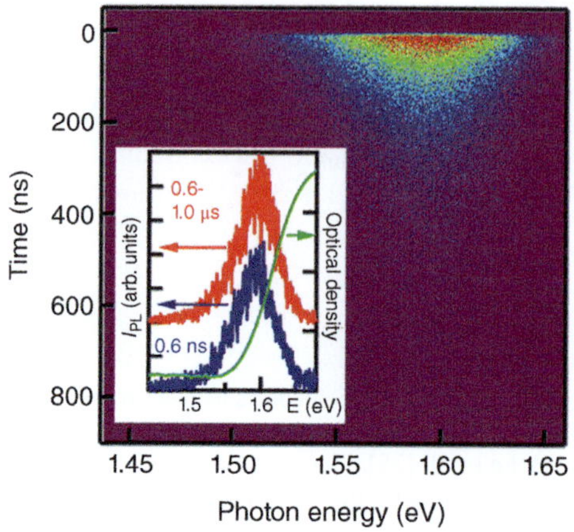

Figure 8.9 Two-dimensional contour image of time-resolved PL spectra of a $MAPbI_3$ film. The inset shows the PL spectra at different delay times. The optical density spectrum is also plotted. Source: Reprinted with permission from Ref. [50]. Copyright 2014, American Chemical Society.

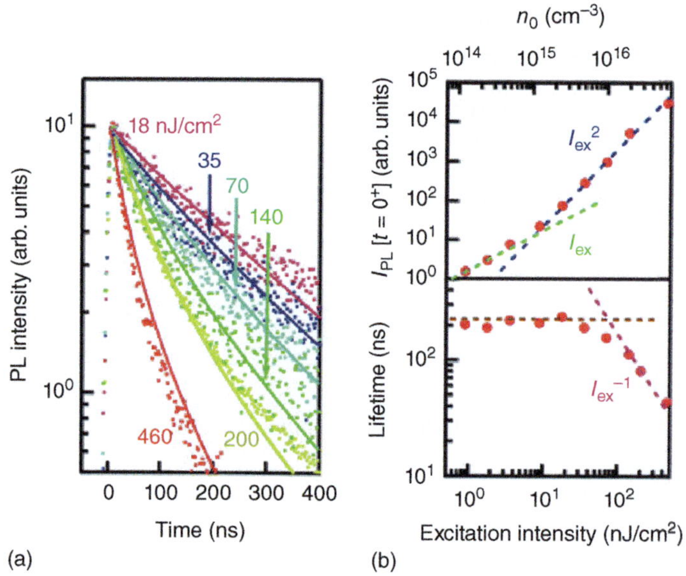

Figure 8.10 (a) PL dynamics of $CH_3NH_3PbI_3$ thin film under different excitation fluence. (b) Excitation fluence (I_{ex}) dependence of the PL intensity and effective PL lifetime. The initial photocarrier density n_0 is shown along the top axis. The dotted lines are for eye guiding. Source: Reprinted with permission from Ref. [50]. Copyright 2014, American Chemical Society.

Figure 8.10a shows the PL decay dynamics under different excitation fluence, where the PL intensities are normalized based on their maximum values [50]. Under weak excitation fluence, the PL dynamics show a slow and nearly monoexponential decay profile, while a fast and nonexponential component appears by intense photoexcitation. Figure 8.10b summarizes the effective PL lifetime and PL intensity immediately after the excitation as a function of excitation fluence. The effective PL lifetime $t_{1/e}$ is defined by $I_{PL}(t_{1/e}) = I_{PL}(0)/e$, where $I_{PL}(t)$ represents the PL decay dynamics. The PL intensity immediately after the excitation shows a quadratic dependence on the excitation intensity at above 20 nJ/cm^2, which evidences that bimolecular recombination dominates the recombination dynamics in this regime.

At excitation intensities below 20 nJ/cm^2, the PL intensity has a linear dependence on the excitation intensity. Thus, the recombination of photoexcited carriers and unintentionally doped carriers is responsible for the dominant light emission mechanism under weak excitation. The photocarrier density at the crossover point ($\sim 1 \times 10^{15}$ cm^{-3}) corresponds to the carrier density resulting from unintentional doping. Considering that the perovskite thin film is fabricated by a solution process (i.e. simple spin coating method), unintentionally doped carrier density, which nearly equals the defect density, is surprisingly low, indicating a unique feature of hybrid perovskites called defect tolerance: defects/impurities are hardly incorporated or do not form deep states that promote the nonradiative recombination processes [51, 52].

The PL lifetime is inversely proportional to the excitation fluence under intense photoexcitation, while it remains nearly the same under weak excitation, indicating that the bimolecular recombination process is dominant under strong excitation. Such power-dependent dynamics are well explained by a simple rate equation that includes single-carrier trapping and two-carrier (bimolecular) radiative recombination [50, 53–55]:

$$\frac{dn}{dt} = -A_n n - Bnp \tag{8.4a}$$

$$\frac{dp}{dt} = -A_p p - Bnp \tag{8.4b}$$

$$I_{PL} \propto Bnp + BN_n p + BN_p n \tag{8.4c}$$

Here n and p are the photoexcited electron and hole densities; A_n, A_p, and B represent the electron trapping rate, the hole trapping rate, and the bimolecular radiative recombination coefficient, respectively (see Figures 8.11a,b). N_n and N_p are the electron and hole densities caused by unintentional doping. In the case that the photogenerated electron and hole densities are always the same (i.e. $n = p$, which strictly holds just after excitation), the aforementioned equations can be simplified to

$$\frac{dn}{dt} = -An - Bn^2 \tag{8.5a}$$

$$I_{PL} \propto Bn^2 + BNn \tag{8.5b}$$

where $A = (A_n + A_p)/2$ and $N = N_n + N_p$. When the photocarrier density (n) is well above the unintentional doping (N), the effective PL lifetime can be approximately written as $t_{1/e} = 1/(A + Bn_0)$, where n_0 is the photocarrier density immediately after the excitation. This expression reproduces well the experimental results shown in Figure 8.10b.

The fact that the aforementioned equations well reproduce the PL dynamics is the evidence that free carriers, not excitons, dominate optical processes in hybrid perovskites because if excitons were dominant, I_{PL} would not increase in proportion to the square of excitation fluence. The free carrier nature of $MAPbI_3$ is consistent

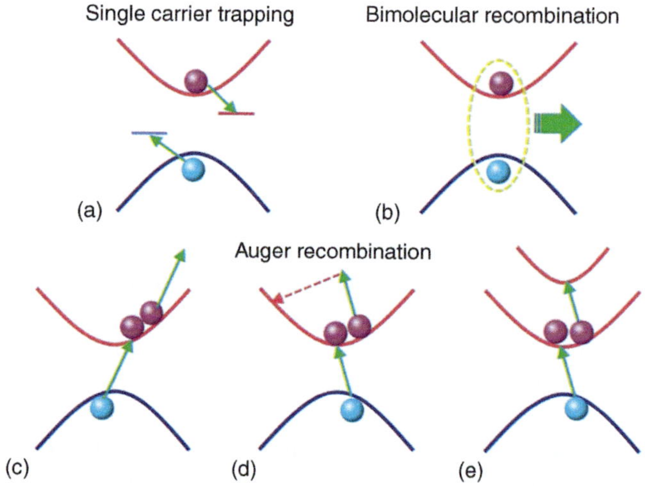

Figure 8.11 Schematic illustration of (a) single-carrier trapping, (b) bimolecular recombination, (c)–(e) three types of Auger recombination involving three-body (electron–electron–hole) processes: (c) direct Auger recombination; (d) indirect Auger recombination involving carrier-phonon interaction ; (e) interband Auger recombination, where an extra electron is excited to second higher-lying band.

with its small exciton binding energy (E_b) compared with room-temperature thermal energy ($k_BT = 26$ meV at 300 K) [23, 26].

From the global fitting of the PL dynamics, we obtain $A = 1.8\times10^7$ s^{-1} and $B = 1.7\times10^{-10}$ cm^3/s for this thin film. Note that A is sensitive to the density of defects or impurities, and thus it differs from sample to sample. On the other hand, B is an intrinsic material parameter, meaning that it is nominally independent of the sample quality. The radiative recombination coefficient B for $MAPbI_3$ is comparable to that in typical direct-gap semiconductors (e.g. GaAs: $B \approx 7.2\times10^{-10}$ cm^3/s). Such a high B coefficient is caused by the direct allowed interband optical transition. Generally, the radiative recombination coefficient (B) can be derived from the absorption coefficient, since the optical absorption and radiative recombination are reverse processes (c.f. van Roosbroeck–Shockley relation). The strong optical absorption coefficient also is consistent with the large B coefficient of hybrid perovskite.

Under much stronger excitation fluence, Auger recombination, that is, a three-body nonradiative recombination involving electron–electron–hole or electron–hole–hole processes appears [21]. In Auger recombination, excess energy generated by the recombination of electrons and hole is used for the high-energy excitation of the extra electron (hole). The high-energy electrons (holes) rapidly fall to the bottom (top) of the conduction (valence) band accompanying phonon emission. When we consider the Auger process, $C_n n^2 p$ and $C_p n p^2$ terms should be added into Eqs. (8.4a) and (8.4b), where C_n and C_p are Auger coefficient for electron–electron–hole and electron–hole–hole processes, respectively. If $n = p$ holds for any time, the rate equation is simply expressed as

$$\frac{dn}{dt} = -An - Bn^2 - Cn^3 \tag{8.6}$$

These carrier recombination processes are schematically illustrated in Figure 8.11. Similar PL dynamics have been reported in other hybrid perovskites such as $MAPbBr_3$. However, the PL dynamics of $MAPbCl_3$ should be different when excitons are dominant (i.e. the exciton binding energy is much larger than the thermal energy at room temperature) [31]. Table 8.1 shows the reported B and C coefficients for $MAPbX_3$.

Next, we focus on the PL dynamics under weak photoexcitation. When the excitation light fluence is small enough so that electron–hole radiative recombination and Auger recombination are negligible, PL lifetime corresponds to the inverse of nonradiative single-carrier trapping rate, A. Because A is determined by the density of deep-level defects and impurities, the PL lifetime under weak excitation can be a good metric for the sample quality.

In the early days of perovskite solar cells, the PL lifetime was as short as ~8 ns [14]. However, it has been rapidly increased as the defect or impurity density of samples was reduced by improving fabrication methods [7–9, 15, 16, 50, 61]; thus, the efficiency of perovskite solar cells improved as well. Currently, the carrier lifetime longer than 30 μs has been reported by time-resolved microwave conductivity [9], suggesting extremely small deep defect or impurity density.

Moreover, such long PL lifetime is quite surprising if we consider that solution-processed $CH_3NH_3PbI_3$ thin films form multigrain structure [4, 62–65]. In many semiconductors, the surface recombination rate is significantly larger compared with that in the bulk region. Thus, the grain structure should shorten the PL lifetime because of the large surface-to-volume ratio. In hybrid perovskites, a small surface recombination rate has been reported [66, 67], leading to the high device performance of perovskite thin-film solar cells.

Table 8.1 Reported B and C coefficients of $MAPbX_3$ measured by PL, pump-probe transient absorption technique (TA), time-resolved microwave spectroscopy (TRMC), and time-resolved THz spectroscopy (THz).

X	B (cm^3/s)	C (cm^6/s)	Methods	References
I	1.7×10^{-10}	—	PL and TA	[50]
	23×10^{-10}	—	TA	[56]
	0.98×10^{-10}	0.68×10^{-28}	TA	[57]
	1.5×10^{-10}	3.4×10^{-28}	TA	[55]
	13×10^{-10}	—	TRMC	[58]
	9.2×10^{-10}	1.3×10^{-28}	THz	[59]
	0.61×10^{-10}	1.6×10^{-28}	THz	[60]
$I_{1-x}Cl_x$	0.87×10^{-10}	0.99×10^{-28}	THz	[59]
Br	4.9×10^{-10}	13.5×10^{-28}	TA	[55]

Related to the long PL lifetime, hybrid perovskite shows different PL spectrum and dynamics in thin films and thick single crystals owing to strong reabsorption of emitted photons (Figure 8.5a) [30, 68–74]. Because of the efficient PL, strong band-to-band emission, small Urbach tail, and long carrier lifetime, high-energy PL is strongly absorbed, which results in the PL red shift for thick single crystals compared with the PL spectrum in thin films. Evaluation of the photon reabsorption effect is important to understand fundamental optical responses for optimal device design. This topic is discussed in the preceding Chapter 9.

8.5 PL Quantum Efficiency

Another indicator of sample quality is PL quantum efficiency. We often consider two types of quantum efficiencies. External quantum efficiency (EQE: η_{ext}) is the ratio of the number of photons emitted outside of the material to the number of photons absorbed to the material (incident photons), whereas internal quantum efficiency (IQE: η_{int}) is the probability that a photogenerated electron and hole recombine radiatively.

EQE can be experimentally obtained by measuring emission intensity and reflectance with high accuracy. The integrating sphere is usually used to collect photons emitted in all directions (Figure 8.12a). The EQE value of a solar cell is directly related to its open-circuit voltage, V_{OC} [76, 77].

$$V_{OC} = V_{OC,ideal} - \frac{k_B T}{q} |\ln \eta_{ext}| \tag{8.7}$$

Here k_B is the Boltzmann constant, T is the cell temperature, and q is the elementary charge. $V_{oc,ideal}$ accounts for the effect of incomplete absorption, thermalization, and solid angle of the solar illumination.

The aforementioned equation can be derived as follows. First, let us consider a solar cell with an external PL efficiency of 100% and ideal open-circuit under quasi-equilibrium condition. At this time, the solar excitation rate and the external light emission rate should be balanced. When the external PL efficiency is η_{ext} ($\neq$1), the quasi-equilibrium condition requires

$$R_{ext} = P_{solar} \times \eta_{ext} \tag{8.8}$$

where R_{ext} is the external PL rate from the solar cell, and P_{solar} is the solar excitation rate. Assuming the PL rate is determined by the bimolecular recombination of electrons and holes, R_{ext} should depend on the product of electron and hole densities, np, that is,

$$np \propto \exp\left(\frac{qV_{OC}}{k_B T}\right) \tag{8.9}$$

By combining this equation with Eq. (8.8), we obtain Eq. (8.7).

The aforementioned equation means that a semiconductor having a high PL efficiency is promising as a solar cell. High PL efficiencies of hybrid perovskites in thin

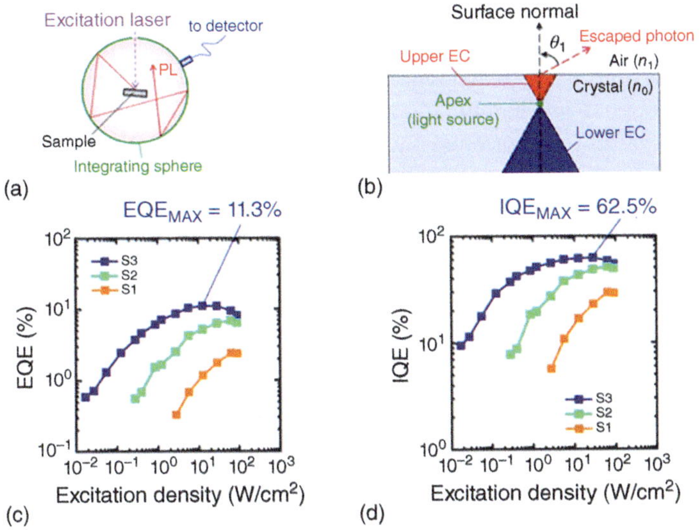

Figure 8.12 (a) Typical experimental setup for measuring PLQE. (b) Schematic illustration of escaping cone. (c), (d) EQE and IQE of PL for $MAPbBr_3$ single crystals fabricated under different growth conditions as a function of excitation fluence. S1, S2, and S3 represent the crystals grown with the MABr concentration of 0.80, 1.00, and 1.25 mol/l, respectively. Source: Reproduced from Ref. [75]. Copyright 2019, AIP Publishing.

films, quantum dots, and crystalline forms have been reported [70–72, 78–82] and are essential for the high performance of perovskite solar cells.

While we can experimentally evaluate EQE, IQE is more appropriate to discuss the optical processes and physics of light emission. For fluorescence particles and chemical molecules, the EQE value is usually close to the value of IQE owing to the large Stokes shift between absorption and emission energies. However, for solid-state light emitters (especially direct band gap semiconductors), EQE is a complicated function of both IQE and light extraction efficiency, because PL photons are partially reabsorbed by the material. As schematically shown in Figure 8.12b, only the PL photons inside the escaping cone can be emitted from the crystal because of the total reflection. Because of these effects, the relation $\eta_{int} \gg \eta_{ext}$ is usually satisfied. The IQE value is determined by a balance between the radiative and nonradiative recombination rates and thus is written by

$$\eta_{int} = \frac{\frac{1}{\tau_{rad}}}{\frac{1}{\tau_{rad}} + \frac{1}{\tau_{nonrad}}} \tag{8.10}$$

where τ_{rad} and τ_{nonrad} are the radiative and nonradiative recombination lifetimes, respectively.

Figures 8.12c,d show the EQE and IQE values of a $MAPbBr_3$ single crystal as a function of excitation intensity measured by omnidirectional PL spectroscopy [75]. Although the PL efficiency changes by fabrication conditions, it increases with excitation intensity under weak excitation. Several crystals were fabricated with different molar ratio of MABr and $PbBr_2$ in the solution for crystal growth. Samples S1, S2, and S3 represent the crystals grown with the MABr concentration of 0.80, 1.00, and 1.25 mol/l, respectively. Here $PbBr_2$ concentration is fixed at 1.00 mol/l. For sample 3 in Figure 8.12d, the maximum IQE reaches 62.5%. The IQE elevation is attributable to the saturation of nonradiative recombination centers. However, in the case of bulk $MAPbBr_3$ crystals, PL lifetime is shortened under strong excitation [69], suggesting that the acceleration of the radiative process mainly occurs in bulk $MAPbBr_3$. Further increment of excitation intensity induces the depression of IQE. A slight PL blue shift was simultaneously observed under intense photoexcitation and can be explained by the band-filling effect [75].

The recombination model quantitatively explains the excitation-intensity dependence of IQE by Eq. (8.6). According to Eqs. (8.6) and (8.10), theoretical IQE (η_{calc}) can be calculated as

$$\eta_{calc}(n_0) = \frac{Bn_0}{A + Bn_0 + Cn_0^2} \tag{8.11}$$

where n_0 is the carrier density at equilibrium under continuous-wave excitation. According to this equation, IQE has a maximum at a certain value, which is consistent with the experimental result [75].

Note that the IQE is significantly enhanced by changing the MABr concentrations in the feed solution as shown in Figure 8.13d, indicating that nonradiative recombination centers correlating with the cation vacancies dominate the nonradiative recombination processes.

8.6 Temperature-Dependent PL

Temperature-dependent PL properties (efficiency, emission energy, linewidth) provide vital information to understand the physical mechanism of the optical response in semiconductors. In many semiconductors, band gap energy (and also PL energy) monotonically decreases with temperature (Varshni's law). However, hybrid halide perovskites display drastically different behavior. One crucial feature is the phase transition. $MAPbX_3$ perovskites undergo cubic-to-tetragonal and tetragonal-to-orthorhombic phase transitions with a decrease in temperature [84, 85]. Figure 8.13 shows the PL spectra of $MAPbI_3$ and $MAPbBr_3$ bulk single crystals measured at various temperatures [83]. The experiments were performed with an objective lens, and only the PL from the near-surface region was collected to minimize the photon reabsorption effect. Lower panel ([c] and [d]) summarizes the temperature dependence of PL peak energy and linewidth. The PL peak positions shift with temperature because of the temperature dependence of the band gap energy. PL energy displays significant "jump" at the tetragonal-to-orthorhombic phase transition temperature. In contrast to the conventional semiconductors whose band gap energy obeys Varshni's law, the PL energy of hybrid perovskites increases with temperature owing to the unique band structure of lead halide perovskites: both conduction and valence bands consist of antibonding orbitals [86].

Unlike the case of PL energy, PL linewidth shows almost no change at the phase transition temperature. PL linewidth is determined by electron–phonon interaction

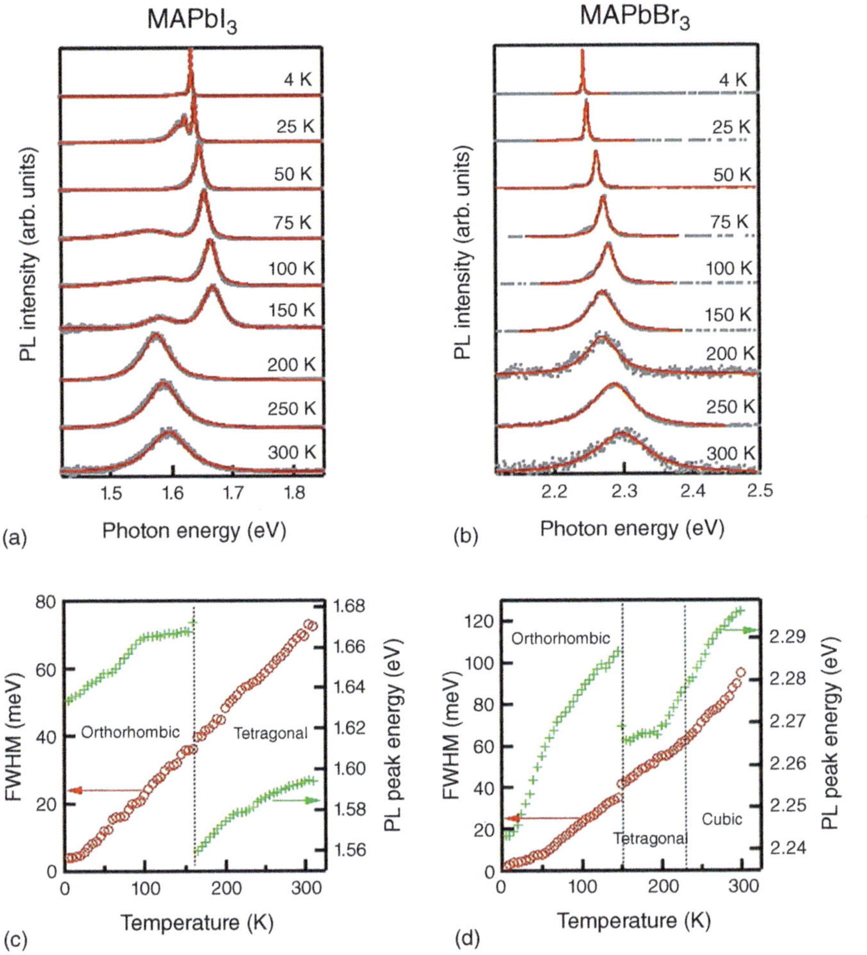

Figure 8.13 PL spectra of (a) $MAPbI_3$ and (b) $MAPbBr_3$ single crystals at various temperatures from 4 to 300 K obtained by PL microscopy. Solid curves represent the fitting results by pseudo-Voigt functions. Temperature dependence of PL energy and linewidth of (c) $MAPbI_3$ and (d) $MAPbBr_3$. Dotted lines indicate the phase transition temperatures. Source: Data Reproduced from [83].

and extrinsic inhomogeneous line broadening. A previous study suggests that the spectral broadening of the exciton is governed by the Fröhlich interaction rather than by charged impurity scattering [87]. The spectral width of the free exciton can be expressed as

$$\Gamma(T) = \Gamma_0 + \frac{\gamma_{LO}}{\exp(E_{LO}/k_B T) - 1} + \gamma_{ac} T \tag{8.12}$$

where Γ_0 is the temperature-independent spectral width, E_{LO} is the phonon energy, γ_{LO} and γ_{ac} are the coupling strengths for LO phonons and acoustic phonons, respectively.

PL linewidths are governed by acoustic phonon interactions below the threshold temperature $T = E_{LO}/k_B$ and by the Fröhlich interaction above this temperature if Γ_0 is negligible.

8.7 Material and Device Characterization by PL Spectroscopy

As demonstrated in Sections 8.2 and 8.3, PL spectroscopy is a powerful and convenient tool to evaluate sample quality by measuring nonradiative decay lifetime. This section provides some examples of material and device characterization using PL spectroscopy.

8.7.1 Degradation and Healing of Hybrid Perovskites

One of the critical issues of device application in hybrid perovskites is their chemical stability. It has been reported that degradation quickly occurs because of exposure to water vapor or photoexcitation, which would be related to the ion migration in hybrid perovskites [88–91]. Conversely, ion migration induces spontaneous defect annihilation or self-healing [92, 93]. Because PL is sensitive to the defects or impurities that promote the nonradiative recombination, we can reveal the degradation and self-healing effects by PL spectroscopy.

Figure 8.14a shows the typical PL decay dynamics of a $CH_3NH_3PbI_3$ thin film under weak excitation where the bimolecular recombination rate (Bn in Eq. (8.5a))

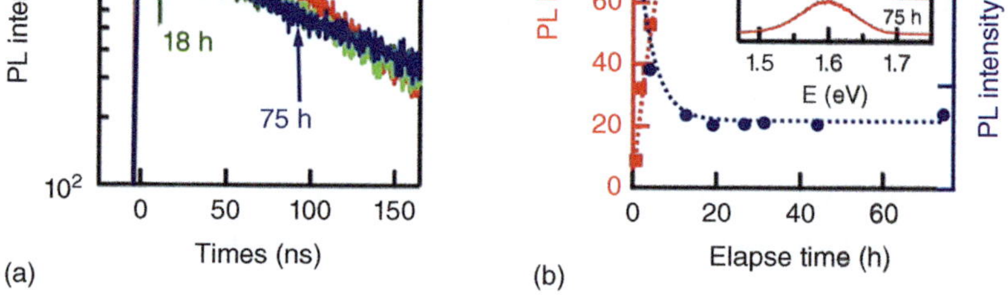

Figure 8.14 (a) Temporal variation of PL dynamics under weak excitation (40 nJ/cm^2) where bimolecular recombination is negligible. (b) PL lifetime and initial PL intensity as a function of elapsed time after fabrication. The dotted curves are fitting results by a double-exponential function. The inset shows the PL spectra at various elapsed times (1.6–75 hours) after fabrication. Source: Reprinted with permission from Ref. [92]. Copyright 2015, American Chemical Society.

is negligibly small compared with the nonradiative single-carrier trapping rate (*A* coefficient) [92]. The PL lifetime lengthens as time elapses, suggesting that the nonradiative recombination rate is reduced. The inset of Figure 8.14b shows that the PL spectrum shape is independent of elapsed time after thin film fabrication, whereas the PL dynamics and intensity change with time. This result indicates that the PL mechanism remains unchanged with elapsed time after fabrication.

Figure 8.14b summarizes the PL lifetime and PL intensity at zero delay time (initial PL intensity) as a function of elapse time. While the PL lifetime becomes longer, the initial PL intensity decreases more rapidly, suggesting that both the density of the nonradiative recombination center and unintentionally doped carriers (N_0) decrease with time, considering that initial PL intensity is proportional to BN_0 under weak excitation according to Eq. (8.5b). These two mechanisms are closely linked to each other, indicating the defect-annihilation effect via room-temperature annealing.

8.7.2 Charge Transfer Mechanism in Perovskite Solar Cell

The carrier dynamics in the device can also be investigated with time-resolved PL measurements. Typical perovskite solar cell is composed of the light-absorbing layer (perovskite), electron-transporting layer (ETL), and hole-transporting layer (HTL). Photocarriers generated in the perovskite layer are separated into ETL and HTL, which is the fundamental mechanism of photovoltage. Therefore, the charge transfer rate dominates the photoconversion efficiency of perovskite solar cells.

The charge transfer process can be estimated by PL quenching due to the addition of ETL (PCBM) and HTL (Spiro-OMeTAD) on the perovskite thin film [15, 16]. The PL lifetime of $MAPbI_3$ thin films with ETL and HTL becomes shorter compared with that with a protective layer (PMMA) [16]. The PL quenching occurs because the photoexcited electrons and holes reach ETL or HTL. Electrons (holes) are extracted from the perovskite layer, and the carrier density is reduced, resulting in the PL reduction. From the analysis of the PL quenching dynamics, carrier diffusion length, as long as 1 μm, has been estimated [15].

Under intense photoexcitation, PL dynamics in perovskite thin films and solar cell devices show more intriguing behavior, providing insights into the carrier injection mechanism at the heterointerface [18]. The PL decay curves of bare $MAPbI_3$ film fabricated on a quartz substrate without any charge transport layers are shown in Figure 8.15a. Equations (8.4a,b) and (8.5) well explain the excitation fluence-dependent PL dynamics; a fast PL decay appears under intense excitation due to electron–hole bimolecular recombination.

In $MAPbI_3$-based solar cell device, which is composed of Spiro-OMeTAD (HTL), TiO_2 (ETL), and perovskite layer, PL shows a fast decay for weak excitation in contrast to the thin film counterpart (see Figures 8.15a,c). The measurement was conducted in a short-circuit condition. Interestingly, PL decay becomes slower with an increase in excitation fluence. Fast PL decay under weak excitation can be explained by hole transfer into the Spiro-OMeTAD layer rather than electron transfer into the TiO_2 layer, which can be inferred from Figure 8.15b. With increasing excitation

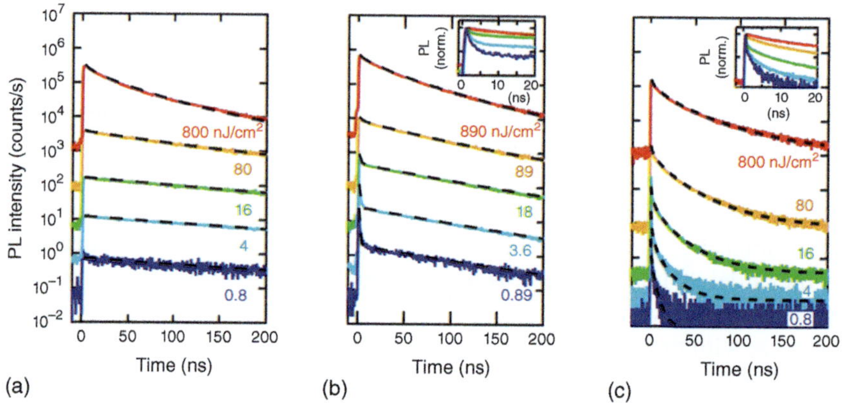

Figure 8.15 Excitation fluence dependence of PL decay curves for (a) $MAPbI_3/SiO_2$, (b) $MAPbI_3/TiO_2$, and (c) the solar cell. The PL from the solar cell was measured under a short-circuit condition. The inset shows the normalized PL decay curves. Dotted lines represent the fitting results. Source: Reprinted with permission from Ref. [18]. Copyright 2017, American Chemical Society.

fluence, suppression of the injection occurs because of unbalanced mobilities as well as the internal electric field across the perovskite layer, which accelerates the charge-carrier separation. Because the mobilities in $MAPbI_3$ are not high enough compared with high-quality solar cell materials such as GaAs, the photogenerated carriers cannot quickly reach the interfaces. Therefore, the distribution of electrons and holes reduces the internal field for separation. For increased photocarrier densities, the internal field weakens, and a slower PL decay is observed. This result might suggest the limitation of photovoltaic performance in concentrator-type solar cells.

8.8 Conclusion

In this chapter, the PL properties of hybrid perovskites were described. The PL origin of $MAPbI_3$ is the bimolecular electron–hole recombination (not exciton), and almost no Stokes shift indicates the lack of influence from the shallow impurity or defect states and small structural change induced by photoexcitation. These features suggest that $MAPbI_3$ is close to a typical inorganic semiconductor from the viewpoint of PL properties rather than an inorganic semiconductor. Furthermore, hybrid perovskites possess unique optical properties even when compared with conventional high-quality inorganic semiconductors. Interestingly, the solution-processed thin film with multigrain structure displays a low nonradiative recombination rate, leading to long carrier lifetime and carrier diffusion length, and high PL efficiency. The bright PL of hybrid perovskites accounts for the high photoconversion efficiency of perovskite solar cells. Apart from the solar cell applications, efficient PL is also promising for photonics applications, such as light-emitting diodes, which operate on the same physical principles in the opposite direction [94–98]. Intensive studies have recently been made for halide perovskite nanomaterials with extremely high PL efficiency, leading to the further device development and opening unique photophysics based on lead halide perovskites.

Acknowledgment

The authors acknowledge support from the Core Research for Evolutional Science and Technology of the Japan Science and Technology Agency (JST-CREST; Grant No. JPMJCR16N3), JSPS KAKENHI (Grant No. 19K03683), and Chiba Iodine Resource Innovation Center (CIRIC).

References

1 Metzger, W.K., Albin, D., Levi, D. et al. (2003). *J. Appl. Phys.* 94: 3549.
2 Tex, D.M., Ihara, T., Akiyama, H. et al. (2015). *Appl. Phys. Lett.* 106: 013905.
3 Tex, D.M., Imaizumi, M., Akiyama, H., and Kanemitsu, Y. (2016). *Sci. Rep.* 6: 38297.
4 Kanemitsu, Y. (2017). *J. Mater. Chem. C* 5: 3427.
5 Kanemitsu, Y., Okano, M., and Yamada, Y. (2018). *ECS J. Solid State Sci. Technol.* 7: R3102.
6 Shockley, W. and Queisser, H.J. (1961). *J. Appl. Phys.* 32: 510.
7 DeQuilettes, D.W., Koch, S., Burke, S. et al. (2016). *ACS Energy Lett.* 1: 438.
8 Noel, N.K., Abate, A., Stranks, S.D. et al. (2014). *ACS Nano* 8: 9815.
9 Brenes, R., Guo, D., Osherov, A. et al. (2017). *Joule* 1: 155.
10 Shi, D., Adinolfi, V., Comin, R. et al. (2015). *Science* 347: 519.
11 Dong, Q., Fang, Y., Shao, Y. et al. (2015). *Science* 347: 967.
12 Even, J., Pedesseau, L., Jancu, J.-M., and Katan, C. (2013). *J. Phys. Chem. Lett.* 4: 2999.
13 Umari, P., Mosconi, E., and De Angelis, F. (2014). *Sci. Rep.* 4: 4467.
14 Motta, C., El-Mellouhi, F., Kais, S. et al. (2015). *Nat. Commun.* 6: 7026.
15 Stranks, S.D., Eperon, G.E., Grancini, G. et al. (2013). *Science* 342: 341.
16 Xing, G., Mathews, N., Sun, S. et al. (2013). *Science* 342: 344.
17 Yamada, Y., Yamada, T., Shimazaki, A. et al. (2016). *J. Phys. Chem. Lett.* 7: 1972.
18 Handa, T., Tex, D.M., Shimazaki, A. et al. (2017). *J. Phys. Chem. Lett.* 8: 954.
19 van Roosbroeck, W. and Shockley, W. (1954). *Phys. Rev.* 94: 1558.
20 Shockley, W. and Read, W.T. Jr. (1952). *Phys. Rev.* 87: 835.
21 Landsberg, P.T. (1992). *Recombination in Semiconductors*. Cambridge University Press.
22 Kanemitsu, Y. and Handa, T. (2018). *Jpn. J. Appl. Phys.* 57: 090101.
23 Yamada, Y., Nakamura, T., Endo, M. et al. (2014). *Appl. Phys. Express* 7: 032302.
24 Shirayama, M., Kadowaki, H., Miyadera, T. et al. (2016). *Phys. Rev. Appl.* 5: 014012.

25 Ledinsky, M., Schönfeldová, T., Holovský, J. et al. (2019). *J. Phys. Chem. Lett.* 10: 1368.

26 Yamada, Y., Nakamura, T., Endo, M. et al. (2015). *IEEE J. Photovoltaics* 5: 401.

27 Yamada, Y., Yamada, T., and Kanemitsu, Y. (2017). *Bull. Chem. Soc. Jpn.* 90: 1129.

28 Ha, S.-T., Shen, C., Zhang, J., and Xiong, Q. (2016). *Nat. Photonics* 10: 115.

29 Yamada, T., Aharen, T., and Kanemitsu, Y. (2019). *Phys. Rev. Mater.* 3: 024601.

30 Yamada, T., Yamada, Y., and Kanemitsu, Y. (2020). *J. Lumin.* 220: 116987.

31 Yamada, T., Aharen, T., and Kanemitsu, Y. (2018). *Phys. Rev. Lett.* 120: 057404.

32 Jang, D.M., Park, K., Kim, D.H. et al. (2015). *Nano Lett.* 15: 5191.

33 McMeekin, D.P., Sadoughi, G., Rehman, W. et al. (2016). *Science* 351: 151.

34 Zarick, H.F., Soetan, N., Erwin, W.R., and Bardhan, R. (2018). *J. Mater. Chem. A* 6: 5507.

35 Hoke, E.T., Slotcavage, D.J., Dohner, E.R. et al. (2015). *Chem. Sci.* 6: 613.

36 Slotcavage, D.J., Karunadasa, H.I., and McGehee, M.D. (2016). *ACS Energy Lett.* 1: 1199.

37 Bischak, C.G., Hetherington, C.L., Wu, H. et al. (2017). *Nano Lett.* 17: 1028.

38 Brennan, M.C., Draguta, S., Kamat, P.V., and Kuno, M. (2018). *ACS Energy Lett.* 3: 204.

39 Tang, X., van den Berg, M., Gu, E. et al. (2018). *Nano Lett.* 18: 2172.

40 Zhang, H., Fu, X., Tang, Y. et al. (2019). *Nat. Commun.* 10: 1088.

41 Sebastian, M., Peters, J.A., Stoumpos, C.C. et al. (2015). *Phys. Rev. B* 92: 235210.

42 Eperon, G.E., Beck, C.E., and Snaith, H.J. (2016). *Mater. Horiz.* 3: 63.

43 Perumal, A., Shendre, S., Li, M. et al. (2016). *Sci. Rep.* 6: 36733.

44 Yamada, Y., Hoyano, M., Oto, K., and Kanemitsu, Y. (2019). *Phys. Status Solidi B* 256: 1800545.

45 Hao, F., Stoumpos, C.C., Cao, D.H. et al. (2014). *Nat. Photonics* 8: 489.

46 Noel, N.K., Stranks, S.D., Abate, A. et al. (2014). *Energy Environ. Sci.* 7: 3061.

47 Hao, F., Stoumpos, C.C., Guo, P. et al. (2015). *J. Am. Chem. Soc.* 137: 11445.

48 Yokoyama, T., Cao, D.H., Stoumpos, C.C. et al. (2016). *J. Phys. Chem. Lett.* 7: 776.

49 Handa, T., Yamada, T., Kubota, H. et al. (2017). *J. Phys. Chem. C* 121: 16158.

50 Yamada, Y., Nakamura, T., Endo, M. et al. (2014). *J. Am. Chem. Soc.* 136: 11610.

51 Yin, W.-J., Shi, T., and Yan, Y. (2014). *Appl. Phys. Lett.* 104: 063903.

52 Brandt, R.E., Stevanović, V., Ginley, D.S., and Buonassisi, T. (2015). *MRS Commun.* 5: 265.

53 Yasuda, H. and Kanemitsu, Y. (2008). *Phys. Rev. B* 77: 193202.

54 Yamada, Y., Yasuda, H., Tayagaki, T., and Kanemitsu, Y. (2009). *Phys. Rev. Lett.* 102: 247401.

55 Yang, Y., Yang, M., Li, Z. et al. (2015). *J. Phys. Chem. Lett.* 6: 4688.

56 Manser, J.S. and Kamat, P.V. (2014). *Nat. Photonics* 8: 737.

57 Yang, Y., Ostrowski, D.P., France, R.M. et al. (2016). *Nat. Photonics* 10: 53.

58 Savenije, T.J., Ponseca, C.S. Jr., Kunneman, L. et al. (2014). *J. Phys. Chem. Lett.* 5: 2189.

59 Wehrenfennig, C., Eperon, G.E., Johnston, M.B. et al. (2014). *Adv. Mater.* 26: 1584.
60 Milot, R.L., Eperon, G.E., Snaith, H.J. et al. (2015). *Adv. Funct. Mater.* 25: 6218.
61 Zhou, H., Chen, Q., Li, G. et al. (2014). *Science* 345: 542.
62 D'Innocenzo, V., Srimath Kandada, A.R., De Bastiani, M. et al. (2014). *J. Am. Chem. Soc.* 136: 17730.
63 Phuong, L.Q., Yamada, Y., Nagai, M. et al. (2016). *J. Phys. Chem. Lett.* 7: 2316.
64 Wehrenfennig, C., Liu, M., Snaith, H.J. et al. (2014). *APL Mater.* 2: 081513.
65 Tahara, H., Endo, M., Wakamiya, A., and Kanemitsu, Y. (2016). *J. Phys. Chem. C* 120: 5347.
66 Yang, Y., Yan, Y., Yang, M. et al. (2015). *Nat. Commun.* 6: 7961.
67 Staub, F., Hempel, H., Hebig, J.-C. et al. (2016). *Phys. Rev. Appl.* 6: 044017.
68 Yamada, Y., Yamada, T., Phuong, L.Q. et al. (2015). *J. Am. Chem. Soc.* 137: 10456.
69 Yamada, T., Yamada, Y., Nishimura, H. et al. (2016). *Adv. Electron. Mater.* 2: 1500290.
70 Pazos-Outón, L.M., Szumilo, M., Lamboll, R. et al. (2016). *Science* 351: 1430.
71 Richter, J.M., Abdi-Jalebi, M., Sadhanala, A. et al. (2016). *Nat. Commun.* 7: 1.
72 Yamada, T., Yamada, Y., Nakaike, Y. et al. (2017). *Phys. Rev. Appl.* 7: 014001.
73 Yamada, Y., Hoyano, M., Akashi, R. et al. (2017). *J. Phys. Chem. Lett.* 8: 5798.
74 Brenes, R., Laitz, M., Jean, J. et al. (2019). *Phys. Rev. Appl.* 12: 014017.
75 Kojima, K., Ikemura, K., Matsumori, K. et al. (2019). *APL Mater.* 7: 071116.
76 Ross, R.T. (1967). *J. Chem. Phys.* 46: 4590.
77 Miller, O.D., Yablonovitch, E., and Kurtz, S.R. (2012). *IEEE J. Photovoltaics* 2: 303.
78 Deschler, F., Price, M., Pathak, S. et al. (2014). *J. Phys. Chem. Lett.* 5: 1421.
79 Tan, Z.-K., Moghaddam, R.S., Lai, M.L. et al. (2014). *Nat. Nanotechnol.* 9: 687.
80 Stranks, S.D., Burlakov, V.M., Leijtens, T. et al. (2014). *Phys. Rev. Appl.* 2: 034007.
81 Sutter-Fella, C.M., Li, Y., Amani, M. et al. (2016). *Nano Lett.* 16: 800.
82 Naphade, R., Zhao, B., Richter, J.M. et al. (2017). *Adv. Mater. Interfaces* 4: 1700562.
83 Nagai, M., Tomioka, T., Ashida, M. et al. (2018). *Phys. Rev. Lett.* 121: 145506.
84 Poglitsch, A. and Weber, D. (1987). *J. Chem. Phys.* 87: 6373.
85 Onoda-Yamamuro, N., Matsuo, T., and Suga, H. (1992). *J. Phys. Chem. Solids* 53: 935.
86 Umebayashi, T., Asai, K., Kondo, T., and Nakao, A. (2003). *Phys. Rev. B* 67: 155405.
87 Wright, A.D., Verdi, C., Milot, R.L. et al. (2016). *Nat. Commun.* 7: 1.
88 Frost, J.M., Butler, K.T., Brivio, F. et al. (2014). *Nano Lett.* 14: 2584.
89 Mosconi, E., Azpiroz, J.M., and De Angelis, F. (2015). *Chem. Mater.* 27: 4885.
90 Wei, W. and Hu, Y.H. (2017). *Int. J. Energy Res.* 41: 1063.
91 McLeod, J.A. and Liu, L. (2018). *J. Phys. Chem. Lett.* 9: 2411.
92 Yamada, Y., Endo, M., Wakamiya, A., and Kanemitsu, Y. (2015). *J. Phys. Chem. Lett.* 6: 482.
93 Nie, W., Blancon, J.-C., Neukirch, A.J. et al. (2016). *Nat. Commun.* 7: 1.

94 Wang, N., Cheng, L., Ge, R. et al. (2016). *Nat. Photonics* 10: 699.
95 Sutherland, B.R. and Sargent, E.H. (2016). *Nat. Photonics* 10: 295.
96 Wang, H.-C., Wang, W., Tang, A.-C. et al. (2017). *Angew. Chem. Int. Ed.* 56: 13650.
97 Chiba, T., Hayashi, Y., Ebe, H. et al. (2018). *Nat. Photonics* 12: 681.
98 Lin, K., Xing, J., Quan, L.N. et al. (2018). *Nature* 562: 245.

9

Role of Grain Boundaries

Jae Sung Yun

University of New South Wales, School of Photovoltaic and Renewable Energy Engineering, Anzac Pde., Kensington 2033, Australia

9.1 Introduction

The precipitation of solid material from liquid solution, so-called "solution process," is a typical process for obtaining high-quality organic–inorganic halide perovskite (OIHP) films on a rigid or flexible substrate. The most conventionally used solution process is a spin-coating process, which results in a polycrystalline thin film and a typical thickness is between 300 and 800 nm for OIHP-based optoelectric devices. These films possess grains with distinct crystal orientations, thus, the formation of grain boundaries (GBs) is inevitable. Interestingly, despite the existence of GBs being often known to degrade the electronic properties, the polycrystalline OIHP-based solar cells have demonstrated outperforming photoconversion efficiency (PCE) compared with their single crystal counterparts.

The definition of a GB is an interface that separates grains with different crystal orientations. GBs are considered as planar defects or 2D defects in the crystal structure and commonly known as deteriorating the electrical performance of semiconductors. GBs are highly disordered or amorphous that usually possess a large number of impurities, thus, creating unique and local electronic states. These local electronic states formed by the GBs are strictly dependent on their distinctive crystallography and chemical bondings. There have been intensive studies on the impact of GBs in commercialized polycrystalline solar cell technologies. For instance, in polycrystalline or multicrystalline silicon, GBs are known to be charged to some extent due to the trapping of majority carriers and create mid-gap states within a bandgap that promote non-radiative recombination. These charged GBs can be passivated using hydrogenation, which reduces the density of dangling bonds at the GBs [1, 2]. In addition, in polycrystalline CdTe or Cu(In,Ga)Se_2 thin film solar cells, the electronic properties of GBs are tuned by a posttreatment or sodium segregation [3].

In polycrystalline OIHP solar cells, various roles and recombination behavior of GBs have been reported [4, 5]. The grain size of OIHP films varies between several

Hybrid Perovskite Solar Cells: Characteristics and Operation,
First Edition. Edited by Hiroyuki Fujiwara.

tenths of nanometers to millimeters depending on processing conditions such as temperature, a solution concentration, processing environment, posttreatment, and additive incorporation [6–10]. These conditions determine not only the size of grain but also the local electronic properties of GBs [6, 7]. In addition, varying the chemical composition such as incorporating excess PbI_2 or I showed significant influence on the GB properties [11]. The composition of the grain interiors (GIs) could be less sensitive toward small changes in overall stoichiometry compared with the GBs. Consequently, examining the roles of GBs in terms of creation of defect state, formation of dangling bond, conductivity, doping concentration, and ion migration behavior are particularly important for the development of OIHP solar cells.

The present chapter will discuss the current status of the research on GBs in the OIHP solar cells. Firstly, Section 9.2 yields a brief introduction to the electronic properties of GBs. In addition, the engineering of the GB properties and their impact on solar cell performance will be discussed. Secondly, Section 9.3 discusses ion migration behavior at the GBs and their impact on the charge separation at the GBs. Lastly, Section 9.4 overviews the role of GBs in respect to the degradation of the materials and device, especially, on the moisture penetration through MA and FA-based OIHPs.

9.2 Role of Grain Boundaries in Device Performance

Since OIHP solar cells first exceeded PCE over 9% in 2012 [12], intensive studies have been performed on the effect of grain size and the role of GBs in terms of charge collection and separation and their influence on the device performance. Up to date, controversy on the role and effect of GBs is still ongoing, and there has been two distant results and conclusions that have been frequently raised: (i) GBs play a benign role or even promote higher charge separation and collection compared with GIs [4, 13] and (ii) non-radiative recombination occurs at GBs so large grain size is advantageous, i.e. lower density of GBs [6, 14]. As stated earlier, the processing conditions that determine the properties of the GBs can substantially manipulate the electrical properties of GBs. Therefore, it is not surprising that many results do not report the same properties of the GBs. Theoretical studies also pose different perspectives. Density functional theory (DFT) calculation performed by Yin et al. indicates that GBs do not generate any gap states, making polycrystalline halide perovskite thin films like the single crystal [15]. On the other hand, a recent time-domain atomistic simulation study pointed out that GBs notably accelerate the electron–hole recombination in $MAPbI_3$ [14]. For inorganic cation halide perovskites, $CsPbI_3$, first-principles DFT calculations showed iodine interstitials in both the positive charge state (Ii^+) and negative charge state (Ii^-). These create deep levels that are energetically more stable at the GBs than those in the bulk [16].

In this section, the evaluation of potential barriers at GBs and their impact on charge separation and collection properties will be discussed first. Then, various approaches to engineer the electronic properties of the GBs will be overviewed as well as their influence on the device performance.

9.2.1 Potential Barrier at GBs and Charge Transport

GBs are regions with dangling bonds, thus, impurities or point defects can effectively be bonded. In OIHP materials, experimental and theoretical results indicate that iodide vacancies can accumulate or be trapped at GBs that changes a local charge state at the GBs [17, 18]. One of the effective techniques to measure the charge state at the GBs is Kelvin probe force microscopy (KPFM), which can map a contact potential difference (CPD) in nanoscale. CPD measures work function between a sample and a tip, thus, any charge state that shifts the work function of the sample surface can be readily revealed in the CPD map. Edri et al. reported that GBs in $CH_3NH_3PbI_{3-x}Cl_x$ have ~40 mV higher CPD compared with GIs [19]. In addition, the manipulation of the GB potential by controlling the PbI_2 concentration in $CH_3NH_3PbI_3$ has been reported. By releasing the organic species during annealing of the film, the higher surface potential at the GBs is changed from 50 mV to lower surface potential around 30 mV [20]. These results provided a strong clue that a charge potential barrier exists across the GBs in OIHP materials. Positive and negative potential barriers formed at GBs are shown in Figure 9.1. Depending on the polarity of segregated species, the barriers can be either positive (downward band bending) or negative (upward band bending).

Figure 9.1c exhibits the case of higher potential barriers at the GBs. The negative charges are trapped at the energy levels within a bandgap, thus, becoming more n-type. From this model, it is explained that the GBs can act as (i) potential barriers for electrons, which impedes their transport between the two grains and thus affects their long-range mobility across the GBs; and (ii) there is a chance that the trapped electrons recombine non-radiatively with free holes in the valence band, thus, enhance non-radiative recombination. When iodide vacancies (V_I^+) reside at the GBs, it is likely that the GBs are positively charged initially. When excess carriers are generated, it is expected that the positively charged V_I^+ becomes electrical neutral by capturing the electrons, which fill the positively charged GBs, removing the potential barrier as described in Figure 9.1d.

In the early phase of perovskite solar cell development, Chen et al. [20] proposed PbI_2 passivation of GBs by thermal annealing and improved efficiency. It is expected that CH_3NH_3I is released from $CH_3NH_3PbI_3$ upon annealing and PbI_2 is left behind at the GBs. Figure 9.1e displays charge separation properties of GBs when the PbI_2 exists at the GBs. PbI_2 has a large bandgap compared with $CH_3NH_3PbI_3$; therefore, an upward band bending is established, which helps to reduce the recombination between the electrons from perovskite and holes from hole transport material (HTM). This study evolved studies on numerous different roles and mechanisms of the existence of PbI_2 in perovskite solar cells [22–24].

As mentioned earlier, KPFM is a powerful tool to visualize a localized potential of the sample surface. A sharp atomic force microscopy probe or tip, typically, 10–30 nm in radius, scans across the sample surface and measures CPD between

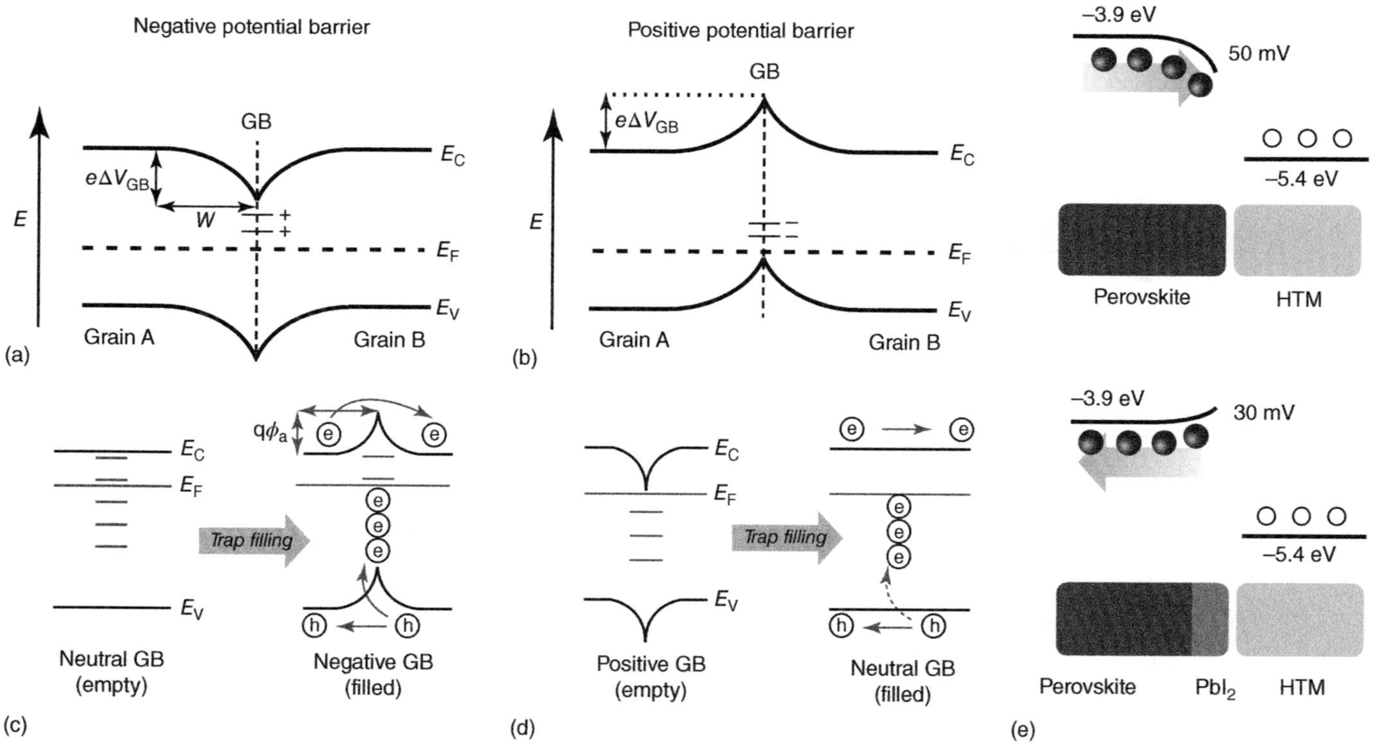

Figure 9.1 Schematics of the electronic band diagram in the vicinity of GBs for (a) a negative and (b) a positive potential barrier. Defect states (arbitrary) at the GB are indicated by a sequence of small dashes. ΔV_{GB} and w denote potential barrier height and barrier width respectively [21]. (c) In typical inorganic solar cells (poly-Si, CdTe), the empty neutral traps at GBs and interfaces when filled with electrons result in a weakened transport due to the potential barrier, and the non-radiative recombination between holes and trapped electrons is strong. (d) In OIHPs, it is likely that the empty traps are positively charged due to accumulated iodide vacancies (V_I^+) at GBs and interfaces. Therefore, when filled with electrons, the traps are neutral, electron transport is relatively unaffected, and non-radiative recombination is weak. (e) The interface of perovskite/HTM is described on the bottom right; the presence of PbI_2 changes the grain to GB bending from downward to upward, which helps to reduce the recombination between the electrons from perovskite and holes from HTM. *Source*: (a–d) Sherkar et al. [21], (e) Chen et al. [20].

the tip and the sample surface. CPD obtained in dark implies a work function between the tip and sample at no other external influences. Under the illumination, a shift of Fermi level occurs due to the excess carriers, and surface photovoltage can be obtained by subtracting CPD measured in dark from CPD measured under the illumination. Comprehensive studies on the potential barrier at GBs and surface photovoltage of $CH_3NH_3PbI_3$ perovskite solar cells are performed by the author. The KPFM measurement setup is shown in Figure 9.2a, and photocurrent measurement setup that allows to map the photocurrent at GBs and GIs is shown in Figure 9.2b. The sample comprises a $CH_3NH_3PbI_3/TiO_2$/FTO/glass heterojunction solar cell. Figure 9.2c shows height profiles of GBs and calculated surface photovoltage profiles obtained by subtracting CPD measured in dark from CPD measured under illumination at various intensities. Surface photovoltage ($CPD_{light} - CPD_{dark}$) can be interpreted as the ability of photo-generated carriers to

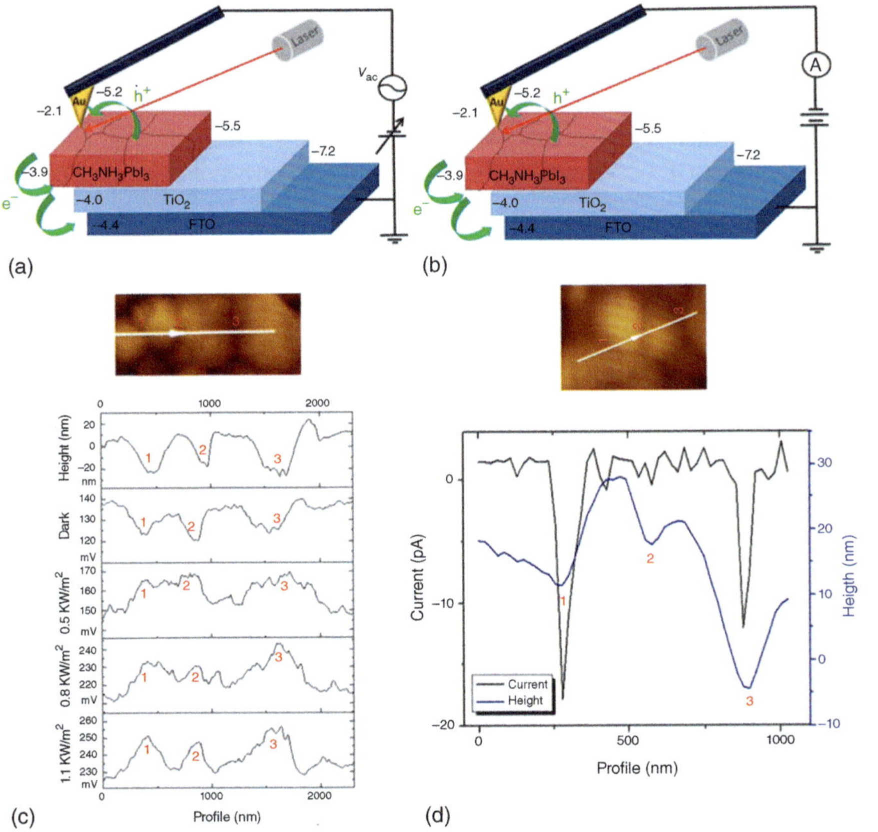

Figure 9.2 Illustration of the (a) KPFM and (b) C-AFM setup. Relevant vacuum energy levels (in eV) for corresponding materials are indicated. (c) Topography map, topography line profile, CPD profile under dark conditions, and surface photovoltage profiles at various intensities. The line profiles were analyzed along the white line indicated in the topography map. Numbers denote locations of GBs. (d) C-AFM current (black) and height (blue) line profiles at 0 V under illumination. *Source*: Yun et al. [4].

reach the top surface before they recombine. In this case, electrons are transported toward an electron transport layer, i.e. TiO_2, and holes are accumulated at the top surface of the perovskite layer. In dark, GBs have ~20 mV lower potential compared with GIs, which indicates that the GBs have downward band bending as described in Figure 9.1d. When illuminated, at the lowest intensity of 0.5 kW/cm^2, surface photovoltage at the GBs 1, 2, and 3 is only a few millivolts higher than that of GIs. As the light intensity increases, surface photovoltage at the GBs is enhanced significantly and the surface photovoltage peaks at the GBs become more prominent with light intensity. This result indicates that the photo-generated carriers are more effectively separated at the GBs compared with GIs. Photocurrent mapping results are shown in Figure 9.2d. From the line profiles of current and height, higher photocurrent (equals to low current in the map) is found at GB1 and GB3, which confirms that GBs act as an efficient charge collector compared with GIs. Interestingly, GB2 does not exhibit the same ways as the other two GBs. This could be responsible for a depth-dependent charge carrier collection properties at GBs [25]. Some of GBs have high resistance at the top surface of the film and act as an impediment to photocurrent collection. In addition, there are lower resistance pathways between grains that exist below the top surface of the film, indicating that there exists a depth-dependent resistance behavior. These results suggest that each GB possesses its unique properties in terms of charge collection properties.

9.2.2 Engineering of GB Properties

As seen from Section 9.2.1, GBs possess their unique electrical properties that are distinct from GIs and even GB-to-GB variation exists. These properties are strongly dependent on fabrication methods and precursor chemicals and conditions. In this section, experiments performed by the author and others on electrical properties of GBs of FA cation-based perovskites as well as multiple cations and mixed halide perovskites will be discussed. Further, its impact on device performance will be examined.

OIHP solar cells made from FA cation showed higher heat tolerance compared with MA cation owing to a lower tendency toward deprotonation [26]. However, the black cubic $FAPbI_3$ perovskite phase (α-phase) is unstable in air and at room temperature so it can convert into the hexagonal yellow non-perovskite $FAPbI_3$ phase (δ-phase). In addition, α-phase $FAPbI_3$ perovskite bandgap below 1.55 eV is of interest for photovoltaics in terms of their close-to-ideal bandgap. Also, high-quality $FAPbI_3$ films that have high coverage, compacity, and uniformity are less prone to degradation and can also perform high efficiency. For this purpose, a small amount of aqueous hydriodic acid (HI) is added (0–0.2 g/ml). $FAPbI_3$ perovskite precursor can control the nucleation and crystallization to form high-quality $FAPbI_3$ films. J–V curves and PCE are displayed in Figure 9.3a,b. As can be seen, the highest performance is achieved when 0.1 g/ml is added. Figure 9.3c–f shows atomic force microscopy (AFM) topography images of the perovskite films using different concentrations of HI additive. Figure 9.3g–j show the corresponding CPD

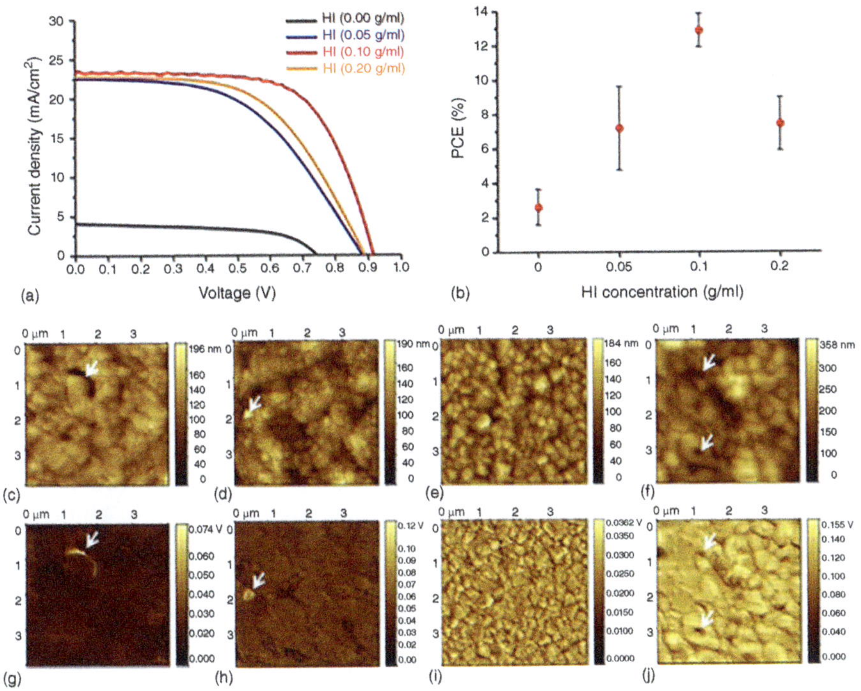

Figure 9.3 (a) $J-V$ curve of $FAPbI_3$ devices fabricated with different HI concentrations and (b) corresponding PCE distribution. AFM topography images of $FAPbI_3$ on bl-TiO_2/FTO glass deposited by gas-assisted spin coating and with HI additive at various concentrations (c) 0, (d) 0.05, (e) 0.10, and (f) 0.20 g/mL. The corresponding KPFM images of the same film are shown in panels (g)–(j). *Source*: (a–c) Kim et al. [27].

images obtained by KPFM. While the individual grains (200–400 nm in size) of films prepared with HI additive $\leq$0.05 g/ml are distinguishable in the topography maps (Figure 9.3c,d), their features are unrecognizable in the CPD maps due to very small variation ($\pm$1 mV) between GBs and GIs (Figure 9.3g,h). This is likely to be due to incompletion of the film formation. In addition, some localized regions (highlighted by white arrows) in Figure 9.3g,h with particularly high CPDs (74 and 120 mV higher) indicate the presence of different phase, which is responsible from the unreacted lead iodides. As the concentration of HI additive reaches 0.1 g/ml, grains become more distinguishable in the CPD maps with the GBs having lower CPD values (Figure 9.3i). There is no irregularly high CPD region when compared with that of the lower concentrations. The CPD at GBs is typically 5–20 mV lower than the CPD within the GIs. Such lower CPD at the GBs is expected to cause a band bending toward GBs that facilitates higher charge collection and separation at GBs. When HI additive concentration $\geq$0.20 g/ml, despite the grain size being larger compared with that of the 0.1 g/ml concentration sample, pin holes are formed in the film, see Figure 9.3f,j for regions highlighted by white arrows. These holes have much (up to 155 mV) lower CPD resulting in nonuniform electrical characteristics of the film and hence final device.

Multiple cation OIHPs have been effectively implemented to improve device performance as well as stability. Especially, triple-cation MA and FA with inorganic cation such as Cs or Rb greatly enhanced thermal stability [28, 29]. The author examined these triple-cation perovskites using KPFM, FAMA ($FA_{0.85}MA_{0.15}PbI_{2.55}Br_{0.45}$), FAMA-Cs ($FA_{0.80}MA_{0.15}Cs_{0.05}PbI_{2.55}Br_{0.45}$), and FAMA-Rb ($FA_{0.80}MA_{0.15}Rb_{0.05}PbI_{2.55}Br_{0.45}$) [30]. Figure 9.4a shows X-ray diffraction (XRD) patterns of these samples showing that the non-perovskite phase at 11.5° is suppressed when Rb or Cs is incorporated. Red arrows in FAMA-Rb present an Rb-related phase. *J–V* curve and performance of each sample are displayed in Figure 9.4b. Figure 9.4c–h displays CPD maps of the FAMA, FAMA-Cs, and FAMA-Rb test structures in

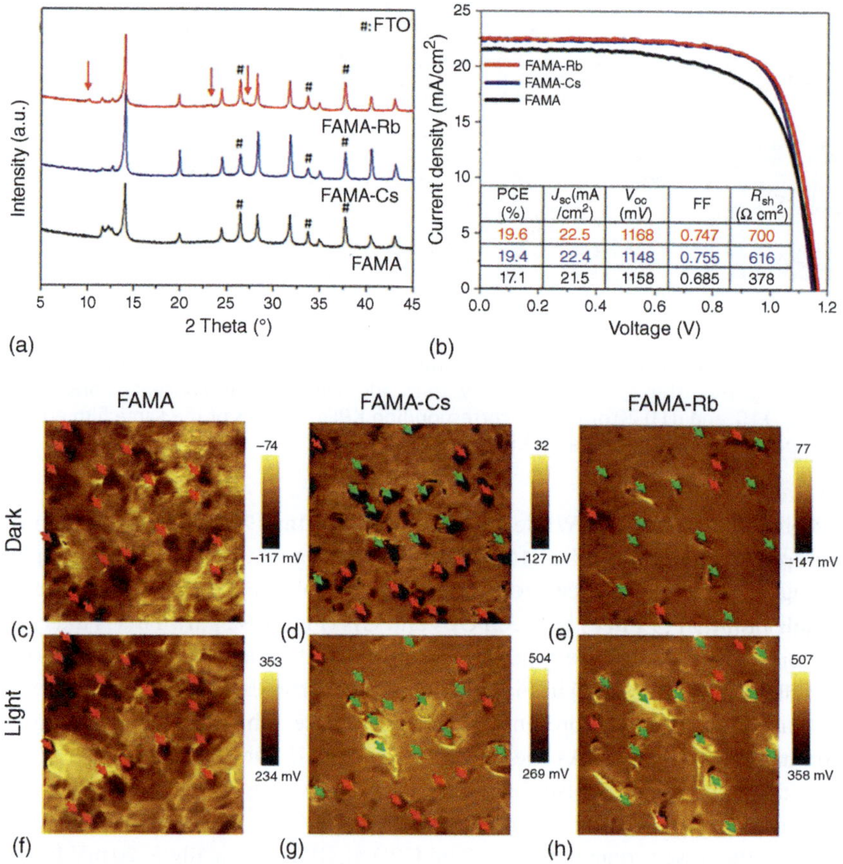

Figure 9.4 (a) XRD patterns of the FAMA, FAMA-Cs, and FAMA-Rb perovskite films. Red arrows indicate an Rb-related phase, and # denotes fluorine-doped SnO. (b) *J–V* curves and performance parameters of the solar cells fabricated with the FAMA, FAMA-Cs, and FAMA-Rb perovskite films. KPFM measurements performed on perovskites/m-TiO_2/c-TiO_2/FTO/glass test structures over an area of $2 \times 2\ \mu m^2$ in the dark and under white light illumination equivalent to 0.6-Sun. CPD maps of (c) FAMA, (d) FAMA-Cs, and (e) FAMA-Rb in the dark. CPD maps of (f) FAMA, (g) FAMA-Cs, and (h) FAMA-Rb under illumination. *Source:* (a–h) Zhang et al. [30].

the dark and under illumination. As can be seen, grains are distinguishable and GBs have lower CPD compared with GIs for the FAMA sample in dark and light (Figure 9.4c,d respectively). On the other hand, FAMA-Cs and FAMA-Rb maps do not show each grain clearly in both dark and light. These imply that the electronic states at GBs have been tuned by substation of these inorganic cations. Additionally, some grains perform differently from the reference FAMA sample. Some grains have relatively lower CPD compare to the surrounding regions (indicated by red and green arrows) in the CPD maps taken in the dark (Figure 9.6c–e). As explained earlier, a variation of CPD or work function difference can be originated from many factors such as carrier concentration, topological defects, impurity segregation, and variation in chemical compositions. Under the illumination, the CPDs at GBs of the FAMA sample increase in Figure 9.4f, showing higher surface photovoltage ($CPD_{light} - CPD_{dark}$) at GBs compared with GIs. In addition, grains with a red arrow in FAMA samples still exhibit relatively lower CPDs to the surroundings in both dark and light. Surface photovoltage at these grains in FAMA sample are similar to other grains indicating that these grains are neither harmful nor beneficial for charge separation. Meanwhile, some of these grains (green arrows) in FAMA-Cs and FAMA-Rb exhibit relatively higher CPD compared with illumination (Figure 9.4g,h) and have higher surface photovoltages than others. These results imply that the photo-generated carriers are more effectively separated at these grains while the GBs do not perform either high or low surface photovoltage compared with the surroundings.

In addition to the inorganic cation incorporated triple-cation perovskites, incorporation of long-changed cations such as phenethylammonium (PEA) significantly alters the electrical properties of GBs. The long-chained cations are known to reduce the dimensionality of the perovskites [32]. Figure 9.5a shows optical properties of perovskite material without and with the incorporation of PEAI into the precusor using UV–vis spectra measurement, which was measured on test structures and results are displayed in Figure 9.5a. It can be seen that both absorption and the optical bandgap (1.46 eV) do not change with the PEAI incorporation, except the absorption of PEAI 10% is slightly lowered. Despite the bandgap remaining almost unchanged, the performance (V_{oc}, J_{sc}, fill factor (FF), and efficiency) of the sample changed greatly (Figure 9.5b). Especially, V_{oc} increases with the PEAI incorporation. Figure 9.5c–g display the CPD maps of samples with different amount of PEAI incorporation measured in the dark. Figure 9.5h–l shows the CPD maps of the same samples under illumination. It is noticeable that the grain size decreases as the amount of PEAI increases. Each grain is clearly distinguishable in all of the CPD maps (Figure 9.5c–l), and the averaged differences between the GB and GI ($CPD_{GB} - CPD_{GI}$) are plotted in the chart in Figure 9.5m for CPD measured in the dark (dark line, calculated from Figure 9.5c–g) and under light (red line, calculated from Figure 9.5h–l). Under the dark condition, the CPD at the GBs is lower than that of GIs. Upon gradually incorporating the PEAI, the CPD at GBs keeps decreasing compared with the GIs; for instance, $CPD_{GB} - CPD_{GI} = -13$ mV without PEAI compared with $CPD_{GB} - CPD_{GI} = -44$ mV with 4.5% PEAI. Such CPD change at the GBs can be related to the formation of a quasi-2D phase, which

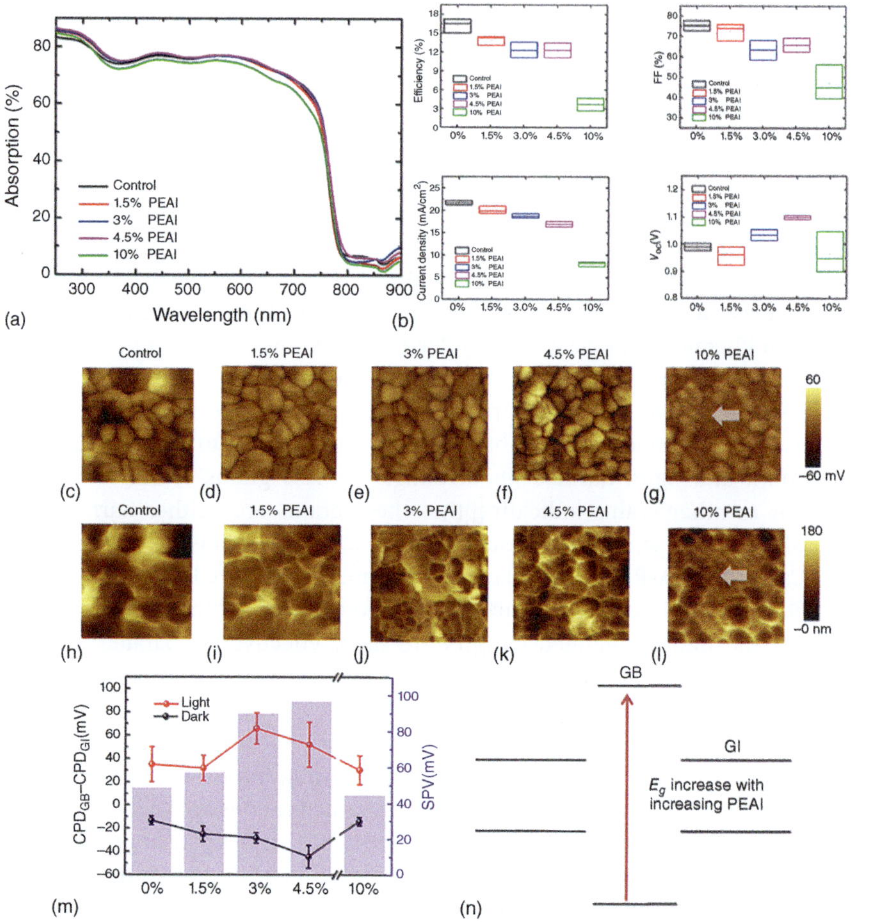

Figure 9.5 (a) Absorption of the test structure FTO/glass/c-TiO_2/meso-TiO_2/$(FAPbI_3)_{0.85}(MAPbBr_3)_{0.15}$ with 0% 1.5%, 3.0%, 4.5%, and 10 mol% of PEAI. (b) Plot of *J*–*V* performance parameters for cells with different levels of PEAI incorporations (0, 1.5, 3.0, 4.5, and 10 mol%). (c)–(g) Spatial CPD maps (1.5 × 1.5 μm^2) of perovskites samples with PEAI incorporated at (c) 0, (d) 1.5, (e) 3.0, (f) 4.5, and (g) 10 mol% measured under dark. (h)–(l) Spatial CPD maps of the same samples measured under illumination. (m) CPD between the GB and GI (CPD_{GB} − CPD_{GI}) in dark and light. Surface photovoltage is indicated with bars. (n) Schematic band diagram illustrating a larger band gap at the grain boundary compared with the GI due to the presence of quasi-2D perovskites. *Source*: (a–n) Lee et al. [33], (c) Quan et al. [32].

has been previously reported by Wang et al. [34]. They reported that plate-like features exist at GBs, which was suspected as a 2D or quasi 2D phase, and form an interface with grains that is a 3D perovskite when *n*-BA is incorporated in $FA_{0.83}Cs_{0.17}Pb(I_{0.6}Br_{0.4})_3$ perovskite. The possible existence of this higher-band-gap quasi-2D (as depicted in Figure 9.5) material has as positive effect preventing recombination of charge carriers that is evident from the fact that V_{oc} linearly increases as the amount of PEAI increases (1.5–4.5%). This is also evident from the fact the surface photovoltage gradually increases from 0 to 4.5% PEAI incorporation as shown in Figure 9.5m. These results evidently indicate the advantageous effects of incorporation of PEAI that lead to higher charge separation at GBs, in turn, improving the V_{oc} of the device. However, this enhancement ceases when PEAI is at 10%. At this concentration, results from both dark and light measurements are very similar to the results for the control film (0% PEAI). This could be due to the presence of an excessive amount of PEAI that inhibits proper formation of the quasi-2D phase at the GBs but rather incurs aggregation of PEAI. This is indicated by the white arrow in Figure 9.5l.

Control of the halide concentration (I_3^-) in the perovskite precursor solution can also tune the potential at GBs. By carefully controlling the excess I_3^- concentration, performance of the device is greatly improved [31]. Figure 9.6a shows the *J–V* curves of the devices with 0 mM of I_3^- (reference sample [PCE = 20.78%]), 10 mM of I_3^- (Target sample [PCE = 22.65%]), and 50 mM of I_3^- (Target sample [PCE = 19.65%]). Figure 9.6d,e display topography and corresponding CPD maps of these samples measured in dark. In the CPD maps, there are some grains that comprise relatively lower CPD compared with the neighboring grains. In the control sample, the percentage of those grains is the highest, while both target and excess samples show relatively fewer numbers of those grains. Such grain-to-grain variation can be visualized in the CPD distribution curves in Figure 9.6b. For CPD distribution, both control and excess samples have a shoulder at around 0.05 and 0.065 V, respectively, while the target sample exhibits a perfect Gaussian curve. This implies that the electrical properties of the target sample surface have the highest homogeneity due to the homogeneous halide distribution or physical orientation of each grain. For each sample, a work function is calculated by subtracting CPD at the peak position from the work function of the tip (−4.0650 eV), and the results are displayed in Figure 9.6b. As can be seen, the work function of the perovskite surface is slightly increased with the addition of I_3^-. Considering that both conduction and valence energy levels have not changed for all of the samples, enhanced work function can be related to a shift of the Fermi energy level toward the valence band, that is, turning the sample surface into more p-type by reducing the disorder or iodide vacancy defects. Figure 9.6c plots the average absolute CPD of the GIs and GBs in each sample. The target sample shows the largest CPD difference between the GI and GB, while the control sample shows the least difference. It is interesting to note that the absolute CPD value at the GBs is comparable in the control and target samples, whereas the CPD at the GIs is higher for the target sample when compared with the control sample. This implies that appropriate incorporation of I_3^- manipulated the electrical properties of GIs, possibly reducing the iodide vacancies. Consequently, the large CPD difference between the GBs and GIs is expected to provide a significant band bending toward the GBs, thus, resulting in a more efficient charge separation at the GBs.

In addition to works performed by the author, there are many other reports on the GB engineering that result in improved performance. Xu et al. attested incorporation of PCBM (Phenyl-C61-butyric acid methyl ester) into perovskite precursor

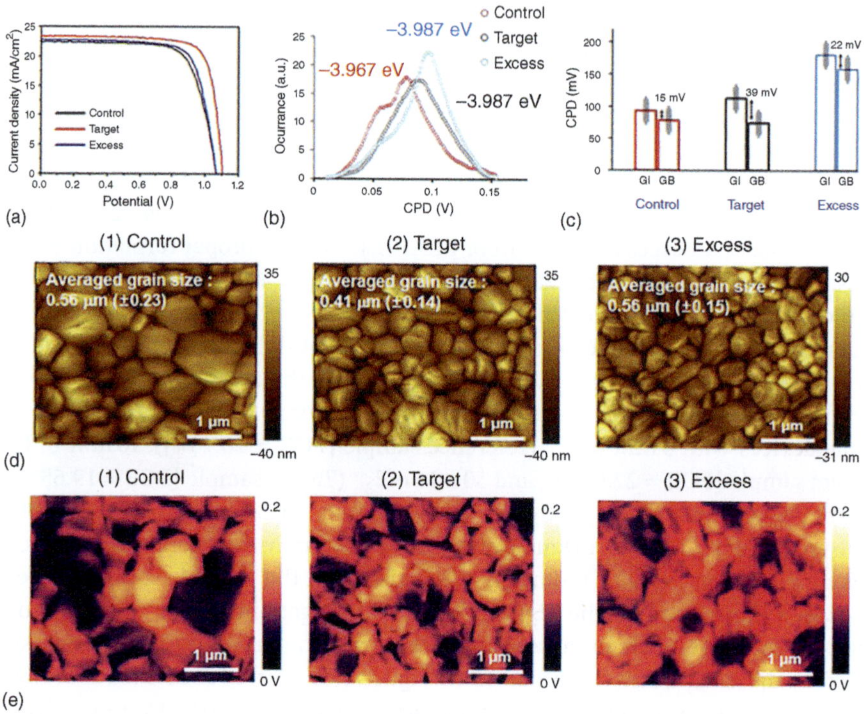

Figure 9.6 (a) $J-V$ curves of the devices with 0 mM of I_3^- (reference sample [PCE = 20.78%]), 10 mM of I_3^- (Target sample [PCE = 22.65%]), and 50 mM of I_3^- (Target sample [PCE = 19.65%]). (b) Average CPD differences between GI and GB of each sample. (c) CPD distribution over-occurrence (256 × 256 pixels over 16 μm^2) with calculated absolute work function of each sample. (d) Topography images with average grain size of each sample. (e) Corresponding CPD images measured in dark for each sample. *Source:* (a–e) Kim et al. [31].

and made perovskite-PCBM hybrid solution [35]. This enables PCBM phase to be aggregated at GBs (Figure 9.7a), which results in improved performance. Electrical properties of GB are probed by conductive atomic force microscopy (C-AFM). As can be seen in Figure 9.7b,c, higher conductivity is found near GBs (Figure 9.7c) when the film is fabricated using perovskite-PCBM hybrid solution. It is explained that the PCBM phase at the GBs formed continuous pathways for electron egress and passivates deep traps at the GBs [35]. Engineering of crystal orientation of GBs is also possible. By pressing polyurethane stamps with hexagonal nanodot arrays on partially dried perovskite (Figure 9.6d), low-angle (mismatch angle below 15%) GBs can be formed [36]. The perovskite film with compression shows a higher current flow (Figure 9.7e,f), and the current increases dramatically. Note the reversed current flows in the inverted device structure, thus, negative bias was applied. Therefore, the darker contrast image in the C-AFM images indicates higher current flow through the perovskite layers. This process improved the efficiency from 16.54 to 19.16%.

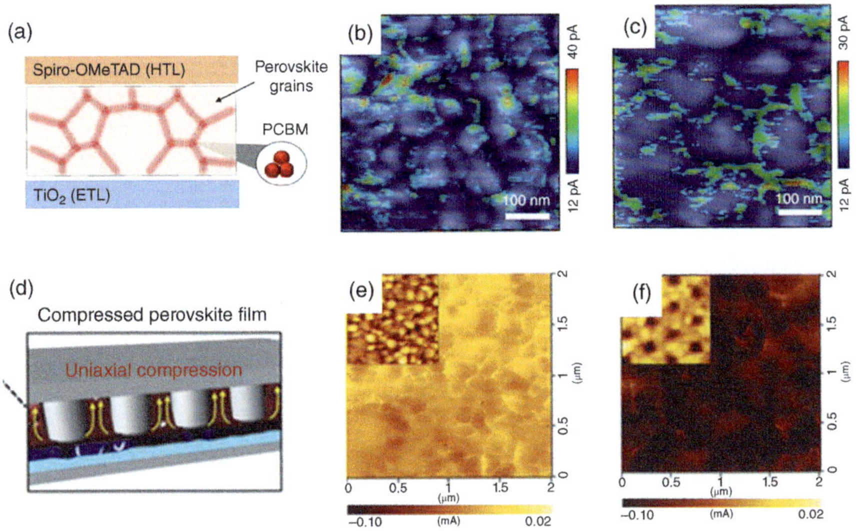

Figure 9.7 (a) PCBM passivation of GBs. (b) Control and (c) PCBM-hybrid film [35]. (d) Schematic of the compression process on the perovskite film with nanoimprinting stamps. C-AFM images of (e) as-prepared and (f) compressed perovskite films [35, 36]. *Source*: (a, b, d) Xu et al. [35], (e) Kim et al. [36].

9.3 Ion Migration Through Grain Boundaries

In OIHP solar cells, hysteresis in light current and voltage (I–V) characteristics has been repeatedly reported and ion migration has been suspected to be one of the main causes [37–41]. Figure 9.8a displays the typical I–V hysteresis behavior of the OIHP solar cells. Usually, a forward bias scan direction from V_{oc} to J_{sc} results in higher efficiency with higher V_{oc} and FF. OIHPs exhibit high ionic conductivity mediated by point defects, typically halide or cation vacancies. Studies have shown that the lead vacancies have relatively higher activation energy for ionic migration compared with cation or halide vacancies so they are suspected to be responsible for the slow response of up to several minutes that may cause the hysteresis and light soaking effect [44]. Interestingly, recent studies reported that GBs serve as ion migration channel [45–47]. Due to the dangling bonds and broken atomic structure of GBs, it is expected that the ion migration activation energy along the GBs is less compared with the bulk as less chemical bonds are reserved [48].

In this section, the ion migration process through the GBs will be evaluated using bias voltage applied KPFM measurement. Then, the consequence of the preferential ion migration through the GBs on the photocurrent hysteresis and charge separation at the GBs will be discussed.

9.3.1 Enhanced Ion Transport at Grain Boundaries

KPFM measurement is capable of detecting surface potentials caused by ionic charges in an ionic conductor [49, 50]. By applying a bias voltage to the AFM tip or

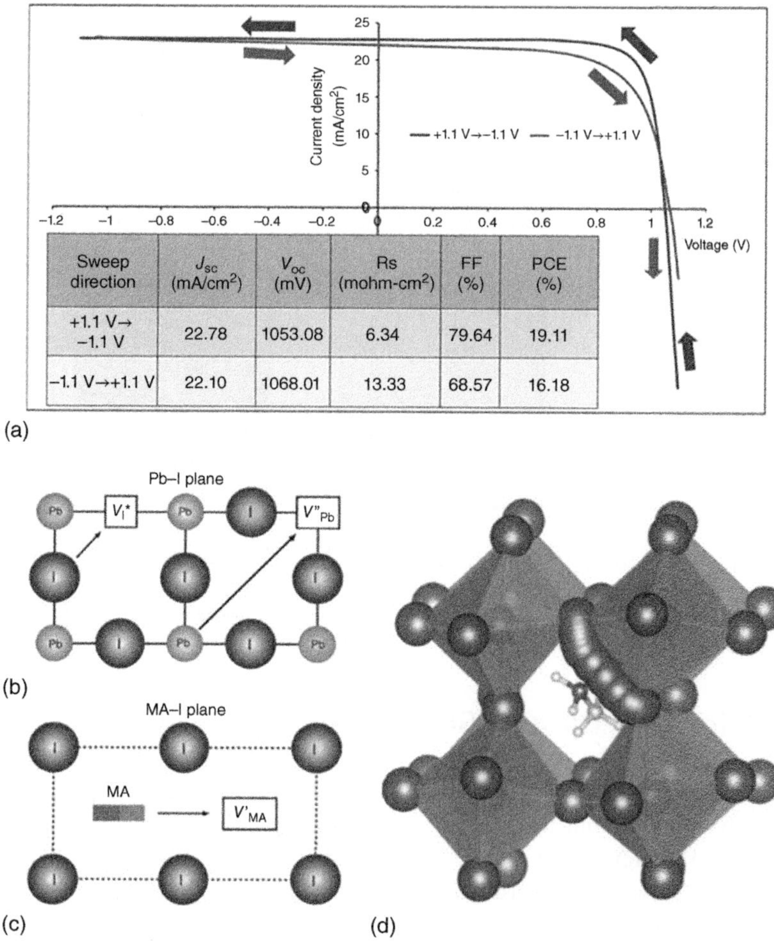

Sweep direction	J_{SC} (mA/cm²)	V_{OC} (mV)	Rs (mohm-cm²)	FF (%)	PCE (%)
+1.1 V→ −1.1 V	22.78	1053.08	6.34	79.64	19.11
−1.1 V→+1.1 V	22.10	1068.01	13.33	68.57	16.18

Figure 9.8 (a) Forward bias to short circuit (FB–SC) and short circuit to forward bias (SC–FB) current–voltage curves measured under AM1.5 simulated sun light for OIHP solar cells. The inset table gives the FB–SC- and SC–FB-determined performance parameters. Schematic illustration of the three ionic transport mechanisms involving conventional vacancy hopping between neighboring positions in $MAPbI_3$: (b) I^- migration along an octahedron edge; Pb^{2+} migration along the diagonal direction <110>; (c) $CH_3NH_3^+$ migration into a neighboring vacant A-site cage involving motion normal to the unit cell face composed of four iodide ions. (d) Illustration of migration path indicating a slightly curved path and local relaxation/tilting of the octahedra. *Source*: (a) Based on Kim et al. [42], (b–d) Eames et al. [43]. Licensed under CC BY-4.0.

the sample, spatially and time-resolved ionic movement can be mapped [46]. In a two-pass KPFM mode, a *dc* bias voltage can be applied to the AFM tip in the first pass to induce ionic motion and the subsequent pass can measure the resulting CPD. Delay time between the first and the second passes is set to several hundred milliseconds, which are equivalent to the ionic motion in halide perovskites [51]. Figure 9.9 shows the CPD maps of $(FAPbI_3)_{0.85}(MAPbBr_3)_{0.15}$ perovskite on TiO_2

measured with different bias voltages applied to the tip. Figure 9.9a,c plot the average CPD values from each map in Figure 9.9b,d, respectively. When positive bias voltages are applied, the average CPD increases. Conversely, the negative voltages decreases the average CPD. Such changes in CPD upon applying the voltages are a clear sign of the movement of ionic defects toward the top surface of

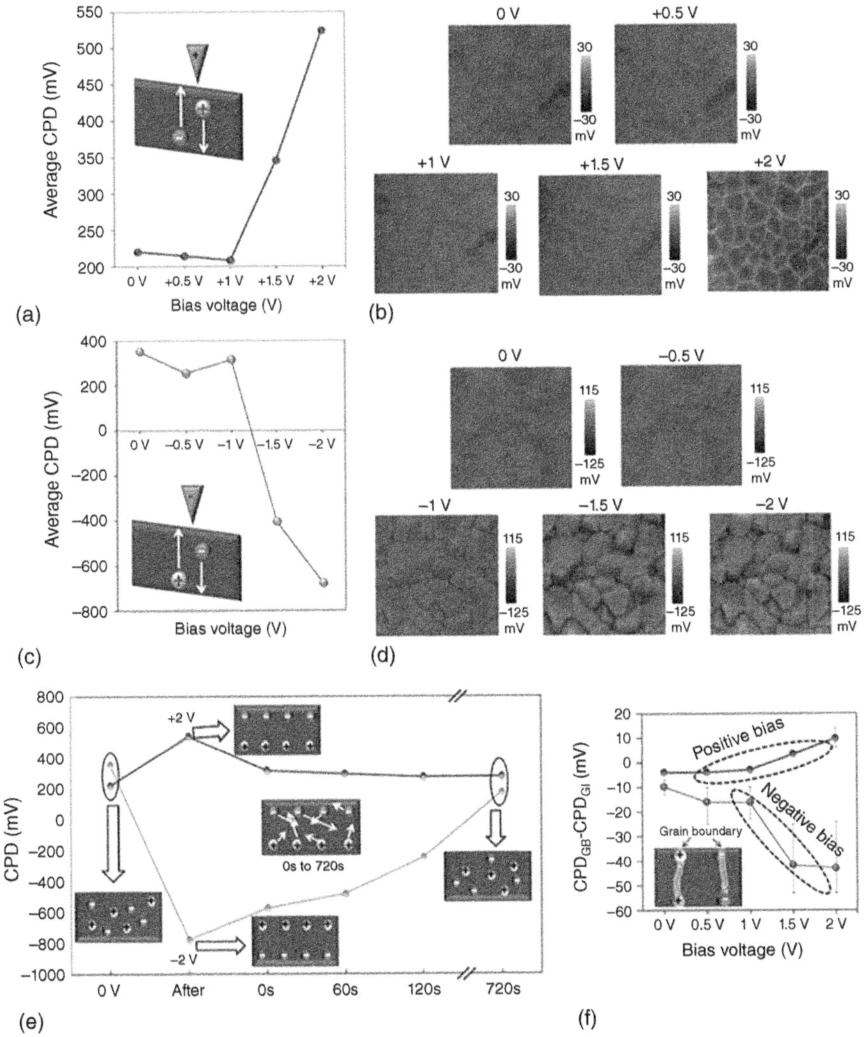

Figure 9.9 CPD spatial maps over an area of 4 μm^2 in the dark after applying (a) positive bias voltages and (c) negative bias voltages. Scale bars are normalized. (b), (d) Plots of average CPDs from each map in (a) and (c), respectively. Inset: Movement of ions with respect to positive and negative bias voltages. (e) Average CPDs as a function of time after the bias voltages have been removed. Inset: Illustration of an arrangement of positive and negative ions. (f) A plot of the differences between CPDs within the GIs and at the GBs (CPD$_{GB}$ − CPD$_{GI}$). Inset shows the movement of positive and negative ions at the GBs. *Source:* (a–e) Yun et al. [46].

the sample that changes the work function of the sample surface. Before applying any stimuli such as light or bias, ions are distributed in thermodynamically favored positions, which is illustrated in Figure 9.9e. When the bias voltage is applied to the tip, ions are redistributed at the surface according to the polarity of the applied bias voltage of the tip, in turn, changing the work function of the sample surface (Figure 9.9e). A positive bias will attract halide ions such as I or Br to the surface (or cation vacancies) and repel cations such as FA or MA (or halide vacancies) and vice versa for negative bias. After removing the voltage, the migrated ions at the top surface are relaxing back to their original position, thus, returning to the original CPD values after around 720 seconds.

Our KPFM maps in Figure 9.9b,d display clear changes of CPD between the GBs and GIs under different bias voltages. Initially, individual grains are distinguishable, GBs have lower CPD of around 15–20 mV compared with the GIs for all samples as shown in Figure 9.9b,d at 0 V. Once the ionic motion takes places by the bias voltage applied to the AFM tip, the contrast between GBs and GIs becomes weaker when the magnitude of the bias voltage is between +0.5 and +1 V. When the magnitude of the bias voltage is equal to and greater than +1.5 V, the contrast between GBs and GIs becomes pronounced. If the bias voltage is positive, CPD at the GBs is higher than the CPD within the GIs, see the brighter GBs in the "+1.5 V" and "+2 V" maps Figure 9.9b. If the bias voltage is negative, the CPD at the GBs is lower than the CPD within the GIs, see the darker contrasted GBs in the "−1.5 V" and "−2 V" maps in Figure 9.9d. Figure 9.9f plots the differences between the average CPDs at the GBs and the average CPDs within GIs ($CPD_{GB} - CPD_{GI}$) from Figure 9.9b,d. This indicates that the CPD change at GBs is more rapid compared with GIs, which clearly suggests that the GBs are the preferred migration paths for the mobile ion in halide perovskites.

9.3.2 Role of GBs for Ion Migration

The impact on the enhanced ion migration at GBs on the charge separation is elucidated by KPFM measurement under the different light conditions. Figure 9.10a shows the CPD maps and the corresponding line profile as the illumination conditions changed over time. The sample used here is $(FAPbI_3)_{0.85}(MAPbBr_3)_{0.15}$/mesostructured TiO_2/compact TiO_2/FTO (fluorine-doped tin oxide). There are a total of five stages involved in terms of charge separation and ion migration:

Stage 1: Initially in the dark, the line profile indicates that the average CPD is around 350 mV. In this configuration, a built-in voltage (V_{bi}) exists in between the perovskite and underneath TiO_2 layer [52]. The direction of the V_{bi} is toward the top surface of the perovskite (Figure 9.10 Stage 1), which can cause the positively charged ions such as cations (or halide vacancies) to migrate toward the top surface. These positively charged ions are known to create energy levels close to the conduction band, thus, making the sample surface more n-type [53]. Since GBs are the channel for ion migration, more ions are segregated at the GBs, thus, GBs have lower (or more n-type) CPD compared with the GIs.

Stage 2: When the light is turned on, photoexcited charge carriers are generated and separated by the V_{bi} formed across the perovskite/TiO_2 interface. As can be seen in Figure 9.10a, the average CPD abruptly increases to ~700 mV due to the accumulation of charge carriers at the top surface. In this case, holes are accumulated at the top surface as the V_{bi} direction is upward. As can be seen from the CPD map, the GBs became indistinguishable within the first four minutes of illumination. Schematic of what is happening during this stage is described in Figure 9.10b. The photo-generated holes at the top surface of the film start to attract negative ions while the photo-generated electrons at the bottom of the film attract positive ions. The movement of these ions is most rapid along the GBs, which makes the GBs indistinguishable as it is under a fast moving stage, thus, CPDs at GBs and GIs are similar during this stage.

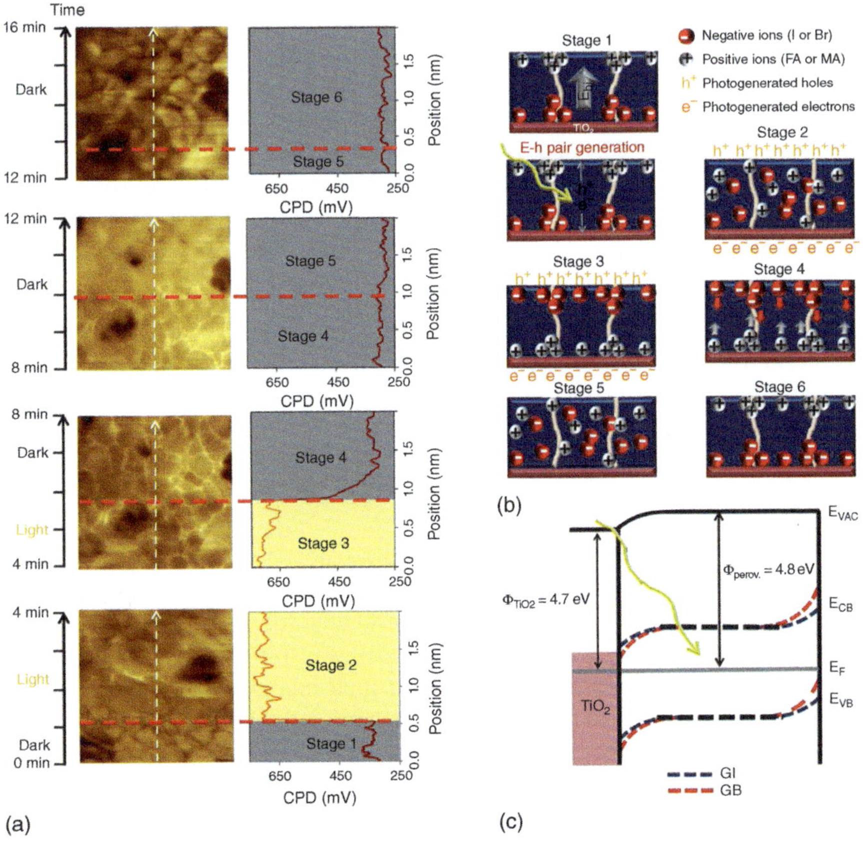

Figure 9.10 Effect of light-induced potential and charge separation. (a) CPD spatial maps and the corresponding CPD line profile (White line) over an area of 4 μm^2 in changing illumination conditions (ON = 0.3 W/m^2; OFF = 0 W/m^2) with time. (b) Schematic illustration of the ion movement based on results in (a). Note that CPD spatial maps are normalized with background subtraction in order to observe a contrast between GBs and GIs, and CPD profile displays absolute values. (c) Energy diagram of the perovskite layer under the influence of built-in potential (V_{bi}). E_{CB} is the conduction band energy, E_{VB} is the valence band energy and E_F is the Fermi level. *Source*: Yun et al. [46].

Stage 3: Continuing of the previous stage, the negative ions (both at the GBs and within the GIs) are at the top surface of the film, while the positive ions are at the bottom. The ions are more concentrated at the GBs, hence the bright outlines observed in the CPD map. Kim et al. [53] reported elemental defects such as MA, Pb, and I forming the shallow level near the band edges suggesting that I vacancy creates shallow-level near conduction band and can act as unintentional n-type while Pb and MA vacancies form an energy level near the valence band and can act as unintentional p-type. Such distribution of ions forms favorable band alignment as shown in Figure 9.10c, and it is more pronounced at the GBs, thus, facilitating the charge separation at the GBs.

Stage 4: The light is turned off at this stage and the CPD gradually decreases due to the absence of photo-generated carriers as well as the pulled ions are now trying to revert back to their original positions. CPD at the GBs remained higher than that of the GIs for ~six minutes after light was turned off due to the higher accumulation of the ions at GBs from the previous stage.

Stage 5: This is an intermediate stage whereby the pulled ions are reverting back to their original positions in the dark. This causes comparable concentrations of ions between the GBs and GIs, making the GBs indistinguishable in this stage.

Stage 6: At this stage, everything is reverted back to the original state. The positive ions (both at the GBs and within the GIs) are back at the top surface of the film, while the negative ions are at the bottom. As in stage 1, the ions are more concentrated at the GBs, hence the darker outlines observed in the CPD map.

As can be seen from earlier results, the GBs in OIHP solar cells have a unique role in terms of ion migration, which has a significant impact on charge separation properties.

9.4 Role of Grain Boundaries in Stability

Halide perovskites are prone to degrade under the moisture, oxygen, high temperature, and UV light. Particularly, multiple reports have suggested that moisture preferentially penetrates through GBs and eventually decomposes the perovskite. Figure 9.11a shows possible decomposition pathway for the $CH_3NH_3PbI_3$ perovskite in the presence of water. When water molecule is introduced **a**, a phase change occurs, hydrogen iodide **b**, the methylammonia **c**, and finally, a yellow solid PbI_2 phase are formed **d** [54]. Leguy et al. [55] reported the formation of hydrate phases in $CH_3NH_3PbI_3$ in the presence of water. It was suggested that the phase transition to the hydrate phase is isotropic on macroscopic scales (i.e. the average film composition is independent of depth), although the phases are likely to separate on a microscopic scale. A high degree of penetration of water molecules into $CH_3NH_3PbI_3$ thin films might arise from diffusion of water molecules along GBs. Wang et al. [56] reported a scaling behavior of grains induced by moisture diffusion through the GBs.

In this section, the moisture penetration through GBs in two different types of perovskites, $MAPbI_3$ and $FAPbI_3$ perovskite, will be discussed. These two perovskites

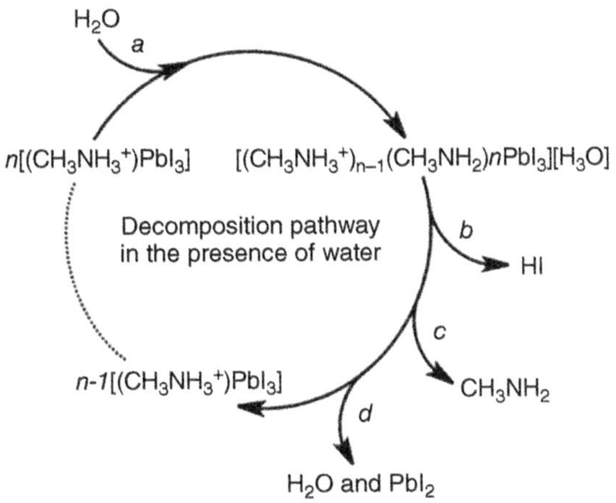

Figure 9.11 Proposed decomposition pathway of $CH_3NH_3PbI_3$, in the presence of water. *Source*: Frost et al. [54].

undergo distinct degradation pathways upon moisture penetration. GBs play important roles during moisture penetration and have a significant influence on the device performance.

9.4.1 $MAPbI_3$ Hydrated Phase at GBs

Leguy et al. [55] studied the formation of hydrate phase ($CH_3NH_3PbI_3{\cdot}H_2O$) in $MAPbI_3$ in the presence of water. Figure 9.12a depicts the forward to reverse bias photocurrent–voltage scans measured at different times during exposure of the device to moist air (77% RH) for three hours followed by exposure to a dry nitrogen stream for approximately five hours. Figure 9.12b shows the corresponding photocurrent–voltage (J–V) curves measured with reverse to forward bias scans. It is evident that both J_{sc} and J_{sc} drop significantly after the device was exposed to moisture for three hours. Following exposure of the device to dry nitrogen for five hours, the J–V curve is almost recovered to very close to its original state in the case of the forward to reverse bias scan. However, this recovery was much less complete for the reverse to forward bias scan, therefore a large J–V hysteresis exists in the device's J–V curves following the hydration–dehydration cycle. Figure 9.12c demonstrates the associated appearance of the monohydrate phase revealed by the XRD pattern that induced the loss in photovoltaic performance seen in Figure 9.12a,b. The XRD measurement of the complete solar cell exposed to moisture demonstrates the formation of $MAPbI_3$ monohydrate (peaks at ~8° and ~10°) in devices. These results clearly suggest that the formation of these hydrate phases can cause a dramatic drop in PCE. From the cross-sectional image in Figure 9.12d, the granular structure of the film with the GBs, it is expected that the efficient penetration of moisture and formation of the hydrate phase occur within the $MAPbI_3$ film along GBs or micro/mesopores, which is illustrated in Figure 9.12e. According to the authors, it is proposed that this partial conversion to $MAPbI_3$ hydrate may cause the isolation of the grains from each other, which would rapidly impede charge carrier transport resulting in increased recombination current at grain interfaces.

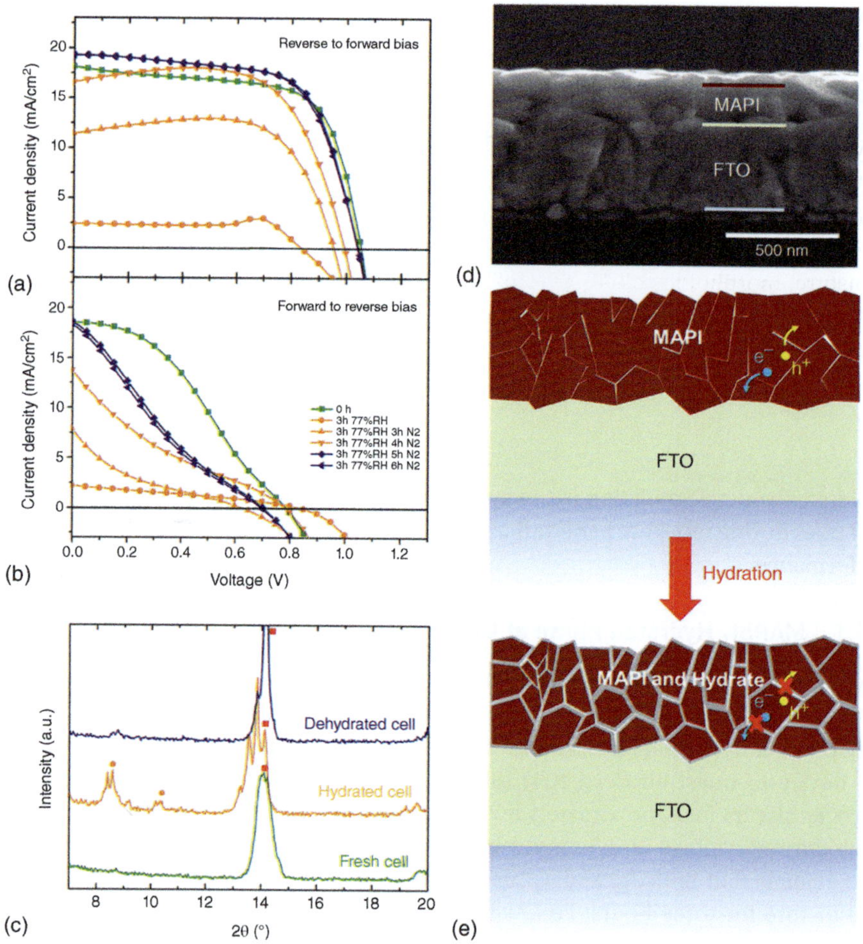

Figure 9.12 (a) and (b) Current–voltage characteristics showing full recovery of a planar heterojunction $MAPbI_3$ solar cell upon exposure to moisture; scanned from reverse to forward bias and from forward to reverse, respectively. (c) XRD patterns of a device before and after hydration showing the emergence of the hydrate specie. Dots are used to tag the main features according to the color of their respective spectra. (d) Cross-sectional scanning electron microscopy (SEM) image of $MAPbI_3$ deposited on a FTO-coated glass slide. (e) Presumed microscopic degradation model of $MAPbI_3$ thin films under partial hydration. *Source*: Leguy et al. [55].

9.4.2 Formation of Non-perovskite Phase at GBs of $FAPbI_3$

As mentioned earlier in Section 9.2.2, the $FAPbI_3$ perovskite has been proposed as more ideal absorber compared with $MAPbI_3$ due to having a more ideal band gap of ~1.45 eV for solar cell performance as well as better thermal stability [57–60]. However, the black cubic $FAPbI_3$ perovskite phase (α-phase) is only stable at relatively high temperature (above 160 °C), and it can convert into the hexagonal yellow non-perovskite $FAPbI_3$ phase (δ-phase) in ambient conditions [55, 61, 62].

Microstructural evolution of the degradation process of $FAPbI_3$ is studied using KPFM by the author [63]. Figure 9.13a–c shows the topography images of $FAPbI_3$ top surface of fresh, initial degradation, and complete degradation. The degraded sample used here is one that was stored at ~50% RH (relative humidity) for three days, which consists of black, pale yellow, and colorless areas. The corresponding CPD maps of the same areas marked in white boxes in the topography are obtained and displayed in Figure 9.13d–f. Figure 9.13a shows the morphology of a fresh sample that has closely packed grains with a size of 300–500 nm. After exposing to the moisture, morphology of the perovskite changes. Figure 9.13b shows some regions where adjacent grains are clustering to form a large grain as marked by red arrows. Within these large grains, round features from the original small grains can still be seen although the GBs have almost disappeared. Further degradation also appears

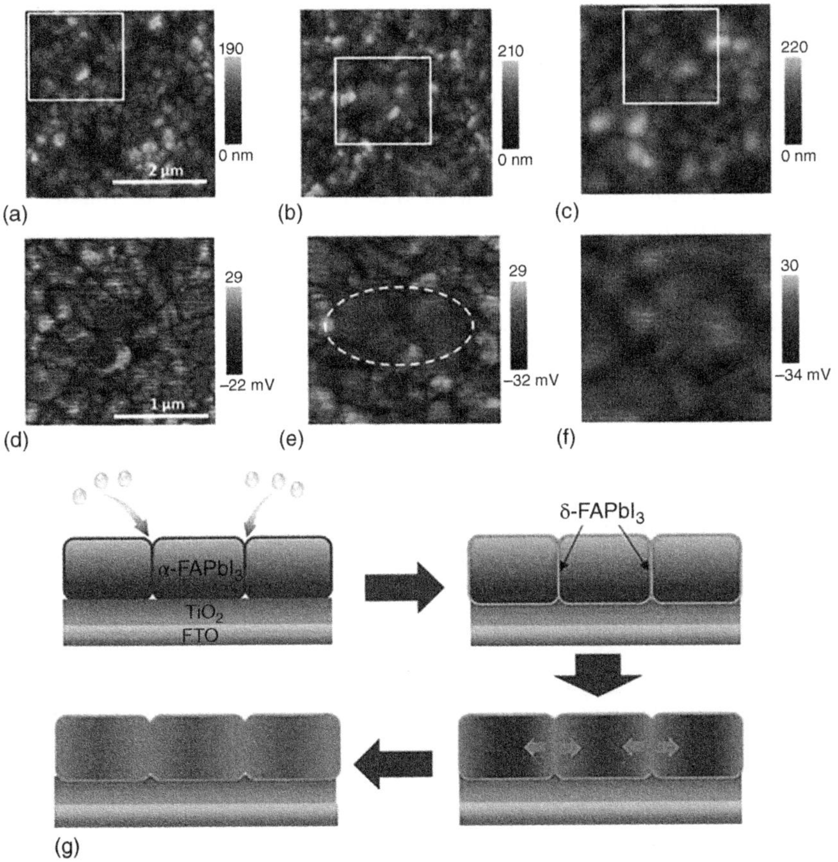

Figure 9.13 (a)–(f) Topography images of a $FAPbI_3$ perovskite film on c-TiO_2/FTO glass. (a) Fresh film when the film is dark black, (b) initial degradation where the film is pale yellow, and (c) with further degraded where the film is colorless, to show the different stages of the degradation process. (d–f) are the corresponding CPD images on the white marked areas in above topography images respectively. (g) Illustrations for the different stages of the degradation process. *Source*: (a–g) Yun et al. [63].

to smooth out the film as shown in the topography map in Figure 9.13c. Finally, the morphology consists of an appearance of several micrometer-sized crystals.

In the corresponding CPD maps, individual grains of fresh films can be clearly identified with 7–18 mV lower CPD at the GBs compared with the GIs as shown in Figure 9.13d. When the film is degraded, the CPD at the GBs is further lowered having darker contrast compared with the GIs. This could be responsible for the phase transition to the non-perovskite phase and a solvation process at the GBs. This observation implies that the electrical property of the GBs changes first, which could be owing to the moisture penetration through GBs. Once the moisture penetrates through the GBs, the transition to the non-perovskite yellow phase could be initiated from the GBs, which is confirmed by XRD pattern of the degraded sample. The cluster outlined by the white dotted line in Figure 9.13e has lower CPD (12–26 mV lower) compared with non-degraded regions. In fact, this value is comparable to the CPD at the GBs that are degraded, which is now 15–32 mV lower compared with the GIs. This could imply that the phase transition progressively proceeds toward the GIs from the GBs. These processes are schematically illustrated in Figure 9.13g. Further degradation continues the phase transition toward the GIs and transforms all the neighboring grains to the non-perovskite phase. Consequently, the grains form a big cluster of non-perovskite crystals as shown in Figure 9.13f. These results are very different from the degradation process of $MAPbI_3$ perovskite where prolonging the exposure time to humidity induced more GBs and steps [64].

References

1 Nickel, N., Johnson, N., and Jackson, W. (1993). *Appl. Phys. Lett.* 62: 3285.
2 Terry, M.L. et al. (2005). *Appl. Phys. Lett.* 86: 172108.
3 Kronik, L., Cahen, D., and Schock, H.W. (1998). *Adv. Mater.* 10: 31.
4 Yun, J.S. et al. (2015). *J. Phys. Chem. Lett.* 6: 875.
5 Castro-Méndez, A.F., Hidalgo, J., and Correa-Baena, J.P. (2019). *Adv. Energy Mater.* 9: 1901489.
6 Nie, W. et al. (2015). *Science* 347: 522.
7 Fei, C. et al. (2017). *Adv. Energy Mater.* 7: 1602017.
8 Euvrard, J., Gunawan, O., and Mitzi, D.B. (2019). *Adv. Energy Mater.* 9: 1902706.
9 Ke, W. et al. (2016). *Adv. Mater.* 28: 5214.
10 Seo, J.Y. et al. (2016). *Adv. Energy Mater.* 6: 1600767.
11 Yang, W.S. et al. (2017). *Science* 356: 1376.
12 Kim, H.-S. et al. (2012). *Sci. Rep.* 2: 591.
13 Yang, M. et al. (2017). *Phys. Chem. Chem. Phys.* 19: 5043.
14 Long, R., Liu, J., and Prezhdo, O.V. (2016). *J. Am. Chem. Soc.* 138: 3884.
15 Yin, W.J., Shi, T., and Yan, Y. (2014). *Adv. Mater.* 26: 4653.
16 Park, J.-S. et al. (2019). *ACS Energy Lett.* 4: 1321.
17 Leijtens, T. et al. (2014). *ACS Nano* 8: 7147.
18 Barboni, D. and De Souza, R.A. (2018). *Energy Environ. Sci.* 11: 3266.
19 Edri, E. et al. (2014). *Nano Lett.* 14: 1000.
20 Chen, Q. et al. (2014). *Nano Lett.* 14: 4158.

21 Sherkar, T.S. et al. (2017). *ACS Energy Lett.* 2: 1214.
22 Kim, Y.C. et al. (2016). *Adv. Energy Mater.* 6: 1502104.
23 Liu, F. et al. (2016). *Adv. Energy Mater.* 6: 1502206.
24 Jacobsson, T.J. et al. (2016). *J. Am. Chem. Soc.* 138: 10331.
25 MacDonald, G.A. et al. (2016). *Energy Environ. Sci.* 9: 3642.
26 Senocrate, A. et al. (2859). *ACS Energy Lett.* 2019: 4.
27 Kim, J. et al. (2016). *J. Phys. Chem. C* 120: 11262.
28 Saliba, M. et al. (2016). *Science* 354: 206.
29 Saliba, M. et al. (2016). *Energy Environ. Sci.* 9: 1989.
30 Zhang, M. et al. (2017). *ACS Energy Lett.* 2: 438.
31 Kim, J. et al. (2020). *J. Am. Chem. Soc.* 142 (13): 6251–6260.
32 Quan, L.N. et al. (2649). *J. Am. Chem. Soc.* 2016: 138.
33 Lee, D.S., Yun, J.S., Kim, J. et al. (2018). *ACS Energy Lett.* 3 (3): 647–654. https://doi.org/10.1021/acsenergylett.8b00121.
34 Wang, Z. et al. (2017). *Nat. Energy* 2: 17135.
35 Xu, J. et al. (2015). *Nat. Commun.* 6: 1.
36 Kim, W. et al. (2018). *Adv. Energy Mater.* 8: 1702369.
37 Zhang, Y. et al. (2015). *Mater. Horiz.* 2: 315.
38 Unger, E. et al. (2014). *Energy Environ. Sci.* 7: 3690.
39 Tress, W. et al. (2015). *Energy Environ. Sci.* 8: 995.
40 Chen, B. et al. (2015). *J. Phys. Chem. Lett.*
41 Shi, J. et al. (2015). *Appl. Phys. Lett.* 107: 163901.
42 Kim, D. et al. (2019). *Nat. Commun.* 10: 1.
43 Eames, C. et al. (2015). *Nat. Commun.* 6: 7497.
44 Azpiroz, J.M. et al. (2015). *Energy Environ. Sci.* 8: 2118.
45 Shao, Y. et al. (2016). *Energy Environ. Sci.* 9: 1752.
46 Yun, J.S. et al. (2016). *Adv. Energy Mater.* 6: 1600330.
47 Xing, J. et al. (2016). *Phys. Chem. Chem. Phys.* 18: 30484.
48 Yuan, Y. and Huang, J. (2016). *Acc. Chem. Res.* 49: 286.
49 Baumgart, C., Helm, M., and Schmidt, H. (2009). *Phys. Rev. B* 80: 085305.
50 Masuda, H. et al. (2017). *Nanoscale* 9: 893.
51 Sanchez, R.S. et al. (2014). *J. Phys. Chem. Lett.* 5: 2357.
52 Jiang, C.-S. et al. (2015). *Nat. Commun.* 6: 1.
53 Kim, J. et al. (2014). *J. Phys. Chem. Lett.* 5: 1312.
54 Frost, J.M. et al. (2014). *Nano Lett.* 14: 2584.
55 Leguy, A.M. et al. (2015). *Chem. Mater.* 27: 3397.
56 Wang, Q. et al. (2017). *Energy Environ. Sci.* 10: 516.
57 Eperon, G.E. et al. (2014). *Energy Environ. Sci.* 7: 982.
58 Pang, S. et al. (2014). *Chem. Mater.* 26: 1485.
59 Lv, S. et al. (2014). *Phys. Chem. Chem. Phys.* 16: 19206.
60 Eperon, G.E. et al. (2014). *J. Phys. Chem. Lett.* 6: 129.
61 Binek, A. et al. (2015). *J. Phys. Chem. Lett.* 6: 1249.
62 Koh, T.M. et al. (2013). *J. Phys. Chem. C* 118: 16458.
63 Yun, J.S. et al. (2018). *Adv. Funct. Mater.* 28: 1705363.
64 Li, D. et al. (2016). *J. Phys. Chem. C* 120: 6363.

10

Roles of Center Cations

Biwas Subedi[1], Juan Zuo[1,2], Marie Solange Tumusange[1], Maxwell M. Junda[1], Kiran Ghimire[1], and Nikolas J. Podraza[1]

[1] *University of Toledo, Department of Physics and Astronomy & Wright Center for Photovoltaics Innovation and Commercialization, 2801 West Bancroft Street, Toledo, OH 43606, USA*
[2] *Xiamen University of Technology, Fujian Provincial Key Laboratory of Functional Materials and Applications, Department of Materials Science and Engineering, Road Ligong No. 600, Xiamen 361005, P. R. China*

10.1 Introduction

All perovskites share the same crystal structure, which includes orthogonal indistinguishable crystal axes of equal length producing a cubic geometric shape; overlapping primitive cubic sub-lattices for each of the three distinct ions; bases of each ion type placed on the vertices, center, and center of the faces of the unit cell (Figure 10.1a); and belonging to the *Pm3m* space group. In the case of hybrid organic–inorganic lead-halide-based perovskites of interest for high-efficiency thin film photovoltaics (PV), the chemical composition is ABX_3, where A is a 1+ organic or inorganic cation, B is a 2+ inorganic cation, and X is a 1− halide anion. The most commonly considered A-cations are methylammonium (CH_3NH_3 or MA), formamidinium (CH_5N_2 or FA), guanidinium (CH_6N_3 or GA), phenylethylammonium ($C_8H_{12}N$ or PEA), cesium (Cs), and rubidium (Rb). B-cations are typically lead (Pb), tin (Sn), or Pb + Sn mixtures. Manganese (Mn), cadmium (Cd), and germanium (Ge) have also been considered for implementation. The X-anion halides are iodine (I), bromine (Br), chlorine (Cl), and their mixtures. In principle, other cations and anions of the required charge can also be used on the appropriate lattice site; however, consideration must be given to ensure that the ionic size and bonding strength results in a sufficiently stable cubic perovskite phase material. Traditionally, the perovskite unit cell is drawn with the A-cation on the vertices of the cube, the B-cation in the center of the cube, and the X-anion in the center of each of the six faces of the cube as shown in Figure 10.1a. The nearest neighbors of A-cations are X-anions, B-cations are X-anions, and X-anions are B-cations. For the hybrid organic–inorganic lead-halide-based perovskite materials of interest for PV, it is common to visually shift the axes of the perovskite unit cell one half unit cell length along each principle direction so that the A-cation is now in the

Hybrid Perovskite Solar Cells: Characteristics and Operation,
First Edition. Edited by Hiroyuki Fujiwara.

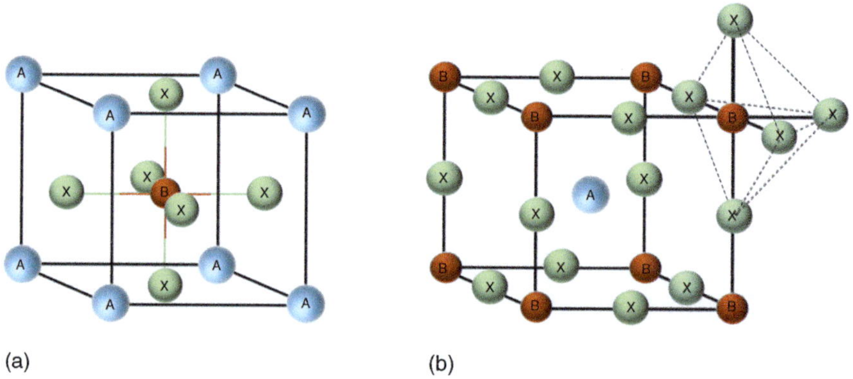

Figure 10.1 Representations of ABX_3 perovskite unit cells centered on the (a) B-cation and (b) A-cation.

"center" with no shifts in the relative ionic positions or nearest neighbors as shown in Figure 10.1b. The A-cation is often referred to as the center cation, and the nature of this ion is of critical importance to ensure the cubic crystal phase structure, producing an environmentally stable perovskite, manipulating the above band gap absorption, and enabling fabrication of high-efficiency thin film PV.

The earliest of this class of perovskite considered for PV was $MAPbI_3$. This composition exhibits a band gap near 1.55 eV, which, although slightly higher than the ideal value for single junction PV, still resulted in very high initial efficiency devices. Unfortunately, these devices had poor stability [1–5]. To address material stability, different combinations of cations and anions were applied with $FAPbI_3$ and other compositions yielding similar band gaps with improved stability [6–20]. Additionally, as the goal of PV research is to improve device efficiency, thin film polycrystalline perovskites have been applied and studied in multiple junction device structures. In these device structures, two or more PV junctions are integrated in mechanically stacked four-terminal devices or monolithically integrated two-terminal devices [21–23]. The goal of these device complications is to combine junctions with different band gap absorbers to collect incident illumination over larger portions of the solar irradiance spectrum.

The A-cation is a substantial contributor to material stability as demonstrated between comparisons of $MAPbI_3$, $FAPbI_3$, $CsPbI_3$, and mixed A-cation perovskites; however, both the B-cation and the fabrication process used also have impacts. Consider $MAPbI_3$ as a baseline material for comparison. Replacing a fraction of Pb with Sn red shifts of the band gap from approximately 1.55 eV to a minimum of 1.20 eV [24–29]. Due to band bowing effects, further increases in Sn content increase the band gap to approximately 1.40 eV. Replacing I with Br or Cl blue shifts the band gap substantially to 2.3 eV when I is replaced with Br and 3.1 eV when replaced with Cl [1, 30]. The choice of A-cation has only a small impact in the band gap energy; however, there are substantial interactions with the nearest neighbor X-anion, which affect the band structure, above band gap electronic transitions, and above band gap optical absorption [30, 31]. The range of chemical possibilities for each ionic site results in complicated interactions in how the crystal structure will behave without yet considering other properties of real thin film materials.

In thermodynamic equilibrium, a material should crystallize into the perovskite crystal structure if that structure yields the lowest Gibbs free energy. However, the perovskite materials for PV are polycrystalline thin films and may be fabricated in nonequilibrium process with nonstoichiometric compositions. This complicates the situation for thermodynamic stability. Gibbs free energy will vary with temperature and pressure. Starting from the desired cubic perovskite phase, reduction in temperature leads to the formation of tetragonal or orthorhombic phases [5, 7, 32, 33]. Increases in pressure cause transitions from the cubic perovskite to tetragonal and orthorhombic phases as well [34]. In $FAPbI_3$, a hexagonal phase is also thermodynamically stable under certain pressure and temperature conditions. For polycrystalline thin films, strain energy and interfacial energy arising from grain boundaries also impact the sample structure which yields the lowest Gibbs free energy. A further complication is that the lowest Gibbs free energy identifies the most stable phase under a given set of conditions; other phases may exist in metastable states. Hysteresis effects in the phase transition cycle from one phase to another and back also enable the existence of nonequilibrium phases. Compositional nonstoichiometry in thin films introduces the possibility of stacked perovskite sheet or two-dimensional perovskite phase structures whereby thin perovskite layers are interspaced with other constituent materials [35]. Only the cubic perovskite phase of all ABX_3 compositions has been demonstrated to be suitable to produce the highest-efficiency thin film solar cells; therefore, only this phase and its stability dependent upon the center cation will be discussed in this chapter. The complexity induced by real materials with grain boundaries, strain effects, phase transition hysteresis, off-stoichiometry, and the presence of nonequilibrium phases all must be considered in the fabrication of the desired cubic perovskite phase.

Even after fabrication of material exhibiting the desired cubic perovskite phase, environmental stability varies depending upon perovskite composition. Decomposition of these perovskites into different phases or phase-segregated mixtures has been observed with exposure to humidity, oxygen, light, heat, and electrical bias [3, 4, 16, 29, 36–42]. With light and electrical bias, ion migration can occur and leads to spatially dependent compositional differences and phase segregation. Exposure to humidity results in incorporation of water molecules into the crystal structure and the formation of water intercalated and hydrated phases. Oxygen exposure alters bonding in the material, most notability in Sn-containing perovskites where the presence of oxygen can shift Sn^{2+} to Sn^{4+}, which destabilizes the crystal structure. Each of these processes may occur simultaneously for perovskite absorber layers in PV; therefore, it is necessary to choose a composition with the desired optoelectronic properties that also exhibits high stability and limits PV device degradation. A-cations exhibit different ion mobilities, which enables fabrication of absorber layers more resistant to phase decomposition toward that goal.

Here we will discuss the role of the center A-cations as well as interplay with B-cations and X-anions on crystal phase structure, perovskite film stability, optoelectronic properties, and solar cell performance. For design of cubic perovskite phase stable materials, empirical tolerance factor criteria will be reviewed [7, 43–47]. Perovskite film stability depending on exposure to humidity, oxygen, light, and bias leading to material decomposition will be considered for $MAPbI_3$ and $FAPbI_3$. Optoelectronic properties including band gaps and band structure will be tracked as functions of A-cation, B-cation, and X-halide compositions. Trends in solar cell performance parameters as functions of A-cation composition conclude this chapter.

10.2 Cubic Perovskite Phase Tolerance Factor

To form a cubic perovskite crystal structure, the appropriate combinations of ionic radii are needed in addition to the balance of cation and anion charges. In a very simplistic view, consider that there is only a range of combinations of ions with the appropriate radii that enable each component ion to fit together into a cubic perovskite crystal structure. These criteria apply to all perovskites, not only the hybrid organic–inorganic lead-halide-based ABX_3 compounds here. Historically, the Goldschmidt tolerance factor is an empirical relationship to determine if combinations of cations and anions yield a given crystal structure [7, 42, 44–46]. This version of the tolerance factor (t) is defined as:

$$t = \frac{r_A + r_X}{\sqrt{2}(r_B + r_X)} \tag{10.1}$$

where r_A is the A-cation radius, r_B is the B-cation radius, and r_X is the X-anion radius. When mixtures of ions occupy the same type of site, the effective ionic site radius is the weighted average of the constituents. For example, in mixed A-cation $FA_{1-x}Cs_xPbI_3$, the effective A-cation radius, r_A, would be equivalent to $(1-x)r_{FA} + xr_{Cs}$. For the organic–inorganic lead halide ABX_3 compounds, the choice of A-cation enables a range of tolerance factors for given ranges of B-cations and X-anions. The most common B-cations are Pb^{2+} and Sn^{2+}, with similar Shannon ionic radii [48] of 119 and 115 pm, respectively. The halides are I^-, Br^-, and Cl^- with Shannon ionic radii of 220, 196, and 181 pm, respectively. The choice of components restricts the window of ionic radii variation to ~3% for the B-cations and ~20% for the X-anions. The ionic radii of FA^+, MA^+, and Cs^+ are 253, 216, and 167 pm, respectively, enabling a variation of ~40% for the A-cation radius with composition. The choice of A-cation enables the largest variation in ionic radius of the constituent ions and the largest impact in the numerator of Eq. (10.1), making the center cation extremely important for determining the tolerance factor and cubic perovskite stability.

Now what physically does the tolerance factor imply and what ranges of t result in a stable cubic perovskite? For classical inorganic perovskites, the cubic perovskite phase forms for $0.9 \le t \le 1$, the orthorhombic phase forms for $0.71 \le t < 0.9$, the hexagonal or tetragonal phase forms at $t > 1$, and other phases form at $t < 0.71$ [43]. Physically, the cubic phase forms when the A- and B-cations have similar sizes, the orthorhombic phase forms when the A-cations are too small to fit within the B-cation sub-lattice, and the hexagonal or tetragonal phase forms when the A-cation is too big to fit within the B-cation sub-lattice. Tolerance factors are empirical relationships, and the ranges require modification for organic–inorganic lead-halide-based ABX_3 compounds. For ABX_3 compounds of interest for PV, the cubic perovskite phase extends from $0.9 \leq t \leq 1.07$, cubic but distorted octahedrally tilted perovskites are found between $0.8 \leq t < 0.9$, orthorhombic phases are found between $0.71 \leq t < 0.8$, the hexagonal or tetragonal phase forms at $t > 1.07$, and other phases form at $t < 0.71$ [7, 43, 45]. These limits are somewhat approximate in that input ionic radii are assumed in the model.

Note that these are approximate ranges and that the calculated values of t depend on assumptions made for the ionic radii. Travis et al. [45] reported a revised system of determining phase structure based on the ionic radii used in tolerance factor calculation and on the octahedral factor (μ) defined as:

$$\mu = \frac{r_B}{r_X} \tag{10.2}$$

which is the ratio of B-cation to X-anion radii. Revised ionic radii account for the interaction between larger, heavier, and less electronegative I, Cl, and Br X-anions and the A- and B-cations. The Shannon ionic radii were developed considering ionic radii of bonding in fluorides and oxides, which are smaller and more electronegative anions. The ionic radii in compounds depend on the other ions, so more realistic values of the radii of specific ions in a specific crystal structure comprised of other ions helps to refine calculated values of the tolerance factor. The octahedral factor $0.41 < \mu < 0.73$ identifies another criterion for formation of the cubic perovskite phase by defining the size of the B-cations assuming a hard sphere geometrical model such that there is not substantial interaction between the X-anions.

An approach based on similar input parameters to the tolerance and octahedral factors is an empirical expression reported by Bartel et al. [47] defining an alternative tolerance factor (τ) as:

$$\tau = \frac{r_X}{r_B} - n_A \left[n_A - \frac{r_A}{r_B \ln \left({}^{r_A}/_{r_B} \right)} \right] \tag{10.3}$$

where n_A is the oxidation state of the A-cation. In this approach, cubic perovskites are formed for $\tau < 4.18$. The center A-cation radius and oxidation state dominate the expression, like the A-cation radius dominating the Goldschmidt tolerance factor.

Each of these criteria is based on empirical relationships and serves as a means of identifying if given ABX_3 compounds are likely to be perovskites. The tolerance factor values depend upon input parameters, such as ionic radii, and will vary with any assumptions made in obtaining those parameters. As such, the limits on perovskite phase or any other phase stability from these tolerance factors should be considered as a guideline. For $FAPbI_3$, the tolerance factor calculated can lie outside that of the original Goldschmidt criteria depending upon assumptions made; however, the cubic perovskite phase can be produced and is quite stable. Alternately, if a compound lies within the stability window of the cubic perovskite phase, the fabrication process employed may not produce the desired perovskite phase. For thin films this can occur due to a variety of reasons including nonstoichiometric composition, substrate-induced strain, and the kinetics of the deposition process. In these cases, identifying compounds with tolerance factors close to that of the cubic perovskite phase is a reasonable way of down-selecting compositions; however, fabrication of material in the desired perovskite phase may still be challenging. ABX_3 compounds with MA, FA, Cs, and mixed A-cations; Sn, Pb, and mixed B-cations; and I, Br, Cl, and mixed X-anions have been fabricated in the perovskite phase as have others with inclusions of GA [18], PEA [49], and Rb [16, 50] A-cations and some Ge and Mn [51] B-cations.

10.3 Thin Film Stability

Aside from crystallization in the cubic perovskite phase, the stability of polycrystalline thin films under external conditions such as humidity and oxygen exposure and light and electrical bias are also considerations. ABX_3 perovskites behave differently depending on composition. Assuming that X-anions are purely I, the roles of Pb, Sn, and mixed B-cations and MA, FA, Cs, and mixed A-cations strongly influence film stability, but each in different ways.

Beginning with $MAPbI_3$, this material will decompose into its constituents over the course of time with exposure to humidity and oxygen, although humidity is likely the more dominant driving force. In the case of humidity exposure, water molecules will intercalate into the $MAPbI_3$ unit cell and ultimately form a monohydrated phase [5, 49, 52]. In both situations, the water molecules are hydrogen bonded to MA, Pb, and I. This interaction of water molecules with the entire perovskite crystal structure lowers energetic barriers for vacancy-driven ion migration to occur, which in turn enables decomposition of the perovskite [3, 38]. This decomposition has been observed to occur initially at the substrate/film and film/ambient interfaces and consists of phase segregation of lead iodide (PbI_2) and methylammonium iodidie (MAI) as shown in Figure 10.2 [36]. Over time the stacked perovskite sheet or two-dimensional perovskite structure occurs along with changes in the optoelectronic response of the film Figure 10.3 [37]. Eventually, the MAI may further decompose into I_2, CH_3I, and NH_3 leaving behind a discontinuous PbI_2 layer.

In the case of $FAPbI_3$, such phase segregation does not occur. Upon exposure to humidity, the cubic perovskite phase of $FAPbI_3$ locally converts to the hexagonal phase starting near grain boundaries as shown in Figure 10.4 [3]. Although the hexagonal phase is not desirable for the fabrication of PV devices as its optoelectronic response is distinct from the cubic perovskite phase [30], the film is stabilized. Beginning from material conversion near grain boundaries, larger cubic phase perovskite grains are converted into smaller hexagonal $FAPbI_3$ grains due to humidity-enhanced ion migration. For $CsPbI_3$, the cubic perovskite phase has been observed to transition to the orthorhombic phase as $FAPbI_3$ transitions to the hexagonal phase or remains stable in the cubic phase depending upon the material preparation method. The orthorhombic $CsPbI_3$ phase is stable at room temperature; however, the cubic perovskite phase has been produced by using rapid cooling from

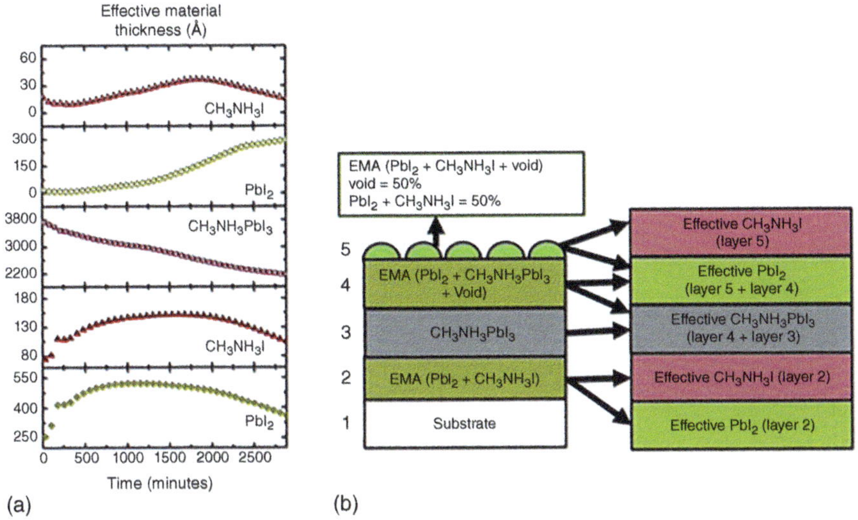

Figure 10.2 Decomposition of a $MAPbI_3$ perovskite into constituent MAI and PbI_2 components at the substrate/film and film/ambient interfaces as analyzed using in situ real-time spectroscopic ellipsometry (a) and the schematic structural model (b). *Source:* Used with permission of IOP Publishing, from Ghimire et al. [36]; permission conveyed through Copyright Clearance Center, Inc.

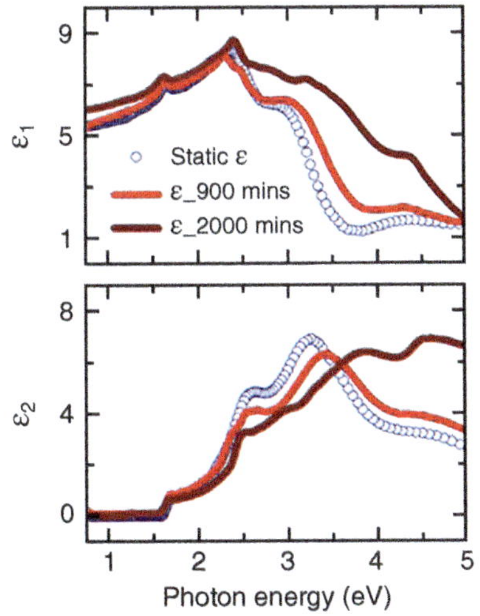

Figure 10.3 Complex optical response in the form of the complex dielectric function ($\varepsilon = \varepsilon_1 + i\varepsilon_2$) spectra showing substantial changes over time (0 or static, 900, 2000 minutes) for a perovskite component in a film undergoing phase decomposition. *Source:* Reprinted, with permission, from Ghimire et al. [37]. ©2016 IEEE.

high temperature fabrication, the inclusion of Eu^{3+} ions, the formation of small well-passivated nanocrystals, interfacial strain, and combining some of these effects [53–56]. Recently, highly stable $FAPbI_3$ has been produced through manipulation of the fabrication process as well [57], although no such humidity-stable $MAPbI_3$ has yet to be produced.

An explanation for lower stability of $MAPbI_3$ relative to even $FAPbI_3$, which converts to the hexagonal phase, relies on the ability of halide X-anions to migrate

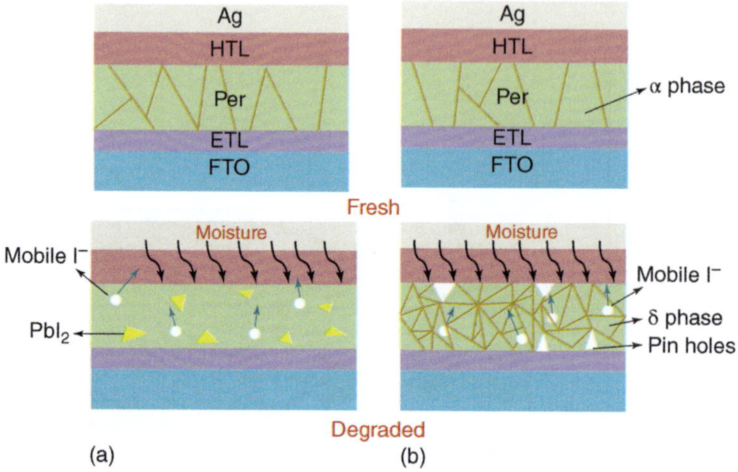

Figure 10.4 Schematics of decomposition of (a) $MAPbI_3$ via the formation of PbI_2 inclusions and (b) $FAPbI_3$ via transition to the hexagonal (δ) phase. *Source*: Reproduced from Ref. Zhang et al. [3] with permission from The Royal Society of Chemistry.

through the material. Incorporation of water molecules into $MAPbI_3$ ultimately results in the formation of PbI_2 inclusions and vacancies (Figure 10.4) [3]. The presence of these vacancies promotes halide X-anion diffusion and migration through the formerly cubic perovskite crystal structure as both the intercalated and monohydrated phases have lower activation energies for X-anion migration than in cubic perovskite $MAPbI_3$. Once substantial X-anion diffusion occurs, the crystal structure may begin to decompose further. In the case of $MAPbI_3$ absorber layers in solar cells, the device structure provides electrical bias as the perovskite is between the hole transfer layer (HTL) and the electron transfer layers (ETL), and photogeneration of charge carriers under illumination further promotes X-anion migration. Similarly, in $FAPbI_3$ X-anion migration occurs; however, the formation of the hexagonal $FAPbI_3$ phase is slower than segregation of PbI_2 from $MAPbI_3$ making both film and PV devices more stable upon exposure to humidity and during device operation, respectively. A direct numerical comparison of operational stability of $FAPbI_3$ and $MAPbI_3$-based PV devices from the literature is difficult since absorber layers are prepared with different methods and incorporated into different device structures. However, in aggregate $FAPbI_3$ devices retain a higher percentage of initial power conversion efficiency for the same operational or storage times (Figure 10.5) [3]. For this comparison, the only difference is the choice of A-cation. Although X-anion mobility is subject to a greater energetic barrier in $MAPbI_3$ than in its water intercalated and monohydrated phases, it is insufficient to prevent perovskite decomposition. For $FAPbI_3$ perovskites, the energetic barrier preventing X-anion diffusion is higher than in $MAPbI_3$ resulting in more stable films even when fabricated under conditions that will lead cubic $FAPbI_3$ to transition to the hexagonal phase. Other $FAPbI_3$ perovskites with sufficiently large grain sizes and stable grain boundary characteristics retain the

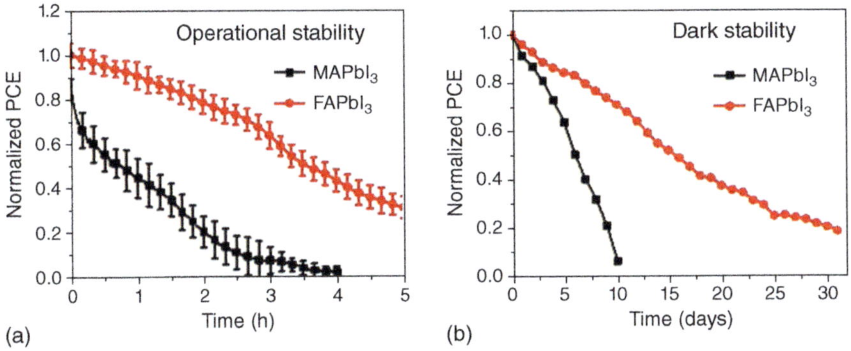

Figure 10.5 Normalized performance in terms of power conversion efficiency (PCE) of similarly fabricated $MAPbI_3$ and $FAPbI_3$ solar cells as functions of (a) storage time and (b) operational time. *Source*: Reproduced from Ref. Zhang et al. [3] with permission from The Royal Society of Chemistry.

cubic perovskite structure even after exposure to humidity and during PV device operation [57].

Another pathway to improve perovskite phase stability is through exploiting configurational entropy by using mixed A-cations and B-cations in films. Gibbs free energy depends on both enthalpy and entropy contributions, with the lowest Gibbs free energy at nonzero temperature dictated by a compromise of low enthalpy and high entropy. For highly ordered material, enthalpy is low, but entropy is also low. For highly disordered material, entropy is high, but enthalpy is also high. Provided that the tolerance factor does not extend beyond that for the desired cubic perovskite phase, substitution of a given ion by another ion of the same charge increases configurational entropy without substantially increasing enthalpy. A variety of examples exist for using mixed A-cations, mixed B-cations, or both in ABX_3 perovskites [7, 10–15, 17–20, 22, 23, 25, 26, 28, 29, 46, 58, 59]. Depending on material fabrication and other characteristics, mixed cations typically produce more environmentally stable films, initially higher efficiency PV devices, and more stable PV devices over time. For two examples, consider the sub band gap absorption induced by defects upon environmental exposure to laboratory ambient air for $FA_{1-x}Cs_xPbI_3$ [59] and $(FASnI_3)_x(MAPbI_3)_{1-x}$ perovskite films [29].

For mixed A-cation $FA_{1-x}Cs_xPbI_3$, sub gap absorption is at a minimum for $x = 0.2$ over a range from $x = 0$ to 0.4 (Figure 10.6a) [59]. Using optical sum rules, the estimated defect density is also the lowest for the $x = 0.2$ composition (Figure 10.6b) [59]. Single-junction solar cells incorporating absorbers fabricated by the same methods showed the highest efficiency for the device incorporating the $FA_{0.8}Cs_{0.2}PbI_3$ perovskite. In this case, the predominately FA A-cations already promote high material stability; however, configurational entropy is increased by limited Cs incorporation to potentially lower Gibbs free energy.

For $(FASnI_3)_x(MAPbI_3)_{1-x}$ perovskites, both A- and B-cations are mixed compositions. The lowest sub band gap absorption upon atmospheric exposure is observed for the $x = 0.5$ and 0.6 compositions (Figure 10.7a,b) [29]. These compositions have

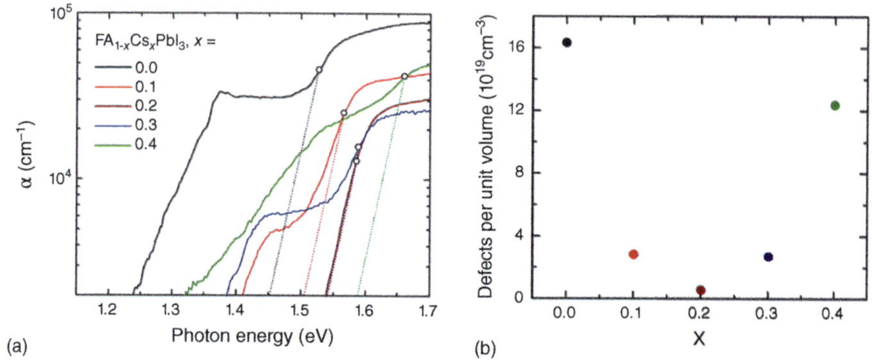

Figure 10.6 (a) Absorption coefficient (α) spectra for $FA_{1-x}Cs_xPbI_3$ thin films exposed to laboratory ambient air showing different absorption below the band gap (band gap energies denoted by ○) as a function of composition. Extrapolated sub gap absorption assuming a 24 meV Urbach tail below the band gap is also shown (dotted lines). *Source*: (a) Reproduced with permission from Subedi et al. [60]. (b) Defect density calculated from sub gap α using optical sum rules. *Source*: Subedi et al. [59].

comparable amounts of FA and MA as well as Sn and Pb. Even though there is not a strong predominance of FA over MA, its presence still helps limit X-anion mobility. The approximately even mixture of the two A-cation species results in high configurational entropy to lower the Gibbs free energy. Similarly, high configurational entropy also occurs through the approximately even mixture of the two B-cation species, further stabilizing these films. Among this composition series, these intermediate compositions exhibit the best environmental stability as reflected in the low defect density estimated from optical sum rules and highest solar cell efficiency. Prior to exposure to laboratory ambient air, $(FASnI_3)_x(MAPbI_3)_{1-x}$ composition perovskites ranging from $x = 0.3$ to 0.8 demonstrated very low Urbach energies, indicative of low defect concentrations, implying that configurational entropy imposed by both mixed A- and B-cations is important to film stability (Figure 10.7c) [29].

The role of Sn in $(FASnI_3)_x(MAPbI_3)_{1-x}$ perovskites leads to another consideration for film stability. The B-cation in perovskites for solar cells is most commonly Pb for wide and intermediate band gap PV or a mixture of Pb and Sn for narrow band gap PV. Sn is particularly sensitive to oxygen as it may shift from the Sn^{2+} configuration necessary for bonding in the cubic perovskite crystal structure to Sn^{4+}. The shift in Sn valence makes the perovskite phase unstable and quickly leads to reduced solar cell efficiency. As an example, the $x = 0.8$ $(FASnI_3)_x(MAPbI_3)_{1-x}$ film exposed to laboratory ambient air exhibits the highest sub band gap absorption and associated defect density among the studied compositions as well as the largest increase in defect density upon exposure (Figure 10.7) [29]. The effect is less pronounced for $x = 0.7$. Both compositions yielded low Urbach energies prior to atmospheric exposure indicating good electronic quality and low defect density; however, the oxygen present in laboratory ambient air caused rapid film decomposition. The sensitivity of high Sn-content perovskites to oxygen is also reflected in PV device performance, as will be discussed later.

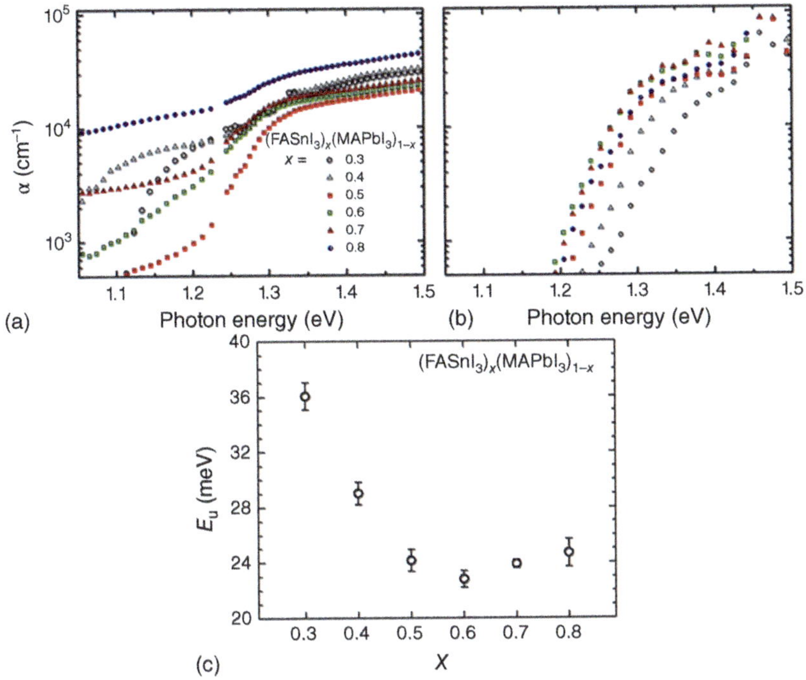

Figure 10.7 Spectra in α for $(FASnI_3)_x(MAPbI_3)_{1-x}$ thin films (a) exposed to laboratory ambient air and (b) without air exposure showing different extents of sub gap absorption induced by air exposure as a function of composition. (c) Urbach energy (E_u) as a function of composition determined from unexposed α. *Source*: Reprinted from Subedi et al. [29], with permission of AIP Publishing.

Overall, the most stable perovskite films at present are produced with substantial or predominate FA based A-cations due to the higher energetic barrier for X-anion migration compared with perovskites fabricated predominately with MA. However, it should be noted that large grain sizes in perovskite films are often attributed to the presence of MA [61–63]. Therefore, mixed FA + MA A-cation perovskites can have large grain sizes with a uniform grain size distribution, which is beneficial for device efficiency and stability [10, 13–15, 17, 19]. Incorporation of mixed A-cations increases configurational entropy to potentially reduce Gibbs free energy and results in more stable perovskites. Although incorporation of Sn is necessary to produce low band gap perovskites, high Sn-content films are highly sensitive to the presence of oxygen.

10.4 Optoelectronic Property Variations

ABX_3 perovskites of interest for PV are in general high electronic quality semiconductors, albeit sometimes not stable, with tunable band gap energies, reasonable mobilities and carrier concentrations, and long minority carrier diffusion lengths.

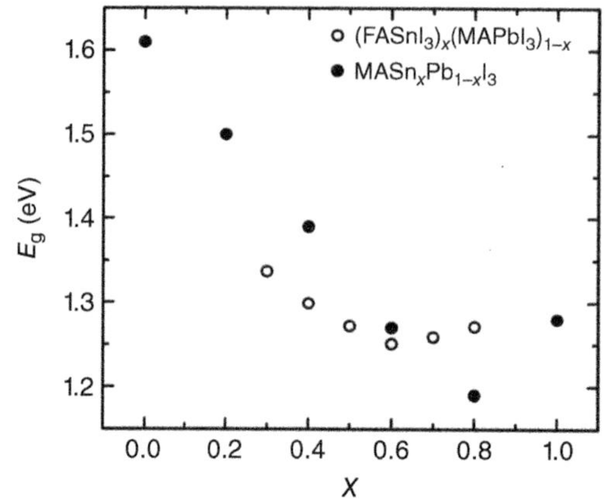

Figure 10.8 Band gap energy (E_g) for $MAPb_{1-x}Sn_xI_3$ [27] and $(FASnI_3)_x(MAPbI_3)_{1-x}$ [29] perovskite thin films as a function of Sn content. Sources: Zhao et al. [27]; Subedi et al. [29].

The compositions of both the B-cation and the X-anion have the strongest impact on the band gap of the perovskite, while the A-cation has a comparatively weaker effect [64]. However, substantial interaction between the A-cation and its nearest neighbor X-anions does impact optical absorption above the band gap, and the volume and stability of the perovskite unit cell stem from the A-cation size making it a substantial contributor to distortions in the unit cell and band structure [30, 31, 64].

The narrowest band gap perovskites suitable for PV involve B-cations based on Sn and mixtures of Sn and Pb (Figure 10.8). When incorporating Sn into the ABX_3 perovskite crystal structure, it must be in the 2+ valence state. As mentioned earlier, perovskites incorporating Sn are very sensitive to oxygen as it can push Sn to the undesirable 4+ state, which will destabilize the perovskite unit cell. The band gap of $MASnI_3$ is approximately 1.3–1.4 eV and lower than that of $MAPbI_3$ at 1.55 eV [25, 27, 65]. The difference in band gap energies between Sn and Pb B-cation perovskites with predominately I X-anions arises from shifting of the valance band maximum more than the conduction band minimum [64, 65]. Replacing Pb with smaller more, electronegative Sn reduces the ionization energy and electron affinity, shifting both the valence band maximum and conduction band minimum upward. However, the upward shift of the s-level contributing to the valence band maximum is larger than the p-level contributing to the conduction band minimum for Sn in the band structure. There is also a contribution to the valence band maximum increase due to increased s–p hybridization between Sn and the X-anion compared with Pb and the X-anion.

The narrowest band gaps near ~1.2 eV, however, are found in perovskites with mixed Sn + Pb B-cations. Band bowing has been observed as a function of B-cation composition in perovskites with fixed X-anions and either fixed or mixed A-cations [10, 24–27, 29]. The lowest gaps are observed at relative Sn to Sn + Pb fractions of approximately $x = 0.5$–0.6 in $(FASnI_3)_x(MAPbI_3)_{1-x}$ [10, 26, 29] and $x = 0.8$ in $MAPb_{1-x}Sn_xI_3$ [25, 27]. The band gap bowing from 1.55 to 1.2 to 1.4 eV for Pb to Pb + Sn to Sn B-cation perovskites is substantial. Due to differences in Pb and Sn atomic orbitals, the valence and conduction band edges are dominated by Sn and

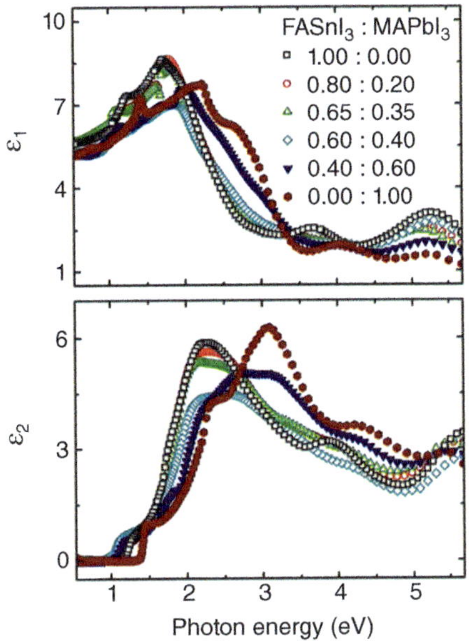

Figure 10.9 Spectra in ε for $(FASnI_3)_x(MAPbI_3)_{1-x}$ perovskite thin films. *Source*: Reproduced from Ghimire et al. [26]. Licensed under CC BY-4.0.

Pb, respectively, in mixed B-cation alloys [65]. As mentioned for the comparison of pure Pb and pure Sn perovskites, both the valence band maxima and conduction band minima shift upward for Sn relative to Pb although the valence band maxima shift is greater. In these alloys, the valence band maximum is dominated by Sn and the conduction band minimum is dominated by Pb. The Sn-dominated valence band maximum is shifted upward relative to the Pb-dominated conduction band minimum, which remains stationary. The band gap decreases. This behavior holds for mixed B-cation alloys with single or mixed A-cations as well. Similar above-gap electronic transitions are observed among these alloy series; however, lower energy transitions increase in strength with increasing Sn content (Figure 10.9) [26].

As mentioned previously, alloying of Sn + Pb perovskites increases configurational entropy to lower the Gibbs free energy and produce a more stable material. Urbach energies are used to describe the exponentially broadened absorption tail in semiconductors below the band gap attributed to defect states within the material; a smaller value of the Urbach energy arises from a narrower absorption tail below the band gap indicating a less defective material. For mixed B-cation $(FASnI_3)_x(MAPbI_3)_{1-x}$ perovskites, Urbach energies characterized prior to any atmospheric exposure show a minimum at $x = 0.6$ with low values observed at $x = 0.5$ and 0.7 indicating that these intermediate alloys exhibit the highest initial electronic quality (Figure 10.7c) [29]. Upon exposure to laboratory ambient air, the sub band gap absorption increases due to widening of the Urbach tail and formation of defect complexes (Figure 10.7a,b). $x = 0.5$ and 0.6 compositions show the least change upon atmospheric exposure illustrating that the appropriate B-cation alloying can produce material more stable against the formation of atmospherically induced defects.

In the past, narrow band gap mixed Sn + Pb perovskites exhibited lower diffusion lengths of ~0.5 μm compared with Pb B-cation perovskites [58]. Although this diffusion length is still quite long, it limits the thickness of the absorber layer in narrow band gap PV. When the absorber layer is approximately greater than 0.6 μm, all photogenerated carriers are not able to be collected and the short circuit current density is limited. This limitation was particularly important for narrow band gap sub cells in tandem PV where larger thicknesses are required to account for the greater penetration depth of long wavelength light. For intermediate and wide band gap perovskites, this effect is less important as the penetration depth is sufficiently small so that all photogenerated charge carriers in a PV absorber layer can be collected. Recently, substantially increased diffusion lengths in Sn + Pb perovskites have been demonstrated through manipulation of processing conditions [22, 23, 58], in turn enabling higher-efficiency single-junction narrow band gap PV efficiency as well as even higher-efficiency tandem devices.

To widen the band gap and produce perovskites suitable for the top junctions in tandem or multijunction PV, the most effective means of increasing the band gap is through modification of the X-anion composition [30, 64, 65]. Comparing the absorption and complex optical response of $MAPbX_3$ with I, Br, and Cl X-anions, the features within the optical properties blue shift with the decreasing mass of the X-anion as I is replaced with Br or Cl. Increased excitonic behavior is observed slightly above the band gap in $MAPbBr_3$ and $MAPbCl_3$ compared with $MAPbI_3$, and the predominant above-gap electronic transition in $MAPbI_3$ blue shifts and splits (Figure 10.10) [30]. For this comparison, the band gap shifts from 1.55 to 2.3 to 3.1 eV as I is replaced with Br and Cl, respectively, in $MAPbX_3$. Mixed X-anions of intermediate compositions show monotonic shifts in the band gap energies [64]. The conduction band minimum is strongly dependent on the p-level of Pb, which shifts upward with the decreasing X-anion size. The valence band maximum shifts

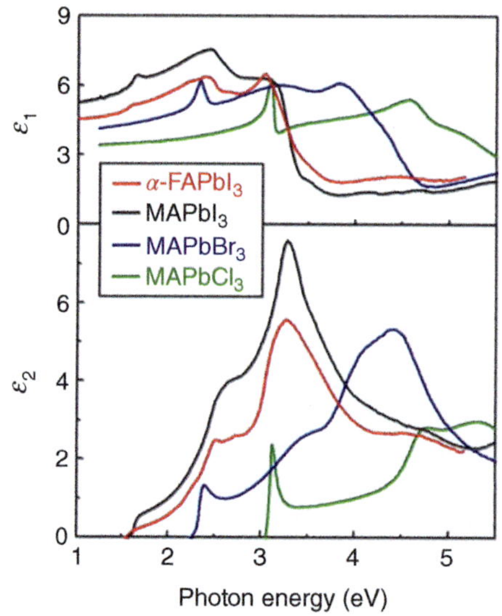

Figure 10.10 Spectra in ε for $MAPbI_3$, $MAPbBr_3$, $MAPbCl_3$, and $FAPbI_3$ perovskites. *Source*: Reprinted from Kato et al. [30], with permission of AIP Publishing.

downward predominately by the larger electronegativity of the smaller X-anions; however, there are weaker upward shifts due to increases in the Pb s-level and s–p hybridization. These upward shifts in the conduction band minimum and net downward shifts in the valence band maximum as I is replaced with Br and then Cl result in the increasingly widening band gaps. Practically, reduction in X-anion size and increased electronegativity shorten the bond lengths between the nearest neighbor B-cations and X-anions resulting in confinement effects. These confinement effects essentially shift electronic transitions in the complex optical response, including the band gap, to higher energies but retain similar relative positions.

The choice of A-cation influences the optoelectronic response of ABX_3 perovskites through relatively small band gap shifts [31, 64] and the magnitude of above band gap optical absorption [30]. In terms of band gap shifting relative to nominally $MABX_3$, the band gap can vary over a range of approximately 1.5 eV depending upon X (I, Br, Cl) and about 0.35 eV depending on B (Pb, Sn, Sn + Pb). Considering $APbI_3$, the band gap can range from 1.72 to 1.55 to 1.50 eV for Cs, MA, and FA A-cations, respectively. Although these changes are measurable and meaningful, the choice of A-cation is not the most effective means of tailoring the band gap. However, the origin of these shifts is related to the strong impact A-cations have on stabilization of the perovskite crystal phase structure. Crystal structure deformation from the ideal perovskite reduces hybridization between the X-anion and B-cation and shifts the valence band maxima downward more than the conduction band minima [64]. Increased A-cation size increases the unit cell volume and shifts the conduction band minima downward due to confinement effects [31]. The choice of A-cation determines the tolerance factor, and the extent of distortions expected with unit cell volume is directly linked to the A-cation size [31]. Increased distortions and unit cell volume increase ionization energy and electron affinity, with greater contributions to each, respectively. $MAPbI_3$ has a tolerance factor well within the stability window for the cubic perovskite phase, so it has limited distortions, low ionization energy, and high hybridization between the X-anion and B-cation. $CsPbI_3$ and $FAPbI_3$ have tolerance factors close to the boundaries of the cubic perovskite phase with a greater degree of distortions and reduced hybridization lowering both the conduction band minima and valence band maxima. FA has a larger ionic radius than MA, increasing the unit cell volume and further shifting the conduction band minimum downward, overall lowering the band gap of $FAPbI_3$ relative to $MAPbI_3$. Cs has a smaller ionic radius than MA, decreasing the unit cell volume and shifting the conduction band minimum upward to result in the higher band gap of $CsPbI_3$ relative to $MAPbI_3$.

The effect of A-cation composition on the above band gap absorption is related to both the A-cation interactions with the X-anion and in turn the impact of the X-anion on the conduction band, which is predominately determined by the B-cation, normally Pb [30]. Increased hydrogen bonding between the A-cation and the X-anion leads to increased anti-coupling behavior. This anti-coupling behavior reduces the strengths of optical transitions above the band gap, thus reducing optical absorption. Again, consider a comparison of the optical response of $FAPbI_3$, $MAPbI_3$, and $CsPbI_3$ (Figure 10.11) [30]. The two nitrogen atoms present in FA

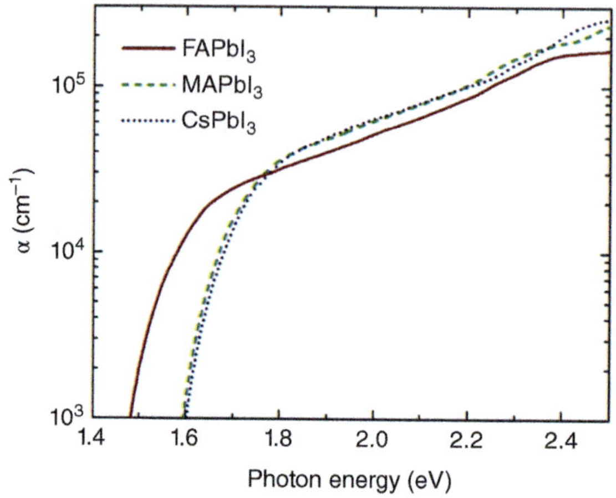

Figure 10.11 Near band gap spectra in α for $FAPbI_3$, $MAPbI_3$, and $CsPbI_3$ perovskites obtained from density functional theory (DFT) calculations. Source: Modified from Kato et al. [30].

result in strong hydrogen bonding to its nearest neighbor X-anion producing strong anti-coupling and reducing the above gap optical absorption relative to $MAPbI_3$. The single nitrogen atom present in MA results in weaker hydrogen bonding to its nearest neighbor X-anion and less anti-coupling. Replacing FA with Cs also reduces the anti-coupling effect.

Returning to sub gap absorption in mixed A-cation perovskites, compositions exhibiting some configurational entropy lower Gibbs free energy and produce a more environmentally stable perovskite among given fabrication conditions. For $FA_{1-x}Cs_xPbI_3$, the lowest atmospherically induced sub gap absorption and deduced defect density occur for $x = 0.2$ [59]. In this situation, there is a monotonic increase in band gap by replacing FA with Cs; however, Cs contents lower and higher to $x = 0.2$ degrade more upon exposure [59]. Here, the stability of the perovskite crystal structure itself is strongly influenced by the A-cation composition.

Overall the complex optical response of ABX_3 perovskites is remarkably similar considering the range of band gap energies possible through compositional tailoring. The band gap of mixed Sn + Pb B-cation perovskites exhibits band bowing, while that of mixed A-cation and mixed X-anion alloys varies monotonically. Above the band gap, critical point transitions generally shift by the same relative amount as the band gap but may exhibit splitting or enhanced excitonic effects depending on the composition. The sub band gap features observed will vary depending mostly upon the initial electronic quality of the film reflected in the Urbach energy or estimated defect density as well as any exposure to humidity or oxygen causing film degradation.

10.5 Solar Cell Performance

The direct impact that the center A-cation has on crystal phase stability, thin film stability, and optoelectronic response transfers to the performance of solar cells with ABX_3 perovskite absorber layers. Unfortunately, due to the wide ranges of deposition processes applied to fabricate a plethora of ABX_3 perovskite compositions, and the continuous improvements in PV device performance parameters reported, it is difficult to make direct comparisons between device performances reported in the literature over time and attribute differences solely to the A-cation. A nonexclusive set of case studies will compare the roles of different single A-cations, mixed A-cations and mixed B-cations, and mixed A-cations for ABX_3 perovskites of different compositions. Other devices incorporating different A-cations will be discussed, and overall considerations for PV device performance dependence on the A-cation will be summarized.

First consider single-junction PV devices with different single A-cation $FAPbI_3$ or $MAPbI_3$ absorbers. These compositions have similar band gaps and have both been implemented in PV devices exceeding 15% power conversion efficiency. A comparative study of solar cell degradation based on these absorber materials indicated the formation of PbI_2 from $MAPbI_3$ and the transition to the hexagonal phase from cubic $FAPbI_3$ when exposed to humidity during operation and storage [3]. The respective decompositions occur on different timescales, leading to $FAPbI_3$-based solar cells retaining higher operational stability and dark stability for hours and days, respectively, after the $MAPbI_3$ device no longer exhibited measurable power conversion efficiency. When perovskites with a substantial fraction of FA A-cations are implemented as absorber layers in solar cells, the devices generally exhibit more stable power conversion efficiency over time with lower hysteresis effects [10, 12–15, 17, 22]. However, it should be noted that the choice of A-cation is not singularly responsible for this behavior. The fabrication process of the perovskite, grain boundary passivation, and solar cell architecture design must be properly implemented, as demonstrated for the case of highly stable mixed FA + MA A-cation solar cells with $MA_{0.7}FA_{0.3}PbI_3$ absorbers [10, 41]. Improvements in the stability of pure $FAPbI_3$ absorbers have also been demonstrated by engineering the perovskite fabrication process, device architecture, and encapsulation [57].

Next consider the behavior of perovskites with mixed Sn + Pb B-cations and with both mixed Sn + Pb B-cations and mixed FA + MA A-cations. In both cases band bowing is observed at an intermediate Sn content; however, depending on the A-cation composition, the B-cation compositions yielding the lowest band gap differ. In the case of $(FASnI_3)_x(MAPbI_3)_{1-x}$ the band gap minimum occurs near $x = 0.5$ or 0.6 [10, 26, 29], while for $FAPb_{1-x}Sn_xI_3$ and $MAPb_{1-x}Sn_xI_3$, a similar band gap minimum occurs near $x = 0.6$ and $x = 0.8$ respectively [25, 27, 66]. The difference in band bowing and the associated minimum band gap with Sn content may arise due to simultaneous interactions between the mixed A-cations and mixed B-cations. More importantly, device performance for $MAPb_{1-x}Sn_xI_3$ at the minimum band gap composition is substantially lower than for $MAPbI_3$ [25, 27, 67], whereas for $(FASnI_3)_x(MAPbI_3)_{1-x}$, comparably high PV performance at the minimum band gap composition is observed [10]. As mentioned previously, the lowest Urbach energies in unexposed $(FASnI_3)_x(MAPbI_3)_{1-x}$ perovskites and defect densities associated with environmental exposure occur with $x = 0.5$ and 0.6 compositions, indicating improved electronic quality of the material and environmental stability [10]. Additionally, highly stable behavior is observed for both low band gap

$(FASnI_3)_{0.6}(MAPbI_3)_{0.4}$ devices and those based upon $(FASnI_3)_{0.6}(MAPbI_{3-3x}Cl_{3x})_{0.4}$ with reported device efficiencies near 18% [10, 15]. In this situation, the presence of a substantial amount of FA may prevent decomposition into PbI_2 while increased configurational entropy via the mixed A-cations and mixed B-cations lowers Gibbs free energy and results in a more stable crystal structure. These characteristics in turn contribute to more stable device performance.

Along that same theme, consider mixed FA and Cs A-cations in $APbI_3$ for solar cell absorber layers. For these $FA_{1-x}Cs_xPbI_3$ compositions, $x = 0.15$–0.2 has been reported to yield the highest efficiency PV performance among series of devices with higher and lower Cs content [7, 8, 11]. Reported device power conversion efficiencies are around 16% for $x = 0.15$ and 0.2 compositions indicating high electronic quality. For thin films exposed to laboratory ambient air, $FA_{0.8}Cs_{0.2}PbI_3$ showed the lowest amount of sub gap absorption indicating the lowest density of atmospherically induced defect states [59]. Stable PV device performance after storage for several hundred hours has been demonstrated for both $FA_{0.8}Cs_{0.2}PbI_3$ and $FA_{0.85}Cs_{0.15}PbI_3$ absorbers [7, 11]. High PV device performance coupled with environmental stability for moderate Cs content reinforces that configurational entropy increase and tolerance factor manipulation by substitution of a fraction of FA by Cs produce a relatively stable material. Furthermore, these $FA_{1-x}Cs_xPbI_3$ compositions serve as a baseline for wider band gap perovskites produced by substituting a fraction of the X-anion I by Br as demonstrated for devices replacing 5% and 30% of I with Br [15].

Variations in A-cations are not limited to the aforementioned compositions. Mixed FA, MA, and Cs A-cation perovskites with a composition of $FA_{0.7}MA_{0.2}Cs_{0.1}Pb(I_{5/6}Br_{1/6})_3$ fabricated with lead thiocyanate $Pb(SCN)_2$ have resulted in 14% efficient devices with limited hysteresis [46]. Perovskites with Sn B-cations have been fabricated with MA, FA, and PEA to produce Pb-free $MA_{0.75}FA_{0.15}PEA_{0.1}Sn(Br_xI_{1-x})_3$ with over 4% efficiency [49]. Replacing the moderate ionic radius MA with the substantially larger ionic radius GA A-cation produces $MA_{1-x}GA_xPbI_3$ perovskites. Solar cells with $MA_{0.86}GA_{0.14}PbI_3$ absorbers demonstrated 19% efficiency and similar compositions yielded >50% of initial power conversion efficiency after over 1000 hours of continuous illumination [18]. Removing MA completely, 20% efficient devices with Rb, Cs, and FA mixed A-cations exhibited high stability after 1000 hours of continuous tracking at the maximum power point [16]. High-efficiency monolithically integrated tandem PV with a wide band gap perovskite top junction and a narrow band gap perovskite bottom junction have been based upon the previously mentioned low band gap $(FASnI_3)_{0.6}(MAPbI_{2.91}Cl_{0.09})_{0.4}$ and wide band gap $FA_{0.8}Cs_{0.2}PbI_{2.1}Br_{0.9}$ compositions [15]. Power conversion efficiency of this tandem device was demonstrated at 21% and retained 85% of its initial efficiency over 80 hours of continuous operation. 23% and 25% efficient two and four terminal tandem devices incorporating triple mixed A-cation $Cs_{0.05}FA_{0.8}MA_{0.15}PbI_{2.55}Br_{0.45}$ top cells and $(FASnI_3)_{0.6}(MAPbI_3)_{0.4}$ bottom cells have also been recently developed [23]. Tuning of the A-cation composition has been a strong contributing factor to achieve these advances in PV performance and device stability.

To review, the choice of A-cation has several roles in the application of thin film polycrystalline perovskite solar cells. Regarding crystal phase stability, combinations of MA, FA, Cs, and other inorganic and organic A-cations can be coupled with appropriate material fabrication processes to produce stable cubic phase perovskites for use as high-efficiency solar cell absorber layers. Appropriate A-cation composition tailoring can be used to manipulate the band gap energy somewhat and alter the above gap optical absorption. Considering above gap optical absorption, the choice of A-cation composition coupled with the B-cation and X-anion can lead to weak absorption at photon energies greater than the band gap energy, which necessitates either thicker absorber layers in solar cells or acceptance of incomplete absorption of incident light in the perovskite. Either situation can limit PV device efficiency, particularly those based on narrow band gap perovskites. Perovskite films incorporating substantial fractions of FA are more environmentally stable, and devices based on these absorbers exhibit lower current–voltage hysteresis and have more stable performance over time. In fact, proper utilization of mixed A-cation wide and narrow gap absorber layers has enabled fabrication of highly efficient and stable tandem PV devices.

References

1 Leguy, A.M., Azarhoosh, P., Alonso, M.I. et al. (2016). *Nanoscale* 8: 6317.
2 Wang, D., Wright, M., Elumalai, N.K., and Uddin, A. (2016). *Sol. Energy Mater. Sol. Cells* 147: 255.
3 Zhang, T., Meng, X., Bai, Y. et al. (2017). *J. Mater. Chem. A* 5: 1103.
4 Song, Z., Shrestha, N., Watthage, S.C. et al. (2018). *J. Phys. Chem. Lett.* 9: 6312.
5 Kye, Y.H., Yu, C.J., Jong, U.G. et al. (2018). *J. Phys. Chem. Lett.* 9: 2196.
6 Seol, D.J., Lee, J.W., and Park, N.G. (2015). *ChemSusChem* 8: 2414.
7 Li, Z., Yang, M., Park, J.-S. et al. (2015). *Chem. Mater.* 28: 284.
8 Yu, Y., Wang, C., Grice, C.R. et al. (2016). *ChemSusChem* 9: 3288.
9 Saliba, M., Matsui, T., Seo, J.Y. et al. (2016). *Energy Environ. Sci.* 9: 1989.
10 Liao, W., Zhao, D., Yu, Y. et al. (2016). *J. Am. Chem. Soc.* 138: 12360.
11 Yi, C., Luo, J., Meloni, S. et al. (2016). *Energy Environ. Sci.* 9: 656.
12 Wang, C., Xiao, C., Yu, Y. et al. (2017). *Adv. Energy Mater.* 7: 1700414.
13 Zhao, D., Yu, Y., Wang, C. et al. (2017). *Nat. Energy* 2: 17018.
14 Xu, G., Bi, P., Wang, S. et al. (2018). *Adv. Funct. Mater.* 28: 1804427.
15 Zhao, D., Chen, C., Wang, C. et al. (2018). *Nat. Energy* 3: 1093.
16 Turren-Cruz, S.-H., Hagfeldt, A., and Saliba, M. (2018). *Science* 362: 449.
17 Li, C., Song, Z., Zhao, D. et al. (2019). *Adv. Energy Mater.* 9: 1803135.
18 Jodlowski, A.D., Roldan-Carmona, C., Grancini, G. et al. (2017). *Nat. Energy* 2: 972.
19 Wang, C., Song, Z., Li, C. et al. (2019). *Adv. Funct. Mater.* 29: 1808801.
20 Prasanna, R., Leijtens, T., Dunfield, S.P. et al. (2019). *Nat. Energy* 4: 939.
21 Eperon, G.E., Leijtens, T., Bush, K.A. et al. (2016). *Science* 354: 861.
22 Lin, R., Xiao, K., Qin, Z. et al. (2019). *Nat. Energy* 4: 864.

23 Tong, J., Song, Z., Kim, D.H. et al. (2019). *Science* 364: 475.

24 Hao, F., Stoumpos, C.C., Chang, R.P., and Kanatzidis, M.G. (2014). *J. Am. Chem. Soc.* 136: 8094.

25 Liu, C., Fan, J., Li, H. et al. (2016). *Sci. Rep.* 6: 35705.

26 Ghimire, K., Zhao, D., Yan, Y., and Podraza, N.J. (2017). *AIP Adv.* 7: 075108.

27 Zhao, B., Abdi-Jalebi, M., Tabachnyk, M. et al. (2017). *Adv. Mater.* 29: 1604744.

28 Lee, S., Ha, T.-J., and Kang, D.-W. (2018). *Mater. Lett.* 227: 311.

29 Subedi, B., Li, C., Junda, M.M. et al. (2020). *J. Chem. Phys.* 52: 064705.

30 Kato, M., Fujiseki, T., Miyadera, T. et al. (2017). *J. Appl. Phys.* 121: 115501.

31 Amat, A., Mosconi, E., Ronca, E. et al. (2014). *Nanoletters* 14: 3608.

32 Whitfield, P.S., Herron, N., Guise, W.E. et al. (2016). *Sci. Rep.* 6: 35685.

33 Lehmann, F., Franz, A., Tobbens, D.M. et al. (2019). *RSC Adv.* 9: 11151.

34 Jiang, S., Luan, Y., Jang, J.I. et al. (2018). *J. Am. Chem. Soc.* 140: 13952.

35 Song, Z., Watthage, S.C., Phillips, A.B. et al. (2015). *Chem. Mater.* 27: 4612.

36 Ghimire, K., Zhao, D., Cimaroli, A. et al. (2016). *J. Phys. D: Appl. Phys.* 49: 405102.

37 Ghimire, K., Zhao, D., Cimaroli, A. et al. (2016). *Proceedings of the 43rd Photovoltaic Specialists Conference*. IEEE 43rd Photovoltaic Specialists Conference, Portland, OR, June 2016, p. 89.

38 Pazoki, M., Wolf, M.J., Edvinsson, T., and Kullgren, J. (2017). *Nano Energy* 38: 537.

39 Jong, U.-G., Yu, C.-J., Ri, G.-C. et al. (2018). *J. Mater. Chem. A* 6: 1067.

40 Song, Z., Wang, C., Phillips, A.B. et al. (2018). *Sustain. Energy Fuels* 2: 2460.

41 Junda, M.M., Zhao, D., Subedi, B. et al. (2019). *J. Phys. D: Appl. Phys.* 52: 175102.

42 Juarez-Perez, E.J., Ono, L.K., Uriarte, I. et al. (2019). *ACS Appl. Mater. Interfaces* 11: 12586.

43 Kieslich, G., Sun, S., and Cheetham, A.K. (2014). *Chem. Sci.* 5: 4712.

44 Kieslich, G., Sun, S., and Cheetham, A.K. (2015). *Chem. Sci.* 6: 3430.

45 Travis, W., Glover, E.N.K., Bronstein, H. et al. (2016). *Chem. Sci.* 7: 4548.

46 Sun, Y., Peng, J., Chen, Y. et al. (2017). *Sci. Rep.* 7: 46193.

47 Bartel, C.J., Sutton, C., Goldsmith, B.R. et al. (2019). *Sci. Adv.* 5: eaav0693.

48 Shannon, R.D. (1976). *Acta Cryst.* 32: 751.

49 Weber, S., Rath, T., Kunert, B. et al. (2019). *Chem. Mon.* 150: 1921.

50 Guo, Y., Li, C., Li, X. et al. (2017). *J. Phys. Chem. C* 121: 12711.

51 Bartesaghi, D., Ray, A., Jiang, J. et al. (2018). *APL Materials* 6: 121106.

52 Zhang, L. and Sit, P.H.L. (2015). *J. Phys. Chem. C* 119: 22370.

53 Liu, D., Yang, C., Bates, M., and Lunt, R.R. (2018). *iScience* 6: 272.

54 Xiang, S., Fu, Z., Li, W. et al. (2018). *ACS Energy Lett.* 3: 1824.

55 Jena, A.K., Kulkarni, A., Sanehira, Y. et al. (2018). *Chem. Mater.* 30: 6668.

56 Steele, J.A., Jin, H., Dovgaliuk, I. et al. (2019). *Science* 365: 679.

57 Pool, V.L., Dou, B., Van Campen, D.G. et al. (2017). *Nat. Commun.* 8: 14075.

58 Yang, Z., Yu, Z., Wei, H. et al. (2019). *Nat. Commun.* 10: 4498.

59 Subedi, B., Guan, L., Yu, Y. et al. (2018). *Sol. Energy Mater. Sol. Cells* 188: 228.

60 Subedi, I., Subedi, B., Uprety, P. et al. (2020). *SPIE Proceedings: Physics, Simulation, and Photonic Engineering of Photovoltaic Devices IX* **11275**, 11275J (2020).

SPIE Physics, Simulation, and Photonic Engineering of Photovoltaic Devices IX in San Francisco, CA, February 2020.

61 Li, C., Zhou, Y., Wang, L. et al. (2017). *Angew. Chem. Int. Ed.* 56: 7674.

62 Tang, S., Deng, Y., Zheng, X. et al. (2017). *Adv. Energy Mater.* 7: 1700302.

63 Yang, M., Li, Z., Reese, M.O. et al. (2017). *Nat. Energy* 2: 17038.

64 Tao, S., Schmidt, I., Brocks, G. et al. (2019). *Nat. Commun.* 10: 2560.

65 Goyal, A., McKechnie, S., Pashov, D. et al. (2018). *Chem. Mater.* 30: 3920.

66 Liu, J., Wang, G., Song, Z. et al. (2017). *J. Mater. Chem. A* 5: 9097.

67 Tavakoli, M.M., Zakeeruddin, S.M., Grätzel, M., and Fan, Z. (2018). *Adv. Mater.* 30: 1705998.

Part II

Hybrid Perovskite Solar Cells

11

Operational Principles of Hybrid Perovskite Solar Cells

Hiroyuki Fujiwara[1], Yoshitsune Kato[1], Yuji Kadoya[1], Yukinori Nishigaki[1], Tomoya Kobayashi[1], Akio Matsushita[2], and Taisuke Matsui[2]

[1] *Gifu University, Department of Electrical, Electronic and Computer Engineering, 1-1 Yanagido, Gifu 501-1193, Japan*
[2] *Panasonic Corporation, Technology Innovation Division, 3-1-1 Yagumo-naka-machi, Moriguchi City, Osaka 570-8501, Japan*

11.1 Introduction

A strong feature of hybrid perovskite solar cells is the simplicity of the device structures. By essentially sandwiching a thin hybrid perovskite layer (~500 nm) between an electron transport layer (ETL) and a hole transport layer (HTL), quite high conversion efficiencies exceeding 20% can be achieved without following complex processing steps [1–13]. This is in sharp contrast with other representative solar cells (i.e. Si, GaAs and $CuInGaSe_2$), in which complicated layered structures must be formed to suppress carrier recombination, particularly at the interfaces [14–17]. Furthermore, high-quality halide perovskite absorbers can be formed at low temperatures (~100 °C) [4] and, in a state-of-the-art hybrid perovskite cell, a remarkably high efficiency of $\eta \sim 25\%$ has been realized [6]. These features make the perovskite devices extremely attractive for both large area and tandem photovoltaic applications.

Quite fortunately, hybrid perovskites, including methylammonium lead iodide ($CH_3NH_3PbI_3$; $MAPbI_3$) and formamidinium lead iodide ($HC(NH_2)_2PbI_3$; $FAPbI_3$), exhibit superior optoelectronic characteristics preferable for photovoltaic devices as described in Part I: namely, an appropriate band gap (E_g) and a high absorption coefficient (Chapter 4), absence of deep traps (Chapter 5), long carrier diffusion length (Chapter 6), and effective carrier extraction from the grain boundary region (Chapter 9). These suitable material properties have enabled an unprecedented rapid development of high-efficiency perovskite solar cells. In particular, it took only 10 years to develop a hybrid perovskite solar cell with an efficiency of 25.2% [18], as opposed to the half century needed for Si solar cells to reach 26.7% efficiency [19, 20]. Quite high conversion efficiencies observed in hybrid perovskite cells can largely be understood by the extreme adaptability of the perovskite thin-film absorbers in simplified solar cell structures.

Hybrid Perovskite Solar Cells: Characteristics and Operation, First Edition. Edited by Hiroyuki Fujiwara.

This chapter reviews the fundamental operational principles and unique characteristics of hybrid perovskite solar cells. In Section 11.2, the basic principles of hybrid perovskite photovoltaics are explained using the band alignments of layered structures as a guiding principle. A more realistic solar cell operation is addressed in Section 11.3 by considering a physical band diagram of semiconducting layers. Section 11.4 further discusses the power loss mechanism in a 20% efficient hybrid perovskite cell, with the goal of further understanding the performance-limiting factors of the device. Finally, Section 11.5 focuses on carrier recombination at interfaces and discusses the crucial factors determining the open-circuit voltage (V_{oc}) of experimental solar cells.

11.2 Operation of Hybrid Perovskite Solar Cells

Hybrid perovskite solar cells have simple structures and operate in a straightforward way. In Section 11.2.1, we see the principle and basic structures of the perovskite devices. In Section 11.2.2, the band alignment of various solar cell component layers is further explained.

11.2.1 Operational Principle and Basic Structures

Figure 11.1 illustrates the operation of a hybrid perovskite solar cell in a simplified picture. Here, the band alignment [i.e. energy positions of valance band maximum (VBM) and conduction band minimum (CBM)] of the solar cell component layers is shown. This solar cell has a basic structure of TCO/ETL/perovskite/HTL/metal, where TCO shows a transparent conductive oxide. For perovskite absorbers with thicknesses of 300–500 nm, a wide variety of perovskite compounds, including $MAPbI_3$, $FAMAPb(I,Br)_3$, and $CsFAPb(I,Br)_3$, have been used (see Figure 4.1). In designing a stable perovskite crystal, the tolerance factor provides a universal guideline (see Figure 3.11). When light is illuminated through the front TCO, the light absorption within the perovskite absorber creates free electrons and holes, which are collected separately in the ETL and HTL (see also Figure 1.1). The choice

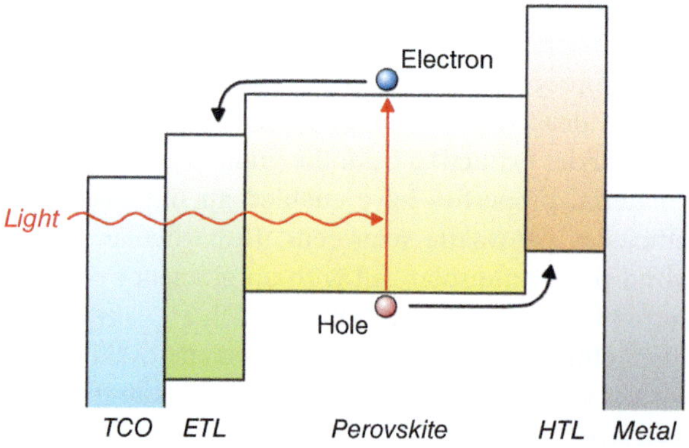

Figure 11.1 Operation of a hybrid perovskite solar cell, explained based on band alignment. In this simplified picture, the electric field in the perovskite absorber layer is neglected. Note that the perovskite layer is simply sandwiched between an ETL and a HTL.

of ETL and HTL is quite important and these layers are selected such that their energy levels match the CBM and VBM [7, 21]. The example in Figure 11.1 shows a preferable band alignment in which electrons and holes are collected without energy barriers.

Figure 11.2 shows the structures of hybrid perovskite solar cells, which can be categorized into (a) n-i-p and (b) p-i-n structures. Here, the meanings of "p," "i," and "n" differ from those of standard solar cells and, in the hybrid-perovskite research field, "p" and "n" denote HTL and ETL, respectively, and not the doped perovskite layers. The "i" represents the perovskite absorber layer, which is generally treated as an intrinsic layer with low carrier concentrations. Note, however, that the carrier type and concentration of the perovskite layers vary substantially with the preparation condition (Figure 6.2) and the resulting solar cell performance also changes [22–25]. Figure 11.2c summarizes the molecular structures of representative organic layers used widely as ETL and HTL.

The perovskite single cell is generally formed on a glass substrate (i.e. a superstrate cell). Structurally, the p-i-n solar cell is simply an inversion of the n-i-p structure, although the selections of ETL and HTL differ due to the difference in their deposition sequences. In particular, a mesoporous TiO_2 (mp-TiO_2) layer is typically used in the n-i-p cell to improve solar cell performance [1, 2, 24]. This layer comprises a continuous structure of anatase TiO_2 with a high porosity of ~60% (Dyesol) [26, 27]. For both n-i-p and p-i-n cells, SnO:F and In_2O_3:Sn (ITO) are widely employed as TCO, whereas Ag and Au rear electrodes are commonly adopted to guarantee efficient backside light reflection [28, 29]. In the standard n-i-p cell, TiO_2 and spiro-OMeTAD (or poly(triaryl amine), PTAA) layers are used as the ETL and HTL, respectively. In the p-i-n cell, a PTAA or poly(3,4-ethylenedioxythiophene): poly(4-styrensulfonate) (PEDOT:PSS) layer is usually employed as the HTL, while (6,6)-phenyl-C_{61}-butyric acid methyl ester (PCBM) combined with a bathocuproine (BCP) interface layer is used as the ETL [30].

In early $MAPbI_3$ research, n-i-p solar cells have been primarily investigated [31–37]. In the cells developed initially, mp-TiO_2 was incorporated almost entirely into the absorber layers [31–36], but as researches progressed, the mp-TiO_2 layer thickness was decreased gradually to ~200 nm. The p-i-n cells shown in Figure 11.2b were developed in a later research stage. One disadvantage of the p-i-n cell structure is the unfavorable parasitic light absorption in the HTL, located at the light illumination side. Specifically, PTAA and NiO layers, often used as the HTL in the p-i-n cell, exhibit strong light absorption in the ultraviolet (UV) region (Figure 11.3), reducing the short-circuit current density (J_{sc}) of the cells [10, 38]. In contrast, for the n-i-p cells, parasitic absorption is suppressed extremely well because (i) TiO_2 shows a high transparency (see Figure 11.3) and the HTL is placed after the perovskite absorber in the light absorption order. The J_{sc} reduction in the p-i-n cell can be avoided by using a PEDOT:PSS HTL, but the incorporation of PEDOT:PSS reduces V_{oc} (see Section 11.5).

In Pb-based perovskite cells, over 20% efficiencies can be obtained in both n-i-p and p-i-n structures [4, 39]. In Sn-based perovskite cells, however, p-i-n cells show substantially higher efficiencies, compared with n-i-p cells (see Table 14.3).

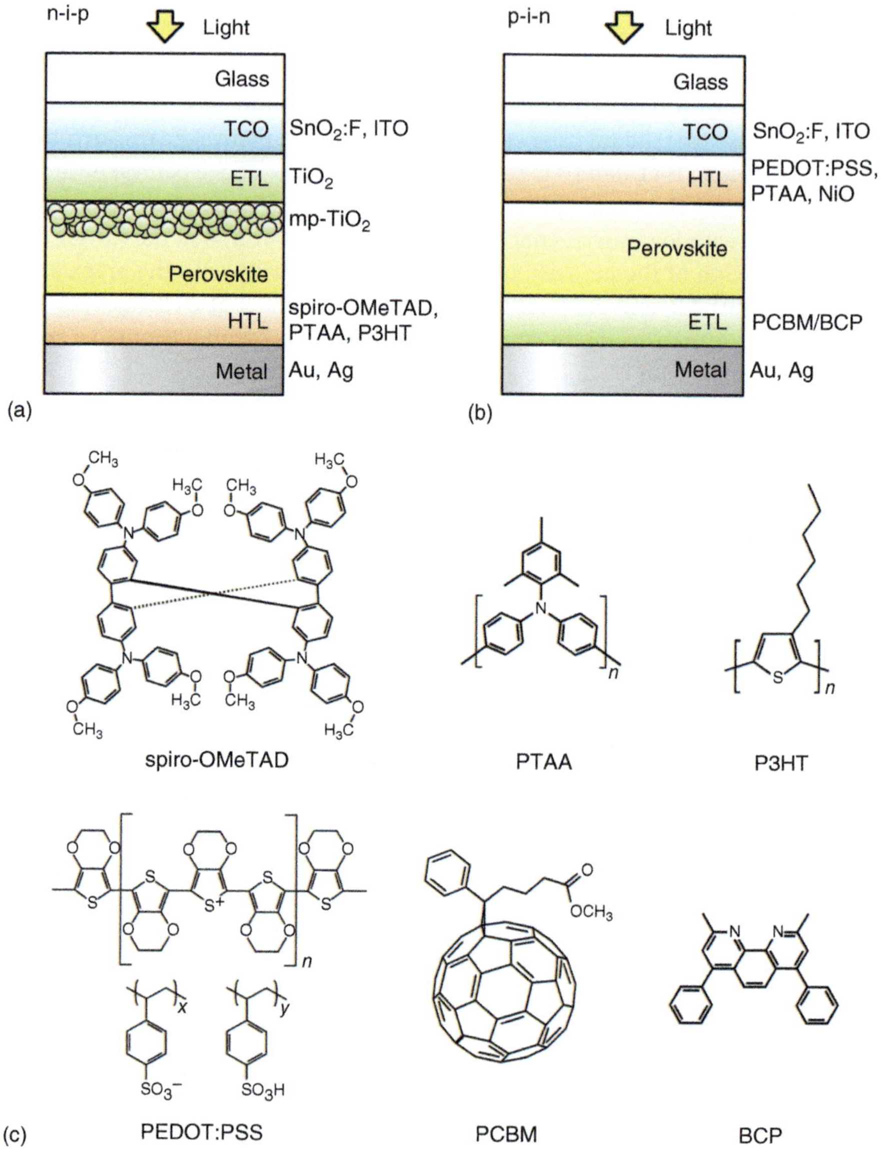

Figure 11.2 Basic structures of hybrid perovskite solar cells in (a) n-i-p and (b) p-i-n configurations and (c) molecular structures of 2,2′,7,7′-tetrakis-(*N*, *N*-di-*p*-methoxyphenylamine) 9,9′-spirobifluorene (spiro-OMeTAD), poly(triaryl amine) (PTAA), poly(3-hexylthiophene) (P3HT), poly(3,4-ethylenedioxythiophene): poly(4-styrensulfonate) (PEDOT:PSS), (6,6)-phenyl-C_{61}-butyric acid methyl ester (PCBM), and bathocuproine (BCP).

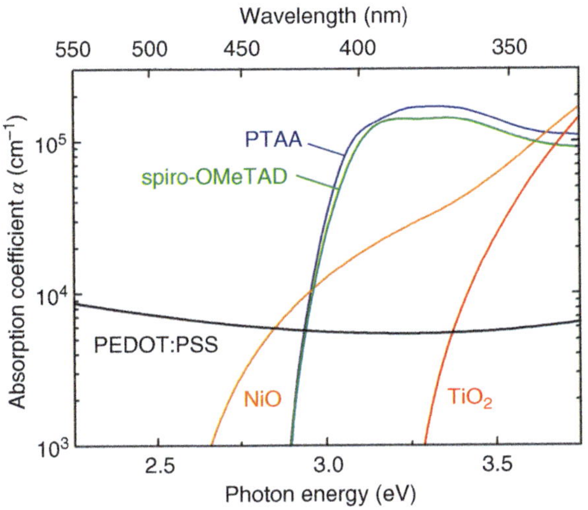

Figure 11.3 Absorption-coefficient spectra of representative ETLs and HTLs, including TiO_2, NiO, spiro-OMeTAD, PTAA, and PEDOT:PSS. The optical spectra calculated from the Tauc–Lorentz model are shown. Except for PTAA, the data were taken from Fujiwara and Collins [29]. PEDOT:PSS shows optical anisotropy [29], and the spectrum in the figure corresponds to that of an ordinary ray (i.e. the electric field of the incident light is parallel to the surface). The background absorption of PEDOT:PSS originates from the strong absorption induced by free carriers.

This difference has been attributed to the generation of trapping sites at the TiO_2/perovskite interface in the n-i-p configuration. In particular, the enhanced SnI_2 formation on the TiO_2 layer in the n-i-p structure, which can be avoided in the p-i-n configuration, has been proposed as a mechanism of high-density trap formation (Section 14.5).

11.2.2 Band Alignment

Figure 11.4 summarizes the band alignments of $MAPbI_3$ solar cells with various ETL and HTL, as reported in Refs. [40–42]. The VBM positions (or ionization potentials) in Figure 11.4 were determined using photoelectron spectroscopy, whereas the CBM positions (or electron affinities) were estimated by adding E_g values to the VBM positions [40]. Note, however, that the absolute energy positions vary slightly with sample surface conditions and reference energy; as a result, there are slight inconsistencies among the reported values [2, 7, 21, 33, 35, 38, 42–45]. In Figure 11.4, the conduction-band edges of all ETLs are slightly lower than that of $MAPbI_3$, while the valence energies of all HTLs are slightly above the $MAPbI_3$ valence band. Accordingly, the operation of a perovskite cell can essentially be expressed using the schematic illustrated in Figure 11.1.

Figure 11.5 summarizes the VBM and CBM positions of various halide perovskites. The VBM positions correspond to those determined by photoelectron spectroscopy [46], while the CBM levels have been derived by adding the E_g values listed in Table 4.3 to the VBM energies. The energy-level diagram for ABX_3

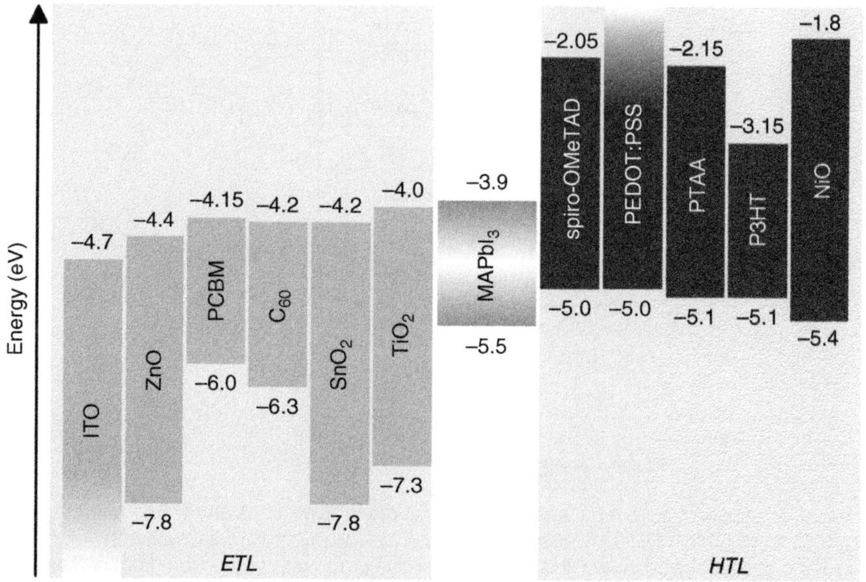

Figure 11.4 Band alignments of $MAPbI_3$ solar cells with various ETL and HTL. The VBM positions (or ionization potentials) and CBM positions (or electron affinities) of the component layers are summarized. The numerical values denote the corresponding energy positions. The energy positions reported in Refs. [40–42] are adopted. Note that there are slight inconsistencies among the reported energy positions. *Source*: Stolterfoht et al. [40]; Kegelmann et al. [41]; Chueh et al. [42].

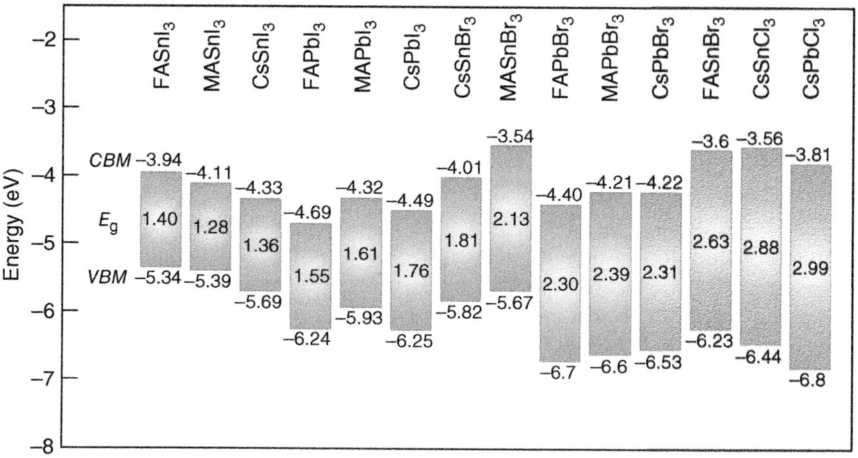

Figure 11.5 Valence band maximum (VBM) and conduction band minimum (CBM) energy positions of various halide perovskites. The E_g of each perovskite is also indicated. The VBM positions correspond to those determined by photoelectron spectroscopy [46], while the CBM levels are derived by adding the E_g values listed in Table 4.3 to the VBM energies. The VBM and CBM positions in the figure differ slightly from those shown in Figures 4.10 and 11.4.

perovskites shows that (i) there are no clear trends in the VBM and CBM with the variation of the A-site center cation, (ii) the VBM is lifted when the divalent B-site cation is changed from Pb^{2+} to Sn^{2+}, and (iii) the VBM shifts downward and CBM moves upward as the X-site halide anion is varied in the order I → Br → Cl (see also Figure 4.10). As shown in Figure 5.6, the A-site cation does not participate in forming the conduction and valence bands and, therefore, the A-site species indirectly influence the VBM and CBM through the crystal structure alteration. The type-II (staggered) band alignment of $ASnI_3$–$APbI_3$ (Figures 4.10 and 14.17) is particularly important for explaining quite strong E_g bowing observed in their alloys (Figures 4.10 and 18.12). When a hybrid perovskite alloy is employed as the absorber, it is important to consider the relative change of the VBM and CBM in designing and selecting ETL and HTL. In particular, higher solar cell performance can be achieved by improving the energy level matching between the perovskite absorber and ETL/HTL [7, 21] (see Section 13.5 and Chapter 14 for more detail).

11.3 Band Diagram of Hybrid Perovskite Solar Cells

The inorganic framework of a hybrid perovskite forms conduction and valence bands (Chapter 5). In a solar cell device, these semiconductor bands are not flat anymore but undergo bending due to the generation of an electric field within the solar cell. The presence of the electric field plays a critical role in collecting photogenerated carriers, particularly under the short-circuit condition (SC). For simplicity, such effects have been neglected completely in Section 11.2. Here, we present an exact model of a perovskite solar cell operation based on the band diagram obtained from device simulation (Section 11.3.1). The band diagrams of experimental perovskite cells are further discussed in Section 11.3.2.

11.3.1 Device Simulation

Figure 11.6 shows the band diagrams of n-i-p $MAPbI_3$ solar cells with a structure of glass/SnO_2:F(500 nm)/TiO_2(150 nm)/$MAPbI_3$(500 nm)/spiro-OMeTAD (50 nm)/Ag, calculated from a one-dimensional device simulator (SCAPS [47–50]). In these simulations, the interface recombination was neglected and, for the $MAPbI_3$ thin layer, a mid-gap trap density of $N_t = 10^{15}\,cm^{-3}$ was assumed to reproduce the experimental solar cell efficiency of ~20%. This assumption is, however, quite hypothetical, as the efficiency of the experimental solar cells is limited essentially by the interface recombination, rather than the bulk recombination (Section 11.5.1). Here, the simulation results of Figure 11.6 are shown to explain the overall operation of $MAPbI_3$ solar cells.

In Figure 11.6a, the band diagram of the n-i-p cell under the SC is shown. The dotted line in Figure 11.6a indicates the Fermi level (E_F) and E_F of the TCO locates inside the conduction band when the carrier concentration is ~$10^{20}\,cm^{-3}$. The E_F of the TiO_2 is close to the conduction band edge, while that of the HTL matches the highest occupied molecular orbital (HOMO) level. It can be seen that the

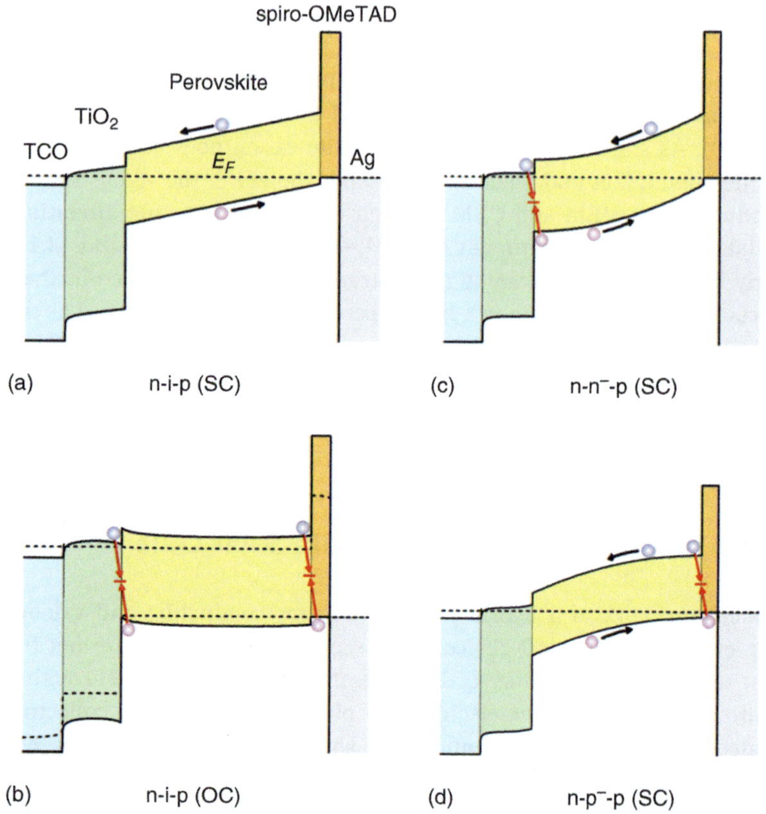

Figure 11.6 Band diagrams of $MAPbI_3$ solar cells with (a) n-i-p (short-circuit condition: SC), (b) n-i-p (open-circuit condition: OC), (c) n-n$^-$-p (SC), and (d) n-p$^-$-p (SC) configurations calculated from a one-dimensional device simulator (SCAPS). A solar cell structure of glass/SnO_2:F(500 nm)/TiO_2(150 nm)/$MAPbI_3$(500 nm)/spiro-OMeTAD (50 nm)/Ag is assumed. In the simulations, a mid-gap trap density of $N_t = 10^{15}$ cm^{-3} is assumed to reproduce an experimental solar cell efficiency of ~20%. In (c) and (d), the simulations were performed assuming n-type and p-type carrier concentrations of 10^{16} cm^{-3} for a $MAPbI_3$ layer. In (b)–(d), hypothetical carrier recombination at the interfaces is illustrated.

conduction and valence bands of the perovskite i-layer are tilted by the uniform electric field present within the i-layer (i.e. the i-layer is depleted completely). In the presence of an electric field, carrier collection is enhanced and, under the SC, the photocarrier collection (i.e. J_{sc}) can reach almost 100% even in the presence of defects (see Section 11.4.2).

We show the corresponding band diagram under the open-circuit condition (OC) in Figure 11.6b. The two dotted lines in Figure 11.6b show the quasi-Fermi levels for electrons and holes. The V_{oc} of the cell can be described by the splitting of these levels (i.e. quasi-Fermi-level splitting). Under the OC, the bands of the perovskite layer become almost flat and the carrier recombination at the interface or within the bulk layer occurs more easily. In other words, the output voltage under this condition (i.e. V_{oc}) is primarily governed by the recombination characteristics, while J_{sc} is less sensitive to the recombination behavior due to the electric field carrier collection. Accordingly, V_{oc} is a good indicator about how significantly carrier recombination occurs within solar cells and the experimental V_{oc} decreases from its theoretical limit (1.32 V for $MAPbI_3$ [16, 51, 52] and 1.28 V for $FAPbI_3$; Table 12.1) with increasing interface and bulk recombinations.

The diffusion length (L_D) is a physical parameter that is inversely related to the recombination (or trap density) in the bulk layer. A significant advantage of hybrid perovskite thin films is very large L_D values (4–23 μm in Table 1.1), which arise from the absence of deep-level traps (Section 5.6) and the resulting long carrier lifetimes (Section 1.2.1). The L_D of the perovskite layers is sufficiently larger than the solar cell absorber thickness (~500 nm); in this case, the recombination process is predominantly governed by interface recombination. In hybrid perovskite cells, therefore, the reduction of interface defects is critical to achieving high efficiency (Section 11.5.1).

The band diagrams in Figure 11.6c,d show cases in which the perovskite layers are n- and p-type, respectively, with low carrier densities of 10^{16} cm^{-3} (i.e. n^- or p^-). As explained in Figure 6.2, the carrier type and concentration of $MAPbI_3$ change with the precursor ratio (i.e. PbI_2/MAI). In the n-type perovskite cell shown in Figure 11.6c, the p–n junction is formed at the $MAPbI_3$/HTL interface and the electric field is concentrated at the HTL interface, while only a weak electric field is present near the ETL as a result of the n–n^- band alignment. In this solar cell, due to the flat band formation at the TiO_2/$MAPbI_3$ interface, the carrier recombination occurs more easily in this region [53, 54]. Moreover, the minority carrier of this solar cell is hole and the conversion efficiency is more influenced by hole mobility. In the n-p^--p cells, the trend is reversed; the carrier recombination in this case tends to occur at the $MAPbI_3$/HTL and the electron mobility influences more. Thus, the formation of an n- or p-type perovskite varies the device operation rather significantly.

11.3.2 Experimental Observation

The electric field (or potential) distributions within hybrid perovskite solar cells can be measured directly by using cross-sectional characterization techniques such as electron beam-induced current (EBIC) and Kelvin probe force microscopy (KPFM). In this section, we discuss the operation of experimental perovskite devices based on EBIC [55, 56] and KPFM [24, 57–59] characterization results.

Figure 11.7a illustrates the EBIC measurement configuration for a planar-type hybrid perovskite cell without mp-TiO_2 [55]. In EBIC, the electron beam of a scanning electron microscope (SEM) is employed to generate the electron–hole pairs and the current flow through the device is characterized. In this technique, therefore, the carrier collection at individual positions can be investigated. Figure 11.7b shows the cross-sectional SEM and EBIC images of the perovskite cell [55]. In the SEM image, a nonuniform $MAPbI_3$ layer fabricated by a spin-coating process can be confirmed. In the corresponding EBIC image, the entire $MAPbI_3$ layer is brightly displayed due to the uniform carrier collection throughout the absorber layer. The result of the EBIC line scan further shows two shoulders near the HTM/$MAPbI_3$

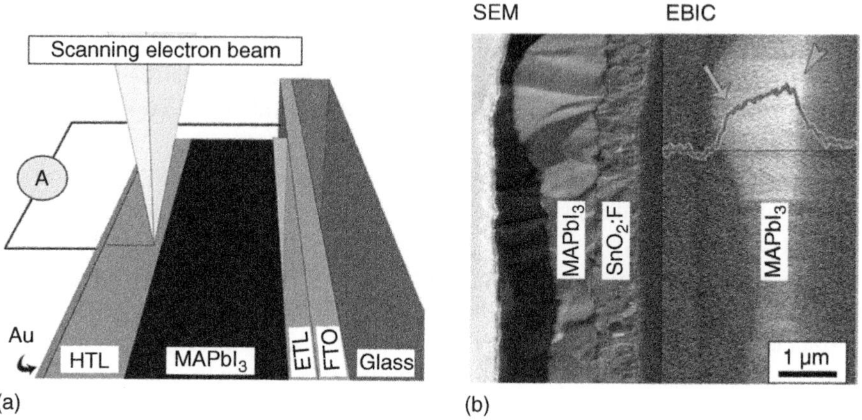

Figure 11.7 (a) EBIC measurement performed for a planar $MAPbI_3$ solar cell with a structure of glass/SnO_2:F(FTO)/TiO_2/$MAPbI_3$/HTL/Au and (b) cross-sectional SEM and EBIC images obtained from the experiment [55]. The $MAPbI_3$ layer formed by a spin-coating process is nonuniform and the bright EBIC image in (b) reveals a uniform carrier collection over the entire $MAPbI_3$ layer, indicating that the perovskite cell operates as a p-i-n device. The result of an EBIC line scan is also indicated. *Source*: Adapted with permission from Edri et al. [55]. Copyright (2014) American Chemical Society.

and $MAPbI_3$/ETL interfaces. Based on this observation, Edri et al. concluded that the perovskite cell operates as a p-i-n device [55, 56].

In contrast, the KPFM characterization of $MAPbI_3$ cells has provided controversial results [24, 57–59]. In particular, Cai et al. performed a KPFM measurement of a 20%-efficient $MAPbI_3$ cell that shows a high V_{oc} of 1.12 V in the standard n-i-p structure [24]. Figure 11.8a shows a profile of the contact potential difference (CPD) in a standard $MAPbI_3$ cell [24]. The CPD profile indicates a change in the electrical potential and the magnitude of the CPD is consistent with the V_{oc} of the device in the same measurement configuration. Note that the electric field is simply given as the change in the voltage with thickness (i.e. $E = V/d$) and thus the CPD slope reflects the electric field. Importantly, in Figure 11.8a, a large potential drop is observed within the $MAPbI_3$-TiO_2 mixed region, indicating that the electric field is concentrated at the $MAPbI_3$-TiO_2 interface. Thus, this high-efficiency cell operates as a simple p–n junction device, whose band profile is represented by that shown in Figure 11.6d. The p–n junction formation is also supported by other KPFM studies [57, 58].

Quite interestingly, in Ref. [24], the large potential drop is always observed within the mesoporous layer, independent of the PbI_2/MAI precursor ratio, and the cell efficiency is rather insensitive to the precursor ratio. In contrast, in planar cells without mp-TiO_2, the location of the potential drop changes with the PbI_2/MAI precursor ratio. When an n-type (PbI_2-rich) $MAPbI_3$ is applied, the CPD drops in the $MAPbI_3$/HTL interface region (Figure 11.8b). In this case, therefore, the solar cell is best described by the n-n^--p type cell shown in Figure 11.6c.

The above results indicate quite important facts that hybrid perovskite cells do not always function as n-i-p (or p-i-n) devices and the detailed operation of the

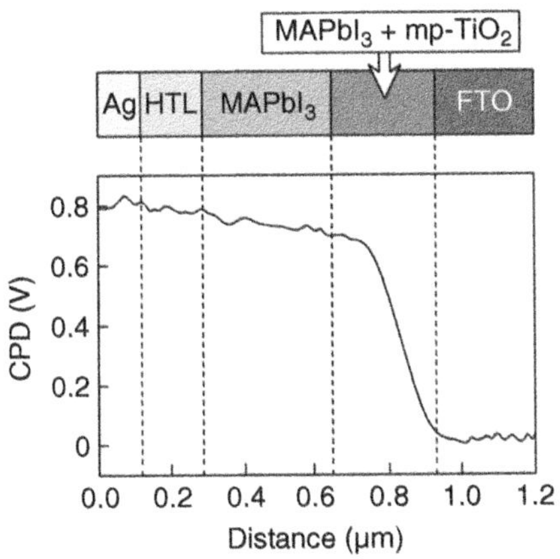

Figure 11.8 Depth profiles of the contact potential difference (CPD) in $MAPbI_3$ solar cells (a) with and (b) without mp-TiO_2, characterized by Kelvin probe force microscopy (KPFM). The solar cell structure is glass/SnO_2:F(FTO)/TiO_2/(mp-TiO_2)/$MAPbI_3$/spiro-OMeTAD. In the figure, the smoothened results from Ref. [24] are shown. The CPD slope reflects the electric field. *Source*: Cai et al. [24].

perovskite cells varies with the solar cell architecture and processing conditions. The interface recombination behavior is also expected to change significantly according to the type of the solar cells (see Figure 11.6).

11.4 Refined Analyses of Hybrid Perovskite Solar Cells

In maximizing solar cell conversion efficiencies, it is critical to understand the carrier generation and recombination dynamics. Although the basic operation of hybrid perovskite devices is rather simple, the actual cell efficiency is determined by a variety of factors, including unfavorable parasitic absorption, carrier recombination, and device resistivity. In this section, we present more refined analyses of how carriers are generated in the perovskite cells (Section 11.4.1) and how the power output of a state-of-the-art high-efficiency cell ($\eta \sim 20\%$) is determined (Section 11.4.2). A Windows-based free software (e-ARC software), which enables the analyses described in this section, is also introduced (Section 11.4.3).

11.4.1 Carrier Generation and Loss

Here, we discuss the spatial distribution of carrier generation and the influence of carrier recombination in an n-i-p $MAPbI_3$ cell. Figure 11.9 shows the optical simulation results obtained from the optical admittance calculation [17, 29] assuming a flat $MAPbI_3$ cell [28]. In Figure 11.9a, the solar cell structure and corresponding internal quantum efficiency (IQE) and internal absorptance (IA) spectra of the cell are shown. The IQE and IA are calculated using the relations $IQE = EQE/(1 - R)$ and $IA = A/(1 - R)$, where EQE, R and A indicate the external quantum efficiency, reflectance, and absorptance of the constituent layers, respectively.

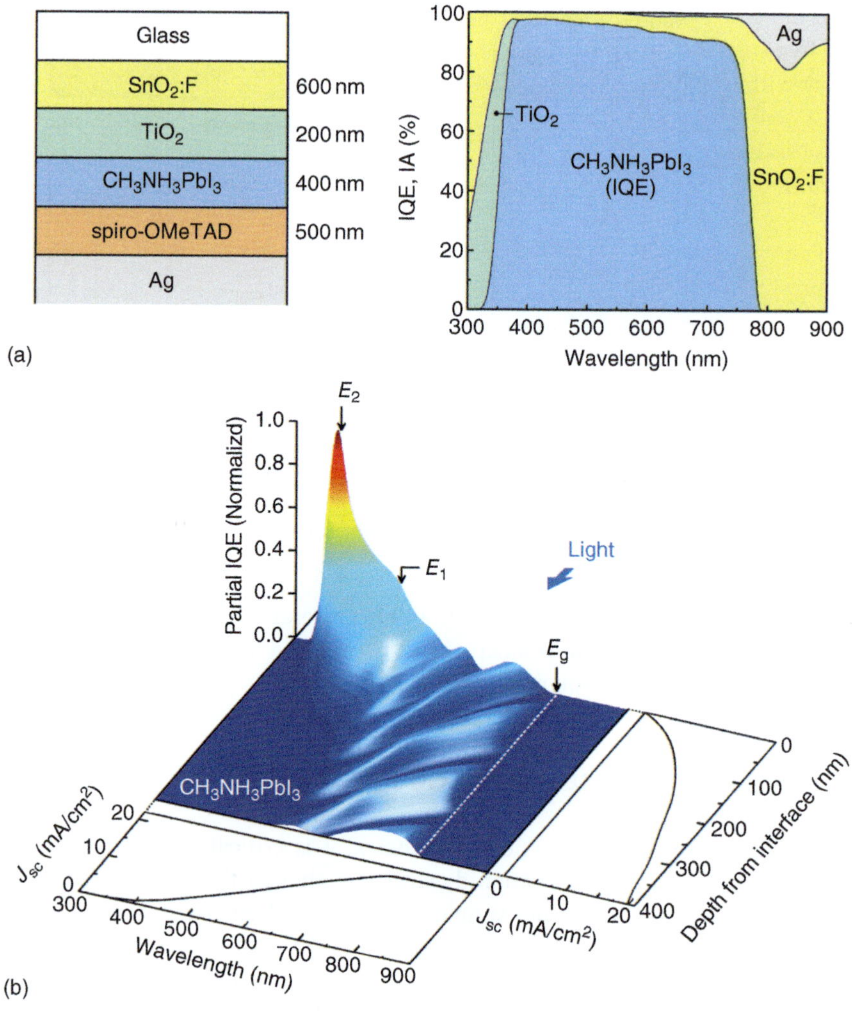

Figure 11.9 (a) $MAPbI_3$ solar cell structure assumed in the optical simulation and calculated internal quantum efficiency (IQE) and internal absorptance (IA) spectra of the component layers and (b) depth-resolved IQE spectra of the $MAPbI_3$ absorber. The arrows in (b) show the energy positions of $E_g(E_0)$, E_1, and E_2 transitions (see Figure 4.4). The integrated J_{sc} values relative to the depth from the TiO_2 interface and λ are also shown. *Source*: Shirayama et al. [28].

The IQE spectrum has a rectangular shape, in which the short-wavelength (λ) response is limited by the light absorption in the SnO_2:F and TiO_2 top layers. The upper bound of the IQE ($\lambda \sim 800$ nm) is simply determined by the E_g of the $MAPbI_3$ layer. Remarkably, the absorption effect of spiro-OMeTAD is almost completely negligible because (i) the E_g of spiro-OMeTAD (2.95 eV) [29] is higher than that of $MAPbI_3$ and (ii) the spiro-OMeTAD layer is placed after $MAPbI_3$ in the order of light absorption.

Note that the maximum IQE in the visible region (λ = 400–750 nm) is limited by the free carrier absorption in the SnO_2:F layer. Free carrier absorption increases strongly with carrier density [60, 61] and, in Figure 11.9, a TCO carrier density of 1.2×10^{20} cm^{-3} is assumed. This density is much lower than those of widely used TEC substrates (5×10^{20} cm^{-3}) [60, 62] and, when a TEC substrate is adopted, the maximum IQE (or EQE) decreases due to enhanced free carrier absorption [54].

In Figure 11.9b, the light absorption in 1-nm-thick $MAPbI_3$ sublayers (partial IQE) is shown. Integration of these partial IQE spectra reproduces the IQE spectrum in Figure 11.9a. The arrows in the figure denote the λ positions of the E_0 (E_g), E_1, and E_2 transitions in $MAPbI_3$ (see Figure 4.4). At $E \geq E_1$, the partial IQE exhibits rapid decay because of the strong light absorption within $MAPbI_3$. In the region $E_g \leq E < E_1$, however, the IQE is low due to the smaller absorption coefficient (α) in this region and photocarriers are generated uniformly throughout the entire $MAPbI_3$ layer with the appearance of the optical interference effect. In this region, the electrons and holes need to travel through the entire $MAPbI_3$ layer and high J_{sc} observed experimentally for hybrid perovskite cells confirms the quite efficient carrier collection within the whole absorber layer [28].

In Figure 11.9b, the integrated J_{sc} relative to the depth and incident light (λ) is also shown. Because of the assumed flat structure, the absolute J_{sc} value is relatively low (~20 mA/cm^2) as compared to the cells with natural textures (~24 mA/cm^2) [1, 2, 16]. The calculation result shows the intense carrier generation near the TiO_2 interface and a gradual increase in J_{sc} with depth. The contribution of J_{sc} at $\lambda \geq 500$ nm accounts for 73% of the total J_{sc}, indicating that the longer λ response is critical to achieving high J_{sc}. In the $MAPbI_3$ cell, the effect of metal-electrode backside reflection is large and the light is absorbed effectively even when the $MAPbI_3$ layer is thin [28]. When a $MAPbI_3$ absorber thickness is varied, J_{sc} saturates at a thickness of 400 nm [28], which is consistent with a general absorber thickness employed in perovskite cell fabrication.

Figure 11.10 further shows the EQE response when interface recombination at the front or rear interface is assumed. In Figure 11.10a, the α spectrum of $MAPbI_3$ (Figure 4.4) and the corresponding penetration depth of light, calculated from $d_p = 1/\alpha$, are indicated. The d_p is very small (<70 nm) in the high-α region ($\lambda < 500$ nm), but increases drastically in the longer λ region. Figure 11.10b shows the depth-resolved light absorption (partial absorptance) of the $MAPbI_3$ layer in the n-i-p cell. Here, the white line indicates the $3d_p$ position, which corresponds to the upper limit of light absorption in the depth direction.

To visualize the influence of carrier recombination, we assume a dead layer in the front or rear interface region [29, 54]. The dead layer represents the carrier recombination region located at the front (rear) interface with a thickness of d_{front} (d_{rear}). In this simulation, all carriers generated in the dead layer are simply lost due to an intense recombination. The dotted lines in Figure 11.10b show the assumed dead layer thickness and the corresponding EQE spectra, obtained by varying d_{front} and d_{rear}, are shown in Figure 11.10c,d, respectively. These EQE spectra were calculated using the e-ARC software [63] (Section 11.4.3).

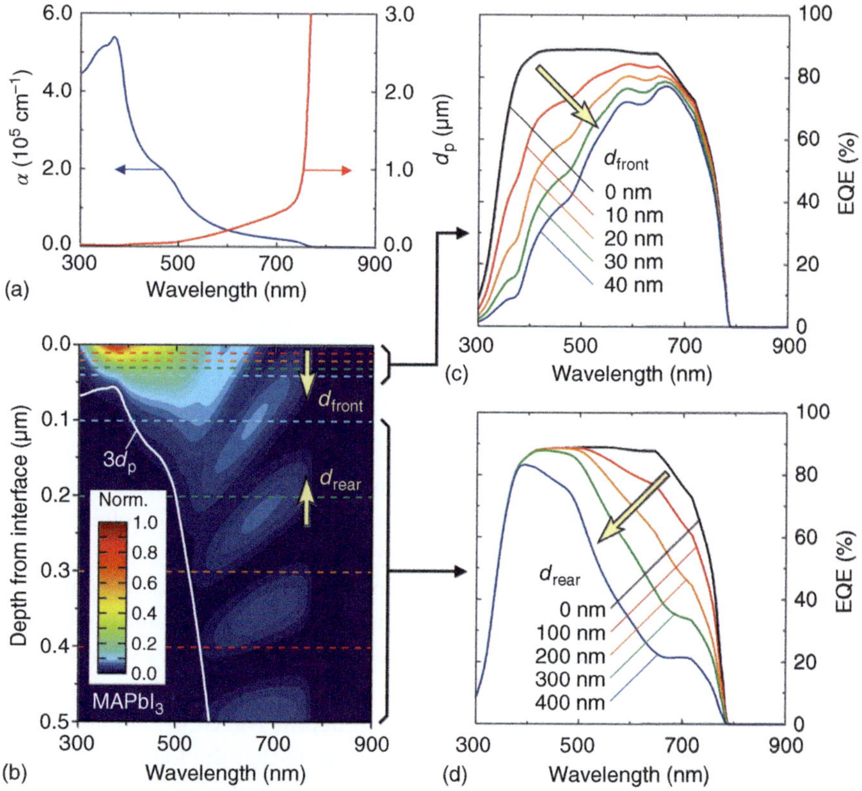

Figure 11.10 (a) α spectrum of $MAPbI_3$ and resulting penetration depth of light ($d_p = 1/\alpha$), (b) depth-resolved absorptance of the $MAPbI_3$ layer in a structure of glass/SnO_2:F(500 nm)/TiO_2(30 nm)/$MAPbI_3$(500 nm)/PTAA(50 nm)/Au, obtained at different depths from the $TiO_2/MAPbI_3$ interface and wavelengths, and EQE spectra calculated for varying (c) front dead layer thicknesses (d_{front} = 0–40 nm) and (d) rear dead layer thicknesses (d_{rear} = 0–400 nm). In (b), the position of $3d_p$ calculated from the results of (a) is shown and the positions of d_{front} and d_{rear} used for the calculations of (c) and (d), respectively, are also indicated. The EQE spectra in (c) and (d) were calculated by using e-ARC software [63].

It can be seen from Figure 11.10c that a slight increase in d_{front} up to 40 nm decreases the short-λ EQE response significantly due to the strong light absorption near the $TiO_2/MAPbI_3$ interface. In contrast, the increase in the rear recombination deteriorates the EQE in the long-λ region, but with a more moderate EQE variation due to uniform carrier generation in this λ region. Consequently, the effects of the front and rear interfaces appear primarily in the short- and long-λ regions, respectively.

In Pb-based perovskite cells, EQE variations in both short-λ [54, 64] and long-λ [3] regions have been observed, which can be interpreted by the change in interface recombination, provided that L_D is sufficiently long. In $MAPbI_3$ cells without HTL (i.e. n-i cells) [65, 66], the absence of an electric field in the rear interface enhances the carrier recombination at the $MAPbI_3$/metal interface, making the EQE spectrum similar to that observed at d_{rear} = 200–300 nm in Figure 11.10d. A large EQE reduction in longer λ has also been confirmed for Sn-based solar cells [67, 68] (see Figure 14.18), due to the significant recombination in the solar cell bottom region. These results demonstrate the usefulness of EQE simulation analysis in finding the dominant carrier recombination region in solar cells. Note, however, that many high-efficiency solar cells show 100% carrier-collection efficiency in the short-circuit condition (see Section 11.4.2).

11.4.2 Power Loss Mechanism

This section presents a detailed power-loss analysis performed for a state-of-the-art perovskite cell ($\eta \sim 20\%$) with a triple-cation-mixed $Cs_{0.05}FA_{0.8}MA_{0.15}Pb(I_{0.95}Br_{0.05})_3$ absorber (see Figure 4.14). Figure 11.11 shows (a) the optical model (structure) of the analyzed cell and (b) the EQE analysis result. This solar cell has an n-i-p structure of glass/ITO/Sb-doped TiO_2(ATO)/TiO_2/mp-TiO_2/perovskite/PTAA/Au. In the analysis, to simplify the simulation, the optical effect of the mp-TiO_2 is neglected while maintaining the total perovskite thickness. The EQE analysis was performed based on the ERS method [17, 28], which performs the EQE analysis by incorporating experimental reflectance spectra (ERS).

As shown in Figure 11.11b, the calculated EQE (red line) shows almost complete agreement with the experimental EQE (open circles), demonstrating a 100% carrier

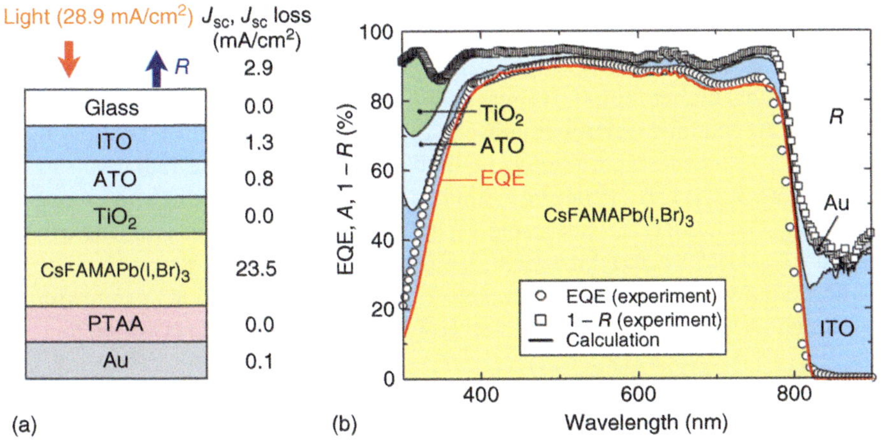

Figure 11.11 (a) Optical model and (b) EQE analysis result for a triple-cation-mixed $Cs_{0.05}FA_{0.8}MA_{0.15}Pb(I_{0.95}Br_{0.05})_3$ solar cell with a 19.95% efficiency. The cell structure is glass/ITO (150 nm)/Sb-doped TiO_2(ATO, 50 nm)/TiO_2(15 nm)/mp-TiO_2(150 nm)/perovskite (500 nm)/PTAA (50 nm)/Au, but the optical effect of mp-TiO_2 is neglected in the optical model of (a). In (b), the open circles (EQE) and squares (R) show experimental results, while the solid lines indicate the calculated absorptance spectra from which the J_{sc} and J_{sc} losses of each layer shown in (a) are deduced. The EQE analysis was performed based on the ERS method [17, 29]. The optical constants of individual layers used for the analysis were obtained from the ellipsometry analyses of separate samples.

collection under the EQE measurement (short-circuit) condition, as observed earlier [28]. Note that a strong electric field is present in the perovskite layer under the short-circuit condition and the photocarriers are collected relatively easily, as discussed in Figure 11.6. In Figure 11.11a, the J_{sc} of the perovskite absorber and the J_{sc} losses induced by the R and A of the component layers are summarized. When 1-sun illumination is converted completely into J_{sc}, we obtain $J_{sc,1sun} = 28.9\,mA/cm^2$ for $E_g = 1.55\,eV$. Among the J_{sc} losses, the R loss is the largest ($2.9\,mA/cm^2$), followed by the TCO loss ($2.1\,mA/cm^2$). Remarkably, the optical losses in the ETL, HTL, and Au metal electrode are almost zero, resulting in a very high optical gain (i.e. $g = J_{sc}/J_{sc,1sun}$) of $g = 81\%$. This optical gain is higher than those of other thin-film devices including $CuInGaSe_2$ and CdTe solar cells [54], but lower than those of crystalline Si devices ($g = 86$–89%) [15].

The above EQE analysis can further be combined with the J–V analysis to determine the precise power loss mechanism in the solar cell. Figure 11.12a shows the J–V curves obtained experimentally (open circles) and analytically (solid lines) for the $CsFAMAPb(I,Br)_3$ cell. When the resistance within solar cells is neglected, the J–V curve is given simply by

$$J = J_0 \left[\exp\left(\frac{qV}{k_B T} \right) - 1 \right] - J_{sc} \tag{11.1}$$

where J_0, q, k_B, and T denote the saturation current density, electron charge, Boltzmann constant, and temperature, respectively [14, 69]. The J–V curve denoted by the blue area in Figure 11.12a is obtained from Shockley-Queisser (SQ) theory (Section 12.2) [70] assuming $E_g = 1.55\,eV$. In this case, two parameters $J_0 = J_{0,SQ}$ and $J_{sc} = J_{sc,SQ}$ can be calculated from the blackbody radiation at 300 K (φ_{BB}) and 1-sun illumination (φ_{1Sun}), respectively (see Section 12.2).

In the SQ calculation, however, the parasitic light absorption is not considered and $J_{sc,SQ}$ is notably higher than the experimental J_{sc}. In the next simulation step (green J–V area), J_{sc} obtained experimentally ($J_{sc,ex}$) is applied for Eq. (11.1) (i.e. $J_{sc} = J_{sc,ex}$, $J_0 = J_{0,SQ}$). From the experimental EQE spectrum and φ_{BB}, the radiative saturation current ($J_{0,rad}$) can further be estimated from Eq. (12.17). For the assumption of $J_0 = J_{0,rad}$, however, the calculated V_{oc} (1.27 V) is much higher than the experimental value (1.08 V). To adjust the theoretical V_{oc} to the experimental V_{oc} ($V_{oc,ex}$), the nonradiative recombination (i.e. defect-induced recombination current) in the experimental cell is considered and the dark current is increased according to

$$J_0 = J_{0,rad} + J_{0,nonrad} \tag{11.2}$$

where $J_{0,nonrad}$ is the nonradiative dark current. The J–V curve obtained by adding the $J_{0,nonrad}$ term is shown as the orange region in Figure 11.12.

In the next analysis, the series resistance (R_s) and shunt resistance (R_{sh}) of the solar cell are considered by the following equation [14]:

$$J = J_0 \left\{ \exp\left[\frac{q(V - R_s J)}{n k_B T} \right] - 1 \right\} - J_{sc} + \frac{V - R_s J}{R_{sh}} \tag{11.3}$$

where n represents the diode factor. The J–V curve with the red region indicates one obtained assuming $R_s = 1.8\,\Omega cm^2$ and $R_{sh} = 1.3\,k\Omega cm^2$ while using $n = 1$ in Eq. (11.3). In the final fitting analysis, n is varied and the experimental J–V

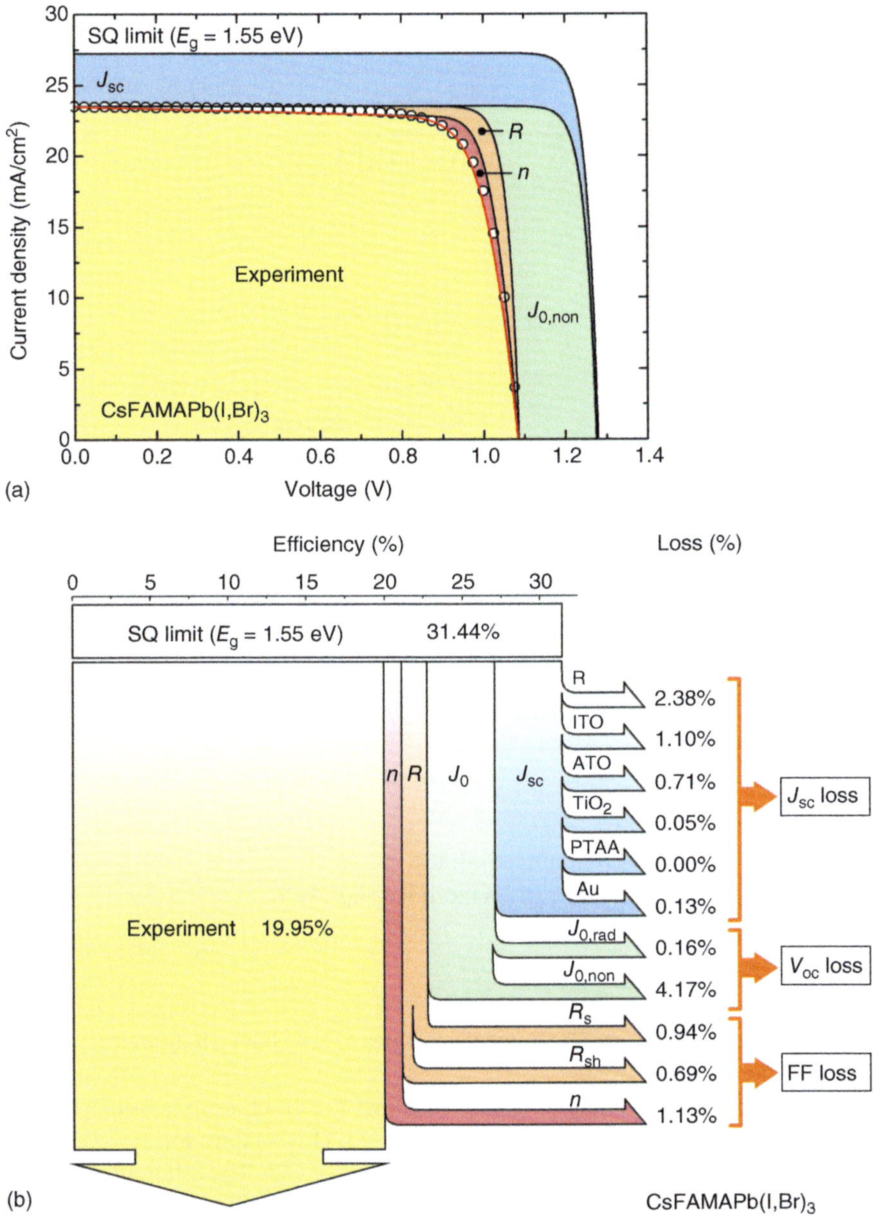

Figure 11.12 (a) J–V curves obtained experimentally (open circles) and analytically (solid lines) and (b) power loss mechanism for a 19.9% efficient $Cs_{0.05}FA_{0.8}MA_{0.15}Pb(I_{0.95}Br_{0.05})_3$ n-i-p cell with a structure of glass/ITO (150 nm)/Sb-doped TiO_2(ATO, 50 nm)/TiO_2(15 nm)/mp-TiO_2(150 nm)/perovskite (500 nm)/PTAA (50 nm)/Au. In (a), the J–V curve of the Shockley and Queisser (SQ) limit (blue color), obtained assuming $E_g = 1.55$ eV, is adjusted to the experimental J–V curve by considering J_{sc} loss (green color), nonradiative recombination ($J_{0,non}$; orange color), R_s and R_{sh} losses (red color), and the loss induced by the diode factor (n; red line). The corresponding power loss contributions are shown in (b). The J_{sc} loss in (b) is determined from the EQE analysis of Figure 11.11b, while the R in the J_{sc} loss denotes reflection loss. The efficiency loss indicated as $J_{0,rad}$ is caused by the reduction in J_{sc} in the experimental cell.

Table 11.1 Characteristics of a $Cs_{0.05}FA_{0.8}MA_{0.15}Pb(I_{0.95}Br_{0.05})_3$ solar cell extracted from the analysis of Figure 11.12.

Cell parameter	Parameter value
$J_{sc,ex}$	23.5 mA/cm^2
$V_{oc,ex}$	1.08 V
FF_{ex}	0.78
η_{ex}	19.95%
$J_{sc,SQ}$	27.2 mA/cm^2
$V_{oc,SQ}$	1.27 V
FF_{SQ}	0.90
$J_{0,SQ}$	9.03×10^{-21} mA/cm^2
η_{SQ}	31.44%
$J_{0,rad}$	1.19×10^{-20} mA/cm^2
$J_{0,nonrad}$ $(n = 1)$	1.34×10^{-17} mA/cm^2
$J_{0,nonrad}$ $(n = 1.6)$	9.31×10^{-11} mA/cm^2
R_s	1.8 Ωcm^2
R_{sh}	1.3 kΩcm^2

The subscript "ex" shows the experimental values, and η_{ex} denotes the experimental cell efficiency. The solar cell parameters extracted from the Shockley–Queisser (SQ) limit assuming $E_g = 1.55$ eV are also shown.

curve (yellow region) is reproduced almost perfectly when $n = 1.6$ (red line in Figure 11.12a). In this case, $J_{0,nonrad}$ is modified slightly by

$$J_{0,nonrad} = \frac{J_{sc}}{\exp\left(\frac{qV_{oc}}{nk_BT}\right) - 1} - J_{0,rad} \tag{11.4}$$

assuming that $V_{oc}/R_{sh} \to 0$. The numerical values of the above analyses are summarized in Table 11.1.

From the step-by-step analyses described above, the exact power loss mechanism in the solar cell can be determined (Figure 11.12b). The SQ limit for $E_g = 1.55$ eV is $\eta = 31.44\%$, where the experimental cell shows $\eta = 19.95\%$. The difference between these efficiencies ($\Delta\eta = 11.5\%$) can be decomposed into the loss contributions arising from J_{sc}, J_0, R (R_s and R_{sh}) and n. Roughly half of the efficiency loss (4.17%) is caused by the dark current generation due to the nonradiative recombination (i.e. $J_{0,nonrad}$). The relatively large loss arising from n ($\Delta\eta = 1.1\%$) can also be interpreted as a result of the recombination within the cell. The second largest loss originates from the J_{sc} reduction due to the light reflection ($\Delta\eta = 2.4\%$), whereas the influences of R_s and R_{sh} are also notable ($\Delta\eta = 1.6\%$ in total). The above analysis results indicate clearly that the 19.9%-efficient $CsFAMAPb(I,Br)_3$ solar cell is limited essentially by the nonradiative recombination and unoptimized high reflectance of the cell. For

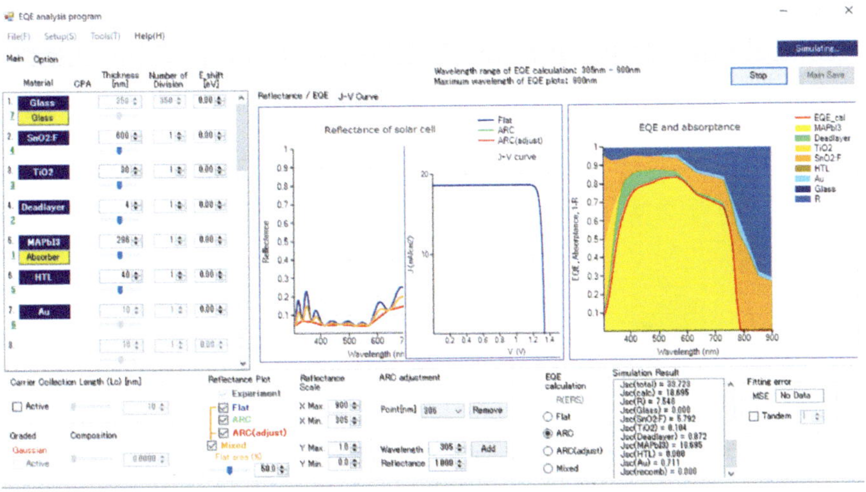

Figure 11.13 Screenshot of e-ARC software, which enables detailed solar cell device simulation. This software can be employed to calculate/simulate the EQE and *J*–*V* curve of solar cells. In particular, complete EQE analyses for single and tandem cells, together with conversion efficiency prediction, can be made from the e-ARC software. This software can be downloaded from a website (https://unit.aist.go.jp/rpd-envene/PV/en/service/e-ARC_en/index_en.html), which can be found using the search term "e-ARC AIST."

this cell, it is most likely that intensive recombination occurs predominantly in the interface regions, as discussed in Section 11.5.

11.4.3 e-ARC Software

Many analyses described in Section 11.4 can be performed based on a Windows-based free computer software (e-ARC software) [63]. Figure 11.13 shows the screenshot of the e-ARC software, which indicates the analysis example for a $MAPbI_3$ cell. This software allows the complete analysis of the optical gain/loss as well as the calculation of the maximum conversion efficiency for designed solar cell structures. In the software, the EQE spectra of various solar cells can be calculated by applying e-ARC method, in which the absorptance of submicron-textured thin film structures can be deduced by applying the antireflection condition [54] (see Figure 12.10). This software also enables an EQE analysis based on the ERS method [17, 54], which was adopted in the EQE analysis of Figure 11.11. The carrier recombination effect within solar cells can explicitly be incorporated into the analysis, as performed in Figure 11.10. The software is quite easy to operate and is recommended for those who wish to gain a better understanding of the carrier generation/loss and theoretical limit of solar cells.

11.5 What Determines V_{oc}?

Minimizing V_{oc} loss is critical to maximizing the conversion efficiency of hybrid perovskite solar cells. In fact, the rapid increase in perovskite cell efficiency

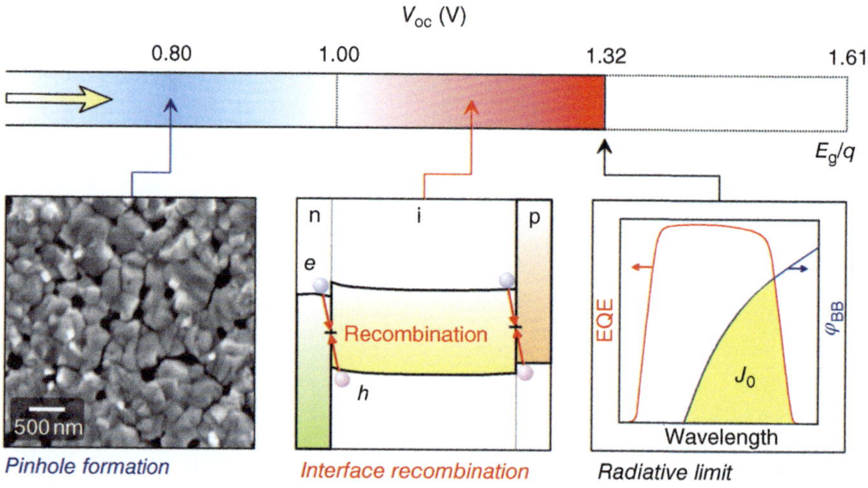

Figure 11.14 V_{oc} of $MAPbI_3$ solar cells determined by three critical factors: (i) pinhole formation (V_{oc} ~ 0.8 V), (ii) nonradiative recombination occurring predominantly at the interfaces (V_{oc} = 1.00–1.32 V), and (iii) radiative limit that determines the theoretical maximum (V_{oc} = 1.32 V). For pinhole formation, an SEM image of a $MAPbI_3$ film, reported in Ref. [71], is shown. The radiative limit is calculated from $J_0 = J_{0,rad}$, determined as a product of the EQE spectrum and blackbody radiation spectrum at 300 K (see Section 12.3 for more details). *Source* (SEM image): Adapted with permission of Liu et al. [71]. Copyright (2016) John Wiley and Sons.

observed over the past decade is primarily attributed to the improvement in V_{oc} (see Figure 1.3). Figure 11.14 illustrates three important factors governing the absolute value of V_{oc}: (i) the radiative limit, which determines the maximum V_{oc}, (ii) the carrier recombination at the ETL and HTL interfaces, and (iii) the formation of pinholes in the perovskite layers. The E_g of $MAPbI_3$ is 1.61 eV (Table 4.3), but V_{oc} corresponding to the E_g (i.e. E_g/q) can never be achieved due to the presence of a high dark current determined by E_g. The radiative V_{oc} limit can be simply calculated assuming $J_0 = J_{0,rad}$, which is obtained as the sum of the product of EQE and φ_{BB} (Section 12.3). For a $MAPbI_3$ solar cell, the theoretical V_{oc} limit is estimated to be V_{oc} = 1.32 V (Table 12.1) [16, 51, 52]. Thus, compared with E_g/q of $MAPbI_3$ (i.e. 1.61 V), the maximum possible V_{oc} is reduced by ~0.3 V.

In high-efficiency $MAPbI_3$ cells, V_{oc} values in the range 1.0–1.26 V have been observed [1, 5, 8–10, 24, 38, 72]. The slightly reduced V_{oc} in these experimental cells, compared with the theoretical limit of 1.32 V, can primarily be attributed to the interface recombination (Section 11.5.1), which increases $J_{0,nonrad}$ in Eq. (11.2). Note that, under the open-circuit condition, the interface recombination becomes more intense due to the flat band formation (see Figure 11.14) and a significant V_{oc} lowering occurs when the interface trap density is high. However, the best experimental V_{oc} of 1.26 V [10] is quite high and is very close to the theoretical limit with the difference of only 60 mV, indicating well-suppressed defect recombination within the cell.

For hybrid perovskite layers formed by spin-coating, the suppression of pinhole formation and the uniform coverage of the underlying layer are critical to improving efficiency [71, 73–76]. In the perovskite devices with pinholes (see Figure 11.14), V_{oc} decreases to the range 0.7–0.9 V [71, 75, 76] and thus the processing needs to be improved in this case. The low V_{oc} could also originate from the increase of the bulk defects in the perovskite layers [77]. Unfortunately, intentional variation of the bulk defect density is difficult for the perovskites and its effect remains ambiguous. Nevertheless, very large L_D with long carrier lifetime observed in hybrid perovskites strongly indicates that the contribution of the bulk defects is minor.

In this section, we first discuss the effect of interface recombination on V_{oc} (Section 11.5.1), in an attempt to interpret the absolute V_{oc} observed in high-efficiency perovskite cells. The latest developments in interface defect passivation in perovskite cells, which has been found to be essential in improving cell efficiencies, are further presented in Section 11.5.2. In the final section (Section 11.5.3), the influence of perovskite grain boundaries on V_{oc} is addressed.

11.5.1 Effect of Interface

Carrier recombination in hybrid perovskite solar cells occurs predominantly at the interfaces [40, 78–82], limiting the V_{oc} of high efficiency cells. The V_{oc} reductions caused by individual interfaces formed with the ETL and HTL have been investigated quantitatively based on photoluminescence (PL) emission measurements [40, 81, 82]. In this method, V_{oc} loss is characterized analytically by observing the PL quantum yield (Q_{PL}) under a 1-sun illumination condition. An important physical concept underlying this technique is the fact that good solar cells are essentially good photo emitters (i.e. light-emitting diodes) [83–85] and thus shows high Q_{PL}. The V_{oc} loss in solar cells can uniquely be related to Q_{PL} by considering the diode equation written simply as

$$J_{rad} = J_{0,rad} \exp\left(\frac{qV}{k_B T}\right) \tag{11.5}$$

where J_{rad} is the radiation current density, which is equivalent to the current contributing to light emission, and $J_{0,rad}$ is the prefactor. For simplicity, the "−1" term in Eq. (11.1) is omitted in Eq. (11.5). By setting $V = V_{oc}$ in Eq. (11.5), we obtain

$$V_{oc} = \frac{k_B T}{q} \ln\left(\frac{J_{rad}}{J_{0,rad}}\right) \tag{11.6}$$

Here, we consider the current density contribution generated by light absorption (J_{ab}) and modify the above equation as follows:

$$V_{oc} = \frac{k_B T}{q} \ln\left(\frac{J_{rad}}{J_{ab}} \cdot \frac{J_{ab}}{J_{0,rad}}\right) = \frac{k_B T}{q}\left[\ln\left(\frac{J_{rad}}{J_{ab}}\right) + \ln\left(\frac{J_{ab}}{J_{0,rad}}\right)\right] \tag{11.7}$$

Now, the ratio of J_{rad}/J_{ab} represents the current density ratio of light emission and absorption and defines Q_{PL} (i.e. $Q_{PL} = J_{rad}/J_{ab}$). Moreover, in solar cells, $J_{ab} = J_{sc}$. Accordingly, from Eq. (11.7), we obtain the following relation [83, 85]:

$$V_{oc} = \frac{k_B T}{q} \ln Q_{PL} + \frac{k_B T}{q} \ln\left(\frac{J_{sc}}{J_{0,rad}}\right) = \frac{k_B T}{q} \ln Q_{PL} + V_{oc,rad} \tag{11.8}$$

where $V_{oc,rad}$ is V_{oc} originating from $J_{0,rad}$, thus the theoretical V_{oc} limit. Note that $Q_{PL} < 1$ and, therefore, the term $k_B T \ln Q_{PL}/q$ shows the negative contribution to V_{oc}. In other words, the detrimental effect of the perovskite interface on V_{oc} can be determined explicitly by measuring Q_{PL}. A significant merit of this method is that V_{oc} reductions induced by the ETL and HTL can be evaluate separately based on the obtained Q_{PL} even for structures that are not solar cells. For example, by measuring the Q_{PL} of a substrate/perovskite/PTAA structure, the V_{oc} reduction due to the perovskite/PTAA interface can be assessed.

Figure 11.15 shows the effect of interface recombination on V_{oc} in CsFAMAPb(I,Br)$_3$, determined analytically from Q_{PL} measurements using Eq. (11.8) [40]. In Figure 11.15, the V_{oc} potential values obtained for each interface formed with the HTL or ETL are summarized. The V_{oc} indicated as "Pero" (1.23 V) is extracted from a substrate (fused silica)/perovskite structure, showing a reference level without a significant influence of interface recombination.

When a perovskite/ETL(HTL) interface is formed, Q_{PL} decreases rather drastically, leading to a notable reduction in the implied V_{oc}. The Q_{PL} reduction by the

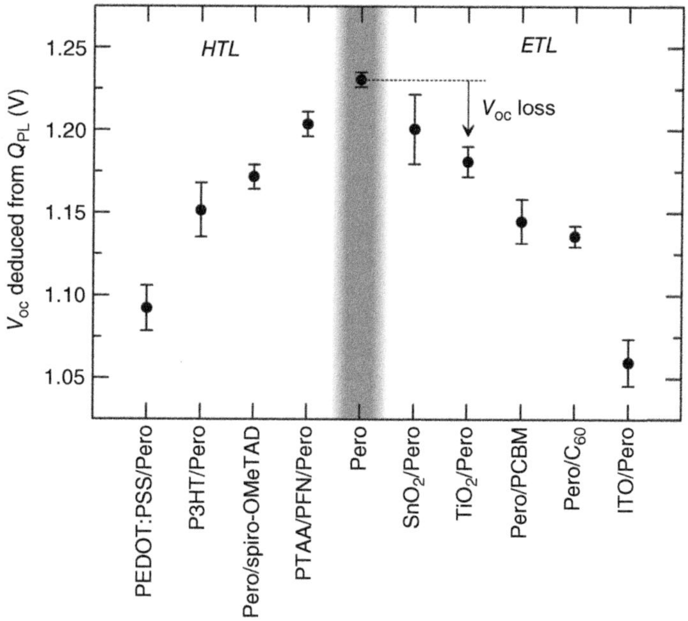

Figure 11.15 V_{oc} potential values obtained for various interfaces formed with HTL or ETL. The V_{oc} (or quasi-Fermi-level splitting) for each interface is deduced from experimental PL quantum yield (Q_{PL}) measurements by applying Eq. (11.8). The Q_{PL} was obtained for CsFAMAPb(I,Br)$_3$ and V_{oc} indicated as "Pero" (1.23 V) refers to the value extracted from a fused-silica substrate/perovskite structure, showing a reference level without a substantial influence of interface recombination. The other results are obtained from substrate/perovskite/ETL(HTL) or substrate/ETL(HTL)/perovskite structures. Note that V_{oc} decreases by the formation of the interface, as indicated by the arrow. In the PTAA sample, a thin PFN-P2 layer is introduced at the PTAA/perovskite interface to improve the perovskite film adhesion. *Source*: Stolterfoht et al. [40].

interface formation is consistent with the general observation that the carrier lifetime is shortened drastically by the ETL (or HTL) interface formation (Figure 6.5b) [86–88]. For the HTL in Figure 11.15, the V_{oc} becomes low when PEDOT:PSS and P3HT layers are adopted, while PTAA provides the best interface with a smaller V_{oc} loss. The V_{oc} characteristic of spiro-OMeTAD is slightly inferior to that of PTAA. For the ETL, the direct formation of the perovskite on ITO increases the V_{oc} loss. The recombination at the PCBM and C_{60} interfaces is also relatively strong, whereas SnO_2 and TiO_2 provide much better interfaces with higher V_{oc}.

The result of Figure 11.15 is of critical importance, as it provides a general guideline for perovskite cell design. In particular, a systematic Q_{PL} measurement shows that the V_{oc} decreases sequentially as the ETL and HTL interfaces are formed. Figure 11.16 shows an example of how V_{oc} is reduced by the formation of ETL and HTL [81]. It can be see that the V_{oc} losses at the ETL and HTL interfaces add up to determine the final V_{oc} of the cell.

The characterization results in Figure 11.15 show that, for the n-i-p configuration, the TiO_2/perovskite/PTAA structure ensures low interface recombination at both

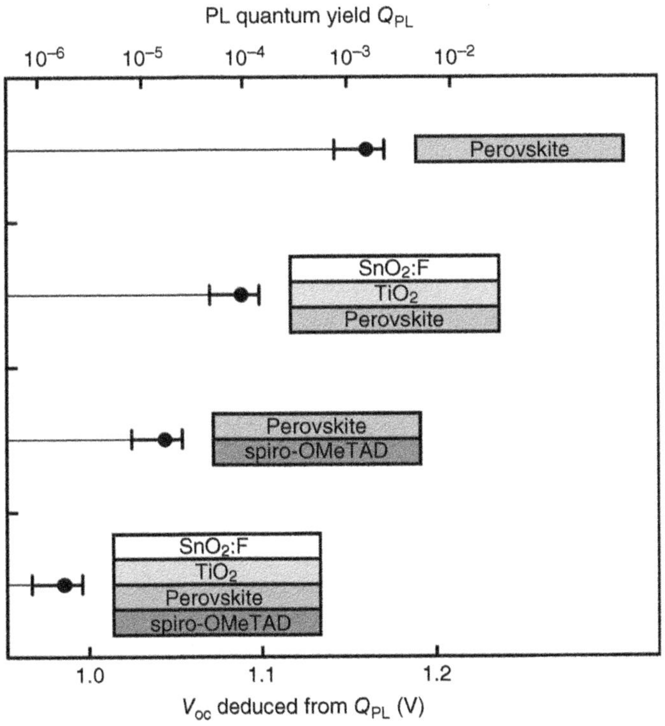

Figure 11.16 V_{oc} deduced from PL quantum yield (Q_{PL}) measurements of different sample (interface) structures. The corresponding Q_{PL} values are also shown. A standalone perovskite ($MAPbI_3$) film was formed on a glass substrate. A lower V_{oc} observed for this sample, compared with the result shown in Figure 11.15, has been suggested to originate from the interface defects formed at the perovskite/glass interface [40]. The V_{oc} decreases sequentially as the ETL and HTL are formed. *Source*: Sarritzu et al. [81]. Licensed under CC BY-4.0.

interfaces, highlighting why this structure has become the standard for hybrid perovskite cells. Moreover, a high V_{oc} of 1.23 V is reported for an n-i-p SnO_2/perovskite/spiro-OMeTAD cell [72], which is consistent with the result of Figure 11.15. For the p-i-n configuration, the application of PEDOT:PSS/perovskite/PCBM leads to a low V_{oc} due to the strong interface recombination occurring at both ETL and HTL interfaces and the PTAA/perovskite/PCBM structure outperforms both PEDOT:PSS and P3HT cells. Note, however, that the overall results shown in Figure 11.15 have been obtained for specific process conditions, and the optimization of layer processing improves the absolute V_{oc} values of Figure 11.15. In fact, a remarkably high V_{oc} of 1.26 V has been reported for a p-i-n PTAA/$MAPbI_3$/PCBM cell [10]. For $MAPbI_3$ cells, $V_{oc,rad} = 1.32$ V, resulting in a simple relation of $V_{oc} = 0.026 \ln Q_{PL} + 1.32$ V from Eq. (11.8). The V_{oc} of 1.26 V observed in the $MAPbI_3$ cell corresponds to $Q_{PL} = 0.1$, indicating that, even in over 20% efficiency perovskite cells, Q_{PL} is rather small and many carriers still recombine nonradiatively at the interfaces.

11.5.2 Effect of Passivation

As we have seen in Section 11.5.1, interface recombination leads to a substantial V_{oc} reduction, limiting the efficiency of perovskite solar cells. Fortunately, the intense carrier recombination at the perovskite interfaces can be suppressed effectively by providing sophisticated passivation layers [89–108]. In fact, when a $MAPbI_3$ layer is terminated by a Lewis-base ligand (tri-*n*-octylphosphine oxide; TOPO), a drastic improvement of Q_{PL} from 0.03 to 0.35 has been reported [78, 79]. An extremely long PL lifetime of 8 μs has also been observed for the TOPO-passivated surface [78] (see Figure 1.6). Accordingly, the passivation of a hybrid perovskite surface (i.e. perovskite rear interface in a cell) has a dramatic impact on suppressing nonradiative recombination, which is presumably induced by vacancies of organic cation and halide atoms [39, 103, 104].

A significant passivation effect has been observed when a wide-bandgap halide (WBH) perovskite layer is introduced at the perovskite/P3HT rear interface (Figure 11.17a) [93]. In this structure, a thin WBH layer (12 nm) consisting of a double-layered perovskite crystal is formed by treating a $FAMAPb(I,Br)_3$ absorber layer with *n*-hexyl trimethlyammonium bromide (HTAB). The P3HT polymer on the WBH is further self-organized along the preferred direction. By introducing the WBH passivation layer, the V_{oc} improves drastically by 227 mV from 0.925 to 1.152 V, while the cell efficiency also increases remarkably from 15.8% to 23.3%. Recall from Figure 11.15 that the perovskite/P3HT interface is particularly defective, compared with the perovskite/PTAA interface, and thus the passivation effect is more drastic. The mechanism for the suppression of nonradiative interface recombination is illustrated in Figure 11.17b. The photoelectron characterization of the layers, however, indicates the formation of small energy barrier for holes at the perovskite/WBH interface [93]. The above study shows the exciting result that high efficiency cells can still be fabricated using an inexpensive P3HT HTL by effectively passivating the perovskite rear interface, while avoiding the use of an expensive spiro-OMeTAD.

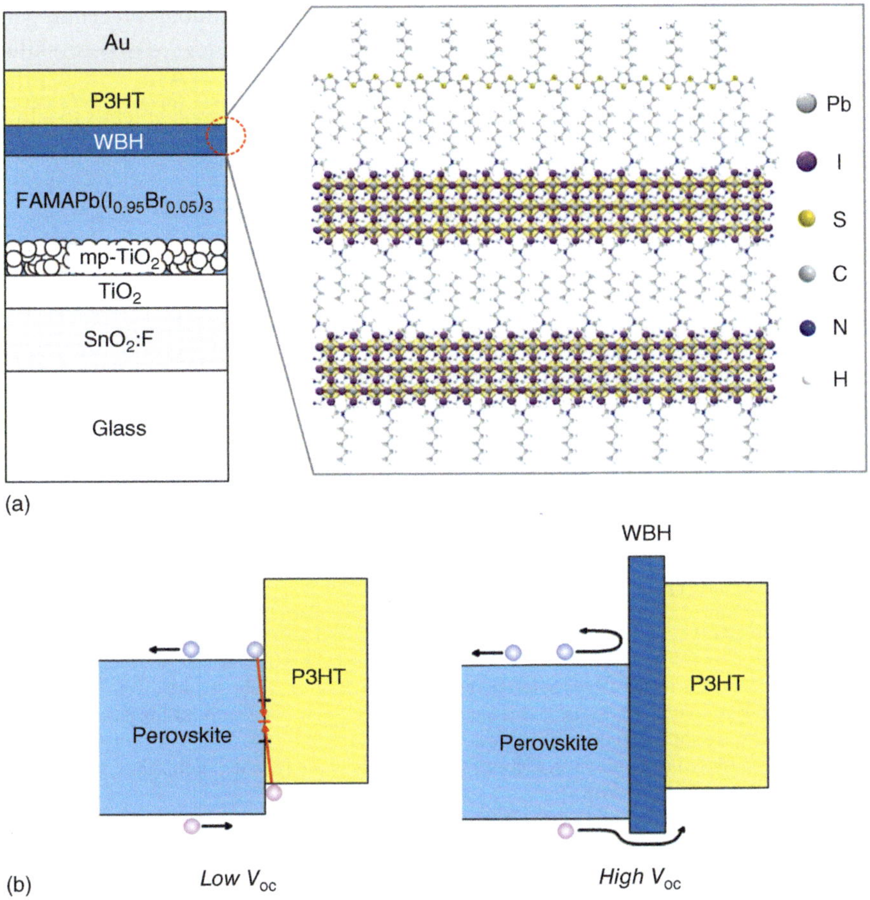

Figure 11.17 (a) Structure of a hybrid-perovskite solar cell with a wide-bandgap halide (WBH) passivation layer introduced at the perovskite/P3HT rear interface in an n-i-p cell and (b) band alignments at the perovskite/P3HT and perovskite/WBH/P3HT interfaces. In this novel structure, a thin WBH layer (12 nm), consisting of a double-layered perovskite crystal, is formed by treating a FAMAPb(I,Br)$_3$ absorber layer with *n*-hexyl trimethlyammonium bromide (HTAB). As shown in (b), the introduction of a wide-gap passivation layer suppresses the interface recombination, which in turn increases V_{oc} dramatically by more than 200 mV. *Source* (Atomic configuration of (a)): Jung et al. [93].

Table 11.2 summarizes the results of different passivation approaches performed for the n-i-p and p-i-n cells. The η and V_{oc} in the table show those obtained by introducing a passivation layer (or reactant) indicated by a square bracket, whereas ΔV_{oc} represents the V_{oc} improvement obtained through the introduction of the passivation layer; $V_{oc} - \Delta V_{oc}$ in the table corresponds to the V_{oc} of the unpassivated cells. The basic concept of passivation involves the introduction of a thin wide-gap perovskite (a 2D planar-type perovskite crystal) [89–93] or an insulating layer. As insulating passivation layers, various organic materials (polymers [95–101], organic salts [102–104], molecules [105, 106]), and inorganic layers (LiF [107] and

Table 11.2 Characteristics of hybrid perovskite solar cells with passivation structures.

Cell interface structure [Passivation layer]	η (%)	V_{oc} (V)	ΔV_{oc} (mV)	References
n-i-p				
mp-TiO_2/FAMAPb$(I,Br)_3$/[**HTAB**][a]/P3HT	23.3	1.15	227	[93]
mp-TiO_2/[**PP**][b]/CsFAMARbPb$(I,Br)_3$/[**PM**][c]/spiro	20.8	1.21	113	[100]
mp-TiO_2/[**PP**][b]/CsFAMARbPb$(I,Br)_3$/spiro	20.4	1.16	70	[101]
SnO_2/FAMAPbI_3/[**PEAI**][d]/spiro	23.5	1.16	56	[104]
mp-TiO_2/CsFAMAPb$(I,Br)_3$/[**PPI**][e]/spiro	20.7	1.15	52	[91]
p-i-n				
PTAA/FAMAPb$(I,Br)_3$/[**CC**][f]/C_{60}	21.0	1.14	110	[103]
PTAA/FAMAPb$(I,Br)_3$/[**GB**][g]/PCBM	21.5	1.21	100	[94]
PTAA/CsFAMAPb$(I,Br)_3$/[**LiF**]/C_{60}	21.6	1.15	35	[107]
PTAA/MAPbI_3/[**PS**][h]/C_{60}	20.3	1.10	30	[95]
PTAA/MAPbI_3/[**BPI**][i]/PCBM	19.5	1.11	30	[89]

The passivation layers (or reactants) incorporated into n-i-p and p-i-n type cells are indicated by square brackets. The conversion efficiency (η) and V_{oc} of the cells with passivation layers are summarized. The ΔV_{oc} in the table indicates the improvement in V_{oc} obtained by incorporating the passivation layer. Thus, $V_{oc} - \Delta V_{oc}$ corresponds to the V_{oc} obtained without the passivation layer. The results are shown in descending order of ΔV_{oc}.

a) Double perovskite layer shown in Figure 11.17, formed by the reaction with *n*-hexyl trimethyl ammonium bromide (HTAB).
b) Layer formed by mixing poly(methyl methacrylate) (PMMA) with PCBM.
c) PMMA: poly(methyl methacrylate).
d) PEAI: phenethylammonium iodide.
e) 2D perovskite layer in the form of PEA_2PbI_4 (PPI); PEA is phenylethylammonium.
f) CC: choline chloride.
g) Widegap layer formed by the reaction with guanidinium bromide (GB).
h) PS: polystyrene.
i) 2D perovskite layer in the form of $(BA)_2PbI_4$ (BPI), formed by the reaction with *n*-butylamine (BA).

Al_2O_3 [108]) have been adopted. In Table 11.2, the largest ΔV_{oc} is obtained for the n-i-p cell shown in Figure 11.17. The passivation effect is smaller in the n-i-p cells with spiro-OMeTAD HTLs ($\Delta V_{oc} \sim 50$ mV). For a p-i-n cell, treatment of the perovskite rear surface by choline chloride [103] or guanidinium bromide [94] is particularly successful with a large ΔV_{oc} exceeding 100 mV.

The results of Table 11.2 demonstrate that the passivation of the perovskite rear surface is a very important approach for achieving high V_{oc}. Note, however, that very high V_{oc} values of 1.23–1.26 V have been reported for the n-i-p and p-i-n cells without the introduction of passivation layers [10, 72]. Accordingly, detailed processing is also of critical importance in improving an absolute V_{oc}. In fact, the V_{oc} increases largely by adopting a process that enhances the ordering of PCBM molecules [109] or by minimizing the HTL doping concentration [72]. Nevertheless, the incorporation of a passivation layer is an effective approach for fabricating higher efficiency cells, enabling more effective control of the nonradiative recombination occurring at the interfaces. Moreover, the introduction of a passivation layer leads to the significant improvements in fill factor (FF), cell stability, and a reduction in unfavorable *J*–*V* hysteresis characteristics [89, 91, 93]. A state-of-the-art passivation technique has also been incorporated in the fabrication of a $FAPbI_3$-based perovskite cell that shows a high efficiency of 24.66% [6].

11.5.3 Effect of Grain Boundary

In general, the grain boundary formation is detrimental for solar cells and the V_{oc} of Si solar cells, for example, increases as grain size becomes larger [110]. Remarkably, even though the grain sizes of hybrid perovskite polycrystals are relatively small (~1 μm), a very high carrier lifetime (~8 μs) [78] and a quite high V_{oc} [10] have been observed. Thus, nonradiative recombination at grain boundaries is suppressed extremely well in hybrid perovskite thin films. This is a characteristic that can be considered as a key factor in achieving >20% efficiencies in polycrystalline devices. In fact, efficient carrier extraction in the grain boundary region has been confirmed experimentally (Chapter 9).

Figure 11.18 shows the influence of average grain size on V_{oc}, FF, and the diode factor n in hybrid perovskite cells with $\eta \sim 20\%$ [6–8, 91, 94, 104, 111]. The n values in Figure 11.18 are extracted from the reported *J*–*V* characteristics by applying Eq. (11.3). Although data shown in Figure 11.18 are obtained from different perovskite materials and structures, the implications of the results are quite clear; (i) V_{oc} is essentially independent of the grain size and remains unchanged unless the grain size is very small (~100 nm) [111, 112] and (ii) FF improves markedly as the perovskite grain size increases, reaching 86% in a large-grain $MAPbI_3$ cell reported in Ref. [8]. The constant V_{oc} observed at different grain sizes indicates that V_{oc} is limited by the surface (interface) recombination and the grain boundary plays a minor role for nonradiative recombination. For the increase of FF with grain size, the improvement of n is mainly responsible and we observe $n = 1.46$ at large grain sizes. The results of Figure 11.18 demonstrate that perovskite polycrystals with larger grains are preferable for improving the cell efficiency.

It should be emphasized that the grain boundary significantly hinders the carrier transport in the perovskite thin films. Indeed, in-plane carrier mobility [~5 $cm^2/(V\,s)$] is much lower than the intragrain mobility [~30 $cm^2/(V\,s)$] (see Figure 6.4). However, when the grain size exceeds the film thickness, there is no grain boundary in the carrier transport direction (i.e. vertical direction) of the cells [111] and thus the detrimental effect of the grain boundary is minimized.

For the exact grain-boundary structures of the perovskite films, however, controversial results have been reported [11, 94, 113]. In an early report [113], it was suggested that the grain boundary is predominantly composed of PbI_2 and that

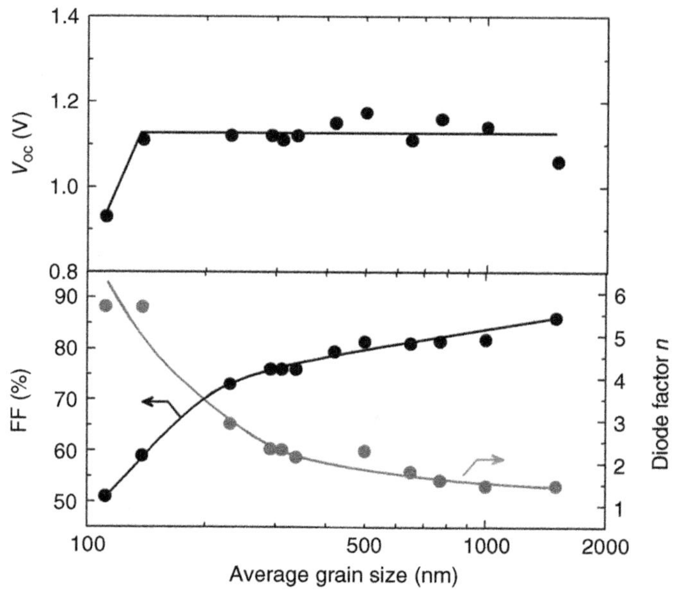

Figure 11.18 Influence of average grain size on V_{oc}, FF, and n in various hybrid perovskite solar cells. Results obtained from different perovskite materials and structures with $\eta \sim 20\%$ are summarized. The n values in the figure have been extracted by analyzing the reported J–V curves using Eq. (11.3). The average size of hybrid perovskite grains was obtained from descriptions or analyses of reported SEM images, which produce average grain sizes of <350 nm [111], 420 nm [91], 500 nm [94], 650 nm [7], 770 nm [104], 1000 nm [6], and 1500 nm [8]. *Source*: Min et al. [6]; Jeon et al. [7]; Chiang and Wu [8]; Cho et al. [91]; Luo et al. [94]; Jiang et al. [104]; Correa-Baena et al. [111].

the perovskite grains are surrounded by thin PbI_2 shells. The perovskite growth under a PbI_2-rich condition is often considered to be beneficial to enhance the grain-boundary passivation through the PbI_2 component. Nevertheless, more recent studies have shown that an excess PbI_2 component is present as isolated PbI_2 grains and does not exist selectively in the grain boundary region [11, 94]. To date, several attempts have been made for "grain-boundary engineering," in which the grain-boundary chemical structures are modified, and the improved solar cell efficiency has been attained (see Chapter 9).

For the p-i-n cells with PCBM ETLs, it has been proposed that PCBM molecules diffuse from the ETL back interface into the perovskite absorber through the grain boundary region, which in turn passivates the grain-boundary trap states [30]. As discussed above, however, the perovskite/PCBM interface shows relatively strong interface recombination and the diffusion of disordered PCBM molecules is expected to enhance the nonradiative recombination by widening the n-type interface region. In fact, when a chemical barrier layer (i.e. passivation layer) is placed at the perovskite/PCBM (or C_{60}) interface, V_{oc} improves effectively (see Table 11.2). Thus, the beneficial role of PCBM diffusion into the grain boundary region is questionable.

References

1 Correa-Baena, J.-P., Saliba, M., Buonassisi, T. et al. (2017). *Science* 358: 739.
2 Jena, A.K., Kulkarni, A., and Miyasaka, T. (2019). *Chem. Rev.* 119: 3036.
3 Yang, W.S., Park, B.-W., Jung, E.H. et al. (2017). *Science* 356: 1376.
4 Saliba, M., Correa-Baena, J.-P., Wolff, C.M. et al. (2018). *Chem. Mater.* 30: 4193.
5 Shin, S.S., Yeom, E.J., Yang, W.S. et al. (2017). *Science* 356: 167.
6 Min, H., Kim, M., Lee, S.-U. et al. (2019). *Science* 366: 749.
7 Jeon, N.J., Na, H., Jung, E.H. et al. (2018). *Nat. Energy* 3: 682.
8 Chiang, C.-H. and Wu, C.-G. (2018). *ACS Nano* 12: 10355.
9 Saliba, M., Matsui, T., Domanski, K. et al. (2016). *Science* 354: 206.
10 Liu, Z., Krückemeier, L., Krogmeier, B. et al. (2019). *ACS Energy Lett.* 4: 110.
11 Jiang, Q., Chu, Z., Wang, P. et al. (2017). *Adv. Mater.* 29: 1703852.
12 Saliba, M., Matsui, T., Seo, J.-Y. et al. (2016). *Energy Environ. Sci.* 9: 1989.
13 Matsui, T., Yamamoto, T., Nishihara, T. et al. (2019). *Adv. Mater.* 31: 1806823.
14 Luque, A. and Hegedus, S. (2011). *Handbook of Photovoltaic Science and Engineering*. West Sussex: Wiley.
15 Nakane, A., Fujimoto, S., and Fujiwara, H. (2017). *J. Appl. Phys.* 122: 203101.
16 Kato, Y., Fujimoto, S., Kozawa, M., and Fujiwara, H. (2019). *Phys. Rev. Appl.* 12: 024039.
17 Hara, T., Maekawa, T., Minoura, S. et al. (2014). *Phys. Rev. Appl.* 2: 034012.
18 National Renewable Energy Laboratory (2020). Best Research-Cell Efficiencies. http://www.nrel.gov/pv/assets/pdfs/best-research-cell-efficiencies.20200406.pdf (accessed 01 April 2021).
19 Green, M.A., Dunlop, E.D., Hohl-Ebinger, J. et al. (2020). *Prog. Photovoltaics Res. Appl.* 28: 629.
20 Yoshikawa, K., Kawasaki, H., Yoshida, W. et al. (2017). *Nat. Energy* 2: 17032.
21 Correa-Baena, J.-P., Steier, L., Tress, W. et al. (2015). *Energy Environ. Sci.* 8: 2928.
22 Wang, Q., Shao, Y., Xie, H. et al. (2014). *Appl. Phys. Lett.* 105: 163508.
23 Emara, J., Schnier, T., Pourdavoud, N. et al. (2016). *Adv. Mater.* 28: 553.
24 Cai, M., Ishida, N., Li, X. et al. (2018). *Joule* 2: 296.
25 Jacobsson, T.J., Correa-Baena, J.-P., Anaraki, E.H. et al. (2016). *J. Am. Chem. Soc.* 138: 10331.
26 Docampo, P., Guldin, S., Stefik, M. et al. (2010). *Adv. Funct. Mater.* 20: 1787.
27 Lin, J., Heo, Y.-U., Nattestad, A. et al. (2015). *Sci. Rep.* 4: 5769.
28 Shirayama, M., Kadowaki, H., Miyadera, T. et al. (2016). *Phys. Rev. Appl.* 5: 014012.
29 Fujiwara, H. and Collins, R.W. (2008). *Spectroscopic Ellipsometry for Photovoltaics: Applications and Optical Data of Solar Cell Materials*, vol. 2. Cham: Springer-Verlag.
30 Shao, Y., Xiao, Z., Bi, C. et al. (2014). *Nat. Commun.* 5: 5784.
31 Kojima, A., Teshima, K., Shirai, Y., and Miyasaka, T. (2009). *J. Am. Chem. Soc.* 131: 6050.
32 Lee, M.M., Teuscher, J., Miyasaka, T. et al. (2012). *Science* 338: 643.

33 Kim, H.-S., Lee, C.-R., Im, J.-H. et al. (2012). *Sci. Rep.* 2: 591.
34 Burschka, J., Pellet, N., Moon, S.-J. et al. (2013). *Nature* 499: 316.
35 Noh, J.H., Im, S.H., Heo, J.H. et al. (2013). *Nano Lett.* 13: 1764.
36 Heo, J.H., Im, S.H., Noh, J.H. et al. (2013). *Nat. Photonics* 7: 486.
37 Liu, M., Johnston, M.B., and Snaith, H.J. (2013). *Nature* 501: 395.
38 You, J., Meng, L., Song, T.-B. et al. (2016). *Nat. Nanotechnol.* 11: 75.
39 Aydin, E., De Bastiani, M., and De Wolf, S. (2019). *Adv. Mater.* 31: 1900428.
40 Stolterfoht, M., Caprioglio, P., Wolff, C.M. et al. (2019). *Energy Environ. Sci.* 12: 2778.
41 Kegelmann, L., Wolff, C.M., Awino, C. et al. (2017). *ACS Appl. Mater. Interfaces* 9: 17245.
42 Chueh, C.-C., Li, C.-Z., and Jen, A.K.-Y. (2015). *Energy Environ. Sci.* 8: 1160.
43 Green, M.A., Ho-Baillie, A., and Snaith, H.J. (2014). *Nat. Photonics* 8: 506.
44 Schulz, P., Edri, E., Kirmayer, S. et al. (2014). *Energy Environ. Sci.* 7: 1377.
45 Olthof, S. (2016). *APL Mater.* 4: 091502.
46 Tao, S., Schmidt, I., Brocks, G. et al. (2019). *Nat. Commun.* 10: 2560.
47 Burgelman, M., Nollet, P., and Degrave, S. (2000). *Thin Solid Films* 361–362: 527.
48 Diekmann, J., Caprioglio, P., Rothhardt, D. et al. (2019). arXiv: 1910.07422.
49 Mandadapu, U., Vedanayakam, S.V., Thyagarajan, K. et al. (2017). *Int. J. Renew. Energy Res.* 7: 1603.
50 Tan, K., Lin, P., Wang, G. et al. (2016). *Solid State Electron.* 126: 75.
51 Tress, W., Marinova, N., Inganäs, O. et al. (2014). *Adv. Energy Mater.* 5: 1400812.
52 Tvingstedt, K., Malinkiewicz, O., Baumann, A. et al. (2014). *Sci. Rep.* 4: 6071.
53 Fujiwara, H., Kato, M., Tamakoshi, M. et al. (2018). *Phys. Status Solidi A* 215: 1700730.
54 Nakane, A., Tampo, H., Tamakoshi, M. et al. (2016). *J. Appl. Phys.* 120: 064505.
55 Edri, E., Kirmayer, S., Henning, A. et al. (2014). *Nano Lett.* 14: 1000.
56 Edri, E., Kirmayer, S., Mukhopadhyay, S. et al. (2014). *Nat. Commun.* 5: 3461.
57 Jiang, C.-S., Yang, M., Zhou, Y. et al. (2015). *Nat. Commun.* 6: 8397.
58 Guerrero, A., Juarez-Perez, E.J., Bisquert, J. et al. (2014). *Appl. Phys. Lett.* 105: 133902.
59 Bergmann, V.W., Weber, S.A.L., Ramos, F.J. et al. (2014). *Nat. Commun.* 5: 5001.
60 Fujiwara, H. and Collins, R.W. (2018). *Spectroscopic Ellipsometry for Photovoltaics: Fundamental Principles and Solar Cell Characterization*, vol. 1. Cham: Springer.
61 Fujiwara, H. and Kondo, M. (2005). *Phys. Rev. B* 71: 075109.
62 Bhachu, D., Waugh, M.R., Zeissler, K. et al. (2011). *Chem. Eur. J.* 17: 11613.
63 This software can be downloaded at the web site of https://unit.aist.go.jp/rpd-envene/PV/en/service/e-ARC_en/index_en.html. This web site can be found using "e-ARC AIST" as key words (accessed 01 April 2021).
64 Tavakoli, M.M., Tress, W., Milić, J.V. et al. (2018). *Energy Environ. Sci.* 11: 3310.
65 Shi, J., Dong, J., Lv, S. et al. (2014). *Appl. Phys. Lett.* 104: 063901.

66 Etgar, L., Gao, P., Xue, Z. et al. (2012). *J. Am. Chem. Soc.* 134: 17396.
67 Ogomi, Y., Morita, A., Tsukamoto, S. et al. (2014). *J. Phys. Chem. Lett.* 5: 1004.
68 Zhao, B., Abdi-Jalebi, M., Tabachnyk, M. et al. (2017). *Adv. Mater.* 29: 1604744.
69 Sze, S.M. (1981). *Physics of Semiconductor Devices*. New York: Wiley-Interscience.
70 Shockley, W. and Queisser, H.J. (1961). *J. Appl. Phys.* 32: 510.
71 Liu, F., Dong, Q., Wong, M.K. et al. (2016). *Adv. Energy Mater.* 6: 1502206.
72 Correa-Baena, J.-P., Tress, W., Domanski, K. et al. (2017). *Energy Environ. Sci.* 10: 1207.
73 Eperon, G.E., Burlakov, V.M., Docampo, P. et al. (2014). *Adv. Funct. Mater.* 24: 151.
74 Wang, Y., Dar, M.I., Ono, L.K. et al. (2019). *Science* 365: 591.
75 Shen, D., Yu, X., Cai, X. et al. (2014). *J. Mater. Chem. A* 2: 20454.
76 Zhang, H., Cheng, J., Lin, F. et al. (2016). *ACS Nano* 10: 1503.
77 Tan, H., Che, F., Wei, M. et al. (2018). *Nat. Commun.* 9: 3100.
78 deQuilettes, D.W., Koch, S., Burke, S. et al. (2016). *ACS Energy Lett.* 1: 438.
79 Braly, I.L., deQuilettes, D.W., Pazos-Outón, L.M. et al. (2018). *Nat. Photonics* 12: 355.
80 Yang, Y., Yang, M., Moore, D.T. et al. (2017). *Nat. Energy* 2: 16207.
81 Sarritzu, V., Sestu, N., Marongiu, D. et al. (2017). *Sci. Rep.* 7: 44629.
82 Wolff, C.M., Caprioglio, P., Stolterfoht, M., and Neher, D. (2019). *Adv. Mater.* 31: 1902762.
83 Rau, U. (2007). *Phys. Rev. B* 76: 085303.
84 Vandewal, K., Tvingstedt, K., Gadisa, A. et al. (2009). *Nat. Mater.* 8: 904.
85 Yao, J., Kirchartz, T., Vezie, M.S. et al. (2015). *Phys. Rev. Appl.* 4: 014020.
86 Xing, G., Mathews, N., Sun, S. et al. (2013). *Science* 342: 344.
87 Stranks, S.D., Eperon, G.E., Grancini, G. et al. (2013). *Science* 342: 341.
88 Hutter, E.M., Eperon, G.E., Stranks, S.D., and Savenije, T.J. (2015). *J. Phys. Chem. Lett.* 6: 3082.
89 Lin, Y., Bai, Y., Fang, Y. et al. (2018). *J. Phys. Chem. Lett.* 9: 654.
90 Chen, P., Bai, Y., Wang, S. et al. (2018). *Adv. Funct. Mater.* 28: 1706923.
91 Cho, K.T., Grancini, G., Lee, Y. et al. (2018). *Energy Environ. Sci.* 11: 952.
92 Cho, Y., Soufiani, A.M., Yun, J.S. et al. (2018). *Adv. Energy Mater.* 8: 1703392.
93 Jung, E.H., Jeon, N.J., Park, E.Y. et al. (2019). *Nature* 567: 511.
94 Luo, D., Yang, W., Wang, Z. et al. (2018). *Science* 360: 1442.
95 Wang, Q., Dong, Q., Li, T. et al. (2016). *Adv. Mater.* 28: 6734.
96 Wolff, C.M., Zu, F., Paulke, A. et al. (2017). *Adv. Mater.* 29: 1700159.
97 Chaudhary, B., Kulkarni, A., Jena, A.K. et al. (2017). *ChemSusChem* 10: 2473.
98 Wang, F., Shimazaki, A., Yang, F. et al. (2017). *J. Phys. Chem. C* 121: 1562.
99 Zuo, L., Guo, H., deQuilettes, D.W. et al. (2017). *Sci. Adv.* 3: e1700106.
100 Peng, J., Khan, J.I., Liu, W. et al. (2018). *Adv. Energy Mater.* 8: 1801208.
101 Peng, J., Wu, Y., Ye, W. et al. (2017). *Energy Environ. Sci.* 10: 1792.
102 Zhao, T., Chueh, C.-C., Chen, Q. et al. (2016). *ACS Energy Lett.* 1: 757.
103 Zheng, X., Chen, B., Dai, J. et al. (2017). *Nat. Energy* 2: 17102.
104 Jiang, Q., Zhao, Y., Zhang, X. et al. (2019). *Nat. Photonics* 13: 460.

105 Wang, F., Geng, W., Zhou, Y. et al. (2016). *Adv. Mater.* 28: 9986.
106 Lin, Y., Shen, L., Dai, J. et al. (2017). *Adv. Mater.* 29: 1604545.
107 Stolterfoht, M., Wolff, C.M., Márquez, J.A. et al. (2018). *Nat. Energy* 3: 847.
108 Koushik, D., Verhees, W.J.H., Kuang, Y. et al. (2017). *Energy Environ. Sci.* 10: 91.
109 Shao, Y., Yuan, Y., and Huang, J. (2016). *Nat. Energy* 1: 15001.
110 Werner, J.H., Dassow, R., Rinke, T.J. et al. (2001). *Thin Solid Films* 383: 95.
111 Correa-Baena, J.-P., Anaya, M., Lozano, G. et al. (2016). *Adv. Mater.* 28: 5031.
112 Saliba, M., Correa-Baena, J.-P., Grätzel, M. et al. (2018). *Angew. Chem. Int. Ed.* 57: 2554.
113 Chen, Q., Zhou, H., Song, T.-B. et al. (2014). *Nano Lett.* 14: 4158.

12

Efficiency Limits of Single and Tandem Solar Cells

Hiroyuki Fujiwara[1], *Yoshitsune Kato*[1], *Masayuki Kozawa*[1], *Akira Terakawa*[2], *and Taisuke Matsui*[3]

[1] *Gifu University, Department of Electrical, Electronic and Computer Engineering, 1-1 Yanagido, Gifu 501-1193, Japan*
[2] *Panasonic Corporation, Energy System Division, 15-2 Nishikiminamimachi, Kaizuka, Osaka 597-0094, Japan*
[3] *Panasonic Corporation, Technology Innovation Division, 3-1-1 Yagumo-naka-machi, Moriguchi City, Osaka 570-8501, Japan*

12.1 Introduction

The theoretical interpretation of solar-cell performances is critical, as it reveals the efficiency-limiting factors of experimental photovoltaic devices [1]. Based on the physics of solar cells, the maximum possible parameters for short-circuit current density (J_{sc}), open-circuit voltage (V_{oc}), and fill factor (FF), as well as the resulting conversion efficiency (η), can be estimated individually [1, 2]. The best-known approach for calculating the solar-cell efficiency limit is that developed by Shockley and Queisser more than half a century ago [2]. Since then, the Shockley–Queisser (SQ) limit has been adopted extensively as an absolute criterion that sets the maximum possible limit of solar-cell efficiency. Remarkably, the SQ limit is estimated by considering only a single physical quantity: the band gap (E_g) of a light absorber at room temperature. Figure 12.1 shows the variation of the SQ limit with E_g, together with the conversion efficiencies of experimental record-efficiency cells [3–7], including methylammonium lead iodide ($CH_3NH_3PbI_3$; $MAPbI_3$) [4] and formamidinium lead iodide [$HC(NH_2)_2PbI_3$; $FAPbI_3$] [5] solar cells. The numerical values of the SQ limit are summarized in Appendix B. It can be seen from Figure 12.1 that the SQ limit exhibits high efficiencies of $\eta \sim 33\%$ in a broad E_g range of 1.1–1.4 eV.

The conversion efficiencies derived from SQ theory are, however, substantially higher than those of the most efficient experimental cells developed to date. In particular, even for the most ideal GaAs cell, the experimental efficiency ($\eta = 29.1\%$ [3]) is substantially inferior to the maximum SQ efficiency of 33.7%. This is, in part, attributed to the overestimated efficiency limit under the SQ model, which is based on highly idealized physical assumptions that can never be achieved in the real world. For example, the theory assumes an infinitely thick light absorber with absolutely zero light reflection (i.e. a perfect light absorber).

Hybrid Perovskite Solar Cells: Characteristics and Operation,
First Edition. Edited by Hiroyuki Fujiwara.

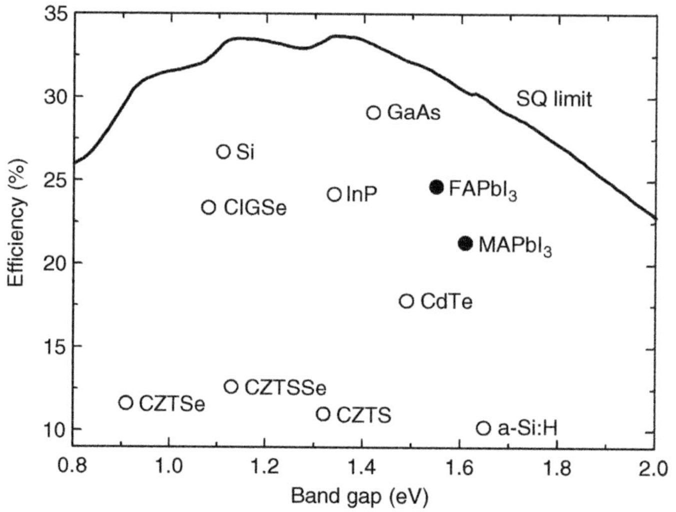

Figure 12.1 Maximum efficiency (η) for different E_g derived from SQ theory (solid line), together with the conversion efficiencies of experimental world record-efficiency cells reported in Refs. [3–7] (circles), including $CuInGaSe_2$ (CIGSe), $Cu_2ZnSn(S,Se)_4$ (CZTSSe), and hydrogenated amorphous silicon (a-Si:H). The efficiencies of the experimental $FAPbI_3$ and $MAPbI_3$ cells are $\eta = 24.66\%$ [5] and 21.3% [4], respectively. For the $FAPbI_3$ cell, a small amount of methylenediammonium dichloride ($MDACl_2$) is incorporated. The numerical values corresponding to the SQ limit are summarized in Appendix B. *Source*: Green et al. [3]; Shin et al. [4]; Min et al. [5]; Green et al. [6]; Lee et al. [7].

A more rigorous approach has been developed to more accurately determine the absolute physical limits of thin-film solar cells, enabling a better understanding of the performance limits of experimental cells [1]. This approach can further be extended to interpret and simulate the performance of perovskite tandem devices. As the tandem solar-cell architecture is complex [8–11], an accurate simulation is of significant importance in designing and improving such devices.

In this chapter, we review all fundamental aspects that are necessary for understanding the theoretical estimation of single- and tandem-cell efficiency limits. We begin by presenting the basic principles of SQ theory (Section 12.2), introducing the optical/physical models adopted in the SQ limit calculation. The latest approach for realistic maximum-efficiency calculation, which has been developed for solar cells in conventional thin-film forms, is explained in Section 12.3. Finally, a rigorous estimation of tandem-cell efficiency is presented in Section 12.4.

12.2 What Is the SQ Limit?

In 1961, Shockley and Queisser developed a theory for predicting the ultimate limit of solar-cell efficiency [2]. This SQ model is based on a simple and straightforward assumption that each photon generates a single electron–hole pair in a p–n junction solar cell at thermal equilibrium. Nevertheless, the understanding of SQ theory is

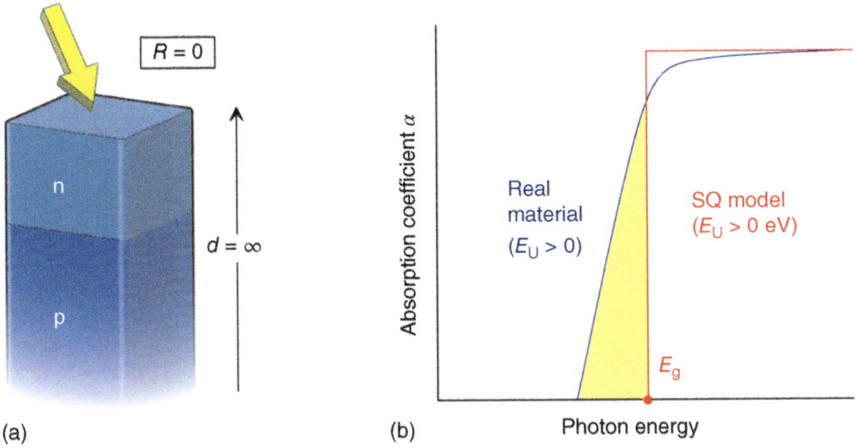

Figure 12.2 (a) Optical model assumed for Shockley–Queisser (SQ) theory and (b) absorption coefficient (α) spectra of the SQ model and a real material. In the SQ model, a p–n homojunction cell with infinite thickness is assumed while forcing the reflectance R of this hypothetical structure to zero (i.e. $R = 0$). For the SQ model, a step-function α spectrum is applied; in contrast, a real material shows finite-tail absorption with nonzero Urbach energy (i.e. $E_U > 0$).

slightly difficult as this model includes a unique physical concept of blackbody radiation. This section provides a step-by-step introduction to SQ theory, which forms a basis for calculating the maximum solar-cell efficiency. Specifically, the physical model assumed in the SQ approach is explained in Section 12.2.1 and the physics of blackbody radiation is fully described in Section 12.2.2. Based on these explanations, the principle of SQ theory (Section 12.2.3) can be understood rather easily.

12.2.1 Physical Model

The solar-cell structure assumed under SQ theory is simple (Figure 12.2a). In this model, a p–n homojunction cell with an infinite thickness is assumed while forcing the reflectance R of this hypothetical structure to zero (i.e. $R = 0$). Figure 12.2b schematically shows the absorption coefficient (α) spectra for the SQ model and a conventional material. Under SQ theory, the α spectrum of the absorber is modeled as a simple step function that shows infinite absorption at the photon energy of $E \geq E_g$. Note that the experimental materials exhibit tail absorption below E_g with Urbach energies of $E_U > 0$ (see Figure 4.17); in contrast, $E_U = 0$ is assumed for SQ theory.

In SQ theory, the current density–voltage (J–V) characteristic is modeled by the following simple diode equation [12, 13]:

$$J = J_0 \left[\exp\left(\frac{qV}{k_B T}\right) - 1\right] - J_{sc} \tag{12.1}$$

where J_0 is the reverse saturation current density (i.e. dark current) and q, k_B, and T are the electron charge, Boltzmann constant, and temperature, respectively.

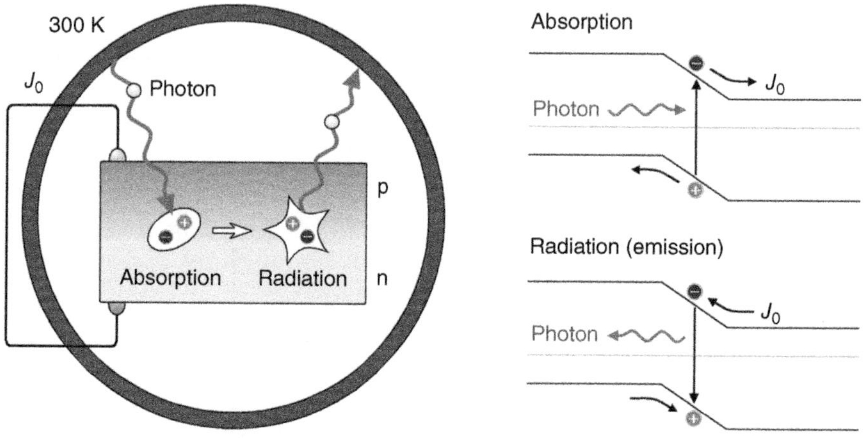

Figure 12.3 Physical model of the SQ model adopted for J_0 calculation. In the SQ model, a p–n homojunction solar cell is surrounded by a spherical blackbody held at 300 K. The absorption of blackbody photons leads to the carrier generation with a current flow of J_0 in the cell. The carrier generation by photon absorption is in equilibrium with the radiative carrier recombination.

In Eq. (12.1), there are only two critical variables (i.e. J_{sc} and J_0) and the solar-cell parameters V_{oc}, FF, and η can be calculated from the J–V curve based on (J_{sc}, J_0). In the SQ limit, J_{sc} is simply estimated by assuming that all electron–hole pairs generated by a 1-sun illumination (φ_{1Sun}) are collected in the energy region $E \geq E_g$. Compared to this simple photocurrent calculation, the estimation of the dark current (J_0) is more complex.

Figure 12.3 shows a physical representation of the SQ model, which is considered for the J_0 calculation. In this approach, a p–n homojunction solar cell is placed in a cavity surrounded by a spherical blackbody held at 300 K. A blackbody is a hypothetical object that absorbs all incoming light, regardless of the wavelength and angle of incidence (i.e. $R = 0$). At elevated temperatures, the blackbody emits electromagnetic waves, which are known as the blackbody radiation. For example, even though graphite is almost completely black, it emits a wide spectrum of blackbody radiation when heated. In the SQ model, the p–n cell is treated as a perfect blackbody in the energy region above E_g, while it is completely transparent at $E < E_g$.

In Figure 12.3, the p–n cell and a spherical blackbody radiator are in thermal equilibrium. If the spherical blackbody radiator is held at 300 K, it emits photons with energies corresponding to those of infrared light (see Section 12.2.2). The generated blackbody photons are absorbed by the solar cell, creating the electron–hole pairs within the absorber. Note that the solar cell in the figure is under the dark condition because it is surrounded completely by the spherical blackbody. In a room covered completely by a thick graphite at 300 K, human eyes cannot detect any light, but blackbody radiation still occurs, generating the dark current in solar cells.

As illustrated in Figure 12.3, the photon absorption in the cell generates carriers with a current flow of J_0. If the direction of the J_0 current is reversed, radiative recombination of carriers occurs, leading to photon radiation (emission). The key assumption of the SQ model is that the rate of photon absorption is exactly equal to the rate of photon emission; therefore, within the cavity, a detailed balance is maintained by the photon exchange, resulting in a complete thermal equilibrium of the system. In other words, in this model, J_0 is ultimately considered as a "virtual current" at $V = 0$, where the current flow generated by photon absorption is counterbalanced by the reverse current flow due to radiative recombination with a net current being $J = 0$. Consequently, under the SQ approach, J_0 is estimated from the absorption of blackbody photons at 300 K. By applying the (J_{sc}, J_0) values obtained from the above-described procedures, the J–V characteristics of a maximum-efficiency cell can be evaluated using Eq. (12.1). This is the fundamental physics of SQ theory.

12.2.2 Blackbody Radiation

Blackbody radiation is governed by a single physical parameter – temperature T. As the physical model of blackbody radiation forms the core of solar-cell limit calculations, the derivation of the Planck's law (http://www.ioa.s.u-tokyo.ac.jp/kisohp/STAFF/nakada/Jugyo/G=Blackbody.ppt) is explained fully in this section.

For this purpose, we consider a blackbody cube with equal sides of length L (i.e. the volume is $V = L^3$). The interior of the cube is empty, but the space inside the cube is filled with blackbody radiation, which is under thermal equilibrium with the blackbody wall maintained at T. The blackbody radiation for a wavelength of λ at T is denoted by $\gamma_{BB}(\lambda, T)$. To describe γ_{BB}, a wavelength space for the photons within the unit volume is defined (Figure 12.4a). This wavelength space differs from the real space and is defined by two factors: λ and the moving direction of the photon (Ω). The $d\Omega$ indicates the solid angle for a traveling photon and the area of the basal plane is $A = \lambda^2\, d\Omega$. When the frustum formed by $d\lambda$ is considered, its volume becomes $\lambda^2\, d\lambda\, d\Omega$. If a photon number density of ρ_n is applied, the number of photons in the frustum is given as $p = \rho_n \lambda^2\, d\lambda\, d\Omega$ per unit volume. In the corresponding real space (Figure 12.4b), on the other hand, the number of photons passing through the small area dS during the time interval dt is expressed as $v = c\,dt\,dS$, where c is the speed of light. Accordingly, the photon number (dN) can be calculated as the product of the

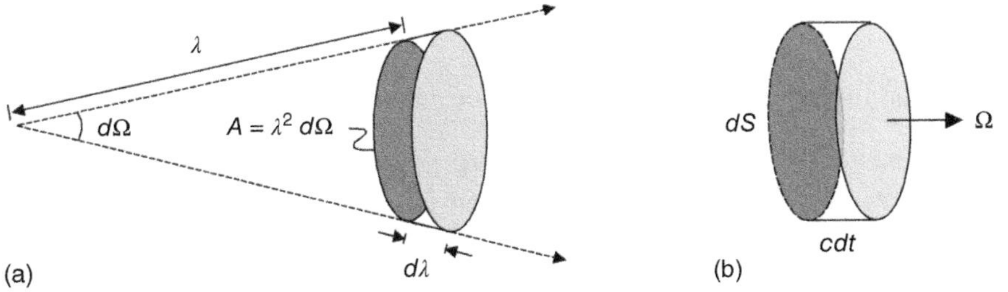

Figure 12.4 Definitions of blackbody radiation in (a) the wavelength space for photons in the unit volume and (b) the real space. The moving direction of photon is Ω, whereas $d\Omega$ indicates the solid angle covered by the path of a traveling photon. In (a), A is the area of the basal plane. The c in (b) shows the speed of light.

real-space volume ($v = c\,dt\,dS$) and the number of photons per unit volume in the wavelength space (i.e. p calculated above):

$$dN = \rho_n c\lambda^2 \; d\lambda \; d\Omega \; dS \; dt \tag{12.2}$$

Because the energy of a single photon is given by $E = hc/\lambda$ (Einstein relation), the energy transferred by dN is $dE = (hc/\lambda)dN$. Thus,

$$dE = \gamma_{BB}(\lambda) \; d\lambda \; d\Omega \; dS \; dt \tag{12.3}$$

where

$$\gamma_{BB}(\lambda) = \rho_n hc^2 \lambda \tag{12.4}$$

To derive ρ_n, we consider the quantum state within the cube ($V = L^3$). For a photon to exist stably within the cube, its wavelength λ in the x, y, and z directions must be exactly the same as L and thus the number of the photon states per volume V is given by

$$\Delta U = 2\Delta\left(\frac{L}{\lambda_x}\right)\Delta\left(\frac{L}{\lambda_y}\right)\Delta\left(\frac{L}{\lambda_z}\right)\cdot\frac{1}{L^3} = \frac{2}{\lambda^6}\Delta\lambda_x\Delta\lambda_y\Delta\lambda_z \tag{12.5}$$

Here, the factor of 2 indicates that the electromagnetic wave can be in two states with different polarizations. By setting $\Delta V_\lambda = \Delta\lambda_x\Delta\lambda_y\Delta\lambda_z$ in Eq. (12.5), we can estimate the density of the quantum state as $\rho_Q = \Delta U/\Delta V_\lambda = 2/\lambda^6$. On the other hand, the quantum-mechanical consideration provides the occupation number of photons in the quantum state of λ as follows (http://www.ioa.s.u-tokyo.ac.jp/kisohp/STAFF/nakada/Jugyo/G=Blackbody.ppt):

$$\Theta = \left[\exp\left(\frac{hc}{\lambda k_B T}\right) - 1\right]^{-1} \tag{12.6}$$

As a result, ρ_n is derived as $\rho_n = \rho_Q\Theta$ and thus γ_{BB} of Eq. (12.4) becomes

$$\gamma_{BB}(\lambda, T) = \frac{2hc^2}{\lambda^5}\left[\exp\left(\frac{hc}{\lambda k_B T}\right) - 1\right]^{-1} \tag{12.7}$$

If hemispherical integration is considered for the blackbody radiation in Figure 12.3, γ_{BB} is multiplied by π (see Section 12.2.3). In this case, the density of blackbody photons (φ_{BB}) per second per meter square can be estimated by dividing γ_{BB} by $E(=hc/\lambda)$:

$$\varphi_{BB}(\lambda, T) = \frac{\pi}{E}\gamma_{BB}(\lambda, T) = \frac{2\pi c}{\lambda^4}\left[\exp\left(\frac{hc}{\lambda k_B T}\right) - 1\right]^{-1} \tag{12.8}$$

Often, a value of φ_{BB} derived for the photon energy is employed (i.e. $\varphi_{BB}(E, T)$). From Eq. (12.3), $\gamma_{BB}(E, T) = \gamma_{BB}(\lambda, T)|d\lambda/dE|$. Since $|d\lambda/dE| = hc/E^2$, we obtain

$$\varphi_{BB}(E, T) = \varphi_{BB}(\lambda, T)\frac{hc}{E^2} = \frac{2\pi E^2}{h^3c^2}\left[\exp\left(\frac{hc}{\lambda k_B T}\right) - 1\right]^{-1} \tag{12.9}$$

Figure 12.5a shows the variation of $\varphi_{BB}(E, T)$ with T, calculated from Eq. (12.9). The spectrum of blackbody photon density changes drastically with T and extends toward the ultraviolet region ($E \sim 3$ eV) as T increases. At 300 K, the φ_{BB} spectrum

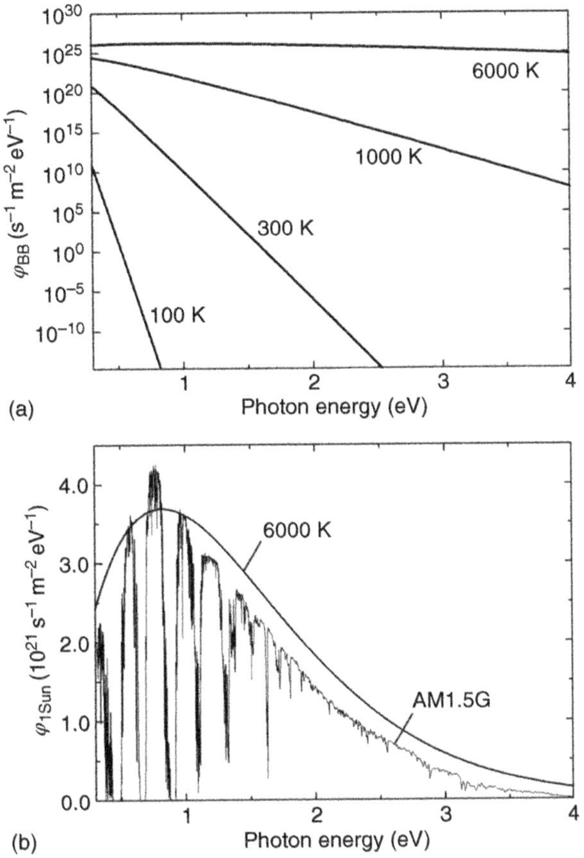

Figure 12.5 (a) Variation in blackbody photon density $\varphi_{BB}(E, T)$ with T, calculated from Eq. (12.9) and (b) photon density of 1-sun illumination (φ_{1Sun}) under the AM1.5G condition, which is compared with blackbody radiation density at $T = 6000$ K.

is nonuniform and φ_{BB} increases in the infrared region ($E \sim 1$ eV); thus, J_0 increases significantly when E_g becomes lower. The spectrum of sunlight can be approximated by setting $T = 6000$ K. In Figure 12.5b, the photon density produced by 1-sun illumination (φ_{1Sun}) under the AM1.5G condition is compared with the blackbody radiation density at $T = 6000$ K and a close match is observed.

12.2.3 SQ Limit

Once the fundamental blackbody concept discussed in Section 12.2.2 is understood, the SQ theoretical-limit calculation becomes straightforward. Figure 12.6a shows the general procedure established under SQ theory for calculating the maximum efficiency. In this method, the optical model is first defined and then used to calculate the external quantum efficiency (EQE). Once the EQE spectrum is known, J_{sc} and J_0 can be calculated from φ_{1Sun} (Figure 12.5b) and φ_{BB} (Figure 12.5a at $T = 300$ K), respectively, as described below. Based on (J_{sc}, J_0), the J–V curve can be calculated from Eq. (12.1), from which all the other solar-cell parameters (i.e. V_{oc}, FF, and η) are deduced.

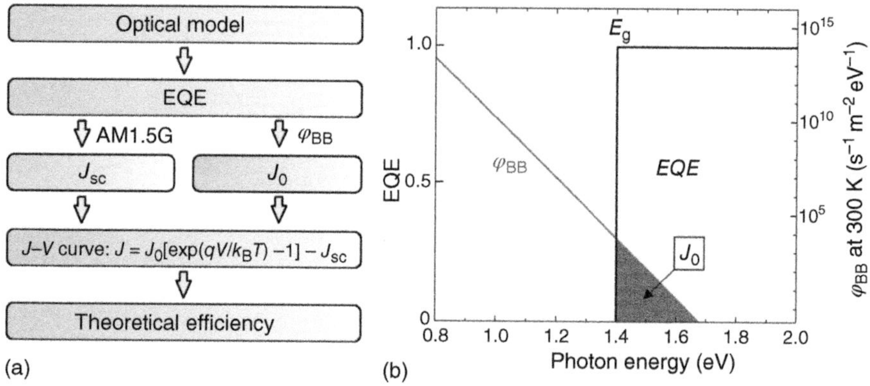

Figure 12.6 (a) Procedure of maximum-efficiency calculation and (b) example of J_0 estimation with $E_g = 1.4$ eV under SQ theory. Here, J_0 is calculated as the product of EQE and φ_{BB} (see Eq. (12.12)). The φ_{BB} spectrum in (b) is consistent with that shown in Figure 12.5a.

Figure 12.6b shows an example of estimating J_0 at $E_g = 1.4$ eV under SQ theory. In the SQ model, (i) an infinite absorber thickness, (ii) a step-function α, and (iii) $R = 0$ are assumed, which lead to the step-function EQE spectrum of $Q_{SQ} = 1$ ($E \geq E_g$) and $Q_{SQ} = 0$ ($E < E_g$). Under SQ theory, J_0 is essentially calculated as an overlap between EQE and φ_{BB} at 300 K (see Figure 12.6b). Accordingly, J_0 increases drastically as E_g decreases.

The actual calculation of J_0 is conducted by considering a hemispherical configuration (Figure 12.7a), in which the blackbody radiation onto a p–n junction solar cell is integrated over a series of solid angles θ and rotation angles ϕ. We consider the blackbody radiation emitted from a small ring-like area, defined by $d\theta$ (gray area in Figure 12.7a). For a hemisphere of radius r, the total gray area on the hemisphere is $dS = 2\pi r \sin\theta L$, where L is the latitudinal distance $L = r\,d\theta$. We further consider the projection of the area dS onto a p–n junction surface with the area of dS' (Figure 12.7b). When the radiation from dS is projected onto dS' at an incident angle of θ, we find a simple relation $dS' = \cos\theta\, dS$.

The total number of blackbody photons entering the p–n solar cell through the area dS' is expressed as

$$\varphi_{BB}(\lambda) = \frac{\lambda}{hc}\int_S \gamma_{BB}(\lambda) dS' \tag{12.10}$$

where the prefactor of $\lambda/hc(=1/E)$ corresponds to the conversion of radiation into photon density, as explained in Eq. (12.8). If $r = 1$, we obtain $dS' = 2\pi \sin\theta \cos\theta\, d\theta$, which can be substituted into Eq. (12.10) to obtain

$$\varphi_{BB}(\lambda) = \frac{\lambda}{hc}\int_0^{2\pi}\int_0^{\pi/2} \gamma_{BB}(\lambda) \sin\theta\, \cos\theta\, d\theta\, d\phi \tag{12.11}$$

In the above equation, θ is integrated over the range 0°–90° and the relation of $\int_0^{2\pi} d\phi = 2\pi$ is used. As $\int_0^{\pi/2} \sin\theta\, \cos\theta\, d\theta = 1/2$, the spherical integrations of $d\theta$ and $d\phi$ produce a value of π (i.e. $2\pi \cdot 1/2$) and $\varphi_{BB}(\lambda)$ in Eq. (12.11) is reduced to that

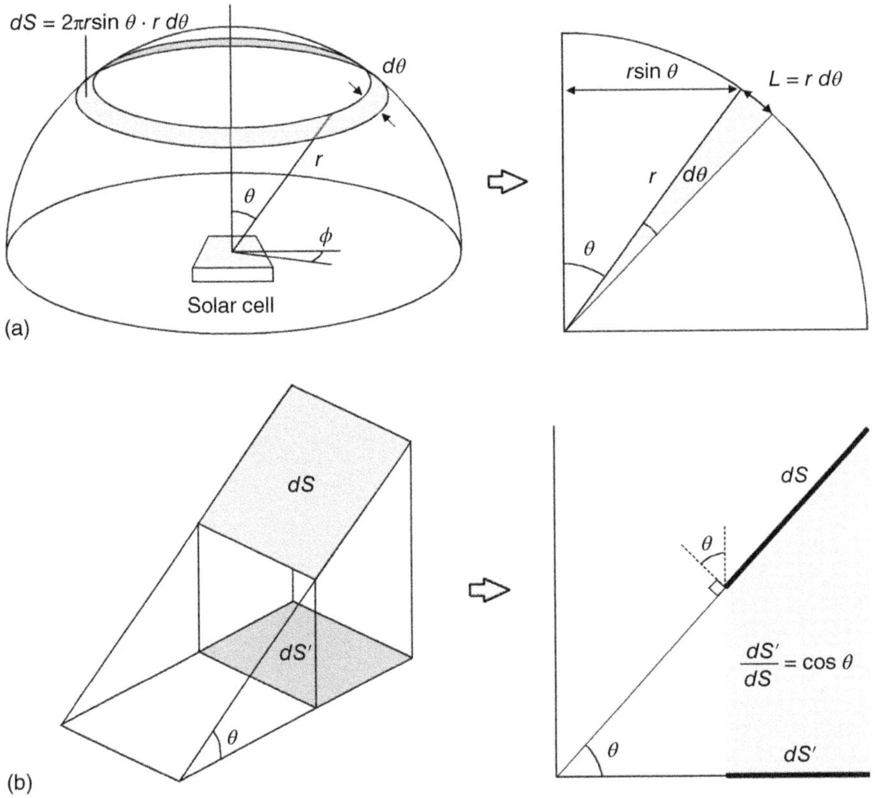

Figure 12.7 (a) Integration method used for the J_0 calculation under SQ theory and (b) projection of area dS onto p–n junction surface with an area of dS'. In (a), θ and ϕ show solid and rotation angles, respectively, whereas r denotes the hemisphere radius. The gray area (dS) indicates the area defined by the variation in $d\theta$.

in Eq. (12.8). Finally, $\varphi_{BB}(\lambda)$ can be integrated over λ to estimate J_0:

$$J_{0,SQ} = q \int Q_{SQ}(\lambda)\varphi_{BB}(\lambda)\, d\lambda \tag{12.12}$$

as illustrated in Figure 12.6b. Under SQ theory, a perfect blackbody solar cell is assumed and the EQE is always unity above E_g, independent of θ and ϕ. On the other hand, J_{sc} of the cell can be calculated by assuming 100% carrier collection at normal incidence:

$$J_{sc,SQ} = q \int Q_{SQ}(\lambda)\varphi_{1Sun}(\lambda)\, d\lambda. \tag{12.13}$$

From the obtained ($J_{0,SQ}$, $J_{sc,SQ}$), the solar-cell parameters can be calculated based on Eq. (12.1).

Figure 12.8 shows the variation of J_{sc}, V_{oc}, and FF with the absorber E_g, estimated from the above SQ approach (solid lines), together with the experimental parameters obtained from record-efficiency solar cells (circles) reported in Refs. [3–7]. The corresponding efficiencies are shown in Figure 12.1. All SQ parameters in Figures 12.1

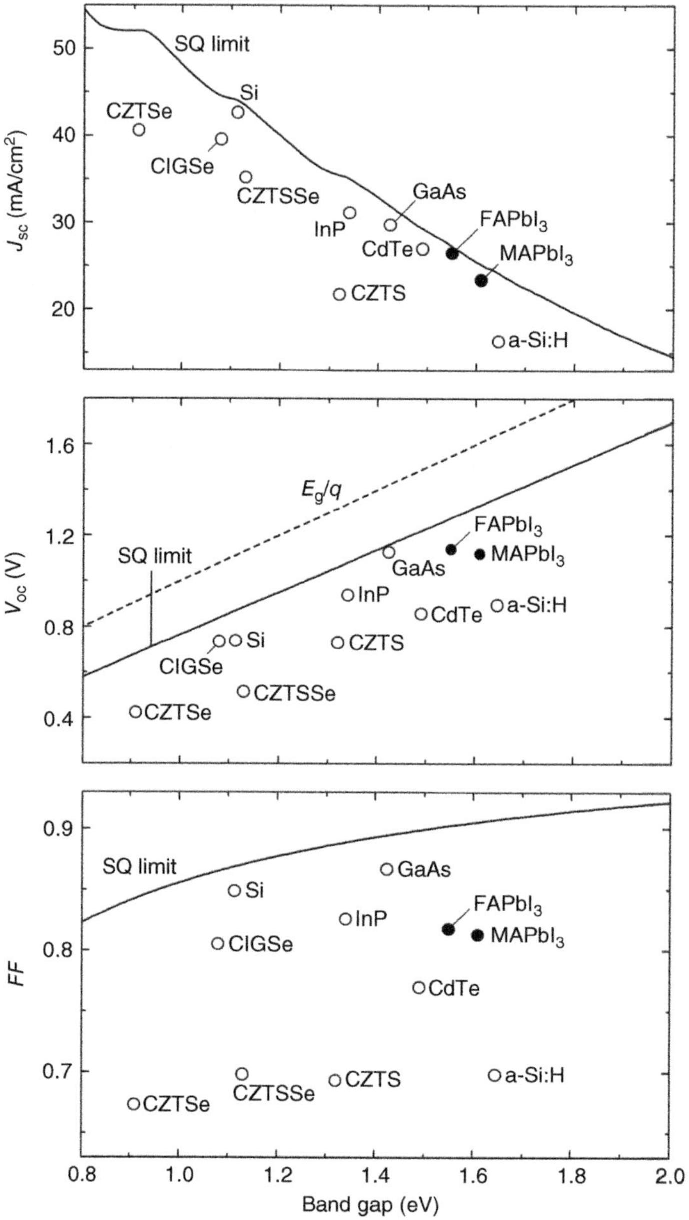

Figure 12.8 Variation in J_{sc}, V_{oc}, and FF with E_g of the absorber, calculated under SQ theory (solid lines), together with the experimental values of record-efficiency solar cells reported in Refs. [3–7] (circles). The corresponding efficiencies are shown in Figure 12.1. The conversion efficiencies of the experimental $FAPbI_3$ and $MAPbI_3$ cells are $\eta = 24.66\%$ [5] and 21.3% [4], respectively. For the $FAPbI_3$ cell, a small amount of methylenediammonium dichloride ($MDACl_2$) has been incorporated. The dotted line shows the V_{oc} derived directly from E_g (i.e. $V_{oc} = E_g/q$). The numerical values corresponding to the SQ limit are summarized in Appendix B.

and 12.8 are summarized in Appendix B. The J_{sc} result is calculated from Eq. (12.13) using φ_{1Sun} shown in Figure 12.5b and J_{sc} decreases as E_g becomes wider. The V_{oc} of the SQ model is calculated directly from $V_{oc} = k_B/q[\ln(J_{sc,SQ}/J_{0,SQ}) + 1]$. Unlike J_{sc}, V_{oc} increases with E_g, primarily because J_0 decreases logarithmically at higher E_g (see Figure 12.6b). The dotted line shows V_{oc} derived directly from E_g (i.e. $V_{oc} = E_g/q$). The V_{oc} obtained under SQ theory is smaller than E_g/q by 0.2–0.3 V, while *FF* remains essentially constant. As a result, the conversion efficiency calculated under SQ theory reaches a maximum at $E_g = 1.34$ eV.

In Figure 12.8, the SQ calculation result reproduces the overall experimental trends well. However, all the experimental solar cells show lower values compared with the SQ results. The reduction from the theoretical J_{sc} can be attributed to (i) the unfavorable light absorption occurring in transparent conductive oxide (TCO) and doped layers of the cells and (ii) the nonzero light reflection in the experimental solar cells, which is neglected completely in the SQ model. The lower V_{oc} in the experimental cells originates from (i) the tail absorption ($E_U > 0$), which reduces the effective E_g and V_{oc}, and (ii) the effect of nonradiative recombination (Section 11.4.2). The reduction of *FF* in the experimental solar cells can be understood by finite series and shunt resistances as well as the increase in the nonideal diode factor, which is neglected in Eq. (12.1), as explained in Section 11.4.2.

12.3 Maximum Efficiencies of Perovskite Single Cells

The SQ limit is determined completely by E_g and does not reflect the optical properties of an actual absorber material. For hybrid perovskite layers fabricated by spin-coating processes, the maximum thickness is ~1 μm [14] and thus the infinite absorber thickness assumed in the SQ model is not valid. Furthermore, the SQ assumptions of $R = 0$ and no tail absorption (see Figure 12.2) cannot be realized in experimental solar cells, contributing to the overestimation of the theoretical limits. Note that $R = 0$ is observed only when the refractive index of the p–n diode is equal to one with no light absorption. From an optical point of view, therefore, the SQ model is highly hypothetical.

In this section, we introduce a new approach that has been developed to calculate more accurate efficiency limits, particularly in thin-film solar cells (Section 12.3.1). In this approach, the calculation of realistic EQE spectra for thin-film solar cells is crucial (Section 12.3.2). In Section 12.3.3, we see the maximum efficiencies of representative solar cells in a 1-μm-thick physical limit. Based on theoretical thin-film limits, the performance-limiting factors of hybrid perovskite devices are further discussed in Section 12.3.4.

12.3.1 Concept of Thin-Film Limit

For thin-film efficiency calculation, a rigorous approach that incorporates all physical and optical aspects of real-world solar cell has been developed by extending SQ theory [1]. In this scheme, a theoretical efficiency limit for a perfectly realizable

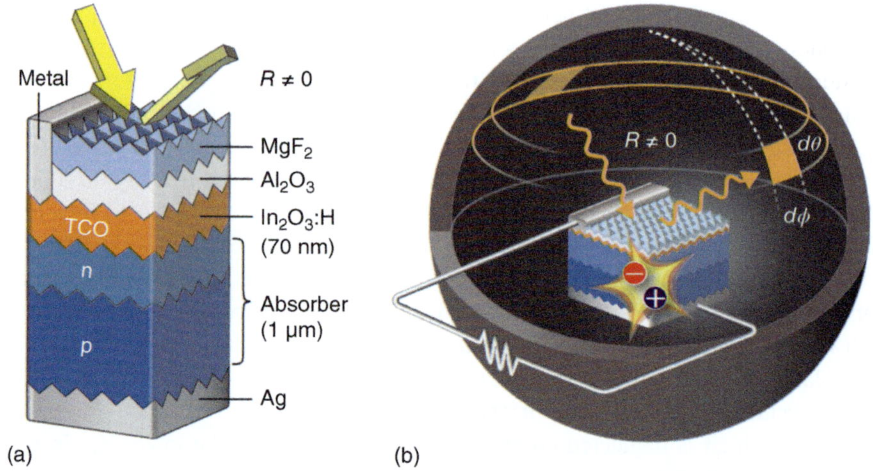

Figure 12.9 (a) Optical model assumed for thin-film limit calculation and (b) physical model of a thin-film device placed in a cavity surrounded by a blackbody radiator. In the thin-film limit calculation, a non-blackbody cell (i.e. $R \neq 0$) is assumed. In the model of (a), MgF_2 and Al_2O_3 are antireflecting coating, whereas In_2O_3:H is a high-mobility TCO material. The area of metal-grid electrode (shadow loss) is assumed to be 5%. *Source*: Kato et al. [1].

thin-film structure in Figure 12.9a is estimated by adopting the true absorption characteristics of the light absorber and constituent layers. In this thin-film concept, a general layer-stacked structure of TCO (70 nm)/absorber (p–n layers: 1 μm)/metal (Ag) is considered. To maximize the light absorption of the thin-film solar cell, the entire cell structure is assumed to be textured. A dual antireflection coating of MgF_2/Al_2O_3 is further incorporated to minimize the front light reflection. In addition, a front metal-grid electrode with a shadow loss of 5% is assumed.

In a thin-film solar cell with a limited absorber thickness, strong optical confinement is crucial for increasing J_{sc}. In the optical calculation of the structure shown in Figure 12.9a, all optical confinement effects due to texture, antireflection coating, and backside reflection are fully incorporated. Moreover, the light reflection of the solar cell is not assumed to be zero ($R \neq 0$) and thus a non-blackbody cell is considered. For solar cells, the presence of TCO is generally problematic because the free carriers (electrons) in the TCO exhibit strong light absorption, which reduces J_{sc} rather significantly [15–19]. In the thin-film model, to suppress unfavorable parasitic absorption in the TCO, a high-mobility TCO layer (H-doped In_2O_3 [20, 21]) with a mobility of $\mu = 100\ \text{cm}^2/(\text{V s})$ and a low carrier concentration of $2 \times 10^{20}\ \text{cm}^{-3}$ is assumed. The use of high-mobility TCO is particularly important in obtaining a sufficiently low TCO series resistance at low carrier concentrations [20]. The carrier concentration in commercial SnO_2:F substrates (TEC substrates) is $5 \times 10^{20}\ \text{cm}^{-3}$ ($\mu = 30\ \text{cm}^2/(\text{V s})$) [20, 22] and the free carrier absorption is substantially stronger. The optical constants of hybrid perovskites can be found in Appendix A and those of the constituent layers can be obtained from Refs. [1, 15].

To estimate the J_0 of a solar cell with the structure of Figure 12.9a, the solar cell is placed in a spherical cavity surrounded by a blackbody radiator

(Figure 12.9b). For a non-blackbody thin-film cell ($R \neq 0$), the optical response is incident-angle-dependent and the optical calculation must be carried out by considering the p- and s-polarized waves (see Figure 4.11a for definitions of these polarizations). Specifically, the incident-angle-dependent EQE spectrum (i.e. $Q(\lambda, \theta)$) is estimated as a simple average of the EQE spectra calculated for the p-polarization $[Q_p(\lambda, \theta)]$ and s-polarization $[Q_s(\lambda, \theta)]$:

$$Q(\lambda, \theta) = [Q_p(\lambda, \theta) + Q_s(\lambda, \theta)]/2 \tag{12.14}$$

In the configuration of Figure 12.9b, $\varphi_{BB}(\lambda)$ in Eq. (12.10) is expressed using the θ-dependent function $\varphi_{BB}(\lambda, \theta)$. Over a small area $d\theta\, d\phi$, we obtain $\varphi_{BB}(\lambda) = \int\int \varphi_{BB}(\lambda, \theta) d\theta\, d\phi$. By inserting this equation and Eq. (12.14) into Eq. (12.12), the blackbody radiation from all the internal surface of the cavity can be integrated using the following exact formula:

$$J_0 = \frac{(1-S)q}{2} \int_0^{\infty} \int_0^{2\pi} \int_0^{\pi/2} [Q_p(\lambda, \theta) + Q_s(\lambda, \theta)] \varphi_{BB}(\lambda, \theta) d\theta\, d\phi\, d\lambda \tag{12.15}$$

where S is the shadow loss of the front metal-grid electrode ($S = 0.05$). On the other hand, the J_{sc} of the solar cell is calculated using the standard equation of

$$J_{sc} = (1-S)q \int Q_{\theta=0}(\lambda) \varphi_{1Sun}(\lambda)\, d\lambda \tag{12.16}$$

where $Q_{\theta=0}(\lambda)$ is the EQE spectrum at $\theta = 0°$. From J_{sc} and J_0 above, the thin-film solar-cell limit can be obtained using Eq. (12.1).

12.3.2 EQE Calculation Method

The EQE spectra $Q_p(\lambda, \theta)$ and $Q_s(\lambda, \theta)$ in Eq. (12.15) can be calculated by applying the antireflection condition (ARC) approach [15–17]. This scheme allows for the incorporation of all possible optical effects generated in textured solar cells. Figure 12.10a shows an example of calculating EQE using the ARC method for a $MAPbI_3$ absorber at $\theta = 0°$. In the ARC method, the R spectrum is first calculated by considering a flat multilayer structure using a conventional optical admittance method [15, 16, 23]. The result obtained assuming this flat structure is indicated as R_{flat} in Figure 12.10a. Next, the minimum R points in the R_{flat} spectrum (open circles) are selected and connected linearly. In the ARC method, the resulting spectrum represents the R spectrum of a textured structure (R_{tex}) that incorporates the widely observed antireflection effect in random textured structures. By adopting this R_{tex} spectrum, the EQE spectrum of a random texture can be approximated based on the optical calculation using a flat optical model [15–17]. This simple approach has been applied successfully in reproducing the optical response of numerous submicron-textured solar cells, including hybrid perovskite [16], CIGSe [17], CZTSSe [16], and CdTe [16]. A free Windows-based software has been released to calculate EQE based on the ARC method (see Figure 11.13) [24].

Figure 12.10b shows an ARC-calculated EQE spectrum plotted on a logarithmic scale. In the visible region, the EQE is almost 100% but is reduced in the ultraviolet region ($\lambda \sim 300$ nm) due to parasitic absorption by the TCO layer.

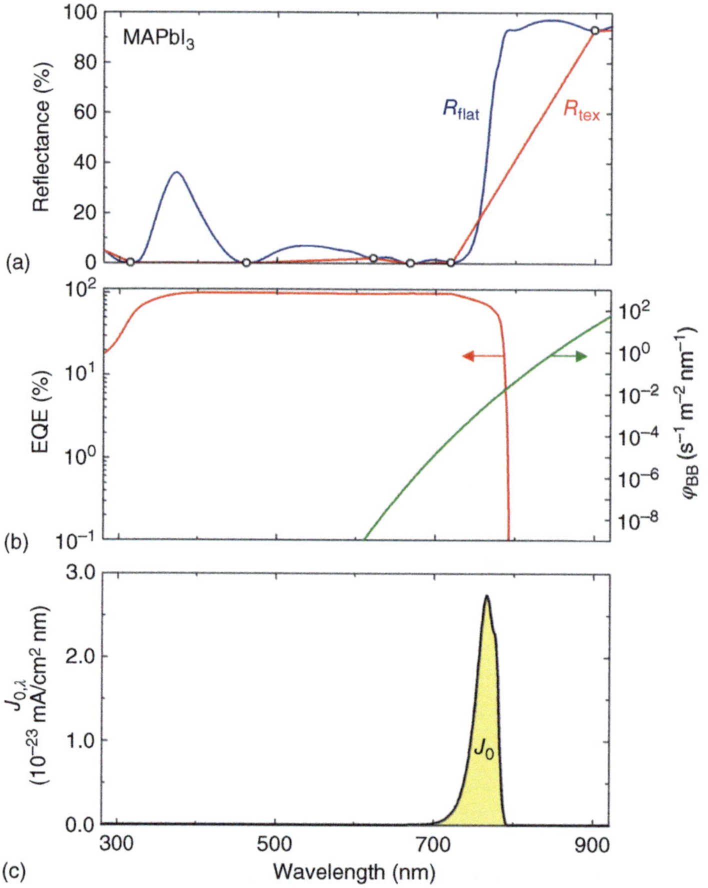

Figure 12.10 Calculation of the EQE spectrum and J_0 for a $MAPbI_3$ absorber at an incident angle of $\theta = 0°$: (a) reflectance spectrum, (b) EQE, and (c) λ-dependent J_0 contribution ($J_{0,\lambda}$). The reflectance and EQE spectra are estimated by applying the ARC method [15–17] (see also Figure 11.13). The R_{flat} and R_{tex} in (a) show the reflectance spectra for flat and textured structures, respectively. The EQE spectrum in (b) is calculated based on R_{tex}. The $J_{0,\lambda}$ in (c) is estimated from the EQE and φ_{BB} in (b) as $J_{0,\lambda} = (1 - S)qQ_{\theta=0}(\lambda)\varphi_{BB}(\lambda)$, where S and $Q_{\theta=0}(\lambda)$ show the shadow loss and EQE spectrum at $\theta = 0°$, respectively.

The figure also shows φ_{BB} at 300 K, indicating the drastic increase in this factor in the longer λ region. In Figure 12.10c, the λ-dependent J_0 contribution ($J_{0,\lambda}$) calculated from $J_{0,\lambda} = (1 - S)qQ_{\theta=0}(\lambda)\varphi_{BB}(\lambda)$ is shown. Note that J_0 is determined by integrating the $J_{0,\lambda}$ spectrum. The result in the figure illustrates the importance of the near-band-edge EQE in determining J_0. In an absorber with extensive tail absorption, the EQE spectrum is extended toward the longer λ region, which in turn increases J_0 and reduces V_{oc} significantly [1].

In the above example, $\theta = 0°$ is assumed and J_0 is calculated as

$$J_0 = (1 - S)q \int Q_{\theta=0}(\lambda)\varphi_{BB}(\lambda)d\lambda \tag{12.17}$$

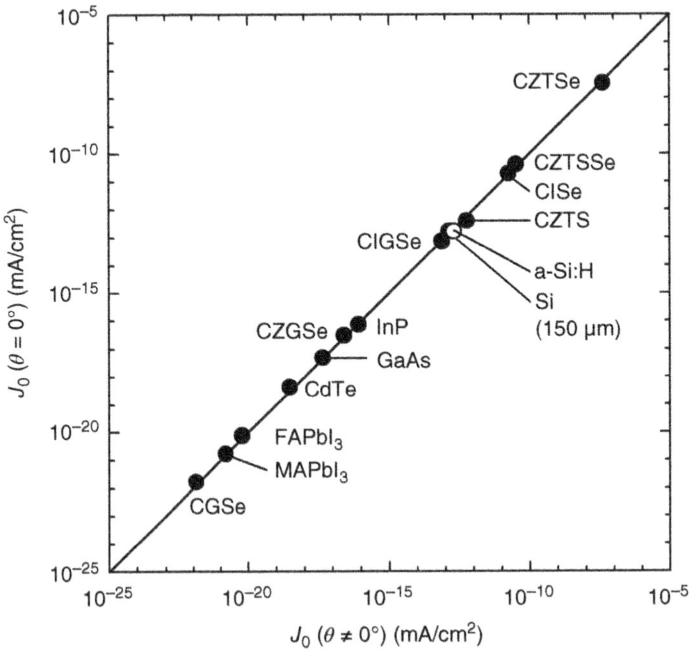

Figure 12.11 Relationship between J_0 values calculated assuming $\theta \neq 0°$ and $\theta = 0°$ [1]. The J_0 values of $\theta \neq 0°$ were evaluated by performing an exact spherical integration of the structure in Figure 12.9b using Eq. (12.15), whereas the J_0 values of $\theta = 0°$ were obtained using Eq. (12.17) assuming a normal incidence for the EQE calculation. *Source*: Kato et al. [1].

This simple approach has been employed widely in the estimation of approximate J_0 values [25–27]. Importantly, it has been confirmed that the J_0 values obtained using the detailed integration of Eq. (12.15) and the simplified calculation of Eq. (12.17) fall into similar ranges (Figure 12.11) [1]. Accordingly, although an exact calculation using Eq. (12.15) is preferable, Eq. (12.17) can still be adopted to estimate an approximate value.

12.3.3 Maximum Efficiencies of Single Solar Cells

Figure 12.12a shows thin-film efficiency limits obtained from the strict J_0 integration performed by Eq. (12.15) assuming 1-μm absorber thickness [1]. Table 12.1 lists the numerical solar-cell parameters derived from these calculations [1]. The calculation results show that the maximum thin-film efficiencies are ≤31%, which are lower than the SQ limits by ~3%. In particular, the realistic maximum efficiencies of $MAPbI_3$ and $FAPbI_3$ solar cells are 28%, while the corresponding SQ limits are 30–31%. As shown in Table 12.1, the estimated maximum V_{oc} values of $MAPbI_3$ and $FAPbI_3$ are 1.32 and 1.28 V, respectively. The V_{oc} limit can also be determined empirically if the experimental EQE spectrum is applied using Eq. (12.17). The maximum V_{oc} of $MAPbI_3$ obtained using this semi-empirical procedure is 1.32–1.33 V [25–27],

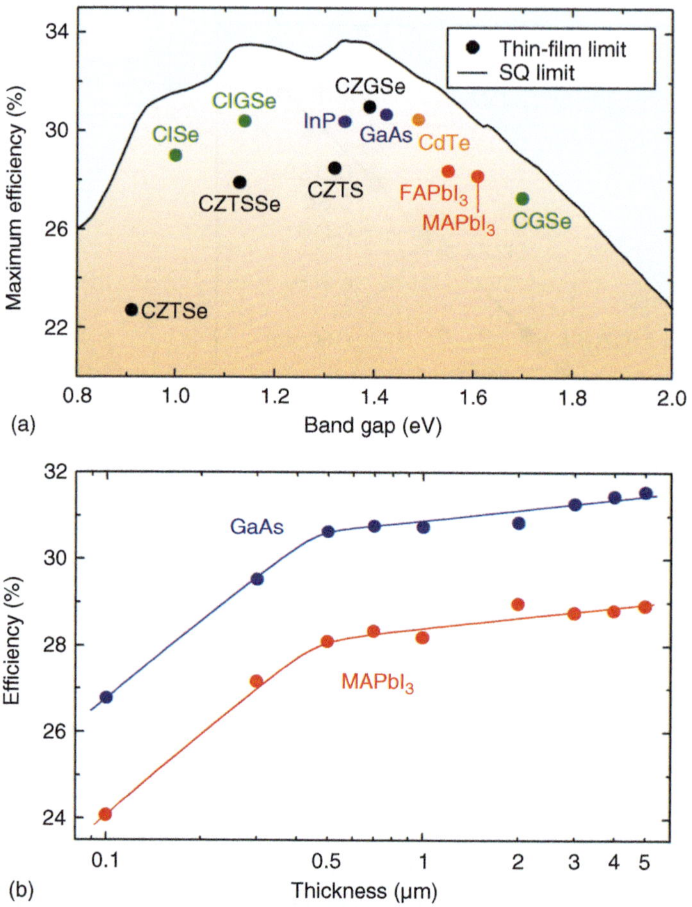

Figure 12.12 (a) Maximum efficiency (η) calculated from the thin-film and SQ limits and (b) thickness dependence of the thin-film limit obtained for GaAs and $MAPbI_3$ solar cells. In (a), the closed circles show the thin-film limits calculated for each absorber, whereas the solid line indicates the SQ limit. For the thin-film limit calculations, a total absorber thickness of 1 μm is assumed. *Source*: Kato et al. [1].

which is consistent with the result of Table 12.1. In the above thin-film limit, the series and shunt resistances are not considered and the obtained maximum *FF* is much higher (~0.9) than the experimental values (Figure 11.18).

In Figure 12.12a, the thin-film limits for CZTS, CZTSe, and CZTSSe are significantly lower than the SQ limits because of the extensive tail absorption in these compounds [28, 29], which lowers the V_{oc} drastically [1]. In fact, the E_U values of these materials are quite high (~80 meV), whereas hybrid perovskites show quite sharp absorption edges with $E_U \sim 15$ meV (Figure 4.18). Accordingly, the maximum V_{oc} values attainable for hybrid perovskites are sufficiently high.

One limitation of the above thin-film limit estimation is that it assumes a substrate-type solar cell as a basic structure, while superstrate-type (i.e. glass substrate type) solar cells are typically adopted for hybrid perovskite single cells

Table 12.1 Maximum conversion efficiencies obtained from the thin-film and SQ limits.

Solar cell	J_{sc}^{max} (mA/cm²)	V_{oc}^{max} (V)	FF^{max}	Thin-film limit (%)	SQ limit(%)
$FAPbI_3$	24.5	1.286	0.903	28.4	31.4
$MAPbI_3$	23.6	1.320	0.905	28.2	30.4
GaAs	30.7	1.121	0.892	30.7	33.1
Si[a)]	39.4	0.861	0.868	29.4	33.3
CIGSe	39.9	0.876	0.870	30.4	33.5
CdTe	28.6	1.189	0.897	30.5	32.2
CZTSSe	45.5	0.723	0.850	27.9	33.5

The maximum J_{sc} (J_{sc}^{max}), V_{oc} (V_{oc}^{max}), and FF (FF^{max}) in this table correspond to those of the thin-film limits indicated by the closed circles in Figure 12.12a. The optical model of Figure 12.9a is adopted for the thin-film limit calculations. For the J_{sc} estimation, a metal-grid electrode with a shadow loss of 5% is assumed.

a) Si absorber thickness is assumed to be 150 μm.

(Figure 11.2). In conventional hybrid perovskite single cells, the shadow loss arising from a front grid electrode can be neglected and J_{sc} could be potentially higher than those listed in Table 12.1. In superstrate perovskite cells, however, the front TCO layer is thicker and the free carrier absorption in the TCO reduces J_{sc}. To estimate the maximum J_{sc}, therefore, more exact optical modeling is necessary. Note that, for hybrid perovskite cells, an almost 100% carrier collection efficiency has been confirmed experimentally under the short-circuit condition (Figure 11.11).

It should be emphasized that, for the SQ efficiency, there is always an ambiguity for the correct choice of material E_g, as experimental E_g changes with the characterization method and definition (see Table 4.1). In contrast, in the derivation of the thin-film limit, E_g is not required and the theoretical efficiency is obtained directly from the thin-film absorption characteristics (i.e. the EQE spectrum). Accordingly, for evaluating the absolute performance limit of a solar cell, the thin-film limit is more appropriate.

Figure 12.12b shows the variation of the $MAPbI_3$ and GaAs thin-film limits with absorber thickness [1]. In both cases, the maximum efficiencies saturate at >0.5 μm. Thus, when the optical confinement is strong, an absorber thickness of 0.5 μm is sufficient. This optimum absorber thickness is consistent with experimental absorber thicknesses adopted typically for the perovskite single cells [4, 5].

12.3.4 Performance-Limiting Factors of Hybrid Perovskite Devices

The potential efficiency calculations for a thin-film configuration further enable the evaluation of the performance-limiting factors in record-efficiency photovoltaic devices. Figure 12.13 compares the calculated maximum efficiencies under the thin-film limit (η_{TF}) with the experimental efficiencies of world record cells (η_{ex}) reported in Refs. [3, 4, 6, 7, 30]. To perform this calculation, the solar-cell structure

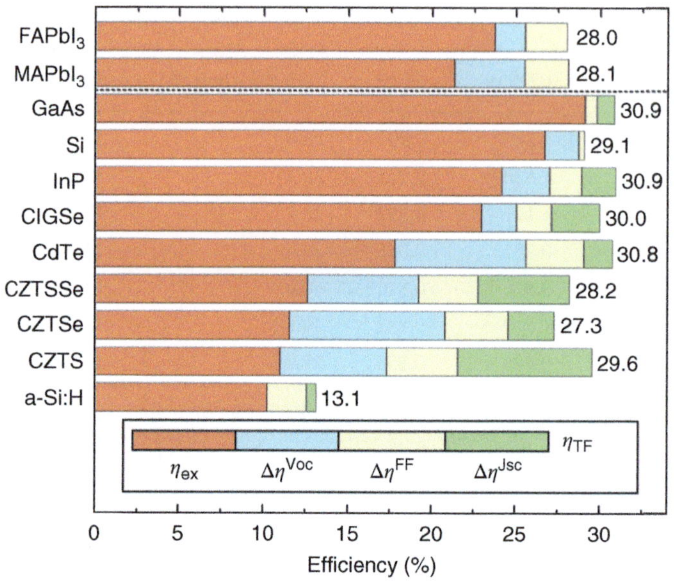

Figure 12.13 Performance-limiting factors of record-efficiency solar cells analyzed by the thin-film limit. The experimental conversion efficiencies (η_{ex}) and the maximum efficiencies estimated from the thin-film limits (η_{TF}) are summarized. The difference between η_{TF} and η_{ex} is further categorized into the efficiency reductions due to the V_{oc} deficit ($\Delta\eta^{Voc}$), J_{sc} deficit ($\Delta\eta^{Jsc}$), and FF deficit ($\Delta\eta^{FF}$). To obtain $FAPbI_3$ and $MAPbI_3$ analysis results, solar-cell parameters reported for $FAMAPb(I,Br)_3$ [30] and $MAPbI_3$ [4] cells are employed. The results for inorganic solar cells are adopted from Ref. [1]. *Source*: Kato et al. [1].

of Figure 12.9a is adopted but its absorber thickness is adjusted to match those of the experimental devices [1]. In Figure 12.13, the difference between the theoretical and experimental efficiencies is further categorized as the efficiency reduction $\Delta\eta$ caused by the V_{oc} deficit ($\Delta\eta^{Voc}$), J_{sc} deficit ($\Delta\eta^{Jsc}$), and FF deficit ($\Delta\eta^{FF}$) relative to the calculated maximum V_{oc}, J_{sc}, and FF values, respectively. The efficiencies of the hybrid perovskite cells employed in the analysis are 23.7% and 21.2% for $FAMAPb(I,Br)_3$ [30] and $MAPbI_3$ cells [4], respectively. The calculation of the $FAMAPb(I,Br)_3$ cell is performed assuming a pure $FAPbI_3$ phase, as the MA and Br contents are very small [31]. In $FAPbI_3$-based solar cells, however, higher record efficiencies of ~25% have been reported [3, 5] and, for these cells, the efficiency deficits are more suppressed.

One of the remarkable features of hybrid perovskite solar cells is the negligible J_{sc} loss due to the very small optical losses in the devices (Section 11.4.2) [16, 18]. In other words, the record-efficiency perovskite cells are limited primarily by V_{oc} and FF losses. Note that the rapid increase in the perovskite cell efficiency observed over the past decade is caused mostly by the improvements of the V_{oc} and FF (see Figure 1.3). The V_{oc} of the current record-efficiency perovskite cells ranges 1.12–1.18 V [3–5, 30], which are lower than the theoretical limits by ΔV_{oc} = 100–200 mV. The relatively large ΔV_{oc} can be interpreted mainly by the presence of nonradiative recombination at the cell interfaces (Figure 11.14). However, a quite high V_{oc} of 1.26 V, with ΔV_{oc} of only 60 mV, has been obtained in a V_{oc}-optimized $MAPbI_3$ p-i-n cell [32], although the J_{sc} of this cell is limited by the parasitic light absorption in the front PTAA layer (see also Figure 11.3).

In contrast, the *FF* values of the record-efficiency cells (80–85%) [3–5, 30] are notably lower than their theoretical counterparts (90%). The reduced *FF* in the experimental cells can be explained by the presence of cell resistance and deterioration of the diode factor (Figure 11.12). Interestingly, *FF* shows a clear correction with the grain size of the perovskite layers, while V_{oc} is independent of the grain size (see Figure 11.18). Consequently, the *FF* loss in the record-efficiency cells can be primarily interpreted by the small grain size (~1 μm) of the polycrystalline absorber.

As shown in Figure 12.13, among all experimental solar cells, the GaAs cell exhibits the highest conversion efficiency [3]. The overall efficiency loss in the GaAs cell is quite small and its experimental efficiency (29.1%) is very close to the thin-film limit (30.9%). Remarkably, for the GaAs cell, $\Delta V_{oc} \sim 0$ and ΔFF is only 2%; in particular, this cell is simply limited by the parasitic light absorption of the front layers [1]. In Si solar cells, in contrast, the V_{oc} deficit is quite large due to the significant Auger recombination in the indirect-transition Si, while the defect-induced nonradiative recombination plays only a minor role [33, 34].

The results of Figure 12.13 show clearly that the performance of all thin-film solar cells is essentially limited by V_{oc} and *FF* losses [1]. In general, V_{oc} can be related to *FF* [35] and a V_{oc} loss tends to increase with a *FF* loss. The $\Delta\eta^{Voc}$ and $\Delta\eta^{FF}$ increase notably in polycrystalline absorbers (i.e. hybrid perovskite, CIGSe, CdTe, and CZT(S)Se) and the formation of larger polycrystalline grains is important in $MAPbI_3$ (see Figure 11.18), CIGSe [36], CdTe [37], and CZTSSe [38] solar cells. Note that all of the chalcogenide-based solar cells (CIGSe and CZTSSe) have rather large values of $\Delta\eta^{Jsc}$ due to the strong parasitic absorption in the Mo back electrode [15–17]. In hybrid perovskite cells, the incorporation of high-reflectivity Ag and Au rear electrodes has been a great advantage in improving J_{sc}.

12.4 Maximum Efficiency of Tandem Cells

To achieve high cell efficiencies of >30%, the use of tandem-cell structures is crucial (Chapter 17). Fortunately, hybrid perovskite cells have been combined successfully with crystalline Si cells to achieve a record 29% efficiency (see Table 17.1). To realize such a high cell efficiency in a tandem-cell configuration, however, a complex multilayer structure is necessary [8–11]. Thus, precise device simulation becomes important not only for determining the maximum potential efficiency but also for designing the solar-cell architecture.

In this section, a rigorous evaluation of the maximum efficiency for fully textured hybrid perovskite/Si tandem cells is demonstrated. As a tandem structure, a $CsFAPb(I,Br)_3$ top cell formed on a Si heterojunction bottom cell is considered. In this section, the optical modeling of the tandem-cell structure (Section 12.4.1) and its calculation method (Section 12.4.2) are introduced. The tandem-cell efficiency limit obtained using the true absorption characteristics of all constituent layers is described in Section 12.4.3. The realistic maximum efficiency of the tandem device is further discussed in the final section (Section 12.4.4).

12.4.1 Optical Model and Assumptions

Figure 12.14a shows an optical model assumed for a two-terminal monolithic $CsFAPb(I,Br)_3$/Si tandem device with a fully textured structure. To derive realistic efficiency limits for the tandem cells, the multilayer structure has been constructed based on experimental perovskite tandem devices [8–10]. In the structure, for a $CsFAPb(I,Br)_3$ top cell, the electron transport layer (ETL) of C_{60} and the hole transport layer (HTL) of spiro-OMeTAD are adopted. The LiF layer acts as a shunt-blocking layer [9], with the LiF also exhibiting a passivation effect (Table 11.2) [40]. On the C_{60} ETL, a thin SnO_2 layer is incorporated to reduce the sputtering damage that occurs during TCO (InZnO: IZO) deposition. For the top cell, a single antireflection layer (MgF_2) is adopted.

The perovskite top cell is formed on an n-i-p type Si heterojunction solar cell with a Si wafer thickness of 260 μm [8]. In the bottom cell, the Si wafer is sandwiched between hydrogenated amorphous silicon (a-Si:H) n-i front and i-p rear layers [41–43]. To construct the tandem cell, the top and bottom cells are combined using recombination junction layers of nano-crystalline Si (nc-Si:H) [8, 10]. The optical constants of all layers in the model can be found in Refs. [15, 39, 44, 45].

The J_{sc} of two-terminal tandem cells is maximized when the J_{sc} values of the top and bottom cells are equal (i.e. the current matching condition). In a perovskite/Si tandem cell, the E_g of the Si bottom cell is fixed (i.e. $E_{g,bottom} = 1.11$ eV [20]), leaving the E_g and thickness of the $CsFAPb(I,Br)_3$ layer as the critical parameters of the tandem-cell design. In this section, we calculate the maximum potential efficiencies using both the true perovskite absorption spectrum and a simplified step-function spectrum and discuss how the results differ.

Figure 12.14b shows the absorption coefficient (α) spectrum of experimental $CsFAPb(I,Br)_3$ with $E_g = 1.65$ eV [39], together with a step-function α spectrum assumed for simplified calculation. Experimentally, the band-edge α of $CsFAPb(I,Br)_3$ is 2×10^4 cm^{-1} (see also Figure 4.14) and the calculation of the light penetration depth ($d_p = 1/\alpha = 500$ nm) immediately gives a rough optimum thickness of the $CsFAPb(I,Br)_3$ top cell. For the step-function α, a constant α of 10^5 cm^{-1} is assumed at $E \geq E_g$.

Note that, when experimental optical properties of hybrid perovskites are modeled by the Tauc–Lorentz model (see Figure 4.4 and Appendix A), the optical functions of their alloys can be approximated by simply shifting the corresponding peak energy positions (Section 4.5.2). Figure 12.14c shows the α spectra of the $CsFAPb(I,Br)_3$ alloys obtained from these energy shift calculations, which show that the α spectrum slide horizontally along the energy scale as E_g changes, a result that has been confirmed experimentally (see Figure 4.15).

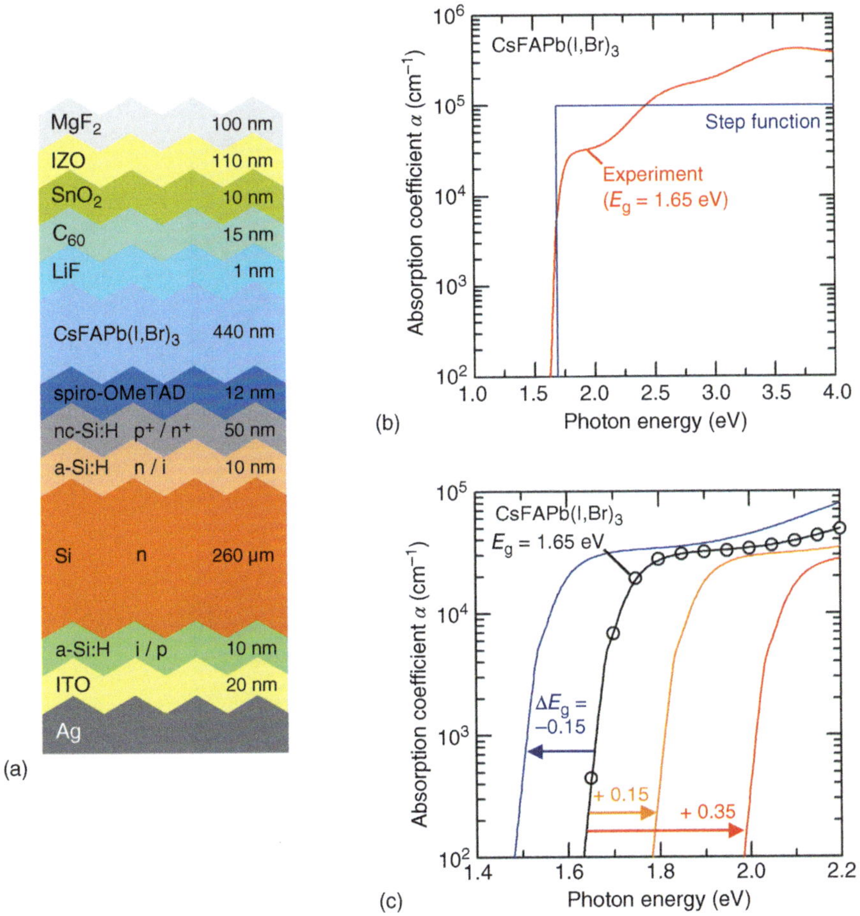

Figure 12.14 (a) Optical model of a two-terminal monolithic $CsFAPb(I,Br)_3$/Si tandem device with a fully textured structure, (b) absorption coefficient (α) spectrum of experimental $CsFAPb(I,Br)_3$ with E_g = 1.65 eV [39] and a step-function α spectrum assumed in the simplified calculation, and (c) α spectra of $CsFAPb(I,Br)_3$ alloys obtained from energy shift calculations based on the Tauc–Lorentz model. The structure and layer thicknesses in (a) are determined based on experimental $CsFAPb(I,Br)_3$/Si tandem devices [8–10]. In (b), the α spectrum of $CsFAPb(I,Br)_3$, described as "CsBr10%(1:2)" in Ref. [39], is shown. In (c), the open circles show the experimental spectrum of (b) and the solid lines indicate the optical functions calculated using the Tauc–Lorentz model. The energy shift value of the spectrum is indicated as ΔE_g.

12.4.2 Calculation of Tandem-Cell EQE Spectra

For fully pyramid-textured solar cells, the optical confinement effect needs to be evaluated accurately. Although a ray-tracing method is frequently applied in the optical calculation of Si textures [46], the computational cost of this method is relatively high and it is relatively difficult to accurately estimate the light absorption within the thin top cell [47]. To overcome these problems, a double-reflection model (DRM),

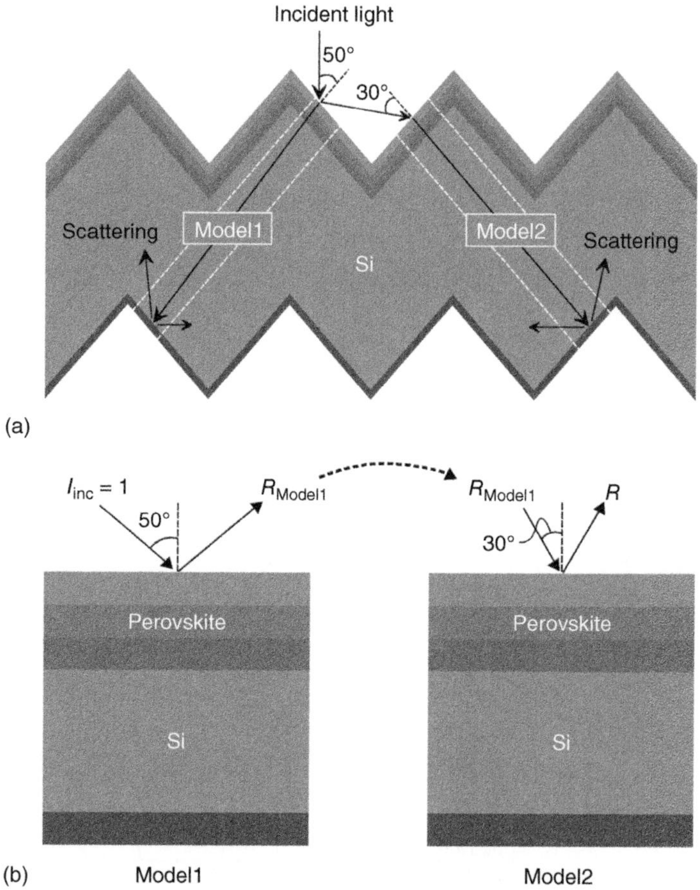

Figure 12.15 Double-reflection model (DRM) [48] employed for the optical calculation of perovskite/Si tandem cells with Si pyramid texture: (a) overall configuration of the two-dimensional optical model consisting of two flat models of Model1 and Model2 and (b) configurations of Model1 and Model2. In the DRM, the light scattering effect that occurs in the Si rear interface is also considered by introducing an optical confinement factor. The I_{inc} and R_{Model1} are the normalized incident light intensity and reflectance of Model1, respectively. Note that R_{Model1} is equivalent to the incident light intensity in the calculation of Model2. *Source*: Modified from Nishigaki et al. [48].

shown in Figure 12.15, has been developed [48]. The essential concept underlying the DRM is to represent the optical absorption in a multilayer structure formed on a Si texture by assuming two flat optical models with different incident angles of $\theta = 50°$ (Model1) and 30° (Model2), as shown in Figure 12.15a. These incident angles are derived simply from the top angle of the texture (80°) [49, 50] and the overall optical response is expressed by combining the calculation results of Model1 and Model2 (Figure 12.15b). Under the DRM, the light scattering effect is also considered by introducing an adjustable parameter (optical confinement factor). The EQE spectra of Si single cells calculated using the DRM show excellent agreement with the experimental EQE spectra [48].

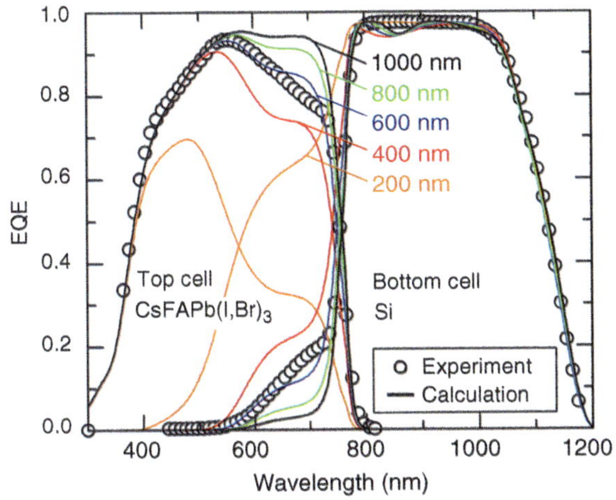

Figure 12.16 EQE spectra of the $CsFAPb(I,Br)_3$ top cell and Si bottom cell for the tandem-cell structure of Figure 12.14a, obtained from the DRM calculations (solid lines), together with the experimental EQE spectra (open circles) reported for a 25.2%-efficiency cell [8]. For the calculation results, the variation of the EQE spectrum with $CsFAPb(I,Br)_3$ thickness (200–1000 nm) is further shown. In the DRM calculation, the optical confinement factor [48] is assumed to be $f = 2$ and the perovskite α spectrum is adjusted to $E_g = 1.59$ eV (i.e. −0.06 eV shift in Figure 12.14c).

Figure 12.16 shows the DRM-calculated EQE spectra of the $CsFAPb(I,Br)_3$ top and Si bottom cells of the tandem structure in Figure 12.14a (solid lines), together with experimental EQE spectra reported for a 25.2% efficiency cell (open circles) [8]. The calculation results further show the variation of the EQE spectrum with the $CsFAPb(I,Br)_3$ top-cell thickness. To perform the EQE calculation, the perovskite α spectrum with $E_g = 1.59$ eV (i.e. −0.06 eV shift in Figure 12.14c) is employed. With increasing the perovskite absorber thickness, the EQE response of the top cell becomes larger, while that of the bottom cell decreases. The calculated EQE spectrum is in excellent agreement when the perovskite thickness is $d = 600$ nm.

12.4.3 Maximum Efficiencies of Tandem Devices

Once the EQE spectra of the top and bottom cells are calculated, their corresponding (J_0, J_{sc}) values can be estimated separately using Eqs. (12.16) and (12.17). By applying a simple diode equation (i.e. Eq. (12.1)), we can further calculate the J–V curve for each cell. From these J–V curves, the J–V curve of the tandem cell is determined [48]:

$$V(J) = V_{top}(J) + V_{bottom}(J) \qquad (12.18)$$

where $V_{top}(J)$ and $V_{bottom}(J)$ are the inverse functions of $J_{top}(V)$ and $J_{bottom}(V)$ calculated for the top and bottom cells, respectively. In estimating $V(J)$ using Eq. (12.18), the current matching condition is also considered; namely, the J_{sc} of the tandem cell is limited by either the top cell or the bottom cell that shows lower J_{sc}. From the

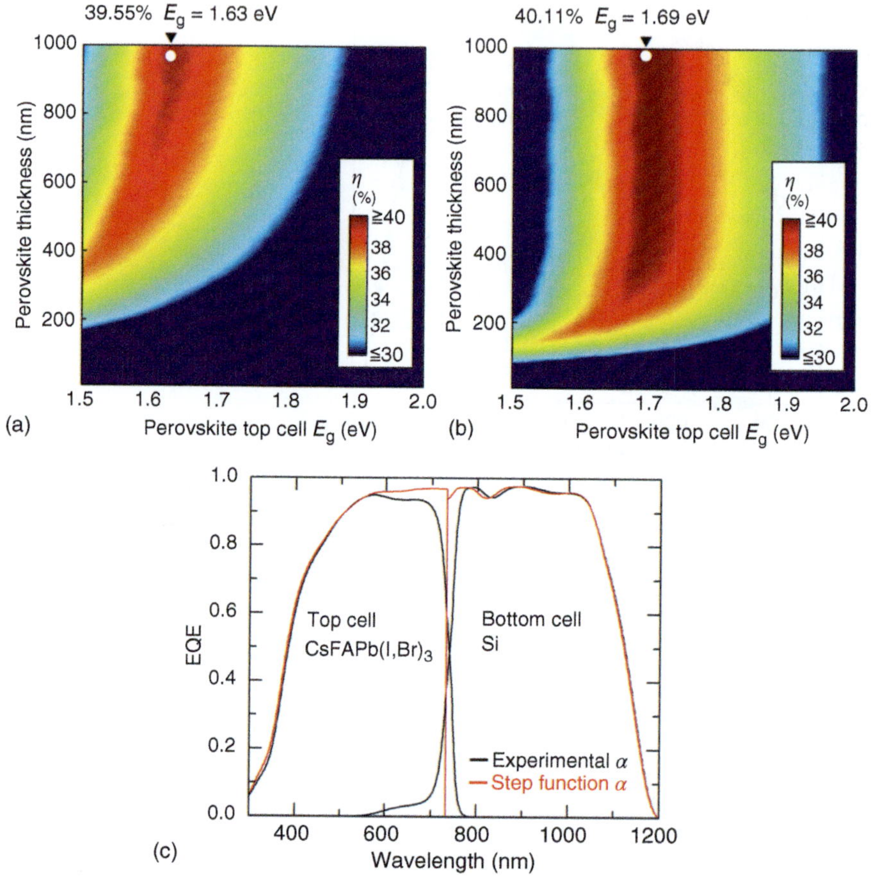

Figure 12.17 $CsFAPb(I,Br)_3$/Si tandem-cell efficiencies calculated by varying the top-cell absorber E_g and thickness assuming (a) the experimental α and (b) the step function α for the $CsFAPb(I,Br)_3$ absorber and (c) EQE spectra of the top and bottom cells obtained under the optimum conditions with the experimental and step-function α spectra. The calculations are performed based on the DRM method [48] with the cell structure shown in Figure 12.14a. The open circles in (a) and (b) show the highest efficiencies obtained for each simulation at $d \leq 1000$ nm. The EQE spectra obtained for these optimum conditions are shown in (c).

$J(V)$ obtained from Eq. (12.18), the theoretical solar-cell parameters and conversion efficiency of the tandem cell are obtained.

Figure 12.17 shows the mapped results of the $CsFAPb(I,Br)_3$/Si tandem-cell efficiency calculated by varying the top-cell E_g and layer thickness in the cell structure of Figure 12.14a. For these calculations, the shadow loss of the front metal-grid electrode is assumed to be zero (i.e. $S = 0$). In Figure 12.17a, the simulation results obtained using the true $CsFAPb(I,Br)_3$ optical properties are shown, whereas the calculation results derived assuming the step-function α are indicated in Figure 12.17b. The optical properties of the top cell absorber modify the maximum cell efficiency rather significantly; specifically, when an exact perovskite α is used, the optimum E_g varies with the top-cell absorber thickness (d) due to limited light absorption. In particular, when the top cell is thin, the J_{sc} of the tandem cell is limited by the top cell and, with decreasing d from 1000 to 400 nm, the optimum top-cell E_g is reduced from E_g = 1.63 to 1.53 eV in an effort to increase the top-cell light absorption. In contrast, when a step-function α is assumed, the light absorption in the top cell becomes unrealistically high, pushing the optimum E_g up to 1.7 eV, essentially independent of the top-cell absorber thickness.

The open circles in Figure 12.17a,b show the highest efficiencies obtained in each simulation with $d \leq 1000$ nm. Although the maximum efficiencies of these calculations are similar ($\eta \sim 40\%$), the optimum E_g differs (i.e. E_g = 1.63 eV for the true perovskite α and E_g = 1.69 eV for the step-function α). The EQE spectra obtained for these optimum conditions are shown in Figure 12.17c, which confirms that the EQE of the top cell becomes much higher when the step-function α is considered due to high α particularly in the band-edge region (see Figure 12.14b).

For the experimental $CsFAPb(I,Br)_3$/Si tandem cell, the optimum top-cell E_g is 1.60 eV [8], with the effective thickness of 600 nm analyzed in Figure 12.16. This experimentally determined optimum E_g is indeed the best value at d = 600 nm in Figure 12.17a and thus the calculation result is consistent with the experimental result. In previous simple optical simulations [51–53], a much wider optimum E_g of ~1.73 eV has been reported. It should be emphasized that V_{oc} loss becomes larger in wide-gap perovskite cells with E_g > 1.65 eV (see Figure 1.10) and a low-E_g perovskite top cell is advantageous in realizing higher efficiencies. Accordingly, the detailed calculation is vital to accurately establish the best tandem-cell design.

In the above analysis, the Si wafer thickness is fixed at 260 μm. When this thickness is varied, however, the optimum top-cell E_g also varies. Figure 12.18 shows the optimum E_g of the perovskite top cell as a function of Si wafer thickness. In Figure 12.18, several results calculated for different top-layer thicknesses in a range of 400–1000 nm are shown. If the Si wafer becomes thin, the J_{sc} of the bottom cell decreases and thus the E_g of the top cell widens to establish the current matching for the lower J_{sc}. In contrast, when the top cell is thick, there is sufficient light

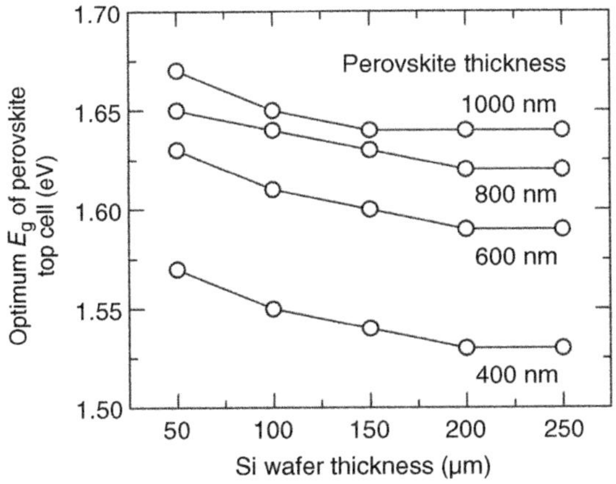

Figure 12.18 Variation of the optimum E_g for the $CsFAPb(I,Br)_3$ perovskite top cell with Si wafer thickness in the $CsFAPb(I,Br)_3$/Si tandem device. Several results calculated for different top-layer thicknesses in a range of 400–1000 nm are shown.

absorption and the E_g of the top cell increases so that the V_{oc} gain becomes larger. In fact, for a very thick perovskite top cell (~1 μm), a wide-gap perovskite with $E_g = 1.68$ eV has been adopted [54]. As shown above, the tandem-cell design varies with the numerous parameters and device simulation is beneficial for efficient optimization of the solar-cell architecture.

12.4.4 Realistic Maximum Efficiency of Tandem Cell

The efficiency limit obtained for the $CsFAPb(I,Br)_3$/Si tandem cell is ~40% (Figure 12.17). In this calculation, however, the critical influences of cell series resistance, shadow loss of the front metal-grid electrode, and Auger recombination in the Si absorber [33, 34] are neglected. In this section, we calculate more realistic potential efficiency by accounting for these effects.

Table 12.2 summarizes the V_{oc}, J_{sc}, *FF*, and conversion efficiency of a $CsFAPb(I,Br)_3$/Si tandem cell obtained under different models (assumptions). The calculation result of Model1 shows the maximum efficiency indicted by the open circle in Figure 12.17a (i.e. $\eta = 39.55\%$). For this tandem cell, the V_{oc} of the top cell is 1.38 V, while that of the bottom cell is 0.82 V. However, this bottom-cell V_{oc} is unrealistically high because it is calculated without considering the strong Auger recombination in the indirect Si absorber. By applying the *J–V* analysis method shown in Figure 11.12, the increase in J_0 in the Si bottom cell due to the Auger recombination was determined based on experimental *J–V* characteristics reported in Ref. [41]. In particular, the J_0 of the Si bottom cell can be expressed as $J_0 = J_{0,rad} + J_{0,Aug}$, where $J_{0,rad}$ indicates the radiative recombination calculated from Eq. (12.17) and $J_{0,Aug}$ is the Auger recombination contribution determined experimentally. The actual values are $J_{0,rad} = 3.1 \times 10^{-13}$ mA/cm^2 and $J_{0,Aug} = 1.3 \times 10^{-11}$ mA/cm^2, indicating that the J_0 of the Si bottom cell is governed by $J_{0,Aug}$. The calculation result obtained by incorporating Auger recombination in the Si cell is shown as Model2 in Table 12.2. In this case, the V_{oc} of the bottom cell decreases to a realistic value of 0.725 V. As a result of the Si Auger recombination, the tandem efficiency is reduced by ~2%.

In Model3, the series resistance of the Si bottom cell ($R_{s,bottom}$) is further considered by applying Eq. (11.3). For this calculation, $R_{s,bottom} = 0.44\ \Omega$cm^2 extracted from an experimental Si heterojunction cell [41] is applied. It can be confirmed that $R_{s,bottom}$ does not affect the overall result notably. In Model4, the effect of R_s in the perovskite top cell ($R_{s,top}$) has been incorporated by assuming $R_{s,top} = 1.8\ \Omega$cm^2 obtained from the analysis of Figure 11.12. In this case, the *FF* decreases slightly, reducing the efficiency to 36.77%. In the final model (Model5), the shadow loss caused by the front metal-grid electrode is considered using $S = 0.023$, as adopted in the experimental perovskite/Si tandem cell [8]. This model results in $\eta = 35.93\%$ with J_{sc} of 19.6 mA/cm^2. In Figure 12.19, the *J–V* characteristics calculated from Model5 are shown. As indicated by this result, a high efficiency of ~36% can be achieved realistically in a hybrid perovskite/Si tandem cell with the practical structure of Figure 12.14a. To realize such a high efficiency, it is necessary to suppress the V_{oc} and *FF* losses in the perovskite top cell, as indicated in Figure 12.13.

Table 12.2 Maximum solar-cell parameters obtained for a $CsFAPb(I,Br)_3$/Si tandem cell under different assumptions.

Model	Additional parameter	V_{oc} (V)	J_{sc} (mA/cm^2)	*FF*	η (%)
1	—	2.20[e)]	20.0	0.895	39.55
2[a)]	$J_{0,\text{bottom}} = 1.3 \times 10^{-11}$ mA/cm^2	2.10[f)]	20.0	0.891	37.63
3[b)]	$R_{s,\text{bottom}} = 0.44\,\Omega\text{cm}^2$	2.10	20.0	0.887	37.46
4[c)]	$R_{s,\text{top}} = 1.8\,\Omega\text{cm}^2$	2.10	20.0	0.871	36.77
5[d)]	$S = 2.3\%$	2.10	19.6	0.871	35.93

The corresponding maximum efficiency (η) is also shown. The results of Model1 correspond to those shown in Figure 12.17a.

a) J_0 of the Si bottom cell is increased to simulate the strong Auger recombination observed in Si cells. The parameter value is extracted from a 24.7%-efficient a-Si:H heterojunction solar cell [41].
b) Series resistance of the Si bottom cell ($R_{s,\text{bottom}}$) is considered. The parameter value is extracted from a 24.7%-efficient Si heterojunction solar cell [41].
c) Series resistance of the $CsFAPb(I,Br)_3$ top cell ($R_{s,\text{top}}$) is considered. The parameter value is adopted from the analysis result of Table 11.1.
d) Shadow loss caused by the front metal-grid electrode is considered. The parameter value is adopted based on an experimental $CsFAPb(I,Br)_3$/Si tandem cell [8].
e) $V_{oc} = 1.38\,\text{V}\,(V_{oc,\text{top}}) + 0.82\,\text{V}\,(V_{oc,\text{bottom}})$.
f) $V_{oc} = 1.38\,\text{V}\,(V_{oc,\text{top}}) + 0.72\,\text{V}\,(V_{oc,\text{bottom}})$.

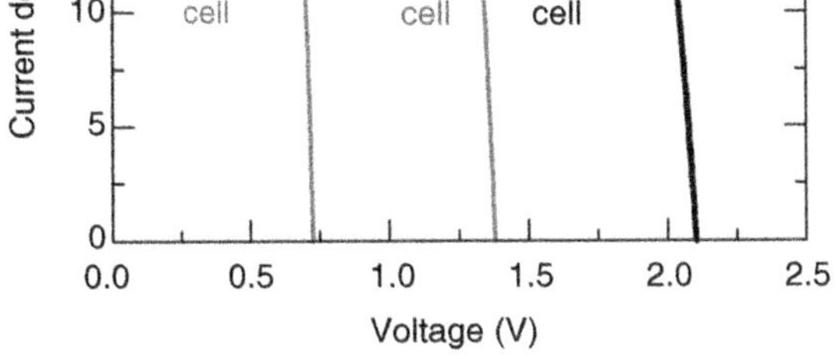

Figure 12.19 *J–V* characteristics of the two-terminal $CsFAPb(I,Br)_3$/Si tandem cell calculated from Model5 in Table 12.2. In Model5, a strong Auger recombination in the Si bottom cell, series resistances of $R_{s,\text{bottom}} = 0.44\,\Omega\text{cm}^2$ and $R_{s,\text{top}} = 1.8\,\Omega\text{cm}^2$, and a shadow loss of 2.3% are assumed, resulting in $\eta = 35.93\%$. The corresponding *J–V* curves of the top and bottom cells are also shown.

12.5 Free Software for Efficiency Limit Calculation

A Windows-based free computer software (e-ARC software) has been developed [24] to perform maximum-efficiency calculations of photovoltaic devices. Using this software, the efficiency limits of single and tandem devices can be calculated by applying Eq. (12.17). From this software, the EQE calculations based on the

ARC method, introduced in this chapter, can also be performed. A more detailed description of the e-ARC software package is given in Section 11.4.3. The package can be downloaded from https://unit.aist.go.jp/rpd-envene/PV/en/service/e-ARC_en/index_en.html, which can be found using the search term "e-ARC AIST."

References

1 Kato, Y., Fujimoto, S., Kozawa, M., and Fujiwara, H. (2019). *Phys. Rev. Appl.* 12: 024039.

2 Shockley, W. and Queisser, H.J. (1961). *J. Appl. Phys.* 32: 510.

3 Green, M.A., Dunlop, E.D., Hohl-Ebinger, J. et al. (2020). *Prog. Photovoltaics Res. Appl.* 28: 629.

4 Shin, S.S., Yeom, E.J., Yang, W.S. et al. (2017). *Science* 356: 167.

5 Min, H., Kim, M., Lee, S.-U. et al. (2019). *Science* 366: 749.

6 Green, M.A., Emery, K., Hishikawa, Y. et al. (2013). *Prog. Photovoltaics Res. Appl.* 21: 1.

7 Lee, Y.S., Gershon, T., Gunawan, O. et al. (2015). *Adv. Energy Mater.* 5: 1401372.

8 Sahli, F., Werner, J., Kamino, B.A. et al. (2018). *Nat. Mater.* 17: 820.

9 Bush, K.A., Palmstrom, A.F., Yu, Z.J. et al. (2017). *Nat. Energy* 2: 17009.

10 Sahli, F., Kamino, B.A., Werner, J. et al. (2018). *Adv. Energy Mater.* 8: 1701609.

11 Jošt, M., Kegelmann, L., Korte, L., and Albrecht, S. (2020). *Adv. Energy Mater.* 10: 1904102.

12 Sze, S.M. (1981). *Physics of Semiconductor Devices*. New York: Wiley.

13 Luque, A. and Hegedus, S. (2011). *Handbook of Photovoltaic Science and Engineering*. West Sussex: Wiley.

14 Correa-Baena, J.-P., Anaya, M., Lozano, G. et al. (2016). *Adv. Mater.* 28: 5031.

15 Fujiwara, H. and Collins, R.W. (2008). *Spectroscopic Ellipsometry for Photovoltaics: Applications and Optical Data of Solar Cell Materials*, vol. 2. Cham: Springer.

16 Nakane, A., Tampo, H., Tamakoshi, M. et al. (2016). *J. Appl. Phys.* 120: 064505.

17 Hara, T., Maekawa, T., Minoura, S. et al. (2014). *Phys. Rev. Appl.* 2: 034012.

18 Shirayama, M., Kadowaki, H., Miyadera, T. et al. (2016). *Phys. Rev. Appl.* 5: 014012.

19 Fujiwara, H., Kato, M., Tamakoshi, M. et al. (2018). *Phys. Status Solidi A* 215: 1700730.

20 Fujiwara, H. and Collins, R.W. (2018). *Spectroscopic Ellipsometry for Photovoltaics: Fundamental Principles and Solar Cell Characterization*, vol. 1. Cham: Springer.

21 Koida, T., Kondo, M., Tsutsumi, K. et al. (2010). *J. Appl. Phys.* 107: 033514.

22 Bhachu, D., Waugh, M.R., Zeissler, K. et al. (2011). *Chem. Eur. J.* 17: 11613.

23 Macleod, H.A. (2010). *Thin-Film Optical Filters*. New York: CRC Press.

24 The e-ARC software has been developed by the research collaboration of Gifu University and AIST. This software can be downloaded at the web site of https://unit.aist.go.jp/rpd-envene/PV/en/service/e-ARC_en/index_en.html. This

web site can be found using "e-ARC AIST" as key words (accessed 01 April 2021).
25 Yao, J., Kirchartz, T., Vezie, M.S. et al. (2015). *Phys. Rev. Appl.* 4: 014020.
26 Tress, W., Marinova, N., Inganäs, O. et al. (2014). *Adv. Energy Mater.* 5: 1400812.
27 Tvingstedt, K., Malinkiewicz, O., Baumann, A. et al. (2014). *Sci. Rep.* 4: 6071.
28 Nishiwaki, M., Nagaya, K., Kato, M. et al. (2018). *Phys. Rev. Mater.* 2: 085404.
29 Nagaya, K., Fujimoto, S., Tampo, H. et al. (2018). *Appl. Phys. Lett.* 113: 093901.
30 Green, M.A., Hishikawa, Y., Dunlop, E.D. et al. (2019). *Prog. Photovoltaics Res. Appl.* 27: 3.
31 Jiang, Q., Chu, Z., Wang, P. et al. (2017). *Adv. Mater.* 29: 1703852.
32 Liu, Z., Krückemeier, L., Krogmeier, B. et al. (2019). *ACS Energy Lett.* 4: 110.
33 Tiedje, T., Yablonovitch, E., Cody, G.D., and Brooks, B.G. (1984). *IEEE Trans. Electron Devices* 31: 711.
34 Green, M.A. (1984). *IEEE Trans. Electron Devices* 31: 671.
35 Green, M.A. (1998). *Solar Cells: Operating Principles, Technology and System Applications*. Kensington: University of New South Wales.
36 Gabor, A.M., Tuttle, J.R., Bode, M.H. et al. (1996). *Sol. Energy Mater. Sol. Cells* 41–42: 247.
37 Ringel, S.A., Smith, A.W., MacDougal, M.H., and Rohatgi, A. (1991). *J. Appl. Phys.* 70: 881.
38 Wang, W., Winkler, M.T., Gunawan, O. et al. (2014). *Adv. Energy Mater.* 4: 1301465.
39 Werner, J., Nogay, G., Sahli, F. et al. (2018). *ACS Energy Lett.* 3: 742.
40 Stolterfoht, M., Wolff, C.M., Márquez, J.A. et al. (2018). *Nat. Energy* 3: 847.
41 Taguchi, M., Yano, A., Tohoda, S. et al. (2014). *IEEE J. Photovoltaics* 4: 96.
42 Taguchi, M., Terakawa, A., Maruyama, E., and Tanaka, M. (2005). *Prog. Photovoltaics Res. Appl.* 13: 481.
43 Fujiwara, H. and Kondo, M. (2007). *J. Appl. Phys.* 101: 054516.
44 Wynands, D., Erber, M., Rentenberger, R. et al. (2012). *Org. Electron.* 13: 885.
45 Holman, Z.C., Filipič, M., Descoeudres, A. et al. (2013). *J. Appl. Phys.* 113: 013107.
46 Baker-Finch, S.C. and McIntosh, K.R. (2011). *Prog. Photovoltaics Res. Appl.* 19: 406.
47 Altazin, S., Stepanova, L., Werner, J. et al. (2018). *Opt. Express* 26: A579.
48 Nishigaki, Y., Nagai, T., Nishiwaki, M. et al. (2020). *Sol. RRL* 4: 1900555.
49 Watanabe, K., Matsuki, N., and Fujiwara, H. (2010). *Appl. Phys. Express* 3: 116604.
50 Matsuki, N. and Fujiwara, H. (2013). *J. Appl. Phys.* 114: 043101.
51 Yu, Z.J., Leilaeioun, M., and Holman, Z. (2016). *Nat. Energy* 1: 16137.
52 Lal, N.N., Dkhissi, Y., Li, W. et al. (2017). *Adv. Energy Mater.* 7: 1602761.
53 Hossain, M.I., Qarony, W., Ma, S. et al. (2019). *Nanomicro Lett.* 11: 58.
54 Hou, Y., Aydin, E., De Bastiani, M. et al. (2020). *Science* 367: 1135.

13

Multi-cation Hybrid Perovskite Solar Cells

Jacob N. Vagott and Juan-Pablo Correa-Baena

Georgia Institute of Technology, School of Materials Science and Engineering, Atlanta, GA 30332-0245, USA

13.1 Introduction

Perovskites are materials with the crystal structure ABX_3. Hybrid organic–inorganic perovskites are those where the A-cation is an organic molecule, the B-cation is typically a divalent metal, and the X anion is a halide. Organic A-cations include methylammonium (MA) and formamidinium (FA). Pb and Sn are commonly used for the B-site cation, and the halogen (X) is usually iodine, bromine, or a combination of the two (known as mixed-halide).

Single-cation perovskites such as $MAPbI_3$, $FAPbI_3$, or $CsPbI_3$, while showing some promising properties for hybrid perovskite solar cells (PSCs), have many downfalls. $MAPbI_3$ has been the most heavily investigated perovskite for PSCs and yet falls short to many of the more recently developed perovskite materials in terms of power conversion efficiency (PCE). $FAPbI_3$ has shown a tendency for the formation of a non-perovskite, photoinactive yellow phase that hampers high efficiencies. The purely inorganic $CsPbI_3$ has shown good emissive properties but is found to be in a photoinactive yellow phase at room temperature [1, 2]. Multi-cation hybrid organic–inorganic perovskites are those that contain one or more cations at either the A-site or B-site. In the typical engineering of these multi-cation materials for PSCs, the A-site is the one that is most frequently considered [3]. The A-site is most commonly filled with a combination of MA, FA, Cs, and Rb. Other cations have also been investigated, including K, guanidinium (GA), ethylammonium (EA), *n*-hexylammonium (HA), and butylammonium (BA) [4–6]. MA has the most ideal radius for the perovskite crystal structure, while FA, GA, EA, and BA are all larger than ideal; in order to account for this, small amounts of MA or another smaller cation can be included in mixtures. Small cations such as Cs, Rb, and K have shown promising improvements of the properties of PSCs. FAMA perovskites have allowed for an incredible rise of efficiencies of PSCs, and small amounts of Cs and Rb show increases in efficiency when added to these perovskites. In these combinations, FA is the most common majority A-cation [3]. Mixing of these cations have resulted in unanticipated, beneficial properties, such as increased stability or a more

Hybrid Perovskite Solar Cells: Characteristics and Operation,
First Edition. Edited by Hiroyuki Fujiwara.

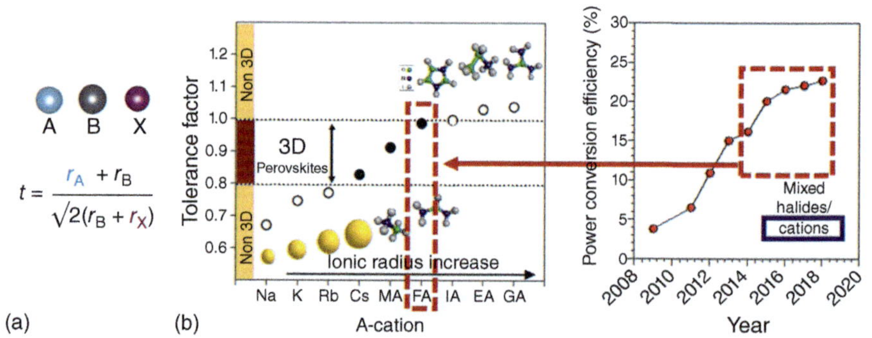

Figure 13.1 Tolerance factor of perovskites: (a) tolerance factor equation and (b) tolerance factor of common perovskites including the regime of most multi-cation perovskites.

ideal bandgap [1]. The resulting perovskite materials often have high absorption coefficients, large charge carrier diffusion lengths, and low trap density [7]. When considering the B-site in multi-cation perovskites, the combinations are more limited. Pb, Sn, and Ge are the most common cations considered for combinations at the B-site. One motivation of manipulating the B-site is to eliminate Pb from the perovskite structure [8, 9]. Multi-cation hybrid PSCs have reached outstanding efficiencies above 23% in 2019, greatly increasing from the 3.8% record reported a decade ago in 2009 [10].

The Goldschmidt tolerance factor (t) equation is shown in Figure 13.1a and allows for the evaluation of a proposed perovskite material. In this equation, r_A, r_B, and r_X represent the ionic radii of the perovskite components: A being the organic cation, B being the metal cation, and X being the halogen anion. To maintain a cubic structure, it can be said that the tolerance factor should be in the range $0.8 < t < 1$. For large organic A-cations, the tolerance factor is often over 1, and the crystal structure is compromised. When the tolerance factor is too low, other structures are formed due to the A-cations' effective radius being too small. Even when the cation is within this range, there can still be lattice strain [11], so making the tolerance factor more ideal through the use of multiple cations will allow for greater stability.

When considering multi-cation perovskites, the tolerance factor equation cannot be used as is. Since there are more than one A-cations present, R_A needs to take this into account. An effective radius and corresponding effective tolerance factor need to be calculated. For this, Eqs. (13.1) and (13.2) can be used for perovskites of composition $A1_xA2_{1-x}BX_3$, which is simply a modified version of the usual perovskite structure but incorporating two A-cations.

$$R_{\text{effective}} = xR_{A1} + (1 - x)R_{A2} \tag{13.1}$$

$$t_{\text{effective}} = \frac{R_{\text{effective}} + R_X}{\sqrt{2}(R_B + R_X)}. \tag{13.2}$$

In these equations, which have been modified from Li et al.'s study on mixed Cs and FA perovskites [12], $R_{\text{effective}}$ represents the effective ionic radius of the combined cations, $t_{\text{effective}}$ is the corresponding tolerance factor, R_{A1} and R_{A2} are the radii of the two A-cations being used, and R_B and R_X represent the radii of the B-cation and halogen, respectively. This formula can also be manipulated to represent triple- or quadruple-cation perovskites, with an adjustment being made to include them in $R_{effective}$. The tolerance from commonly considered A-cations can be seen in Figure 13.1b. The most efficient multi-cation perovskites have been FA-based, so it is worth noting the regime in which most multi-cation perovskites lie. While FA perovskite is at the upper threshold for tolerance factor, the smaller cations that are used in cation cascades lower the effective tolerance factor.

13.2 Types of A-Site Multi-cation Hybrid Perovskite Solar Cells

13.2.1 Pb-Based Multi-cation Hybrid Perovskite Solar Cells

Since 2015, the highest efficiency PSCs to date have been Pb-based multi-cation hybrid organic–inorganic PSCs (Figure 13.1b). As the PCE of modern PSCs are getting closer to their theoretical limits, it is important to understand the effect that each component of the device has on the overall performance. Multi-cation perovskites have shown their superiority over single-cation perovskites, but it is necessary to know the contribution of each cation to determine the optimum compositions. Equally important is the selection of the charge transporting layers, which will be covered later on in this chapter. Due to the large size of the FA molecule, adding in smaller cations like MA, Cs, and Rb have an obvious effect on the overall stability of the device since the effective radius and therefore Goldschmidt tolerance factor are decreased to a more ideal value. While the exact impact of each cation has yet to be determined, there have been great strides to do so in recent years. Idealization of the tolerance factor can be achieved through compositional engineering of the present cations, so that the effective radius is with the 0.9–1.0 range for ideal cubic structure. It has been suggested that the chief benefit of cation mixing is the suppression of carrier recombination centers [11, 13].

Recognizing the benefits of including both MA and FA organic cations has led to a rapid increase in PCE, and interest in adding more cations to the mix has also been successful. FAMA perovskites are majority-FA while having higher stability than pure FA perovskites due to the added MA lowering the effective cationic radius of the FA perovskite. FAMA-based perovskites are typically >80% FA. Only a small amount of MA is needed in order to restore balance to the perovskite crystal structure distorted by the large size of FA molecules; the MA is acting as a crystallizer to the FA. Improved thermal and structural stability in FAMA perovskites is an added benefit over pure MA and FA perovskites. The first report of mixing MA and FA cations for PSCs was in 2014 by Pellet et al. [14]. This study focused on providing a proof-of-concept for mixing cations for enhanced properties in PSCs. It was found that $FA_{0.4}MA_{0.6}PbI_3$ had the highest PCE of the considered combinations at 14.9% PCE. In 2015, Jeon et al. investigated the properties of mixed FAMA PSCs by incorporating varying amounts of $MAPbI_3$ into $FAPbBr_3$ [15]. The range that was observed was 0–30% $MAPbI_3$. If the amount of $MAPbI_3$ present was too small (e.g. 5%), the surface was found to be considerably rough. Increasing this percentage to 15% allowed for a relatively smooth surface. For 15% $MAPbI_3$, substantial increases in stability and crystallinity of the perovskite were observed. Accounting for hysteresis in previous reports (where 19.3% was reported from a reverse scan), this was the highest PCE for PSCs at the time. This gave ample motivation to build off the multi-cation FAMA perovskite composition for further PCE improvement. In 2015, Yang et al. used FAMA perovskite to create the first device with a maximum efficiency of over 20% [16].

While FAMA Pb-based PSCs have shown incredible efficiencies when compared to purely MA PSCs, there is still room to improve overall stability of the device. This is typically done by introducing a small cation such as Cs or Rb. Introducing Cs to FAMA Pb-based perovskites has proven to be an attractive method to increase thermal and structural stability of the perovskite and in turn increase the efficiency of the PSC. In 2014, Choi et al. demonstrated the effect of adding Cs cations to $MAPbI_3$ perovskite [17]. Cs was added in small increments to determine the optimal concentration of Cs so that the optimal properties from $MAPbI_3$ could be retained. Choi et al. theorized that since Cs and MA are similar in size, that the Cs atoms could effectively dope the perovskite film by substituting in place of MA cations. The perovskite $Cs_{0.1}MA_{0.9}PbI_3$ showed a 7.68% PCE, up from 5.51% in the control device ($MAPbI_3$). Both J_{sc} and V_{oc} showed promising rises when compared to the control device. While the long-term stability of the PSCs in this study was considered, it was not an accurate representation of the effect of Cs on stability due to the poor long-term stability of the electrode and hole transporting material (HTM) used in the device architecture (Al and poly(3,4-ethylenedioxythiophene) polystyrene sulfonate (PEDOT:PSS), respectively). Ma and Li considered the effect of Cs on $MAPbI_3$ in a more recent study in 2018 [18]. Instead of focusing on incorporating Cs in the bulk perovskite, only the (1 0 0) surface was manipulated. Previous research by Niu et al. showed that the preferential orientation of perovskite could be manipulated by mixing cations [19]. It was found that when increasing the Cs being incorporated into the perovskite, the decomposition energy was raised, indicating increased thermal stability [18]. With Cs incorporation, a certain degree of tilt in the PbI_6 octahedra was observed. By changing the Cs percentage, it was seen that the tilt orientation was dependent on the degree of Cs-doping. The findings from this study suggest that the bandgap of the perovskite with structure $Cs_xMA_{1-x}PbI_3$ can be tuned through increasing Cs percentage. Both thermal stability and optical absorption coefficient were found to increase with Cs-doping of the (1 0 0) surface [18].

Shortly after Choi et al.'s initial investigation into the effect of Cs on $MAPbI_3$, Lee et al. published their study on the effect of Cs incorporation in $FAPbI_3$ in 2015 [13]. The optimum percentage of Cs in $FAPbI_3$ was also found to be 10%. The PCE was found to have increased when compared to the control device (using $FAPbI_3$) as a result of suppressed charge recombination in the Cs-doped devices [13]. Humidity/moisture stability and photostability were also found to have increased in the Cs-containing perovskites. The champion device had an efficiency of 19.01%

(reverse scan), with an average PCE of 16.5%. Yi et al. continued to investigate the effect of Cs incorporation in FA-based perovskite in 2016, when mixed-halide perovskite was used [11]. The perovskite structure $Cs_xFA_{1-x}PbBr_{0.16}I_{2.84}$ was used for maximum PCE in this study, and an optimized composition of $Cs_{0.2}FA_{0.8}PbBr_{0.16}I_{2.84}$ was determined; however, $Cs_xFA_{x-1}PbI_3$ was also analyzed. The devices fabricated using the optimized perovskite were found to have negligible hysteresis and improved long-term stability. While 20% Cs was found to maximize PCE, it is worth noting that maximum stability was found at 40–60% Cs [11]. J_{sc} was found to slightly increase when a device with perovskite composition $Cs_{0.2}FA_{0.8}PbI_3$ was compared to a control device with composition $FAPbI_3$, but the V_{oc} and FF increases were found to benefit most from Cs incorporation. When compared to the control, the V_{oc} and FF increased by 22.8% and 16.9%, respectively.

Looking at the Goldschmidt tolerance factor of a perovskite composition has shown to give valuable insight into whether it will form a stable photoactive phase. While purely inorganic Cs-based perovskites have tolerance factors below the accepted threshold, the effective ionic radius can be altered by utilizing multiple cations. FA-based perovskite, for example, is within the tolerance factor acceptable range, but it is nearing the top. By altering the effective radius by incorporating both Cs and FA into the perovskite structure, the tolerance factor becomes more ideal. This is supported by the work of Li et al. who showed that the tolerance factor of $Cs_xFA_{1-x}PbI_3$ can be tuned by altering the percentage of Cs present [12]. It was found that a structure of $Cs_{0.15}FA_{0.85}PbI_3$ had an ideal tolerance factor that allowed for high stability and photovoltaic performance. With a more ideal ionic radius, it makes sense that mixed CsFA perovskites are more likely to be stable and photoactive at room temperature. Due to the large size difference in Cs and FA molecules, much more so than the difference in MA and FA, FA should be more stable when mixed with Cs than with MA [12, 13]. Li et al. observed that even if the tolerance factor is acceptable, too high of lattice mismatch due to large differences in atomic radii can have an impact on the overall stability of the perovskite; this held true for large amounts of Cs being incorporated into the FA-based perovskite structure [12]. Phase separation occurs at room temperature in these perovskite films, inducing a photoinactive phase.

Triple-cation perovskites have many promising features. In studies investigating triple-cation perovskites, MA, FA, and Cs are the three most commonly combined cations. The high efficiencies found in FAMA perovskites have been built upon by incorporating Cs. When incorporating any amount of Cs into the perovskite, the effective tolerance factor decreases, so it is intuitive that the stability of the devices will increase. Adding multiple cations increases the complexity of the perovskite, which is unideal given the drive for commercialization. This increase in complexity can be justified in the improvements to stability that resulted from preliminary research. Triple-cation perovskites were first introduced in 2016 by Saliba et al. [1]. The perovskites considered in this study had the empirical form $Cs_x(FA_{0.83}MA_{0.17})_{1-x}Pb(Br_{0.17}I_{0.83})_3$, where the ratios of MA/FA and I/Br were determined for the best known structure for high-efficiency PSCs at the time.

Rb, which has a slightly smaller ionic radius than that of neighbor alkali Cs, has also been considered for perovskite application. First introduced in 2016 by Saliba et al. Rb inclusion in PSCs has shown promising results when introduced as a complementing cation [20]. Due to the small ionic radius of Rb, it does not have a promising Goldschmidt tolerance factor for stand-alone Rb perovskites. Pure Rb perovskites have also not exhibited sustainable stabilities when tested experimentally. Saliba et al. considered four cation combinations in which the tolerance factor showed that it should exhibit a relatively stable photoactive black phase: FARb, CsFARb, FAMARb, and CsFAMARb. Of these combinations, FARb was found to exhibit the lowest PCE when optimized at 14%. While it does lower the effective ionic radius of FARb, there is greater lattice mismatch due to the large difference in size between Rb and FA. The triple-cation CsFARb and FAMARb fared much better and resulted in efficiencies as high as 19.3% and 19.2%, respectively. The most notable result of this study was the outstanding performance of the quadruple-cation CsFAMARb, with PCEs of up to 21.6%. The effect of Rb on FA-based perovskites was considered further in a study by Zhang et al. [21]. Moisture stability was substantially increased in $FA_{0.95}Rb_{0.5}PbI_3$ when compared to $FAPbI_3$, as was the overall PCE. Over a four-week storage in dark lighting and ambient air, the FARb PSCs maintained 90% of their original efficiencies, while the $FAPbI_3$ showed considerably lower PCEs. Zhang et al. also looked at the effect of Rb incorporation in $MAPbI_3$, although it was found that there were no substantial improvements to stability or PCE.

Another cation that has shown promise in increasing the PCE and reducing the hysteresis in PSCs is K [22–24]. While there is uncertainty surrounding its full incorporation in the perovskite structure [23], it has been shown to fully eliminate hysteresis and improve device performance [24]. In 2017, Bu et al. introduced the “quadruple-cation” perovskite $Cs_{0.05}K_x(FA_{0.85}MA_{0.15})Pb(Br_{0.15}I_{0.85})_3$, or CsFAMAK perovskite [24]. It was seen that the grain size of the perovskite increased drastically, up to double that of the control devices. The increased properties were suspected to be due to a decrease in defect density and longer carrier lifetimes with the incorporation of K. With a PCE of over 20%, the overall impact of K on the perovskite layer was clearly positive. The CsFAMAK perovskite also showed outstanding stability, with no loss in PCE after 1000 hours stored under ambient air. In 2018, Abdi-Jelebi et al. found that introducing K to the triple-cation $(Cs_{0.06}FA_{0.79}MA_{0.15})Pb(Br_{0.15}I_{0.85})_3$ perovskite resulted in inhibition of ion migration and increased stability; however, full incorporation of K into the perovskite structure is doubted [25].

13.2.2 Sn-Based Multi-cation Hybrid Perovskite Solar Cells

Sn-based perovskite materials are desirable in part due to their lack of inclusion of Pb. Sn-based PSCs are also inherently more stable than Pb-based cells [26]. Multi-cation hybrid perovskites based around Sn have seen increased efficiencies in recent years but still lag behind Pb-based PSCs. The highest Sn-based PSC made thus far reached 9.6% PCE [27]. This device was realized by introducing the large-cation guanidinium (GA) to Sn-based perovskites. $FAGASnI_3$ was the structure that showed the highest efficiency and also had notably high stability [7]. One of the main concerns with incorporating Sn in perovskites is the tendency for the oxidation of Sn from the Sn^{2+} to Sn^{4+} state. There is also concern over the often-short carrier lifetimes, which in turn limit diffusion lengths [28].

Using multiple cations in fully Pb-free $ASnX_3$ perovskites has not been thoroughly investigated, although in recent years there has been some work done in the area. In 2016, Liu et al. investigated the effect of Cs substitution in primarily $MAPb_xSn_{1-x}I_3$ [29]. As a part of this study, 100% Sn perovskite was considered. The findings showed very low PCE for devices with this perovskite layer, only achieving 0.33% PCE. While both V_{oc} and J_{sc} were very low, a V_{oc} of only 0.20 V severely limited the devices. In 2017, Zhao et al. fabricated Pb-free perovskites with the structure $FA_xMA_{1-x}SnI_3$, which resulted in moderate success [30]. After optimization, a champion device with the composition $FA_{0.75}MA_{0.25}SnI_3$ exhibited a PCE of 8.12%. This increase in efficiency resulted from the reduced charge carrier recombination and improved V_{oc} of the devices.

The effect of adding Cs in $FASnI_3$ perovskite shows a surprising effect on the bandgap. With increasing Cs, the bandgap is lowered for $Cs_xFA_{1-x}SnI_3$ perovskites. While Pb-based CsFA perovskites have a distorted, tetragonal structure, this is not the case for $Cs_xFA_{1-x}SnI_3$ perovskites [28]. Instead of distortion in the structure, the Sn-based perovskites show an isotropic contraction that decreases the bond length between the Sn and the halogen. This decreased bond length increases the valence band maximum and shows cause for a decreased bandgap [28].

13.3 Cation Selection in Mixed-Cation Hybrid Perovskite Solar Cells

13.3.1 Organic A-Cations

MA has been the most extensively considered A-site cation; however, MA-based perovskites (e.g. $MAPbI_3$) are known to be unstable when exposed to moisture, heat, and light. One reason that MA-based perovskites have received so much attention is due to their close-to-ideal tolerance factors. For example, $MAPbI_3$ has a tolerance factor of 0.91, which is very close to the middle of the generally accepted 0.8–1.0 range. Single-crystal $MAPbI_3$ perovskites, however, have reached a PCE of over 21% [31]. MA is useful as a constituent in multi-cation FA-based perovskites. In FAMA perovskites, the MA cation acts as a stabilizer of the photoactive FA. MA-based perovskites have a high bandgap when compared to that of FA- and Cs-based perovskites. The most common majority cation in multi-cation hybrid perovskites is FA, although it is not feasible in pure, single-cation perovskites. In single-cation perovskites such as $FAPbI_3$, FA forms a photoactive black phase as well as a non-perovskite photoinactive yellow phase (δ-phase) [32]. Transition occurs to this yellow phase when the perovskite is held at room temperature. This phase is detrimental to the overall PCE of the solar cell, and ways of preventing this phase have been a target of recent research. It has been shown that this yellow phase can be reduced or eliminated by the inclusion of another cation, effectively stabilizing the FA. When considering the integrity of the perovskite structure, it is important to note the size of the molecule that is filling the A-site. The FA molecule is larger than MA, resulting in a somewhat distorted perovskite structure. This has been counterbalanced by incorporating MA into FA-based multi-cation perovskites.

EA has been shown to be suitable for high-performance PSCs [4]. In this study by Gholiour et al. it was experimentally shown that large amounts of EA could be introduced into perovskites without becoming detrimental to the film quality. The atomic radius of EA is approximately 2.3 angstroms, which makes it larger than MA but smaller than FA [33, 34]. This shows that perovskites with the formula $EA_xMA_{1-x}PbI_3$ have tolerance factors ranging between 0.83 and 0.94, which is within the acceptable range. It has been shown that adding EA to MA-based perovskites does result in increased stability [33].

BA has been introduced into CsFA perovskite by Wang et al. [6]. This added cation results in the formation of two-dimensional (2D) perovskite platelets that suppress non-radiative charge recombination. The champion BACsFA PSC by Wang et al. had a PCE of 19.5%. The addition of the BA cation resulted in enhanced stability under simulated sunlight. Plate-like 2D-phase crystallites form between the 3D perovskite grains when BA is introduced to the CsFA perovskite, which greatly improves crystallinity. The multi-cation perovskite $BA_xMA_{1-x}PbI_3$ can be seen in Figure 13.2a. The 2D nature of the perovskite can be seen, and there is significant elongation of the perovskite cage. Other cations that form 2D perovskites are GA and HA [36]. Fu et al. synthesized perovskite mixtures of $MAHA_2Pb_2I_7$ (Figure 13.2b) and $GAHA_2Pb_2I_7$. Like $BAMAPbI_3$, the perovskite lattice has been elongated and distorted.

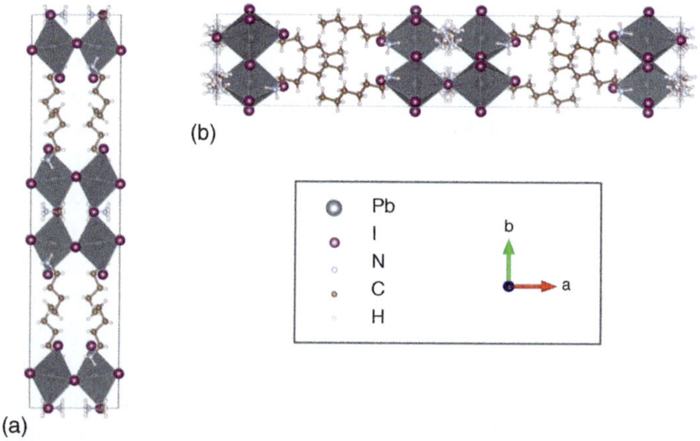

Figure 13.2 Crystal structures for (a) $BAMAPbI_3$ [35] and (b) $MAHA_2Pb_2I_7$ [36]. Structures were generated using VESTA. *Source:* (a) Based on Stoumpos et al. [35]. (b) Based on Fu et al. [36].

13.3.2 Inorganic A-Cations

Introducing an inorganic molecule such as Cs or Rb has shown to increase the performance of hybrid PSCs as well as to aid in the crystallization process [20, 37]. These molecules provide stabilization to the perovskite material when included in small amounts, and recently, quadruple-cation perovskites, which include small amounts of both Cs and Rb, have shown promise. When both Cs and Rb are incorporated into the PSC, there has been shown to be better charge transport, lower defect densities, and longer charge carrier lifetimes [38]. The synergy between the Cs and the Rb when utilized for devices has been noted, although the exact nature of their effect on the properties of the cells is still being investigated [32, 38]. When both Cs and Rb are present, the radiative recombination within the perovskite has shown to decrease by about 60% [38].

Cs is a much smaller atom than MA or FA. When combined with MA to form $MAPbI_3$, it has been shown to increase the stability of the typical $MAPBI_3$ perovskite [1]. Alone, Cs-based perovskites have a low Goldschmidt tolerance factor that is very close to the accepted threshold. This can be counterbalanced by incorporating some amounts of larger cations like MA or FA into Cs-based perovskite structures. Cs has been shown to aid in the crystallization of black phase FA in these triple-cation perovskites [1]. Cs has been shown to integrate more easily into FAMA perovskite than Rb, which results in increased perovskite grain size [38]. Perovskites have shown reduced number and depth of trap states when Cs is present [39]. Both thermal stability and structural stability have been shown to increase with incorporation of a small amount of Cs [40]. The small size of Cs suggests that there may be difficulty with integrating within the perovskite structure; however, it has been confirmed by multiple researchers that this is not the case. Purely inorganic $CsPbI_3$ perovskite is not viable for PSCs because it forms a photoinactive δ-phase, or yellow phase, at room temperature [12]. Since the tolerance factor for $CsPbI_3$ is very low, the idea that it will not successfully maintain a photoactive phase at room temperature is supported. In addition, the photoactive black phase is only stable at very high temperatures (>300 °C). On the other hand, the useful properties that Cs attributes when incorporated into perovskite absorber layers make utilization of Cs in multi-cation PSCs very attractive.

Rb as a cation in PSCs has become a topic of discussion since 2016 [20]. The purely inorganic, single—cation perovskite $RbPbI_3$ has no photoactive phase at room temperature. Unlike Cs, even when raised to high temperatures, there is no photoactive black phase that can be established, since the perovskite experiences irreversible melting. Rb has recently shown promise for being included in multi-cation perovskites. Mixed-cation perovskites including Rb have shown encouraging PCEs. Rb has been shown to stabilize the black phase of FA when included in FA-based perovskites, although it is not suitable to produce pure single-cation perovskites (e.g. $RbPbI_3$). Rb is smaller than FA, so it helps to balance out the effective ionic radius, which corresponds to a more ideal Goldschmidt tolerance factor. Introduction of the Rb cation has shown to increase the inherent conductivity of the perovskite, which in turn increases the fill factor (FF) [37].

CsFAMARb quadruple-cation perovskite has shown promising yet surprising improvements over FAMA perovskites and triple-cation CsFAMA perovskites. When introduced to CsFAMA perovskite, it has been shown to increase V_{oc} and photostability [22]. At high Rb concentrations, yellow-phase $RbPbI_3$ is detected through X-ray diffraction (XRD), suggesting phase segregation. Introducing Rb directly into the perovskite structure (e.g. FAMARb perovskite) has been difficult to realize due to the inherent size of the Rb cation; however, when Cs is also introduced, the perovskite is much more stable [39]. CsFAMARb perovskites have been shown to have high PCE, up to 21.6%. In 2017, a study by Yadav et al. showed that the increased performance of multi-cation perovskites involving Rb could be a result of increased open circuit voltage (V_{OC}), higher FF, increased recombination rate, and less interfacial reactive sites [37]. Hu et al. suggests that Rb results in the formation of multiple photoinactive side phases. It can also be seen in XRD measurements that the (1 1 0) reflection angle shifts lower with increasing amounts of Rb [40], which supports the idea that it is not fully incorporated into the perovskite lattice.

Introduction of Cs and Rb have been known to be beneficial in increasing the performance of the perovskite absorber layer. Understanding how these cations influence the perovskite is crucial to further success and was investigated by Dang et al. [32]. Even though high-efficiency devices have been made using FAMA perovskites, regions of MAI and FABr agglomeration have been observed [32]. Cs and Rb additives have been successful at keeping phase segregation such as this from occurring. These two small cations have been shown to alter the solidification of the perovskite and encourage a homogenous film of the photoactive black phase.

Aside from Cs and Rb, introduction of K in perovskites has also been investigated. While K has been suggested to incorporate itself into perovskite structures [41], there are reports that discuss why this may not be the case. Kubicki et al. challenges the idea that there is K incorporation in perovskite lattices [23]. Using nuclear magnetic resonance (NMR) spectroscopy, the ^{39}K spectra supports the idea that there is not K incorporation in many of the previously reported perovskites. While it is still under debate whether there is full acceptance of the K cation into the perovskite structure, there is a mutual agreement that the addition of K increasing the efficiency of PSCs. K acts as a passivation agent when introduced to perovskites and increases stability in the resulting PSCs (Figure 13.3d) [22]. Perovskites with K-doping experienced very minimal decrease in PCE after 1000 hours, which was not the case for the other multi-cation perovskites tested. Doping with K has shown to significantly reduce hysteresis of PSCs [24–26]. Photoluminescence (PL) has been used to compare the photovoltaic performances of FAMA, CsFAMA, and CsFAMAK perovskites [24]. As seen in Figure 13.3a, the PL intensity of FAMA is the lowest, although very close to that of CsFAMA. The CsFAMAK perovskite, however, has a much higher PL intensity. This suggests that there are less defects present in the CsFAMAK film [24]. Furthermore, PCE was found to increase with each added cation, and there was also found to be less hysteresis for CsFAMAK perovskites. It is also worth noting that an increase in the concentration of K in the perovskite resulted in an increasing wavelength at the PL peak (Figure 13.3c).

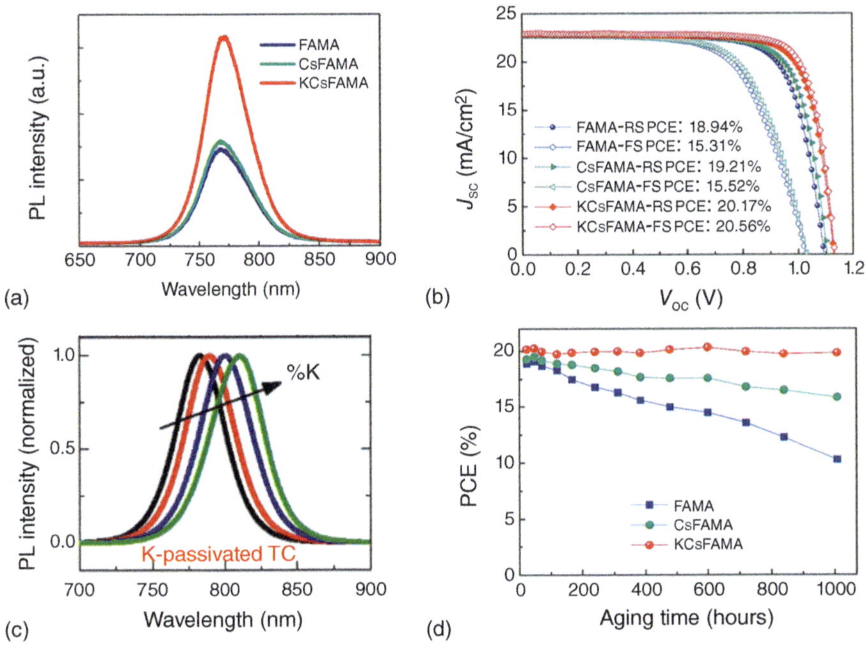

Figure 13.3 PL and PCE measurements for K-containing perovskites. (a) PL comparison of FAMA, CsFAMA, and CsFAMAK perovskites [24]. (b) *J*–*V* curves for FAMA, CsFAMA, and CsFAMAK perovskites [24]. (c) PL measurements for various K concentrations [22]. (d) PCE over time stability measurements [24]. *Source*: (a, b, d) Bu et al. [24]. (c) Abdi-Jalebi et al. [22].

13.4 Fabrication of Mixed-Cation Hybrid Perovskite Solar Cells

13.4.1 Traditional Fabrication Approach

The highest efficiency PSCs to date have been fabricated using solution processing, and spin-coating is currently the preferred method of depositing perovskite layers for PSCs. This method allows for high-efficiency devices at a low cost and aside from an initial learning curve, deposition is not difficult. Spin-coating lends itself well to multi-cation PSC applications, since the correct stoichiometry precursors can easily be prepared and deposited. For example, in order to deposit the triple-cation perovskite $Cs_xFA_yMA_{1-x-y}PbBr_zI_{3-z}$, stock solutions of CsI, MABr, FAI, PbI_2, and $PbBr_2$ can be prepared and then combined for the desired stoichiometry. The prepared perovskite stock solution is spin-coated, and high-quality multi-cation perovskites are deposited. When introducing a new cation, such as K or Rb, the atom is paired with a halogen and the stoichiometry is calculated for ideal formation of perovskite. In spin-coating of perovskites, there are inevitable defects, and the film may not be as homogeneous as with vapor-processed perovskite films. Interfacial defects are also present [42], which lowers the overall PCE of the device. The grain size at the surface of solution-processed perovskite films tends to be small, which typically results in more surface defects than large-grain perovskite. With solution processing of perovskite films, low-temperature (less than 120 °C) thermal annealing is necessary. Among the downsides to solution processing of perovskites is the lack of scalability, with record efficiency devices having been fabricated on small-scale cells less than 1 cm^2. With spin-coating, the uniformity drastically decreases for high-area substrates, and this has been a consistent hurdle for commercialization of PSCs.

The one-step spin-coating method for perovskite films was first reported by Kojima et al. [10]. At this time, perovskites for photovoltaic applications were just beginning to be studied. A precursor solution was applied to the surface of TiO_2, which was the electron transport layer (ETL) of the cell, in a dilute amount. Spin-coating of this solution and subsequent drying resulted in the transition to the black perovskite phase. Two-step spin-coating was introduced by Im et al. [43]. A comparison of one-step vs. two-step spin-coating methods can be seen in Figure 13.4.

Two-step spin-coating involves the preliminary step of depositing BX_2, where B is typically Pb and X is a halogen, before adding the organic iodide in a second step. This process can also be adapted for multi-cation perovskites (Figure 13.4a) using the desired ratio of organics in solvent during the second step.

Among popular solution processing techniques for perovskite films are one-step and two-step spin-coating [46]. Dimethylformamide (DMF) and dimethyl sulfoxide (DMSO) are commonly used solvents for this process in a 4 : 1 ratio [46]. A perovskite solution of desired stoichiometry is crafted using this solvent mixture. The antisolvent method is perhaps the most widely accepted perovskite deposition technique and involves dropping an antisolvent such as chlorobenzene on the perovskite surface during spin-coating. Other antisolvents include IPA, DCM [47], and toluene. Once the antisolvent has been dripped onto the perovskite and the spinning has completed, the film is immediately annealed and the perovskite transitions to a reflective black film. Figure 13.4b shows the spin-coating process using toluene as an antisolvent.

13.4.2 Emerging Fabrication Technologies

Recently, researchers have considered alternatives to the traditional spin-coating deposition techniques for perovskites. One substantial benefit of alternative processes includes that the solvents involved in spin-coating are no longer required, reducing the overall toxicity of the fabrication process. The two main types of vapor deposition for solar cell applications are chemical vapor deposition (CVD) and physical vapor deposition (PVD). A form of CVD known as atomic layer deposition (ALD) has also been considered as a method of depositing perovskite materials.

CVD of perovskites is a beneficial method of depositing perovskite due to the ability to deposit highly conformal films without the need for toxic solvents. CVD may allow for large-scale deposition of perovskite materials, which would alleviate some of the concern with industrializing PSCs. Unfortunately, CVD processes also require high vacuum, which will likely hamper large-scale attempts. In a typical

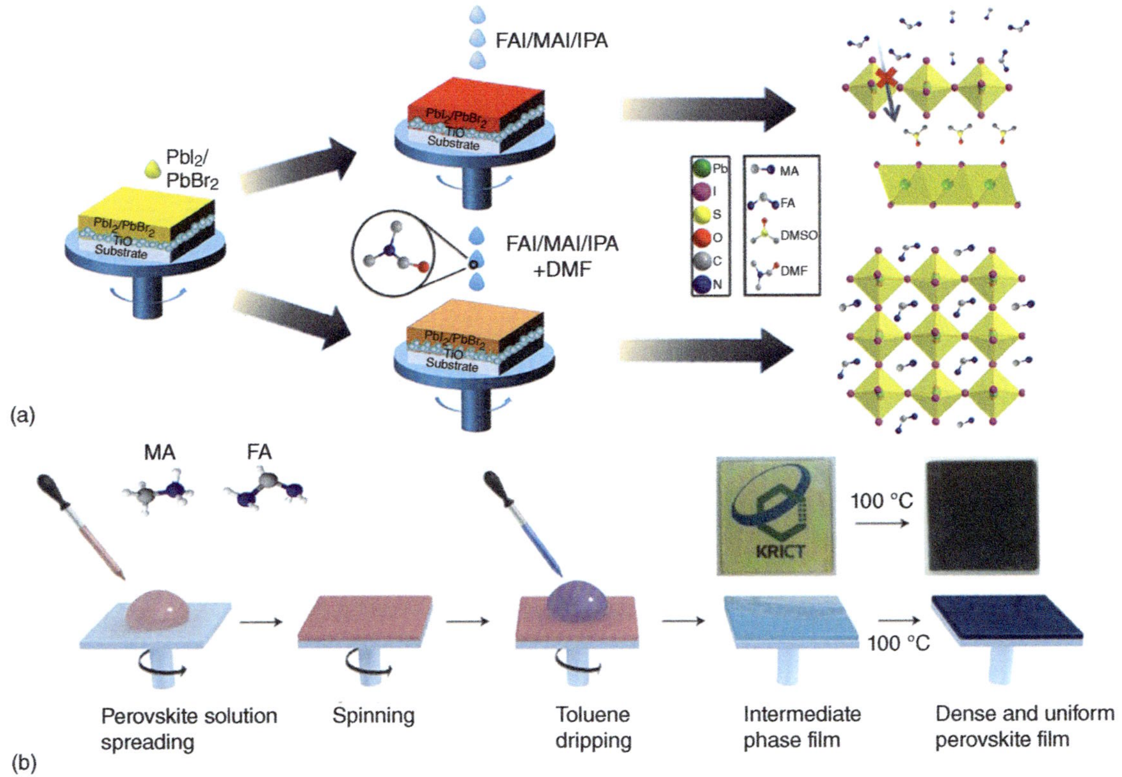

Figure 13.4 A comparison of one-step and two-step deposition methods for multi-cation perovskite. (a) Two-step deposition for multi-cation perovskites [44]. (b) Antisolvent method (one-step) deposition of FAMA perovskite using toluene as antisolvent [45]. *Source*: (a) Wu et al. [44]. (b) Jeon et al. [45].

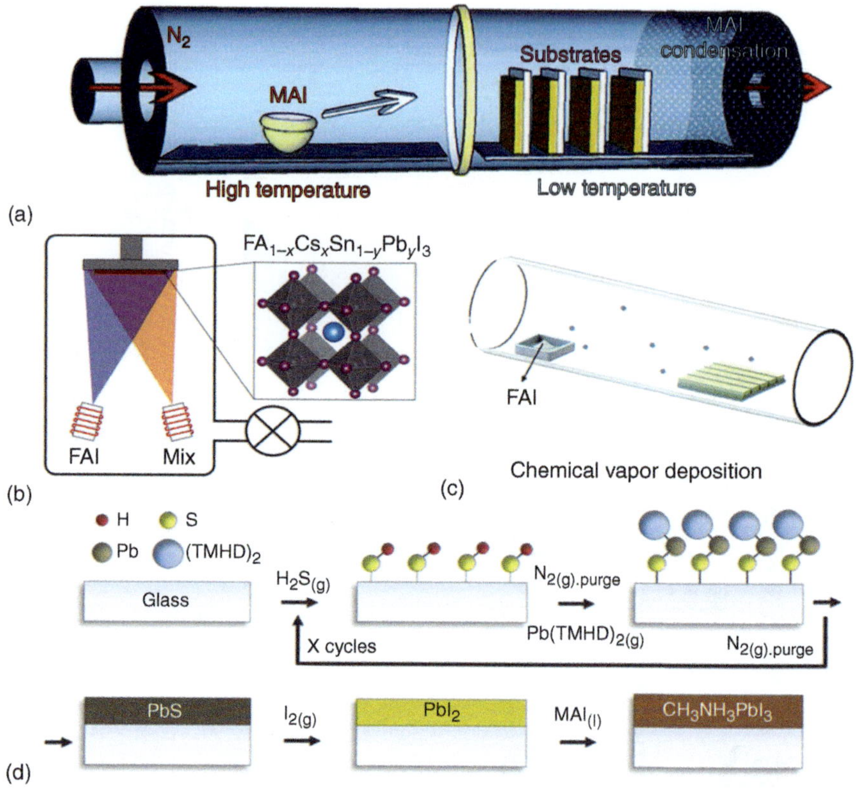

Figure 13.5 Vapor deposition approaches for depositing perovskites. (a) The HCVD designed by Leyden et al. and used in the deposition of perovskites for PSCs [48]. (b) Co-evaporation process for deposition of FA-based multi-cation perovskites [50]. (c) The HCVD design utilized by Qiu et al., which involved using co-evaporation (PVD) of the metal halide precursors before conversion to perovskite by CVD of FAI [51]. (d) Process of ALD-assisted perovskite deposition by Sutherland et al. [52]. *Source*: (a) Leyden et al. [48]. Licensed under CC BY 3.0. (b) Ball et al. [50]. Licensed under CC BY 4.0. (c) Qiu et al. [51]. (d) Sutherland et al. [52].

CVD process, precursors are introduced into the chamber together and the final thickness of the thin film is determined primarily by reaction time. CVD processes have been used to convert PbI_2 films into the perovskite $MAPbI_3$ [48, 49]. Due to the need for evaporation of the organic precursor MAI, the process is considered hybrid chemical vapor deposition (HCVD). An example of one of these HCVD tools can be seen in Figure 13.5a.

$FA_xMA_{1-x}PbI_3$ has also been deposited using a combination of spin-coating and CVD [49]. The resulting device showed proof that CVD could result in high-quality devices with an efficiency of 15.86% and negligible current hysteresis. In this study, similar to past attempts at using CVD in the deposition of perovskites, the PbI_2 film was first prepared by spin-coating. The PbI_2 film was then converted to $FA_xMA_{1-x}PbI_3$ by CVD process using FAI and MAI as precursors. Incorporating Cs has also been successfully achieved by a HCVD process [51]. In this study, a large-area substrate (91.8 cm^2) was used to show the potential of CVD processes to allow for large-scale deposition of perovskites (Figure 13.5c). $Cs_{0.1}FA_{0.9}PbI_{2.9}Br_{0.1}$ was deposited using this system, and an efficiency of 9.34% was achieved on the 91.8 cm^2 substrate [51]; this provides a proof-of-concept that efficient, large-scale perovskites could become possible through HCVD reactions. Since double-cation hybrid perovskites have been able to be fabricated in this manner, it provides a direction for research of more multi-cation perovskites. Perovskite bandgap tuning is critical to its successful photovoltaic (PV) applications, and the tuning capabilities of these combinations must also be considered. It has been shown that the growth time directly impacts the optical and electrical properties of mixed-cation perovskites formed by CVD [51]. Perovskite grain size has shown to increase with increasing growth time. Increased reaction/growth time results in less dense and less compact perovskite films with many pinholes; however, if the growth time is too short, the perovskite has a significantly shorter lifetime [51].

Conversely, ALD, which is a sub-type of CVD, is more useful for fine-thickness applications. ALD involves a series of repeated self-limiting reaction steps, where a precursor is introduced and saturates the substrate; the next reactant is then introduced, allowing the deposition of a monolayer of the desired material. ALD for perovskites was first attempted in a 2015 publication by Sutherland et al. [52]. A PbS film was deposited by ALD using $Pb(TMHD)_2$ and H_2S as precursors, which could then be converted to PbI_2 by exposing it to iodine vapor in a nitrogen-sealed container. The conversion to perovskite is completed by submerging the PbI_2 film in an MAI bath. The full process can be seen in Figure 13.5d. Another attempt at using ALD for perovskites was in the work done by Popov et al. [53]. Popov et al. investigated SnI_4 as an iodine precursor, so that PbI_2 films can be deposited directly from the precursors $Pb(btsa)_2$ and SnI_4. One potential issue with this method is that there are small amounts of Sn left in the film.

PVD typically involves the deposition of thin films by evaporating the precursor(s) onto a substrate. Co-evaporation is the PVD process of evaporating precursors at the same time so that the desired material is formed on the surface of the substrate. Electrical contacts (e.g. gold or silver) can be deposited onto the top of PSCs by PVD. This provides a conformal coating of the desired contact material, and masks are easily integrated so that only the needed area is covered. Evaporation can also be used to deposit the perovskite itself. It has also been shown possible to deposit multi-cation perovskites using co-evaporation [50]. Ball et al. shows that it is possible to fabricate mixed A-site and B-site perovskites using co-evaporation; deposition of the low-bandgap $Cs_xFA_{1-x}Sn_{1-y}Pb_yI_3$ perovskite was demonstrated (Figure 13.5b).

13.5 Charge Transport Materials

Efficient charge transport materials are critical to achieve highly efficient PSCs. PSCs contain two charge transport layers: a hole transport layer (HTL) and an ETL. A great number of these charge transport layers have been created and used in state-of-the-art solar cells. In selecting materials for these layers, the configuration of the device is important. Planar n-i-p devices are those where the ETL is deposited before the perovskite and the HTL afterward, and p-i-n (inverted) devices have the HTL below the perovskite and the ETL above.

An HTL is responsible for the efficient hole extraction from the perovskite absorber layer to the respective contact (depending on device configuration). It is important that the HTL fully covers the perovskite layers so that there is no contact with the electrode or conducting oxide layer. When selecting a HTM, stability is critical. The material must be able to withstand temperatures up to the required processing temperatures and must be photochemically stable. Hole mobility must be acceptable for efficient charge transport, and the highest occupied molecular orbital energy must be slightly higher than that of the perovskite valence band edge in order to extract the holes from the perovskite [54]. The current HTM of choice for high-efficiency n-i-p PSCs is spiro-OMeTAD, which has been used in most of the recent record-breaking PSCs [54, 55]. It is common to employ a buffer layer between contacts and charge transport layers. For example, MoO_3 has been shown to enhance the performance of PSCs as a buffer layer between spiro-OMeTAD and the electrode [56].

The HTM used depends on the configuration of the device. In n-i-p devices, the HTL is deposited onto the perovskite layer. For this type of PSC, the most commonly used HTM is spiro-OMeTAD, which has been used in most record-breaking devices since its adoption. Spiro-OMeTAD is commonly doped to increase performance. For example, bi-dopants bis(trifluoromethane)sulfonimide lithium salt (Li-TFSI) and 4-*tert*-butylpyridine are commonly used. Doping is unfavorable in any instance of solar cell production due to the added cost and complexity, but it does drastically increase performance in many cases. While shown to increase PCE in some cases, there have also been reports about the downsides that these dopants result in, such as degradation of the perovskite, decreased stability, and increased hysteresis [57–59].

When considering p-i-n devices, HTMs do not need to withstand high temperatures since processes typically stay under 100 °C. Common HTMs for p-i-n devices include PEDOT:PSS, poly[bis(4-phenyl)(2,4,6-trimethyl-phenyl)amine] (PTAA), and nickel oxide (NiO), among others [60]. The HTL for the first device that with efficiencies over 20% was PTAA [16]. PTAA has shown to be a promising HTM, although it does have a low inherent carrier mobility [16, 60]. In most cases of high-efficiency devices using PTAA, the polymer is doped with F4-TCNQ to increase conductivity [60]. Li-TFSI and 4-*tert*-butylpyridine are also effective at doping PTAA. As mentioned previously, there is motivation to develop highly efficient PSCs with as little dopants as possible. Recently, high-efficiency PSCs with non-doped PTAA have been fabricated, with efficiencies approaching 20% [60].

The ETL is responsible for extracting photogenerated electrons from the perovskite layer and delivering them to the electrical contact. Metal oxides are commonly employed as ETLs due to their high electron mobility and robustness [59].

ETLs have shown to be relatively easy to deposit by solution processing, spray pyrolysis, and ALD.

TiO_2 is the most common ETL for high-performing PSCs currently. Mesoporous TiO_2 has shown to be efficient at the extraction and transportation of electrons and has led to the further development of other metal oxides for ETL applications [59, 61]. The first PSC to reach 20% efficiency was using mesoporous TiO_2 as the ETL [16]. Compact TiO_2 is commonly used for n-i-p devices and has resulted in very high-efficiency devices [62]. Unfortunately, one drawback of using TiO_2 as a HTL is its tendency to decompose the perovskite layer and result in poor device stability [63]. SnO_2 has been considered a potential alternative for high-efficiency devices, and it has shown high effectiveness as an electron transporter [64]. A comparison of TiO_2 and SnO_2 can be seen in Figure 13.6.

The bandgap alignment of the ETL and perovskite layers is unideal for TiO_2, which may result in less efficient charge selection. SnO_2 has shown promise as a solution to this problem for multi-cation perovskites. As shown in Figure 13.6b, for single-cation perovskite $MAPbI_3$, bandgap alignment is more ideal for TiO_2 than SnO_2; however, this is not the case for multi-cation perovskites. Since the highest efficiencies have come from multiple cation perovskites, it is important to reconsider the ETL to maximize efficiencies with the new compositions. SnO_2 has shown more ideal bandgap alignment (Figure 13.6c), reduced hysteresis in J–V curves (Figure 13.6e), and higher efficiencies when compared to devices with TiO_2. SnO_2 can be easily deposited by ALD, making it a strong candidate as an ETL in modern multi-cation PSCs. The transient absorption measurements in Figure 13.6d, which was performed at 750 nm, the peak of the photobleach of the perovskite [64], effectively shows the relative charge carrier population. While both decay, it is much greater for SnO_2, meaning that there is faster charge movement in the PSCs with SnO_2 as the electron transporting layer.

Doping of ETLs has shown to significantly increase the electronic properties, particularly of metal oxides. P-type doping has also shown to decrease surface defects and allow for a better energy-level alignment [61]. Li-doping of TiO_2, for example, has been shown to have a positive impact on the PCE of the device [65]. The amount of electron traps was decreased substantially, and the FF of the PSC was improved when compared to the undoped control device. Aside from doping, other methods of increasing the efficiency of an ETL is to include a buffer layer or form a bilayer of two materials with synergetic behavior. Bathocuproine (BCP) is an example of a commonly utilized buffer layer material due to its ability to influence electrical properties at the interface of the ETL and electrode [63]. When forming bilayers for ETLs, the same material or different materials may be used. For example, a compact TiO_2 layer can be used in conjunction with the mesoporous layer [16]. C_{60} is an ETL which results in improved UV stability but is difficult to deposit correctly, so utilizing it as a bilayer with another electron transporting material (ETM) has shown to be beneficial; for example, performance increases when used with an ultrathin TiO_x layer [47]. Bilayers such as this have also shown to increase the cohesion of the device resulting in less charge recombination.

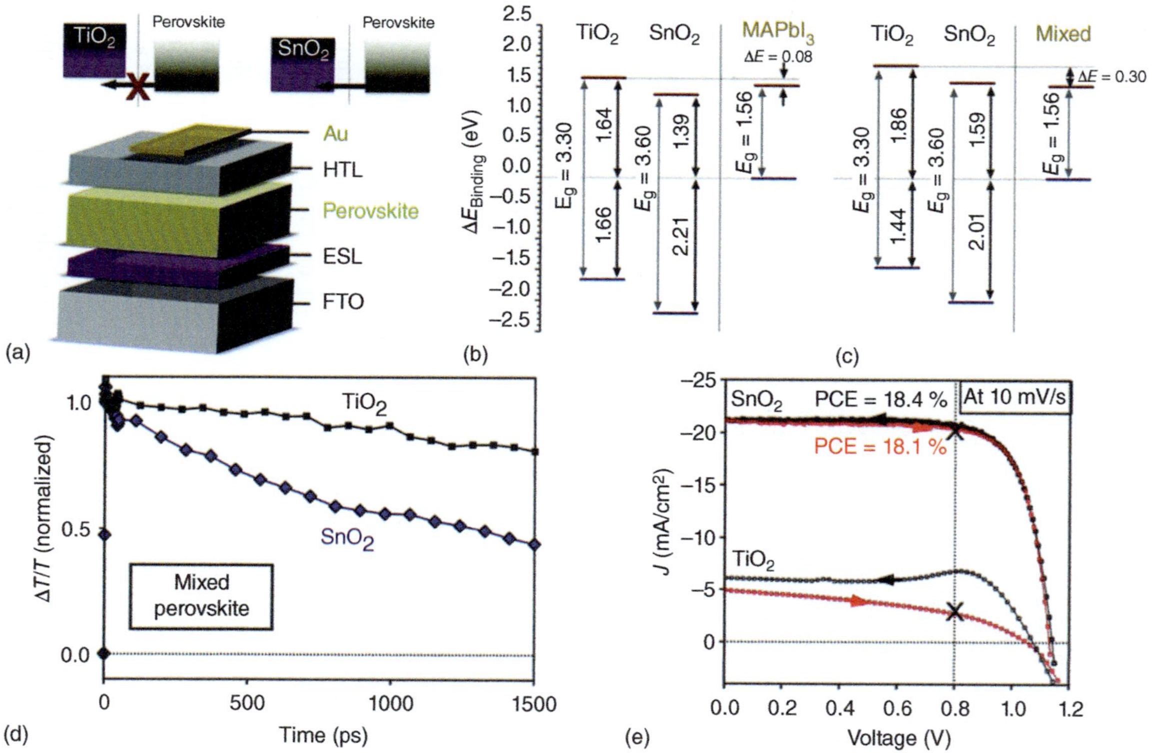

Figure 13.6 Comparison of TiO_2 and SnO_2 electron transport layers [64]. (a) Example device stack and alignment comparison using TiO_2 or SnO_2 as the ETL. (b) The bandgap alignment for $MAPbI_3$ perovskite. (c) The bandgap alignment for a typical mixed-cation perovskite. (d) Transient absorption measurements for PSCs with both TiO_2 and SnO_2 under short-circuit condition. (e) J–V curves for PSCs with either TiO_2 or SnO_2, where reverse scan is denoted with a black arrow and forward scan with a red arrow. *Source*: Correa-Baena et al. [64].

13.6 Surface Passivation

Effective defect passivation is crucial when fabricating high-efficiency solar cells. This section focuses on advancements in surface passivation for mixed A-cation PSCs. There is a known relationship between V_{oc} and recombination of defects in the bulk or the interfaces of the perovskite absorber layer [66]. By decreasing the recombination of defects, the V_{oc} can be raised and in turn result in more efficient cells. Defects are most likely to form at surfaces, so effective surface passivation in PSCs is very important. The focus of this section will be on advancements in passivation for multi-cation PSCs. There has been successful reduction of surface defects using various methods. The 2D perovskites with wide bandgaps and super-thin insulated materials have been used in the past, but newer passivation materials used for multi-cation PSCs include quaternary ammonium halides (QAHs) [67, 68], ammonium salts [69], rubidium halides [25], and potassium halides [25, 66]. A comparison of the effects of QAH and ammonium salt passivation methods can be seen in Figure 13.7. Other methods of surface passivation include using an excess of PbI_2 in the perovskite layer, controlled doping of the perovskite layer [70], and adding small cations to the perovskite lattice have also shown to have passivating effects on PSCs [22, 71].

Adding excess PbI_2 to PSCs has shown to result in surface passivation and allow for higher PCEs. Although easily implemented, the exact mechanism with which excess PbI_2 aids in passivation has only recently begun to be understood. In 2018, Park et al. investigated the impact of the excess PbI_2 on device performance [72]. In particular, the effect on charge diffusion length and film formation properties were considered. They found that with excess PbI_2, there was a lower defect concentration due to decreased halide vacancies. It was also seen that there was formation of α-perovskite crystals, which could also be a reason for increased PCE. More work needs to be done in order to fully understand the mechanism behind increased PCE in devices with excess PbI_2.

It has been proposed that defect passivation is possible through the addition of smaller cations to the perovskite lattice; for example, K and Rb [22]. In this study by Abdi-Jalebi, it was shown that K allows for better luminescent properties than passivation using Rb, while both showed improvements over non-passivated cells. When K and Rb were added to the triple-cation perovskite $Cs_{0.06}FA_{0.79}MA_{0.15}PbBr_{0.45}I_{2.55}$, it was seen that K selectively binds to Br and Rb binds to I. Since KBr has a higher solubility than RbI in common perovskite processing solvents (DMF/DMSO), a higher amount of K can be added to the perovskite, therefore allowing greater passivation effects. K-passivated perovskite films also showed a much greater resistance to humidity when compared to the Rb-passivated films.

The addition of potassium iodide to the perovskite precursor for multi-cation perovskites has shown to be beneficial to the overall PCE of the device. Considering triple-cation perovskites, Abdi-Jalebi et al. show that when a small amount of potassium iodide is added, the PCE can increase substantially [25]. With a 0.1 potassium iodide fraction, the PCE of the triple-cation perovskite $Cs_xFA_yMA_{1-x-y}PbBr_{0.45}I_{2.55}$ rose from 17.3% in the control device to 21.5% with

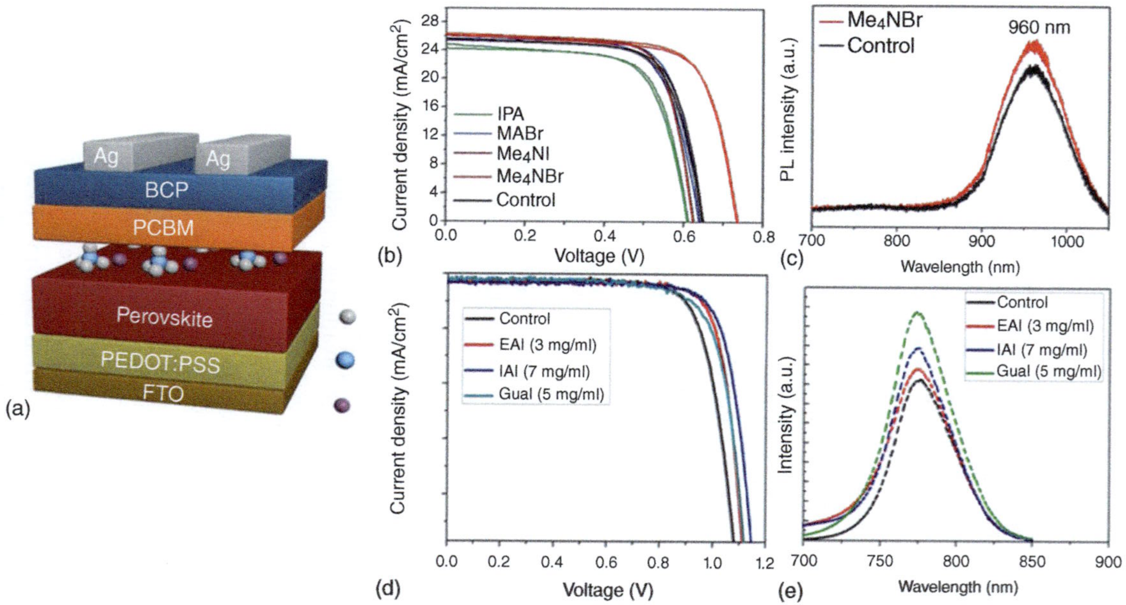

Figure 13.7 Comparison of surface passivation techniques on PCE and PL. (a) The device architecture used for surface passivation in Ref. [67]. (b) *J–V* curves for various surface passivation methods [67]. (c) PL spectra for PSCs with Me_4NBr surface passivation [67]. (d) *J–V* curves for ammonium salt-passivated PSCs [69]. (e) PL spectra for PSCs with ammonium salt passivation [69]. *Source*: (a–c) Du et al. [67]. (d, e) Alharbi et al. [69]. Licensed Under CC BY 4.0.

passivation. The FF also increased from 73% to 79%. The maximum V_{oc} that was achieved was 1.17 V. Changing the halide ratios in the starting triple-cation perovskite maintained the passivating effects, although a lower maximum PCE was reached. For $Cs_xFA_yMA_{1-x-y}PbBr_{1.8}I_{1.2}$, the V_{oc} increased from 1.12 to 1.23 V, but due to the lower J_{sc} compared to $Cs_xFA_yMA_{1-x-y}PbBr_{0.45}I_{2.55}$, the overall efficiency was lower at 17.5% for the passivated device. For both perovskites, the bandgap decreased by 2–3 eV when passivated. Yang et al. also investigated the effects of potassium iodide when incorporated into PSCs. Targeting grain boundaries and interfacial defects, potassium iodide was used within the perovskite layer in order to increase grain size and decrease the contact potential distribution at grain boundaries. The crystallinity of the perovskite layer was improved, which signifies a decrease in defects. Potassium iodide showed to be effective at reducing the hysteresis in multi-cation perovskites, with the hysteresis factor reducing from 0.13 to 0.03 in the work of Yang et al. [42]. Another notable result from this work was that the long-term stability of the device was found to increase from less than 200 hours to over 1000 hours (τ_{80}).

xsAmmonium salts that have been considered for PSC surface passivation include ethylammonium, imidazolium, guanidinium iodide [69], cetyl trimethyl ammonium bromide [73], and phenethylammonium iodide (PEAI) [66, 74]. Alharbi et al. reported on the effects of ethylammonium, imidazolium, and guanidinium iodide, finding that all three increased the PCE of the devices when compared to the un-passivated control. In comparing the PL spectra for these ammonium salts, guanidinium iodide shows the greatest peak intensity, while the PL peak intensity did increase for all passivated PSCs (Figure 13.7e). It is likely that there was greater non-radiative recombination in the control when compared to the passivated PSCs. The maximum PCE achieved for each was 22.3, 22.1, and 21.0% for ethylammonium, imidazolium, and guanidinium iodide, respectively, and the J–V curves can be seen in Figure 13.7d [69]. It is also noteworthy that in this study, a 3% excess of PbI_2 was also used in the preparation of the perovskite layer. It was found that the ammonium salts formed a buffer layer at the interfaces between the perovskite layer and the transport layers, which led to increased V_{oc} in passivated devices. Imidazolium-passivated devices' V_{oc} rose as high as 70 mV when compared to the control, which lead to an increase in the PCE. FF was shown to increase for the ethylammonium and imidazolium, reaching as high as 81% on the champion device. Jiang et al. studied the effect of organic halide salt PEAI as a passivation agent for PSCs [66]. They investigated the effects of adding PEAI as a passivating agent to FAMA perovskites and found positive effects on the overall efficiency of the device, fabricating a champion device with a certified 23.32% efficiency. While it has been reported that the 2D perovskite PEA_2PbI_4 can be used for surface passivation, it was shown that the organic halide salt had a much greater passivation effect. PEAI as an additive allows for an increase in V_{oc} primarily due to reduction of defect recombination on the surface of the perovskite absorber layer. Jiang et al. were able to achieve a V_{oc} as high as 1.18 V, which is 94.4% of the Shockley–Queisser limit of 1.25 V. Zhuang et al. were also able to achieve a 1.18 V V_{oc} using PEAI passivation and had a champion device with efficiency of 22.98%, or roughly a 14% increase compared to the control device [74]. The effects of different functional groups attached to PEAI was investigated by Zhuang et al. and it was determined that the $-CH_3O$ group lead to improvements over the control device, such as an increased PL lifetime, in addition to the increased V_{oc} and PCE. It is also noteworthy that hysteresis is greatly reduced in PSCs that have been treated with PEAI passivation [66, 74].

N-Butylammonium iodide (*n*-BAI) has been investigated for passivation of perovskite layers in PSCs [71]. Mahmud et al. used the quadruple-cation $Cs_{0.07}FA_{0.765}MA_{0.135}Rb_{0.03}PbBr_{0.45}I_{2.55}$ in an n-i-p device for their study. Using *n*-BAI, a champion device with PCE of 22.77% (stabilized 22.60%) was synthesized. Notably, the passivated device had negligible hysteresis and much less non-radiative recombination at both perovskite interfaces. The V_{oc} and FF were enhanced, leading to the increased PCE. The proposed reason for the high-performing nature of the device is that there are discontinuous 2D perovskite layers that form on at the interfaces between the perovskite layer and the respective charge transport layer. The 2D perovskite layer passivates the interface, allowing localized charge extraction through un-passivated regions [71].

QAHs have been investigated for use as surface passivators due to their ability to decrease charged defects in hybrid perovskites [67, 68]. QAHs have a structure of $NR_r{}^+X^-$ with R being either an alkyl or an aryl group and X a halide [68]. Zheng et al. showed that QAH passivation reduces trap density and increases carrier lifetime, which allows for a higher V_{oc} compared to non-passivated devices. While Pb–Sn perovskites allow for the reduction in the amount of Pb in a cell, the instability of Sn is an obstacle which has yet to be overcome. In such devices, it is crucial to have good surface passivation. The device structure used by Du et al. for examining the effects of various passivating agents can be seen in Figure 13.7a. PEDOT:PSS was used as the HTL and phenyl-C61-butyric acid methyl ester ($PC_{61}BM$) the ETL, with BCP acting as a hole blocking layer, and a Pb–Sn perovskite composition was used. Me_4NBr is a QAH showing a significant increase in PCE (Figure 13.7b), so PL measurements were performed (Figure 13.7c). In comparison to the control, a non-passivated device with otherwise identical architecture, there is an increase in PL intensity, suggesting less non-radiative recombination. It was found that Me_4NBr increases V_{oc} and FF [67]. The PCE of the cells rose by 18% when compared to a device without the Me_4NBr surface passivation, with a champion device of 14.56% efficiency. Passivation by Me_4NBr was found to be, at least in part, a decrease in carrier recombination and charge trap-state density.

Doping of perovskites has shown some promise for surface passivation. Noel et al. examined the effect of p-doping of the interfaces of the charge transport layers and the perovskite layer [70]. A molybdenum tris(dithiolene) complex was used to dope the perovskite layer surface due to its electron-withdrawing nature, and the shift in work function indicated that p-doping had occurred. The p-doped interface resulted in improved charge extraction and therefore higher PCEs. The champion-passivated device had a PCE of 21.2%, which showed a 7.33% increase over the highest PCE control device. While the V_{oc} increased from 1.16 to 1.18 V for the champion-passivated device, the J_{sc} decreased from 22.9 to 22.7 mA/cm^2. It was also shown that the FF increased from 74% to 79% for the doped perovskite devices.

13.7 Mixed B-Cation Hybrid Organic–Inorganic Perovskite Solar Cells

Most high-efficiency devices have arisen from mixing A-site cations to form new perovskite combinations; however, another form of mixed-cation PSCs are those where the perovskite has multiple cations in the B-site. Due to the high toxicity of Pb, it is widely considered favorable to decrease the amount of Pb in PSCs. One method of decreasing the amount of Pb in the perovskite layer is to include some amount of Sn. It is also worth noting that with the structure ABX_3, where A is an organic cation, B is Pb_xSn_{1-x}, and X is a halide, the bandgap becomes lower and more ideal [75–77].

The first mixed B-cation PSCs were fabricated in 2014 by Ogomi et al. [75]. For the perovskite layer, $MAPb_xSn_{1-x}I_3$ was used, and the maximum efficiency was attained for $MAPb_{0.5}Sn_{0.5}I_3$ at 4.18%. Due to the low stability of Sn^{2+}, which easily reduces to the unfavorable Sn^{4+}, some amount of Pb must be present in the PSCs. The reported V_{oc} was particularly low at 0.42 V and the FF was only 50%. Ogomi et al. reported that bandgap tuning was possible, which lead to further investigation of the practicality of mixed Pb–Sn PSCs. Advancement was made quickly, and within a few months, a device with the perovskite $MAPb_{0.75}Sn_{0.25}I_3$ was reported by Hao et al. with a PCE of 7.37% [76]. The V_{oc} was greatly increased from the previous work with Pb–Sn PSCs at ~0.738 V. The FF was also closer to that of the control device ($MAPbI_3$ with FF of 68%) at 64%. The J_{sc} of the optimized perovskite composition was improved from the control devices at 15.82 mA/cm^2. Zuo et al. developed the first 10+% Pb–Sn PSC using the perovskite $MAPb_xSn_{1-x}Cl_yI_{3-y}$, where the optimum ratio of Pb/Sn was found to be 85 : 15 [77]. The reported V_{oc} for the 15% Sn perovskite was 0.76 V, the J_{sc} was 19.1 mA/cm^2, and the FF was 66%. The V_{oc} and FF are comparable to that of the previous work by Hao et al. but the overall efficiencies of the devices were increased due to the higher J_{sc} in Zuo et al.'s devices. In the beginning years of investigating Pb–Sn PSCs, it was found that these devices commonly have large amounts of V_{oc} loss, and this is still one of the main hurdles in developing highly efficient mixed B-cation perovskites today.

In 2016, Liao et al. was able to achieve 15% efficiency in a Pb–Sn PSC for the first time [78]. The champion device had an efficiency of 15.08% with a V_{oc} of 0.795 V, J_{sc} of 26.86 mA/cm^2, and FF of 70.6%. The composition of the optimized perovskite layer was $FA_{0.6}MA_{0.4}Pb_{0.4}Sn_{0.6}I_3$. The method of using $MAPbI_3$ and $FASnI_3$ to fabricate a new perovskite precursor allowed for the precise control over molar ratio. The lowest bandgap mixture was found to be approximately 1.2 eV at 60% $FASnI_3$ and led Liao et al. to focus on this composition for measured devices. Hysteresis was also found to be greatly reduced in the fabricated devices. By increasing the thickness of the devices to 620 nm, the group was later able to fabricate devices with a champion efficiency of 17.6%. The V_{oc}, J_{sc}, and FF of the champion device were found to be 0.853 V, 28.5 mA/cm^2, and 72.5%, respectively. This improvement in V_{oc} was substantial to increasing the PCE. The authors implemented their newly designed perovskite layer into an all-perovskite four-terminal tandem solar cell and reported an overall PCE of ~21%.

High J_{sc} values have been found for more recent attempts at Pb–Sn perovskites, which have allowed higher PCEs. Recently, it has become the focus of researchers to minimize voltage loss in Pb–Sn perovskites. In 2018, Kapil et al. suggested using an n-type interface with spike structure between the perovskite and the electron transporting layers could reduce voltage loss [79]. The V_{oc} was able to be improved to 0.75 with a maximum efficiency of 17.6%. The J_{sc} for this highest performing device was 30.56 mA/cm^2 and the FF was 76%. Around the same time, Xu et al. also attempted to decrease the voltage loss in Pb–Sn PSCs by including an ultrathin bulk-heterojunction (BHJ) organic semiconductor (PBDB-T:ITIC) buffer layer at the interface between the HTL and the perovskite absorber layer [80]. This improved interface proved to increase the PCE of the devices by providing energy alignment in the device and passivating defects at the perovskite surface. High stability in comparison with past Pb–Sn was realized with this new layer and a champion device was fabricated with a PCE of 18.03%. The passivating effect due to the buffer layer allowed for greater carrier mobility and a high V_{oc}; the champion device was recorded at 0.86 V. The champion device also had a reported J_{sc} value of 27.92 mA/cm^2 and FF of 75.1%. Xu et al.'s PBDB-T:ITIC was shown to have effectively negated the voltage loss in Pb–Sn PSCs, allowing for further increase in PCE.

13.8 Basic Characterization of Mixed-Cation Hybrid Perovskite Solar Cells

The most important metrics in understanding photovoltaic performance are V_{oc}, J_{sc}, FF, and PCE. The PCE of a solar cell can typically be calculated as follows [81]:

$$\eta = \frac{V_{oc} J_{sc} \mathrm{FF}}{P_{in}} \tag{13.3}$$

P_{in} is the power being applied to the solar cell, which is 1 kW/m^2 and η is PCE [81]. Currently, the highest efficiency PSC fabricated has been reported at 25.2%, while the highest published is 23.56% (Figure 13.8f,g). Most high-efficiency cells have been small-area cells since solution processing allows for more control over deposition on a small area. The *J–V* curve of the perovskite, measured by various tools such as a DC voltage/current generator, allows for the determination of V_{oc}, J_{sc}, and ultimately PCE. The *J–V* curve of the highest published efficiency device can be seen in Figure 13.8f.

Due in part to the hysteretic behavior found in the current density vs. voltage measurements, careful characterization is necessary to fully understand PSCs and ensure that the results are reproducible [58]. It has been determined that ion migration is a primary cause of hysteresis, which is found in varying degrees for different perovskite compositions [83]. This hysteresis was first reported by Snaith et al. [84]. External quantum efficiency (EQE) is another metric for efficiency of PSCs. Figure 13.8g shows the EQE spectra for the current record efficiency PSC. As a measure of the electrical properties of a device, EQE is the ratio of number of collected electron–hole pairs to the number of impinging photons [85]. The purpose

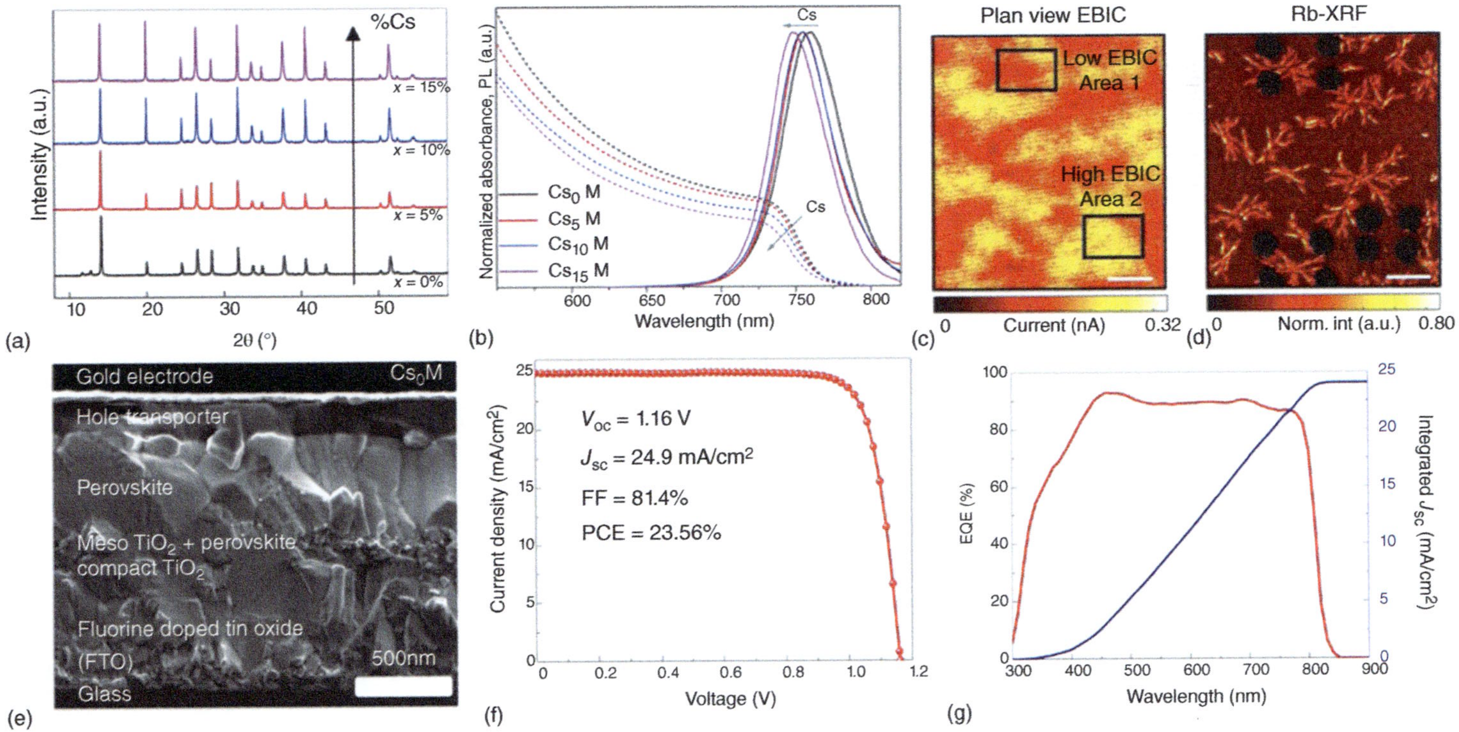

Figure 13.8 Characterization methods for analyzing PSCs. (a) XRD spectra for increasing Cs concentration in triple-cation perovskites [1]. (b) Absorption and PL spectra for increasing Cs concentration [1]. (c) Plan-view EBIC profile of a 5% Rb-I/Br PSC [82]. (d) Rb-XRF of a 5% Rb-I/Br PSC [82]. (e) SEM cross section [1]. (f) The J–V curve for the current highest efficiency PSC [66]. (g) EQE for the current highest efficiency PSC [66]. *Source*: (c, d) Correa-Baena et al. [82]. (e) Saliba et al. [1]. (f, g) Jiang et al. [66].

of EQE is to understand how efficiently a device is at absorbing photons and extracting charges. The EQE of a solar cell can be measured by applying an AC disturbance in light intensity and measuring the response of the cell and is usually measured at short circuit. J_{sc} can be determined from EQE spectra, and EQE is useful for approximating the quality of PSCs.

Several characterization techniques are necessary to fully understand the behavior and characteristics of perovskites. Among the most critical as the development of multi-cation perovskites increases is XRD, which is the primary method of analyzing the crystalline characteristics of PSCs [86]. It is possible to determine if the added cation has been successfully incorporated into the perovskite by looking at the diffraction peaks with and without the additional cation. XRD spectra for a triple-cation perovskite can be seen in Figure 13.8a, and it supports Cs incorporation in the perovskite lattice. There is also suppression of the photoinactive yellow phase and reduction in PbI_2 peak with increasing Cs concentration [13]. There is some limitation in analysis using XRD due to preferential orientation in perovskite crystals. This limitation is not present in using the advanced characterization techniques of grazing incident small-angle X-ray scatting (GISAXS) and grazing incident wide-angle X-ray scatting (GIWAXS) [86, 87]. These two methods can be used to study film morphology of perovskites and the relationship of precursor and perovskite crystal orientation/formation [87]. Information from these measurements allows for investigation into degradation and crystallization processes, which are necessary for understanding and predicting long-term device stability of a given precursor/composition combination.

As absorbers, PL and time-resolved photoluminescence (TRPL) are very helpful in understanding the quality of perovskite films. PL and TRPL are valuable in considering different combinations of cations for multi-cation PSCs. In general, the higher the PL intensity for a given perovskite, the less defects there are in the film. The efficiency of surface passivation in decreasing the amount of surface defects can be observed in this way. PL allows for a determination of the absorption peaks of perovskite over a certain range of wavelength and for the measurement of carrier recombination lifetimes [88]. In PL measurements, if the wavelength of the peak shifts to lower wavelengths, this is called a blue shift; likewise, when the wavelength of the peak shifts to higher wavelengths, this is known as a red shift. PL spectra for triple-cation perovskites with varying concentrations of Cs can be seen in Figure 13.8b. The PL and absorption spectra are blue shifted with increasing Cs concentration. TRPL is important for examining the carrier extraction at the interfaces between the charge transport layers and the perovskite layer. PL analysis can be used on complete and incomplete devices, allowing for a good tool for analyzing limitations of PSCs.

In order to get an understanding of what is happening at the ground boundaries of mixed-cation perovskites, scanning electron microscopy (SEM) can be used. In SEM imaging, different phases can be seen or a cross section of the device can be obtained. For example, the characteristic rods of PbI_2 can be differentiated from the more circular grains of perovskite in $APbI_3$ perovskites, where A is any combination of cations. SEM allows valuable insight into the grain size of the perovskite. Being able to determine grain size is important since it has shown to have a large impact on the overall efficiency of the device [86]. An example device cross section for a FAMA perovskite can be seen in Figure 13.8e. This device shows the traditional ETL of high-efficiency devices in TiO_2, for which there is a compact layer and mesoporous layer. Through SEM, it can be seen that there is a mixed mesoporous TiO_2/perovskite region present in the device. Atomic force microscopy (AFM) is also useful for analyzing surface characteristics of PSCs. Aside from grain size, these methods can be used to look at the surfaces of the perovskite and charge transport layers for defects and uniformity. Both methods also allow for the determination of layer uniformity and surface roughness. In considering long-term stability of PSCs, knowing the surface morphology of the cell is important [87]. The interfaces of the charge transport layers and the perovskite can give insight into the presence of deep traps or other defects within the cell which have an impact on the ability to efficiently extract charges from the perovskite absorber.

Nanoscale X-ray fluorescence (n-XRF) is a synchrotron-based advanced characterization technique that is very useful in determining the roles of new cations being introduced in new perovskite compositions [82]. n-XRF allows for the mapping of elemental composition of a sample, and for multi-cation perovskites, it allows for distinguishing between components of a cation cascade. Electron beam-induced current (EBIC) measurements allow for analysis of current variation in the sample. The EBIC profile in Figure 13.8c shows that there is some fluctuation in current in different parts of the sample. The sample being analyzed here is a FAMA perovskite with 5% RbI additive. When combined with n-XRF data for the same region of the sample (Figure 13.8d), it is possible to see the correlation between charge collection and localized Rb concentration. There is a negative correlation here since the current is lower in areas of high Rb content. This suggests that the agglomeration of Rb negatively impacts the performance of the PSC.

References

1 Saliba, M., Matsui, T., Seo, J. et al. (2016). Cesium-containing triple cation perovskite solar cells: improved stability, reproducibility and high efficiency. *Energy Environ. Sci.* 9 (6): 1989–1997.

2 Bekenstein, Y., Koscher, B.A., Eaton, S.W. et al. (2015). Highly luminescent colloidal nanoplates of perovskite cesium lead halide and their oriented assemblies. *J. Am. Chem. Soc.* 137 (51): 16008–16011.

3 Saliba, M. (2019). Polyelemental, multicomponent perovskite semiconductor libraries through combinatorial screening. *Adv. Energy Mater.* 9 (25): 1–8.

4 Gholipour, S., Morteza Ali, A., Correa-Baena, J.P. et al. (2017). Globularity-selected large molecules for a new generation of multication perovskites. *Adv. Mater.* 29 (38).

5 De Marco, N., Zhou, H., Chen, Q. et al. (2016). Guanidinium: a route to enhanced carrier lifetime and open-circuit voltage in hybrid perovskite solar cells. *Nano Lett.* 16 (2): 1009–1016.

6 Wang, Z., Lin, Q., Chmiel, F.P. et al. (2017). Efficient ambient-air-stable solar cells with 2D-3D heterostructured butylammonium-caesium-formamidinium lead halide perovskites. *Nat. Energy* 2 (9): 1–10.

7 Prochowicz, D., Tavakoli, M.M., Kalam, A. et al. (2019). Influence of A-site cations on the open-circuit voltage of efficient perovskite solar cells: a case of rubidium and guanidinium additives. *J. Mater. Chem. A* 7 (14): 8218–8225.

8 Chu, L., Ahmad, W., Liu, W. et al. (2019). Lead-free halide double perovskite materials: a new superstar toward green and stable optoelectronic applications. *Nano-Micro Lett.* 11 (1): 1–18.

9 Chen, M., Ju, M., Garces, H.F. et al. (2019). Highly stable and efficient all-inorganic lead-free perovskite solar cells with native-oxide passivation. *Nat. Commun.* 10 (1): 1–8.

10 Kojima, A., Teshima, K., Shirai, Y., and Miyasaka, T. (2009). Organometal halide perovskites as visible-light sensitizers for photovoltaic cells. *J. Am. Chem. Soc.* 131 (17): 6050–6051.

11 Yi, C., Luo, J., Meloni, S. et al. (2016). Entropic stabilization of mixed A-cation ABX_3 metal halide perovskites for high performance perovskite solar cells. *Energy Environ. Sci.* 9 (2): 656–662.

12 Li, Z., Yang, M., Park, J.S. et al. (2016). Stabilizing perovskite structures by tuning tolerance factor: formation of formamidinium and cesium lead iodide solid-state alloys. *Chem. Mater.* 28 (1): 284–292.

13 Lee, J.W., Kim, D.H., Kim, H.S. et al. (2015). Formamidinium and cesium hybridization for photo- and moisture-stable perovskite solar cell. *Adv. Energy Mater.* 5 (20).

14 Pellet, N., Gao, P., Gregori, G. et al. (2014). Mixed-organic-cation perovskite photovoltaics for enhanced solar-light harvesting. *Angew. Chem. Int. Ed.* 53 (12): 3151–3157.

15 Jeon, N.J., Noh, J.H., Yang, W.S. et al. (2015). Compositional engineering of perovskite materials for high-performance solar cells. *Nature* 517 (7535): 476–480.

16 Yang, W.S., Noh, J.H., Jeon, N.J. et al. (2015). High-performance photovoltaic perovskite layers fabricated through intramolecular exchange. *Science* 348 (6240): 1234–1237.

17 Choi, H., Jeong, J., Kim, H. et al. (2014). Cesium-doped methylammonium lead iodide perovskite light absorber for hybrid solar cells. *Nano Energy* 7: 80–85.

18 Ma, X.X. and Li, Z.S. (2018). Substituting Cs for MA on the surface of $MAPbI_3$ perovskite: a first-principles study. *Comput. Mater. Sci.* 150: 411–417.

19 Niu, G., Yu, H., Li, J. et al. (2016). Controlled orientation of perovskite films through mixed cations toward high performance perovskite solar cells. *Nano Energy* 27: 87–94.

20 Saliba, M., Masui, T., Domanski, K. et al. (2016). Incorporation of rubidium cations into perovskite solar cells improves photovoltaic performance. *Science* 354 (6309): 206–209.

21 Zhang, M., Yun, J.S., Ma, Q. et al. (2017). High-efficiency rubidium-incorporated perovskite solar cells by gas quenching. *ACS Energy Lett.* 2 (2): 438–444.

22 Abdi-Jalebi, M., Andaji-Garmaroudi, Z., Pearson, A.J. et al. (2018). Potassium-and rubidium-passivated alloyed perovskite films: optoelectronic properties and moisture stability. *ACS Energy Lett.* 3 (11): 2671–2678.

23 Kubicki, D.J., Prochowicz, D., Hofstetter, A. et al. (2018). Phase segregation in potassium-doped lead halide perovskites from ^{39}K solid-state NMR at 21.1 T. *J. Am. Chem. Soc.* 140 (23): 7232–7238.

24 Bu, T., Liu, X., Zhou, Y. et al. (2017). A novel quadruple-cation absorber for universal hysteresis elimination for high efficiency and stable perovskite solar cells. *Energy Environ. Sci.* 10 (12): 2509–2515.

25 Abdi-Jalebi, M., Andaji-Garmaroudi, Z., Cacovich, S. et al. (2018). Maximizing and stabilizing luminescence from halide perovskites with potassium passivation. *Nat. Publ. Gr.* 555 (7697): 497–501.

26 Ke, W. and Kanatzidis, M.G. (2019). Prospects for low-toxicity lead-free perovskite solar cells. *Nat. Commun.* 10 (1): 1–4.

27 Jokar, E., Chien, C.H., Tsai, C.M. et al. (2019). Robust tin-based perovskite solar cells with hybrid organic cations to attain efficiency approaching 10%. *Adv. Mater.* 31 (2): 1804835.

28 Prasanna, R., T. Leijtens, A. Gold-Parker et al. (2018). Compositional engineering of tin-lead halide perovskites for efficient and stable low band gap solar cells. *2018 IEEE 7th World Conference on Photovoltaic Energy Conversion, WCPEC 2018 – A Joint Conference of 45th IEEE PVSC, 28th PVSEC 34th EU PVSEC*, no. Figure 1, pp. 1718–1720.

29 Liu, X., Yang, Z., Chueh, C. et al. (2016). Improved efficiency and stability of Pb-Sn binary perovskite solar cells by Cs substitution. *J. Mater. Chem. A* 4 (46): 17939–17945.

30 Zhao, Z., Gu, F., Li, Y. et al. (2017). Mixed-organic-cation tin iodide for lead-free perovskite solar cells with an efficiency of 8.12%. *Adv. Sci.* 4 (11).

31 Chen, Z., Turedi, B., Alsalloum, A.Y. et al. (2019). Single-crystal $MAPbI_3$ perovskite solar cells exceeding 21% power conversion efficiency. *ACS Energy Lett.* 4 (6): 1258–1259.

32 Dang, H.X., Wang, K., Ghasemi, M. et al. (2019). Multi-cation synergy suppresses phase segregation in mixed-halide perovskites. *Joule* 3 (7): 1746–1764.

33 Liu, D., Li, Q., and Wu, K. (2019). Ethylammonium as an alternative cation for efficient perovskite solar cells from first-principles calculations. *RSC Adv.* 9 (13): 7356–7361.

34 Hsu, H.L., Chang, C.C., Chen, C.P. et al. (2015). High-performance and high-durability perovskite photovoltaic devices prepared using ethylammonium iodide as an additive. *J. Mater. Chem. A* 3 (17): 9271–9277.

35 Stoumpos, C.C., Cao, D.H., Clark, D.J. et al. (2016). Ruddlesden-popper hybrid lead iodide perovskite 2D homologous semiconductors. *Chem. Mater.* 28 (8): 2852–2867.

36 Fu, Y., Hautzinger, M.P., Luo, Z. et al. (2019). Incorporating large A cations into lead iodide perovskite cages: relaxed Goldschmidt tolerance factor and impact on exciton–phonon interaction. *ACS Cent. Sci.* 5 (8): 1377–1386.

37 Yadav, P., Dar, M.I., Arora, N. et al. (2017). The role of rubidium in multiple-cation-based high-efficiency perovskite solar cells. *Adv. Mater.* 29 (40): 1–7.

38 Solanki, A., Yadav, P., Turren-Cruz, S.H. et al. (2019). Cation influence on carrier dynamics in perovskite solar cells. *Nano Energy* 58: 604–611.

39 Hu, Y., Hutter, E.M., Rieder, P. et al. (2018). Understanding the role of cesium and rubidium additives in perovskite solar cells: trap states, charge transport, and recombination. *Adv. Energy Mater.* 8 (16): 1–11.

40 Hu, Y., Aygüler, M.F., Petrus, M.L. et al. (2017). Impact of rubidium and cesium cations on the moisture stability of multiple-cation mixed-halide perovskites. *ACS Energy Lett.* 2 (10): 2212–2218.

41 Nam, J.K., Chai, S.U., Cha, W. et al. (2017). Potassium incorporation for enhanced performance and stability of fully inorganic cesium lead halide perovskite solar cells. *Nano Lett.* 17 (3): 2028–2033.

42 Yang, Y., Wu, L., Hao, X. et al. (2019). Beneficial effects of potassium iodide incorporation on grain boundaries and interfaces of perovskite solar cells. *RSC Adv.* 9 (49): 28561–28568.

43 Im, J.H., Kim, H.S., and Park, N.G. (2014). Morphology-photovoltaic property correlation in perovskite solar cells: one-step versus two-step deposition of $CH_3NH_3PbI_3$. *APL Mater.* 2 (8).

44 Wu, J., Xu, X., Zhao, Y. et al. (2017). DMF as an additive in a two-step spin-coating method for 20% conversion efficiency in perovskite solar cells. *ACS Appl. Mater. Interfaces* 9 (32): 26937–26947.

45 Jeon, N.J., Noh, J.H., Kim, Y.C. et al. (2014). Solvent engineering for high-performance inorganic-organic hybrid perovskite solar cells. *Nat. Mater.* 13 (9): 897–903.

46 Bi, D., Tress, W., Dar, M.I. et al. (2016). Efficient luminescent solar cells based on tailored mixed-cation perovskites. *Sci. Adv.* 2 (1).

47 Konstantakou, M., Perganti, D., Falaras, P., and Stergiopoulos, T. (2017). Anti-solvent crystallization strategies for highly efficient perovskite solar cells. *Crystals* 7 (10): 1–21.

48 Leyden, M.R., Ono, L.K., Raga, S.R. et al. (2014). High performance perovskite solar cells by hybrid chemical vapor deposition. *J. Mater. Chem. A* 2 (44): 18742–18745.

49 Kim, J., Kim, H.P., Teridi, M.A.M. et al. (2016). Bandgap tuning of mixed organic cation utilizing chemical vapor deposition process. *Sci. Rep.* 6: 1–11.

50 Ball, J.M., Buizza, L., Sansom, H.C. et al. (2019). Dual-source coevaporation of low-bandgap $FA_{1-x}Cs_xSn_{1-y}PbyI_3$ perovskites for photovoltaics. *ACS Energy Lett.* 4 (11): 2748–2756.

51 Qiu, L., He, S., Jiang, Y. et al. (2019). Hybrid chemical vapor deposition enables scalable and stable Cs-FA mixed cation perovskite solar modules with a designated area of 91.8 cm^2 approaching 10% efficiency. *J. Mater. Chem. A* 7 (12): 6920–6929.

52 Sutherland, B.R., Hoogland, S., Adachi, M.M. et al. (2015). Perovskite thin films via atomic layer deposition. *Adv. Mater.* 27 (1): 53–58.

53 Popov, G., Mattinen, M., Hatanpää, T. et al. (2019). Atomic layer deposition of PbI_2 thin films. *Chem. Mater.* 31 (3): 1101–1109.

54 Urieta-Mora, J., García-Benito, I., Molina-Ontoria, A., and Martín, N. (2018). Hole transporting materials for perovskite solar cells: a chemical approach. *Chem. Soc. Rev.* 47 (23): 8541–8571.

55 Li, X., Bi, D., Yi, C. et al. (2016). A vacuum flash-assisted solution process for high-efficiency large-area perovskite solar cells. *Science* 353 (6294): 58–62.

56 Jiang, K., Wu, F., Zhang, G. et al. (2019). Efficient perovskite solar cells based on dopant-free spiro-OMeTAD processed with halogen-free green solvent. *Sol. RRL* 3 (5): 1900061.

57 Lee, I., Yun, J.H., Son, H.J., and Kim, T.S. (2017). Accelerated degradation due to weakened adhesion from Li-TFSI additives in perovskite solar cells. *ACS Appl. Mater. Interfaces* 9 (8): 7029–7035.

58 Luo, J., Xia, J., Yang, H. et al. (2018). Toward high-efficiency, hysteresis-less, stable perovskite solar cells: unusual doping of a hole-transporting material using a fluorine-containing hydrophobic Lewis acid. *Energy Environ. Sci.* 11 (8): 2035–2045.

59 Zheng, S., Wang, G., Liu, T. et al. (2019). Materials and structures for the electron transport layer of efficient and stable perovskite solar cells. *Sci. China Chem.* 62 (7): 800–809.

60 Zhao, Q., Wu, R., Zhang, Z. et al. (2019). Achieving efficient inverted planar perovskite solar cells with nondoped PTAA as a hole transport layer. *Org. Electron.* 71: 106–112.

61 Zhou, Y., Li, X., and Lin, H. (2019). To be higher and stronger – metal oxide electron transport materials for perovskite solar cells. *Small* 16 (15): 1902579, 1–21.

62 Möllmann, A., Gedamu, D., Vivo, P. et al. (2019). Highly compact TiO_2 films by spray pyrolysis and application in perovskite solar cells. *Adv. Eng. Mater.* 21 (4): 1–8.

63 Zhao, P., Han, M., Yin, W. et al. (2018). Insulated interlayer for efficient and photostable electron-transport-layer-free perovskite solar cells. *ACS Appl. Mater. Interfaces* 10 (12): 10132–10140.

64 Correa-Baena, J.P., Steier, L., Tress, W. et al. (2015). Highly efficient planar perovskite solar cells through band alignment engineering. *Energy Environ. Sci.* 8 (10): 2928–2934.

65 Liu, D., Li, S., Zhang, P. et al. (2017). Efficient planar heterojunction perovskite solar cells with Li-doped compact TiO_2 layer. *Nano Energy* 31: 462–468.

66 Jiang, Q., Zhao, Y., Zhang, X. et al. (2019). Surface passivation of perovskite film for efficient solar cells. *Nat. Photonics* 13 (7): 460–466.

67 Du, X., Qiu, R., Zou, T. et al. (2019). Enhanced uniformity and stability of Pb–Sn perovskite solar cells via Me_4NBr passivation. *Adv. Mater. Interfaces* 6 (14): 1–7.

68 Chen, B., Rudd, P.N., Yang, S. et al. (2019). Imperfections and their passivation in halide perovskite solar cells. *Chem. Soc. Rev.* 48 (14): 3842–3867.

69 Alharbi, E.A., Alyamani, A.Y., Kubicki, D.J. et al. (2019). Atomic-level passivation mechanism of ammonium salts enabling highly efficient perovskite solar cells. *Nat. Commun.* 10 (1): 1–9.

70 Noel, N.K., Habisreutinger, S.H., Pellaroque, A. et al. (2019). Interfacial charge-transfer doping of metal halide perovskites for high performance photovoltaics. *Energy Environ. Sci.* 12 (10): 3063–3073.

71 Mahmud, M.A., Duong, T., Yin, Y. et al. (2020). Double-sided surface passivation of 3D perovskite film for high-efficiency mixed-dimensional perovskite solar cells. *Adv. Funct. Mater.* 30 (7): 1–11.

72 Park, B., Kedem, N., Kulbak, M. et al. (2018). Understanding how excess lead iodide precursor improves halide perovskite solar cells performance. *Nat. Commun.* 9 (1): 1–8.

73 Wang, S., Zhu, Y., Wang, C., and Ma, R. (2019). Interface modification by a multifunctional ammonium salt for high performance and stable planar perovskite solar cells. *J. Mater. Chem. A* 7 (19): 11867–11876.

74 Zhuang, J., Mao, P., Luan, Y. et al. (2019). Interfacial passivation for perovskite solar cells: the effects of the functional group in phenethylammonium iodide. *ACS Energy Lett.* 4 (12): 2913–2921.

75 Ogomi, Y., Morita, A., Tsukamoto, S. et al. (2014). $CH_3NH_3SnxPb_{(1-x)}I_3$ perovskite solar cells covering up to 1060 nm. *J. Phys. Chem. Lett.* 5 (6): 1004–1011.

76 Hao, F., Stoumpos, C.C., Chang, R.P.H., and Kanatzidis, M.G. (2014). Anomalous band gap behavior in mixed Sn and Pb perovskites enables broadening of absorption spectrum in solar cells. *J. Am. Chem. Soc.* 136 (22): 8094–8099.

77 Zuo, F., Williams, S.T., Liang, P.W. et al. (2014). Binary-metal perovskites toward high-performance planar-heterojunction hybrid solar cells. *Adv. Mater.* 26 (37): 6454–6460.

78 Liao, W., Zhao, D., Yu, Y. et al. (2016). Fabrication of efficient low-bandgap perovskite solar cells by combining formamidinium tin iodide with methylammonium lead iodide. *J. Am. Chem. Soc.* 138 (38): 12360–12363.

79 Kapil, G., Ripolles, T.S., Hamada, K. et al. (2018). Highly efficient 17.6% tin-lead mixed perovskite solar cells realized through spike structure. *Nano Lett.* 18 (6): 3600–3607.

80 Xu, G., Bi, P., Wang, S. et al. (2018). Integrating ultrathin bulk-heterojunction organic semiconductor intermediary for high-performance low-bandgap perovskite solar cells with low energy loss. *Adv. Funct. Mater.* 28 (42): 1–8.

81 PVEducation. (2010) Solar Cell Efficiency. https://www.pveducation.org/pvcdrom/solar-cell-operation/solar-cell-efficiency (accessed: 18 March 2020)

82 Correa-Baena, J.P., Luo, Y., Brenner, T.M. et al. (2019). Homogenized halides and alkali cation segregation in alloyed organic-inorganic perovskites. *Science* 363 (6427): 627–631.

83 Kang, D.H. and Park, N.G. (2019). On the current–voltage hysteresis in perovskite solar cells: dependence on perovskite composition and methods to remove hysteresis. *Adv. Mater.* 31 (34): 1–23.

84 Snaith, H.J., Abate, A., Ball, J.M. et al. (2014). Anomalous hysteresis in perovskite solar cells. *J. Phys. Chem. Lett.* 5 (9): 1511–1515.

85 Ravishankar, S., Aranda, C., Boix, P.P. et al. (2018). Effects of frequency dependence of the external quantum efficiency of perovskite solar cells. *J. Phys. Chem. Lett.* 9 (11): 3099–3104.
86 Huang, Y.C., Tsao, C., Cho, Y. et al. (2015). Insight into evolution, processing and performance of multi-length-scale structures in planar heterojunction perovskite solar cells. *Sci. Rep.* 5: 1–11.
87 Schlipf, J. and Müller-Buschbaum, P. (2017). Structure of organometal halide perovskite films as determined with grazing-incidence X-ray scattering methods. *Adv. Energy Mater.* 7 (16): 1–10.
88 Baloch, A.A.B., Alharbi, F.H., Grancini, G. et al. (2018). Analysis of photocarrier dynamics at interfaces in perovskite solar cells by time-resolved photoluminescence. *J. Phys. Chem. C* 122 (47): 26805–26815.

14

Tin Halide Perovskite Solar Cells

Gaurav Kapil and Shuzi Hayase

The University of Electro-Communications, i-Powered Energy System Research Center (i-PERC), 1-5-1 Chofugaoka, Chofu, Tokyo 182-8585, Japan

14.1 Introduction

Over the last decade, we have witnessed a huge growth in the field of perovskite solar cells (PSCs). It was predicted at the beginning of the research that these solid-state hybrid solar cells based on organometal trihalide perovskite absorbers are the potential candidate for highly efficient low-cost solar power [1]. Undoubtedly, the predictions made at that time are now seemed to be realized with power conversion efficiencies near to 25% and are now on the verge of surpassing the single-crystalline Si-based solar cells. The idea of the implementation of perovskites as an absorber was motivated by the dye-sensitized solar cells (DSSCs). In a DSSC, there are three main components, a mesoporous n-type semiconductor TiO_2, a light-absorbing material (dye), and a redox electrolyte. The first demonstration of PSC in a peer-reviewed journal was done by Miyasaka and coworkers [2] in 2009, a perovskite $CH_3NH_3PbI_3$ was used as a light-absorber employing the conventional iodide/triiodide redox couple. Later, Park and coworkers [3] showed improvements in the efficiencies by exhibiting that these materials have a high absorption coefficient and can be adsorbed on thinner layers of TiO_2 compared to conventional DSSCs. Their results further indicated the need for thinner layers (<3.6 μm) to enhance efficiency. It was also reported that the redox electrolyte was responsible for the fast (within a few minutes) decompositions of $CH_3NH_3PbI_3$. To address this problem, collaborative efforts were done and an organic hole conductor 2,2′(7,7′)-tetrakis-[*N*,*N*-di-4-methoxyphenyl-amino]9,9′-spirobifluorene (spiro-OMeTAD) was incorporated leading to power conversion efficiencies more than 10% [4, 5]. The hole conductor is soluble in nonpolar organic solvents (such as chlorobenzene) which can form a thin dry film after evaporation of the solvent and can be deposited on the surface of perovskite (soluble in polar solvents such as dimethylformamide) film and therefore perovskite absorber layer cannot be dissolved by hole conductor, unlike redox electrolyte.

Hybrid Perovskite Solar Cells: Characteristics and Operation,
First Edition. Edited by Hiroyuki Fujiwara.

14.1.1 Device Structure and Operating Principle

The first PSC structure fabricated was in n-i-p configuration as shown in Figure 14.1a, where the mesoporous TiO_2 layer was used to provide sufficient internal surface area to adsorb perovskites to enable enough sunlight harvesting possible. Subsequently, with the ongoing rapid development in PSC research now, there are broadly two categories of device structures, n-i-p or p-i-n configuration, as shown in Figure 14.1a–c, wherein the nominally intrinsic (i) perovskite layer is sandwiched between an n-type electron transport layer (ETL) and a p-type hole transport layer (HTL). The perovskite layer is generally assumed to possess intrinsic nature; however, this cannot be strictly true, it depends on the position of the Fermi level that further complicates the intrinsic nature of perovskite. However, the perovskite layer usually shows less charge carrier concentrations compared to ETL and HTL, and therefore, it may have less impact on the built-in electric field of the solar cell, although this also might not always be the case. Figure 14.1d shows the general operating phenomenon involved in the PSCs [6]. The photoexcitation process (1) in the perovskite layer causes the formation of excitons and carriers which are then transferred to their respective transport layers (electron: e^- to ETL and hole: h^+ to HTL) (2 and 3). In some PSCs, such as $MAPbI_3$, carriers are generated directly by the photon harvesting. These processes are very fast and take place in the range of 10^{-12} to 10^{-9} seconds. It is worth mentioning that the processes 2 and 3 are governed by the built-in electric field as illustrated in Figure 14.1e, which further depends on the carrier concentration of ETL and HTL. Therefore, the processes 1, 2, and 3 are the desired processes for an efficient solar cell. The undesired processes marked as 4, 5, 6, and 7 are known as exciton annihilation. The processes 4, 5, and 6 are assigned to the nonradiative recombination occurring at the

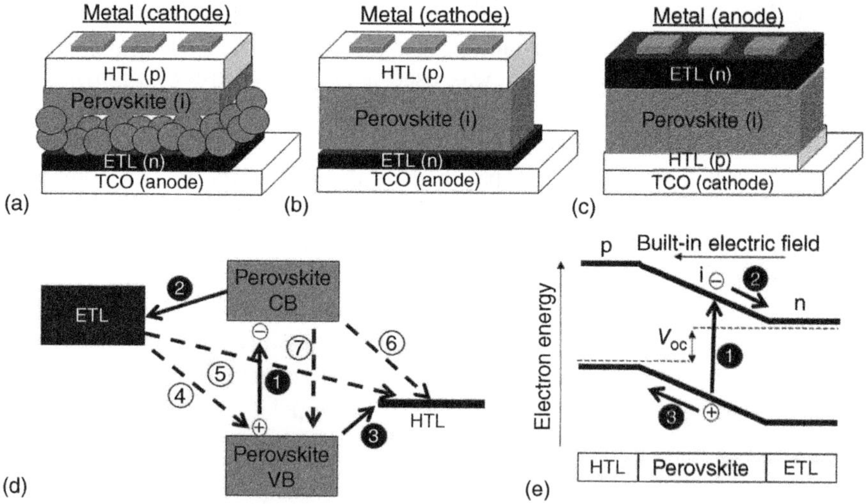

Figure 14.1 (a) n-i-p with mesoporous layer, (b) planar n-i-p, and (c) planar p-i-n device configurations used in PSC fabrication. (d) Different charge transfer processes in the PSC. (e) Approximate band alignment in the PSC.

interface, while process 7 is radiative recombination known as photoluminescence (PL). In the working state of the PSC, the annihilation processes compete with the charge extraction and thus expected to control the overall photovoltaic conversion efficiency.

14.1.2 Crystal Structure

"Perovskite" is the name given to the materials that exhibit the same crystal structure as calcium titanate, that is, ABX_3 (Figure 14.2a). There are hundreds of materials that have perovskite structure and show different properties such as thermoelectric, antiferromagnetic, piezoelectric, insulating, semiconducting, conducting, and superconducting. Conventionally, mostly researched perovskites are oxides and have been prepared by utilizing solid-state synthesis employing high-temperature sintering (>1300 K). Another kind of perovskites, halide perovskites are prepared by drying the precursor salt solutions relatively at a lower temperature, display interesting semiconducting properties, and are highly useful in printable electronic applications. Solar cells based on halide perovskites have attracted researchers all around the world in the last decade owing to their fascinating electronic properties such as tunable bandgap, high absorption coefficient, ambipolar charge transport, long charge carrier diffusion length. In a typical halide perovskite ABX_3, A is a monovalent cation such as Cs^+, CH_6N^+ (MA^+), $C_2H_5N_2^+$ (FA^+), B is bivalent cation such as Pb^{2+}, Sn^{2+}, Ge^{2+}, and X is halide anion such as Cl^-, Br^-, I^-. Therefore, by keeping either of them or by sharing the different elements at A, B, or X position, the different bandgap of perovskites can be obtained which further can be used to harvest the sunlight in different wavelength regions. In the case of perovskites, a geometrical parameter called as Goldschmidt tolerance factor, $\tau = (r_A + r_B)/\{\sqrt{2}(r_B + r_X)\}$ where r_A, r_B, and r_X are the radii of the corresponding ions, indicates for the stability and formability of

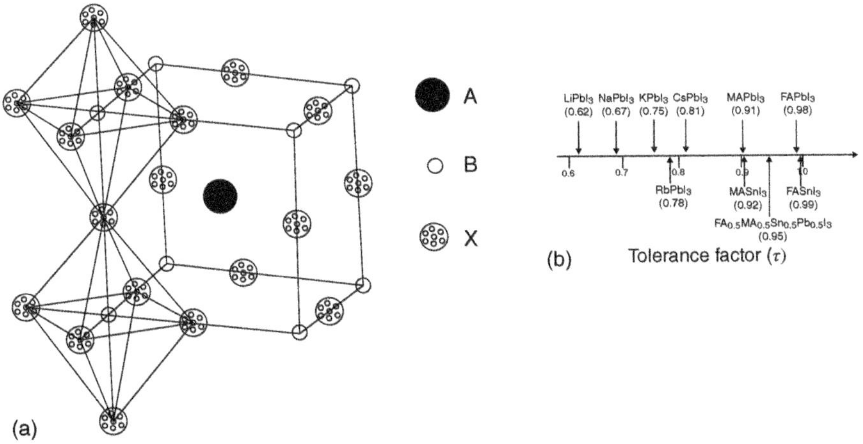

Figure 14.2 (a) A perovskite ABX_3 crystal structure. (b) Tolerance factor of some perovskite configurations.

perovskites structure. Empirically, a tolerance factor of 1 result in an ideal cubic structure, and structures showing values less than 1 tend to be distorted into a low symmetry. In general, the 3D perovskite structure is favored for the values of τ between 0.8 and 1. It is important to note that many perovskite structures fulfill the condition for τ but still exhibit the non-perovskite phase. Hence, there is another necessary geometrical parameter that needs to be satisfied is known as an octahedral factor, $\mu = r_B/r_X$, which determines whether the B cation will sit in the octahedral formed by X anions. To satisfy the condition, the preferred value should be between 0.4 and 0.9 [7]. However, it should be noted that τ and μ are not the only factors to determine the formability and stability of a perovskite structure, there are various nongeometric factors such as phase stability, moisture stability, oxygen stability, and many other. Overall, it is correct to say that both the tolerance factor and octahedral factor are necessary but not sufficient conditions for ABX_3 halide perovskite formability. But still, the geometrical factors can provide useful information for selecting the appropriate A, B, and X ions to guess the stable perovskite structure. Figure 14.2b illustrates some of the perovskite structures along with their tolerance factor calculated from the ionic radii of Li^+(76 pm), Na^+(102 pm), K^+(138 pm), Rb^+(152 pm), Cs^+(167 pm), MA^+(217 pm), FA^+(253 pm), Pb^{2+}(119 pm), Sn^{2+}(118 pm), I^-(220 pm). The octahedral factor for all the perovskites is between 0.4 and 0.9. $LiPbI_3$(0.62), $NaPbI_3$(0.67), $KPbI_3$(0.75), and $RbPbI_3$(0.78) cannot form the perovskite structure, whereas $CsPbI_3$(0.81), $MAPbI_3$(0.91), $FAPbI_3$(0.98), $FASnI_3$(0.99) appear to have perovskite structure. As mentioned earlier, geometrical stability cannot guarantee the formability as well, for example, in the case of $FAPbI_3$ phase stability at room temperature, and for $FASnI_3$ oxygen, stability is the concern, although both the structures are geometrically more stable. In this regard, to attain phase stability, the alloying of different atoms has been reported to be one of the possible solutions to this kind of issue while fulfilling the condition for τ and μ [8, 9]. Other than phase stability, the alloying of the atoms also improves the stability against oxygen as reported earlier [10].

14.2 Tin Perovskite Solar Cells

The highest efficiency of the PSC that has been published so far is 24.66%, adopting a bandgap of around 1.48 eV [11]. As we can see the majority of the high-efficiency PSCs are fabricated by incorporating the Pb, which is a highly toxic metal. Besides, the optical bandgaps >1.45 eV have been reported in almost all kinds of efficient perovskite structures, where Pb only sits at B position. According to the Shockley–Queisser (SQ) limit, a bandgap of 1.34 eV is the ideal value for a single-junction solar cell to achieve the maximum theoretical efficiency. Therefore, the bandgap of only Pb containing ("only Pb" refers to the positioning at B only) perovskites is far from the ideal value. Fortunately, the solution to the toxicity as well as bandgap can be answered by a metal that belongs to the same group 14 of the periodic table and located just above Pb, i.e. tin (Sn). There are now many reports showing the PSCs fabricated by $ASnI_3$ perovskites (with A = Cs, MA, and FA) exhibiting the bandgaps of the order of 1.2–1.4 eV [12, 13]. Surprisingly, similar, or even lower, bandgaps could be obtained by sharing the B position between Sn and Pb due to anomalous bandgap effect, notably increasing the air stability of the compounds and the corresponding PSCs along with the enhancement in the efficiencies. Also, the solar cells based on Sn or mixed Sn–Pb could reach up to the approximately 33% power conversion efficiency (PCE), apart from the efficient single-junction solar cells these can be implemented in tandem solar cell technology. Tandem solar cells based on perovskites are now highly researched as they have the potential to reach a higher theoretical maximum of more than 40%, therefore Sn and Sn–Pb mixed solar cells (known as narrow bandgap perovskites) can be coupled with wideband perovskites to attain higher efficiencies. In short, the ideal bandgap of Sn perovskites, less toxicity, and the possibility of utilizing them in the tandem solar cell as a narrow bandgap cell gives them strong advantages compared to Pb counterparts.

14.2.1 Intrinsic Properties

The initial work on Sn-based perovskite was started in the mid-1990s by Mitzi and coworkers, they demonstrated the existence of inherent high hole mobilities [14] in the Sn-based systems. In the work, a field-effect transistor was fabricated using layered tin perovskites which were showing the formation of p-type transistor characteristics, it was further shown that by increasing the thickness of perovskite layers a 3D $MASnI_3$ could be obtained which exhibited metal-like conductivity. It was evaluated by using extended Huckel band structure calculation using a basis set consisting of Sn 5s and 5p and I 5s and 5p valence orbitals that the metallic behavior in tin halide systems arises due to the dispersion of the Sn 5s band (hybridized with I 5p) along the ⟨111⟩ direction of the cubic Brillouin zone. As a result, the conduction band in these systems consisting of Sn 5p overlaps with Sn 5s bands near the R point ($[\frac{1}{2}, \frac{1}{2}, \frac{1}{2}]2\pi/a$), with Fermi energy level between the two bands. The high conductivity has been noticed in many of the research works, which is attributed to the high carrier concentrations and high charge carrier mobility within these Sn halide perovskites [13, 15–17]. Table 14.1 illustrates the intrinsic carrier concentrations in some of the halide perovskites along with the commercially used c-Si [19] and GaAs [18] photovoltaic materials for the parallel analysis. In the case of primitive Pb perovskite thin films such as $MAPbI_3$, carrier densities are usually reported in the range of 10^{13}–10^{15}, whereas the values are 2–5 orders higher in the case of Sn-based perovskites. It can be observed that the values of carrier densities in halide perovskites are quite higher than the commercially used photovoltaics, for c-Si values are in the order of $10^9\ cm^{-3}$ and for GaAs, the values are around $10^6\ cm^{-3}$, which hints for the possibility of higher charge carrier recombination in the halide perovskites and more specifically the recombination within Sn perovskites could be high. As mentioned

Table 14.1 Physical properties of different photovoltaic materials.

Material	Carrier density (cm^{-3})	Mobility (e^-, h^+) ($cm^2/(V\,s)$)	Lifetime, τ (ns)	Diffusion length, LD (μm)	References
GaAs	1.79×10^6	8500, 400	1000	148.26, 32.16	[18]
c-Si	1.45×10^9	1500, 450	500×10^3	1392.66, 762.79	[19]
$MAPbI_3$	10^{13}–10^{15}	1.68, 2.16	2000	2.89, 3.28	[13]
$CsSnI_3$	10^{14}–10^{19}	—, 585	0.05–19.7	—, 0.86–5.45	[13]
$MASnI_3$	10^{14}–10^{19}	—, 1.6–170	0.2–10.3	—, 0.03–0.55	[13]
$FASnI_3$	5.83×10^{17}	—, 22	0.2–6.8	—, 0.104–0.620	[15]
$FA_{0.75}MA_{0.25}SnI_3$	9.11×10^{17}	—, 12.49	3.46	—, 0.350	[16]
$FA_{0.75}MA_{0.25}Sn_{0.95}Ge_{0.05}I_3$	1.69×10^{15}	—, 98.27	5.04	—, 1.186	[16]
$FA_{0.45}Cs_{0.05}MA_{0.5}Sn_{0.5}Pb_{0.5}I_3$	1×10^{16}	0.65, >3.25	140 ± 9	0.49 ± 0.10, $>1.08 \pm 0.10$	[20]
0.03 mol% CdI_2 + $FA_{0.45}Cs_{0.05}MA_{0.5}Sn_{0.5}Pb_{0.5}I_3$	4.2×10^{15}	1.98, >3.86	1450 ± 160	2.72 ± 0.15, $>3.80 \pm 0.20$	[20]

earlier [14] that the intrinsic p-type behavior is dominant in Sn halide perovskites, the similar observations are reported by many other research groups when they have evaluated the charge carrier mobilities and found that the majority carriers are holes, whereas Pb halide perovskite exhibits ambipolar charge transport.

14.2.2 Carrier Lifetime and Diffusion Length

The charge carrier properties such as doping densities, carrier lifetimes, mobilities, and diffusion lengths are crucial in determining solar cell performances. It can be noticed from Table 14.1 that the halide perovskite show relatively higher values of carrier concentration than commercially used c-Si and GaAs photovoltaic materials, which can increase the possibility of bimolecular carrier recombination possibility (Langevin theory) and hence low photogeneration current. In general, the charge-carrier recombination involves the monomolecular and higher order processes represented by the following rate equation,

$$dn/dt = G - k_1 n - k_2 n^2 - k_3 n^3 = G - nR_T(n) \tag{14.1}$$

where G is the charge-density generation rate, k_1 the monomolecular charge-recombination rate, k_2 the bimolecular electron–hole recombination rate constant, and k_3 the Auger recombination rate constant. The Langevin model assumes the recombination will occur once an electron and a hole bounded by coulombic attraction in a radius (r_c) which is assumed to be larger than their mean free path

(l_c). Therefore, proposes a relation between the bimolecular recombination rate constant k_2 and the charge-carrier mobility μ,

$$k_2/\mu = e/(\varepsilon_0 \varepsilon_r) \tag{14.2}$$

The Pb halide perovskites exhibit this ratio of $(k_2/\mu)_{evaluated}$ four orders of magnitude less than predicted from the Langevin theory, $(k_2/\mu)_{predicted}$ [21–23]. The diffusion length of 100 nm was predicted for $MAPbI_{3-x}Cl_x$ calculated according to the Langevin model which is less than the diffusion length of 3 μm evaluated experimentally. Such effects are likely to arise from the spatial separation of opposite charge carriers within the Pb halide perovskite crystals. In other words, the carrier lifetime is longer which is comparable to GaAs as shown in Table 14.1. The low value of the ratio of bimolecular recombination rate constant to the charge carrier mobility was evaluated in the case of $MASnI_3$ as well [24]; however, experimentally low values of carrier lifetime were observed, also listed in Table 14.1. This deviation from the behavior can be explained by the high value of monomolecular charge recombination rate (k_1), a value of $8 \times 10^9\ s^{-1}$ has been reported, which is almost 3 orders of magnitude higher than the values typically found for $MAPbI_3$. It is important to mention that monomolecular recombination originates from trap-assisted recombination, and in the case of Sn perovskites Sn^{2+} undergoes self-oxidation to form energetically more stable Sn^{4+} which increases the p-type doping densities within these Sn perovskites in the range of 10^{16}–$10^{18}\ cm^{-3}$, acting as trapping centers. Therefore, this high self-doping of all the Sn halide perovskites (Table 14.1) in the presence of Sn^{4+} leads to fast recombination and hence short carrier lifetimes and diffusion lengths. At this stage of discussion, it is worth mentioning that the reduction of these trap centers can increase the diffusion length as predicted by Noel et al. also [24]. For a doping density of $10^{18}\ cm^{-3}$, the diffusion length of 30 nm was calculated which was increased to 1 μm with the decrease in doping density to $10^{15}\ cm^{-3}$. Recently, also shown that $FA_{0.75}MA_{0.25}SnI_3$ ($9.11 \times 10^{17}\ cm^{-3}$) perovskite thin films when used for solar cells showed a diffusion length of 350 nm which was increased in the case of $FA_{0.75}MA_{0.25}Sn_{0.95}Ge_{0.05}I_3$ ($1.69 \times 10^{15}\ cm^{-3}$) to 1.186 μm [16].

14.3 The Status of Sn Perovskite Solar Cells

Tin halide PSCs have seen a rapid increase in research interest owing to their less toxic nature compared to Pb counterparts. However, they have broadly two main bottlenecks that need to be overcome for the development of efficient and stable devices. The first one is the rapid oxidation of Sn^{2+} into Sn^{4+}, without encapsulation and reducing additives tin halide perovskite decomposes to tin oxide under ambient conditions. The second one is the fast crystallization of the tin halide perovskite that leads to the poor morphology of thin film which is a rough film containing a lot of pinholes and small flower-like grains.

Therefore, it is crucial to develop strategies that can address the issues mentioned above for tin halide perovskites to improve the performance and stability of solar

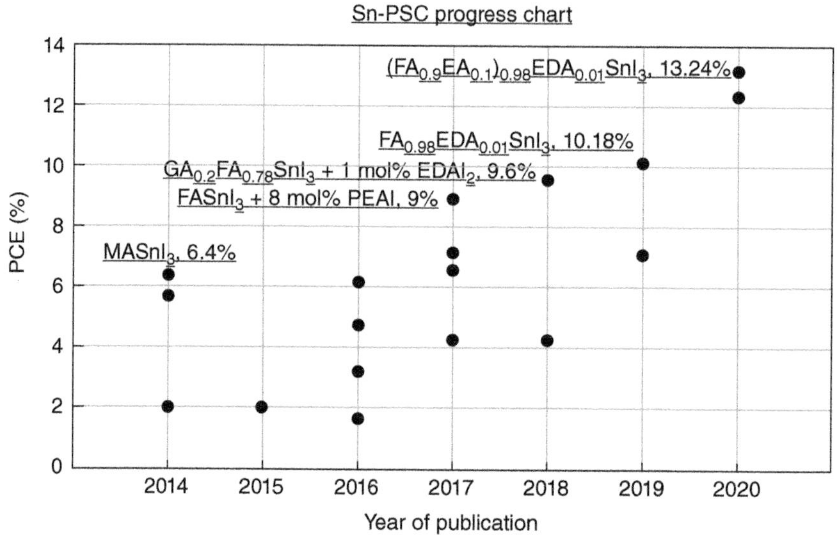

Figure 14.3 Progress in the efficiency of Sn halide perovskites.

cells. The status of the Sn-based PSCs has been summarized in Figure 14.3 and Table 14.2.

14.3.1 Different Type of Sn Perovskite Solar Cells

14.3.1.1 $CsSnI_3$

$CsSnI_3$ is a 3D orthorhombic perovskite which is a p-type conductor with a bandgap of 1.3 eV (Figure 14.4a). It is one of the initial perovskite materials known to be implemented in the DSSCs. It was used as a hole transport material and not as an absorber layer because of its dominant p-type semiconductor behavior [41]. However, because of its favorable bandgap, high absorption coefficient (10^{-4} cm^{-1}), and low exciton binding energy (18 meV), it is a potential candidate for the lead-free absorber layer. Also, it is correct to say that it was the initially known perovskites that were incorporated as the absorber layer in PSCs. Therefore, $CsSnI_3$ was the first "eco-friendly" materials used as an absorber in a solar cell configuration by Chen et al. [42]. The absorber possesses a low exciton binding energy (18 meV) implying that most of the photogenerated carriers are free and need small driving energy to separate the excitons toward the contacts. $CsSnI_3$ was deposited by evaporation on the sputtered Ti (100 nm)/Au (20 nm) metal contacts and on the top was contacted by a transparent indium tin oxide (ITO) electrode. It was reported that the photovoltaic effect was mainly driven by the built-in field between Au/$CsSnI_3$ junction leading to an efficiency of 0.9% with a V_{oc} of 0.42 V. $CsSnI_3$ can absorb the solar light up to the infra-red wavelength region because of its narrow bandgap and thus has higher theoretical photocurrent density (34 mA/cm^2) compared to the known Pb perovskite materials. However, $CsSnI_3$ is prone to form intrinsic defects associated with Sn-cation vacancies resulting in metallic conductivity. Therefore,

Table 14.2 Photovoltaic parameters of different kind of Sn-based PSCs including the highest reported values.

Perovskite	Bandgap (eV)	Structure type	J_{sc} (mA/cm^2)	V_{oc} (V)	FF (%)	PCE (%)	Publication year	References
$MASnI_3$	1.23	n-i-p	16.8	0.88	42	6.4	2014	[24]
$MASnIBr_2$	1.75	n-i-p	11.74	0.82	57	5.73	2014	[25]
$MASnI_3$ + 15%$C_2H_{10}N_2I_2$	1.3	n-i-p	24.28	0.428	63.72	6.63	2017	[26]
$MASnI_3$	1.3	n-i-p	22.91	0.486	64	7.13	2019	[27]
$MASnI_3$	1.3	p-i-n	12.1	0.377	36.6	1.7	2016	[28]
$MASnI_3$	1.3	p-i-n	17.8	0.595	29.8	3.2	2017	[29]
$CsSnI_3$	1.3	n-i-p	22.7	0.24	37	2.02	2014	[30]
$CsSnI_3$	1.3	n-i-p	18.5	0.44	52.9	4.3	2018	[31]
$CsSn_{0.5}Ge_{0.5}I_3$	1.5	p-i-n	18.61	0.63	60.6	7.11	2019	[32]
$FASnI_3$	1.41	n-i-p	24.45	0.238	36	2.1	2015	[33]
$FASnI_3$	1.4	n-i-p	23.7	0.32	63	4.8	2016	[34]
$FASnI_3$ + 15 mol%$EDAI_2$	1.41	n-i-p	22.54	0.48	65.96	7.14	2017	[35]
$FASnI_3$ + 15 mol%$EDAI_2$	1.41	n-i-p	22.54	0.46	69.74	7.23	2017	[35]
$FASnI_3$	1.41	p-i-n	22.07	0.47	60.67	6.22	2016	[36]
$FASnI_3$ + 8 mol%PEAI	1.41	p-i-n	24.1	0.525	71	9	2018	[37]
$GA_{0.2}FA_{0.78}SnI_3$ + 1 mol% $EDAI_2$	1.5	p-i-n	21.2	0.619	72.9	9.6	2018	[38]
$FA_{0.98}EDA_{0.01}SnI_3$	1.38	p-i-n	23.09	0.6	73	10.18	2019	[15]
$(FA_{0.9}EA_{0.1})_{0.98}EDA_{0.01}SnI_3$	1.42	p-i-n	20.32	0.84	78	13.24	2020	[39]
$PEA_{0.15}FA_{0.85}SnI_3$ + 5 mol% NH_4SCN	1.39	p-i-n	17.4	0.94	75	12.4	2020	[40]

in 2014, Kumar et al. [30] explored that the addition of SnF_2 is very effective in controlling the metallic conductivity of these materials. Sn^{2+} vacancies can be monitored through the measurement of hole carrier density that can be measured by Hall measurement. As shown in Figure 14.5a, the pristine $CsSnI_3$ films exhibited a very high carrier density $\sim 10^{19}$ cm^{-3} that decreases with the concentration of SnF_2. The cross-sectional image of the PSC fabricated in the work is shown in Figure 14.4c, a PCE of 3×10^{-4}% with pristine $CsSnI_3$ was dramatically improved to 2.02% in the case of 20%SnF_2-$CsSnI_3$ based absorber PSC (Figure 14.5b). This kind

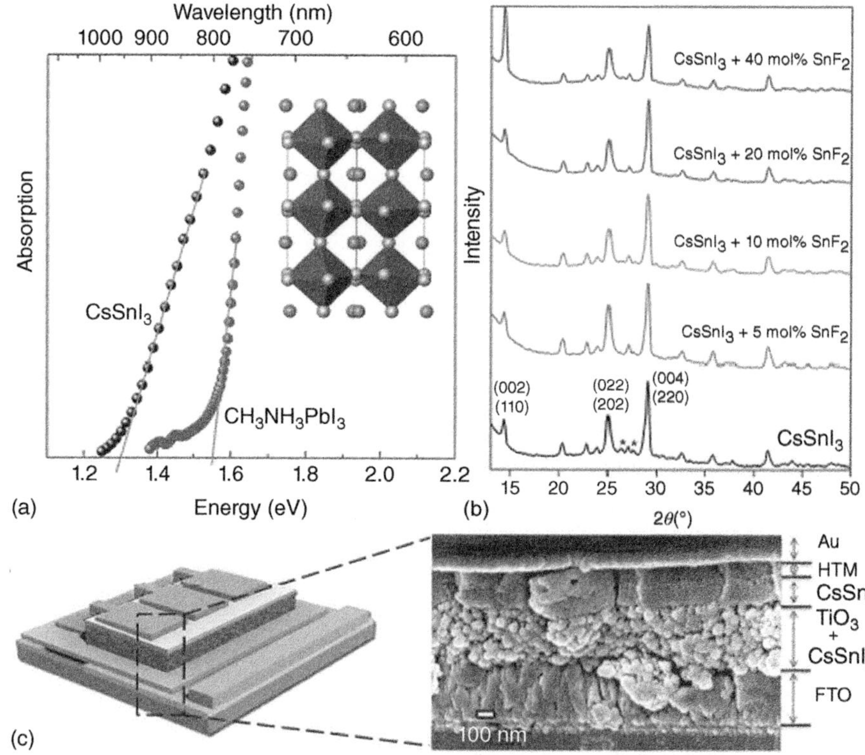

Figure 14.4 (a) Absorption spectra of $CsSnI_3$ and $MAPbI_3$. (b) XRD spectra of the $CsSnI_3$ thin films with the addition of different concentrations of SnF_2. (c) Cross-sectional scanning electron microscope (SEM) image of the PSC fabricated. *Source*: Reproduced with permission from Kumar et al. [30]. Copyright 2014, Wiley-VCH.

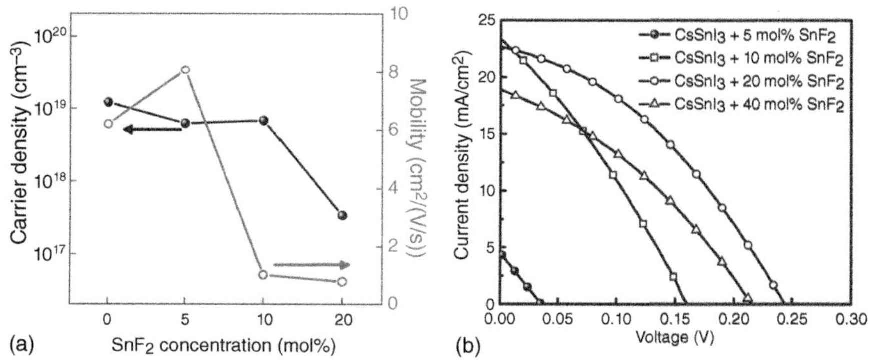

Figure 14.5 (a) Change in carrier density with the addition of SnF_2 in $CsSnI_3$ perovskite thin films. (b) IV curve of the devices prepared. *Source*: Kumar et al. [30].

of approach of the addition of SnF_2 to realize the improved photovoltaic behavior in Sn halide perovskite is now most effective and now used in almost every kind of Sn halide perovskite to be implemented in PSCs. The X-ray diffraction (XRD) patterns of pristine $CsSnI_3$ and with SnF_2 (5, 10, 20, and 40 mol%) are shown in Figure 14.4b, all the reflections observed in XRD pattern show the orthorhombic (*Pnam*) structure. The addition of SnF_2 did not change the lattice parameters of the films and unwanted phases confirming that F^- is not substituting I^- in the perovskite which could be due to the much smaller radii of F (1.33 Å) compared to I (2.20 Å). However, the addition of SnF_2 eliminates the weak unidentified reflections attributed to the yellow nonperovskite polymorph of $CsSnI_3$. In some reports, other SnX_2 (X = Cl, Br) has also been reported in removing the Sn vacancies [31].

14.3.1.2 $MASnI_3$

The first attempt at using $MASnI_3$ ($CH_3NH_3SnI_3$) as the light-absorbing material to fabricate the solution-processed solid-state photovoltaic device was done by Hao et al. [25].

$MASnI_3$ is a direct-gap semiconductor (1.3 eV) with an absorption onset of 950 nm, adopts the perovskite structure type, crystallizing in the pseudocubic space group *P4mm* at ambient conditions. It has a high symmetry phase (α-phase) at room temperature, unlike $CH_3NH_3PbI_3$ that has a lower symmetry (β-phase). The conductivity of these compounds is $\sim 5 \times 10^{-2}$ S/cm corresponding to a Seebeck coefficient of ~ -60 μV/K (n-type). The carrier concentrations reported are $\sim 1 \times 10^{14}$ cm^{-3} with high electron mobility on the order of ~ 2000 $cm^2/(V\,s)$ which is comparable or even superior to most traditional semiconductors such as Si, $CuInSe_2$, and CdTe. However, these materials also show inherent high doping density $\sim 1 \times 10^{19}$ cm^{-3} leading to a strong p-type character and a metallic behavior due to the presence of Sn^{4+}. In the work, different perovskites such as $CH_3NH_3SnI_{3-x}Br_x$ were demonstrated. It can be observed from the UV absorption that bandgap can be varied by changing the ratio of I/Br anions and the bandgaps in between 1.3 and 2.15 eV are obtained. These materials were used in PSC using regular n-i-p structure, PCEs of 5.23% 5.48%, 5.73%, and 4.27% were reported for the first time in case of $CH_3NH_3SnI_3$, $CH_3NH_3SnI_2Br$, $CH_3NH_3SnIBr_2$, and $CH_3NH_3SnBr_3$ respectively.

Another important initial research work in the $MASnI_3$ was reported by Noel et al. [24]. According to the research, there was a broad spread in the performance of $MASnI_3$ possibly due to fast oxidation issues especially when devices were measured in the ambient atmosphere compared to the Pb counterparts where high reproducibility can be observed. In the work, a PCE of 6.4% with remarkably high V_{oc} of 0.88 V under 1-sun illumination was reported. The fundamental energy loss in a solar cell is the difference between bandgap and an open-circuit voltage obtained, which is 0.35 eV for c-Si and 0.27 eV for GaAs. And in the case of $MASnI_3$ (1.23 eV), this loss is 0.35 eV which is surprisingly high despite the low diffusion length (30 nm) estimated in the work. But, the low value of V_{oc} loss obtained suggesting that when the full potential of $MASnI_3$ is exploited, the PCEs could reach near to c-Si and GaAs. Also, if the background concentration of holes can be decreased then the diffusion length could approach 1 μm which can lead to the use of these materials in the planar configuration as well. According to the research work, the pivotal issue is the stabilization of these materials by suppressing the oxidation of Sn within the crystal, thus reducing the doping levels of the material, and enabling long term stable solar cell operation. Moreover, if the issues are solved, then we may see the performance of Sn perovskites surpassing the PCEs of Pb-based PSCs in the future with less toxicity concern.

14.3.1.3 $FASnI_3$

The room temperature resistivity and Hall effect measurements have shown that $FASnI_3$ has large resistivity (11.8 Ωcm) and lower mobility (103 $cm^2/(V\,s)$) compared to MA^+ (0.5 Ωcm, 2320 $cm^2/(V\,s)$) and Cs^+ (0.9 Ωcm, 585 $cm^2/(V\,s)$) analogues [13]. It has been shown that the antibonding coupling between Sn-s and I-p is weaker in $FASnI_3$ than in $MASnI_3$ owing to the larger ionic radius of FA^+, resulting in higher formation energies of Sn vacancies in $FASnI_3$, hence lower doping density and carrier concentrations [43]. $FASnI_3$ shows a bandgap of 1.41 eV which is higher compared to $MASnI_3$ (1.23–1.3 eV) and $CsSnI_3$(1.3 eV). According to the XRD data, $FASnI_3$ exists in a single-phase unlike its Pb counterpart $FAPbI_3$ which exists in two phases black and yellow (the yellow phase restricts the application in solar cells). Therefore, $FASnI_3$ exhibits the orthorhombic (*Amm*2) crystal structure with unit cell dimensions of $a = 6.3096(15)$, $b = 8.9298(21)$, and $c = 9.0622(35)$. The first PSC based on $FASnI_3$ absorber was demonstrated by Koh et al. [33] with a PCE of 2.10%, a regular n-i-p structure utilizing mesoporous TiO_2 was fabricated. Another initial work utilizing $FASnI_3$ was reported by Lee et al. [34], they demonstrated that rapid crystallization of perovskite and phase segregation due to excess use of SnF_2 could be the reason for the poor performance of $FASnI_3$-based PSCs. The morphology of the $FASnI_3$ thin film was crucial in determining the overall PCE of the device. To deposit dense and uniform films, the antisolvent dripping method was used. At first, SnI_2(1 M), FAI(1 M), and SnF_2(0.1 M) were dissolved in *N*,*N*-dimethylformamide (DMF) and spin-coated on the top of mesoporous TiO_2 followed by heating at 60 °C for 10 minutes leading to poor morphology and coverage as shown in Figure 14.6a. The same morphology was obtained even with the use of antisolvent dripping (toluene [TL] was dropped on the top of the spinning substrate covered with a solution containing SnI_2(1 M), FAI(1 M) and SnF_2(0.1 M) dissolved in DMF) (Figure 14.6b). The uniformity and coverage of the $FASnI_3$ surface layer were significantly improved when a mixed solvent of DMF/dimethyl sulfoxide (DMSO) (4 : 1 volume ratio) was used (Figure 14.6c). It has been reported that the addition of DMSO forms the SnI2·3DMSO intermediate phase which retards the crystallization of FAI and SnI_2 by holding the component during spin coating, thereby producing a film with much better quality [44]. However, it was noticed that there was the formation of plate-like aggregates on the film resulting from excess SnF_2 which can be removed by the introduction of pyrazine into the solvent-engineered solution leading to a very smooth and dense $FASnI_3$ perovskite layer without any plate-like aggregates (Figure 14.6d). The efforts in the research finally led to the PCE of 4.8% with device stability of more than 100 days, maintaining 98% of the initial efficiency. In the next stage of development of $FASnI_3$-based PSCs it was observed that in the

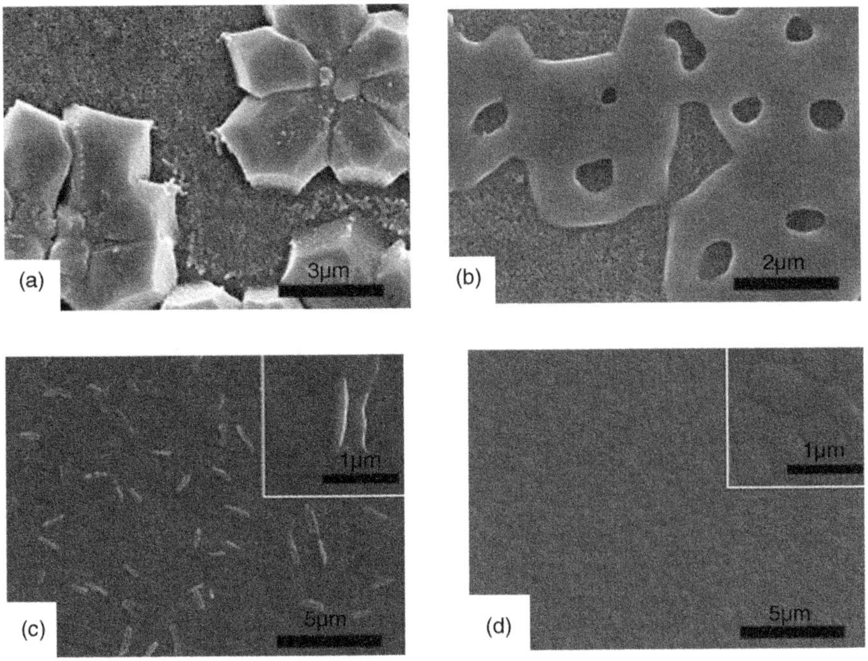

Figure 14.6 SEM image of $FASnI_3$ thin films prepared by spin coating, (a) precursor solution (DMF) containing SnF_2 without using antisolvent, (b) precursor solution (DMF) containing SnF_2 using toluene antisolvent, (c) precursor solution (DMF:DMSO = 4 : 1) containing SnF_2 using Toluene antisolvent, and (d) precursor solution (DMF:DMSO = 4 : 1) containing SnF_2 and pyrazine using toluene antisolvent. *Source*: Reproduced with permission from Hao et al. [44]. Copyright 2016, American Chemical Society.

devices with n-i-p configuration, spiro-OMeTAD is used as HTL which contains lithium (Li) and/or cobalt (Co) salts that damage the Sn-based perovskite films and lead to poor performances. Additionally, it was observed that surface traps can be increased due to the formation of Ti-O-Sn-I on the TiO_2 surface [45]. Therefore, to circumvent these issues, Liao et al. [36] demonstrated an inverted p-i-n PSC and device was fabricated on PEDOT: PSS as bottom HTL with C60 (Fullerene) as ETL (Figure 14.7), hence the corrosive elements such as TiO_2 and spiro-OMeTAD for $FASnI_3$ were avoided and in the work, an improved PCE of 6.22% was obtained.

14.3.1.4 $FA_xMA_{1-x}SnI_3$

In an ABX_3 perovskite, the sharing of A, B, or X by different atoms is known as composition engineering. The composition engineering has proven to be effective in enhancing the performances of the Pb-based solar cells [8, 9]. The same strategy was first implemented for Sn-based PSCs as well by Zhao et al. [46] to evaluate the change in optical as well as solar cell properties. Different perovskites were prepared by changing the ratio of FA/MA in $FA_xMA_{1-x}SnI_3$ as shown in Figure 14.8. The XRD data (Figure 14.8a) show the contraction of the crystal lattice by incorporation of MA^+ in $FASnI_3$ which is obvious because of the ionic radius of MA^+ (2.17 Å) is smaller than FA^+ (2.53 Å). It was also observed that the bandgap changes with the

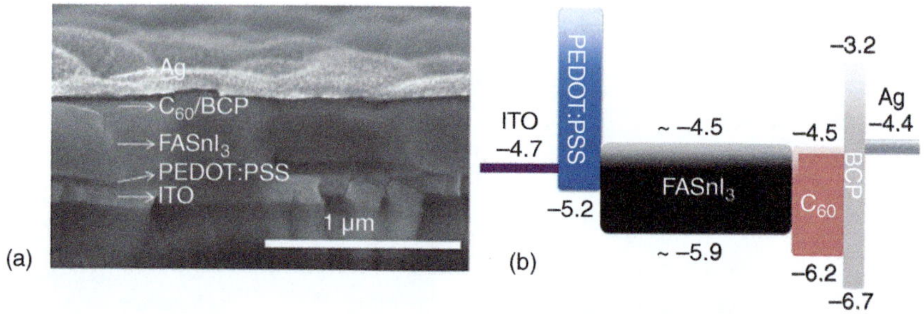

Figure 14.7 (a) Cross-sectional SEM image of the PSC. (b) Energy diagram showing the energy levels of different elements of the PSC. *Source*: Reproduced with permission from Liao et al. [36]. Copyright 2015, Wiley-VCH.

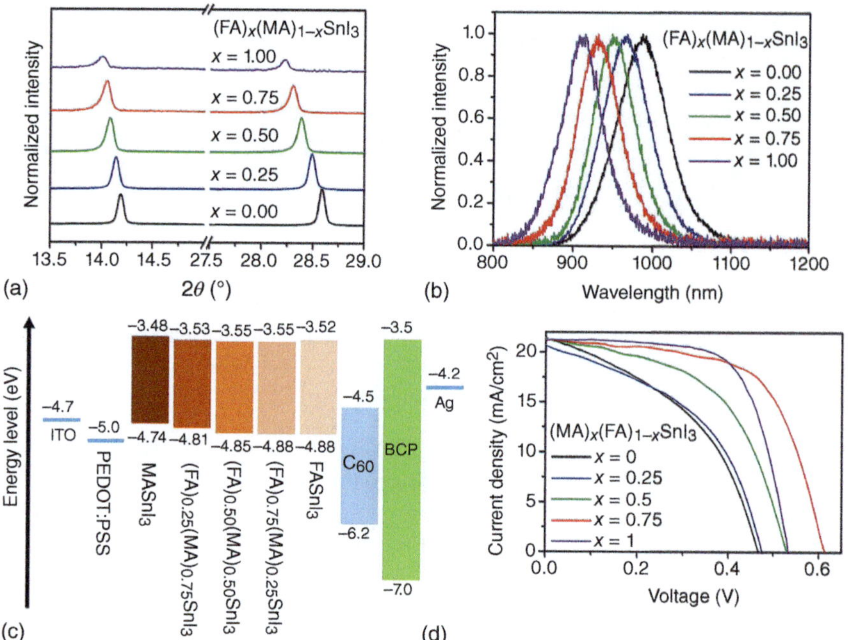

Figure 14.8 (a) XRD spectra, (b) photoluminescence (PL), (c) energy diagram of the PSCs, and (d) IV curve of the PSCs fabricated by employing $FA_xMA_{1-x}SnI_3$ thin films. *Source*: Zhao et al. [46]. Licensed Under CC BY 4.0.

different ratios of FA/MA, for $FASnI_3$ bandgap, is around 1.36 eV and it changes to 1.26 eV for $MASnI_3$ which can be seen through the shift in PL peaks (Figure 14.8b) toward the infrared region with the incorporation of MA ion. The solar cell performances show that the perovskite $FA_{0.75}MA_{0.25}SnI_3$ exhibited the best PCE of 8.12% with a high V_{oc} of 0.61 V (Figure 14.8d) compared to $FASnI_3$ (6.60%) and $MASnI_3$ (4.29%). This difference in solar cell performance was attributed to the lower charge recombination in the case of $FA_{0.75}MA_{0.25}SnI_3$, which can be further related to the improved surface morphology of the films with the incorporation of FA and MA in

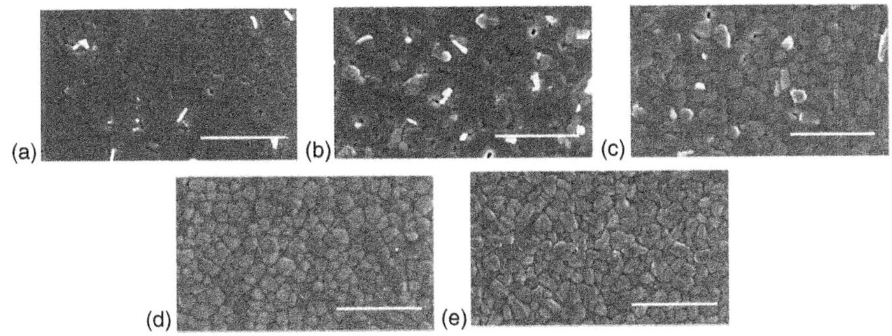

Figure 14.9 SEM image of $FA_xMA_{1-x}SnI_3$ thin films with (a) $x = 0$, (b) $x = 0.25$, (c) $x = 0.50$, (d) $x = 0.75$, and (e) $x = 1$. *Source*: Reproduced with permission from Zhao et al. [46]. Copyright 2017, Wiley-VCH.

A position of the perovskite lattice (Figure 14.9). The $MASnI_3$ thin films manifest a continuous surface morphology without much obvious grain boundaries. The surface morphology starts changing to a dense and with grain-like appearance with the addition of FA^+. At lower FA contents ($x = 0.00$, 0.25, and 0.50) films show a lot of pinholes, and some white aggregates are observed which can be ascribed to the phase separation induced by SnF_2 leading to poor fill factor (FF) and PCE of the solar cells. However, with higher FA content ($x = 0.75$ and 1.00), uniform and dense films are obtained that contribute to the superior device performances.

14.3.1.5 2D/3D $FASnI_3$

The positioning of a small amount of bulkier organic cation at A position improves the stability of Pb-based PSCs. Therefore, a similar effect can be observed in the case of Sn-based PSCs as well. Therefore, Shao et al. [37] added a small amount of phenethyl ammonium ($PEA^+ = C_6H_5(CH_2)_2NH_3^+$) iodide to fabricate the solar cells. The main research achievement in the work was that the 2D–3D mixed $FASnI_3$ leads to suppression in the formation of the Sn vacancies and therefore the background carrier density. It was demonstrated that the addition of a very small amount (0.08 M) of layered tin perovskite led to superior crystallinity and a well-defined orientation of the $FASnI_3$ crystals, as shown in Figure 14.10. It was shown that the five dominant peaks at angles of 14.0°(100), 24.4°(120)/(102), 28.22°(200), 31.65°(122), 40.37°(222) for reference $FASnI_3$ crystal reduces to only two highly oriented peaks of 14.0°(100) and 28.22°(200), as well as a weaker peak at 42.9°(300). This observation is further confirmed by the grazing incidence X-ray scattering data as shown in Figure 14.10d,e. Therefore, the extended ordering and packing of crustal planes and suppress background carrier density due to the addition of 2D tin perovskite led to a high PCE of 9%.

14.3.1.6 Sn–Ge mixed PSCs

Introduction of Ge at B position and the PSC performances for $CsGeI_3$ (1.63 eV), $MAGeI_3$ (2.0 eV), and $FAGeI_3$ (2.35 eV) were reported by Krishnamoorthy et al. [47]. The purpose of the work was to evaluate the performances of Pb-free PSCs.

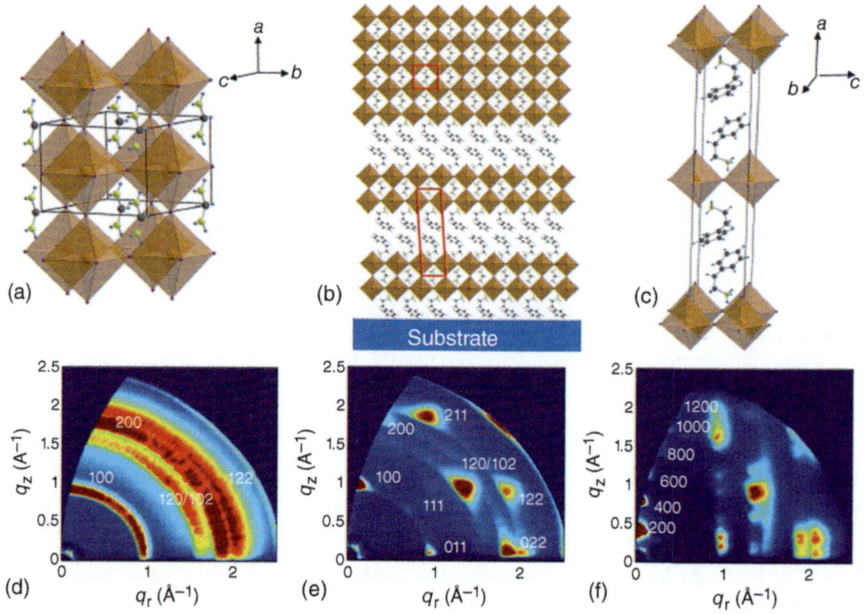

Figure 14.10 Schematic crystal structure of (a) $FASnI_3$, (b) 2D/3D mixture (2D = 0.08 M), and (c) 2D with their grazing incidence X-ray scattering data (d)–(f), respectively. *Source:* Shao et al. [37].

Unfortunately, the device performances were not very high compared to Sn-based PSCs and Pb-based PSCs. However, the research showed prospects of utilizing Ge in PSCs. It was reported in the work that the formation of Ge^{4+} due to rapid oxidation of Ge^{2+} was responsible for the poor device performances. This property of rapid oxidation of Ge was cashed in by Ito et al. [48] when they prepared $FA_{0.75}MA_{0.25}Sn_{0.95}Ge_{0.05}I_3$ which gave an improved PCE of 6.92% compared to non-Ge perovskite $FA_{0.75}MA_{0.25}SnI_3$ (3.31%). The improvement in PCE was attributed to the surface passivation by Ge atoms which forms a thin GeO_2 layer on the surface of the perovskites. As the amount of Ge added was only 5 mol%, it was surprisingly noticed that in the bulk of the perovskite is the $FA_{0.75}MA_{0.25}SnI_3$ structure, and hence Ge addition was effectively passivating the perovskite that further suppressing the oxidation of Sn^{2+} or the formation of Sn^{4+}. The role of Ge addition in improving the stability and performance was demonstrated by Chen et al. [32] also, an all-inorganic $CsSn_{0.5}Ge_{0.5}I_3$ was reported. A promising efficiency of 7.10% was reported with less than 10% decay of the PCE when kept for 500 hours under the N_2 atmosphere. The improved performance was attributed to the formation of the oxide layer due to oxidation of Ge^{2+} to Ge^{4+} which is the similar observation to Ito and coworkers As shown in Table 14.1 as well, the addition of Ge reduces the carrier concentration by order of 2 [16] and also increases the carrier mobility which further increases the diffusion length. These improvements in physical properties are the result of passivation by an oxide layer which reduces the formation of Sn vacancies.

14.3.2 Strategies to Improve the Efficiency

It can be concluded from the abovementioned performances of the Sn perovskites that control in the morphology and reducing the chance of oxidation of Sn^{2+} are the two key factors that need to be taken care of to further improve the PCEs. Therefore, the following are some of the recent strategies being implemented to achieve higher and stable Sn-based PSCs.

14.3.2.1 Film Fabrication Methods

The spin-coating is a commonly used method to fabricate the thin films of perovskite owing to its low cost and easy fabrication process. There are broadly two methods which are popularly used, one-step method and two-step method. In one-step method, first, the precursor was prepared by mixing the basic components, AX and BX_2 to get ABX_3 precursor solution, in DMF and DMSO. The precursor solution is then spin-coated followed by dropping an anti-solvent such as chlorobenzene (Ph-Cl), toluene (TL), or diethyl ether (DE) to get the thin film which can be used for further processing accordingly. However, the two-step method is a sequential deposition method, at first, usually, BX_2 only is dissolved in DMF and/or DMSO which is spin-coated to form the BX_2 thin layer and then a second layer was coated on the top consisting of AX layer (dissolved in isopropyl alcohol [IPA]). The final film ABX_3 then forms after the reaction between AX and BX_2 layers. These two methods are widely used for Pb-based PSCs, but in the case of Sn-based PSCs, the sequential deposition method is difficult to process because of the high solubility of SnI_2 in IPA [49]. There are various other methods reported that can be used for preparing films such as thermal evaporation [28] and vapor-assisted solution methods [50]. Another approach of film fabrication is hot antisolvent treatment (HAT) reported by Liu et al. [51], the hot anti-solvent Ph-Cl was dripped for the film formation. The HAT method increases the density of nucleation sites, which can assist in improving the coverage of the perovskite film. In their research, they further annealed the films in the DMSO vapor atmosphere to control the crystallization process and to obtain the films with large grain size. This method of HAT and solvent vapor annealing (SVA) led to a PCE of 7.2% compared to the solar cell prepared without HAT and SVA method (5.7%). Recently, Li et al. [27] demonstrated a new method of fabrication to deposit dense and uniform films of $MASnI_3$ known as cation exchange method which is a two-step process. At first, a hydrazinium tin iodide ($N_2H_5SnI_3$, $HASnI_3$) thin film was fabricated by spin coating the $HSnI_3$ solution followed by annealing at 80 °C for five minutes, then the film was annealed in the atmosphere of MA vapor to form $MASnI_3$ through an organic cation displacement reaction. In this process, the oxidation of Sn^{2+} can be suppressed by the reducing agent hydrazine generated inside the film during the displacement reaction. The films were fabricated on the mesoporous TiO_2 and a PCE of 7.13% was obtained.

14.3.2.2 Use of Reducing Agents

In the Sn halide perovskite film fabrication, the oxidation of Sn^{2+} to Sn^{4+} is very difficult to avoid even if the films are prepared in the N_2 atmosphere. A small amount of oxygen is enough to deteriorate the perovskite structure. The oxygen present in the solvents such as DMF and DMSO is very difficult to remove, apart from this, the SnI_2 purity is also crucial because the amount of SnI_4 present can finally affect the solar cell performances. The addition of various reducing agents can suppress the Sn^{4+} amount either by donating electrons before oxidation of Sn^{2+} or can help to reduce Sn^{4+} to Sn^{2+} ($Sn^{2+} \rightleftharpoons Sn^{4+} + 2e^-$). Therefore, many researchers have used various kinds of reducing agents such as hypophosphorous acid (HPA) [52] and hydrazine (N_2H_4) [53]. The HPA is used as an additive in perovskite precursor solution facilitating the nucleation rate in perovskite as well as acts as a reducing agent to avoid oxidation of Sn^{2+}. In the case of application of N_2H_4, perovskite annealing is done in the vapor atmosphere of N_2H_4 leading to PCE improvements. Similarly, in another work, the hydrazinium chloride (N_2H_5Cl) is used as an additive in perovskite precursors [54]. It is explored that apart from the reducing nature of N_2H_5Cl, its addition forms an intermediate fused template phase of perovskite that acts as a seed and encourages the growth of larger perovskite grains. Recently, Gu et al. [55] have shown that Sn metal can be used as the additive in perovskite precursor, it is reported that metallic Sn is a better reducing agent than Sn^{2+} and therefore redox reaction $Sn^{4+} + Sn \rightarrow 2Sn^{2+}$ is more favorable. As shown in Figure 14.11, the oxidized solution of $FASnI_3$ (red color indicates the formation of Sn^{4+} in the precursor solution) can be changed back to the original $FASnI_3$ (yellow color) solution containing less amount of Sn^{4+}. This kind of strategy is very helpful and in reported work, it enhanced the performance of the devices prepared from oxidized solution (0.09%) to 6.19% with the films prepared from the solution mixed with Sn powder for five hours.

14.3.2.3 Doping Effect of Large Organic Cations

The use of large organic cation in improving stability and efficiency has been mentioned in the earlier section (*2D–3D $FASnI_3$*). In this process, a very interesting

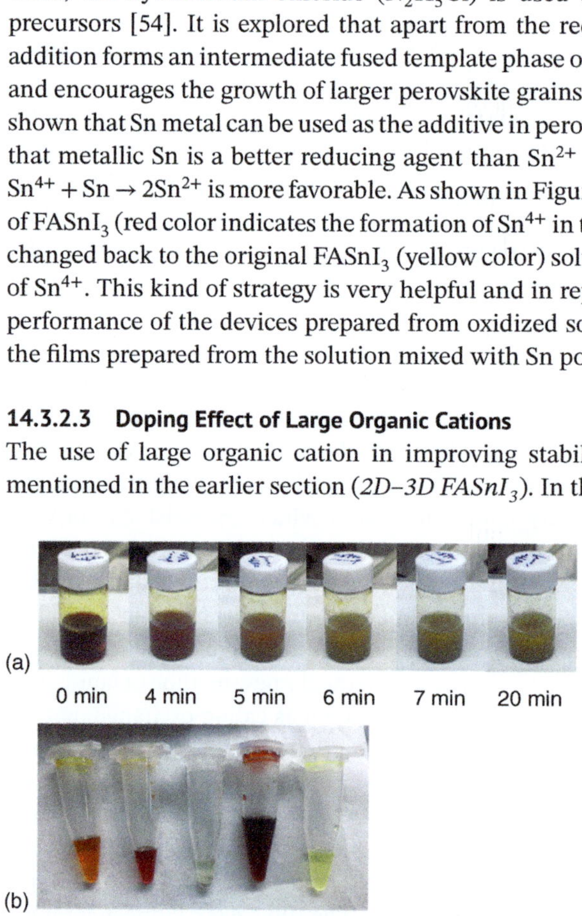

Figure 14.11 (a) Photographs of $FASnI_3$ precursor solutions (0.8 mol/l in DMSO) with different stirring time in ambient air after adding Sn powder. (b) Photographs of different solutions in DMSO. From left to right: 0.1 M SnI_4, 0.1 M SnI_4 + FAI, 0.1 M SnI_4 + FAI + Sn, 0.8 M SnI_2 (99%) + FAI + 10% SnF_2, and 0.8 M SnI_2 (99%) + FAI + 10% SnF_2 + Sn after filtration. *Source*: Reproduced with permission from Gu et al. [55]. Copyright 2018, Wiley-VCH.

cation ethylene diammonium diiodide ($EDAI_2$) has been reported by Ke et al. [26, 35]. According to the work, ethylenediamine (EDA) insertion into the lattice led to the dramatic increase in the air stability and photoelectric properties of tin-based absorbers. A high PCE of 7.14% was reported which retained 96% of its initial value over 1000 hours with encapsulation. Also, it was mentioned that bandgap of $FASnI_3$ can be increased by the introduction of EDA into the lattice; however, interesting information was that with the increase in the concentration of $EDAI_2$ to 25 mol%, the XRD of the $EDAI_2$ added $FASnI_3$ still exhibits 3D perovskite unlike 2D–3D $FASnI_3$ formation as described earlier. However, there is an increase in the bandgap with $EDAI_2$ which is attributed to the formation of Schottky defects in the 3D structure. Following this interesting work, Jokar et al. [56] reported the stoichiometric doping effect of $EDAI_2$ on solar cell performances and stability. The fast crystallization in tin-based perovskite is a critical problem for getting uniform and pinhole-free films. The small addition (1 mol%) of $EDAI_2$ helped to fabricate the pinhole-free films by slowing down the crystallization process. The devices show the increase in PCE to 8.9% (PCE measured on the first day was 6–7%) after they were stored in a glove box for more than two months. This kind of phenomenon might be due to the slow decrease in surface defects and the relaxation of crystal strain in the presence of $EDAI_2$. Moreover, this could be due to conformational rearrangement or slow migration of bulky EDA^+ cation into the perovskite crystal. In another recent report by Kamarudin et al. [15] shows that these bulky EDA can be used to passivate the films as well which led to a dramatic enhancement in PCEs of Sn halide perovskites to more than 10%. EDA is a Lewis base that bonds the undercoordinated Sn, passivates the dangling bonds and defects, resulting in suppressed charge carrier recombination that further enhances the physical properties of the perovskite films. The passivation effect can be observed also by the emergence of relatively less amount of Sn^{4+} on the surface of the perovskite film as confirmed by X-ray photoelectron spectroscopy (XPS). Figure 14.12a,b shows the IV curve of the solar cells fabricated, it shows that with the passivation using EDA (0.01, 0.05, and 0.1 mM), the solar cell PCEs have improved and further improved to higher values after one-week storage in a glove box which is the same observation as Jokar et al. [56]. It is explored that the charge recombination lifetime (τ_{rec}) increases (Figure 14.12d) resulting in increased diffusion length after EDA passivation (Figure 14.12f). Also, the overall charge collection efficiency evaluated from charge transport time (τ_{tr}) (Figure 14.12c) and recombination lifetime (τ_{rec}), enhances with the EDA passivation (Figure 14.12e) which could be attributed to the change of carrier type from p-type to n-type with EDA passivation that increases the electron collection possibility in p-i-n structure.

14.3.2.4 Device Engineering and Lattice Relaxation

So far, we have noticed that by reducing the formation of Sn vacancies within the Sn halide perovskites the solar cell PCEs of Sn-based PSCs can be improved. Another important criterion is the energy level matching at the interface of absorber/HTL and absorber/ETL that becomes important for the facile electron and hole transport. In general, we have limited options for the HTL and ETL materials, and most

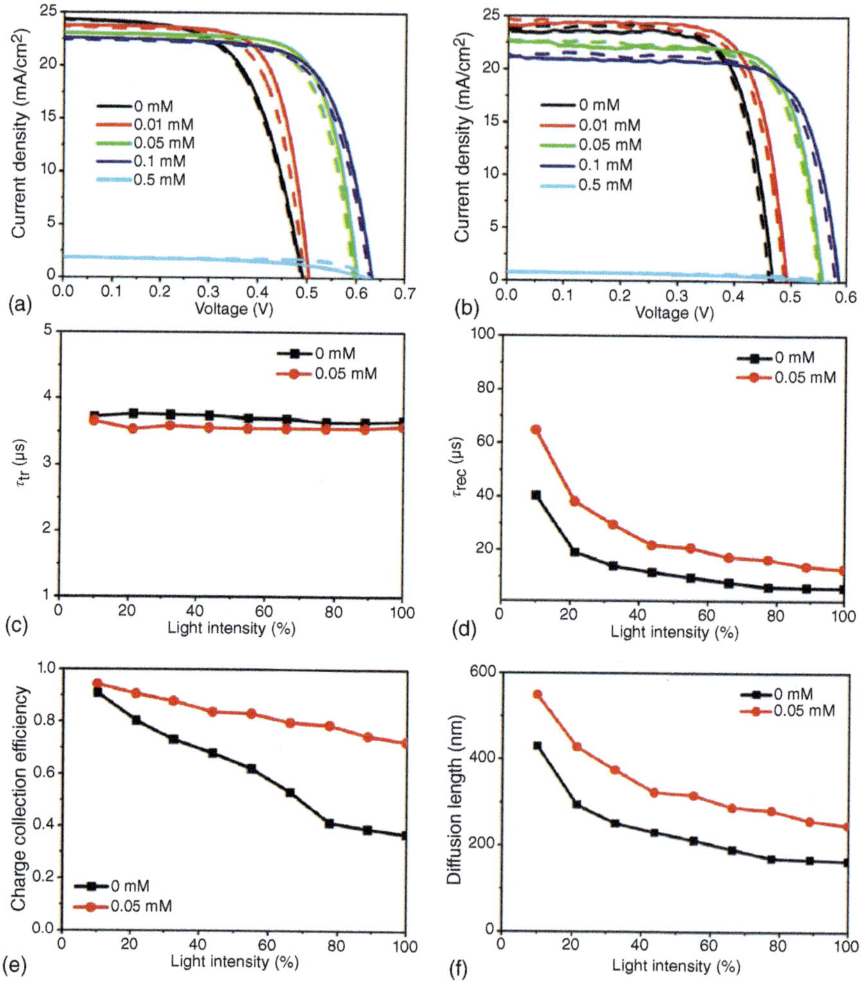

Figure 14.12 Light *J*–*V* plot of champion $FA_{0.98}EDA_{0.01}SnI_3$ with different concentrations of edamine for (a) fresh samples and (b) samples after seven days of storage in the glovebox. (c) Plot of transport lifetime as a function of light intensity between 0 and 0.05 mM edamine-passivated devices obtained from IMPS measurements. (d) Plot of recombination lifetime as a function of light intensity of 0 and 0.05 mM edamine-passivated devices obtained from intensity modulated voltage spectroscopy (IMVS) measurements. (e) Charge collection efficiency plot against light intensity comparing the reference tin perovskite solar cell device and edamine-passivated device. (f) Plot of diffusion length as a function of light intensity of 0 and 0.05 mM edamine-passivated devices. *Source*: Kamarudin et al. [15].

commonly used are PEDOT: PSS with highest occupied molecular orbital (HOMO) ~−5.1 eV and C60 with lowest unoccupied molecular orbital (LUMO)~−4.5 eV. Therefore, it becomes necessary that we should tune the bandgap and energy levels of the absorber layer according to the ETL and HTL in such a way that LUMO of absorber would be shallower than LUMO of the ETL, and HOMO of the absorber would be deeper than HOMO of the HTL. The best case would be the minimum difference in these energy levels provided that effective charge transport is occurring that finally controls the photovoltaic parameters such as V_{oc}, J_{sc}, and FF. The bandgap tuning as mentioned earlier can be achieved by sharing the A position in ABX_3 perovskite [8]. Therefore, recently there is some research work focusing on these aspects while fulfilling the purpose of reduction of Sn vacancies. Jokar et al. [38] demonstrated the addition of bigger non-polar organic cation guanidinium (GA^+) into the $FASnI_3$ crystal structure in the presence of 1% $EDAI_2$, the best device efficiency was obtained for $GA_{0.2}FA_{0.8}SnI_3$–1%$EDAI_2$, a PCE of 8.5% was obtained which increased to 9.6% after storing in a glove box for 2000 hours. It is mentioned in the report that with the use of such non-polar molecules the resistance against moisture can be improved for the perovskites that can also increase the stability of the PSCs. This kind of strategy has been reported recently by Nishimura et al. also [39] they reported the incorporation of ethylamine (EA^+) into the perovskite, GeI_2 doped $(FA_{1-x}EA_x)_{0.98}EDA_{0.01}SnI_3$. In the work, the highest PCE of 13.24% is reported so far in the field of Sn-based PSCs by using $(FA_{0.9}EA_{0.1})_{0.98}EDA_{0.01}SnI_3$, a V_{oc} of 0.84 V is obtained which is also among the highest values of V_{oc} reported so far (Figure 14.13a). The highest PCE parameters (Figure 14.13a,b) obtained can be attributed to the implementation of reducing the surface traps by EDA passivation in the same manner as mentioned by Kamarudin et al. [15], the strain relaxation by addition of EA^+ [39], and an improved energy level (Figure 14.13c) matching with ETL and HTL. It can be observed that carrier concentration decreases with the addition of EA^+ with a noticeable increase in mobility (Figure 14.13d,e), this effect is due to the decrease in Sn vacancies as mentioned earlier also. Further, XPS analysis concludes that the reference sample without EA is not only deficient of iodide but also shows higher oxygen content at the surface (Figure 14.13f,g). Therefore, with the EA incorporation, the conditions have been improved increasing solar cell performances. Recently, there is another report where energy alignment was achieved with the use of a proper selection of ETL layer, indene-C60 bisadduct (ICBA) is used which possesses the shallower LUMO energy level ~−3.7 eV that matches well with the perovskite, $PEA_{0.15}FA_{0.85}SnI_3$ (where $PEA^+ = C_6H_5CH_2CH_2NH_3^+$) with additive 5 mol% NH_4SCN (LUMO ~ −3.69 eV) [40]. A high PCE of 12.4% is reported with the highest value of V_{oc} (0.94 V) so far for the absorber layer with the bandgap of 1.42 eV.

14.4 Sn–Pb Perovskite Solar Cells

The highest efficiencies possible for single-junction solar cells are decided on the bases of the SQ limit. Figure 14.14 illustrates the current situation of some of the widely researched PSCs. For comparison, commercially used c-Si (PCE-26.1%) and GaAs (PCE-27.8%) are also marked in the plot. We can see that the peak of the highest efficiency possible approximately lies in between 1.1 and 1.4 eV, written as "High PCE Zone." Fortunately, Sn and Sn–Pb mixed perovskite absorber's bandgap falls in this zone, which highly motivates the researchers all around the world to improve the performances of the solar cells based on these materials, and according to the SQ, diagram PCEs more than 30% are possible (Figure 14.15). The rapid growth in

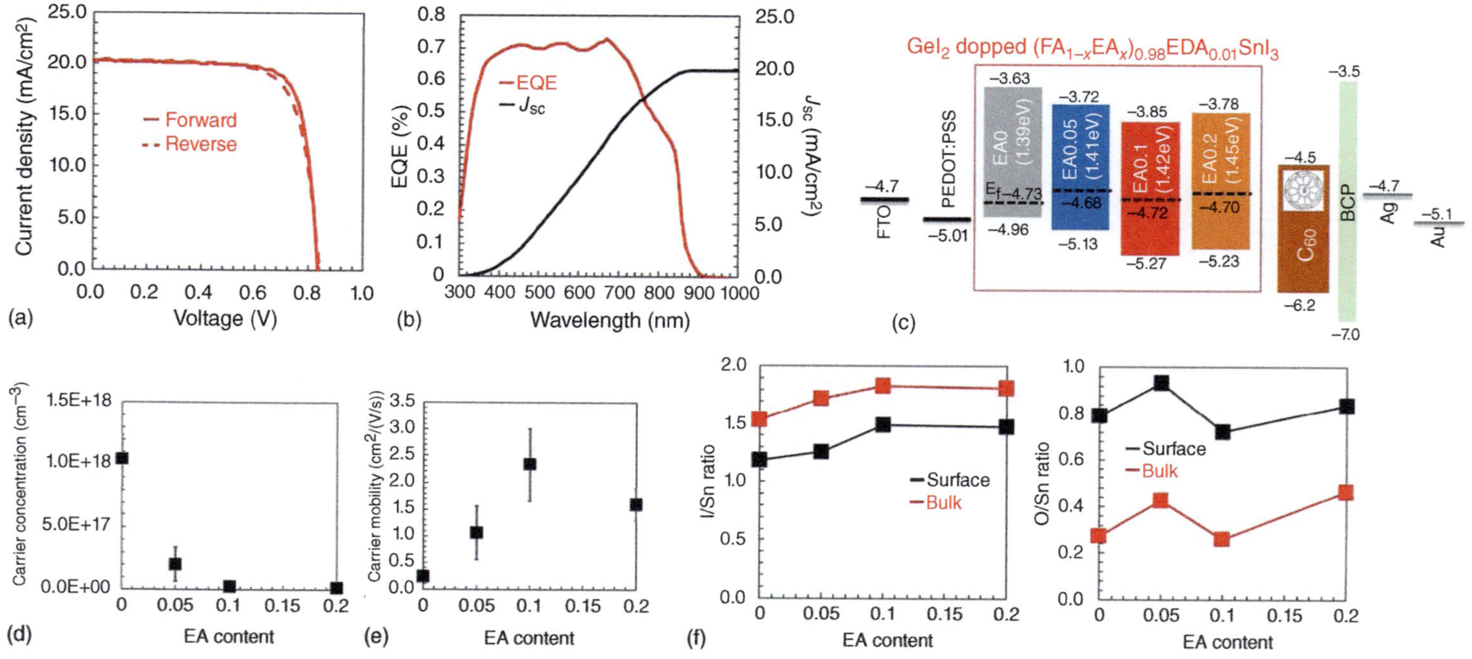

Figure 14.13 (a) IV curve, (b) IPCE of the PSC prepared by $(FA_{0.9}EA_{0.1})_{0.98}EDA_{0.01}SnI_3$, (c) energy diagram of the PSCs fabricated, (d) carrier concentration vs. EA content, (e) carrier mobility vs. EA content, (f) I/Sn ratio vs. EA content, and (g) O/Sn ratio vs. EA of the perovskite thin films. *Source*: Nishimura et al. [39].

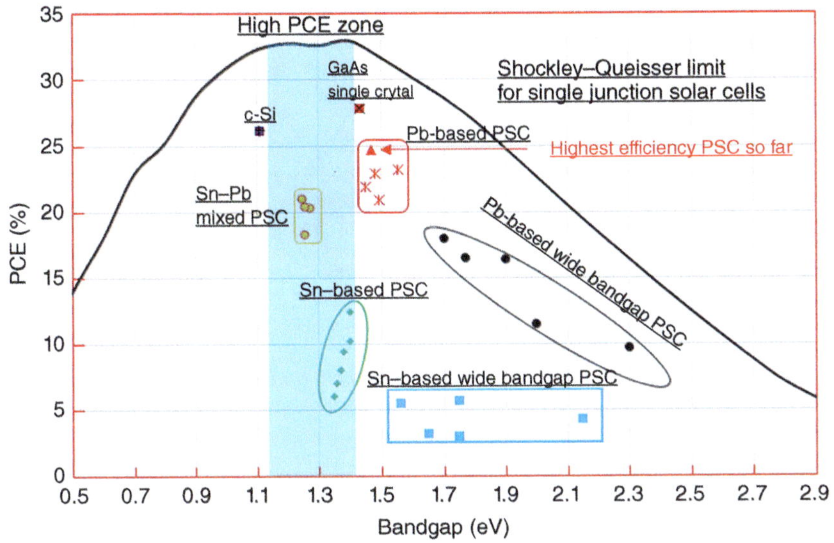

Figure 14.14 Current status of the different kinds of photovoltaic materials used.

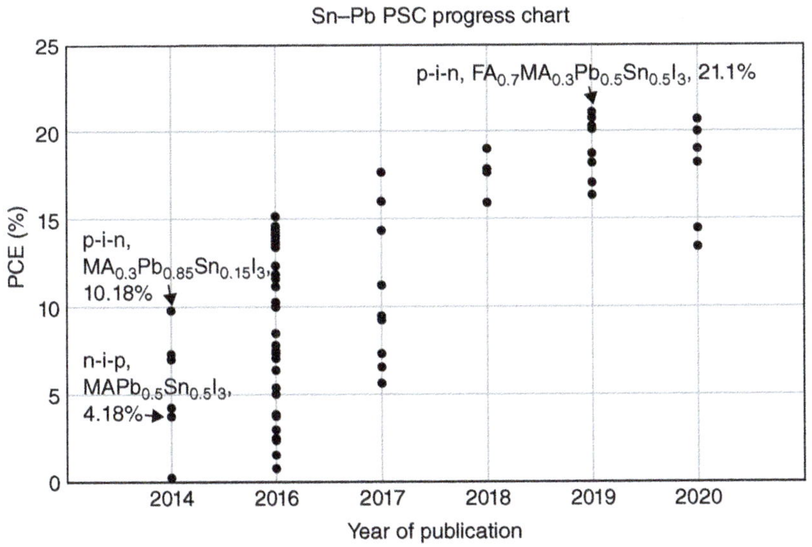

Figure 14.15 Progress in the efficiency of Sn–Pb mixed halide perovskites.

efficiency of PSCs in one decade from 3.8% [2] in 2009 to 24.66% [11] in 2019 gives strong intuition that in the coming years there is a huge possibility that their performances can cross that of c-Si or possibly GaAs also. And this possibility is more for the materials whose bandgap lies in the range of 1.1–1.4 eV, and up to now highest PCE for PSCs is obtained by Pb-based PSCs where most of the material's bandgap lies in the range of 1.5–1.6 eV. Therefore, research in the field of Sn and Sn–Pb becomes important from a high PCE point of view as well. Another most important advantage

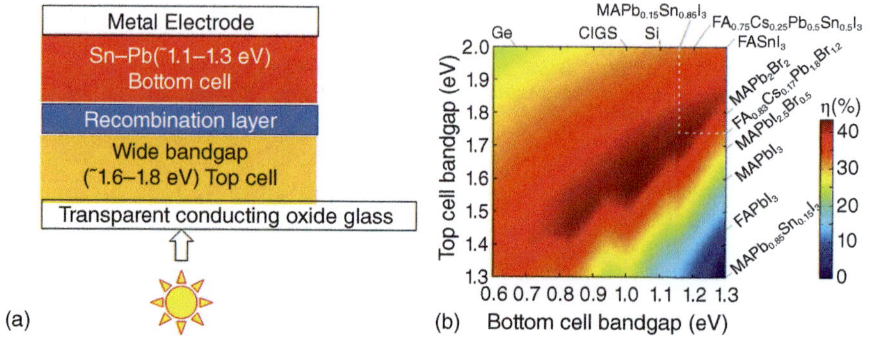

Figure 14.16 (a) Schematic showing the tandem solar cell. (b) Color contour showing the maximum possible PCEs that can be achieved by a tandem solar cell. *Source*: Anaya et al. [57].

of Sn–Pb PSCs is their potentiality as a narrow bandgap solar cell in the fabrication of tandem solar cells. Tandem solar cell technology is a proven method to break the SQ limit, they reduce the losses coming from thermalization by implementing different semiconductor layers with different bandgaps as illustrated in Figure 14.16. Therefore, a higher PCE can be obtained by utilizing the tandem solar cells in comparison to single-junction solar cells. A typical color contour plot is shown in Figure 14.16b suggest that a PCE near to 40% can be achieved by coupling of a "wide-bandgap cell" and "narrow bandgap cell" [57]. In a two-junction tandem, a "wide-bandgap cell" absorbs high-energy photons but allows lower energy photons to be absorbed in the "narrow bandgap cell." Therefore, the higher energy photon absorption results in a high voltage in the wide bandgap cell while low-energy photons are absorbed in the narrow bandgap cells to obtain the same photocurrent as wide bandgap cell. The main purpose of using a tandem of two cells is to achieve a PCE higher than that of the individual solar cell.

14.4.1 Anomalous Bandgap of SnPb (The Bowing Effect)

Sn–Pb mixed halide perovskites exhibit the lower bandgap compared to their parental products. This kind of behavior is known as the bowing effect and the reason behind this is still under debate. However, there are enough experimental proofs to confirm this effect. Ogomi et al. [10] and Hao et al. [58] evaluated the bandgaps of $MASn_xPb_{1-x}I_3$ perovskite by changing the ratio of Sn/Pb and realized first time the anomalous bandgap behavior in SnPb mixed systems, as shown in Figure 14.17a,b. There are some simulation-based studies as well to see the bowing effect that could answer the possible reason behind the phenomenon involved. Density functional theory (DFT), hybrid DFT, and the many-body quasiparticle self-consistent GW (QSGW) were employed [59]. It was explored that in the mixed Sn–Pb systems, the valence band maximum (VBM) is derived from the interactions between Sn-s and I-p orbitals, whereas the conduction band minimum (CBM) is obtained from the Pb-p and I-p orbitals as shown in Figure 14.17c,d, resulting

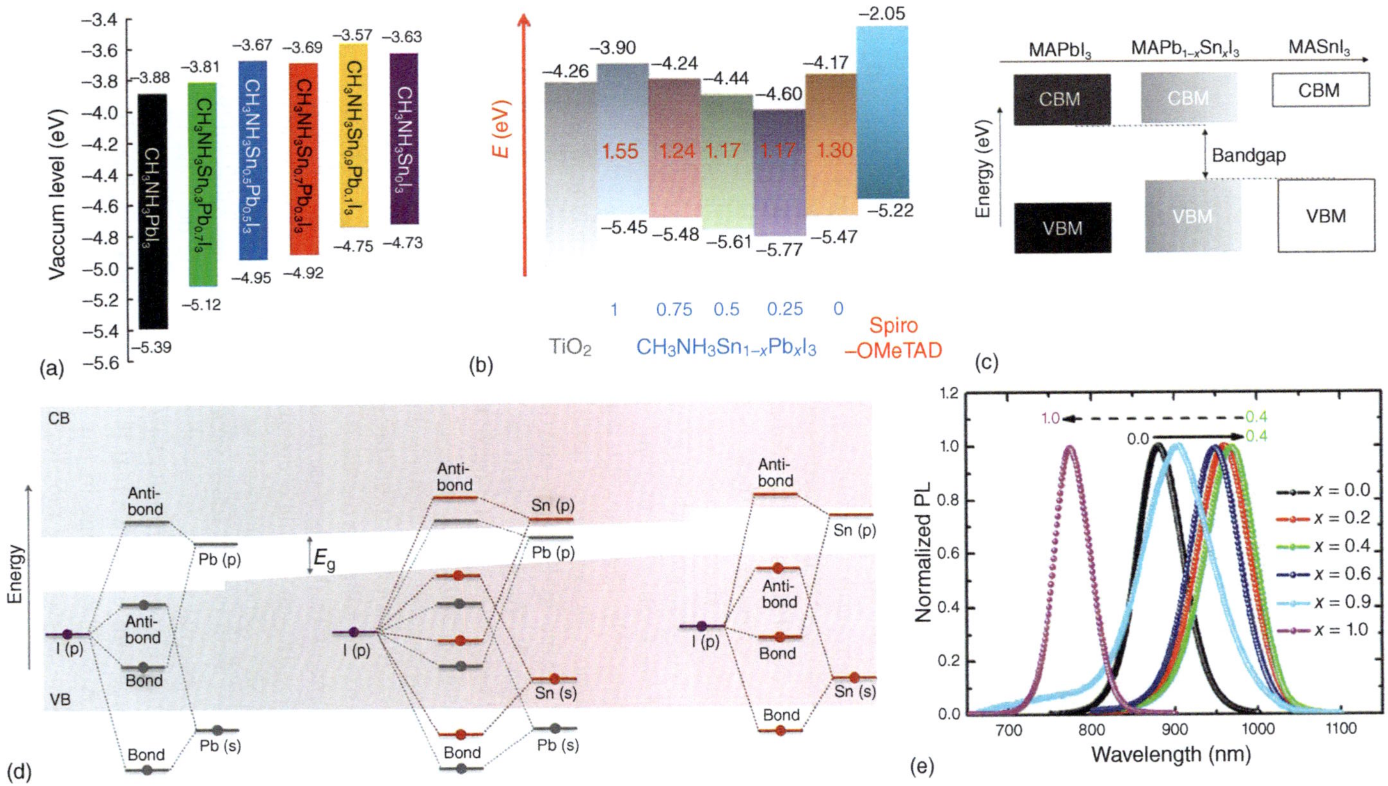

Figure 14.17 (a) Energy diagram of different Sn–Pb mixed perovskite films. *Source*: Ogomi et al. [10]. (b) Energy diagram of $MASn_{1-x}Pb_xI_3$. *Source*: Hao et al. [58]. (c) Schematic showing the bowing effect after mixing Sn and Pb perovskites. (d) Molecular orbital diagram, from the left to right, the Pb, Sn–Pb, and Sn perovskites. *Source*: (c, d) Goyal et al. [59]. (e) PL spectra of $(FASnI_3)_{1-x}(MAPbI_3)_x$. *Source*: Liao et al. [60].

into the lowering of bandgap compared to either of the Sn and Pb compounds. These simulation results match well with the experimental results demonstrated by Hayase and coworkers [10], band energy levels of SnPb perovskites were evaluated using XPS and optical absorbance spectroscopies showing the upward shift in VBM and CBM with the increase in Sn content. However, in the work of Hao et al. [58], there was a tendency of the first downward shift of both VBM and CBM and then upward with the increase in Sn content. Despite these differences in change of energy levels, all the studies clearly show the decrease in bandgap for mixed Sn–Pb mixed systems at some specific compositions such as Yanfa-Yan and coworkers [60] showed this behavior for $(FASnI_3)_{1-x}(MAPbI_3)_x$ as shown in PL spectra measured (Figure 14.17e). It was illustrated that because of the bowing effect, the PL emission peak position of Sn–Pb mixed perovskites exhibits as redshift compared to the pure Sn and Pb compounds.

14.4.2 Physical Properties

14.4.2.1 Intrinsic Carrier Concentration

Unlike Sn, the Sn–Pb mixed systems have superior physical properties and have been evaluated by many research groups [29, 60–63]. Stoumpos et al. [13] evaluated the structural and optoelectronic properties of Sn and SnPb mixed perovskites. They synthesized the perovskite crystals by solution routes employing a reducing agent (H_3PO_2), and it was explored that these perovskites exhibit the n-type behavior due to inhibition of Sn^{4+}, whereas these materials show p-type behavior when synthesized via solid-state reaction without reducing agents. The electrons mobilities measured for $MASnI_3$, $CsSnI_3$, $FASnI_3$, and $MASn_{0.5}Pb_{0.5}I_3$ are 2320, 536, 103, and 270 $cm^2/(V\,s)$, respectively. Therefore, it is correct to say that Sn–Pb mixed systems also show the change in physical properties based on the formation of Sn vacancies and Sn^{4+}, and most of the Sn–Pb perovskite thin films exhibit p-type semiconductor characteristics. Recently, Yang et al. [20] explored an interesting study, they measured the physical properties of thin films of $FA_{0.45}Cs_{0.05}MA_{0.5}Sn_{0.5}Pb_{0.5}I_3$ with and without Cd^{2+} as listed in Table 14.1. It is explored that $FA_{0.45}Cs_{0.05}MA_{0.5}Sn_{0.5}Pb_{0.5}I_3$ have intrinsic p-type nature with higher values of hole mobility (>3.25 $cm^2/(V\,s)$) compared to electron mobility (0.65 $cm^2/(V\,s)$). However, with the addition of Cd^{2+} the electron mobility increased to 1.98 $cm^2/(V\,s)$ and hole mobility to >3.80 $cm^2/(V\,s)$ along with the decrease in intrinsic carrier concentration by one order (1×10^{16}–4.2×10^{15} cm^{-3}). The decrease in carrier concentration is attributed to the decrease in Sn vacancies, the authors explained that the Cd^{2+} ions because of their small ionic radii can fill those vacancies to de-dope Sn perovskites and hence reduce the internal strain. In another research, the decrease in intrinsic carrier concentration by two to three order (10^{17}–10^{15} cm^{-3}) was obtained by the addition of Br anion into the lattice [64] that resulted in an increase in mobility and hence PCE. A similar change in physical properties is observed with the addition of Cl anion as well, the mobility for perovskite $(FASnI_3)_{0.6}(MAPbI_3)_{0.4}$ (27 ± 2 $cm^2/(V\,s)$) increased to 40 ± 5 $cm^2/(V\,s)$ with the addition of 2.5 mol% of Cl^- [65]. In another approach [66], the addition of Sn metal was done to reduce the formation of Sn^{4+} as mentioned earlier also [55]. It was explored the perovskite

$FA_{0.7}MA_{0.3}Sn_{0.5}Pb_{0.5}I_3$ exhibits a high hole concentration (1.4×10^{16} cm^{-3}), also the density of hole traps were higher than the density of electron traps due to the formation of Sn vacancies. By the addition of Sn metal, there was a decrease in both hole trap density (2.14×10^{16} to 1.06×10^{16} cm^{-3}) and hole carrier concentration (1.4×10^{16} cm^{-3} to 0.54×10^{16} cm^{-3}) which finally led to increasing in overall solar cell performance. Hence, these above discussions reveal that similar approaches as in the case of pure Sn-based perovskite are required, the Sn vacancies need to be reduced to improve the physical properties and PCEs.

14.4.2.2 Carrier Lifetime and Diffusion Length

Sn–Pb mixed systems show a superior carrier lifetime compared to Sn perovskites but inferior to that of Pb perovskites [60]. The Sn vacancies act as charge trapping centers which are in higher concentration in Sn perovskites than Sn–Pb mixed perovskites owing to the presence of Pb^{2+} ion which is less sensitive to oxygen and thus resulting in improved carrier lifetimes. Therefore, the role of Sn vacancies become important for carrier lifetime as well. Long carrier lifetime and diffusion length are necessary for any solar cell to achieve high PCEs. As illustrated in Table 14.1, the charge carrier lifetime for $FA_{0.45}Cs_{0.05}MA_{0.5}Sn_{0.5}Pb_{0.5}I_3$ (140 ± 9 ns) was increased to 1450 ± 160 ns along with the increase in diffusion length of both the charge carriers to more than 1 μm due to the reduction in Sn vacancies by the de-doping effect of Cd^{2+}, and therefore, the PCE in the research work has increased from 14.1% to 20.3% [20]. The carrier lifetime for $(FASnI_3)_{0.6}(MAPbI_3)_{0.4}$ thin films can be varied based on the thickness of the films [61]. The carrier lifetime of 15, 96, 255, and 226 ns were obtained for 400, 510, 620, and 1010 nm, respectively. This difference in a lifetime can be attributed to the difference in the crystallinity and grain size of the films with the change in thickness. The carrier lifetime has been reported to be affected by the addition of Cs^+ into the lattice as well [67], carrier lifetime of $FA_{0.5}MA_{0.5}Sn_{0.5}Pb_{0.5}I_3$ can be improved from ~60.08 to 100.85 ns with $Cs_{0.025}FA_{0.475}MA_{0.5}Sn_{0.5}Pb_{0.5}I_3$. The improvement in carrier lifetime is attributed to the reduction in strain energy of the crystal lattice which led to an increase in PCE to 20.4%. There are some other reports where surface passivation of the $Cs_{0.1}MA_{0.2}FA_{0.7}Pb_{0.5}Sn_{0.5}I_3$ by phenethylammonium iodide (PEAI) ($PEA^+ = C_6H_5(CH_2)_2NH_3^+$) can increase the carrier lifetime from 32 to 210 ns [68]. Similarly, the addition of bigger cations such as Guanidinium ($CH_6N_3^+$) into the perovskite $(FASnI_3)_{0.6}(MAPbI_3)_{0.4}$ has been reported to increase the carrier lifetime from ~139 to 1232 ns [69] that is comparable to pure Pb-based PSCs which led to increasing in PCEs to 20.4%.

14.5 The Status of Sn–Pb Perovskite Solar Cells

Sn–Pb mixed perovskites can harvest the sunlight up to the infrared region. The bandgap for Sn–Pb mixed systems has been reported to be between 1.17 and 1.3 eV [10, 58] which corresponds to the absorption from 950 to 1050 nm. This wavelength range is according to the ratio of Sn/Pb while keeping the elements at A and X position of ABX_3 same. However, it is not true that Sn–Pb systems can only have narrow bandgaps, Sn–Pb perovskites with large bandgaps can also be obtained

Table 14.3 Photovoltaic parameters of different kind of Sn–Pb mixed PSCs including the highest reported values.

Perovskite	Bandgap (eV)	Structure type	V_{oc} (V)	J_{sc} (mA/cm^2)	FF (%)	PCE (%)	Year	References
$MASn_{0.5}Pb_{0.5}I_3$	1.28	n-i-p	0.42	20.4	50	4.2	2014	[10]
$MASn_{0.25}Pb_{0.75}I_3$	1.24	n-i-p	0.728	15.7	64.01	7.37	2014	[58]
$MASn_{0.75}Pb_{0.25}I_{1-x}Cl_x$	1.25	n-i-p	0.457	19.63	57.2	5.13	2016	[63]
$MASn_{0.62}Pb_{0.38}I_3$	1.3	n-i-p	0.796	25.5	69.2	14	2018	[73]
$MASn_{0.15}Pb_{0.85}I_3$	1.4	p-i-n	0.77	19.5	67	10.1	2014	[71]
$Cs_{0.1}MA_{0.9}Sn_{0.25}Pb_{0.75}I_3$	1.39	p-i-n	0.84	22.8	76	14.55	2016	[74]
$FA_{0.6}MA_{0.4}Sn_{0.6}Pb_{0.4}I_3$	1.25	p-i-n	0.853	28.7	72.5	17.6	2017	[61]
$FA_{0.6}MA_{0.4}Sn_{0.6}Pb_{0.4}I_3$	1.25	p-i-n	0.83	27.3	80	18.2	2017	[75]
$Cs_{0.025}FA_{0.475}MA_{0.5}Sn_{0.5}Pb_{0.5}I_3$	1.27	p-i-n	0.81	33.14	76	20.4	2019	[67]
$FA_{0.6}MA_{0.4}Sn_{0.6}Pb_{0.4}I_3$	1.25	p-i-n	0.834	30.4	80.8	20.5	2019	[69]
$FA_{0.7}MA_{0.3}Pb_{0.5}Sn_{0.5}I_3$	1.22	p-i-n	0.831	31.4	80.8	21.1	2019	[69]

by choosing the different composition of A and X [70]. However, as mentioned before, the "High PCE Zone" falls in between 1.1 and 1.4 eV; therefore, most of the Sn–Pb-based study has been focused on the preparation of low bandgap perovskites. At the beginning of the Sn–Pb research, these systems were suffering from high V_{oc} loss [10, 58], which can be attributed to the use of regular n-i-p device structure that employs the porous semiconductor TiO_2, and according to one of the reports [45], the SnI_2 coating can form the trapping centers on the surface of TiO_2 which may lead to the decrease in performances of Sn–Pb mixed systems. Therefore, Sn–Pb solar cells with inverted p-i-n configurations were fabricated first time by Jen and coworkers [71, 72] which are now used in almost every Sn–Pb PSCs reported these days. By employing the inverted p-i-n structures, now the V_{oc} loss of 0.384 eV [64] can be achieved which is comparable to that of highly efficient c-Si (0.35 eV) and GaAs (0.27 eV)-based solar cells. Despite the low V_{oc} loss, the Sn–Pb-based PSCs are still lagging their Pb counterparts which are because of the low external quantum efficiencies in these materials, especially in the infrared region. Therefore, the recent research trend is to increase the photocurrents obtained by these materials by various approaches such as by increasing the thickness of the absorber layers, by increasing the diffusion length, and most important is by decreasing the Sn vacancies. The status of the Sn–Pb mixed PSCs has been summarized in Figure 14.15 and Table 14.3.

14.5.1 Different Types of Sn–Pb Perovskite Solar Cells

14.5.1.1 First Kind of Sn–Pb PSC absorber: $MASn_xPb_{1-x}I_3$

$MASn_xPb_{1-x}I_3$ crystal structures fall in the category of tetragonal *I4cm* space group for the less amount of Sn ($0.5 \geq x \geq 0$) and with the increase in Sn content, the perovskite structures show the tetragonal *P4mm* space group [58, 71]. These are the first kind of Sn–Pb mixed PSCs fabricated using a regular n-i-p device structure as shown in Figures 14.18a–c by Ogomi et al. [10], a PCE of 4.18% was reported for the $MASn_{0.5}Pb_{0.5}I_3$ with a J_{sc} of 20.04 mA/cm^2 (with photon harvesting until 1060 nm), V_{oc} of 0.42 V and FF of 0.50. They also highlighted the low V_{oc} loss (<0.5 V) possibilities in PSCs and concluded that in the case of Sn–Pb PSCs this value could be achieved with the encapsulation of the device and proper selection of HTL. Hao et al. [58] also fabricated the solar cells using $MASn_xPb_{1-x}I_3$ as an absorber in n-i-p device structure using Spiro-OMETAD as HTL (Figure 14.18d,e) and demonstrated the PCEs of 5.44%, 3.74%, 7.27%,7.37% and 8.31% for $MASnI_3$, $MASn_{0.75}Pb_{0.25}I_3$, $MASn_{0.5}Pb_{0.5}I_3$, $MASn_{0.25}Pb_{0.75}I_3$ and $MAPbI_3$, respectively. Recently, Tavakoli et al. [73] deposited the $MASn_xPb_{1-x}I_3$ by using a chemical vapor deposition method. In the work, they prepared the perovskite films by employing the reaction between Sn/Pb binary alloy metal and methylammonium iodide (MAI). At first, MAI reacts with Pb/Sn to form $(Pb/Sn)I_2$, and then $(Pb/Sn)I_2$ further reacts to form $CH_3NH_3(Pb/Sn)I_3$. The PCEs of 14.04%, 14.21%, and 4.62% were reported for $MASn_{0.62}Pb_{0.38}I_3$, $MAPbI_3$, and $MASnI_3$, respectively, as shown in Figure 14.18f,g using the same deposition method. The PCE obtained in the work is also the highest value reported in n-i-p device structure for Sn–Pb mixed systems. Figure 14.18h shows the cross-sectional image of the device prepared with planar n-i-p configuration. The p-i-n configurations were also implemented for these perovskites. The first device prepared in this configuration showed the PCE of 10.10% for $MASn_{0.15}Pb_{0.85}I_3$; however, fewer PCEs were reported for other Sn/Pb ratios containing solar cells [71]. These initial results were very encouraging as the PCEs obtained were higher than devices prepared in n-i-p configuration. However, it can be observed that the films were exhibiting many pinholes with poor surface coverage, which is the problem of fast crystallization like pure Sn perovskites. Therefore, to improve the perovskite films surface coverage in planar p-i-n configuration, Zhu et al. [76] demonstrated the use of mixed solvents of DMF and DMSO to prepare the perovskite precursor along with the use of the antisolvent dripping method as mentioned in Section 14.3.2. It was explored that films prepared by precursor dissolved in DMF-only exhibit the high roughness (65 nm) with small grain size and many pinholes. However, the addition of DMSO retards the crystallization of perovskite which resulted in the formation of pinhole-free smooth films with the roughness of 12 nm (DMF:DMSO = 1 : 1) and 100% surface coverage. As a result, this research work demonstrated a high PCE of 15.2% for $MASn_{0.25}Pb_{0.75}I_3$ perovskite-based PSCs.

14.5.1.2 Multi Cation Sn–Pb Perovskites: (FA, MA, Cs) (Sn, Pb) $(I, Br, Cl)_3$

The mixing of monovalent cations has been the key to improve the stability and PCE of the Pb-based PSCs. The same approach has been implemented in Sn–Pb

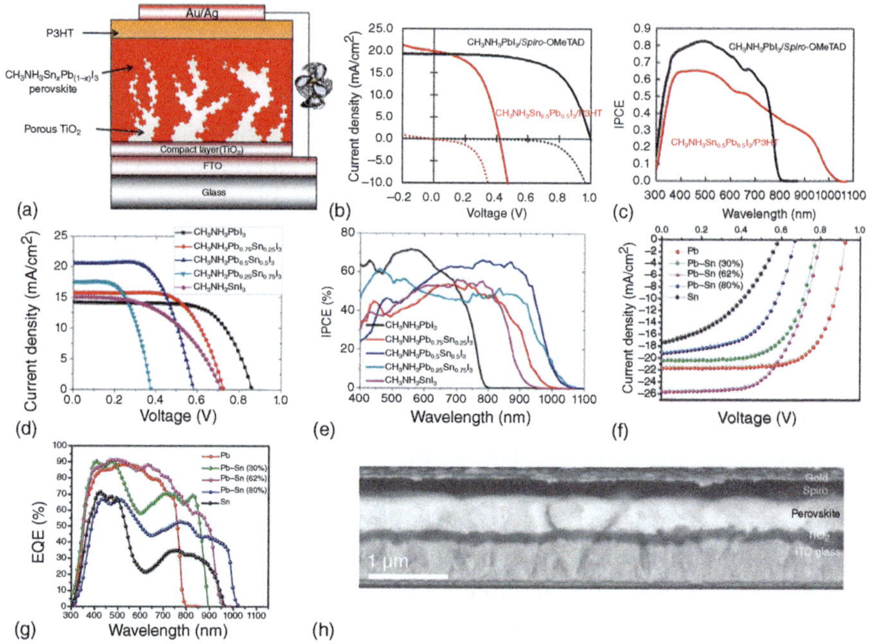

Figure 14.18 (a) Schematic showing the conventional n-i-p PSC fabricated, (b) IV curve, and (c) IPCE spectrum of the PSCs. *Source*: (a–c) Ogomi et al. [10]. (d) IV curve and (e) IPCE spectrum of $MASn_{1-x}Pb_xI_3$-based PSCs. *Source*: (d, e) Hao et al. [58]. (f) IV curve and (g) IPCE spectrum of different PSCs fabricated by alloying. *Source*: (f, g) Liu et al. [74]. (h) Cross-sectional SEM image of the PSC fabricated by the alloying method. *Source*: Reproduced with permission from Tavakoli et al. [73]. Copyright 2018, Wiley-VCH.

mixed perovskites as well. Liu et al. [77] reported the perovskite $FAPb_{1-x}Sn_xI_3$ to explore the effect of FA in Sn–Pb-based PSCs for the first time. The inclusion of FA^+ in Pb perovskites improves the thermal stability as well as broadens the absorption region compared to MA only based Pb perovskites. However, $FAPbI_3$ forms the yellow non-perovskite phase at room temperature which can be stabilized by using a mixed perovskite such as $FA_{1-x}MA_xPbI_{3-x}Br_x$. Interestingly, the substitution of Pb by Sn also stabilizes the perovskite structure, and hence, a PCE of over 10% was reported with $FAPb_{0.5}Sn_{0.5}I_3$, covering an absorption spectrum up to 965 nm. Further, the substitution of FA^+ with Cs metal cation is known to be retarding the crystallization of perovskite leading to the better quality of films with higher stability [74, 75]. The substitution of MA by Cs in $MASn_xPb_{1-x}I_3$ also shows the same behavior, and an improvement in PCE and stability has been noticed. Leijtens et al. [78] showed that the PSC based on a perovskite $FA_{0.75}Cs_{0.25}Sn_{0.5}Pb_{0.5}I_3$ exhibits the air stability of 150 hours at 85 °C which was the longest air stability of any Sn–Pb PSC reported by the time. Recently, the operational stability of 1000 hours in N_2 atmosphere and 1000 hours of shelf-life stability in ambient air at 85 °C has been reported for unencapsulated PSCs fabricated by using such perovskite materials [79]. The cation mixing of MA and FA at A position has also been reported ($FA_{1-a}MA_aSn_xPb_{1-x}I_3$), and so far the solar cells fabricated by this kind of perovskite material ($FA_{0.7}MA_{0.3}Sn_{0.5}Pb_{0.5}I_3$) hold the highest PCE (21.1%) record in Sn–Pb mixed systems [66]. The first-ever report on using such MA and FA cationic mixture in Sn–Pb perovskites showed a PCE of 15.08% with $FA_{0.6}MA_{0.4}Sn_{0.6}Pb_{0.4}I_3$ [60] perovskites. Later, the same perovskite materials with the improvements in crystallinity and carrier lifetime by increasing the perovskite layer thickness reached a high PCE of 17.6% [61], which was a remarkable work and after the report, many research papers came on utilizing the $FA_{1-a}MA_aSn_xPb_{1-x}I_3$ perovskite materials with high PCEs. Moreover, with the ongoing research to improve the PCEs, there are various efforts in decreasing the V_{oc} loss also. For that purpose, the substitution of I anion by Br anion has been reported [64], the small amount of Br inclusion in the lattice leads to a slight increase in the bandgap; however, the prominent effect is the decrease in the reverse saturation current leading to increase in the V_{oc}. Also, to reduce the bulk recombination within the perovskite the addition of Cl anion has been reported in Pb-based PSCs, the same effect is observed in the Sn–Pb mixed systems as well with the $FA_{0.6}MA_{0.4}Sn_{0.6}Pb_{0.4}I_xCl_{3-x}$ perovskite-based PSCs, PCE was improved from 17.29% to 18.4% with Cl anion addition [65]. The addition of three cations (FA, MA, Cs) ion into the lattice of Pb perovskites has been extensively studied by several groups to increase the efficiency, stability, and reproducibility [8]. Hence, in the case of Sn–Pb perovskites as well the role of the addition of three cations has been reported recently by a few groups [20, 67]. The perovskite material $Cs_{0.025}FA_{0.475}MA_{0.5}Sn_{0.5}Pb_{0.5}I_3$ and $Cs_{0.05}FA_{0.45}MA_{0.5}Sn_{0.5}Pb_{0.5}I_3$-based PSCs showed the PCEs of 20.4% and 20.2%, respectively. The small addition of Cs cation into the perovskite leads to the decrease in the crystal lattice strain and trap densities resulting in overall improved PCEs.

14.5.2 Strategies to Improve the Efficiency

The Sn–Pb perovskites are derived from the Sn perovskites; therefore, they have the same kind of issues and solutions to them, even though Sn–Pb mixed perovskites are more stable than Sn perovskites. In the development of Sn–Pb PSCs, the focus has been on decreasing the trap densities, both at the interface and bulk of the perovskite films to increase the carrier lifetime and diffusion lengths by various strategies.

14.5.2.1 Use of Additives

Like Sn perovskites, the addition of SnF_2 in a small amount is necessary to reduce the formation of Sn vacancies [30]. However, the SnF_2 alone is not enough to reduce the formation of Sn^{4+}; therefore, the addition of a small amount of Cl and Br anion has been reported to achieve the better performance of the solar cells as mentioned in the previous section also. Another kind of additives which led to the improvement in Sn–Pb PSCs are thiocyanates (SCN^-) salts such as lead thiocyanate $Pb(SCN)_2$ [60], methylammonium thiocyanate (MASCN) [80], and guanidinium thiocyanate (GuaSCN) [69]. All these additives were added into the perovskite precursor solution to fabricate the films. $Pb(SCN)_2$ addition has been reported to obtain high-quality perovskite films with large and uniform grains. The perovskite films with large grains are beneficial for the solar cell performances owing to the reduced grain boundary scattering, extended carrier lifetime, and improved crystal quality. The addition of MASCN also demonstrated the same effect as shown in Figure 14.19a. The addition of GuASCN has also been reported to improve the optoelectronic properties of $(FASnI_3)_{0.6}(MAPbI_3)_{0.4}$ by reducing the defect density, increasing the carrier lifetime to >1 μs and diffusion length to 2.5 μm. The GuASCN can form two-dimensional (2D) structures at gain boundaries and can passivate the defects formed at the grain boundaries leading to suppression in Sn vacancies and enhanced solar cell characteristics. Some of the other approaches involve the addition of fluoro-alkyl substituted fullerenes (DF-C_{60}) within the bulk of the Sn–Pb perovskite to passivate the bulk generated traps [81] as shown in Figure 14.19b. The additive can be used to passivate the defects and can work as ETL as well. The additive can be mixed with perovskite precursor forming a bulk heterojunction (BHJ) of additive with perovskite or can be added to the antisolvent to fabricate a graded heterojunction (GHJ) type solar cell. The outcome of this research work was the high V_{oc} of 0.89 V (~92% of SQ limit) for a perovskite with a bandgap of 1.22 eV. Recently, the addition of tin metal powder in perovskite precursor has been reported [66], similar to the work reported in Sn-based PSCs [55]. The Sn metal acts as a reducing agent, thus reduces the concentration of Sn^{4+} by reducing it to Sn^{2+}. As a result, the trap densities were reduced, and carrier diffusion length increased up to 3 μm leading to the highest PCE of 21.1% so far in Sn–Pb mixed PSCs.

14.5.2.2 Device Engineering

To achieve better performances from a perovskite absorber the electrons and holes need to be extracted efficiently. In recent years, surface passivation has been in focus to improve the interfacial charge transfer in all kinds of PSCs. Interfaces play a more dominant role when it comes to Sn and Sn–Pb mixed PSCs because of their fast tendency to form SnO_2 and iodide-deficient surface [39] resulting in a nonstochiometric perovskite surface that needs to be reduced for the better charge transfer. Recently, the passivation by two-dimensional (2D) organic perovskites has been adopted in Sn–Pb mixed systems as well. PEA^+ ligands have been used [68] in the same way as mentioned for DF-C60 (GHJ) in the previous section (Figure 14.19c). The perovskite films prepared by dropping PEAI mixed in antisolvent exhibited better carrier lifetime owing to the reduced surface and bulk traps. The carrier lifetime is measured from both the film and glass sides as shown in Figure 14.19d, which confirmed that PEAI addition in antisolvent is passivating both the surface and bulk defects. Similarly, guanidinium bromide (CH_6N_3Br; GABr) has been demonstrated to passivate the surface traps [83] leading to a FF of 83%. This posttreatment by GABr resulted in more favorable band alignment at the interfaces, enabling better charge extraction. The energy level matching at the interfaces is very crucial for an efficient solar cell, some efforts on energy alignment on HTL side have been demonstrated for Sn–Pb PSC [82], as shown in Figure 14.19e. It is reported that by doping generally used HTL (PEDOT: PSS) with a perfluorinated ionomer (PFI) the HOMO energy of HTL can be shifted from −5.02 to −5.19 eV resulting in better hole collections that led to increase in V_{oc} over 50 mV. However, recently, it is

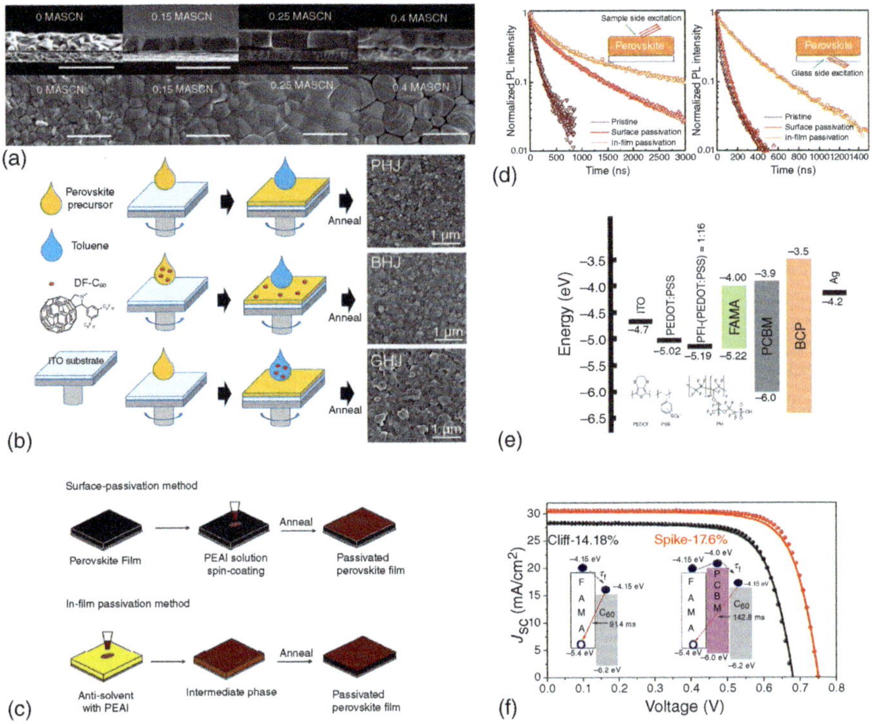

Figure 14.19 (a) Change in surface morphology with different amounts of MASCN additive. *Source*: Reproduced with permission from Lian et al. [80]. Copyright 2019, Wiley-VCH. (b) Thin films prepared by using different techniques. *Source*: Reproduced with permission from Rajagopal et al. [81]. Copyright 2017, American Chemical Society. (c) Thin films prepared by using surface-passivation method and in-film passivation method, (d) PL spectra measured from the sample side and glass side. *Source*: (c, d) Wei et al. [68]. (e) Energy diagram showing the role of energy level alignment. *Source*: Tang et al. [82]. (f) Energy diagram showing the effect of conduction band offset in Sn–Pb mixed PSCs. *Source*: Kapil et al. [62].

also reported that a conduction band offset (CBO) between the absorber layer and ETL might be more effective to increase the PSC performance [62], as shown in Figure 14.19f. A bilayer of PCBM/C60 (ETL) is used which resulted in increased carrier lifetime, increased recombination lifetime (91.4–142.8 ms) that finally led to increased V_{oc} (0.75 V) and PCE (17.6%). Some other recent efforts in device engineering are related to the light management in the PSCs to increase the photon harvesting by the absorber layer, which has been reported by few groups leading to short circuit current density of more than 30 mA/cm^2. The use of transparent conducting oxide (TCO) glass with less free carrier absorption is important for the Sn–Pb mixed PSCs [67], due to which incident photon-to-electron conversion efficiency (IPCE) ~80% at 900 nm is achieved with a record J_{sc} of 33.14 mA/cm^2 so far. Also, by increasing the absorber layer ($FA_{0.7}MA_{0.3}Sn_{0.5}Pb_{0.5}I_3$ perovskite) thickness to 860 nm [66], the J_{sc} > 32 mA/cm^2 has been reported that led to overall best PCE of 21.1% so far in Sn–Pb mixed PSCs. As mentioned earlier, high carrier diffusion length (3 μm) is the key to better charge collection in thicker layers.

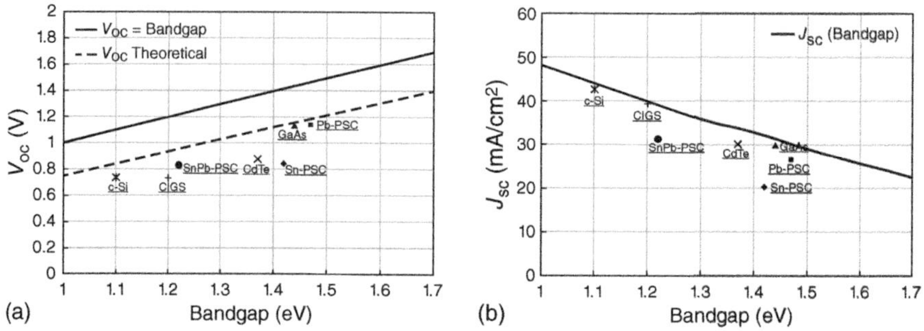

Figure 14.20 (a) Diagram showing V_{oc} deficit and (b) J_{sc} deficit in different photovoltaic technologies.

14.6 Conclusion and Outlook

This thorough study on Sn and Sn–Pb mixed PSCs has highlighted the main research findings so far in the research field. Different kinds of perovskite crystal structures utilized in the fabrication of the PSCs are discussed with their advantages and disadvantages. In this study, we mentioned that there are two major problems of Sn and Sn–Pb mixed perovskites; first, the fast crystallization of Sn perovskite that can lead to poor film morphology; and second, the formation of Sn vacancy owing to the fast oxidation of Sn^{2+} to Sn^{4+}. Several methods that have been used to address these problems are discussed in detail. We have summarized the V_{oc} and J_{sc} of some of the commercially available solar cells along with that of the best Sn, Sn–Pb, Pb PSCs in Figure 14.20a. The plot shows that the V_{oc} loss for best reported Sn, Sn–Pb, and Pb PSCs are 0.58, 0.39, and 0.33 V, respectively. Hence, according to the data Pb-based PSC is almost near to the V_{oc} decided by the SQ limit. However, for Sn and Sn–Pb-based PSCs still V_{oc} near to 1.12 and 1 V, respectively are possible. Therefore, several methods such as interface and bulk defect passivation, device engineering, and with the use of proper perovskite absorber these mentioned V_{oc} values can be achieved. Similarly, from Figure 14.20b, we can conclude that maximum (assuming 100% photon harvest) J_{sc} of ~32.5 mA/cm² for Sn and ~39.5 mA/cm² for Sn–Pb mixed PSCs are achievable. All the necessary steps for getting the high V_{oc} apply to J_{sc} as well, besides, the light absorption can be enhanced by increasing thickness of the perovskite absorber layer with diffusion length comparable to thickness.

References

1 Snaith, H.J. (2013). *Phys. Chem. Lett.* 4: 3623.

2 Kojima, A., Teshima, K., Shirai, Y., and Miyasaka, T. (2009). *J. Am. Chem. Soc.* 131: 6050.

3 Im, J.H., Lee, C.R., Lee, J.W. et al. (2011). *Nanoscale.* 3: 4088.

4 Lee, M.M., Teuscher, J., Miyasaka, T. et al. (2012). *Science* 338: 643.

5 Kim, H.S., Lee, C.R., Im, J.H. et al. (2012). *Sci. Rep.* 2: 1.
6 Marchioro, A., Teuscher, J., Friedrich, D. et al. (2014). *Nat. Photonics* 8: 250.
7 Li, C., Lu, X., Ding, W. et al. (2008). *Acta Crystallogr., Sect. B: Struct. Sci.* 64: 702.
8 Saliba, M., Matsui, T., Seo, J.-Y. et al. (2016). *Energy Environ. Sci.* 9: 1989.
9 Saliba, M., Matsui, T., Domanski, K. et al. (2016). *Science* 354: 206.
10 Ogomi, Y., Morita, A., Tsukamoto, S. et al. (2014). *J. Phys. Chem. Lett.* 5: 1004.
11 Min, H., Kim, M., Lee, S.U. et al. (2019). *Science* 366: 749.
12 Kapil, G., Ohta, T., Koyanagi, T. et al. (2017). *J. Phys. Chem. C* 121: 13092.
13 Stoumpos, C.C., Malliakas, C.D., and Kanatzidis, M.G. (2013). *Inorg. Chem.* 52: 9019.
14 Kagan, C.R., Mitzi, D.B., and Dimitrakopoulos, C.D. (1999). *Science* 286: 945.
15 Kamarudin, M.A., Hirotani, D., Wang, Z. et al. (2019). *J. Phys. Chem. Lett.* 10: 5277.
16 Ng, C.H., Nishimura, K., Ito, N. et al. (2019). *Nano Energy* 58: 130.
17 Ng, C.H., Hamada, K., Kapil, G. et al. (2020). *J. Mater. Chem. A* 8: 2962.
18 Gruginskie, N., Cappelluti, F., Bauhuis, G.J. et al. (2020). *Prog. Photovoltaics Res. Appl.* 28: 266.
19 Andreani, L.C., Bozzola, A., Kowalczewski, P. et al. (2019). *Adv. Phys. X* 4: 1548305.
20 Yang, Z., Yu, Z., Wei, H. et al. (2019). *Nat. Commun.* 10: 1.
21 Wehrenfennig, C., Liu, M., Snaith, H.J. et al. (2014). *Energy Environ. Sci.* 7: 2269.
22 Johnston, M.B. and Herz, L.M. (2016). *Acc. Chem. Res.* 49: 146.
23 Wehrenfennig, C., Eperon, G.E., Johnston, M.B. et al. (2014). *Adv. Mater.* 26: 1584.
24 Noel, N.K., Stranks, S.D., Abate, A. et al. (2014). *Energy Environ. Sci.* 7: 3061.
25 Hao, F., Stoumpos, C.C., Cao, D.H. et al. (2014). *Nat. Photonics* 8: 489.
26 Ke, W., Stoumpos, C.C., Spanopoulos, I. et al. (2017). *J. Am. Chem. Soc.* 139: 14800.
27 Li, F., Zhang, C., Huang, J.H. et al. (2019). *Angew. Chem. Int. Ed.* 58: 6688.
28 Yu, Y., Zhao, D., Grice, C.R. et al. (2016). *RSC Adv.* 6: 90248.
29 Zhao, B., Abdi-Jalebi, M., Tabachnyk, M. et al. (2017). *Adv. Mater.* 29: 1604744.
30 Kumar, M.H., Dharani, S., Leong, W.L. et al. (2014). *Adv. Mater.* 26: 7122.
31 Heo, J.H., Kim, J., Kim, H. et al. (2018). *J. Phys. Chem. Lett.* 9: 6024.
32 Chen, M., Ju, M.G., Garces, H.F. et al. (2019). *Nat. Commun.* 10: 1.
33 Koh, T.M., Krishnamoorthy, T., Yantara, N. et al. (2015). *J. Mater. Chem. A* 3: 14996.
34 Lee, S.J., Shin, S.S., Kim, Y.C. et al. (2016). *J. Am. Chem. Soc.* 138: 3974.
35 Ke, W., Stoumpos, C.C., Zhu, M. et al. (2017). *Sci. Adv.* 3: 1.
36 Liao, W., Zhao, D., Yu, Y. et al. (2016). *Adv. Mater.* 28: 9333.
37 Shao, S., Liu, J., Portale, G. et al. (2018). *Adv. Energy Mater.* 8: 1702019.
38 Jokar, E., Chien, C.H., Tsai, C.M. et al. (2019). *Adv. Mater.* 31: 1.
39 Nishimura, K., Kamarudin, M.A., Hirotani, D. et al. (2020). *Nano Energy* 74: 104858.
40 Jiang, X., Wang, F., Wei, Q. et al. (2020). *Nat. Commun.* 11: 1.

41 Chung, I., Lee, B., He, J. et al. (2012). *Nature* 485: 486.
42 Chen, Z., Wang, J.J., Ren, Y. et al. (2012). *Appl. Phys. Lett.* 101: 093901.
43 Shi, T., Zhang, H.S., Meng, W. et al. (2017). *J. Mater. Chem. A* 5: 15124.
44 Hao, F., Stoumpos, C.C., Guo, P. et al. (2015). *J. Am. Chem. Soc.* 137: 11445.
45 Hayase, S. (2017). *Electrochemistry* 85: 222.
46 Zhao, Z., Gu, F., Li, Y. et al. (2017). *Adv. Sci.* 4: 1700204.
47 Krishnamoorthy, T., Ding, H., Yan, C. et al. (2015). *J. Mater. Chem. A* 3: 23829.
48 Ito, N., Kamarudin, M.A., Hirotani, D. et al. (2018). *J. Phys. Chem. Lett.* 9: 1682.
49 Konstantakou, M. and Stergiopoulos, T. (2017). *J. Mater. Chem. A* 5: 11518.
50 Yokoyama, T., Song, T.B., Cao, D.H. et al. (2017). *ACS Energy Lett.* 2: 22.
51 Liu, J., Ozaki, M., Yakumaru, S. et al. (2018). *Angew. Chem. Int. Ed.* 57: 13221.
52 Li, W., Li, J., Li, J. et al. (2016). *J. Mater. Chem. A* 4: 17104.
53 Song, T.B., Yokoyama, T., Stoumpos, C.C. et al. (2017). *J. Am. Chem. Soc.* 139: 836.
54 Kayesh, M.E., Chowdhury, T.H., Matsuishi, K. et al. (2018). *ACS Energy Lett.* 3: 1584.
55 Gu, F., Ye, S., Zhao, Z. et al. (2018). *Sol. RRL* 2: 1800136.
56 Jokar, E., Chien, C.H., Fathi, A. et al. (2018). *Energy Environ. Sci.* 11: 2353.
57 Anaya, M., Lozano, G., Calvo, M.E., and Míguez, H. (2017). *Joule* 1: 769.
58 Hao, F., Stoumpos, C.C., Chang, R.P.H., and Kanatzidis, M.G. (2014). *J. Am. Chem. Soc.* 136: 8094.
59 Goyal, A., McKechnie, S., Pashov, D. et al. (2018). *Chem. Mater.* 30: 3920.
60 Liao, W., Zhao, D., Yu, Y. et al. (2016). *J. Am. Chem. Soc.* 138: 12360.
61 Zhao, D., Yu, Y., Wang, C. et al. (2017). *Nat. Energy* 2: 1.
62 Kapil, G., Ripolles, T.S., Hamada, K. et al. (2018). *Nanoletters* 18: 3600.
63 Tsai, C.M., Wu, H.P., Chang, S.T. et al. (2016). *ACS Energy Lett.* 1: 1086.
64 Li, C., Song, Z., Zhao, D. et al. (2018). *Adv. Energy Mater.* 9: 1803135.
65 Zhao, D., Chen, C., Wang, C. et al. (2018). *Nat. Energy* 3: 1093.
66 Lin, R., Xiao, K., Qin, Z. et al. (2019). *Nat. Energy* 4: 864.
67 Kapil, G., Bessho, T., Ng, C.H. et al. (2019). *ACS Energy Lett.* 4: 1991.
68 Wei, M., Xiao, K., Walters, G. et al. (2020). *Adv. Mater.* 32: 1.
69 Tong, J., Song, Z., Kim, D.H. et al. (2019). *Science* 364: 475.
70 Yang, Z., Rajagopal, A., Jo, S.B. et al. (2016). *Nano Lett.* 16: 7739.
71 Zuo, F., Williams, S.T., Liang, P.W. et al. (2014). *Adv. Mater.* 26: 6454.
72 Yang, Z., Rajagopal, A., Chueh, C.C. et al. (2016). *Adv. Mater.* 28 (40): 8990–8997.
73 Tavakoli, M.M., Zakeeruddin, S.M., Grätzel, M., and Fan, Z. (2018). *Adv. Mater.* 30: 1.
74 Liu, X., Yang, Z., Chueh, C.C. et al. (2016). *J. Mater. Chem. A* 4: 17939.
75 Prasanna, R., Gold-Parker, A., Leijtens, T. et al. (2017). *J. Am. Chem. Soc.* 139: 11117.
76 Zhu, H.L., Xiao, J., Mao, J. et al. (2017). *Adv. Funct. Mater.* 27: 1605469.
77 Liu, J., Wang, G., Song, Z. et al. (2017). *J. Mater. Chem. A* 5: 9097.
78 Leijtens, T., Prasanna, R., Bush, K.A. et al. (2018). *Sustain Energy Fuels* 2: 2450.

79 Prasanna, R., Leijtens, T., Dunfield, S.P. et al. (2019). *Nat. Energy* 4: 939.

80 Lian, X., Chen, J., Zhang, Y. et al. (2019). *Adv. Funct. Mater.* 29: 1.

81 Rajagopal, A., Liang, P.W., Chueh, C.C. et al. (2017). *ACS Energy Lett.* 2: 2531.

82 Tang, H., Shang, Y., Zhou, W. et al. (2019). *Sol. RRL* 3: 1.

83 Jayawardena, K.D.G.I., Bandara, R.M.I., Monti, M. et al. (2020). *J. Mater. Chem. A.* 8: 693.

15

Stability of Hybrid Perovskite Solar Cells

Seigo Ito

Department of Materials and Synchrotron Radiation Engineering, Graduate School of Engineering, University of Hyogo, 2167 Shosha, Himeji, Hyogo 671-2280, Japan

15.1 Introduction: Trigger of the Degradation

Photovoltaic panels have three requirements for commercialization: "high efficiency," "low cost," and "high stability." Although organo-lead-halide perovskite solar cells (PSCs) have a high efficiency of 25.2% [1, 2] and a low cost [3], the stability is insufficient, which is the most important issue. This organo-lead-halide perovskite crystal is quite weak against "water," "heat," and "light," because it is just a salt similar to a soap ("savon" in French). Moreover, one of the components (methyl ammonium cation ($CH_3NH_3^+$)) can react with water and polar sorbents, easily [4–6]. In the initial stage of this research field (in 2014), the PSCs ($CH_3NH_3PbI_3$, written as $MAPbI_3$) deteriorated in one day of sunlight irradiation without sealing [7]. Afterwards, the stability was improved drastically with the efforts of scientists in this field. However, understanding of the test conditions is very complicated due to the combination of "***water***," "***heat***," and "***light***." Specifically, the deterioration by "light" can occur, resulting in issues with "***applied voltage***," "***crystal defect***," and "***ion migration***" simultaneously in the PSCs (Figure 15.1).

Another problem in this stability issue leads to the significantly less reproducibility in the fabrication of PSCs. In spite of producing PSCs with the same structure, the stability can differ based on the fabricating laboratory and scientists. The fabrication window is quite narrow due to the effects of moisture, temperature, material purity, and handling ability. Since the resulting devices are different from each other, simple comparisons between the published results and the understanding of the inside of the cells are quite difficult.

The deteriorations of perovskite crystal can occur by the defects in or on the crystals [8–11]. The defects can be in the crystal with a void in the crystal elements (A^+, Pb^{2+}, or X^-) or contamination (water, Li^+ dopants from HTM, and so on) inside or on the surface of the crystal grain. The void can occur by the non-stoichiometry of elements ratio. The defects itself can migrate with light and applied voltage. Finally, the perovskite crystal can decompose and be inactivated for photovoltaics. The pure

Hybrid Perovskite Solar Cells: Characteristics and Operation,
First Edition. Edited by Hiroyuki Fujiwara.

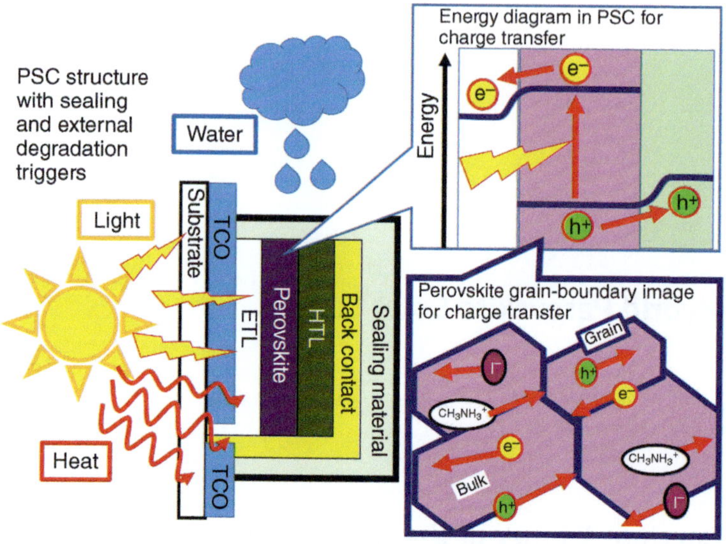

Figure 15.1 Illustration of deterioration factors in perovskite solar cells (PSCs). "ETL" and "HLT" are electron and hole transporting layers, respectively.

materials for perovskite source and mixed cation/anion perovskite crystals can help to reduce the occurrence of defects.

Habisreutinger et al. (Snaith group, Oxford University) have summarized the stability tests in a review paper with a big data table [12], which is one of the best efforts as of the time of this paper's publication to understand the stability issue. In the table, they have categorized the results using the terms of "system and structure," "humidity (RH) (%) of atmosphere," "light intensity (mW/cm^2) with UV and heat (°C) at MPP," "time (h)," "initial PCE (%) with decrease rate (%/100 h), total decrease (%) and control degradation (%)." The terms of "System" and "Structure" show the method to improve the stability of PSCs. Changing structure using new materials, the stability of PSCs can be improved. The degradation points are (i) interface between electron-transportation layer (ETL) and perovskite, (ii) perovskite itself, and (iii) interface between perovskite and hole-transportation layer (HTL). The ETL and HTL can also deteriorate, but the weakest point in PSCs is the perovskite crystal. Using new materials for ETL and HLT, the stability of PSCs has been improved by passivating the interface. For the real commercialization in the field of photovoltaic energy resources, the stability against "humidity (RH) (%) of atmosphere" was tested intensively, because the weakest point of metal-lead-halide perovskite crystal is water. The next issue was "light intensity (mW/cm^2) with UV and heat (°C) at MPP," which can be applied on PSCs at the same time. Of course, the stability duration from initial *Eff*. (PCE) (%) is quite important with "decrease rate (%/100 h), total decrease (%) and control degradation (%)."

Additionally, the condition of stability test is important. One of the most "soft" conditions is "shelf stability," which means that the PSCs were kept on laboratories shelves under dark conditions, and that the photovoltaic parameters were measured at intervals. Of course, this ability of "shelf stability" cannot work for the utilizations of sunlight energy conversion. However, the results have been important at the initial stage of the PSCs research [13], because the first PSCs deteriorated in minutes just after assembling without photovoltaic measurement [14].

Now, the degradation "trigger" and "cause and efforts" were listed as below:

- *Triggers*: Water/heat/light (bias voltage)
- *Causes and efforts*: Crystal quality by changing materials/water stability/methyl ammonium (MA) free/defects and grain passivation/ion migration/2-D perovskite/interface degradation/sealing effect.

The topics of causes and efforts for the degradation of PSCs above are mixed up. These factors inside of PSCs make it complex and obscure understanding of the stability issue. Moreover, since the number of papers about PSCs has been over 3000 per year (2018–2020, verified by SCOPUS, Elsevier), listing of all of publications is impossible. Hence, although there are a lot of missing points in this manuscript due to the high-speed progress of stability issue in the world, this topic has been presented in order to assist understanding for the readers.

15.2 Crystal Quality for Stable Perovskite Solar Cells

In order to fabricate stable PSCs, preparation of stable perovskite crystal is quite important. "Perovskite" is a crystal, which has a cubic structure as shown in Figure 15.2. The original compound found by L.A. Perovski (1792–1856) was metal oxides [15], such as $CaTiO_3$, which contains three components of Ca^{2+}, Ti^{4+}, and O^{2-}. For the perovskite crystal of organo-lead halide, it is composed of A-site mono-valence cation (A^+), B-site two-valence lead cation (B^{2+}), and X-site mono valence anion (X^-). At A, B, and X sites, organo cations, lead cation (Pb^{2+}), and halide anions can be located, respectively. It can be written as ABX_3 or $APbX_3$. The most famous combination of $APbX_3$ for PSCs is MA lead iodide (written as "$MAPbI_3$" or "$CH_3NH_3PbI_3$"), which has been published in the initial stage of PSC research [13, 16]. In order to consider the possible combinations of these components, tolerance factor (t) should be considered $[(r_B + r_X)\sqrt{2} = t(r_A + r_X)]$, which can be derived using the two equations in Figure 15.2. If t value is too small

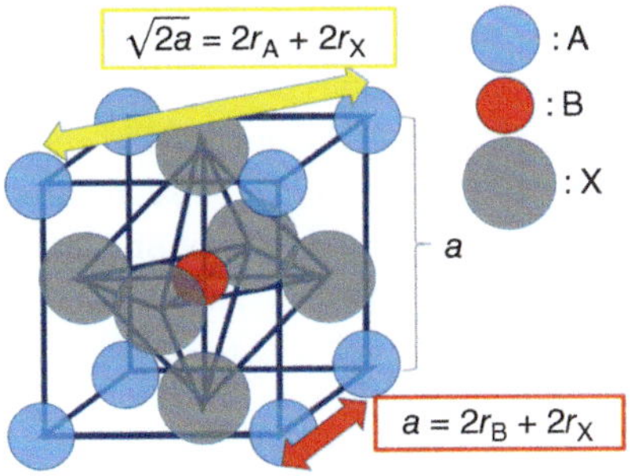

Figure 15.2 Unit structure of organo-metal-halide perovskite crystal.

or too large from a range ($0.8 < t < 1.0$) [17], the perovskite crystal can become unstable, and it can change to another crystal phase, easily. For example, $MAPbI_3$ (or $CH_3NH_3PbI_3$) consisted of $CH_3NH_3^+$ ($r = 0.180$ nm), Pb^{2+} ($r = 0.119$ nm), and I^- ($r = 0.220$ nm), to be $t = 0.8345$. In the case of $CaTiO_3$ (Ca^{2+}: $r = 0.100$ nm; Ti^{4+}: $r = 0.061$ nm; O^{2-}: $r = 0.140$ nm) and $CsPbI_3$ (Cs^+: $r = 0.167$ nm; Pb^{2+}: $r = 0.119$ nm; I^-: $r = 0.220$ nm), $t = 0.844$ and 0.807, respectively. Hence, perovskite crystal can be stable with t around 0.83.

"A^+" can be methyl ammonium ion ($CH_3NH_3^+$: MA^+), formamidinium ion (NH_2-CH-NH_2^+: FA^+), cesium ion (Cs^+), rubidium ion (Rb^+), potassium ion (K^+), and so on. Na^+ and Li^+ are too small to be in the perovskite crystal. Li^+ ion has been utilized as the dopant in spiro-OMeTAD (organic HTL) [13, 16]. It was known that this Li^+ ion can decompose the $MAPbI_3$ perovskite crystal and deteriorate the photovoltaic parameters. Investigating dopant-free HTM has been an important topic for the stability issue [18, 19]. Fr^+ is the radioactive element, which cannot be used normally for material scientists. Other large organo cations (e.g. $C(NH_2)_3^+$ [guanidium]: GA^+) can be squeezed at the A site between MA^+ and FA^+ [20]. Since such organic cations (except for MA^+ and FA^+) are too large to be in the range of tolerance factor (t), the utilization of such single large cation cannot stabilize the perovskite crystal.

Pb^{2+} at "B-site" can be changed to another two-valence cation, such as tin cation (Sn^{2+}), germanium cation (Ga^{2+}), and so on. Pb^{2+} can perform the highest efficiency as PSCs with 25.2% [1, 2]. However, due to the toxicity, it has been thought that Pb^{2+} should be removed from PSCs. Hence, another element has been considered as an alternate. However, the photoenergy conversion efficiencies were not so much as Pb^{2+}, for example, 12.4% [21] and 13.24% [22] using Sn^{2+}, and 7.1% using Ge^{2+}–Sn^{2+} (as $CsSn_{0.5}Ge_{0.5}I_3$) [23]. Although Sn metal and SnO_2 (as Sn^{4+}) are quite safe and stable materials to be used tableware and heat ray-reflecting glass plates (transparent conducting oxide, called "Low-E"), respectively, it was reported that Sn^{2+} is more toxic than Pb^{2+} [24]. Anyway, for the utilization of PSCs as a large field PV system, these toxic materials should be encapsulated perfectly so as not to contaminate food and human body [25]. Afterwards, as long as the regulation can control the lead leakage, it appears that utilization of lead can be continued for the PSCs, as lead batteries in automobiles.

Similarly, the elements at "X" site can also be changed. However, the number of candidates is quite limited for the "X" site mono anions, which are Cl^-, I^-, Br^-, and SCN^-. At^- is the radioactive element, and we cannot use it. F^- is too small to be in the perovskite crystal. Cl^- can be in the perovskite crystal as pure $MAPbCl_3$ [26]. However, Cl^- ion is too small to be in a perovskite crystal with I^- ion. In this case, Cl^- can be eliminated from $MAPbI_3$ crystal [27]. Interestingly, the mixed halide crystal (e.g. $MAPb(I,Br)_3$) can be segregated to $MAPbI_3$ and $MAPbBr_3$ by light irradiation [28]. Only a limited ratio of Br^- can be present in the perovskite crystal with I^- while maintaining stability. The stability of perovskite crystal was improved with mixed halogen at the "X" site as $MAPb(Br_xI_{1-x})_3$ [29].

Basically, the fabrication of perfect crystal is very difficult. Diamond and silicon single crystal also can have defects due to the crystal tension inside. Therefore, in order to reduce the crystal tension inside, mixing several element at A- and X-sites has been quite effective, resulting in high-stability and high-efficiency PSCs. As 3-cations and 2-halides perovskite, $Cs_{0.05}(MA_{0.17}FA_{0.83})_{0.95}Pb(I_{0.83}Br_{0.17})_3$ (as the material ratio in precursor solution) exhibited a performance of 21.17% conversion efficiency and 250 hours stability at room temperature under light [30]. As 4-cations and 2-halides perovskite, $FA_{0.9025}MA_{0.1805}Cs_{0.045}Rb_{0.05}Pb_{1.1913}I_{2.9855}Br_{0.5776}$ (as the material ratio in precursor solution) showed improved performance of 21.8% conversion efficiency and 500 hours stability at 85 °C under full-sun illumination [17]. For the perovskite crystal in the published best-efficiency PSCs (PCE: 23.32% and 23.2%), MA^+ and FA^+ at the A site have been used with I^- and Br^- at the X site, as $(FAPbI_3)_{0.97}(MAPbBr_3)_{0.03}$ [31] and $(FAPbI_3)_{0.95}(MAPbBr_3)_{0.05}$ [32], which have relatively high stability at 85 °C for 500 hours and at 60 °C for 300 hours, respectively.

15.3 Water-Stable and MA-Free Perovskites

$CH_3NH_3PbI_3$ ($MAPbI_3$) perovskite crystal is quite weak against water. One of the important points is the decomposition of $CH_3NH_3^+$ (MA^+) ion in the $CH_3NH_3PbI_3$ crystal due to the equilibrium with water as [4],

$$CH_3NH_3^+ + H_2O \Leftrightarrow CH_3NH_2 + H_3O^+ \tag{15.1}$$

This Eq. (15.1) can lead the decomposition of $CH_3NH_3PbI_3$ crystal as,

$$CH_3NH_3PbI_3 + H_2O \Leftrightarrow CH_3NH_2 + Pb^{2+} + 3I^- + H_3O^+ \tag{15.2}$$

The boiling point of CH_3NH_2 is 266.83 K [33]. Hence, the CH_3NH_2 molecule can transition to a gas at room temperature. Moreover, the emitted I^- can react with the emitted H_3O^+ in Eq. (15.2) above, as,

$$I^- + H_3O^+ \Leftrightarrow HI + H_2O \tag{15.3}$$

The boiling point of HI is 237.60 K [33], which can also transition to a gas at room temperature. The emitted H_2O in Eq. (15.3) can be recycled into the reaction of Eq. (15.1). Hence, H_2O can work as the catalyst to decompose the $CH_3NH_3PbI_3$ perovskite crystal, resulting in PbI_2 [7].

Therefore, it can be thought that MA^+-free ABX_3 perovskites, such as $FAPbI_3$ or $CsPbI_3$, would be more stable than $MAPbI_3$, because MA^+ is the decomposing center of the perovskite crystal due to the equilibrium equations as above. Actually, the proton from MA^+ (Eq. (15.1)) can react with ZnO EHL, which can accelerate the decomposition of $MAPbI_3$ perovskite crystal to PbI_2. Since $FAPbI_3$ does not have an equilibrium reaction as above, perovskite is stable with ZnO ETL [34]. Hence, as MA-free PSCs, $FAPbI_3$ and $CsPbI_3$ have been investigated. However, the photoactive black phases (α, β, and γ) of $FAPbI_3$ and $CsPbI_3$ can transform immediately into the undesired non-photoactive yellow phase (δ) by water in an ambient atmosphere [35]. Hence, the black-phase stability has been the most important target for MA-free $FAPbI_3$ and $CsPbI_3$-based PSCs [36].

For FA-based PSCs, in order to improve the black-phase stability as MA-free perovskite crystal, $FAPbI_3$ and $CsPbI_3$ were mixed to be stable $FA_{0.9}Cs_{0.1}PbI_3$ PSCs (with 19.0% PCE) for 200 hours stability under continuous white light (sulfur lamp, ≈100 mW/cm^2) illumination (relative humidity [RH] <50%, temperature [T] <65 °C) by N. G. Park group (SKKU, Korea). The devices were encapsulated under N_2 atmosphere using UV-curable sealant [37]. By S. I. Seok group (UNIST), methylenediammonium dichloride (MDACl2) (Figure 15.3a) was mixed with $FAPbI_3$ to create very stable PSCs as unencapsulated devices which retained more than 90% of their initial PCE even after annealing for 20 hours at 150 °C in air (the certified PCE was 23.7%) [38]

For Cs-based PSCs, which can be inorganic perovskite crystal, sodium (Na) doping to $CsPbI_3$ was improved to be non-encapsulated stable device showing no performance degradation after 70 days of storage in air atmosphere [39]. And, using sulfobetaine zwitterions (Figure 15.3b) at the grain boundary of α-$CsPbI_3$, the device became thermally stable and maintained the original efficiency at 90 °C for 20 hours [40]. And, using choline iodide (Figure 15.3c), the black β-$CsPbI_3$ was stabilized as solar cells with 18.4% PCE for 500 hours under light (100 mW/cm^2 intensity) [41]. Additionally, adding chlorine (Cl^-) doping, the stability of black γ-$CsPbI_3$ solar cells was improved with efficiency beyond 16% over 60 days under AM 1.5 light irradiation [42].

It was known that 2-D crystal was more stable than 3-D crystal using amino-alkyl compounds of organo-lead-halide perovskite crystal [43] (Figure 15.4). Reducing the dimension from 3-D to 2-D, the α-$CsPbX_3$ perovskites became stable under ambient air atmosphere without sealing over 40 days [44].

Many efforts have been performed for the creation of stable and high-efficacy MA-free PSCs. But, due to the limitation of this book chapter, all of the publications cannot be listed (please read a good reviewing paper of this field [36, 45]). It is recommended for readers to catch up the latest progress by internet searching for the terms of "MA free perovskite solar cells."

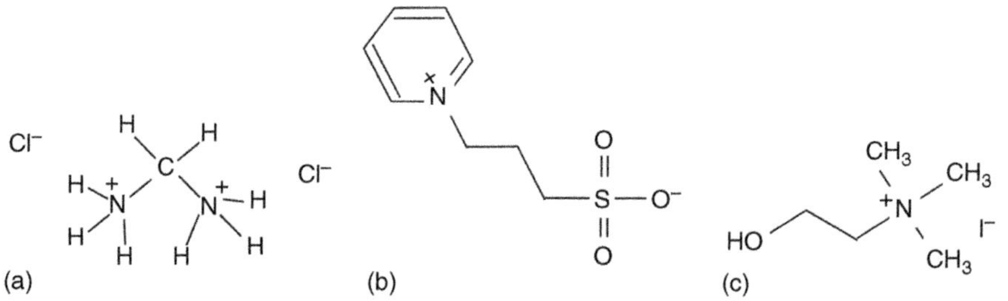

Figure 15.3 Chemical structures of (a) methylenediamine dihydrochloride, (b) sulfobetaine zwitterions, and (c) choline iodide.

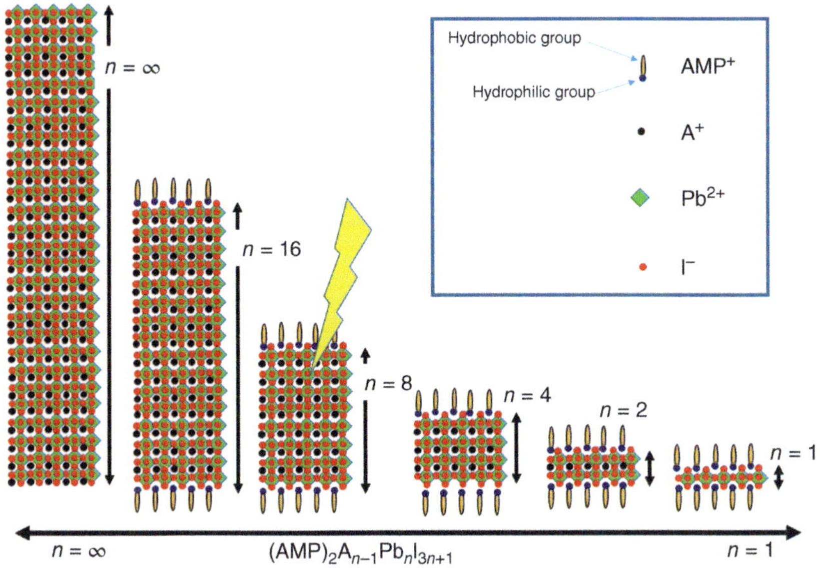

Figure 15.4 Image of reducing dimension from 3-D ($n = \infty$) to 2-D crystal (AMP, amphiphilic molecule).

15.4 Defects and Grain-Surface Ion Migration, and Passivation (Including 2-D Crystal)

The ABX_3 perovskite crystal is very similar to soap, which is a very soft material. A computational chemist (Prof. Aron Walsh) showed a presentation slide as, "the ABX_3 perovskite crystal is soft like a jelly" in a conference (EMRS-2015), and wrote, "The hybrid perovskites are structurally soft materials" [8]. Actually, when you fabricate a perovskite single crystal, you can squeeze it in your fingers, and you can feel the softness of perovskite single crystal like a jelly. Since this perovskite thin-film crystal can be fabricated by spin coating deposition, the crystal cannot be a single crystal and should have considerable defects present. In the ABX_3 perovskite crystal, ions can migrate easily via defects at the grain surface of perovskite crystal [9]. Although ion migration through grain boundaries might dominate, bulk migration can be still large and cannot be completely excluded [10].

By the experiments, the ion migration has been observed by applying voltage and light irradiation on the film [46]. After such ion migration, the ABX_3 perovskite crystal can be changed to non-photoactive material. This light-induced ion migration can be explained using the fact that light irradiation on perovskite crystal can produce zero-valence iodide (I_0) at the defects and that the I_0 can be transformed to I_2 molecule, which was found from light irradiation on perovskite crystal [11]. This light-induced I_2 molecule is not an ion material without ion bonding in perovskite crystal, which will decompose the perovskite crystal structure.

In order to eliminate such surface and grain boundary defects for the stability, they can be passivated using organic or inorganic materials (Figure 15.5a). For the

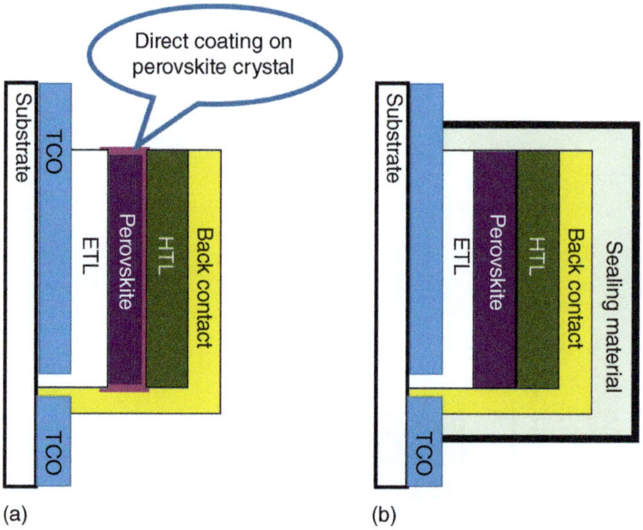

Figure 15.5 (a) Direct coating on perovskite crystal and (b) outer-shell sealing.

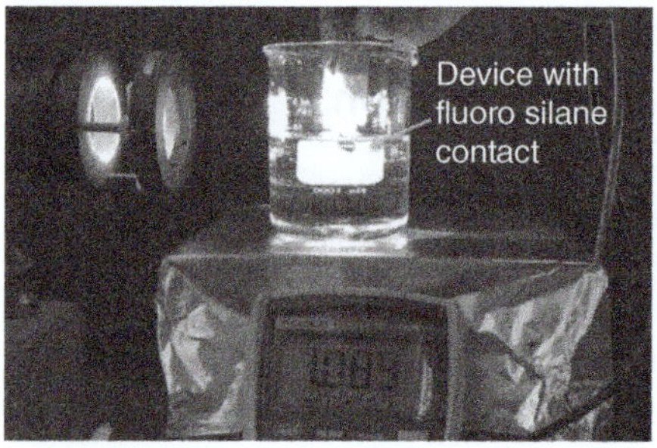

Figure 15.6 Photograph of direct coting perovskite solar cells (as Figure 15.5a) working in water without outer sealing. *Source*: Wang et al. [48].

surface passivation on perovskite crystal directory, the utilized organic compounds are adamantanes [47], polystyrene and fluoro-silane [48], polyethylene-imine [49], PCBM [50], and so on. The utilized inorganic compounds are Al_2O_3 [51, 52], amorphous SiO_2 [53], and so on. These surface passivating materials can work also as the surface sealing materials to prevent water exposure. In some case, it can produce a PV device that can work in water (Figure 15.6) [48].

For the grain-boundary passivation, two-anchored connecting molecules have been applied [53–55] (Figure 15.7a) or a polymer (poly(bithiophene imide) [PBTI]) [56]. Another method is to use amino-alkyl or amino-phenyl amphiphilic molecules [57, 58] (Figure 15.7b). Amino group on molecule can be anchored on the perovskite crystal surface, and alkyl croup on molecule can create a hydrophobic layer between the perovskite crystal grains. Each passivation can work to elongate the lifetime of PSCs.

Since the organo-lead-halide perovskite crystal is ionic, it can be stabilized using double-anchoring or amphiphilic molecules to be "2-D crystal" (Figure 15.4) [43],

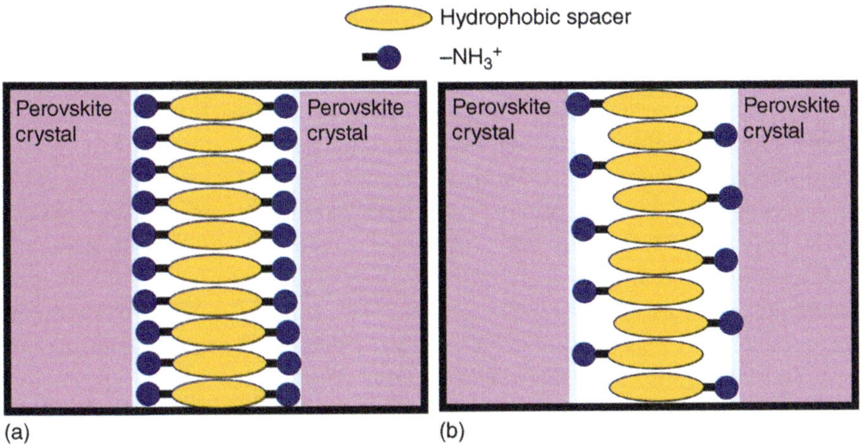

Figure 15.7 Images of grain passivation using (a) two-anchored connecting molecules and (b) amino-alkyl (or phenyl) amphiphilic molecules.

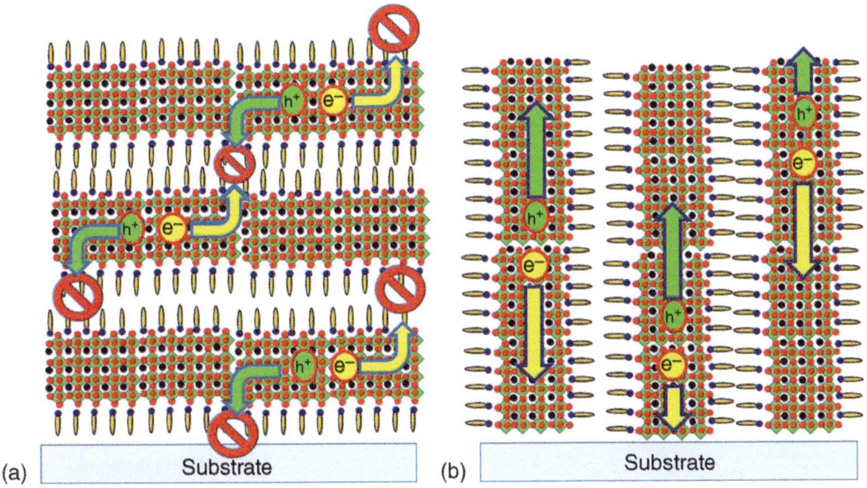

Figure 15.8 2-D perovskite crystal solar cells of (a) face-on and (b) edge-on structure for solar cells.

which can be really close to "soap-structured crystal." Due to the electrically inert long alkyl chain, however, the carrier transportation in 2-D perovskite crystals has been limited (Figure 15.8a). But the stability of 2-D perovskite crystal is quite important. In order to eliminate the hindrance against the carrier transportation by long alkyl chain, the disk of 2-D perovskite crystal was set perpendicular against substrate (Figure 15.8b), resulting in the high-efficiency and high-stability PSCs [59]. As another possibility, the stable 2-D perovskite crystal was coated on 3-D perovskite crystal to create more stable PSCs, in order to protect 3-D perovskite crystal from water (as shown in Figure 15.8a) [60].

15.5 Degradation at Interface with Metal Oxides

Perovskite crystals can deteriorate at the interface with metal-oxide charge transfer materials, such as TiO_2, SnO_2, ZnO, and NiO. Specially, TiO_2 has the functions of photo- and thermal-catalysts. Just by passivating the TiO_2 surface by Sb_2S_3, the stability of PSC was significantly improved (Figure 15.9) [61].

Snaith group (Oxford University, UK) reported the UV effect of perovskite deterioration using TiO_2 ETL [62]. Afterward, Oxford PV (a venture company working with Snaith group) has shifted their R&D target of the PSCs for the commercialization as, (a) TiO_2 ETL/organic HTL [62] → (b) SnO_2 ETL/organic HTL [63] → (c) NiO HTL/organic ETL (PCBM) [64] → (d) organic HTL/organic ETL (C60) [65] (Figure 15.10). However, the best-efficiency PSCs with high stability are still using TiO_2 ETL by the group of Prof. Seo and Prof. Seok, KRICT (as Figure 15.10a) [1, 2]. This topic needs further investigations to be understood.

15.6 Porous Carbon Electrode to Be Very Stable Multiporous-Layered-Electrode Perovskite Solar Cells (MPLE-PSC)

One of the significant strategies to enhance the stability of PSCs is to use carbon electrodes as the back contact, because Au and Ag can be dissolved (or reacted) with

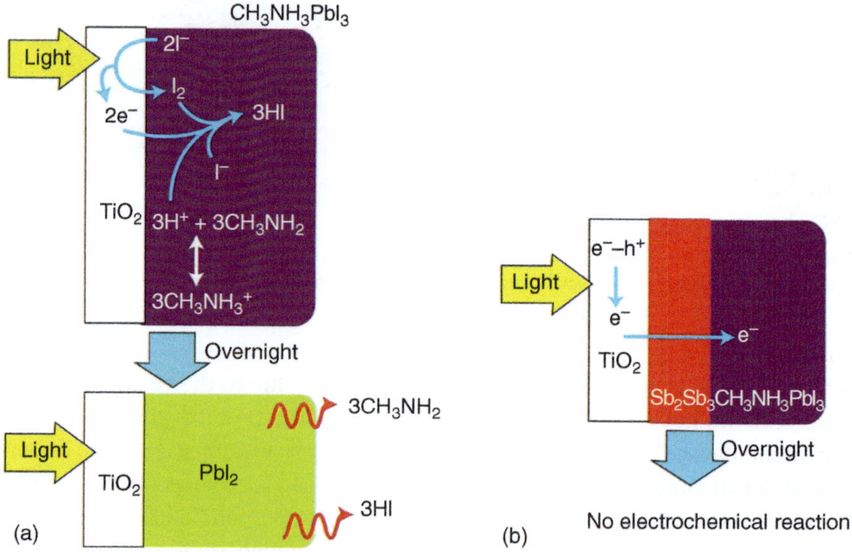

Figure 15.9 Degradation scheme of $CH_3NH_3PbI_3$ perovskite solar cells during light exposure test: (a) ⟨$TiO_2/CH_3NH_3PbI_3$⟩ and (b) ⟨$TiO_2/Sb_3S_3/CH_3NH_3PbI_3$⟩. *Source*: Modified from Ito et al. [61].

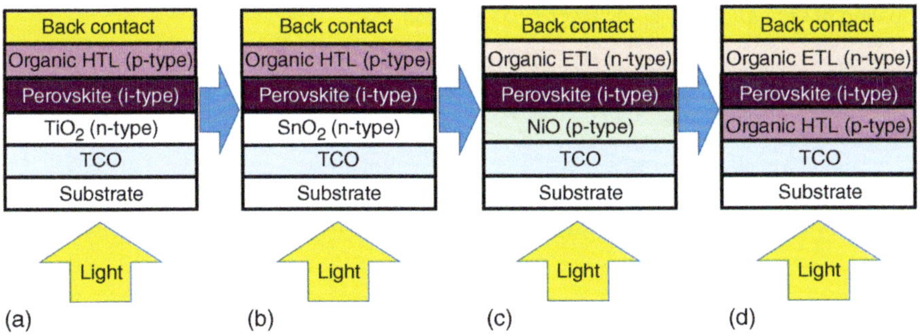

Figure 15.10 Image of R&D progress of Oxford University (Snaith group) with Oxford PV (UK). (a) ⟨substrate/TCO/TiO_2/perovskite/organic HTL/back contact⟩; (b) ⟨substrate/TCO /SnO_2/perovskite/organic HTL/back contact⟩; (c) ⟨substrate/TCO/NiO/perovskite/organic ETL/back contact⟩; (d) ⟨substrate/TCO/organic HTL/perovskite/organic ETL/back contact⟩.

halides in perovskite crystals [66, 67]. Actually, fully printed monolithic-porous-layered-electrode perovskite solar cells (MPLE-PSCs) (Figure 15.11), which was investigated by Hongwei Han group (Huazhong University of Science and Technology, China), have quite high light stability under 1-sun irradiation (AM1.5G) over 1000 hours without sealing [68]. The thermal stability was confirmed at 100 °C over 4000 hours with sealing [69]. This MPLE-PSC has sufficient stability for commercialization, and the best published efficacy has been 17.02% [70]. Now, Wonder Solar (a venture company related with Hongwei Han's group) is producing the perovskite photovoltaic modules (http://www.wondersolar.cn).

15.7 Damp Heat Tests

In order to improve the device stability against water, elimination of water from the perovskite crystal is quite important. However, perfect sealing to block the penetration of water from the outside is quite difficult. Hence, some stability tests of PSCs have been performed in ultra-dry argon-filled glove box [17, 29], resulting in the very high-stability PSCs under 1-sun solar simulator irradiation (AM1.5G) with high temperature (85 °C).

In order to protect perovskite crystal from water, there are two ideas: one is to coat the perovskite crystal, directly (Figure 15.5a) and the other is to coat the device as the outer-shell sealing (Figure 15.5b). Direct coating of perovskite crystal (Figure 15.5a) had been performed using fluoro-silane. With optimized surface coatings, the device can survive in water without outer-shell sealing (Figure 15.6) [48]. The device outer-shell sealing (Figure 15.5b) is quite important for commercialization. Most of the highly stable cells are protected by sealing glue. Dump heat (thermal) tests have been performed for the PSCs with good outer-shell sealing materials (Table 15.1) [71–80]. The targeting test is "IEC 61215 damp heat test" at 85 °C with 85%RH for 1000 hours. In Table 15.1, one of the papers passed the IEC 61215 test using a good sealing material [73].

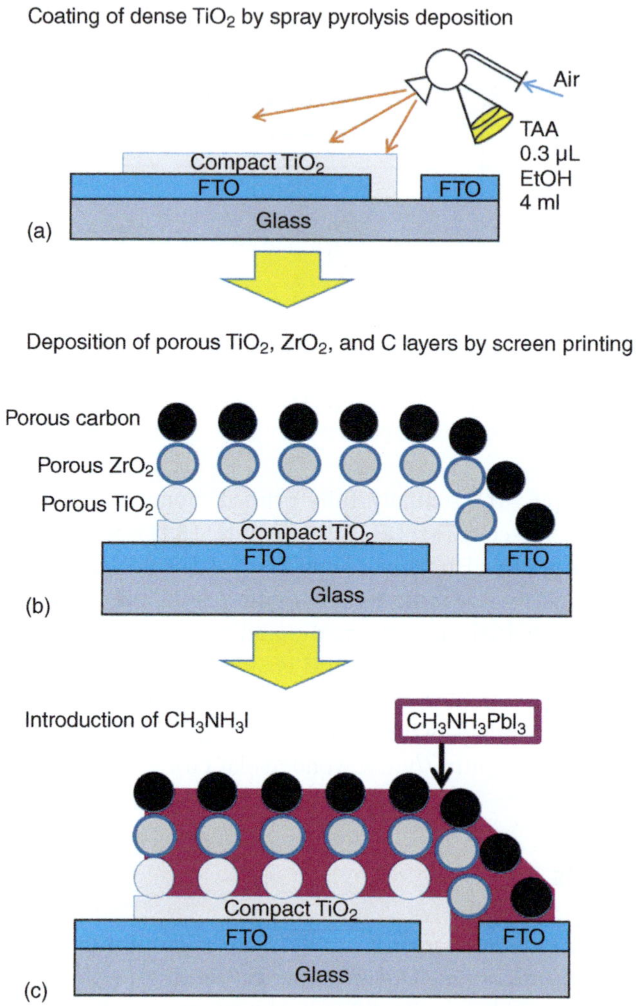

Figure 15.11 Fabrication scheme of monolithic-porous-layered-electrode perovskite solar cells (MPLE-PSC): (a) compact TiO_2 deposition; (b) porous-layered-electrode deposition; (c) perovskite crystal insertion.

15.8 Conclusion

The PSCs had been weak against water and drastically deteriorated during the stability test, because the organic lead halide perovskite crystal has phenomena similar to "soap," as stated above. Due to the hard work by scientists, at present, the PSCs have performed at a high efficiency of 25.2%, and the PSCs have finally passed 85 °C-85%RH dump heat test [73]. Moreover, the 150 °C thermal stability has also been confirmed [38].

Table 15.1 Summery of dump-heat (thermal) tests on perovskite solar cells.

Material	Structure	Sealing material	Temperature (°C)	Humidity (%RH)	Duration (hours)	Starting efficiency (%)	Keeping ratio from the initial efficiency (%)	References
Glass/FTO/c-TiO_2/$CsPbI_2Br$/P3HT/Au	Thin films	No information	85	85	240	6.7	90	[71]
Glass/ITO/SnO_2/$MAFAPbI_3$/P3CT-BN/P3HT/Au	Thin films	P3CT-BN (PVK surface coating)	60	60–70	386	19.23	79	[72]
Glass/FTO/TiO_2/$Cs_{0.05}FA_{0.8}MA_{0.15}Pb(I_{0.85}Br_{0.15})_3$/PTAA/Au	Thin films	Polyisobutylene (PIB)-based polymer sealant (HelioSeal PVS101, KÖMMERLIN, Berlin, Germany)	85	85	1800	13.7	100 (no degradation)	[73]
Glass/FTO/TiO_2/ZrO_2-$(5\text{-AVA})_{0.05}(MA)PbI_3$/carbon	MPLE	Glass frit	85	85	50	8.21	109	[74]
Glass/FTO/NiO/Aa-$MAPbI_3$/TEACl/PC61BM/PEI/Ag	Thin films	A UV curable epoxy resin/a butyl rubber paste at the edge of the top sealing glass	85	85	1300	20	80	[75]

(continued)

Table 15.1 (Continued)

Material	Structure	Sealing material	Temperature (°C)	Humidity (%RH)	Duration (hours)	Starting efficiency (%)	Keeping ratio from the initial efficiency (%)	References
Glass/FTO/TiO_2/$FAPbI_{3-x}Br_x$/PEDOT:PSS/ITO/glass	Thin films	Edge-encapsulated	85	85	1000 (plus 1 sun)	16.4	94.8	[76]
Glass/ITO/E-NiO_x/$MAPbI_3$/PCBM/Ti(Nb)O_x/Ag	Thin films	UV glue	85	85	500	18.49	90	[77]
Glass/ITO/NiO_x/$Cs_{0.17}FA_{0.83}Pb(Br_{0.17}I_{0.83})_3$/LiF/$C_{60}$/$SnO_2$/ITO/Ag	Thin films	ENLIGHT polyolefin	85	85	1000	**11–14**	99	[78]
Glass/FTO/TiO_2/$FAPbI_3$/PTAA/gold	Thin films	PIB edge-sealed	85	85	110	8.8	90	[79]
Glass/FTO/TiO_2/$MAPbI_3$/Spiro-OMeTAD(with Li)/gold	Thin films	Kapton adhesive sheet and UV curable glue (Threebond)	40–50	95	104	13.6 (1 cm^2 area)	73	[80]

"MPLE" in the line of "Structures" means "monolithic-porous-layered-electrode."

Now, due to the toxicity of lead element, a method to block the leakage of lead to circumstance is important [24]. Yabin Qi's group (OIST, Japan) showed the possibility of leakage-blocking ability of self-healing glue on PSCs submodules [81]. A lot of progress is still required for the future commercialization of PSCs.

References

1 NREL. Best Research-Cell Efficiency Chart (by NRLE, USA). https://www.nrel.gov/pv/cell-efficiency.html (accessed 02 April 2021).
2 Green, M.A., Dunlop, E.D., Hohl-Ebinger, J. et al. (2020). *Prog. Photovoltaics Res. Appl.* 28: 3.
3 Cai, M., Wu, Y., Chen, H. et al. (2016). *Adv. Sci.*: 1600269.
4 Ouellette, R.J. (1997). *Organic Chemistry: A Brief Introduction*, 2e. Pearson Education, US. ISBN-10: 0138419337 and ISBN-13: 978-0138419332.
5 Ito, S., Kanaya, S., Nishino, H. et al. (2015). *Photonics* 2: 1043.
6 Kim, B.J., Kim, D.H., Kwon, S.L. et al. (2016). *Nat. Commun.* 7: 11735.
7 Ito, S., Tanaka, S., Vahlman, H. et al. (2014). *ChemPhysChem* 15: 1194.
8 Frost, J.M., Butler, K.T., and Walsh, A. (2014). *APL Mater.* 2: 081506.
9 Eames, C., Frost, J.M., Barnes, P.R.F. et al. (2015). *Nat. Commun.* 6: 7497.
10 Yuan, Y. and Hu, J. (2016). *Acc. Chem. Res.* 49: 286.
11 Motti, S.G., Meggiolaro, D., Barker, A.J. et al. (2019). *Nat. Photonics* 13: 532.
12 Habisreutinger, S.N., McMeekin, D.P., Snaith, H.J., and Nicholas, R.J. (2016). *APL Mater.* 4: 091503.
13 Kim, H.S., Lee, C.R., Im, J.H. et al. (2012). *Sci. Rep.* 2: 591.
14 Kojima, A., Teshima, K., Shirai, Y., and Miyasaka, T. (2009). *J. Am. Chem. Soc.* 131: 6050.
15 Katz, E.A. (2020). *Helvetica* 103: e2000061.
16 Lee, M.M., Teuscher, J., Miyasaka, T. et al. (2012). *Science* 338: 643.
17 Saliba, M., Matsui, T., Domanski, K. et al. (2016). *Science* 354: 206.
18 Kim, G.-W., Kang, G., Kim, J. et al. (2016). *Energy Environ. Sci.* 9: 2326.
19 Liu, J., Wu, Y., Qin, C. et al. (2014). *Energy Environ. Sci.* 7: 2963.
20 Jodlowski, A.D., Roldán-Carmona, C., Grancini, G. et al. (2017). *Nat. Energy* 2: 972.
21 Jiang, X., Wang, F., Wei, Q. et al. (2020). *Nat. Commun.* 11: 1245.
22 Nishimura, K., Kamarudin, M.A., Hirotani, D. et al. (2020). *Nano Energy* 74: 104858.
23 Chen, M., Ju, M.G., Garces, H.F. et al. (2019). *Nat. Commun.* 10: 16.
24 Babayigit, A., Thanh, D.D., Ethirajan, A. et al. (2016). *Sci. Rep.* 6: 18721.
25 Babayigit, A., Ethirajan, A., Muller, M., and Conings, B. (2016). *Nat. Mater.* 15: 247.
26 Yuan, Z., Huang, W., Ma, S. et al. (2019). *J. Mater. Chem. C* 7: 5442.
27 Zhao, Y. and Zhu, K. (2014). *J. Phys. Chem. C* 118: 9412.
28 Yoon, S.J., Draguta, S., Manser, J.S. et al. (2016). *ACS Energy Lett.* 1: 290.
29 Noh, J.H., Im, S.H., Heo, J.H. et al. (2013). *Nano Lett.* 13: 1764.

30 Saliba, M., Matsui, T., Seo, J.Y. et al. (2016). *Energy Environ. Sci.* 9: 1989.
31 Jiang, Q., Zhao, Y., Zhang, X. et al. (2019). *Nat. Photonics* 13: 460.
32 Jeon, N.J., Na, H., Jung, E.H. et al. (2018). *Nat. Energy* 3: 682.
33 Lide, D.R. (ed.) (Editor in chief) (1994). *CRC Handbook of Chemistry and Physics*, 75e. CRC Press Inc.
34 Song, J., Hu, W., Wang, X.-F. et al. (2016). *J. Mater. Chem. A* 4: 8435.
35 He, X., Guo, P., Wu, J. et al. (2017). *Sol. Energy* 157: 853.
36 Qiu, Z., Li, N., Huang, Z. et al. (2020). *Small Methods* 4: 1900877.
37 Lee, J.-W., Kim, D.-H., Kim, H.-S. et al. (2015). *Adv. Energy Mater.* 5: 1501310.
38 Min, H., Kim, M., Lee, S.-U. et al. (2019). *Science* 366: 749.
39 Xiang, S., Li, W., Wei, Y. et al. (2019). *Science* 15: 156.
40 Wang, Q., Zheng, X., Deng, Y. et al. (2017). *Joule* 1: 371.
41 Wang, Y., Dar, M.I., Ono, L.K. et al. (2019). *Science* 365: 591.
42 Wang, K., Jin, Z., Liang, L. et al. (2019). *Nano Energy* 58: 175.
43 Mitzi, D.B., Wnag, S., Feild, C.A. et al. (1995). *Science* 267: 1473.
44 Jiang, Y., Yuan, J., Ni, Y. et al. (2018). *Joule* 2: 1356.
45 Tai, Q., Tang, K.-C., and Yan, F. (2019). *Energy Environ. Sci.* 12: 2375.
46 Xiao, Z., Yuan, Y., Shao, Y. et al. (2015). *Nat. Mater.* 14: 193.
47 Tavakoli, M.M., Bi, D., Pan, L. et al. (2018). *Adv. Energy Mater.* 8: 1800275.
48 Wang, Q., Dong, Q., Li, T. et al. (2016). *Adv. Mater.* 28: 6734.
49 Yun, J.H., Lee, I., Kim, T.-S. et al. (2015). *J. Mater. Chem. A* 3: 22176.
50 Bai, Y., Dong, Q., Shao, Y. et al. (2016). *Nat. Commun.* 7: 12806.
51 Dong, X., Fang, X., Lv, M. et al. (2015). *J. Mater. Chem. A* 3: 5360.
52 Niu, G., Li, W., Meng, F. et al. (2014). *J. Mater. Chem. A* 2: 705.
53 Kanda, H., Shibayama, N., Huckaba, A.J. et al. (2020). *Energy Environ. Sci.* 13: 1222.
54 Chen, J., Lee, D., and Park, N.-G. (2017). *ACS Appl. Mater. Interfaces* 9: 36338.
55 Liu, C., Huang, Z., Hu, X. et al. (2018). *ACS Appl. Mater. Interfaces* 10: 1909.
56 Chen, W., Wang, Y., Pang, G. et al. (2019). *Adv. Funct. Mater.* 29: 1808855.
57 Jung, M., Shin, T.J., Seo, J. et al. (2018). *Energy Environ. Sci.* 11: 2188.
58 Lee, D.S., Yun, J.S., Kim, J. et al. (2018). *ACS Energy Lett.* 3: 647.
59 Yang, Y., Liu, C., Mahata, A. et al. (2020). *Energy Environ. Sci.* 13: 3093.
60 Lin, Y., Bai, Y., Fang, Y. et al. (2018). *J. Phys. Chem. Lett.* 9: 654.
61 Ito, S., Tanaka, S., Manabe, L., and Nishino, H. (2014). *J. Phys. Chem. C* 118: 16995.
62 Leijtens, T., Eperon, G.E., Pathak, S. et al. (2013). *Nat. Commun.* 4: 2885.
63 Habisreutinger, S.N., Wenger, B., Snaith, H.J., and Nicholas, R.J. (2017). *ACS Energy Lett.* 2: 622.
64 Bai, S., Da, P., Li, C. et al. (2019). *Nature* 571: 245.
65 Lin, Y.-H., Sakai, N., Da, P. et al. (2020). *Science* 369: 96.
66 Domanski, K., Correa-Baena, J.-P., Mine, N. et al. (2016). *ACS Nano* 10: 6306.
67 Xie, F.X., Zhang, D., Su, H. et al. (2015). *ACS Nano* 9: 639.
68 Mei, A., Li, X., Liu, L. et al. (2014). *Science* 345: 295.
69 Baranwal, A.K., Kanda, H., Shibayama, N. et al. (2019). *Energy Technol.* 7: 245.
70 Liu, S., Huang, W., Liao, P. et al. (2017). *J. Mater. Chem. A* 5: 22952.

71 Ma, Q., Huang, S., Chen, S. et al. (2017). *J. Phys. Chem. C* 121: 19642.
72 Zhang, W., Wan, L., Fu, S. et al. (2020). *J. Mater. Chem. A* 8: 6546.
73 Shi, L., Bucknall, M.P., Young, T.L. et al. (2020). *Science* 368: 1328.
74 Emami, S., Martins, J., Ivanou, D., and Mendes, A. (2020). *J. Mater. Chem. A* 8: 2654.
75 Hsiao, K.-C., Jao, M.-H., Tian, K.-Y. et al. (2020). *Sol. RRL* 4: 2000197.
76 Heo, J.H., Choi, Y.K., Koh, C.W. et al. (2019). *Adv. Mater. Technol.* 4: 1800390.
77 He, J., Bi, E., Tang, W. et al. (2018). *Sol. RRL* 2: 1800004.
78 Cheacharoen, R., Boyd, C.C., Burkhard, G.F. et al. (2018). *Sustain. Energy Fuels* 2: 2398.
79 Shi, L., Young, T.L., Kim, J. et al. (2017). *ACS Appl. Mater. Interfaces* 9: 25073.
80 Liu, X., Bu, T., Li, J. et al. (2018). *Nano Energy* 44: 34.
81 Jiang, Y., Qiu, L., Juarez-Perez, E.J. et al. (2019). *Nat. Energy* 4: 585.

16

Hysteresis in *J*–*V* Characteristics

Wolfgang Tress

Zurich University of Applied Sciences, Institute of Computational Physics, School of Engineering, Wildbachstr. 21, 8401 Winterthur, Switzerland

16.1 Introduction and Definitions: What Do We Mean by Hysteresis?

Several review papers on hysteresis exist already mainly following the same scheme of extensively listing reported results on origins and suggested elimination strategies of hysteresis [1–4]. I myself contributed to a book chapter on this topic recently as well [5]. To create something new, I tried to write this chapter in hand from a different perspective. Instead of reviewing what has been reported, I invite the reader to undertake the journey of exploring and understanding the hysteresis phenomenon in perovskite solar cells (PSCs) by themselves guided by this book chapter.

Prior to jumping into PSCs, which are topic of this book, we want to start with some general points regarding the phenomenon hysteresis. One of the most prominent examples in physics, where we observe hysteresis, is found in the field of magnetism. In a ferromagnetic material, the magnetic flux density B lags behind the magnetization force H, giving rise to a hysteresis loop when B is plotted vs. H. A ferroelectric material behaves analogously, when plotting the displacement field $D = \varepsilon_0 E + P(E, \ldots)$ vs. the electric field E (ε_0 is the vacuum permittivity), where remnant polarization P can be observed (Figure 16.1a). The general scheme behind these effects is a delayed response of the material upon an external force. This leads to remnant effects and therefore a dependence of the measured state on the history of the sample.

Generally speaking we can distinguish two major reasons behind hysteresis:

1. A certain threshold force is required to change/switch one quantity such as the magnetic flux density.
2. The response time of the material/device is continuously lagging behind while measuring the desired quantity and simultaneously sweeping the relevant force, e.g. the electric field acting on charge carriers by applying a voltage.

The first leads to a rather rate-independent hysteresis (here labeled as *type I*) meaning that the hysteresis does not depend on how fast the force is swept but

Hybrid Perovskite Solar Cells: Characteristics and Operation,
First Edition. Edited by Hiroyuki Fujiwara.

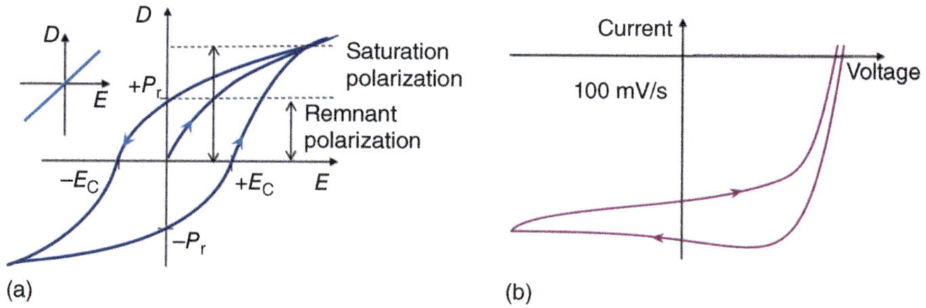

Figure 16.1 Examples of hysteresis. (a) Hysteresis loop of a ferroelectric material, where displacement field D is plotted vs. electric field E and a remnant polarization P_r is observed, when $E = 0$. The inset shows a dielectric material, where $D = \varepsilon_r\varepsilon_0 E$. (b) Example of a hysteresis in the current–voltage (*JV*) curve of a perovskite solar cell measured at a voltage sweep rate of 100 mV/s.

mainly on the history in the sense of previous (extreme) forces, the sample has seen (Figure 16.1a). The second leads to a rate-dependent hysteresis (*type II*), meaning that the material or system is not characterized by a single hysteresis. Instead, despite the same scan range of the force, different hysteresis curves exist dependent on the sweep rate. In this case it is essential to state the conditions not only before the measurement, but also during the measurement, i.e. the sweep rate. Ideally, in such a system the hysteresis vanishes when the sweep rate of the force becomes infinitely low.

Now we want to move to hysteresis in the current–voltage (*JV*) curve of electronic devices such as solar cells. In contrast to the aforementioned examples, we would not expect hysteresis in a *JV* curve because we would implicitly assume that the measured current density J is a unique function of the applied voltage V with the resistance as quotient of both: $R = V/(JA)$ (A is area). Therefore, we would not expect a hysteresis of *type I*. On the other hand, the *JV* curve is obtained by sweeping the voltage, which is a dynamic measurement, whose result could depend on the voltage sweep rate, if the response time of the system is low. Usually, we would assume that the voltage sweep is much slower than any electronic process in our system, and we would measure a steady-state situation, i.e. a unique *JV* curve. Such a *JV* curve for a solar cell under standardized illumination and ambient conditions allows us to directly determine the maximum output power of the solar cell by finding the maximum power point (MPP), where the electrical power (density) $P = VJ$ is maximum. Knowing P and the light intensity, we can determine the power-conversion efficiency (PCE), the main figure of merit of a solar cell. However, if the *JV* curve unprecedentedly shows hysteresis (Figure 16.1b), we measure two curves, one for scanning the voltage from short circuit to open circuit (called *forward scan*) and one vice versa (labeled *backward* or *reverse scan*). When determining the PCE as just described, this implies two values for the PCE, which is obviously one value too much for a solar cell. After the report of a severe *JV* hysteresis under common measurement conditions in 2014 [6], a big debate arose on whether reported PCE values for PSCs can be trusted.

To pragmatically resolve this ambiguity in PCE, researchers started to report the hysteresis, i.e. both *JV* curves, and to calculate an average of the two PCE values. Furthermore, one attempted to quantify the hysteresis with various indices [5, 7], which should provide means of judging the uncertainty of the reported PCE values. However, these attempts neglect the fact that the hysteresis is of *type II*, i.e. rate-dependent (will be discussed in detail in Section 16.3), meaning that we do not have only two *JV* curves (forward and reverse) but a whole set of curves, when we vary the voltage sweep rate. Consequently, the hysteresis index becomes a function of this rate.

As the rate-dependent hysteresis is indicative of a non-steady-state situation at each voltage during the *JV* measurement, a simple solution to reach more reliable PCE values would be to determine the PCE not from *JV* sweeps but from a transient measurement of the PCE, where the applied voltage is roughly kept constant. This approach became standard for PSCs [8] and is usually implemented either by:

- Applying a constant voltage at the MPP determined from a preceding *JV* sweep and monitoring the transient current, coined "stabilized power output" [6].
- Dynamically tracking the MPP and monitoring the output power.

In both cases, a convergence of the power is required, whose value is then reported. Usually the measurement is performed for a couple of minutes. This approach re-established trust in reported PCE values. Furthermore, optimization of the device architecture and, in particular, the contact layers has lead to a reduction of the hysteresis, often making this issue less relevant for determining the PCE of state-of-the-art solar cells.

However, the hysteresis is not completely tamed, and investigations on its origin are still ongoing. Furthermore, it is still unclear to which extent hysteresis and its origins are related to long-term instabilities of PSCs. Therefore, a profound understanding of the hysteresis is desirable.

The goal of this chapter is to address the hysteresis in a systematic way combining theoretical aspects with experimental evidence on the macro- and nanoscale. We will proceed step by step:

- To understand hysteresis, we need to understand the *JV* curve of PSCs in general (Section 16.2).
- To understand the phenomenon hysteresis, we need to characterize the device properly and unravel parameters that govern it (Section 16.3).
- To propose microscopic origins, we need to develop a consistent picture between theory and experiment (Section 16.4).

In the final section (Section 16.5), we will summarize the current knowledge and address open questions.

16.2 The *JV* Curve of a Solar Cell: What Does It Tell?

This chapter provides a general brief review on the *JV* curve of a solar cell, independent of whether there is hysteresis or not. Understanding the factors dominating the

JV curve will allow us to better grasp the effects that underlie hysteresis. We want to address

- The fundamental solar cell structure
- The photocurrent/short-circuit current
- The photovoltage/open-circuit voltage
- The ideal *JV* curve under illumination
- The ideal *JV* curve in the dark
- The voltage dependence of the photocurrent
- The series and parallel resistance
- The meaning of the fill factor

A **solar cell** is an active device. Under illumination it provides electrical power without any externally applied voltage. This is very different to e.g. a photoresistor as visualized in Figure 16.2. In the convention we are using, power can be generated in the fourth quadrant, where a positive voltage comes along with a negative current. To make such a situation feasible, the photo-generated electron–hole pairs in our semiconductor absorber have to be spatially separated and collected at two different electrodes. This is only possible if the device exhibits an asymmetric structure. For example, just having one homogeneous photoactive absorber material would not be sufficient to exploit the photoelectric effect in a solar cell, but could work as a photoresistor, a device whose resistance can be modified by light, which changes the charge carrier density (Figure 16.2a). In a solar cell, however, the photo-generated charges have to flow to their respective electrode without any additional external driving force. This can happen, e.g. in a structure, where an electric field is built into the absorber, such as is the case in a p-i-n solar cell. Such a built-in field can also be caused by a built-in potential that originates from the work function difference of e.g. the metal contacts. However, such a field is not necessarily required. If electron and hole are able to diffuse efficiently and independently in the absorber, so-called selective contacts are sufficient to construct a solar cell using a single and homogeneous absorber material (Figure 16.2b). Such selective contacts can be realized by using wide-gap semiconductors with suitable energy levels to generate a band diagram as depicted in Figure 16.2c.

Any efficient perovskite solar cell exhibits such selective layers, which are often called electron or hole transporting layer. Despite this name, their major function is actually to block the "other" charge carrier type.

Having realized such a structure, we can examine the **photocurrent** density under short circuit (J_{sc}). The short-circuit current flows when both contacts are connected with a piece of wire, yielding $V = 0\,\mathrm{V}$ (Figure 16.2b). J_{sc} is a function of the absorption efficiency and the charge carrier collection efficiency (under short circuit). The former depends on how many photons are absorbed in the absorber material and the latter on how many of these photons are converted into electron–hole pairs that are collected at the electrodes. For optimized PSCs, both quantum efficiencies are rather high. If the collection efficiency is not 100%, photo-generated charges are lost. The only possible loss process in steady state is recombination of electrons with holes, which brings us to the next point:

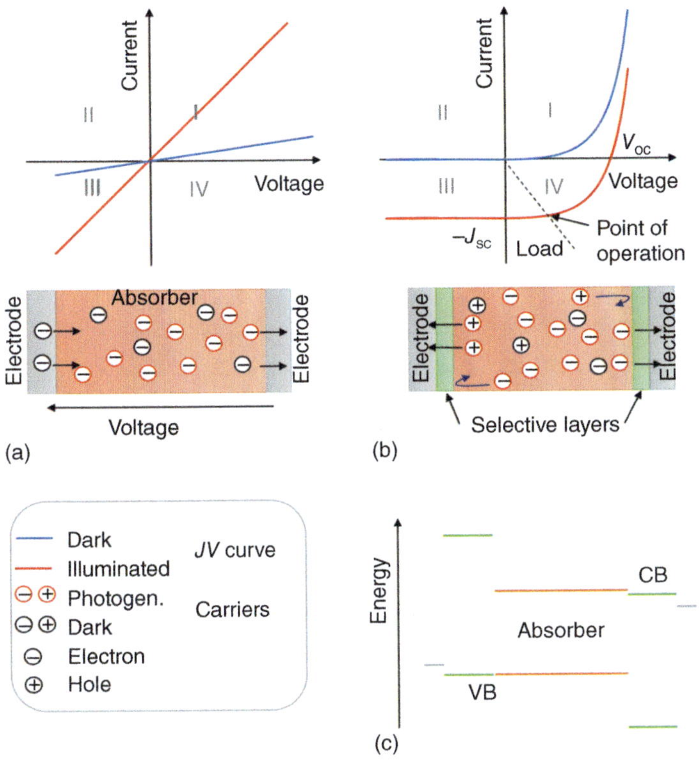

Figure 16.2 Solar cell and photoresistor. (a) Photoresistor as a passive device, where absorption enhances the conductivity by increasing the density of charge carriers (for simplicity, only electrons shown). (b) Solar cell as an active device, being able to provide electrical power in the fourth quadrant, due to a selective collection of electrons and holes at the respective contact. The dashed line shows the negative *JV* curve of a resistive load. (c) Possible energy diagram of a solar cell with wide-gap charge-selective layers. Energy levels (VB, valance band; CB, conduction band, gray work functions of electrodes) are shown vs. a common vacuum level for the individual layers.

The **open-circuit voltage** V_{oc}. If we do not connect any load to the solar cell, but just keep the contacts open, we do not extract any electron or hole. Therefore, all photo-generated charges have to recombine. However, we measure a voltage between the two contacts. Where does this voltage originate from? Coming back to the idea of the previously discussed driving force (drift due to electric field, diffusion due to selectivity), we might imagine that electrons and holes are driven to different electrodes and accumulate there. This accumulated charge might repel further charges and can be measured as an external voltage. This intuitive picture is correct, though a bit simplified and contains the risk of wrong conclusions, such as that the open-circuit voltage equals the built-in potential or is limited by it. To be precise, though a bit less intuitive at first glance, we should talk about electrochemical potentials (quasi-Fermi levels), which we address only very roughly here: Under light we increase both electron (n) and hole (p) densities, as we generate electron–hole pairs (cartoon in Figure 16.2b). Therefore, their chemical

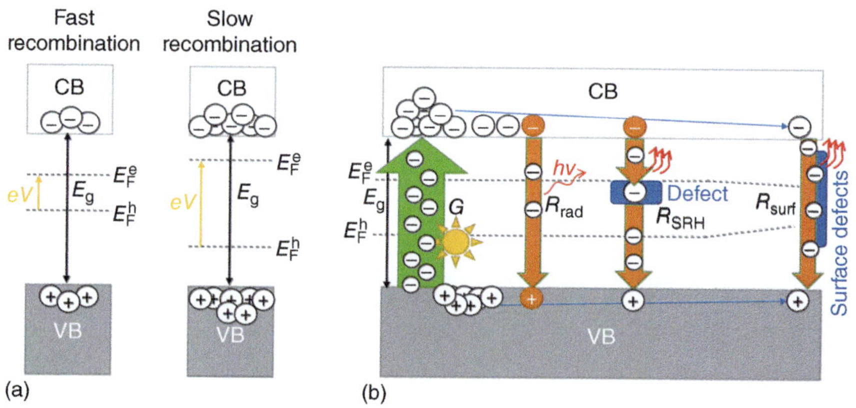

Figure 16.3 Open-circuit voltage and recombination. (a) Slower recombination leads to higher charge carrier densities and therefore higher voltage *V*. (b) Visualization of different recombination mechanisms and the fact that optical generation *G* equals the overall recombination $R_{rad} + R_{SRH} + R_{surf}$. SRH stands for Shockley–Read–Hall recombination at defects and the arrow bundle indicates generation of heat.

potentials rise until a point is reached, where the concentrations are sufficiently high to lead to a recombination rate (R) that is equal to the photo-generation rate (G) of charge carriers. If we express the recombination rate by $R = \beta np$, which should equal G, we directly see that a higher light intensity (higher G) leads to higher np (β independent of np). Therefore, we can very generally conclude that lower recombination rate constants (here β) lead to higher np for a given G. As the chemical potentials of electrons and holes scale with (the logarithm of) n and p, respectively, the open-circuit voltage becomes higher for reduced recombination, reaching in the ideal case the thermodynamic limit (around 0.25…0.35 V below the band gap divided by elementary charge, see e.g. in Ref. [9]). Figure 16.3a visualizes how the externally measured voltage relates to the difference of the electrochemical potential of the charge carriers at their respective electrode. Beyond the mentioned radiative recombination $R = \beta np$, recombination at defects and at surfaces/interfaces prevails, which commonly dominates the overall recombination and can lead to gradients in the electrochemical potentials, also under open circuit (Figure 16.3b, details in Refs. [10, 11]). After having understood the concept of recombination and electrochemical potentials, we can now explain the **Ideal *JV* curve under illumination**. Short (0 load) and open circuit (infinite load) are extreme cases the solar cell can operate in. Connecting a load with a finite resistance, the solar cell will operate at a current that is between 0 and $-J_{sc}$ and a voltage that is between 0 and V_{oc}, dependent on the resistance and the J–V relation (*JV* curve) of the solar cell (dashed line in Figure 16.2b). This *JV* curve could have various shapes, most intuitively monotonically increasing with voltage, as we were discussing that buildup of the voltage is related to enhanced charge carrier densities, which increase the overall recombination rate and thus decrease the photocurrent. Looking into the (Boltzmann approximated Fermi–Dirac) statistics for charges in their bands, it turns out that there is an exponential relationship between charge carrier densities (and therefore recombination) and the chemical potential (and therefore voltage). This means that when increasing the voltage from 0 to V_{oc}, we enhance the recombination exponentially with voltage, leading to a reduction of the absolute current under illumination and an exponential *JV* curve (red curve in Figure 16.2b). Having understood this relation between current and voltage, we can also address the question of the *JV* curve in the dark.

In the dark, there are no photo-generated carriers, hence no short-circuit current density or open-circuit voltage unequal to 0 can be measured. However, if voltage is applied actively by connecting the solar cell to a voltage source, electrochemical potentials move and charge carrier concentrations increase exponentially with voltage, leading to a current that increases exponentially with voltage (blue line in Figure 16.2b), resembling a classical Shockley diode equation:

$$J_{\text{dark}}(V) = J_0\left(e^{\frac{eV}{nkT}} - 1\right) \tag{16.1}$$

with $\frac{kT}{e}$ being the thermal voltage and n the ideality factor. The blocking behavior under negative voltage is a direct consequence of the selectivity of the contacts. As this exponential increase in recombination current depends only on the voltage and not on the illumination, we can directly relate the ideal *JV* curve under illumination with the dark *JV* curve by just subtracting the J_{sc} from the dark *JV* curve, referred to as superposition principle (Figure 16.4a):

$$J_{\text{illu}}(V) = J_0\left(e^{\frac{eV}{nkT}} - 1\right) - J_{sc} \tag{16.2}$$

So far we just discussed that the *JV* curve is completely defined by recombination current. In a real device, however, the current can be subjected to ohmic losses. In other words, the charge carrier collection efficiency might depend on the voltage, meaning that J_{sc} in Eq. (16.2) has to be replaced by a **photocurrent** J_{photo} **that depends on the voltage**:

$$J_{\text{illu}}(V) = J_0\left(e^{\frac{eV}{nkT}} - 1\right) - J_{\text{photo}}(V) \tag{16.3}$$

This can happen if charge carrier diffusivities and mobilities are low. In that case, an electric field originating from the built-in potential can support charge carrier extraction. As this field is more and more compensated with enhanced voltage, the photocurrent decreases toward open circuit and the curve leaves the ideal diode curve (Figure 16.4b). In extreme cases the curve can show features as depicted in Figure 16.4c and called s-shape/s-kink. Such s-shaped *JV* curves can also result from non-optimum properties of the selective contacts. If these contacts are not ideal for the charge carrier to collect, meaning that their conductivity is low or energy levels are misaligned causing charge extraction barriers, photocurrent extraction is strongly hampered in a highly nonlinear way [12, 13]. Piled-up charges become prone to recombination and only voltages much smaller than V_{oc} can support sufficient extraction of charges.

Apart from these processes related to the absorber and its interfaces, there are further parasitic effects that influence the *JV* curve, both in the dark and under illumination. Lateral transport through the electrode, wiring, etc. introduce a **resistance** that is in series to the solar cell (R_S). This series resistance does not influence

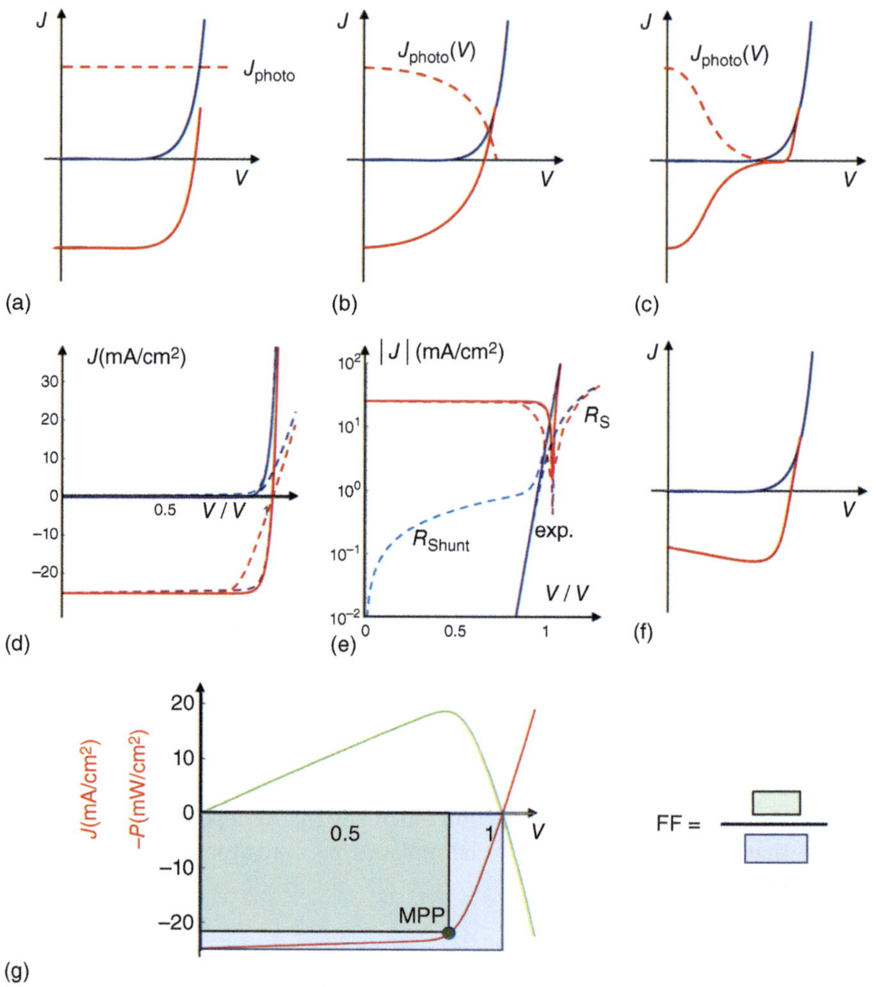

Figure 16.4 The *JV* curve. (a) Ideal situation with a constant photocurrent J_{photo}. (b) Photocurrent that decreases with voltage due to charge collection issues. (c) More extreme case of (b), which causes s-shape, often due to hindered charge extraction at the contacts. (d) Effect of series resistance ($R_S = 5\,\Omega\,cm^2$, blue, red dashed) and parallel (shunt) resistance ($R_{Shunt} = 1\,k\Omega\,cm^2$, cyan, magenta dashed). (e) Data of (d) in logarithmic current scale. (f) "Bump" (g) output power (green) and definition of the FF.

the discussed physics of the solar cell, but modifies the appearance of the externally measured *JV* curve. For a given current, a certain fraction of the voltage drops over this resistance and the "intrinsic" *JV* curve could be revealed by correcting the voltage:

$$V_{\text{intrinsic}}(J) = V(J) - V_{R_s}(J) = V(J) - JAR_s \tag{16.4}$$

The influence of a series resistance is shown with the dashed red/blue lines in Figure 16.4d,e. Note that it does not modify V_{oc} because current does not flow under open circuit. Leakage currents give rise to a parallel or shunt resistance, often visible close to J_{sc} and mainly in the dark curve (magenta/cyan dashed lines in Figure 16.4d,e).

All these resistive effects influence the *JV* curve, including the voltage range between 0 V and V_{oc}. To maximize the PCE, the device should be operated at a current/voltage pair, where the output power $P = JV$ becomes maximum, as stated in Section 16.1. P can be visualized by a rectangular area as shown in Figure 16.4g. The closer this area is to the product of V_{oc} and J_{sc}, i.e. the more "rectangular" the *JV* curve, the higher is P. The fill factor (FF) quantifies the ratio of this rectangle to $V_{oc} \times J_{sc}$. Based on the discussions earlier, we can conclude that a rectangular shape of the *JV* curve cannot be approached infinitely closely. The closest match is the exponential behavior in the recombination-limited case with $n = 1$, which dependent on the V_{oc} can exceed 90% in $MAPbI_3$ solar cells. In experiment FFs lower than the recombination limit are very common due to the discussed charge-transport limitations and parasitic resistances. In PSCs, it looks like the contact layers with a high selectivity also introduce an obstacle for collecting the desired charge carrier. This limits the fill factor commonly to around 80% for the highly efficient devices.

It is hard to imagine a situation as in Figure 16.4f, where the FF would be even >100%. Such "bump" curves, which are observed for dye sensitized solar cells (DSSC) [14, 15] and PSCs [16], are a clear indication that these curves do not represent a steady-state situation but must be part of a hysteresis curve.

For more details regarding different recombination mechanisms and correct measurements of PCE and *JV* curves, the reader is referred to my previous book chapters and reviews [9–11].

16.3 Characteristics of Hysteresis: What Does It Depend on?

Having understood what determines the *JV* curve in general, we can attempt to investigate the hysteresis. We will approach the phenomenon experimentally step by step without having any predefined hypothesis on its microscopic origin in mind. This is a wise approach for such a complex problem and does not make us mix up cause and effect. It ensures that we will not neglect indirect interactions between charges and electric field in the solar cell stack.

We start with a measured hysteresis curve, shown in Figure 16.1b. Obviously, V_{oc}, J_{sc}, FF, and hence PCE are lower for the curve obtained from the forward scan. The first question we want to ask is whether the hysteresis depends on the **scan rate**. We have already anticipated the answer when introducing the two types of hysteresis in Section 16.1. And, indeed, Figure 16.5a confirms a strong rate dependence. For a certain intermediate rate (here around 1000 mV/s), the hysteresis is largest and reduces for lower and higher scan rates. This trend is strong evidence for a process that reacts slowly on the applied bias voltage. If the scan rate is sufficiently low, this process reaches a steady state at each voltage during the scan, and the hysteresis should vanish. If the scan rate is sufficiently high, the slow process cannot react and a quasi-steady-state *JV* curve is measured, which is not unique but depends on the

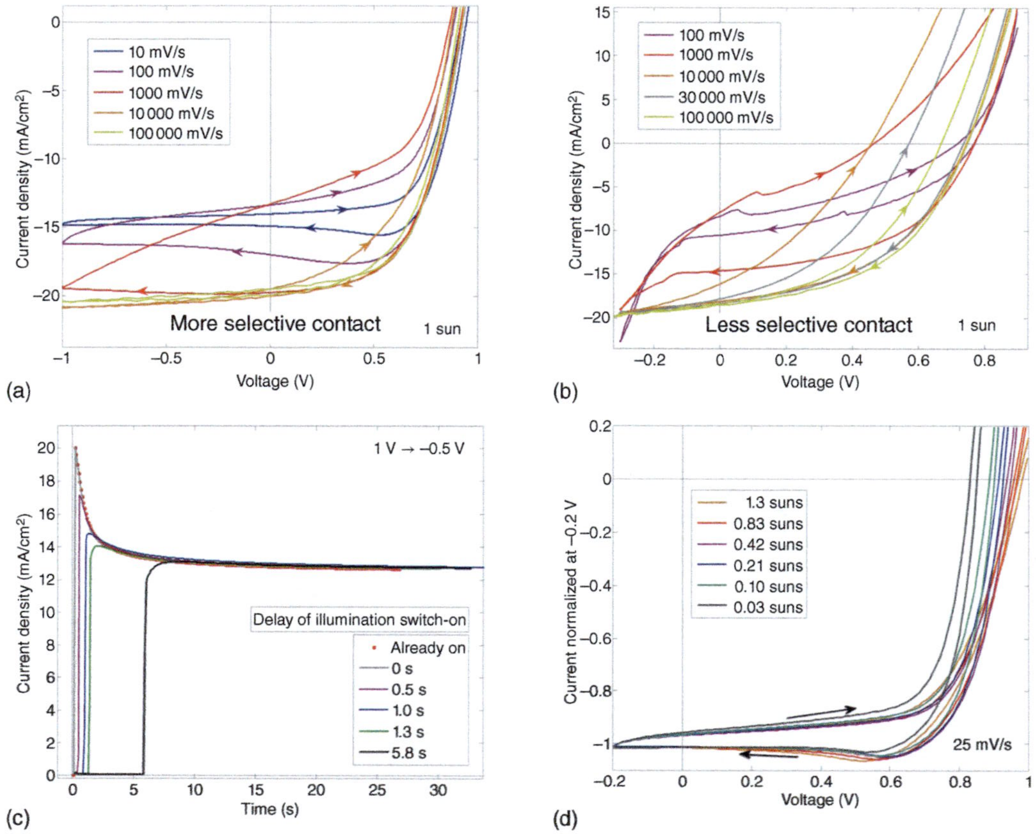

Figure 16.5 Factors influencing hysteresis. (a) Scan rate for device with more selective contact, where mainly the FF is affected. (b) Scan rate for device with less selective contact (no compact TiO_2 layer), where also V_{oc} is affected. (c) Transient response in the dark and (partially) under light. (d) Light-intensity-dependent normalized *JV* curves. *Source*: Tress et al. [16].

state directly before the measurement (**"poling," preconditioning**). This behavior has severe practical consequences:

- It is not straightforward to make claims on a hysteresis-free *JV* curve/device.
- Even if the hysteresis is negligible at a certain scan rate, it is not guaranteed that the *JV* curve is a unique steady-state curve.
- When comparing two devices at one selected scan rate, a reduced hysteresis does not imply that the origin of the hysteresis got really reduced. It might just be that the rate for the strongest response shifted.

The transient response can nicely be visualized by performing a step-wise voltage sweep and recording the transients for every voltage step as shown in Figure 16.6a [17, 19, 20].

Although we can already conclude that there is a slow transient process, we leave discussions on its origin for Section 16.4 and proceed with systematically characterizing further parameters.

We continue with the influence of **illumination**, as the *JV* curves in Figure 16.5a are measured under illumination. We make use of transient currents again, but this time applying a big voltage step and monitoring the current response. Figure 16.5c shows that it takes a couple of seconds until the current is stabilized. If we repeat the experiment with the device being for part of the time in the dark, it turns out that the response (detectable under light) remains unmodified. Therefore, we can conclude that this slow transient process happens the same way in the dark. To go one step further, we can have a look at the dependence on the light intensity, which can be nicely visualized by plotting normalized *JV* curves (Figure 16.5d). Also in this case, the hysteresis remains completely independent of light intensity, indicating that the slow process is not directly influenced by absorption of light and/or photo-generated charge carriers. This observation is consistent with the fact that hysteresis is also observed in dark *JV* curves. Applying voltages to symmetric devices allows them to develop an asymmetry in their *JV* curve giving rise to diode characteristics dependent on the sign of the poling voltage [21], which confirms the importance of the applied voltage independent of illumination.

A further important parameter is the **temperature** [22]. It has been found that it influences the hysteresis in a similar way as the scan rate, meaning that the hysteresis is the highest for a certain temperature and decreases for higher and lower temperatures [23, 24]. Therefore, identical hysteresis curves can be obtained when simultaneously decreasing the temperature and voltage sweep rate, as shown in Figure 16.7b for a photodetector [26]. This is indicative of a temperature-activated process. At low temperatures, it cannot follow and the hysteresis vanishes, whereas at high temperature it can follow the voltage change completely and the hysteresis vanishes again. For intermediate temperatures, the hysteresis is most pronounced. This temperature dependence of the hysteresis [29] or the transient response [30] has been used to determine an activation energy of the slow process. For devices, which seem to be rather hysteresis-free for common voltage sweep rates at room temperature, hysteresis can occur at lower temperatures [24, 31].

There is a plethora of metal-halide perovskite compositions and device architectures. Almost all of these devices show hysteresis to some extent under

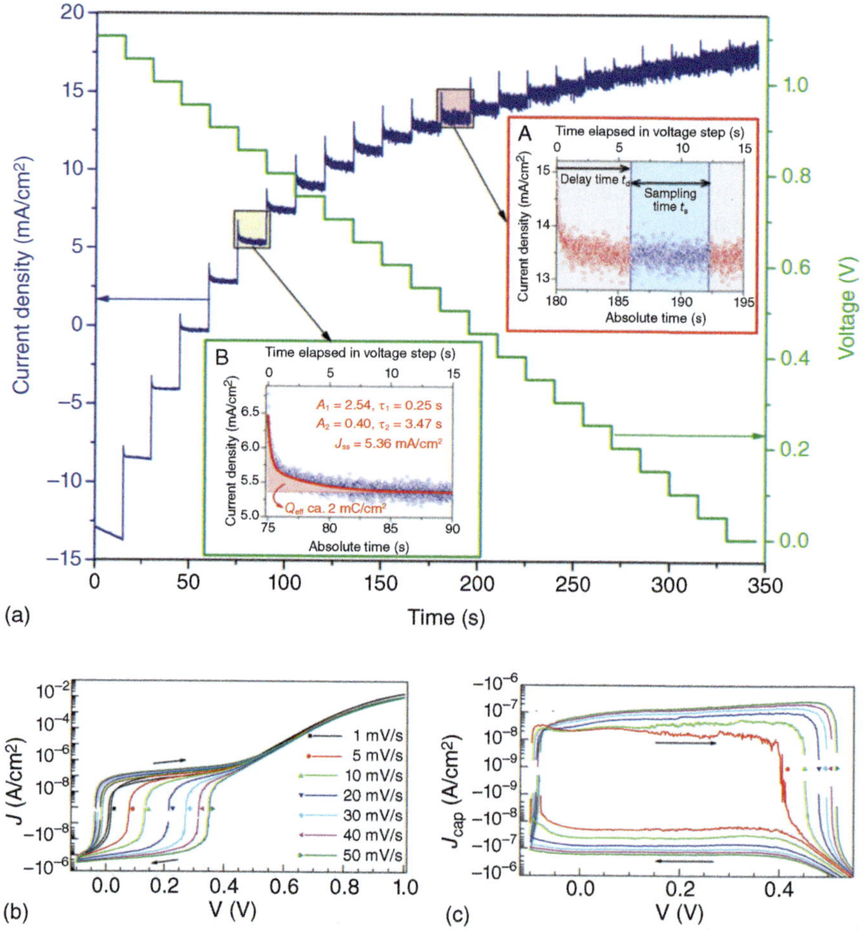

Figure 16.6 Non-capacitive and capacitive effects. (a) Transient current response (blue) during a staircase voltage sweep (green), revealing an initially higher charge collection efficiency after increasing the voltage, which looks similar to a classical capacitive response, where the displaced charge would be gigantic (mC/cm^2). *Source*: (a) Christoforo et al. [17]. Licensed under CC BY 4.0. (b, c) Capacitive dark currents in the order of 10…100 nA/cm^2. J_{cap} in (c) is obtained from J in (b) by subtracting the approximate steady-state current. *Source*: (b) Almora et al. [18].

certain conditions. For example, decreasing the temperature or increasing the scan rate makes a hysteresis appear in devices labeled hysteresis-free (example of a p-i-n device in Figure 16.7c). Therefore, hysteresis is something very prominent in PSCs. However, e.g. in p-i-n small-molecule organic solar cells based on zinc phthalocyanine:C_{60}, hysteresis could not be detected for common scan rates (<100 V/s) (Figure 16.7d).

Nevertheless, already in the early days it has been observed that hysteresis can be tuned by the **solar cell architecture**, the adjacent charge transport layers, and also the perovskite itself. A common trend seems to be that hysteresis

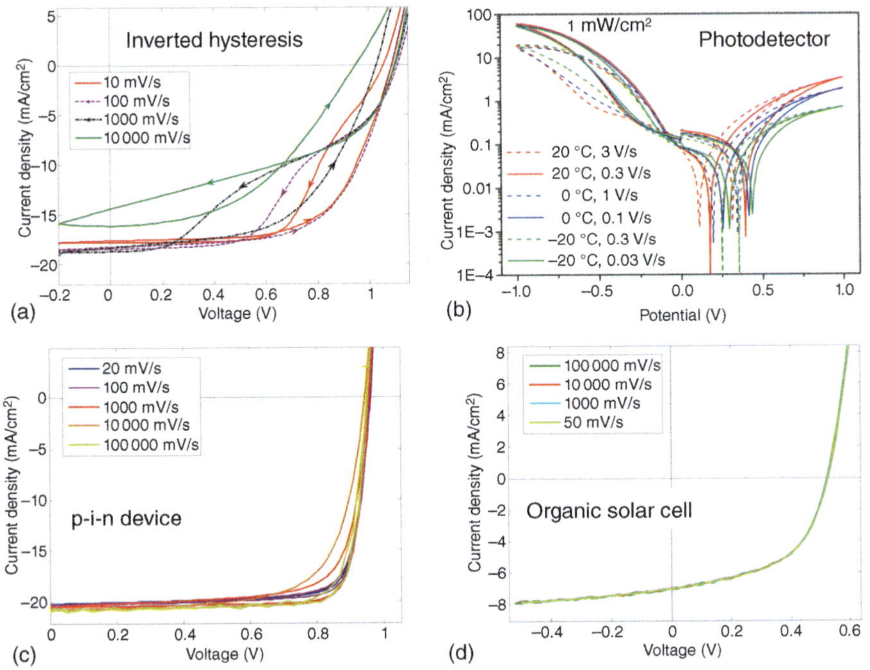

Figure 16.7 Hysteresis in different solar cells. (a) Inverted hysteresis, as e.g. observed at instable mixed perovskites. *Source*: Adapted from Tress et al. [25]. Copyright 2016 WILEY-VCH Verlag GmbH & Co. KGaA, Weinheim. (b) Temperature and rate dependence of hysteresis of a photodetector. *Source*: Adapted from Domanski et al. [26]. Copyright 2015 WILEY-VCH Verlag GmbH & Co. KGaA, Weinheim. (c) Planar p-i-n device with architecture presented in Ref. [27]. *Source*: Based on Chiang et al. [27]. (d) Organic solar cell ZnPc:C_{60} with structure shown in Ref. [28]. *Source*: Based on Tress et al. [28].

decreases when changing the electron transport layer from planar TiO_2 to planar SnO_2 or mesoporous TiO_2 [20, 32]. Modifications of the oxide contacts using e.g. phenyl-C61-butyric acid methyl ester (PCBM) reduce hysteresis as well [33, 34]. Modifications or elimination of the p-side contact seems to influence the hysteresis less according to my experience but exchanging spiro-MeOTAD by CuI leads to reduced hysteresis [35]. When flipping the layer sequence (from n-i-p to p-i-n) and using organic contact layers, the hysteresis vanishes at common scan rates [36–38]. However, also in this architecture hysteresis can be prominent, e.g. when a NiO_x hole transport layer is used or at higher scan rates (Figure 16.7c). Also in the inverted structure, PCBM in between perovskite and C_{60} has turned out to be beneficial in reducing hysteresis [39].

Having a look at the **perovskite layer itself**, also some trends have been observed: For n-i-p devices with a mesoporous TiO_2 electrode, the hysteresis becomes more prominent with reduced grain size [20, 40]. In this case, the increased hysteresis is accompanied by a reduced FF. On the other hand, PSCs that are vacuum-deposited are almost hysteresis-free independent of architecture despite small grains in the film [41]. The composition of the perovskite itself independent of the morphology might be relevant and materials present at grain boundaries such as PbI_2 seem to influence hysteresis as well [42, 43]. In addition, additives can modify the hysteresis, where e.g. PCBM [44], carbon nanoparticles [45], or potassium iodide [46] have been found to reduce hysteresis.

The *JV* hystereses in the figures presented so far have one feature in common despite the different shape and differently affecting J_{sc}, FF, and V_{oc}: The reverse scan yields higher PCE values than the forward scan. However, this feature is not to be generalized and there are devices that show the opposite behavior, also called *inverted* hysteresis (Figure 16.7a) [25]. It is even possible that the same device can change from normal to inverted hysteresis upon light soaking or dependent on the scan rate. Furthermore, the *JV* curves from forward and reverse scan can cross (several times) between 0 V and V_{oc}.

From the phenomenological characterization of hysteresis and experimental variation of the contact and absorber layers in PSCs, we can conclude:

- Hysteresis in PSCs is universal but not unique: It depends on measurement conditions (scan rate, temperature, preconditioning). Therefore, it is ambiguous to talk about "hysteresis-less" devices, when comparing only two *JV* curves.
- Hysteresis is triggered by changes in voltage due to a slow transient process in the PSC. It is most pronounced when scan rate and the dynamics of this slow process are on the same timescale.
- Factors in the device design influencing the appearance of hysteresis are various: A single culprit or a magic modification that universally removes hysteresis cannot be identified.
- Properties of the contact layers have the strongest influence on hysteresis.
- Hysteresis occurring on slow timescales of minutes can lead to a smooth transition to reversible [47] and irreversible degradation phenomena.

In Section 16.4 we want to finally address possible reasons for such a slow response and identify the most likely one that consistently describes the experimental data.

16.4 Mechanistic and Microscopic Origin of Hysteresis: What Changes Slowly?

In this section, we want to go the reverse way compared with Section 16.3. We start with hypotheses based on microscopic theories and scrutinize whether they can describe the "slow process" and its consequences.

We start with **capacitive effects**. As we are performing a voltage sweep, we are applying a continuously changing voltage, characterized by the sweep rate dV/dt. The changing electric field results in a displacement current $J = \varepsilon_0 \varepsilon_r \frac{dE}{dt}$, which can be expressed as a capacitive current:

$$J_{cap} = \frac{dQ}{dt} = C\frac{dV}{dt} \tag{16.5}$$

Such a capacitive current occurs in the dark and under illumination. To judge whether it plays a role in the hysteresis, we calculate this current assuming a realistic "geometric" capacitance of 50 nF/cm^2 (plate capacitor model: $C = \frac{\varepsilon_0 \varepsilon_r}{d}$ with thickness $d = 500$ nm and $\varepsilon_r \approx 30$). We make the case more realistic by considering a finite resistance that the current faces when charging the capacitor ($R = 100\,\Omega\,\text{cm}^2$, unrealistically high for a series R_s/A in Eq. (16.4)).

Figure 16.8a shows the results with the following characteristic features: As obvious from Eq. (16.5), the capacitive current is independent of voltage for a constant scan rate $\frac{dV}{dt}$ and proportional to the scan rate, changing sign with the sign of the sweep rate. The current-limiting effect of R is visible when modifying the sweep rate. Comparing these data with the hysteresis in Figure 16.5, we observe that the features look different and that the values of C required for the current to show up in the hysteresis plot (e.g. several mA/cm^2 at 100 mV/s) would be gigantic (Figure 16.8b shows a case for 1 V/s and 1000 × larger capacitance). Therefore, we can already conclude that the type of hysteresis seen in Figure 16.5 is not caused by capacitive effects. Nevertheless, also capacitive effects can be observed in PSCs, e.g. in µs transients [48], when using high scan rates, or when investigating dark currents, where the steady-state currents around 0 V are much lower than under solar illumination. This is e.g. shown in Figure 16.6b,c from Ref. [18]. Note that the currents have similar positive and negative values in the order of 10–100 nA/cm^2, compatible with a capacitance value of 3 µF/cm^2. This value is realistic for an accumulation capacitance of e.g. ionic charges at the interface between charge transport layers and perovskite. Figure 16.8c shows that such capacitive effects are expected to become visible in the *JV* curve under solar illumination for scan rates in the kV/s range. Capacitive effects have been observed in other solar cell technologies as well, e.g. in Si [49] or CdTe [50].

In addition to this textbook-like displacement current, further capacitive/charge accumulation effects might be considered to explain the hysteresis. Those are related to storage of electrons or holes as a function of prebias and their collection as a function of sweep rate. Such an effect was proposed to explain the already mentioned "bump" in the *JV* curve [51]. However, absence of the bump in the dark and the peak current under light always being smaller than the maximum obtainable photocurrent in the PSCs do not advocate for such an effect.

Naturally, the hysteresis comes along with "capacitive" effects seen in impedance spectroscopy, yielding results of gigantic and photoinduced capacitances [52, 53]. However, these results are unphysical when interpreted literally. They cannot be taken as a support of capacitive currents because they are just a manifestation of the same slow process in another measurement. Therefore, these capacitances have to be seen as *apparent capacitances*, not directly originating from charge stored somewhere in the device [54–57].

This brings us to **non-capacitive origins** of hysteresis, which are the dominating ones for the common hysteresis observed at low scan rates (V/s to mV/s) and under light [16, 19, 58, 59]. (Most of the effects assigned to capacitive contributions in Ref. [60] are assigned to the non-capacitive category in this book chapter.)

The already mentioned poling effects on symmetrical devices, which become p–n or n–p diodes dependent on prebias [21], are strong evidence that electrical properties can be modified by applying a voltage and remain for a certain time. A

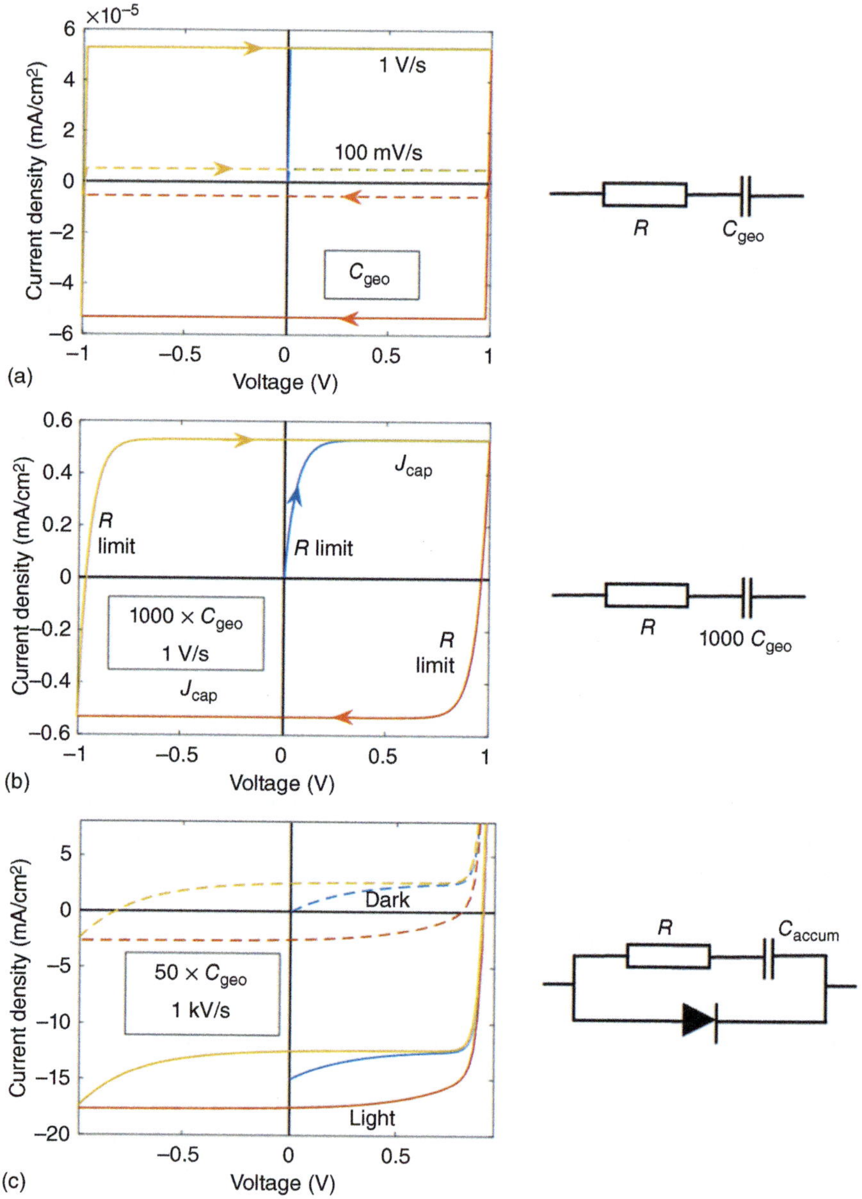

Figure 16.8 Demonstration of capacitive effects. (a) Calculated current vs. voltage for forward and backward scan with typical scan rates considering the geometric capacitance. (b) 1000 × higher capacitance and 1 V/s scan rate to reach about $J_{cap} \approx 1\,\text{mA/cm}^2$. (c) Capacitive effects superimposed on diode *JV* curves. The photocurrent is 15 mA/cm². *R* is 100 Ω cm² and the scan rate 1 kV/s.

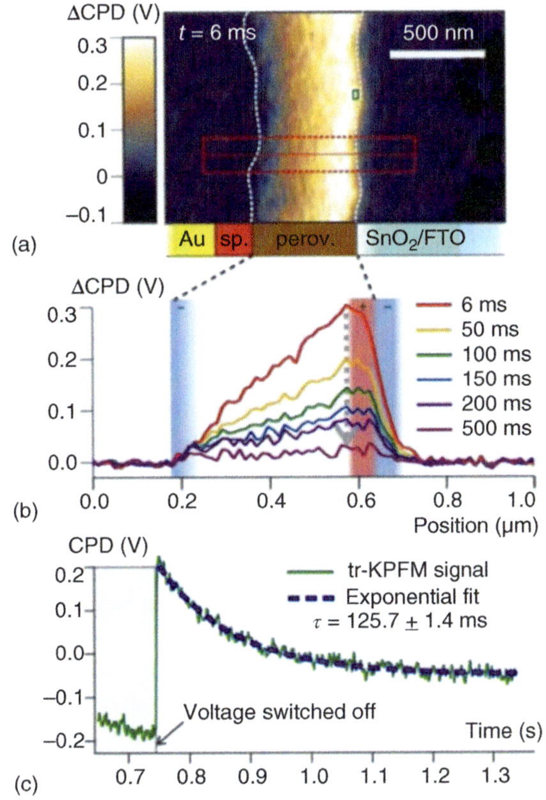

Figure 16.9 Transient KPFM measurements of perovskite solar cells upon switching the applied voltage from 0.5 to 0 V. (a) Contact potential map 6 ms after switching the voltage. (b) Evolution of the potential with time. (c) Contact potential transient at the SnO_2 interface. *Source*: From Weber et al. [61]. Published by The Royal Society of Chemistry.

direct proof of slowly changing electric fields upon a voltage step was obtained by Kelvin probe force microscopy (KPFM) measurements on cross sections of solar cells (Figure 16.9) [61].

Recalling the theory on the *JV* curve in Section 16.2, we understand that the hysteresis has to be a result of modified charge carrier collection and recombination dependent on the history (prebias and scan rate) of the device ($J_{photo}(V)$ in Eq. (16.3) becomes $J_{photo}[V$, scan rate, prebias]). In other words, the hysteresis (and the associated giant capacitance) appears when probed by electron–hole currents, in particular under illumination, whereas the underlying microscopic mechanism governs the timescale and less the magnitude of the effect.

Already in the first paper [6] dedicated to the *JV* hysteresis, three origins for the remnant electric field in the perovskite have been proposed:

1. Purely electronic origin due to
   a. *Deep electronic traps*: Charges trapped predominantly at interfaces would cause remnant space charge and affect the *JV* curve. So far, no clear evidence has been collected for this hypothesis.
   b. *Electron/hole transport*: Sometimes insufficient charge collection due to imbalanced electron/hole transport is discussed as reason for the hysteresis [37, 62, 63]. As electron/hole dynamics happen on much faster timescales, we exclude this effect as the underlying mechanism.

2. Ferroelectrics, meaning that an electric polarization remains after switching off the electric field (cf. Figure 16.1a); or in other words, the existence of a spontaneous polarization that can be switched by an external electric field: Perovskites can be ferroelectric, either by displacement of lattice sites leading to a polar distortion [64] or by oriented dipolar cations such as MA [65]. For $MAPbI_3$, the latter is unlikely due to a ps scale randomization of the MA orientation [29, 66]. Regarding the former, the absence of a (detectable) second harmonic signal was taken as an indication for a centrosymmetric, nonpolar structure of all phases in a recent study [67]. Later, second harmonic generation has been reported for tetragonal $MAPbI_3$ [68].

   The fact that hysteresis can be as prominent in Cs or FA-based perovskites shows that the orientation of MA can be excluded as a major reason. Here, we leave the discussions on the existence of ferroelectric domains at room temperature to the experts. Some identified them [69] related to tetragonal-to-cubic phase transitions [70, 71], others exclude their existence, despite theoretical predictions [72] or mainly by alternative interpretations of the microscopic measurements [73–75]. Some studies clearly rule out a ferroelectric origin of the hysteresis [76]. Here, we just conclude that we cannot see clear indications that ferroelectric domains influence the *JV* curve or are even required to explain efficient charge separation and transport in metal-halide perovskites as hypothesized in Refs. [77–80]. Only for completely unrealistically high recombination rate constants, ferroelectric polarization was shown to influence the FF in a simulation study [81].
3. *Mobile ions*: Metal halide perovskites are ionic conductors [82–84], meaning that defect ions such as halide vacancies are mobile and tend to accumulate at the surface of the perovskite to screen the electric field [85], which confirms the intuitive pictures brought up for hysteresis [16]. The double layer formed can be seen in the capacitance under low voltage and in the dark. Ion diffusion has been identified in a Warburg feature [86] or by transient capacitance measurements [87] yielding diffusion coefficients in a range of 10^{-12}–10^{-8} cm2/s. Theoretical work in combination with temperature-dependent experiments revealing activation energies for mobile ions provides strong evidence for halide vacancy migration being the response-time-determining process [29, 30, 88], although the error bars and parasitic influences on experimentally determined activations energies hardly allow for a meaningful comparison [89]. The picture of mobile ions being the source of changed charge-carrier collection and recombination probabilities is well supported and developed by numerical device simulation [90–94] that is capable of reproducing the major features of hysteresis.

There are several studies explicitly comparing ferroelectric and ionic origins of hysteresis, concluding that the ionic origin is the more likely one [76, 95].

Now we examine whether and how mobile ions allow for a consistent explanation of aforementioned experimental observations. The basic concept is sketched in Figure 16.10, indicating that for each applied voltage, mobile ions equilibrate in a way to reduce/minimize the electric field in the perovskite (situation at short circuit shown assuming an electric built-in potential). Consequently, when sweeping

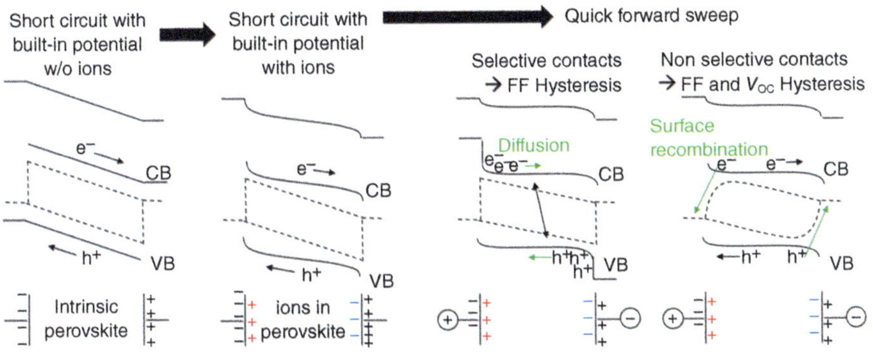

Figure 16.10 Band diagrams showing the flow of photo-generated electrons and holes in their respective bands. Mobile ions partially screen the electric field in the perovskite by accumulating at interfaces. A forward sweep in voltage faster than the response time of the ions leads to an unfavorable electric field for charge extraction. Dependent on the selectivity of the contacts, not only FF but V_{oc} is affected as well (compare with Figure 16.5a,b). *Source*: Tress [89].

or switching to a lower voltage ("backward"), the generated electric field is beneficial for drift-assisted charge collection. When switching or sweeping to a higher voltage ("forward"), as shown in the right panels of Figure 16.10, the electric field opposes charge collection at the designated electrodes. Once the ions have equilibrated, this effect diminishes and the current stabilizes, as seen in the transients in Figure 16.6. We discuss along the following list whether this explanation is fully consistent with experimental observations:

- *Rate dependence of hysteresis (Figure* 16.5*a)*: Yes, hysteresis when measured in a *JV* loop becomes largest, when the response time (i.e. transit time according to the sketches in Figure 16.10) is in the time range the *JV* scan takes. For slower scans, ions can follow and hysteresis reduces; for much faster scans, ions cannot follow and hysteresis reduces as well. If forward and backward curves are not measured in a loop but after equilibration at the starting voltage, hysteresis would not vanish at higher scan rates.
- *Temperature dependence (Figure* 16.7*b)*: Yes, it is analogous to the scan rate. Lower temperatures decrease ion mobility and hysteresis might become more or less apparent dependent on the scan rate. A simultaneous reduction of temperature and scan rate can lead to the same hysteresis (Figure 16.7b). In other words, when decreasing the temperature, the scan rate, where hysteresis is maximum, becomes lower [26].
- *Light-intensity independence (Figure* 16.5*c,d)*: Yes, because the ion distributions are mainly influenced by the voltage and the effect on the relative charge collection efficiency is independent of illumination intensity. If the ionic conductivity depends on illumination as discussed in Ref. [85] and found in Ref. [96], the hysteresis might change. How pronounced this change is expected to be should be clarified by device simulation.

- *Lower FF in forward scan (Figure* 16.5*a)*: Yes, it fits the idea that the ions screen the electric field and a forward scan, where voltage is applied that opposes charge collection, leads to a lower FF (Figure 16.10).
- *Lower V_{oc} in forward scan (Figure* 16.5*b)*: Yes. In this case, it is not sufficient to only discuss modified charge collection, but also changed recombination probabilities. The most straightforward example is surface recombination, where charges are lost at the wrong electrode (cf. Figure 16.3b). Ion distributions caused by a positive prebias induce an electric field that increases the selectivity of the device, whereas ions equilibrated at low or negative voltage decrease the selectivity (Figure 16.10 right panel).
- *Modified forward (dark) currents*: Yes, ions accumulated at interfaces to charge transport layers together with electronic counter charge on the electrodes, create interface dipoles, which modify charge injection properties. The lower or even negative the prebias, the more positive (negative) is the ion concentration at the hole (electron)-injecting contact, hindering forward current.
- *Grain size and composition dependence*: This is a complex problem and trends are not unique (cf. Section 16.3). Theoretical descriptions of ion motion are commonly limited to single crystals. Whether grain boundaries impede ion migration or whether ion migration is facilitated by grain boundaries remains an open question. Furthermore, grain boundaries affect electron/hole transport as well [40]. Thus, it is not surprising that the hysteresis is influenced by the morphology albeit predictions are not possible.
- *Inverted hysteresis (Figure* 16.7*a)*: Yes, more complex shapes of hysteresis can be explained. It was proposed that in case of inverted hysteresis, charge extraction at the contacts is the bottleneck due to extraction barriers [25] or reversed band bending [97]. In such a case, lower prebias is beneficial as accumulated ions improve charge extraction by the enhanced electric field close to the contact(s).
- *Timescale and quantities*: So far we have been talking about a “slow” process without clearly specifying the timescale. And we have been discussing migrating ions without talking about their concentrations. We analyze these parameters with Table 16.1, which shows that a broad range of diffusion coefficients and ion concentrations are reported. We can conclude that from various transient measurements, also on non-solar-cell architectures, a mobile ionic species with a diffusion coefficient in the range of 10^{-9}–10^{-8} cm^2/s has been identified and attributed to iodine vacancies [84]. Such a diffusion coefficient would result in hysteresis at scan rates in the order of 100 V/s, where capacitive and non-capacitive contributions are relevant as shown by Neukom et al. [101], who simulated not only the *JV* hysteresis but also a set of complementary experiments. This hysteresis, which also becomes apparent for nominally “hysteresis-free” devices, is consistent with the idea that the underlying mechanism is ion migration in the perovskite. However, to reproduce the common hysteresis at scan rates in the mV/s regime using device simulations, the ion mobility has to be decreased by 3–4 orders of magnitude. Furthermore, the ion densities are set in the order of 10^{18}–10^{19} cm^{-3} to provide sufficient space charge to screen the electric field and

show influence on the charge collection efficiency. Measured values, however, are more in the 10^{16}–10^{17} cm^{-3} regime. One explanation would be that ions in "hysteresis-devices" are much slower as proposed by Bertoluzzi et al. [99]. However, this is unlikely as also in these devices the "fast" contribution can be seen, e.g. in impedance spectroscopy. Another explanation would be that there is an additional slower mobile ionic species, as e.g. observed in Futscher et al. [87], where, however, the charge of the fast species is negative, which is in contradiction to the consensus that it should be a halide vacancy (positive). Furthermore, the explanation of a slower species is problematic as the slower species should be present also in hysteresis-less devices. Nevertheless, the simulations are capable of reproducing the hysteresis in the mV/s range very well, as long as slow mobile ions are assumed. As these simulations describe the effect of this slow process, not the dynamics itself, we would conclude that compared with hysteresis-less devices, either the electric field in the perovskite is much lower than predicted from these simulations, as it e.g. drops on a low-mobility transport layer, or the mobility of the ionic charge is strongly reduced, e.g. because these ions get stuck at interfaces.

- *Interface dependence*: An obvious question is why do we extensively discuss mobile ions in the perovskite as underlying reason for the hysteresis, knowing that p-i-n devices with organic charge transport layers hardly show hysteresis? One can counter this question by saying that hysteresis in oxide-based n-i-p devices can be widely tuned by the perovskite composition itself without modifying the contacts. The proper answer to this apparent paradox is that hysteresis is a product of an interplay between ion motion, setting the timescale, and the "reaction" of electronic charges, expressing the result in the *JV* curve. That is why it is not astonishing that certain electronic and ionic properties need to be fulfilled when simulating the hysteresis with numerical device simulators, such as existence of mobile ions and surface traps [90]/surface recombination [93] or mobile ions paired with extraction barriers [104] if recombination is not sufficiently strong. This interplay is summarized in Figure 16.11 from Ref. [89]. The consequence is that there are many handles to tune hysteresis, both on the ionic and electronic material and device properties. Nevertheless, this interplay cannot explain the different timescales, where halide vacancy migration is expected to happen compared with the ion mobilities required to reproduce hysteresis (cf. Table 16.1). This discrepancy could indeed be due to the influence of the charge transport layers. Recent device simulations have examined the role of charge transport layers and found that their properties indeed affect hysteresis. It was reported that the timescale decreases slightly with the product of doping concentration and dielectric constant [103], however, not to an extent to describe the difference between TiO_2 and p-i-n devices. Another simulation study showed that energetic barriers at the interface to the electron transport layer tune the appearance of hysteresis without modifying ionic parameters [104]. This is another example, demonstrating the importance of the electronic–ionic interplay sketched in Figure 16.11 and its subtle nuances that are yet to be discovered.

Table 16.1 Parameters for mobile ions in $MAPbI_3$, deduced by various experimental and simulation approaches.

Method	Scan rate[a)]	Diffusion coefficient (cm^2/s)	Mobility at 300 K ($cm^2/(V\,s)$)	Concentration (cm^{-3})	Mobile species[b)]	References	
Transient current, dark		2.4×10^{-8}	9.3×10^{-7}	5×10^{16}[c)]	I^- (measured)	Yang et al. [84]	Measurement
Electrochemical impedance spectroscopy (EIS), Warburg fit light only		1.4×10^{-12}	5.4×10^{-11}		MA^+	Bag et al. [86]	Measurement
Transient capacitance		3×10^{-9}		1×10^{15}	I^-	Futscher et al. [87]	Measurement
		3×10^{-12}		1.3×10^{16}	MA^+ tetragonal grain boundary (GB)		Measurement
		7×10^{-12}		1.3×10^{16}	MA^+ cubic GB		Measurement
		2×10^{-12}		5×10^{15}	MA^+ cubic		Measurement
Photoluminescence (PL) migration		1×10^{-9}	5×10^{-8}		I^-	Li et al. [98]	Measurement
Transients[d)]		8×10^{-9}	3×10^{-7}	1×10^{18}		Bertoluzzi et al. [99]	Measurement
Charge upon voltage step				4×10^{17}		Bertoluzzi et al. [100]	Measurement
DD simulated transients + hysteresis	100 V/s	1×10^{-9}/ 3×10^{-16}	5×10^{-8}/ 1×10^{-14}	5×10^{17}/ 5×10^{17}	Cation/anion	Neukom et al. [101]	Fitting simulated
DD simulated	100 kV/s	2×10^{-6}	1×10^{-4}	1×10^{18}		van Reenen et al. [90]	Fitting simulated
Estimated	1 V/s	2×10^{-11}	1×10^{-9}				Fitting simulated

hysteresis to experiment

DD hysteresis fit	1 V/s 250 mV/s	1×10^{-11} 3.7×10^{-12}	5×10^{-10} 1.4×10^{-10}	1.6×10^{19}	Cation	Richardson et al. [91]
DD fit to transients dark current		7.5×10^{-13}	3×10^{-11}	1.6×10^{19}	Cation	O'Kane et al. [102]
DD hysteresis fit	40 mV/s	1×10^{-12}	5×10^{-11}	1×10^{19}	Cation	Calado et al. [93]
Only DD simulated, parameters from prev. publ.	75 mV/s	1×10^{-13}	5×10^{-12}	1.6×10^{19}	Cation	Courtier et al. [103]
Only DD simulated, parameters from prev. publ.	20 mV/s	1×10^{-13}/ 1×10^{-13}	5×10^{-12}/ 5×10^{-12}	1×10^{18}/ 1×10^{18}	Anion/cation	Sun et al. [104]
Only DD simulated, parameters from prev. publ.	240 mV/s	2×10^{-12}	1×10^{-10}	5×10^{17}		Yang et al. [105]

DD stands for drift-diffusion device simulation.

a) Order of magnitude, where hysteresis is pronounced.

b) Commonly assigned with reference to preceding literature such as density functional theory (DFT) calculations.

c) Deduced from the published conductivity value.

d) Mixed perovskite composition.

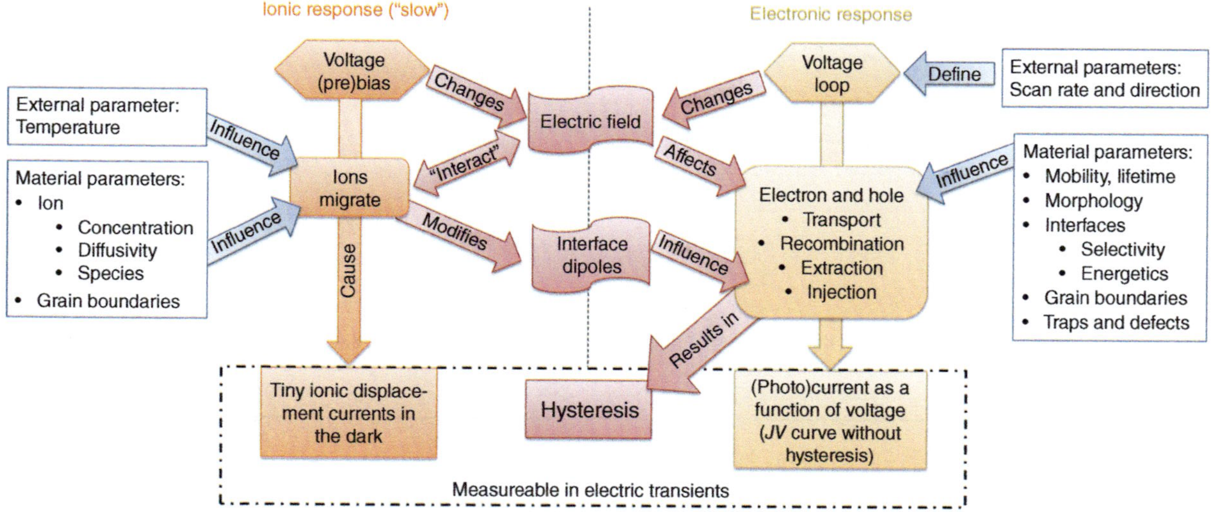

Figure 16.11 Interplay of ionic and electronic response focusing on the electrostatic effect of the mobile ions. The *JV* hysteresis seen in Figures 16.1b and 16.5 is a result of slowly modified electron/hole currents. *Source*: Tress [89].

16.5 Issues with Hysteresis: How to Tune/Avoid/Suppress?

The brief literature overview in Section 16.3 showed that many factors influence hysteresis. In Section 16.5, we saw that this is not surprising as many properties of the material and the device stack influence the "outcome" of the mixed ionic electronic conductivity. Therefore, it is not astonishing that the literature does not offer a conclusive strategy to suppress hysteresis. Based on earlier explanations, we can conclude the following: Measures that are beneficial for the PCE are the best candidates for decreasing hysteresis as well [92], because

- "Passivation" measures reduce recombination losses at interfaces to the contact layers and hence increase contact selective. This makes the device less prone to surface recombination and thus less sensitive to changes in the electrostatics by mobile ions.
- "Good" morphology, i.e. films with low defect densities and high overall charge carrier mobilities, allows for a solar cell, working efficiently in a diffusion-driven regime. Thus, charge collection does not rely on the electric field and its modification by moving ions.
- Conductive charge transport layers minimize voltage losses during current flow and avoid accumulation of electronic charge. Therefore, efficient charge extraction does not rely on an electric field, which could be compromised by mobile ions.

Measures targeting directly the mobile ions by decreasing their mobility or density are less straightforward, as it is expected that they would mainly shift the timescale where hysteresis is observable. Furthermore, as the ions are intrinsic defects, their equilibrium concentration is given by thermodynamics, and a certain concentration is therefore unavoidable. Due to their low formation energies and a potential dependence on illumination, they are dynamically generated and eliminated [106]. Extrinsic ions such as Li (dopant of the hole-transport material spiro-MeOTAD) have been reported to contribute to hysteresis as well [107]. Therefore, their trajectories should be controlled, also with respect to the long-term device stability.

16.6 Conclusion and Open Questions

Hysteresis in the *JV* curve of PSCs is a rather universal feature that is mostly of non-capacitive origin. Instead, collection and extraction efficiencies of photo-generated charge carriers and thus the fraction of recombining charges become a function of the history, i.e. prebias and voltage sweep rate. Hysteresis is more pronounced the more important an electric field is for charge extraction, such as for low-conductivity active and charge transport layers. The microscopic process causing this delayed response of the electric field are mobile ions that contribute to the electrostatic picture of the device. Ferroelectric effects are less likely to play a role for hysteresis, independent of the fact that discussions on ferroelectric domains in metal-halide perovskites are still ongoing.

Mobile ions result mainly from intrinsic crystal defects that are thermodynamically unavoidable and proven to be migrating, both experimentally and theoretically. Their effect on the electric field and thus the *JV* hysteresis has been confirmed by transient measurements of the electric field, which is consistent to device simulations that reproduce hysteresis assuming mobile ions and competing charge extraction and recombination rates.

Beyond these mainly qualitative coincidences, questions remain open, e.g. when comparing measured ion diffusion coefficients with those required to reproduce hysteresis curves in device simulation. The same holds for ion concentrations and a clear assignment of a chemical species to an effect observed in the *JV* curve. Therefore, further work has to be carried out to disentangle the exact interplay of ionic and electronic charge carriers in PSCs. This will lead to a deeper understanding of which parameter changed when observing different hysteresis curves. The obtained insights will allow for a better control, targeted tuning, and even exploitation of the hysteresis phenomenon in future device generations.

References

1 Kang, D.-H. and Park, N.-G. (2019). On the current–voltage hysteresis in perovskite solar cells: dependence on perovskite composition and methods to remove hysteresis. *Adv. Mater.* 31 (34): 1805214.

2 Elumalai, N.K. and Uddin, A. (2016). Hysteresis in organic-inorganic hybrid perovskite solar cells. *Sol. Energy Mater. Sol. Cells* 157: 476–509.

3 Chen, B., Yang, M., Priya, S., and Zhu, K. (2016). Origin of J–V hysteresis in perovskite solar cells. *J. Phys. Chem. Lett.* 7 (5): 905–917.

4 Li, C., Guerrero, A., Zhong, Y., and Huettner, S. (2017). Origins and mechanisms of hysteresis in organometal halide perovskites. *J. Phys. Condens. Matter* 29 (19): 193001.

5 Unger, E.L., Czudek, A., Kim, H.-S., and Tress, W. (2020). Current-voltage analysis: lessons learned from hysteresis. In: *Characterization Techniques for Perovskite Solar Cell Materials*, Chapter 4 (eds. M. Pazoki, A. Hagfeldt and T. Edvinsson), 81–108. Elsevier.

6 Snaith, H.J., Abate, A., Ball, J.M. et al. (2014). Anomalous hysteresis in perovskite solar cells. *J. Phys. Chem. Lett.* 5 (9): 1511–1515.

7 Habisreutinger, S.N., Noel, N.K., and Snaith, H.J. (2018). Hysteresis index: a figure without merit for quantifying hysteresis in perovskite solar cells. *ACS Energy Lett.* 3: 2472–2476.

8 Christians, J.A., Manser, J.S., and Kamat, P.V. (2015). Best practices in perovskite solar cell efficiency measurements. Avoiding the error of making bad cells look good. *J. Phys. Chem. Lett.* 6 (5): 852–857.

9 Tress, W. (2016). Maximum efficiency and open-circuit voltage of perovskite solar cells. In: *Organic-Inorganic Halide Perovskite Photovoltaics* (eds. N.-G. Park, M. Grätzel and T. Miyasaka), 53–77. Springer International Publishing.

10 Tress, W. (2017). Perovskite solar cells on the way to their radiative efficiency limit – insights into a success story of high open-circuit voltage and low recombination. *Adv. Energy Mater.* 7 (14): 1602358.

11 Tress, W. (2014). *Organic Solar Cells*. Cham: Springer International Publishing.

12 Tress, W., Leo, K., and Riede, M. (2011). Influence of hole-transport layers and donor materials on open-circuit voltage and shape of I–V curves of organic solar cells. *Adv. Funct. Mater.* 21 (11): 2140–2149.

13 Tress, W. and Inganäs, O. (2013). Simple experimental test to distinguish extraction and injection barriers at the electrodes of (organic) solar cells with S-shaped current–voltage characteristics. *Sol. Energy Mater. Sol. Cells* 117: 599–603.

14 Contreras, L., Idígoras, J., Todinova, A. et al. (2016). Specific cation interactions as the cause of slow dynamics and hysteresis in dye and perovskite solar cells: a small-perturbation study. *Phys. Chem. Chem. Phys.* 18 (45): 31033–31042.

15 Sarker, S., Seo, H.W., Jin, Y.-K. et al. (2015). On the hysteresis of current density-voltage curves of dye-sensitized solar cells. *Electrochim. Acta* 182: 493–499.

16 Tress, W., Marinova, N., Moehl, T. et al. (2015). Understanding the rate-dependent J–V hysteresis, slow time component, and aging in $CH_3NH_3PbI_3$ perovskite solar cells: the role of a compensated electric field. *Energy Environ. Sci.* 8 (3): 995–1004.

17 Christoforo, M.G., Hoke, E.T., McGehee, M.D., and Unger, E.L. (2015). Transient response of organo-metal-halide solar cells analyzed by time-resolved current-voltage measurements. *Photonics* 2 (4): 1101–1115.

18 Almora, O., Zarazua, I., Mas-Marza, E. et al. (2015). Capacitive dark currents, hysteresis, and electrode polarization in lead halide perovskite solar cells. *J. Phys. Chem. Lett.* 6 (9): 1645–1652.

19 Unger, E.L., Hoke, E.T., Bailie, C.D. et al. (2014). Hysteresis and transient behavior in current–voltage measurements of hybrid-perovskite absorber solar cells. *Energy Environ. Sci.* 7 (11): 3690–3698.

20 Kim, H.-S. and Park, N.-G. (2014). Parameters affecting I–V hysteresis of $CH_3NH_3PbI_3$ perovskite solar cells: effects of perovskite crystal size and mesoporous TiO_2 layer. *J. Phys. Chem. Lett.* 5 (17): 2927–2934.

21 Xiao, Z., Yuan, Y., Shao, Y. et al. (2015). Giant switchable photovoltaic effect in organometal trihalide perovskite devices. *Nat. Mater.* 14 (2): 193–198.

22 Ono, L.K., Raga, S.R., Wang, S. et al. (2014). Temperature-dependent hysteresis effects on perovskite-based solar cells. *J. Mater. Chem. A* 3: 9074–9080.

23 Jacobsson, T.J., Tress, W., Correa-Baena, J.-P. et al. (2016). Room temperature as a goldilocks environment for $CH_3NH_3PbI_3$ perovskite solar cells: the importance of temperature on device performance. *J. Phys. Chem. C* 120 (21): 11382–11393.

24 Levine, I., Nayak, P.K., Wang, J.T.-W. et al. (2016). Interface-dependent ion migration/accumulation controls hysteresis in $MAPbI_3$ solar cells. *J. Phys. Chem. C* 120 (30): 16399–16411.

25 Tress, W., Correa Baena, J.P., Saliba, M. et al. (2016). Inverted current–voltage hysteresis in mixed perovskite solar cells: polarization, energy barriers, and defect recombination. *Adv. Energy Mater.* 6 (19): 1600396.
26 Domanski, K., Tress, W., Moehl, T. et al. (2015). Working principles of perovskite photodetectors: analyzing the interplay between photoconductivity and voltage-driven energy-level alignment. *Adv. Funct. Mater.* 25 (44): 6936–6947.
27 Chiang, C.-H., Tseng, Z.-L., and Wu, C.-G. (2014). Planar heterojunction perovskite/$PC_{71}BM$ solar cells with enhanced open-circuit voltage via a (2/1)-step spin-coating process. *J. Mater. Chem. A* 2 (38): 15897–15903.
28 Tress, W., Leo, K., and Riede, M. (2011). Effect of concentration gradients in ZnPc:C_{60} bulk heterojunction organic solar cells. *Sol. Energy Mater. Sol. Cells* 95 (11): 2981–2986.
29 Meloni, S., Moehl, T., Tress, W. et al. (2016). Ionic polarization-induced current-voltage hysteresis in $CH_3NH_3PbX_3$ perovskite solar cells. *Nat. Commun.* 7: 10334.
30 Eames, C., Frost, J.M., Barnes, P.R.F. et al. (2015). Ionic transport in hybrid lead iodide perovskite solar cells. *Nat. Commun.* 6: 7497.
31 Bryant, D., Wheeler, S., O'Regan, B.C. et al. (2015). Observable hysteresis at low temperature in "hysteresis free" organic–inorganic lead halide perovskite solar cells. *J. Phys. Chem. Lett.* 6 (16): 3190–3194.
32 Baena, J.P.C., Steier, L., Tress, W. et al. (2015). Highly efficient planar perovskite solar cells through band alignment engineering. *Energy Environ. Sci.* 8 (10): 2928–2934.
33 Patel, J.B., Wong-Leung, J., Reenen, S.V. et al. (2017). Influence of interface morphology on hysteresis in vapor-deposited perovskite solar cells. *Adv. Electron. Mater.* 3 (2): 1600470.
34 Kegelmann, L., Wolff, C.M., Awino, C. et al. (2017). It takes two to tango – double-layer selective contacts in perovskite solar cells for improved device performance and reduced hysteresis. *ACS Appl. Mater. Interfaces* 9 (20): 17245–17255.
35 Sepalage, G.A., Meyer, S., Pascoe, A. et al. (2015). Copper(I) iodide as hole-conductor in planar perovskite solar cells: probing the origin of J–V hysteresis. *Adv. Funct. Mater.* 25 (35): 5650–5661.
36 Kim, H.-S., Jang, I.-H., Ahn, N. et al. (2015). Control of I–V hysteresis in $CH_3NH_3PbI_3$ perovskite solar cell. *J. Phys. Chem. Lett.* 6 (22): 4633–4639.
37 Heo, J.H., Han, H.J., Kim, D. et al. (2015). Hysteresis-less inverted $CH_3NH_3PbI_3$ planar perovskite hybrid solar cells with 18.1% power conversion efficiency. *Energy Environ. Sci.* 8 (5): 1602–1608.
38 Wu, C.-G., Chiang, C.-H., Tseng, Z.-L. et al. (2015). High efficiency stable inverted perovskite solar cells without current hysteresis. *Energy Environ. Sci.* 8 (9): 2725–2733.
39 Shao, Y., Xiao, Z., Bi, C. et al. (2014). Origin and elimination of photocurrent hysteresis by fullerene passivation in $CH_3NH_3PbI_3$ planar heterojunction solar cells. *Nat. Commun.* 5: 5784.

40 Correa-Baena, J.-P., Anaya, M., Lozano, G. et al. (2016). Unbroken perovskite: interplay of morphology, electro-optical properties, and ionic movement. *Adv. Mater.* 28 (25): 5031–5037.

41 Momblona, C., Gil-Escrig, L., Bandiello, E. et al. (2016). Efficient vacuum deposited p-i-n and n-i-p perovskite solar cells employing doped charge transport layers. *Energy Environ. Sci.* 9 (11): 3456–3463.

42 Lee, J.-W., Kim, S.-G., Bae, S.-H. et al. (2017). The interplay between trap density and hysteresis in planar heterojunction perovskite solar cells. *Nano Lett.* 17 (7): 4270–4276.

43 Yu, H., Lu, H., Xie, F. et al. (2016). Native defect-induced hysteresis behavior in organolead iodide perovskite solar cells. *Adv. Funct. Mater.* 26 (9): 1411–1419.

44 Xu, J., Buin, A., Ip, A.H. et al. (2015). Perovskite–fullerene hybrid materials suppress hysteresis in planar diodes. *Nat. Commun.* 6 (1): 1–8.

45 Yavari, M., Mazloum-Ardakani, M., Gholipour, S. et al. (2018). Carbon nanoparticles in high-performance perovskite solar cells. *Adv. Energy Mater.* 8 (12): 1702719.

46 Son, D.-Y., Kim, S.-G., Seo, J.-Y. et al. (2018). Universal approach toward hysteresis-free perovskite solar cell via defect engineering. *J. Am. Chem. Soc.* 140 (4): 1358–1364.

47 Domanski, K., Roose, B., Matsui, T. et al. (2017). Migration of cations induces reversible performance losses over day/night cycling in perovskite solar cells. *Energy Environ. Sci.* 10 (2): 604–613.

48 Wang, Z.S., Ebadi, F., Carlsen, B. et al. Transient photovoltage measurements on perovskite solar cells with varied defect concentrations and inhomogeneous recombination rates. *Small Methods* 4: 2000290.

49 Herman, M., Jankovec, M., and Topič, M. (2013). Optimisation of the I–V measurement scan time through dynamic modelling of solar cells. *IET Renew. Power Gener.* 7 (1): 63–70.

50 Albin, D.S. and del Cueto, J.A. (2010). Effect of hysteresis on measurements of thin-film cell performance. *Reliability of Photovoltaic Cells, Modules, Components, and Systems III*, 7773, 77730N.

51 Ravishankar, S., Almora, O., Echeverría-Arrondo, C. et al. (2017). Surface polarization model for the dynamic hysteresis of perovskite solar cells. *J. Phys. Chem. Lett.* 8 (5): 915–921.

52 Cojocaru, L., Uchida, S., Jayaweera, P.V.V. et al. (2015). Origin of the hysteresis in I–V curves for planar structure perovskite solar cells rationalized with a surface boundary-induced capacitance model. *Chem. Lett.* 44 (12): 1750–1752.

53 Juarez-Perez, E.J., Sanchez, R.S., Badia, L. et al. (2014). Photoinduced giant dielectric constant in lead halide perovskite solar cells. *J. Phys. Chem. Lett.* 5 (13): 2390–2394.

54 Ebadi, F., Taghavinia, N., Mohammadpour, R. et al. (2019). Origin of apparent light-enhanced and negative capacitance in perovskite solar cells. *Nat. Commun.* 10 (1): 1574.

55 Pockett, A., Eperon, G.E., Sakai, N. et al. (2017). Microseconds, milliseconds and seconds: deconvoluting the dynamic behaviour of planar perovskite solar cells. *Phys. Chem. Chem. Phys.* 19 (8): 5959–5970.

56 Moia, D., Gelmetti, I., Calado, P. et al. (2019). Ionic-to-electronic current amplification in hybrid perovskite solar cells: ionically gated transistor-interface circuit model explains hysteresis and impedance of mixed conducting devices. *Energy Environ. Sci.* 12 (4): 1296–1308.

57 Jacobs, D.A., Shen, H., Pfeffer, F. et al. (2018). The two faces of capacitance: new interpretations for electrical impedance measurements of perovskite solar cells and their relation to hysteresis. *J. Appl. Phys.* 124 (22): 225702.

58 Almora, O., Aranda, C., Zarazua, I. et al. (2016). Noncapacitive hysteresis in perovskite solar cells at room temperature. *ACS Energy Lett.* 1: 209–215.

59 Chen, B., Yang, M., Zheng, X. et al. (2015). Impact of capacitive effect and ion migration on the hysteretic behavior of perovskite solar cells. *J. Phys. Chem. Lett.* 6 (23): 4693–4700.

60 Garcia-Belmonte, G. and Bisquert, J. (2016). Distinction between capacitive and noncapacitive hysteretic currents in operation and degradation of perovskite solar cells. *ACS Energy Lett.* 1: 683–688.

61 Weber, S.A.L., Hermes, I.M., Turren-Cruz, S.-H. et al. (2018). How the formation of interfacial charge causes hysteresis in perovskite solar cell. *Energy Environ. Sci.* 11 (9): 2404–2413.

62 Jeon, N.J., Noh, J.H., Kim, Y.C. et al. (2014). Solvent engineering for high-performance inorganic–organic hybrid perovskite solar cells. *Nat. Mater.* 13 (9): 897–903.

63 Heo, J.H., Song, D.H., Han, H.J. et al. (2015). Planar $CH_3NH_3PbI_3$ perovskite solar cells with constant 17.2% average power conversion efficiency irrespective of the scan rate. *Adv. Mater.* 27 (22): 3424–3430.

64 Choi, T., Lee, S., Choi, Y.J. et al. (2009). Switchable ferroelectric diode and photovoltaic effect in $BiFeO_3$. *Science* 324 (5923): 63–66.

65 Leguy, A.M.A., Frost, J.M., McMahon, A.P. et al. (2015). The dynamics of methylammonium ions in hybrid organic-inorganic perovskite solar cells. *Nat. Commun.* 6: 7124.

66 Quarti, C., Mosconi, E., and De Angelis, F. (2014). Interplay of orientational order and electronic structure in methylammonium lead iodide: implications for solar cell operation. *Chem. Mater.* 26 (22): 6557–6569.

67 Sharada, G., Mahale, P., Kore, B.P. et al. (2016). Is $CH_3NH_3PbI_3$ polar? *J. Phys. Chem. Lett.* 7 (13): 2412–2419.

68 Rakita, Y., Bar-Elli, O., Meirzadeh, E. et al. (2017). Tetragonal $CH_3NH_3PbI_3$ is ferroelectric. *Proc. Natl. Acad. Sci. U.S.A.* 114 (28): E5504–E5512.

69 Kutes, Y., Ye, L., Zhou, Y. et al. (2014). Direct observation of ferroelectric domains in solution-processed $CH_3NH_3PbI_3$ perovskite thin films. *J. Phys. Chem. Lett.* 5 (19): 3335–3339.

70 Röhm, H., Leonhard, T., Hoffmann, M.J., and Colsmann, A. (2017). Ferroelectric domains in methylammonium lead iodide perovskite thin-films. *Energy Environ. Sci.* 10 (4): 950–955.

71 Vorpahl, S.M., Giridharagopal, R., Eperon, G.E. et al. (2018). Orientation of ferroelectric domains and disappearance upon heating methylammonium lead triiodide perovskite from tetragonal to cubic phase. *ACS Appl. Energy Mater.* 1 (4): 1534–1539.

72 Fan, Z., Xiao, J., Sun, K. et al. (2015). Ferroelectricity of $CH_3NH_3PbI_3$ perovskite. *J. Phys. Chem. Lett.* 6 (7): 1155–1161.

73 Liu, Y., Collins, L., Proksch, R. et al. (2018). Chemical nature of ferroelastic twin domains in $CH_3NH_3PbI_3$ perovskite. *Nat. Mater.* 17 (11): 1013–1019.

74 Sajedi Alvar, M., Kumar, M., Blom, P.W.M. et al. (2017). Absence of ferroelectricity in methylammonium lead iodide perovskite. *AIP Adv.* 7 (9): 095110.

75 Gómez, A., Wang, Q., Goñi, A.R. et al. (2019). Ferroelectricity-free lead halide perovskites. *Energy Environ. Sci.* 12 (8): 2537–2547.

76 Beilsten-Edmands, J., Eperon, G.E., Johnson, R.D. et al. (2015). Non-ferroelectric nature of the conductance hysteresis in $CH_3NH_3PbI_3$ perovskite-based photovoltaic devices. *Appl. Phys. Lett.* 106 (17): 173502.

77 Frost, J.M., Butler, K.T., and Walsh, A. (2014). Molecular ferroelectric contributions to anomalous hysteresis in hybrid perovskite solar cells. *APL Mater.* 2 (8): 081506.

78 Pecchia, A., Gentilini, D., Rossi, D. et al. (2016). Role of ferroelectric nanodomains in the transport properties of perovskite solar cells. *Nano Lett.* 16 (2): 988–992.

79 Rossi, D., Pecchia, A., der Maur, M.A. et al. (2018). On the importance of ferroelectric domains for the performance of perovskite solar cells. *Nano Energy* 48: 20–26.

80 Wei, J., Zhao, Y., Li, H. et al. (2014). Hysteresis analysis based on the ferroelectric effect in hybrid perovskite solar cells. *J. Phys. Chem. Lett.* 5 (21): 3937–3945.

81 Sherkar, S. and T., and Koster, L.J.A. (2016). Can ferroelectric polarization explain the high performance of hybrid halide perovskite solar cells? *Phys. Chem. Chem. Phys.* 18 (1): 331–338.

82 Mizusaki, J., Arai, K., and Fueki, K. (1983). Ionic conduction of the perovskite-type halides. *Solid State Ionics* 11 (3): 203–211.

83 Maeda, M., Hattori, M., Hotta, A., and Suzuki, I. (1997). Dielectric studies on $CH_3NH_3PbX_3$ (X = Cl and Br) single cystals. *J. Phys. Soc. Jpn.* 66 (5): 1508–1511.

84 Yang, T.-Y., Gregori, G., Pellet, N. et al. (2015). The significance of ion conduction in a hybrid organic–inorganic lead-iodide-based perovskite photosensitizer. *Angew. Chem. Int. Ed.* 127 (27): 8016–8021.

85 Belisle, R.A., Nguyen, W.H., Bowring, A.R. et al. (2017). Interpretation of inverted photocurrent transients in organic lead halide perovskite solar cells: proof of the field screening by mobile ions and determination of the space charge layer widths. *Energy Environ. Sci.* 10 (1): 192–204.

86 Bag, M., Renna, L.A., Adhikari, R.Y. et al. (2015). Kinetics of ion transport in perovskite active layers and its implications for active layer stability. *J. Am. Chem. Soc.* 137 (40): 13130–13137.

87 Futscher, M.H., Lee, J.M., McGovern, L. et al. (2019). Quantification of ion migration in $CH_3NH_3PbI_3$ perovskite solar cells by transient capacitance measurements. *Mater. Horiz.* 6 (7): 1497–1503.

88 Azpiroz, J.M., Mosconi, E., Bisquert, J., and Angelis, F.D. (2015). Defect migration in methylammonium lead iodide and its role in perovskite solar cell operation. *Energy Environ. Sci.* 8 (7): 2118–2127.

89 Tress, W. (2017). Metal halide perovskites as mixed electronic–ionic conductors: challenges and opportunities – from hysteresis to memristivity. *J. Phys. Chem. Lett.* 8 (13): 3106–3114.

90 van Reenen, S., Kemerink, M., and Snaith, H.J. (2015). Modeling anomalous hysteresis in perovskite solar cells. *J. Phys. Chem. Lett.* 6 (19): 3808–3814.

91 Richardson, G., O'Kane, S.E.J., Niemann, R.G. et al. (2016). Can slow-moving ions explain hysteresis in the current–voltage curves of perovskite solar cells? *Energy Environ. Sci.* 9 (4): 1476–1485.

92 Neukom, M.T., Züfle, S., Knapp, E. et al. (2017). Why perovskite solar cells with high efficiency show small IV-curve hysteresis. *Sol. Energy Mater. Sol. Cells* 169: 159–166.

93 Calado, P., Telford, A.M., Bryant, D. et al. (2016). Evidence for ion migration in hybrid perovskite solar cells with minimal hysteresis. *Nat. Commun.* 7: 13831.

94 Jacobs, D.A., Wu, Y., Shen, H. et al. (2017). Hysteresis phenomena in perovskite solar cells: the many and varied effects of ionic accumulation. *Phys. Chem. Chem. Phys.* 19 (4): 3094–3103.

95 Seol, D., Jeong, A., Han, M.H. et al. (2017). Origin of hysteresis in $CH_3NH_3PbI_3$ perovskite thin films. *Adv. Funct. Mater.* 27 (37): 1701924.

96 Kim, G.Y., Senocrate, A., Yang, T.-Y. et al. (2018). Large tunable photoeffect on ion conduction in halide perovskites and implications for photodecomposition. *Nat. Mater.* 17 (5): 445–449.

97 Shen, H., Jacobs, D.A., Wu, Y. et al. (2017). Inverted hysteresis in $CH_3NH_3PbI_3$ solar cells: role of stoichiometry and band alignment. *J. Phys. Chem. Lett.* 8 (12): 2672–2680.

98 Li, C., Guerrero, A., Zhong, Y. et al. (2017). Real-time observation of iodide ion migration in methylammonium lead halide perovskites. *Small* 13 (42): 1701711.

99 Bertoluzzi, L., Belisle, R.A., Bush, K.A. et al. (2018). In situ measurement of electric-field screening in hysteresis-free $PTAA/FA_{0.83}Cs_{0.17}Pb(I_{0.83}Br_{0.17})_3/C_{60}$ perovskite solar cells gives an ion mobility of $\sim 3 \times 10^{-7}$ $cm^2/(V\ s)$, 2 orders of magnitude faster than reported for metal-oxide-contacted perovskite cells with hysteresis. *J. Am. Chem. Soc.* 140 (40): 12775–12784.

100 Bertoluzzi, L., Boyd, C.C., Rolston, N. et al. (2020). Mobile ion concentration measurement and open-access band diagram simulation platform for halide perovskite solar cells. *Joule* 4 (1): 109–127.

101 Neukom, M.T., Schiller, A., Züfle, S. et al. (2019). Consistent device simulation model describing perovskite solar cells in steady-state, transient, and frequency domain. *ACS Appl. Mater. Interfaces* 11 (26): 23320–23328.

102 O'Kane, S.E.J., Richardson, G., Pockett, A. et al. (2017). Measurement and modelling of dark current decay transients in perovskite solar cells. *J. Mater. Chem. C* 5 (2): 452–462.

103 Courtier, N.E., Cave, J.M., Foster, J.M. et al. (2019). How transport layer properties affect perovskite solar cell performance: insights from a coupled charge transport/ion migration model. *Energy Environ. Sci.* 12 (1): 396–409.

104 Sun, Z., Kang, Y., Wang, G. et al. (2020). Impact of interface energy alignment on the dynamic current–voltage response of perovskite solar cells. *J. Phys. Chem. C* 124 (24): 12912–12921.

105 Yang, C., Shan, X., and Xie, T. (2020). Insights of hysteresis behaviors in perovskite solar cells from a mixed drift-diffusion model coupled with recombination. *Photonics* 7 (3): 47.

106 Ceratti, D.R., Rakita, Y., Cremonesi, L. et al. (2018). Self-healing inside $APbBr_3$ halide perovskite crystals. *Adv. Mater.* 30 (10): 1706273.

107 Li, Z., Xiao, C., Yang, Y. et al. (2017). Extrinsic ion migration in perovskite solar cells. *Energy Environ. Sci.* 10: 1234–1242.

17

Perovskite-Based Tandem Solar Cells

Klaus Jäger[1] and Steve Albrecht[2]

[1]Helmholtz-Zentrum Berlin für Materialien und Energie, Department of Optics for Solar Energy, Albert-Einstein-Straße 16, 12489 Berlin, Germany
[2]Helmholtz-Zentrum Berlin für Materialien und Energie, Young Investigator Group Perovskite Tandem Solar Cells, Kekuléstraße 5, 12489 Berlin, Germany

17.1 Introduction

As thoroughly discussed in the previous chapters, perovskite solar cells made tremendous progress in recent years. However, as for all single-junction solar cells, their power conversion efficiency (PCE) is limited by the detailed balance limit [1], as discussed in Chapter 12. Figure 17.1a shows the two major loss mechanisms of *single-junction* solar cells. The solar cell is based on a semiconductor, which is transparent for photons with energies below the semiconductor bandgap. This loss is called *below-bandgap loss*. For silicon (Si), with a bandgap of 1.1 eV, corresponding to a wavelength of 1127 nm, this loss accounts for around 19% of the total solar irradiance. On the other hand, the maximum energy that the solar cell can utilize per absorbed photon is slightly lower than the bandgap. Therefore, the usable energy is considerably below the available energy from the photons, as illustrated in Figure 17.1a, which is called *thermalization loss* and accounts for at least 33% of the available energy for 1.1 eV semiconductor bandgap.

The thermalization loss can be reduced considerably when combining more or two solar cells in *multi-junction solar cells*. *Tandem solar cells*, where two solar cells are combined, are constructed such that incident sunlight first hits the *top cell* with the higher bandgap. The fraction of the solar spectrum with energies below the top cell bandgap can pass the top cell and can be absorbed by the bottom cell. Because the high energy photons are utilized by the top cell, the overall thermalization losses can be reduced considerably, as illustrated in Figure 17.1b.

Figure 17.2 shows absorption spectra of the absorbers in (a) a state-of-the-art silicon solar cell with pyramidal textures on both sides and (b) a state-of-the art monolithic[1] perovskite-silicon tandem solar cell with a flat perovskite top cell and a pyramidally back-textured silicon cell. We see that in (b) the absorption is split

1 The concept of monolithic tandem solar cells is described in Section 17.2.1.

Hybrid Perovskite Solar Cells: Characteristics and Operation,
First Edition. Edited by Hiroyuki Fujiwara.

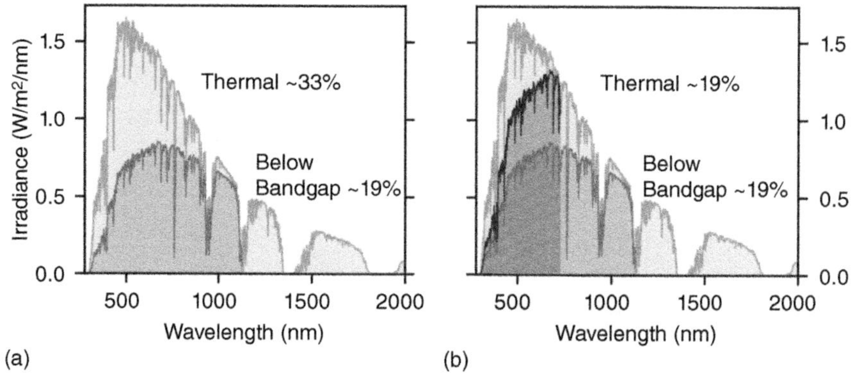

Figure 17.1 The AM1.5g solar spectrum [2] and the two most important losses for (a) single-junction and (b) tandem solar cells. As bandgap for the single-junction cell and the bottom cell of the tandem cell, 1.1 eV (1127 nm) is chosen, which corresponds to the bandgap of silicon. As top-cell bandgap, 1.71 eV are chosen, which is the optimal bandgap for series-connected two-terminal devices. *Source*: A.S.T.M.S. G173-03 [2].

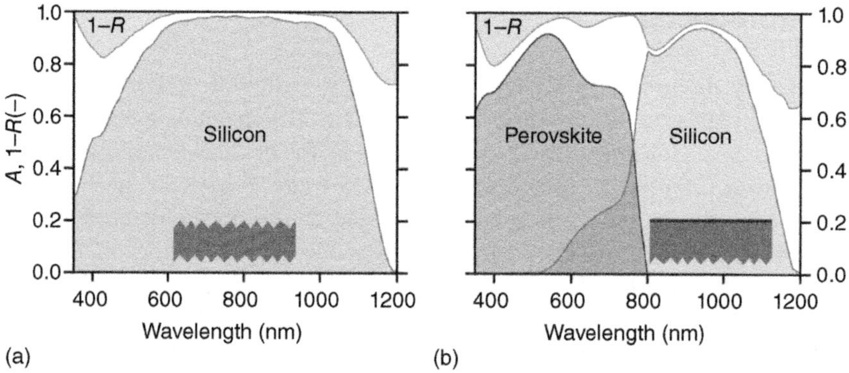

Figure 17.2 Example of (a) a simulated absorption spectrum of the silicon absorber in a state-of-the-art silicon heterojunction (SHJ) solar cell and $1-R$ spectrum of this cell with pyramidal textures on both sides. Example of (b) simulated absorption spectra of the perovskite and silicon absorbers in a state-of-the-art silicon monolithic perovskite-silicon tandem solar cell based on a back-textured SHJ cell and $1-R$ spectrum of this tandem cell. The detailed layer stack of the tandem cell is given in Ref. [3]. The SHJ cell in (a) is based on the same design but with an additional pyramidal texture on front and a 100 nm thick indium tin oxide (ITO) layer as front electrode. All simulations were performed with the package GENPRO4 [4], which is based on the net-radiation method and ray tracing. *Source*: Jäger et al. [3].

between the two sub cells. Short-wavelength light is absorbed in the top cell, while long-wavelength light traverses the top cell and is absorbed in the bottom cell. The pyramidal texture at the back increases the average light path and therefore increases the absorption in the silicon wafer for wavelength longer than ~900 nm. Because of manufacturing issues, most perovskite top cells still are flat. This leads to interferences, which are visible in Figure 17.2b. Using textures is one way of *light management*, which will be discussed in Section 17.5.

In this chapter we will discuss tandem solar cells with a perovskite-based top cell. In Section 17.2 we discuss different types (or architectures) of tandem solar cells and review the general efficiency limits for tandem solar cells in Section 17.3. In Section 17.4 we discuss, why perovskites are well-suited for tandem solar cells. Section 17.5 is devoted to two-experimental results of perovskite-based tandem solar cells. In Section 17.6 we briefly discuss energy yield calculations before we conclude the chapter with the outlook and guidelines in Section 17.7. Perovskite–perovskite tandem solar cells are discussed in Chapter 18.

17.2 Architectures of Tandem Solar Cells

Tandem solar cells can be manufactured in various different architectures, as illustrated in Figure 17.3. For all these concepts light is considered to arrive at the

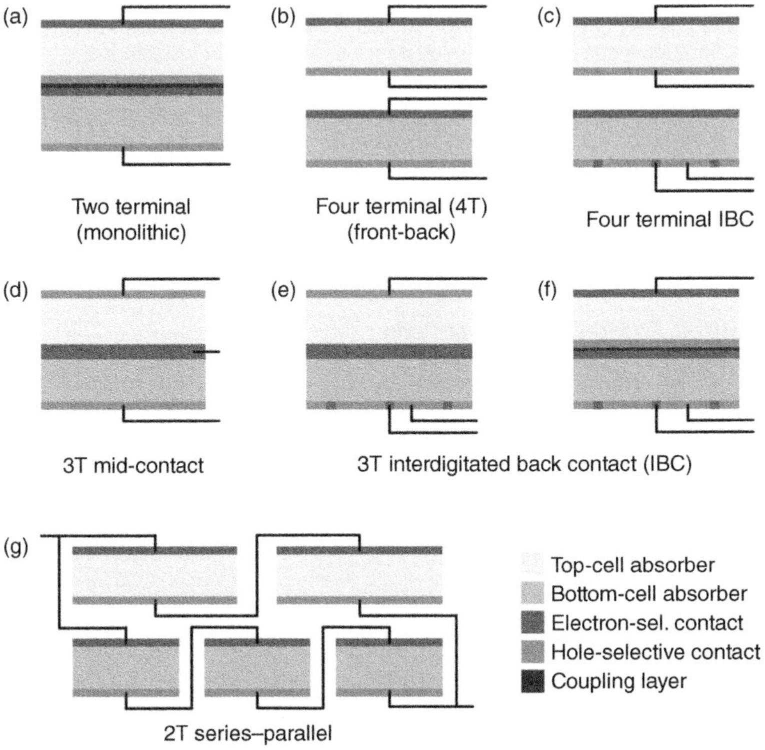

Figure 17.3 Illustrating different architectures for tandem solar cells. For all, light is illuminating the cell from top. The most common concepts are (a) monolithic two-terminal (2T) and (b)–(c) four-terminal (4T) cells. (d)–(f) Also three-terminal (3T) concepts are investigated, where we distinguish between (d) mid-contacted cells and (e)–(f) cells with interdigitated back contacts (IBC). (g) Also, series-parallel connected tandem cells a considered. For all concepts light arrives on the cells from top. *Source*: Based on Rienäcker et al. [5].

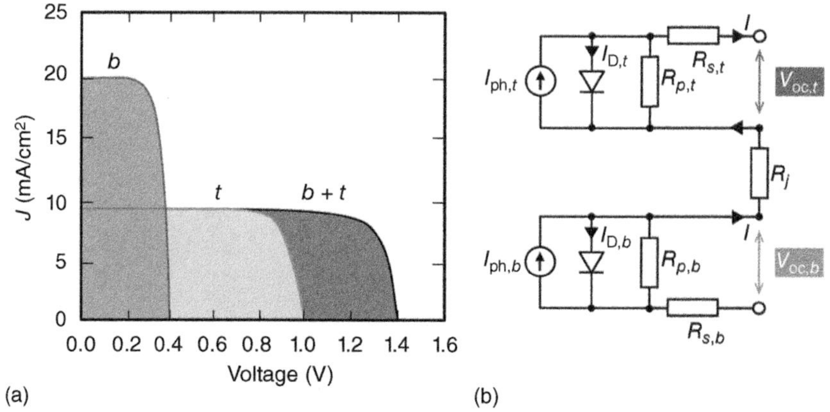

Figure 17.4 (a) Current density (J) – voltage (V) characteristics for single-junction solar cells under AM1.5 illumination for cells with a large bandgap "t" and a small bandgap "b" as well as the J–V characteristics of a monolithic series-connected tandem solar cell consisting of these two subcells "t+b." (b) The equivalent circuit of a monolithic series-connected tandem solar cell, where each subcell consists of a photocurrent source I_{ph}, a diode with diode/dark current I_D, series resistance R_s and parallel resistance R_p. Further, the resistance R_j at the junction between the two subcells is considered. *Source*: Modified from Smets et al. [6].

solar cell from top, hence, it reaches first the solar cell with the higher bandgap. The most common concepts are monolithic two-terminal (2T) tandem solar cells, shown in Figure 17.3a, and four-terminal (4T) tandems solar cells Figure 17.3b,c.

17.2.1 Monolithic Two-Terminal Solar Cells

In monolithic two-terminal cells the two subcells are electronically connected in series. To understand, how such a solar cell works, we look at Figures 17.4 and 17.5, where we follow Ref. [6]. Figure 17.4a shows current density (J) – voltage (V) characteristics for single-junction solar cells under AM1.5 illumination for cells with a large bandgap "t" and a small bandgap "b" as well as the J–V characteristics of a monolithic series-connected tandem solar cell consisting of these two subcells "b+t."

To understand why the combined J–V curve looks as plotted in the graph, we take a look at Figure 17.4b, which shows an equivalent circuit of the tandem cell: in *series connection*, the voltages add up and the same current flows through both cells – it is determined by the subcell generating the lower current. This important property of monolithic tandem cells is called *current matching* and negatively affects the overall performance if the currents generated in the two subcells differ significantly. We will discuss this topic in more detail in Section 17.5.

Figure 17.5 sketches a band diagram of a tandem solar cell (a) in the dark and (b) under illumination. In (a) the situation is drawn under thermal equilibrium, meaning that the Fermi energy E_F is constant across the whole device. Here we assume that the two solar cells are homojunctions, hence their respective

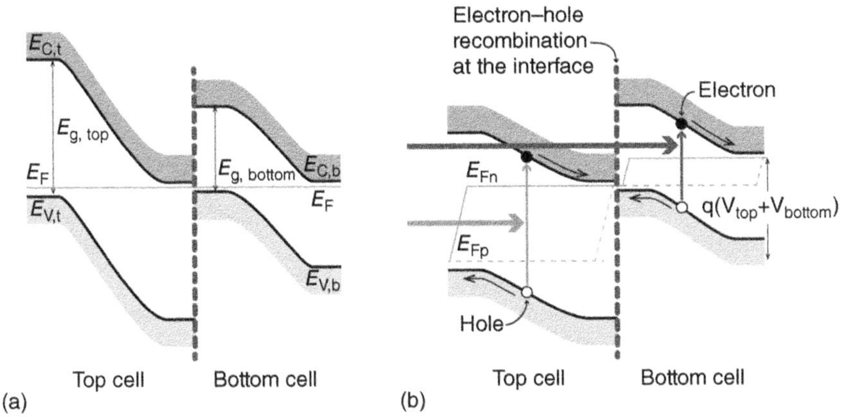

Figure 17.5 Schematic band diagram of tandem solar cells (a) in the dark at thermal equilibrium and (b) under illumination at the maximum power point (MPP). In (a) the energy levels of the valence band E_V and conduction band E_C are depicted for both top and bottom cells. Further, the Fermi energy E_F and the bandgaps E_g are depicted. In (b), quasi-Fermi level splitting is observed in both subcells, with the quasi Fermi levels E_{Fn} and E_{Fp} for electrons and holes, respectively. In the ideal case the voltages add up without voltage loss across the interface, where q denotes the elementary charge. Therefore, efficient recombination of electrons and holes at the interface between the two subcells is required. *Source*: Modified from Smets et al. [6].

bandgaps are constant across the whole junction. In real devices, often different materials with different bandgaps will be used, hence the band diagram looks more complicated.

Under illumination, high-energy (short-wavelength) photons are absorbed by the top cell while low-energy (long wavelength) photons can pass through the top cell and are absorbed in the bottom cell, as illustrated in Figure 17.5b. In both subcells the Fermi level separates into quasi-Fermi levels for electrons and holes, respectively, as shown in Figure 17.5b. In the ideal case the two voltages add up, which is also depicted. Therefore, it is necessary, that electrons and holes can combine efficiently at the interface between the top and bottom cells. This can be either done with tunnel junctions or with a (metallic) recombination layer. Both concepts are discussed in Section 17.5.

17.2.2 Four-Terminal Tandem Solar Cells

Figures 17.3b,c sketch four-terminal (4T) tandem solar cells – the other common tandem-cell concept. As in monolithic 2T solar cells, light passes first through the high-bandgap top cell and then through the low-bandgap bottom cell. In contrast to 2T cells, here each subcell has its own electric connections and can be operated independently from the other subcell. Therefore, both subcells can be operated at their respective maximum power points (MPPs) – no matter how the light is distributed between the two subcells.

This makes 4T cells much less sensitive to changes in the spectrum of the light, which illuminates the cell. Hence, the overall energy yield in principle can be higher than with 2T solar cells. However, for the bottom contact of the top cell and for the top contact of the bottom cell two additional contacting layers need to be established for an efficient lateral transport of the charge carriers. Such contacts can be transparent conducting oxide (TCO) layers such as indium tin oxide, zinc oxide or tin oxide. For wafer-based silicon solar cells, also a metal grating system can be used. In any case, optical losses because of parasitic absorption and/or shading by metal fingers will be larger than for 2T devices, where only vertical transport of charge carriers is required between the two subcells. If the bottom cell is an interdigitated back-contact cell, as sketched in Figure 17.3c, these losses can be slightly reduced.

Further, every module has four connections, therefore the number of cables in a photovoltaic (PV) system consisting of 4T modules roughly doubles, which increases the balance of system (BOS) costs of the PV system.

Several numerical studies compared the energy yield of photovoltaic systems based on two-terminal and four-terminal tandem solar cells with a perovskite top cell – we will review a few of them in Section 17.6.

17.2.3 Other Concepts

As discussed above, the major drawback of two-terminal tandem solar cells is current mismatch, which means that the overall performance goes down when the current densities generated in the top and bottom cells differ. On the other hand, four-terminal cells suffer from higher optical losses because of the contacting layers on bottom of the top cell and on top of the bottom cell.

Figure 17.3d–f shows sketches of *three-terminal* (3T) tandem solar cells, which aim to combine the advantages of the two concepts by adding a third terminal to a 2T cell [5]. The third terminal matches the currents of the two subcells and therefore overcomes the performance losses due to current mismatch. The two subcells are monolithically connected. Therefore, optical losses are less a problem than in 4T cells.

Figure 17.3d illustrates a mid-contacted three-terminal solar cell, where the third terminal connects to the back of the top cell and the front of the bottom cell. However, also this concept needs a contacting layer which ensures sufficient lateral transport. If a TCO is used, lateral transport can be controlled by the layer thickness as well as by the conductivity – both measures increase parasitic absorption.

If the bottom cell is designed as interdigitated-back-contacted cell with both the electron- and hole-selective contacts at the back, the optical losses are not larger than in 2T cells. Such 3T cells can be designed as two anti-series connected diodes with a common ground (Figure 17.3e) or as series-connected diodes with a coupling layer (Figure 17.3f). The power output of such 3T solar cells can be almost as high as for 4T solar cells [5].

Another concept is the *series–parallel connected tandem solar cell*, which is sketched in Figure 17.3g [7]. When two cells are connected in parallel, their voltages have to be matched for proper operation. To achieve this in a tandem configuration, where two types of cells with different bandgaps are used, n_1 top cells and n_2 bottom cells are connected in series, respectively, such that their string voltages roughly match, $n_1 \times V_1 \approx n_2 \times V_2$. These two strings are then connected in parallel.

The cell voltage only depends weakly (logarithmically) on the illumination, hence the voltages of the two strings are roughly independent from the illumination. Therefore, series–parallel connected tandem solar cells are much less susceptible to the illumination conditions than monolithic two-terminal cells. As a consequence, the bandgaps of the top and bottom cell materials are much less constraint than for monolithic 2T cells. Further, PV modules based on this concept would only have two connections which is an advantage with respect to 3T and 4T concepts. However, they would still be sensitive toward spectral changes, which naturally occur for outdoor illumination as discussed in Section 17.6.

17.2.4 Bifacial Solar Cells

At the end of this section we briefly discuss the concept of *bifacial solar cells* and solar modules, which not only can utilize light that is impinging on the front side of the solar module, but also light impinging at the solar module back. Depending on the reflectivity of the ground, this can significantly increase the energy yield over the year. For example, a study by Tillmann et al. showed that more than 10% of all the photon power incident on the solar module arrives at the back side for standard PV-field configurations, when the reflectivity of the ground is assumed to be 30% [8]. Because of the rise of new technologies for silicon solar cells, such as the *passivated emitter rear contact* concept discussed in Section 17.5.1, bifacial solar cells are expected to have a market share of around 50% by 2029 [9].

Bifacial tandem solar cells in a four-terminal configuration can be used just as standard monofacial cells, because the two subcells operate separately from each other. Hence, the additional light on the back side will only increase the current density of the bottom cell, shifting the MPP of the bottom cell with respect to a monofacial cell. In contrast, for monolithic tandem solar cells the additional current density at the bottom requires a change of design to achieve current matching, such that the current density in the top cell increases. We can achieve this by increasing the thickness of the absorber layer in the top cell, by using top cell materials with lower bandgaps or by using more advanced light management techniques, as discussed in Section 17.5. Finally, because of changing illumination conditions (position of the sun, clouds, etc.) the ratio between front and back-side illumination can vary throughout the day and the year.

Energy yield analyses, explained in Section 17.6, can help to find the best band gap/thickness combination for the top cell (under the assumption that the bottom cell parameters are fixed, as e.g. for silicon solar cells). Further, these analyses can be used to assess estimate the differences in energy yield for 4T and 2T tandem cells both in monolithic and bifacial configurations.

17.3 Efficiency Limits of Multi-Junction Solar Cells

In this section we review the theoretical efficiency limits for four- and two-terminal tandem solar cells and briefly look at cells with more than two junctions. The theoretical efficiency limit for single-junction solar cells was established by the seminal work of Shockley and Queisser in 1961 using the detailed balance approach [1, 10]. For four-terminal cells the efficiencies of the two subcells can be maximized independently; only the bandgap of the top cell affects the short-circuit current density of the bottom cell. For two-terminal cells, the derivation is more complex because the two subcells are electrically connected.

17.3.1 Efficiency Limit for Four-Terminal Tandem Solar Cells

For four-terminal cells we can follow the derivation by De Vos from 1980 Ref. [11]. The overall efficiency η_{4T} is given by

$$\eta_{4T}(E_t, E_b) = \eta_t(E_t) + \eta_b(E_t, E_b) = \frac{p_t(E_t) + p_b(E_t, E_b)}{p_{in}} \tag{17.1}$$

where E_t and E_b denote the bandgap energies of the top and bottom cells, respectively, and p_t and p_b denote the power densities at the respective MPPs. Finally, p_{in} denotes the incident power density, which is 1000 W/m^2 for the standardized AM1.5g solar spectrum, which we use for this derivation [2]. For the whole section we assume $E_t > E_b$.

The maximum power density for the top cell is given by

$$p_t = \max_V [V \cdot J_t(V)] \tag{17.2}$$

where the maximum can be found numerically. The current density – voltage characteristic is based on the *Shockley equation*, which describes an ideal diode,

$$J_t(V) = J_t^0 \left[\exp\left(\frac{qV}{kT_e}\right) - 1\right] - J_t^{ph} \tag{17.3}$$

with the dark current density J_t^0, the photocurrent density J_t^{ph},the elementary charge q, the Boltzmann constant k and the temperature of the environment of the solar cell T_e, which is set to 25 °C for solar cell standard testing conditions (STCs) [12]. To calculate the dark current density, we use the *detailed balance assumption*, which assumes that the solar cell behaves as a black body for wavelengths up to the bandgap and is fully transparent for longer wavelengths and that the black body radiation is driven by radiative recombination in the solar cell,

$$\begin{aligned} J_t^0 &= 2q \int_{2\pi} \frac{2hc^2}{\lambda^5} \left[\exp\frac{hc}{\lambda k T_e} - 1\right]^{-1} \cos\theta \, d\Omega \, d\lambda \\ &= 2\pi q \int_0^{\lambda_t} \frac{2hc^2}{\lambda^5} \left[\exp\frac{hc}{\lambda k T_e} - 1\right]^{-1} d\lambda \end{aligned} \tag{17.4}$$

with the Planck constant h, the speed of light c and the solid angle element $d\Omega = \sin\theta d\theta \, d\phi$. The factor 2 arises from the fact that the solar cell is assumed to emit from both front and back sides. The factor $\cos\theta$ arises from Lambert's cosine law of emission. The wavelength λ_t is given by $\lambda_t = hc/E_t$. More details can be found in Ref. [6], chapter 10.

For the bottom cell Eqs. (17.2)–(17.4) hold as well, the subscript t needs just to be only replaced by b.

The photocurrent density for the top cell is given by

$$J_t^{ph} = q \int_0^{\lambda_t} \Phi_{ph,\lambda} d\lambda + \frac{1}{2} J_t^0 \tag{17.5}$$

where the first term is the photocurrent density due to solar illumination with the spectral photon flux $\Phi_{ph,\lambda}$ and the second term arises from the blackbody emission from the bottom cell, cut off at the bandgap of the top cell (the bottom cell also emits blackbody radiation between λ_t and λ_b, but this cannot be utilized by the top cell).

For the bottom cell, we find

$$J_b^{ph} = q \int_{\lambda_t}^{\lambda_b} \Phi_{ph,\lambda} d\lambda + J_{LC} + \frac{1}{2} J_t^0 \tag{17.6}$$

where the first term is the photocurrent density due to solar illumination, which passes through the top cell with the spectral photon flux $\Phi_{ph,\lambda}$ and the last term arises from the blackbody emission from the top cell.

The second term J_{LC} in Eq. (17.6) accounts for *luminescence coupling* (LC, also called *luminecent coupling*): in a high-quality top cell the part of the generated current $J_t^{ph} - J_t(V_t)$, which is not utilized electrically, can lead to radiative recombination and hence to emission of light by the top cell. A fraction of this light reaches the bottom cell, where it can be absorbed. A simple expression for the additional current density generated in the bottom cell is given by [13, 14]

$$J_{LC} = \eta_{LC}[J_t^{ph} - J_t(V_t)] \tag{17.7}$$

with the luminescent-coupling efficiency η_{LC}. No further conversion factors are needed, because one electron–hole pair in the top cell, which recombines radiatively, leads to the emission of one photon, which may generate one electron–hole pair in the bottom cell. η_{LC} might also depend on the bias voltage V_t of the top cell, which, however, is out of the scope of this book [13]. In principle also LC from the bottom cell into the top cell should be considered. However, as we may safely assume that the light is emitted with energies close to the bandgap, light emitted by the bottom cell has energies below the top-cell bandgap and cannot be utilized.

Figures 17.6a,b shows the theoretical efficiency limit for four-terminal cells as a function of the top- and bottom-cell bandgaps. For 4T cells the efficiencies of the top and bottom subcells can be maximized independently of each other for each combination of bandgaps. Without luminescence coupling, shown in Figure 17.6a, the maximum PCE is 45.2% with top and bottom bandgap energies of 1.73 and 0.94 eV, respectively. This value is significantly higher than the detailed balance limit for single junction solar cells, which is slightly above 30%. For 4T cells, maximal luminescence coupling ($\eta_{LC} = 100\%$) only would change the picture slightly, as shown in Figure 17.6b, where the maximum efficiency is 45.4% at 1.73 eV top- and 0.95 eV bottom-cell bandgaps. The little effect of LC can be understood as follows: in a 4T cell, the top cell operates at its MPP with a current density J_t close to J_t^{ph}. Hence, $J_t^{ph} - J_t$ and subsequently J_{LC} are very small.

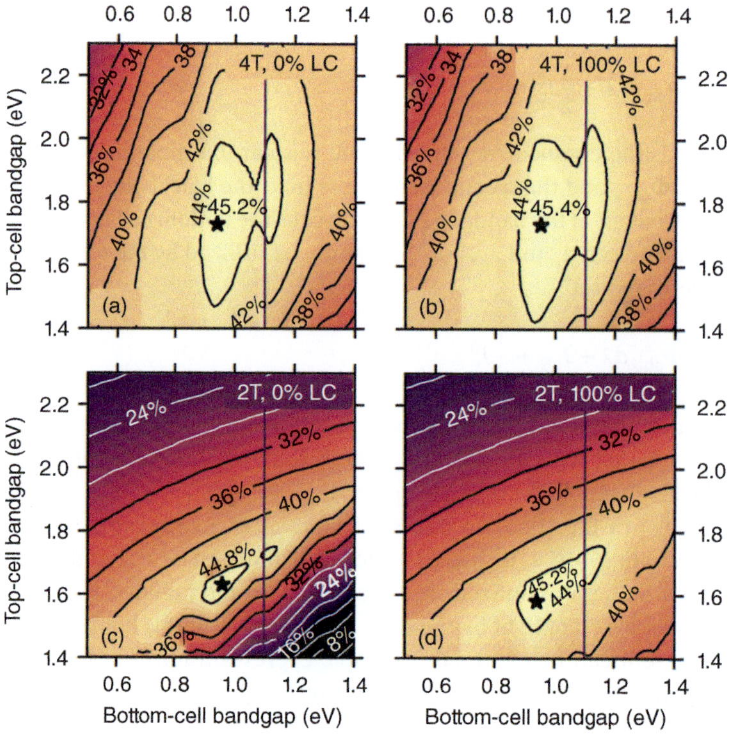

Figure 17.6 The theoretical efficiency limits for tandem solar cells as functions of the top and bottom cell bandgap energies. Results are shown for four-terminal cells (a) without and (b) with ideal luminescence coupling (LC efficiency $\eta_{LC} = 100\%$): and for two-terminal cells (c) without and (d) with 100% LC. In all plots the maximum is marked by a star; the purple line denotes a bottom-cell bandgap of $E_b = 1.1$ eV, for which section plots are shown in Figure 17.7.

17.3.2 Efficiency Limit for Two-Terminal Tandem Solar Cells

For two-terminal cells, the two subcells cannot be treated independently because of their electrical series connection. As a consequence, the same current density flows through both subcells. Hence, we write

$$\eta_{2T}(E_t, E_b) = \frac{p_{2T}(E_t, E_b)}{p_{in}} \tag{17.8}$$

where the maximum power density is given by

$$p_{2T} = \max_J [V_t(J) \cdot J + V_b(J) \cdot J] \tag{17.9}$$

In contrast to Eq. (17.2), here the voltages V have to be treated as dependent variables – hence the *Shockley equation* (17.3) must be inverted. Therefore, identifying the maximum power is numerically more involved than for 4T solar cells. The dark current density and the photocurrent densities can be calculated as in the 4T case using Eqs. (17.4)–(17.6).

Figure 17.6c shows the theoretical efficiency limit for two-terminal tandem solar cells with no and 100% luminescence coupling. Without LC, the maximum PCE is 44.8% with top and bottom bandgap energies of 1.63 and 0.96 eV, respectively. While this limit is only slightly below that of 4T cells, the sweet spot is much smaller for this architecture. Especially, the top-cell bandgap strongly affects the overall efficiency: at higher top-cell bandgaps, the top photocurrent density decreases, increasing the potential photocurrent density of the bottom cell. However, this increase cannot be utilized because of current matching. On the other hand, 1.63 eV is closer to the bandgap of high-performance perovskites than 1.73 eV, which is the optimum for 4T cells. If 100% LC was occurring, as shown in Figure 17.6d, the maximum would be 45.2% at 1.58 eV top- and 0.94 eV bottom-cell bandgaps. For top-cell bandgaps below the maximum, Figure 17.6d looks almost like for 4T cells, meaning that luminescence coupling can (partially) compensate for the disadvantages of 2T cells.

Figure 17.7 shows the efficiency limit for 2T and 4T tandem solar cells as functions of the top-cell bandgap for a bottom-cell bandgap of 1.1 eV, which corresponds to the bandgap of silicon – the semiconductor on which most solar cells are based on. While the top-cell bandgap hardly affects the overall efficiency for 4T cells, 2T cells without luminescence coupling are very susceptible to changes of the top-cell bandgap with a very narrow maximum. This also makes them more sensitive to spectral changes in the illumination, which naturally occur throughout the course of time. We will discuss this topic in Section 17.6. For ideal luminescence coupling (η_{LC} = 100%), the maximum efficiency is almost as high as for 4T cells, when the top-cell bandgap is below the maximum: in this regime the 2T-cells are bottom-cell limited, which can be partially compensated by the additional current from luminescence coupling. However, for larger wavelength, the 2T cells are top-cell limited and luminescence coupling has no added value. As discussed above, luminescence coupling can only improve 4T cells slightly.

Experimentally, luminescence coupling is not a relevant effect in perovskite-based tandem solar cells yet. For perovskite solar cells, external luminescence quantum efficiencies of up to 15% were reported as of summer 2020 [15–20]. For III–V multi-junction solar cells, where the strongest LC is observed, efficiencies of 33% were reported [21].

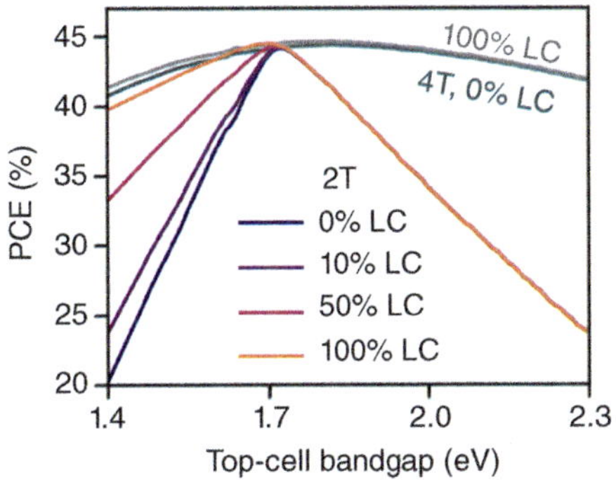

Figure 17.7 Efficiency limits for 2T and 4T tandem solar cells as a function of the top-cell bandgap for different luminescence coupling efficiencies, when the bottom-cell bandgap is 1.1 eV, which corresponds to the bandgap of silicon.

Note that the theoretical limit for silicon solar cells (29.4%) is a few per cent below the detailed balance limit for 1.1 eV [22]. Silicon is an indirect semiconductor, therefore recombination happens mainly via Auger processes, while radiative recombination, which is assumed in the Shockley–Queisser limit, hardly occurs.

17.3.3 Efficiency Limit for Cells with More Junctions

As we have seen, tandem solar cells can yield higher PCEs than single-junction solar cells, because the top cell with the higher bandgap can utilize the energy of high-energy photons more effectively. The efficiency can be increased even more by building solar cells with more than two junctions.

Figure 17.8 shows the bandgap utilization and the efficiency limits for cells with one, two, three, four, and infinitely many junctions according to data from Martí and Araújo [23]. If the sunlight is not concentrated, a cell consisting of one, two, three, or four junctions would yield a PCE of 32.5%, 44.3%, 50.1%, and 54.0%, respectively. In the limit of infinitely many junctions 65.4% of the incident sunlight would be converted into electricity. The additional gain decreases with every additional junction the optimal number of junctions therefore finally depends on the economics: is the cost of additional junctions outweighed by the gain in performance. Note that the efficiency limit for tandem cells (44.3%) slightly differs from the limits presented in Figure 17.7. The reason for this is probably different spectral data for the AM1.5 spectrum as well as other numerical effects.

17.4 Perovskites as Tandem Solar Cell Materials

The development of tandem solar cells based on mature commercial solar cell technologies such as *silicon* (Si) or *copper indium gallium selenide* (CIGS) poses several requirements on the properties of the partnering subcell. The most obvious one is a suitable complementary bandgap (see Figure 17.6) with high absorption for photons with energies exceeding the bandgap and very low absorption for photons with energies below the band gap. The material should with a very sharp absorption onset, which reflects low energetic disorder, i.e. a low Urbach energy. In addition, to achieve high power conversion efficiencies (PCEs), two-terminal cells require a suitable electrical connection of the subcells and compatible processing of the layer stack. Moreover, general and economical aspects must be considered, for example material abundance, ease of processing, compatibility with existing fabrication lines, market readiness and cost effectiveness.

Because of their easily tunable bandgap through compositional engineering, *metal-halide perovskites* are prominent candidates for perfectly adjustable top cells, with bottom cells made from Si, CIGS, or low band-gap perovskites with Sn–Pb compositions. Besides the tunable bandgap, metal-halide perovskites also absorb strongly at energies above the bandgap and weakly below the bandgap energy [24]. Therefore, already a perovskite layer of around 500 nm thickness

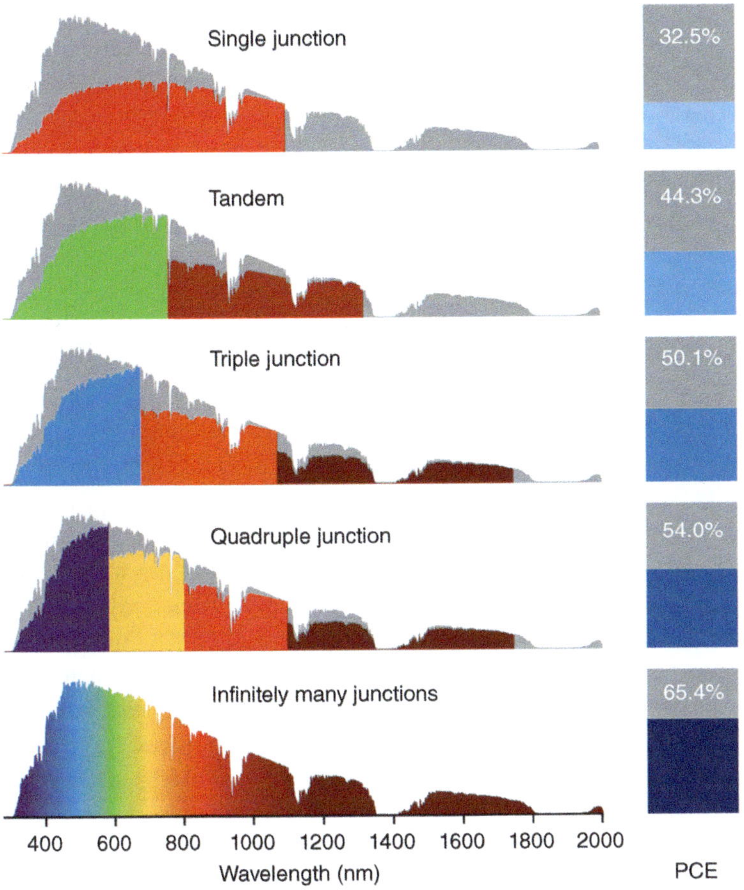

Figure 17.8 Spectral utilization for solar cells with different numbers of junctions and corresponding power-conversion efficiencies (PCE) according to the detailed balance limit. The data are taken from Ref. [23] for non-concentrated sunlight. Optimal bandgaps for single junction: 1.13 eV; tandem: 0.94 and 1.64 eV; triple junction: 0.71, 1.16, and 1.83 eV; quadruple junction: 0.71, 1.13, 1.55, and 2.13 eV. For maximum concentration, the efficiency limits would be higher, approaching the thermodynamic limit of 86% for infinitely many junctions. *Source*: Data from Martí and Araújo [23].

absorbs a large fraction of the incident light during a single pass and enables high transmission of near-infrared light to the bottom cell. In addition, a variety of deposition techniques can be applied to yield perovskite films with high opto-electronic quality: the different deposition techniques include solution processes like spin coating, inkjet printing and blade or slot-die coating as well as vacuum processes like thermal co-evaporation or a hybrid process combining different steps and processes [25]. The variety of different deposition techniques allows to cover the underlying subcell conformally, independent of the nano roughness of e.g. CIGS or the pyramidal microtexture of typical silicon solar cells. The defect tolerance of metal-halide perovskites enables high-efficiency solar cells at manageable requirements on material purity and process as well as equipment costs, especially with the low processing temperatures needed. Because of this combination, highly efficient tandem modules at reasonable costs will be available in the near future [26, 27].

The bandgap of perovskites can be tuned ranging from 1.2 to 2.3 eV by appropriately choosing the monovalent cations (methyl ammonium [MA], Cs, formamidinium [FA]), anions (I, Br, Cl), metals (Pb, Sn), and mixtures thereof [28]. On the one hand, increasing the bandgap leads to a blue-shifted absorption edge and hence a lower photogenerated current, but on the other hand it leads to an increased open circuit voltage in solar cells. The optimal trade-off between current and voltage depends on the architecture (i.e. two-terminal or four-terminal, as discussed in Section 17.3). However, a substantial scientific challenge is to enable good electrical quality and photostability over a wide range of perovskite compositions and subsequently the bandgap. In this context, the V_{oc}-to-bandgap loss is an important metric to quantify the electrical quality of the perovskite absorber.

One part of the V_{oc}-to-bandgap loss consists of unavoidable radiative recombination of free charges in the absorber layer, when it is in equilibrium with its surrounding. In this *detailed balance limit*, the solar cell emits black body radiation according to its temperature and band gap, as discussed in Section 17.3. Additionally, non-radiative losses caused by trap-assisted recombination in the perovskite bulk and primarily at interfaces significantly reduce the V_{oc}s reported in experiment as compared to the detailed balance limit (see e.g. [6], section 7.3). A specific property of metal-halide perovskites, which can be related to the V_{oc}-to-bandgap loss is the crystalline-phase instability observed for certain compositions.

Figure 17.9 shows experimentally measured open circuit voltages as a function of the material bandgap and reveals two main features: In general, the V_{oc} follows the bandgap between 1.45 and 1.65 eV with high V_{oc} values well above 90% of the respective detailed balance limit. However, for bandgaps above 1.65 eV, the reported V_{oc} values are not increasing linearly with bandgap. In this bandgap range, which is of particular interest for tandem applications, the anion site of the perovskites is typically occupied by a mixture of iodine (I) and bromium (Br). In particular high Br contents were earlier reported to lead to phase instability [34], which induces additional traps and therefore nonradiative recombination centers. A reduction of these trap-induced recombination requires additional efforts in compositional engineering [35, 36] and recently some publications have reported high V_{oc} values for that bandgap region [29, 30]. In addition to adjusting the bulk properties of the perovskite, the charge-selective contacts, which are well suited for the range between 1.45 and 1.65 eV may induce losses at higher bandgaps, which for example are caused by energetic misalignment. Thus, one of the main challenges for approaching the theoretical tandem efficiency limits with perovskite top cells is a reduction of non-radiative losses and the improvement of phase stability at higher bandgaps, to reach open circuit voltages above 90% of the value from the respective detailed balance limit.

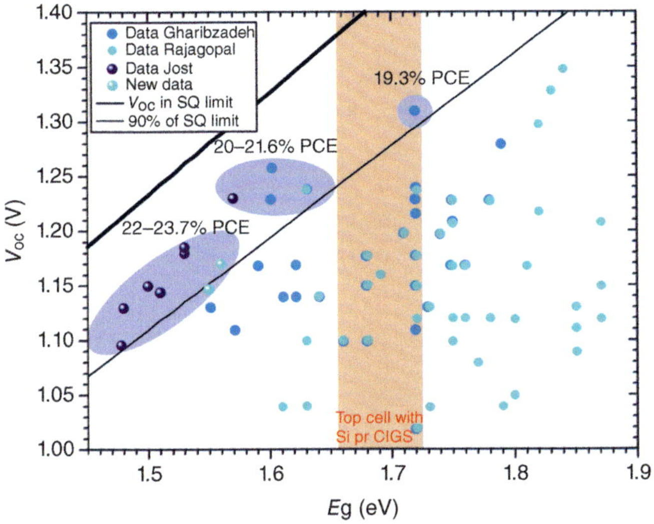

Figure 17.9 Experimentally reported open-circuit voltages plotted as function of perovskite band-gap. Data are collected from Gharibzadeh et al Ref. [29], Rajagopal et al. [30], and Jošt et al. [31]. In addition, new data for high efficiency results are included [32, 33]. As comparison, the detailed balance (SQ) limit and 90% of it are shown as well. The round colored areas denote efficiencies reached with the corresponding compositions (band gaps) and the band gap regime for optimized top cells with silicon and CIGS (assuming a band gap of 1.1 eV for both) bottom-cells is also indicated. *Sources*: Data from Gharibzadeh et al. [29]; Rajagopal et al. [30]; Jošt et al. [31]; Zheng et al. [32]; Zhu et al. [33].

17.5 Experimental Results on Perovskite-Based Tandem Solar Cells

The first reports on four-terminal perovskite/silicon [37] and perovskite/CIGS [38] tandem solar cells were published by the end of 2014. In 2015, first reports on two-terminal tandem cells followed for both technologies [39, 40]. Since then, the two tandem technologies underwent a fascinating improvement within a short timescale. With PCEs below 14% for perovskite/silicon (both 2T and 4T) and 18% or 11% for perovskite/CIGS for 4T or 2T tandems, respectively, all reported first tandem PCEs were well below those of best single-junctions, which is the benchmark for any tandem technology. In early 2020, the certified PCE for 1 cm^2 active area devices improved to 29.2% and 24.2% for perovskite/silicon and perovskite/CIGS, respectively (see Figure 17.10, Tables 17.1 and 17.2).

For 4T tandems, best reported (not-certified) values as of July 2020 are 28.2% and 25.9% for bottom cells made from silicon [51] and CIGS [70], respectively. Especially the 4T tandems improved with rather linear slopes between 2017 and 2019 with an annual absolute PCE gain of ~0.5% and ~1.4% for silicon and CIGS, respectively. So far, most 2T tandems were reported with silicon as a bottom cell and a steady PCE improvement over time was reported, which is also visualized by the

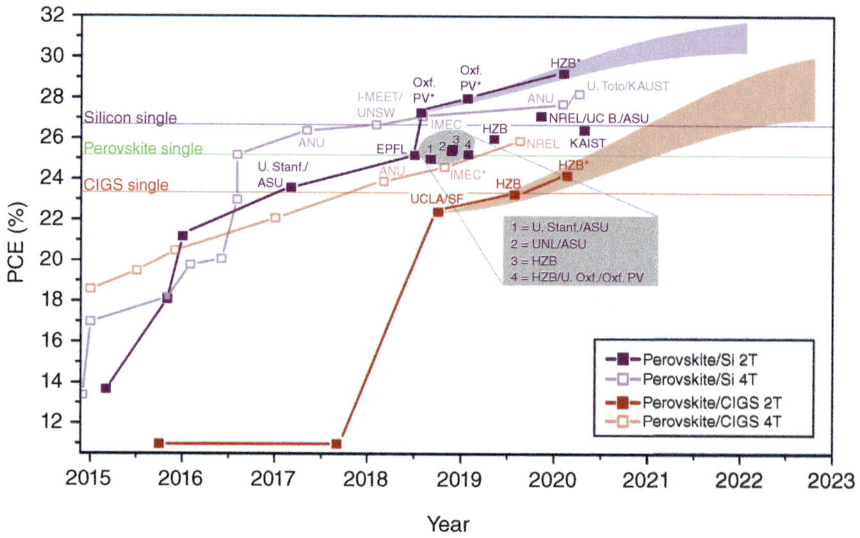

Figure 17.10 Efficiency evolution for perovskite-based tandem solar cells in two-terminal (2T) and four-terminal (4T) configuration with Silicon (Si) or Cu(In,Ga)Se_2 (CIGS) bottom cells until mid 2020. Data are extracted from press releases (marked with *) or published papers and the abbreviation indicates the company or university/institute reporting the value. The colored areas indicate a possible efficiency evolution scenario for perovskite-based 2T tandems with CIGS (red) and silicon (purple). *Source*: Based on Bush et al. [41].

high number of datapoints in Figure 17.10. Recently, a mismatch occurred between record values reported in press releases and scientific publications. Therefore, Figure 17.10 includes numerous "non-record" values to show the evolution of the status of publications. Although the two different tandem technologies discussed in this chapter have comparably high efficiency potentials (compare Tables 17.1 and 17.2), the development for perovskite/CIGS, especially the 2T architecture, was delayed and it took three years after the first publication to improve the PCE well above 20%. Thus, it seems that this design is more challenging: as a matter of fact, the achieved PCE values and the number of published papers is yet behind the ones of perovskite/silicon solar cells. Note that the 2T tandem layer stacks with multiple layers processed on top of each other and the series connection restrict the processing and operation conditions.

As discussed earlier, for 2T tandem cells both subcells are connected in series and the electrical connection between them is formed by a p/n tunnel or recombination junction. These recombination junctions are either formed by highly doped silicon layers [42] or by degenerately (very highly) doped transparent conductive oxides such as indium tin oxide (ITO) [49]. In each of the two subcells, one electron-hole pair is generated. One of the two charge carriers can be extracted through the external contact of this subcell – an electron leaving through the "negative" terminal of the 2T tandem cell and the hole recombining with an external electron at the "positive" contact. The other two charge carriers must recombine at the junction of the two sub cells to provide overall charge neutrality. In other words, one electron is

Table 17.1 Latest results for perovskite/silicon tandem solar cells with 2T or 4T configurations.

References	Abbreviations used in Figure 17.10	Published (Y/M)	Area (cm^2)	J_{SC} (mA/cm^2)	FF (%)	V_{oc} (V)	PCE (%)	MPP (%)	E_g Pero (eV)
			2T perovskite/silicon						
		Recent efficiency evolution with silicon heterojunction bottom cells							
Sahli [42]	EPFL	2018/06	1.42	19.5	73.2	1.79	25.5	25.2[a)]	1.60
Bush [43]	U. Stanf/ASU	2018/08	1.0	18.4	77	1.77	25.0		1.63
Chen [44]	UNL/ASU	2019/01	0.42	17.8	79.4	1.80	25.4		1.64
Jošt [45]	HZB	2018/10	0.77	18.5	78.5	1.76	25.5		1.63
Mazzarella [46]	HZB/U.Oxf./Ox PV	2019/02	1.10	19.0	74.6	1.79	25.4	25.2[a)]	1.63
Köhnen [47]	HZB	2019/05	0.77	19.2	76.6	1.77	26.0	26.0	1.63
Oxford PV [48]	Ox PV[b)]	2018/12	1.03	19.8	78.7	1.80	28.0[a)]		—
Xu/Boyd [49]	NREL/UC.B./ASU	2020/03	1.0	19.1	75.3	1.89	27.13	27.04	1.67
Al-Ashouri/ Köhnen [50]	HZB[b)]	2020/01	1.06	19.2	79.4	1.90		29.15[a)]	1.68
Kim [51]	KAIST	2020/04	1.00	18.8	76.4	1.82		26.20[a)]	1.68
			Results on fully textured bottom cells						
Sahli [42]	EPFL	2018/06	1.42	19.5	73.2	1.79	25.5	25.2[a)]	1.60
Nogay [52]		2019/03	1.42	19.5	74.7	1.74	25.4	25.1	—
Lamanna [53]		2020/02	1.43	18.8	77.5	1.80	26.3	25.9	1.64
Chen [54]		2020/01	0.42	19.2	74.4	1.82	26.2	26.1	1.65[c)]
Hou/Aydin [55]		2020/03	0.83	19.1	73.4	1.79		25.7[a)]	1.68

(continued)

Table 17.1 (Continued)

References	Abbreviations used in Figure 17.10	Published (Y/M)	Area (cm^2)	J_{SC} (mA/cm^2)	FF (%)	V_{oc} (V)	PCE (%)	MPP (%)	E_g Pero (eV)
		Results on non-silicon-heterojunction bottom cells							
Wu [56]		2017/10	1	17.6	73.8	1.75	22.8	22.5	1.62
Zheng [57]		2018/10	16	16.2	78	1.74	21.9	21.8	1.58
Shen [58]		2018/12	1	17.8	78.1	1.76	24.5	24.1	1.63
Zheng [59]		2019/10	4	16.5	81.0	1.73	23.1	23.0	1.61
References	Abbreviations used in Figure 17.10	Published (Y/M)	Areas (cm^2)	J_{SC} (mA/cm^2)	FF (%)	V_{oc} (V)	PCE (%)	Combined (%)	
			4T perovskite/silicon[d)]						
			Recent efficiency evolution						
Duong [35]	ANU	2017/04	0.3/4	19.4/18.8	70/80	1.13/0.69	16.0/10.4	26.4	1.73
Ramírez Quiroz [60]	i-MEET/UNSW	2018/01	0.1/1	21.0/17.7	74/80	1.10/0.67	17.1/9.6	26.7	1.57
Jaysankar [61]	IMEC	2018/12	0.13/4	15.4/24.1	73/81	1.22/0.68	13.8/13.3	27.1	1.72
Duong [62]	ANU	2020/01	0.21/4	18.0/19.6	78/78	1.2/0.69	17.1/10.7	27.7	1.72
Duong [62]		2020/01	1/1	17.5/18.6	76/80	1.2/0.68	16.1/10.7	26.2	1.72
Chen [63]	U. Toronto/KAUST	2020/03	0.05/3	22.3/17.2	78/77	1.12/0.7	19.0/9.2	28.2	1.63

2T monolithic solar cells are listed in three different categories: (i) recent efficiency evolution with mostly silicon-heterojunction bottom cells which is so far the major design route in terms of publication number, (ii) results on fully textured wafers and (iii) results on other (non-silicon-heterojunction) bottom cells.

a) Certified values.
b) Values taken from press releases.
c) Value for band-gap taken from PL emission maximum.
d) For 4T results, the first value refers to the perovskite top cell and the second value to the silicon bottom cell.

Table 17.2 Results of notable perovskite/CIGS tandem solar cells.

References	**Abbreviations used in Figure 17.10**	**Published (Y/M)**	**Area (cm^2)**	**J_{SC} (mA/cm^2)**	**FF (%)**	**V_{oc} (V)**	**PCE (%)**	**MPP (%)**	**E_g Pero (eV)**
				2T perovskite/CIGS					
Han [64]	UCLA/SF	2018/08	0.04	17.3	73	1.77	22.43[a]		1.59
Jost [65]		2019/01	0.8	18.0	76	1.58	21.6	21.6	1.63
Al-Ashouri [66]	HZB	2019/10	1.04	19.2	72	1.68		23.26[a]	1.63
HZB [67]	HZB[b]	2020/01	1.05	18.8	72	1.77		24.16[a]	1.68
References	Abbreviations used in Figure 17.10	Published (Y/M)	Areas (cm^2)	J_{SC} (mA/cm^2)	FF (%)	V_{oc} (V)	PCE (%)	Combined (%)	
				4T perovskite/CIGS[c]					
Shen [68]	ANU	2018/01	0.3/0.5	22.2/13	70/0.72	1.13/0.62	18.1/5.8	23.9	1.62
IMEC [69]	IMEC[b]	2018/09						24.6	1.72
Hoe Kim [70]	NREL	2019/07	0.06/0.42	19.6/15.6	77/79	1.13/0.71	17.1/8.8	25.9	1.68

Certified values are labeled with[a], first value refers to perovskite top cell, second value to CIGS bottom cell.
a) Certified values.
b) Values taken from press releases.
c) For 4T results, the first value refers to the perovskite top cell and the second value to the CIGS bottom cell.
Source: Based on Han et al. [64].

entering the tandem cell via the "positive" terminal, gains a voltage in the first sub cell, changes into the other sub cell via the recombination junction (losing a bit of its energy), gains more voltage in the second sub cell and leaves the solar cell at the "negative" terminal.

The scenario is also illustrated in Figure 17.5b. In order to enable the sketched scenario without significant losses, the recombination layer must provide the following three main properties: (i) very good selectivity for the respective charge carriers, (ii) favorable energetic alignment for lossless photo-voltage addition, and (iii) low series resistance. Furthermore, it must also provide high optical transparency for the bottom-cell response in the near infrared (NIR) region.

As for all solar cells, the absorption of the incident sunlight can be improved using various *light management* techniques [71, 72]. Especially for 2T tandem solar cells current matching at highest possible values is of utmost importance. First, measures to increase the *in-coupling* of light into the solar cell, which is to reduce the reflective losses at the front side, and secondly *light trapping* techniques, which aim to increase the absorption in the absorber layer usually with textures that scatter the incident light and hence lead to an increase of the average path length of the light in the absorber. Textures or nanotextures also can be used for better in-coupling, however here also planar antireflective layers or layer stacks can be used. In tandem solar cells, the optics of intermediate layers in-between the top and bottom cells can be used to redistribute and light between top and bottom cells [46, 72].

In Sections 17.5.1 and 17.5.2, we discuss the developments in engineering and improved scientific understanding for perovskite/silicon and perovskite/CIGS tandem solar cells, respectively.

17.5.1 Perovskite/Silicon Tandem Solar Cells

Crystalline silicon photovoltaics is the leading technology in the global PV market, with 95% market share from total module production in 2018 [9]. Therefore, out of the different possible bottom cell technologies for perovskite-based tandem solar cells, silicon receives by far the highest attention resembling in the highest tandem efficiency realized so far and the highest number of published reports, as shown in Table 17.1 and Figure 17.10. Most silicon solar cells consist of a diffused homojunction and an aluminium back surface field (Al BSF) for enhanced charge extraction. In the improved *passivated emitter rear contact* (PERC) design, most of the back-side contact is passivated with a dielectric, such as silicon oxide and or silicon nitride, to reduce surface recombination losses. Only local electrical contacts are formed. For PERC cells, efficiencies up to 24% in commercial wafer size are reported [73]. For passivated emitter and rear locally diffused (PERL) cells with *passivated emitter and rear locally diffused* design, the long-time efficiency record has been 25.0% [74]. Single-junction silicon solar cells with even higher PCE values can be fabricated by implementing full-area contacts, which are both passivating and selective. One example for such contacts are ultrathin silicon oxides, which are combined with doped polycrystalline layers. Devices with such layers are called polycrystalline silicon on oxide (POLO) or TOPCon and they have demonstrated efficiencies up to

26.1% (with interdigitated back contact [IBC] contacts) [75] and 25.8% (with double sided contacts) [76], respectively. Another concept are silicon heterojunction (SHJ) solar cells, where two ultrathin intrinsic and doped hydrogenated amorphous silicon (a-Si;H) layers are deposited onto both sides of the solar cell. SHJ cells reach the highest efficiencies for silicon single cells of 25.1% and 26.7% for double sided [77] and IBC [78] contacts schemes, respectively. For further reading on high efficiency silicon single junction solar cells, we refer to reviews by Battaglia et al. [79] and Hermle et al. [80].

In this section, we present the development of perovskite/silicon tandem solar cells until mid 2020; first for 2T monolithic and then for 4T architectures. The 2T tandem section includes a summary of the efficiency improvement, the development of certain device structures such as the bottom and top cell design and electrode polarities as well as implementation of textured surfaces and light management strategies. The main focus here is directed towards the majority of reported results for perovskite/silicon tandem cells that are shown in purple color in Figure 17.10 with rather similar device design being described in Figure 17.11 and the corresponding performance data also found in Table 17.1. In addition, we highlight results on textured wafers and other (non-silicon-heterojunction) bottom cells, both categories are also listed in Table 17.1. For more information on the efficiency evolution of perovskite/silicon tandem solar cells, we refer the reader to several review papers by Werner et al. [81], Tai et al. [82], Torabi et al. [83] or, most recently, by Song et al. [84], Hu et al. [85], Jost et al. [31], and the progress report by Shen et al. [86].

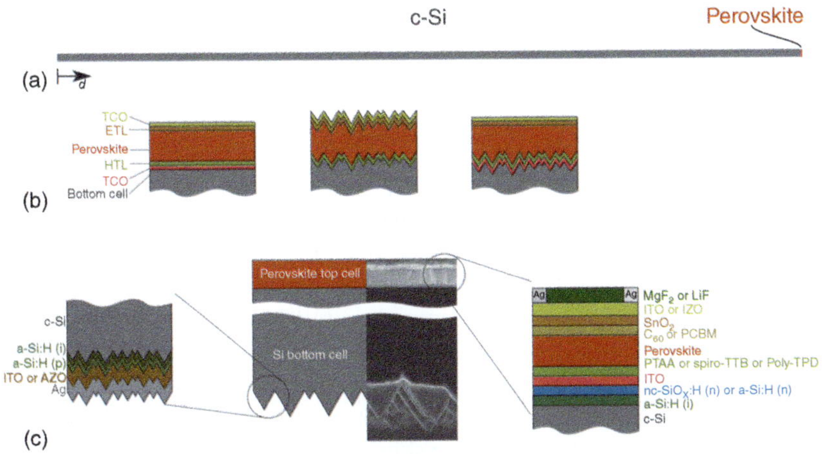

Figure 17.11 Schematic depiction of the 2T perovskite/silicon tandem device architecture: (a) size comparison of the bottom and top cell assuming a wafer thickness of 250 µm and a 500 nm thick top cell; (b) the front side texture and respectively the top cell processing options, reported so far (see Table 17.1, section results on fully textured bottom cells); (c) specific device design and layer stack utilized for most reports in Table 17.1 that enable highest PCE values well above 25%, SEM images and a zoom-in to top and bottom cell layers are shown as well. Sources: Sahli et al. [42]; Nogay et al. [52].

The first perovskite/silicon 2T tandem solar cell was realized on a silicon homojunction bottom cell processed with classical high temperature diffusion process [39]. The tandem cell reached a PCE of 13.7%, strongly limited by its low J_{SC} of 11.5 mA/cm^2 which was limited by the highly absorptive, doped Spiro-OMeTAD film present in the front contact stack. Thus, the overall parasitic absorption equivalent to a loss in photogeneration current density was over 8 mA/cm^2 strongly limiting the PCE.

Since then, many researchers focused on SHJ solar cells as bottom cells. The advantage of these cells is their high efficiency, which is reached because of very high V_{oc} values with records up to 750 mV [87]. In addition, the optical loss arising from the front side selective contact made from a-Si:H does not limit tandem cell operation, because this fraction of the incident light is already absorbed by the perovskite top cell. Besides the high efficiency, typical SHJ cells already feature a TCO, which fully covers the front side. This TCO layer can serve as the bottom contact for the perovskite top cell, just as in single-junction perovskite solar cells fabricated on glass. This allows relatively simple process integration. Another aspect is the flexibility of the polarity that is offered by the use of SHJ: the bottom cells can be utilized with both polarities and such also the polarity of the perovskite top-cell can be selected as well which was a key step toward >25% for perovskite/silicon tandem cells, as described below.

The first 2T perovskite/SHJ tandem solar cell was demonstrated in 2015 by Albrecht et al. [88] with 19.9% *JV*-scan efficiency and – due to hysteresis – 18.1% efficiency from stabilized MPP-tracks efficiency, as introduced in Chapter 16. The SHJ utilized as bottom cell was made from an *n*-type, double-side polished float zone (FZ) c-Si wafer with a front emitter fabricated from thin layers of intrinsic and p-doped hydrogenated amorphous silicon (a-Si:H). The p/n recombination junction interconnecting the two sub-cells was formed between the p-doped (*p*)a-Si:H film and a layer of ITO. ITO is a well-established electrode material for perovskite single-junction cells. The perovskite cell was then processed at temperatures below 120 °C in order to avoid degradation of the SHJ cell: The ITO was covered by an atomic layer deposition (ALD)-processed SnO_x film, to form the electronic contact of the perovskite sub-cell. The $FAMAPbI_3$ perovskite absorber ($E_g \sim 1.57$ eV) was 600 nm thick being processed by spin-coating. The hole selective contact and top contact stack was Spiro-OMeTAD and thermally evaporated MoO_x buffer combined with sputtered ITO. Finally, a LiF film was evaporated on top of the front ITO layer as an antireflective coating. Later that year, Werner et al. showed a strongly improved efficiency in a similar device [89]: They used an almost identical SHJ cell, but with slightly different TCOs and another electron selective (fullerene C_{60} instead of ALD SnO_2) contact for the perovskite top cell to reach a PCE from MPP tracks of 21.2%.

The three papers discussed above are plotted as the first three unlabeled dark purple datapoints in Figure 17.10. In 2017, a key achievement toward higher efficiencies followed by Bush et al. [41]: they switched the tandem polarity from n-i-p top cells with front-emitter SHJ bottom cells to rear-junction SHJ and p-i-n top cells. With that, the parasitic absorption in the n-type front contact layer stack could be reduced significantly by thinner and less absorbing films. This inverted design has been used ever since and resulted in most of the results summarized in Table 17.1 exceeding 25% PCE.

The above-mentioned tandem design with rear-junction SHJ and p-i-n perovskite top cells utilized for most tandem reports with efficiency above 25% from Table 17.1 is schematically represented in Figure 17.11. Figure 17.11c displays the general monolithic 2T tandem design with a zoom into the back side and the top cell. Typically, the SHJ back side is fabricated on textured wafers by passivating the wafer with intrinsic (*i*)a-Si:H, utilizing (*p*)a-Si:H as selective contact and implementing a low absorptive TCO such as ITO or aluminium-doped zinc oxide (AZO) between the selective contact and the reflecting silver (Ag) back electrode contact. Some tandem designs use silica nanoparticles between the TCO and silver layers to enhance the NIR absorption in the bottom cell and leave approximately 5% of the area uncovered for direct electrical contact between the TCO and the Ag [41, 54]. On the planar (or polished) front side, (*i*)a-Si:H forms the passivated, selective contact together with *n*-doped (*n*)a-Si:H [42] or (*n*)nc-SiO_x:H [46, 47] and the subcells are interconnected by TCOs such as ITO or tunnel junctions formed by doped silicon layers [47, 52]. The hole-selective contact of the perovskite top cells is typically formed by organic molecules such as Spiro-TTB or polymers such as poly(triaryl amine) (PTAA) [42, 49] and the perovskite is processed on top. Spin coating is still used for most of the reported perovskite absorbers in the tandems with compositions such as the "triple cation" $FAMACsPb(I_xBr_{1-x})_3$ or double cations such $FACs(I_xBr_{1-x})_3$. Band gaps of these compositions utilized so far range from 1.63 to 1.68 eV and a constant increase in band gap toward 1.68 eV was found in the development of 2T perovskite/SHJ tandems as shown in Table 17.1. The semitransparent n-type top cell stack is typically formed by fullerenes or derivatives such as C_{60} or phenyl-C_{61}-butyric acid methyl ester (PCBM), ALD SnO_2 as buffer layer, TCOs sputtered at low temperature such as ITO or indium zinc oxide (IZO) and anti-reflection coating such as LiF or MgF_2. Metal fingers or grids are typically made from evaporated Ag. This tandem layer design and material stack enabled reported PCEs up to 29.15% [50]. Future improvements might be possible – next to the optical optimization – by reduction of non-radiative recombination for these perovskite top-cells (see also Figure 17.9) in order to improve the top cell V_{oc} and fill factor (FF) and with that the tandem PCE. In addition, Figure 17.11b also visualizes the yet reported strategies to cover textured silicon wafers by either complete conformal coverage [42, 52] or utilization of thick absorber layers in combination with mild textures to fully cover the texture height. Both being explained in more detail below. Figure 17.11 also includes in (a) a scale bar to highlight the size difference between the bottom and top cells.

Usually, single-junction silicon solar cells are made from textured silicon wafers to effectively reduce reflection and to extend the light path within the absorber layer (light trapping). If light trapping happens via the rear side of the silicon bottom cell only, the perovskite top cell can be processed on the flat front side of the Si wafer without constraints. Therefore, standard texturing processes can be used on the silicon rear side, and indeed, most of today's high efficiency perovskite/SHJ tandem cells are processed on rear side textured wafers with beneficial light trapping in the NIR – data are presented in Table 17.1. Optical simulations, which are briefly discussed in Section 17.6.2, can help to estimate the effect of different light-trapping measures. The optical simulation studies by Jošt et al. [90] and Santbergen et al. [91] of 2T perovskite/SHJ tandems with such a random pyramid textured rear side showed a gain around ~0.7 mA/cm^2, if a rear side texture is present, which was also demonstrated experimentally by Werner et al. [92]. For reducing reflective losses, the front side has to be textured as well. Jost et al. carried out sophisticated optical simulations of perovskite/SHJ tandems on flat, rear side and both side textured silicon wafers, also considering an optimization of the perovskite band gap [45]. They find a difference of ~0.5 mA/cm^2 in photocurrent density between optimized rear side texture and a double side textured cells, which can result in 1% absolute efficiency gain. For monocrystalline silicon, random pyramid textures with a height and base length of a few micrometers are typically used [42]. However, such feature sizes are incompatible with spin-coating, which is typically applied for high efficiency lab-based solar cells. Here, typical thicknesses of around 500 nm and well below 100 nm for the absorber and charge selective layers are utilized, respectively. Therefore, usually additional chemical and/or mechanical polishing is required to planarize the Si surface and enable perovskite spin coating atop [42].

Despite the potential advantages not only in optical properties, but also in terms of process compatibility to standard silicon cell technology, only few results on implementing perovskite cells on bottom cells with random pyramid textures on both wafer sides are reported. As illustrated in Figure 17.11b, 2T perovskite/silicon tandem solar cells with front side textures can be implemented in two ways: (i) a fully conformal deposition of the top cell layers or (ii) utilizing a mild texture and a thick perovskite absorber to fully cover the texture height with the perovskite absorber layer. The first strategy of fully conformal coverage was realized by Sahli et al. in 2018 [42]: To achieve a conformal coating of the perovskite film, they used a hybrid process consisting of first co-evaporating the inorganic precursor species PbI_2 and CsBr to form a porous but conformal film on the texture. Then, this film was infiltrated by a spin-coated solution of FAI and FABr perovskite precursors and annealed at 150 °C to form the $Cs_xFA_{1-x}Pb(I,Br)_3$ perovskite film. Interestingly, spin coating the organic components on top of the textured and conformally covered (with PbI_2 and CsBr) wafer yielded a conformal perovskite film. The final tandem cell has a certified MPP-tracked PCE of 25.2%. Indeed, the reflection losses are minimal between 400 and 1100 nm and the external quantum efficiency (EQE) of the fully textured tandem reported by Nogay et al. shows low parasitic losses as well as a good current matching between the sub-cells [52]. The later strategy of utilizing a thick absorber to fully cover the texture height one is reported by Chen et al. [54] and Hou et al. [55]: utilizing a mild sub-micron to micron texture and a thick perovskite absorber from blade- or spin-coating enabled the complete coverage of perovskite films that showed a rather planarized perovskite top surface on top of the textured silicon wafer. Remarkable monolithic tandem cell efficiencies of 25.7% and 26.1% were realized by the teams of Chen and Hou/Aydin, respectively. Another interesting approach is mechanical stacking into 2T monolithically tandem designs as shown by Ramirez Quiroz et al. [78] and Lamanna et al. [53]. Ramirez Quiroz et al. used a planar bottom cell and a conductive adhesive to mechanically and electrical connect both sub cells realizing remarkable fill factors above 80% [78]. Lamanna et al. utilized a fully textured bottom cell by contacting the back electrode of the perovskite top cell with the texturized and metalized front contact of the silicon bottom cell [53]. As of July 2020, the utilization of front side texturing has not enabled highest photocurrents compared to front side flat wafers (compare e.g. the results by Köhnen et al. [47] and Sahli et al. [42]), but it was shown that double side textured wafers, i.e. the standard in silicon PV manufacturing, can be utilized for highly efficient tandems as well and that these could have a better optical efficiency limit.

The first silicon/perovskite tandem solar cell based on an n-type Si homojunction bottom cell with a standard diffused junction was built by Mailoa and coworkers in 2015 [39]. Soon after, Werner et al. demonstrated 16.3% efficient cells using ZnSnO as recombination layer, a stack of sputtered, compact TiO_2 and mesoporous TiO_2 as electron transport layer (ETL), and $MAPbI_3$ as top cell absorber [93]. Just as in the cell by Mailoa, they used Spiro-OMeTAD as hole-selective layer, which suffered from parasitic absorption at wavelengths shorter than 400 nm limiting the photocurrent density. In 2017, We and coworkers manufactured a tandem cell with a silicon homojunction bottom cell featuring passivating Al_2O_3 and SiN_x films with local contact openings and Cr/Pd/Ag point contacts, covered on the front with an ITO electrode; hence their passivation/point contact schemes was very similar to those in industrial PERC/passivated emitter rear totally diffused (PERT) Si cells. The top cell was similar to Werner's device; the PCE was 22.5% [56]. In 2018, Zheng et al. built tandem solar cells with 16 cm^2 device area with an interconnecting layer of SnO_2 processed by spin coating from a colloidal precursor and annealing at 150 °C. For this device, a reverse scan PCE of 17.6% and an MPP-tracked PCE of 17.1% were demonstrated, whereas on 4 cm^2 a PCE of 21.0% under reverse scanning and a stabilized PCE of 20.5% was shown [94]. Later that year, the same group demonstrated a 16 cm^2 cell with 21.8% PCE; here $(FAPbI_3)_{0.83}(MAPbBr_3)_{0.17}$ was used as perovskite [57]. Shen and coworkers applied a different approach by implementing an interconnecting layer stack consisting of ~54 nm ALD-grown compact TiO_2, followed by a mesoporous TiO_2 layer, on two different high-temperature processed front p/n junctions. Their n-type c-Si cells had either a diffused front (p+/n)c-Si homojunction, or a (p+)polysilicon/silicon oxide/(*n*)c-Si passivated front junction (being called POLO or TOPCon contact). The perovskite composition in the top cell was $CsRbFAMAPbI_{3-x}Br_x$ with 1.63 eV bandgap and the front contact was a Spiro-OMeTAD/MoO_x buffer/sputtered ITO stack. With an additional antireflection foil, the 1 cm^2 cell with passivated polysilicon contact reached an MPP-tracked efficiency of 24.1% whereas the homojunction-based tandem had a PCE of 22.9% [58]. In 2019, Zheng et al. presented a 2T perovskite/silicon-homojunction tandem solar cell with 23.0% MPP efficiency on a 4 cm^2 cell. It incorporated a down-shifting phosphor in the polydimethylsiloxan (PDMS) antireflective film, which absorbs UV photons and re-emitting green light leading to a small improvement of the integrated photocurrent improvement in external quantum efficiency spectra of around 0.3 mA/cm^2 [59].

In 4T perovskite/silicon tandem cells both subcells are processed individually and mechanically stacked on top of each other. A selection of relevant results is listed at the end of Table 17.1. We summarize here the latest developments in efficiency improvements only. 4T tandem cells suffer from less restrictions on the layer integration and enjoy more freedom toward the perovskite top cell design than 2T cells. Free choice of polarity and layer arrangement (substrate vs. superstrate) enabled a continuous efficiency improvement but also more spread and diverse device designs. This can also be seen by the various perovskite top cell band gaps utilized for recent high efficiency designs ranging from 1.57 to 1.72 eV. Typically, the active layer sizes are not matched for the bottom and top cells; Table 17.1 therefore lists the areas of both cells. For unmatched areas the small semitransparent top cell is measured first and then a larger top cell-like device with similar optical properties (smaller devices have reduced losses from serial losses) is used as a filter for the measurement of the larger bottom cell (due to the rather large diffusion length in crystalline silicon, the bottom cell suffers from edge effects at smaller active areas).

The first report on 4T perovskite/silicon tandem cells with efficiency close to the best silicon single junction was presented by Duong et al. in 2017 [35]. Using a "quadruple" cation composition with MA, FA, Cs, and Rb stabilized the top cell absorber at 1.73 eV bandgap and therefore enhanced efficiency, and reduced both hysteresis and degradation under operation. The mesoporous top cell with PTAA instead of Spiro-OMeTAD as hole-transporting layer had an enhanced NIR transmission and allowed to realize a 26.4% tandem solar cell. In early 2018, Ramirez Quiroz et al. achieved a tandem efficiency of 26.7% using a planar p-i-n top cell with a $CH_3NH_3PbI_3$ absorber (1.57 eV bandgap) and silver nanowires as transparent top contact [60]. In late 2018, Jaysankar et al. improved the 4T tandem further by using a CsFAPbIBr perovskite composition with 1.72 eV band gap [61]. Further, they passivated the perovskite surface by a sub-nm layer of atomic-layer-deposited Al_2O_3 to further enhance the open-circuit voltage of the top cell. Combining this planar n-i-p top cell with an IBC silicon bottom cell, a PCE of 27.1% was reached. In early 2020, Duong and coworkers achieved a new efficiency record of 27.7% tandem efficiency with a size-unmatched IBC bottom cell and 26.2% for 1 cm^2 size-matched cell using a PERL bottom cell [62]. They improved their mesoporous n-i-p cell by passivating the surface of the quadruple cation perovskite with a large cation *n*-butylammonium bromide to enhance both the V_{oc} and also slightly the FF. As of July 2020, the highest 4T tandem efficiency combining perovskite and silicon is 28.2% and was reached using a triple cation perovskite composition with 1.63 eV band gap [63]. They used urea as additive to the perovskite, which enhanced the carrier diffusion length and therefore allowed to increase the perovskite thickness, leading to increased absorption and an unprecedented photocurrent density exceeding 22 mA/cm^2 in the semitransparent top cell and 39.5 mA/cm^2 in the 4T tandem.

After discussing the efficiency and device-development of 2T and 4T perovskite/silicon tandem solar cells, we now will briefly look at more practical properties such as operational stability and upscaling toward larger. To evaluate the operational stability, long-term MPP tracking is used and the most stable perovskite single cells reach T80 stabilities (the time when they are degraded to 80% of the initial performance) of >1000 hours [95, 96]. The stability under MPP tracking depends on the measurement conditions, such as device temperature, spectral illumination condition such as with or without UV irradiation, and the atmospheric conditions such as oxygen and moisture levels. As good practice, writing these conditions directly into the MPP tracking figures has evolved. However, long-term MPP tracking of tandem solar cells is only rarely reported in the literature and yet non-standardized. Tandem devices without encapsulation were tested in ambient conditions and retained 90% after 61 hours [42] and 92% after 100 hours [54], as reported by Sahli and Chen, respectively. Encapsulated tandem devices on the other hand by Hou et al. were light-soaked under 400 hours MPP load at 40 °C but showed no degradation [55].

In 2019, Kamino et al. reported on a perovskite/SHJ tandem cell with 57.4 cm^2 aperture area and 22.6% PCE [97]. However, for commercial applications, at least the full wafer formats M2 or better M6 need to be developed, with areas of around 244 and 274 cm^2, respectively. The continuous trend toward larger wafers might also force the required area for top cells to M12 format, which is 21 cm × 21 cm = 441 cm^2 area. It is very important to demonstrate homogeneous top-cell layer coverage on this area, which can be potentially achieved by blade coating as used by Chen et al. [54] or vacuum-based methods [98]. In addition, also low-temperature curing grids and grip pastes with minimal contact resistance to the TCO need to be developed and implemented. First results on this aspect were reported by Kamino et al. [99]. For further reading, we highlight reviews on large area processing for perovskite single junctions that were e.g. published by Chen et al. [100].

17.5.2 Perovskite-Chalcogenide Tandem Solar Cells

Besides crystalline silicon, also CIGS [$Cu(In,Ga)Se_2$] or copper indium selenide (CIS) [$CuInSe_2$] are suitable materials for the bottom cells of tandem cells with a perovskite top cell. In 2019, 1.6 GWp of PV modules based on CIGS or CIS were produced, which is around 1.2% of the totally produced PV modules [101]. Because of the direct bandgap of CIGS, the absorbers can be made much thinner than in Si solar cells. Hence, this technology enables low material and energy consumption. In other words, CIGS solar cells are a thin-film technology and therefore would allow true thin-film tandem cells, which can be fabricated as light-weight modules with a high power-to-weight ratio, flexibility and radiation hardness for space applications [102]. The bandgap of CIGS can be tuned by varying the In:Ga ratio between ~1.7 eV for $CuGaSe_2$ and ~1.0 eV for $CuInSe_2$. Comparing the best CIGS single-junction cell with 23.25% PCE [103] to the best front-side-planar silicon solar cell with 22.1% PCE [104] shows that CIGS might have an obvious advantage.

In this section we discuss the efficiency evolution of 2T and 4T perovskite/CIGS tandems solar cells until mid 2020 and conclude with a short paragraph on stability and upscaling. The number of published reports on perovskite/CIGS tandem solar cells is significantly lower than for perovskite/silicon tandem solar cells, which can be seen by the number of data points in Figure 17.10 and when comparing the

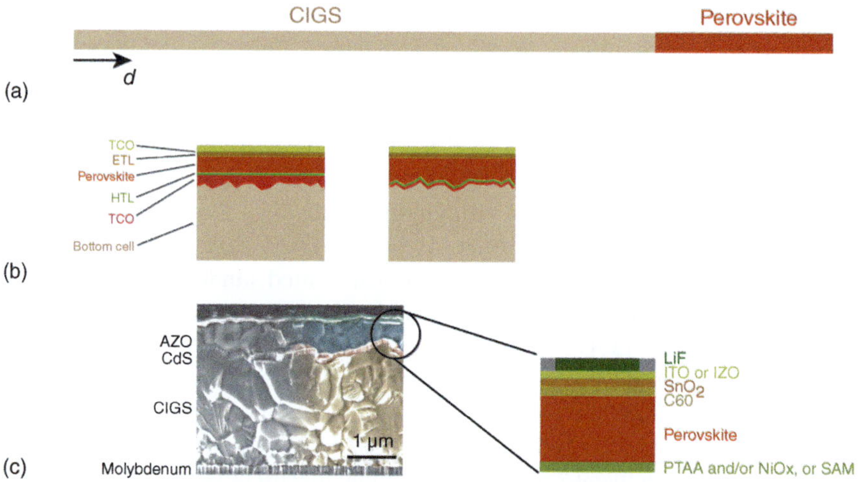

Figure 17.12 Schematic depiction of the 2T perovskite/CIGS tandem device architecture: (a) size comparison of the bottom and top cell assuming a 2–3 μm thick CIGS bottom cell and a 500 nm thick perovskite top cell; (b) the front side texture and respectively the top cell processing options, reported so far; (c) specific device design and layer stack utilized for the few reports in Table 17.2, SEM images for top and bottom cell layers are shown as well. *Source*: Based on Han et al. [64].

numbers of entries in Tables 17.1 and 17.2. Only few reports on 2T perovskite/CIGS solar cells have been published. This is likely because of the roughness of the CIGS bottom cell surface, as we will discuss below.

Figure 17.12a shows a thickness comparison of a 2–3 μm thick CIGS bottom cell and a 500 nm thick perovskite top cell. In contrast to wafer-based silicon tandem solar cells, here a true thin film tandem with an overall thickness below 5 μm can be built. Figure 17.12c shows the typical device structure of CIGS bottom cells: it consists of a substrate, a molybdenum (Mo) back contact, the p-type CIGS absorber, an n-type CdS absorber and a zinc oxide (ZnO) front contact. Here, the CIGS and CdS layers form a p–n junction. Due to this defined polarity in the typical substrate configuration, only the p-i-n perovskite top cell configuration has been used in 2T tandem devices so far. As a consequence, the top cell designs used for CIGS or SHJ bottom cells were recently merged and p-i-n perovskite top cells were implemented in both designs (compare Figures 17.11 and 17.12). Luckily, this restriction is not crucial, because optical losses in a p-i-n top cell configuration are much smaller than in n-i-p designs, as explained in Section 17.5.1.

Figure 17.12b illustrates two examples of monolithic integration for 2T perovskite/CIGS tandem solar cells by either using polishing of the intermediate TCO (left) or conformal coverage of the charge selective layers (right). Both will be described below. Overall, only a few reports on monolithic perovskite/CIGS tandem solar cells can be found in the literature, with best ones summarized in Table 17.2. Reasons for this low number are – amongst others – polarity restrictions and processing restrictions. Most importantly, the deposition of CIGS absorber results in a relatively rough surface, in the range of $\sigma_{RMS} = 50–200$ nm, with surface features of high aspect ratio [65]. While the typical perovskite absorber thickness of around 500 nm is sufficient to fully cover the rough CIGS surface, high-performance hole selective layer for the perovskite top cell, commonly deposited via solution processing, cannot cover the rough CIGS surface. This very likely leads to a shunted top cell and makes fabrication of a monolithic perovskite/CIGS tandem very challenging. Consequently, only three reports on 2T perovskite/CIGS cells with efficiencies higher than 20% were published by July 2020; they are summarized in Table 17.2.

Todorov et al. reported the first 2T perovskite/CIGS tandem device in 2015 with 11% PCE [40]. This tandem cell was based on a bottom cell being processed from solution instead of typical sputtering or co-evaporation processes, thus reducing the surface roughness. As a hole selective layer between top and bottom cells a thick poly(3,4-ethylenedioxythiophene):polystyrene sulfonate (PEDOT:PSS) layer was used, which smoothened the bottom-cell surface even further. While this layer protected the top cell from shunting, the overall tandem cell performance might have suffered from reduced V_{oc} because of high recombination losses at the PEDOT:PSS/perovskite interface [105] and an ultrathin semitransparent metal front contact.

In 2018, Han and coworkers presented a top-cell integration, where the front surface is polished to obtain a flat surface [64], similar to what is done for perovskite/silicon tandems. They sputtered a thick ITO recombination layers on top of the CIGS bottom cell and polished it to reduce roughness as shown in Figure 17.12b. With that, the spin-coating of perovskite and very thin hole selective layers could be implemented. Consequently, a PCE of 20.8% on 0.52 cm^2 was obtained and a certified PCE of 22.4% on 0.042 cm^2 (see also Table 17.2). This tandem shows a high V_{oc} of 1.78 V, which is the highest value reported for this tandem technology, as of July 2020. Because of the thick ITO recombination layer, the free-carrier absorption increases in the NIR, reducing the generated current density in the bottom CIGS sub-cell. Thus, the overall tandem J_{SC} of only 17.3 mA/cm^2 is rather moderate.

Another approach was reported by Jošt et al. in early 2019: instead of polishing the CIGS cell to obtain a flat surface, they conformally deposited a hole transport layer directly on top of the rough as-grown CIGS surface [65]. 10 nm of ALD grown NiO_x were sufficient to form a conformal, closed layer. In addition, a spin-coated PTAA polymer layer reduced interface recombination between the conformal NiO_x and the perovskite, even though, the polymer itself might not have been covered fully conformal in a dense film, leaving some of the NiO_x uncovered. This enabled 21.6% PCE on an area of 0.78 cm^2. The current density of this cell was 18.0 mA/cm^2 and therefore higher than by Han et al., despite a strong current mismatch between the sub-cells. This high current density could be achieved by low parasitic absorption in the recombination layer and the rough internal interfaces, which reduced internal reflection. However, this tandem device has a rather low V_{oc} below 1.6 V.

Al-Ashouri et al. developed the idea of conformally deposited layers even further and used a self-assembled monolayer (SAM) acting as the hole selective layer [66]. This ultrathin SAM layer binds to oxide surfaces (e.g. ITO, ZnO) and thus connects the tandem recombination contact with the perovskite top-cell absorber. Despite being only a single molecular layer, the strong binding affinity and self-assembly ensures closed layers even on rough surfaces. The almost lossless SAM/perovskite interface additionally resulted in an improved V_{oc}, which, combined with improved light management, yielded a certified current record PCE for perovskite/CIGS tandems of 23.3% on more than 1 cm^2 area (see also Figure 17.10).

In January 2020, a team from Helmholtz-Zentrum Berlin (HZB) manufactured a perovskite/CIGS tandem solar cell with a certified efficiency of 24.16% [106]. This result was realized by improving the V_{oc} to the value by Han et al. Based on these results, the future efficiency development is sketched in Figure 17.10. As the efficiency potential for perovskite/CIGS tandem is highly comparable to perovskite/silicon tandems [31], it is assumed that the efficiency will quickly increase to well above 26% for perovskite/CIGS tandem and could ultimately catch up with perovskite/silicon tandems above 30%.

Now we discuss developments on four-terminal perovskite/CIGS solar cells starting with the first cell that was significantly outperforming the best CIGS single junction: Shen et al. reported in early 2018 a combined tandem efficiency of 23.9% by using a quadruple-cation perovskite composition with 1.62 eV band gap as also utilized for their 2T perovskite/silicon tandems [58, 68]. In addition, they also report on a MA-free perovskite compoisiton with 1.75 eV band gap that is characterized by higher stability toward oxygen atmosphere. It is supposed that MA cations and oxygen molecules interact and enhance the surface recombination, which is identified as main driver for oxygen-induced degradation [68]. A further enhancement toward efficiency was reported by IMEC in September the same year: They improved the NIR transmittance of the perovskite top cell by adding optical coupling layers and by optimizing the transparent electrodes. In addition, the perovskite itself was optimized toward a 1.72 eV band gap which is close to the fundamental optimum for higher tandem efficiency [69]. The latest report with the highest 4T tandem efficiency of 25.9% was announced by Hoe Kim and coworkers: using additives such as the large organic cation phenethylammonium iodide (PEAI) and lead thiocyanate ($Pb(SCN)_2$) improved the perovskite film quality with enhanced crystallinity, reduced formation of excess PbI_2, a lower defect density and also higher charge carrier mobility as well as lifetime [70]. This enabled a highly performing perovskite top cell with around 20% and a combined tandem efficiency of 25.9%, the yet highest reported value for any kind of perovskite/CIGS tandem solar cells.

The results in this section highlight the versatile potential of perovskite/CIGS tandem solar cells. Although less results and lower performance was reported in comparison to perovskite/silicon tandems, Jošt et al. recently compared the efficiency potential between silicon and CIGS and found that comparable efficiency metrics above 33% could be achieved [31] with realistic material properties and device designs. Therefore, we forecast a strong efficiency improvement for perovskite/CIGS tandem solar cells toward 30% within the next two to three years.

At the end of this section, we briefly summarize the more practical properties such as operational stability and upscaling toward larger areas for perovskite/CIGS tandems although the number of reports is yet limited. Larger area perovskite mini modules and their integration into tandem devices were reported by Jaysankar et al. [107]. When scaling the semi-transparent from 0.13 toward 4 cm^2 (seven series connected sub cells) and 16 cm^2 the efficiency drops from 16.7% to 14.8% and to 12.0%, respectively. Combining the 4 cm^2 mini module with a CIGS bottom cell, an area matched tandem efficiency of 21.3% was achieved [107]. For perovskite/CIGS tandems, module integration by series connection of sub-cells and the deposition on the m^2 area are two important challenges toward commercialization. Another important aspect for outdoor tandem modules is operational stability. Long term MPP tracking was reported by Han et al.: non-encapsulated tandems retained 88% of the initial performance after 500 hours under continuous one-sun illumination at 30 °C ambient conditions. After storing the tandems in dark for 12 hours, it recovered to 93% of its initial performance [64]. The perovskite-specific dark recovering might induce higher operational stability under outdoor conditions [108], ensuring high energy yields over a long tandem module operation time scale.

17.6 Energy Yield Calculations

The performance of solar cells is usually measured under *standard testing conditions* (STC) [12], hence the solar cells are illuminated according to the AM1.5g spectrum [2] with a total radiant power density of 1000 W/m^2 at 25 °C.

A tandem solar cell, which outperforms a comparable single-junction solar cell under STC, does not necessarily perform better than the reference cell under realistic conditions. Especially for monolithic two-terminal cells, where current matching is required for a high overall photocurrent density, spectral changes may be detrimental to the solar-cell performance. Further, changes in temperature affect the performance of solar cells. Therefore, STC are often not sufficient for reliably assessing the potential of tandem solar cells.

A better assessment can be achieved by estimating the *energy yield* (EY), which is the electric energy a photovoltaic system is expected to generate in a given amount of time, often a year. While the performance of a solar cell under STC can be measured in laboratories, the energy yield only can be measured in outdoor testing facilities. In addition, measuring a solar cell in such a facility throughout the course of a year will not directly allow to interpolate the solar cell performance to the future because of variabilities of weather.

Therefore, the energy yield often is assessed using numerical methods. For estimating the annual energy yield, the power $P(t)$ needs to be calculated throughout the year and the power must be integrated over this time. Figure 17.13 shows a schematic flowchart for calculating the power at a given timestamp.

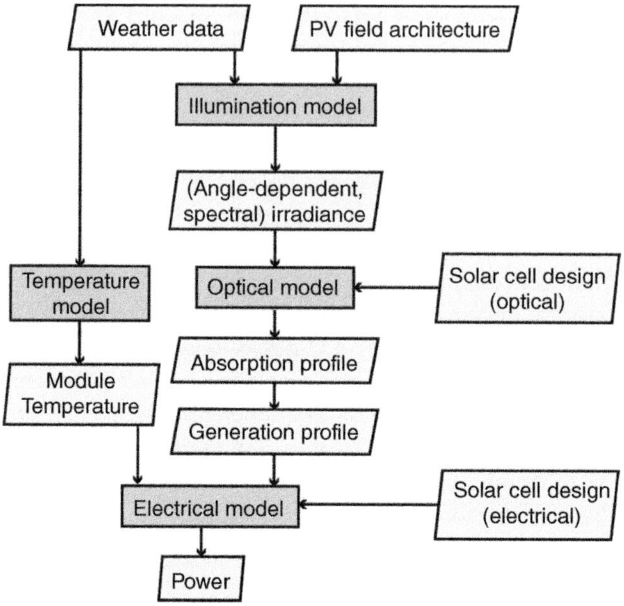

Figure 17.13 Flowchart for calculating the power output of a solar module.

17.6.1 Illumination Model

First, the irradiance onto the module needs to be calculated. It depends, amongst others, on the position of the Sun in the Sky, the direct normal irradiance (DNI) of the sunlight, the diffuse horizontal irradiance (DHI) irradiance of the light from the sky and the orientation of the module. These data are provided by many institutions, such as national weather services or the National Renewable Energy Laboratory (NREL) in the USA. Some sources provide data for many years, here the annual energy yield would be calculated by averaging over several years. Others, such as NREL, provide data for a *typical meteorological year* (TMY). For tandem solar cells, it is especially relevant to have spectral data, because this determines how the incoming light is distributed between the two subcells. Currently, most databases do not provide these data, however. If no spectral data is available the spectral distribution of the light can also be described by the *average photon energy* (APE), which for example was used by Futscher and Ehrler [109]. A high APE value corresponds to a blue-rich spectrum, while a red-rich spectrum leads to a low APE. While some models consider a singular module with no shadowing effects by other modules [110], other models consider the PV module to be in a large PV field, therefore taking shadowing effects into account [8]. The most complex models also consider the surrounding, including detailed shadowing by buildings at a specific location [111].

17.6.2 Optical Model

When the irradiance is determined, it can be used to calculate the spectral intensity profile throughout the absorber layer $I_\lambda(z)$, where z is the vertical position within the absorber. A detailed *optical model* should take all layers of the solar cell into account; it is relevant to know the thickness and the wavelength-dependent complex refractive index data of every layer. Especially in the quickly evolving field of perovskite solar cells, where material compositions of the different layers change often, this task is very relevant and often requires an individual characterization of the layers in the respective labs. Further, textures of the different surfaces and interfaces in the solar cell strongly affect the optics. Taking them into account in an appropriate manner always increases the simulation time and is not always possible. While large pyramids, which e.g. are commonly present in silicon solar cells can be simulated with approaches such as ray tracing [4, 112], periodic structures with dimensions in the range of the wavelength need rigorous approaches for reliable results, for example the finite elements method [113, 114].

From the intensity profile the absorption in the absorber layer is given by

$$A(\lambda) = \frac{I_\lambda(0) - I_\lambda(d)}{I_\lambda(0)} \tag{17.10}$$

where d is the thickness of the absorber. If light traverses the absorption layer only once, the intensity decays according to the *Lambert–Beer law*,

$$I(z) = I(0)\exp(-\alpha z) \tag{17.11}$$

with the absorption coefficient α. However, this relation does not hold if light is reflected from the back into the absorber, if interference is relevant or if the device contains scattering structures.

For a rough estimate, simple optical approaches can be used. For example, every solar cell absorber layer can be assumed to absorb all the light with energies larger or equal then the bandgap: then, all the light reaching the solar cell with energies larger than the top-cell bandgap is considered to be absorbed there while longer wavelength can reach the bottom cell. In a more realistic but still idealistic approach, the absorption in the weakly absorbing region close to the bandgap can be estimated by the Tiedje–Yablonovitch limit [115] or by an approach by Green [116]. However, these approaches only will yield a layer absorptivity A but no depths-dependent information.

For detailed electrical simulations, we need the generation rate $G(z)$ (in cm^{-3}), which tells us, how many electron-hole pairs are generated at depths z. Starting from the *spectral* intensity $I_\lambda(z)$, we can obtain the spectral generation rate by

$$G_\lambda(z) = -\eta_G \frac{\lambda}{hc} \frac{dI_\lambda(z)}{dz} \tag{17.12}$$

where hc/λ is the photon energy at wavelength λ and η_G is the generation quantum efficiency, usually assumed equal to unity. The generation rate is then obtained via integration

$$G(z) = \int_0^\infty G_\lambda(z) d\lambda \tag{17.13}$$

Simpler electrical models or estimates do not require generation profiles, but only the *short-circuit current density*. When we assume that every absorbed photon leads to the generation of an electron–hole pair, we obtain the *maximum achievable current density*,

$$J_{ph} = q \int_0^\infty A(\lambda)\Phi_{ph,\lambda} d\lambda \tag{17.14}$$

where $\Phi_{ph,\lambda}$ denotes the spectral photon flux impinging on the solar modes. For standard optical modeling it is obtained from the AM1.5g spectrum at normal incidence [2]; for detailed energy yield analyses we obtain it from an illumination model as discussed in Section 17.6.1.

Note that the generation rate and the current density need to be calculated for the absorbers in the top cell and the bottom cell.

17.6.3 Electrical Model

The final step in the chain of models sketched in Figure 17.13 is the electrical model. A very rigorous method for electrical modeling is to calculate the depth profiles of the electron concentration $n(z)$ and hole concentration $p(z)$, from which the electrical potentials and hence the band diagram of the device can be calculated, which is the base for the current density – voltage characteristic. All this is explained in detail for example in Reference [6], Chapters 6–8. The calculations are based on three sets of equations:

- The *continuity equation* ensures the conservation of charge, not only when electrons and holes flow through the solar cell, but also when electron-hole pairs are generated as described in Section 17.6.2, or when they recombine via recombination mechanisms such as radiative recombination, defect-driven Shockley-Read-Hall recombination or Auger recombination, which is relevant in indirect semiconductors. More information on generation and recombination is for example given in Ref. [6], chapter 7.
- The *transport equation* relates the current density to gradients in potential and charge density. Here the current density $J(z)$ is driven by diffusion, which is caused by concentration differences of the charge carriers at different positions z in the semiconductor, and by drift due to electric fields.
- The *Poisson equation* relates the electrical potential and hence the electrical field to the sum of all charge densities.

Using these equations, the electron and hole concentrations across a solar cell can be calculated and the current density – voltage characteristic can be extracted. Usually, these equations cannot be solved analytically but require a numerical approach. Next to commercial programs, several academic groups have developed solvers to study the device physics of solar cells, e.g. [117, 118]. While these programs are well-suited to simulate single-junction solar cells, they often have great difficulties in covering the physics happening at the recombination contact of monolithic tandem solar cells, where the top and bottom cells are electrically connected to each other. Further, this kind of simulations is relatively expensive in terms of computational capacity and takes relatively long, which makes them

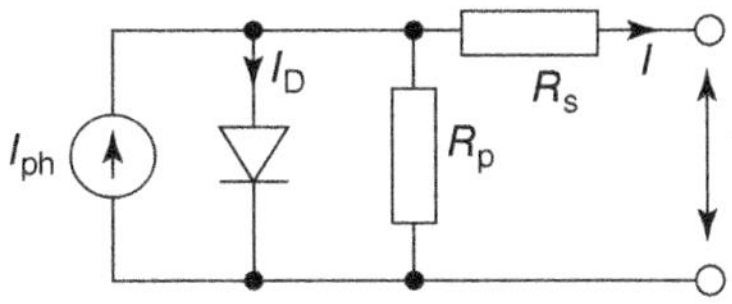

Figure 17.14 The equivalent circuit according to the one-diode model.

less suited for energy yield analyses, where the simulations must be conducted for many time steps.

A much simpler analytical approach is to electrically describe the solar cell via a *one-diode model*, where we consider the solar cell to consist of a photocurrent source I_{ph}, a diode, a parallel (shunt) resistance R_p and a serial resistance R_s, as illustrated in Figure 17.14. Mathematically, the current density – voltage characteristic of this circuit is given by

$$I(V) = I_0 \left\{ \exp\left[\frac{q(V - IR_s)}{mkT}\right] - 1 \right\} + \frac{V - IR_s}{R_p} - I_{ph} \tag{17.15}$$

with the dark current I_0, the diode ideality factor m, the unit charge q, the Boltzmann constant k, and the temperature T. For $R_s = 0$, $R_p \to \infty$, and $m = 1$ the equation reduces to the *Shockley equation* (17.3), which describes an ideal diode. We used it in our derivation of tandem solar cell efficiency limits in Section 17.3. For some types of solar cells, such as silicon solar cells, models consisting of a single diode cannot describe the solar-cell behavior accurately. Then, two-diode models can be used where two diodes are assumed to be in parallel: one is an ideal diode with ideality factor $m = 1$, the other one has a different ideality factor (see e.g. Ref. [6], chapter 9): With such models, monolithic two-terminal solar cells can be modeled as sketched in Figure 17.4b. One-diode models have been used by several authors for their energy yield analysis of tandem solar cells [110, 119].

Each of the approaches described above leads to a current density – voltage characteristic, for which the MPP needs to be determined. The maximum power can then be used for the energy yield calculation.

In an even simpler approach, the external solar cell parameters of experimentally measured single-junction solar cells are taken as starting points: the open-circuit voltage $V_{oc,\,ref}$, the short-circuit current density J_{sc}, and the fill factor FF. Using optical simulations, a maximum achievable current density J_{ph} is determined for each subcell and taken as new value for J_{sc} – for monolithic tandem solar cells the lower one of these two photocurrents is also used for the other subcell in order to mimic current matching. Then, the open-circuit voltage is determined by accounting for the logarithmic relationship between current density and voltage,

$$V_{oc} = V_{oc,ref} + \frac{kT}{q} \ln \frac{J_{ph}}{J_{sc}} \tag{17.16}$$

Finally, the FF is assumed to stay constant. Assuming this approach, a power density is determined for the two subcells and the two power densities are added. This approach was for example used for layer thickness optimizations of monolithic tandem solar cells [45, 113, 120].

17.6.4 Temperature Model

The electrical performance is not only determined by the irradiance which reaches the solar cell, as described in Section 17.6.1, but also by the solar cell temperature. The temperature affects many electronic processes in the solar cell, such as recombination rates and – as a consequence – the voltage. For detailed energy-yield analyses and electrical modeling, knowledge of the solar-cell temperature, which might be significantly hotter than the surface of the solar module, is therefore very important.

The temperature of a solar cell within the solar module is determined by several factors: temperature of the surrounding, irradiance on the module, wind at the module – which is also dependent on the installation height of the module. Some models that can be used to estimate the module temperature are described in Ref. [6], section 20.3.

The effect of temperature on the performance of silicon PV modules is often approximated linearly, where the change per degree Kelvin of a parameter such as the module's PCE, is given as a *temperature coefficient*. Temperature coefficients are published in datasheets of silicon PV modules. For perovskite solar cells a temperature coefficient of −0.17% per K between 25 °C and 85 °C was published in 2020 [121], which is comparable with silicon solar cells. For perovskite-based tandem solar cells no temperature coefficients have been determined yet. However, as both perovskite and silicon solar cells show similar temperature coefficients, it is likely that perovskite-silicon tandem solar cells behave analogously.

17.6.5 Energy Yield Calculation

Above we described the models that are required for an energy yield analysis:

- An illumination model to determine the amount of radiation – and hence the wavelength-dependent photon flux – hitting the solar module. For tandem modules, where the two cells are susceptible for different wavelength ranges, the illumination model should provide spectral information.
- An optical model that calculates, how much light is absorbed in the absorber layers of the subcells, with the output of the illumination model as input.
- An electrical model, which calculates the current density – voltage characteristic and subsequently the maximum power output at the given illumination and temperature.
- Finally, a temperature model that determines the temperature of the solar module, which is relevant for accurate electrical modeling.

All these models must be performed sequentially for every time step in the period, for which the energy yield should be calculated. Some models described above are very time consuming, such as the finite element method e.g. for optical modeling or numerical electrical models based on the continuity, transport, and *Poisson equations*. It is not practical to perform them for every time step. Therefore, for energy yield calculations we need to fall back on simpler and faster models. Alternatively, we can use *surrogate modeling methods* [122], where we perform the rigorous calculations only for a few points. An example for such methods is the *reduced basis method* [123], which can be applied to the finite elements method [124, 125].

Some authors use energy yield calculations for optimizing different aspects of tandem solar cells, e.g. to study the effect of bandgap and layer stack changes on the performance of 2T- and 4T solar cells [119, 126].

17.7 Conclusions and Outlook

In this chapter we have shown that perovskite-based tandem solar cells with a perovskite top cell and a crystalline silicon or CIGS bottom cell are a very interesting concept because of their high efficiency limits around 45% and more practical limits around 34% for both bottom-cell materials. Both tandem technologies will break the 30% efficiency barrier within the early 2020s with a sooner breakthrough for perovskite/silicon tandem solar cells. Hence, they offer a huge efficiency potential as compared to any single junction concept.

Both material combinations can be either realized as two-terminal tandem solar cells, were the two subcells are electrically connected in series or four-terminal cells, where the two subcells are electrically independent of each other. In addition, also three-terminal configurations with e.g. interdigitated back contact (IBC) bottom cells become more relevant.

To get closer to the fundamental efficiency limit, more fundamental understanding of the working principle of tandem solar cells, and of the optimal device designs such as two-, three-, or four-terminal configuration is necessary. Therefore both optical and electrical simulations can help to further optimize the designs and choice of materials used for certain charge selective and contact layers.

As of 2020, the major limitations for the two-terminal devices are engineering problems, such as processing restrictions e.g. the maximum allowed temperature or solvents used, device polarity matching with p-i-n or n-i-p topcells, and surface texture/roughness of the utilized bottom cells. In addition, also band-gap optimized perovskite compositions and charge selective contact combinations to enable highest passivation and with that radiative efficiency and open-circuit voltages. Also, the fill factor of these tandems solar cells lacks behind theoretical predictions. Further optimization of the series resistance losses of the tandem contact design but also more understanding on the internal electrical field distribution might help to improve these parameters further.

In addition, also more practical issues must be considered. Even though scaling toward larger areas already is demonstrated, module integration with typical thin-film series connection of module sub-areas and larger area contact design is not yet developed far enough. Further, standard testing protocols for perovskite-based tandem solar cells need to be defined by the research community, especially accelerated testing protocols under elevated temperature and/or humidity (damp heat) and temperature cycling. In addition, also true outdoor measurements with day/night cycling need to be performed and reported for perovskite-based tandem solar cells for better comparison.

Looking a decade ahead, also triple- and potentially quadruple-junctions solar cells with perovskite sub-cells will become more important as they have even higher theoretical efficiency limits than tandems.

In conclusion, multi-junction solar cells with perovskite subcells are likely to play a crucial role in next-generation low cost solar modules with efficiencies exceeding 30%.

Acknowledgments

Preparing this chapter would not have been possible without the support of many. First of all, we would like to thank Hiroyuki Fujiwara for giving us the opportunity to contribute to this book with a chapter. Further, we would like to thank Amran Al-Ashouri, Eike Köhnen and Philipp Tockhorn for supporting us in the preparation of the manuscript. We would like to acknowledge Robert Wenisch, Peter Tillmann and Lars Korte for critically reviewing parts of the manuscript and Christian Kaufmann for providing us with information on CIGS solar cells.

References

1 Shockley, W. and Queisser, H.J. (1961). Detailed balance limit of efficiency of p–n junction solar cells. *J. Appl. Phys.* 32: 510–519.

2 A.S.T.M.S. G173-03 (2008). *Standard Tables for Reference Solar Spectral Irradiances: Direct Normal and Hemispherical on 37 Tilted Surface*. West Conshohocken, PA: ASTM International.

3 K. Jäger, M. Werth, L. Mazzarella, et al. (2017). Numerical optical optimization of perovskite-silicon tandem solar cells, in *33rd European Photovoltaic Solar Energy Conference and Exhibition*. Amsterdam, the Netherlands, 1057–1060 (25–29 September 2017).

4 Santbergen, R., Meguro, T., Suezaki, T. et al. (2017). GenPro4 optical model for solar cell simulation and its application to multijunction solar cells. *IEEE J. Photovolt.* 7: 919–926.

5 Rienäcker, M., Warren, E.L., Schnabel, M. et al. (2019). Back-contacted bottom cells with three terminals: maximizing power extraction from current-mismatched tandem cells. *Prog. Photovoltaics Res. Appl.* 27: 410–423.

6 Smets, A.H.M., Jäger, K., Isabella, O. et al. (2016). *Solar Energy: The Physics and Engineering of Photovoltaic Conversion Technologies and Systems*. UIT Cambridge.

7 M. Stocks, Y.X. Loo and N.N. Lal (2016). Free the Bandgap! Series-Parallel Connection of Tandem Cells, in *Proceedings of the 32nd European Photovoltaic Solar Energy Conference and Exhibition*. Munich, Germany, 223–227 (21–24 June 2016).

8 Tillmann, P., Jäger, K., and Becker, C. (2020). Minimising levelised cost of electricity of bifacial solar panel arrays using Bayesian optimisation. *Sustain. Energy Fuels* 4: 254–264.
9 (2019). *International Technology Roadmap for Photovoltaic (ITRPV) – 2018 Results*, 10e. Frankfurt am Main, Germany: VDMA.
10 Guillemoles, J.-F., Kirchartz, T., Cahen, D., and Rau, U. (2019). Guide for the perplexed to the Shockley–Queisser model for solar cells. *Nat. Photonics* 13: 501–505.
11 Vos, A.D. (1980). Detailed balance limit of the efficiency of tandem solar cells. *J. Phys. D: Appl. Phys.* 13: 839.
12 IEC:60904–3: *Photovoltaic devices – Part 3: Measurement principles for terrestrial photovoltaic (PV) solar devices with reference spectral irradiance data*, 2008.
13 Jia, J., Miao, Y., Kang, Y. et al. (2015). Bias-dependence of luminescent coupling efficiency in multijunction solar cells. *Opt. Express* 23: A219.
14 M. Mazouchi, J. Jia, Y. Huo, et al. 2015. Luminescent coupling effects in double-junction solar cells, in *2015 IEEE 42nd Photovoltaic Specialist Conference (PVSC)*. New Orleans, LA, USA (14–19 June 2015).
15 Liu, Z., Krückemeier, L., Krogmeier, B. et al. (2018). Open-circuit voltages exceeding 1.26 V in planar methylammonium lead iodide perovskite solar cells. *ACS Energy Lett.* 4: 110–117.
16 Ban, M., Zou, Y., Rivett, J.P.H. et al. (2018). Solution-processed perovskite light emitting diodes with efficiency exceeding 15% through additive-controlled nanostructure tailoring. *Nat. Commun.* 9: 9.
17 Zou, W., Li, R., Zhang, S. et al. (2018). Minimising efficiency roll-off in high-brightness perovskite light-emitting diodes. *Nat. Commun.* 9: 2.
18 Yang, X., Zhang, X., Deng, J. et al. (2018). Efficient green light-emitting diodes based on quasi-two-dimensional composition and phase engineered perovskite with surface passivation. *Nat. Commun.* 9: 2.
19 Xiao, Z., Kerner, R.A., Zhao, L. et al. (2017). Efficient perovskite light-emitting diodes featuring nanometre-sized crystallites. *Nat. Photonics* 11: 108–115.
20 Cho, H., Jeong, S.-H., Park, M.-H. et al. (2015). Overcoming the electroluminescence efficiency limitations of perovskite light-emitting diodes. *Science* 350: 1222–1225.
21 Walker, A.W., Höhn, O., Micha, D.N. et al. (2015). Impact of photon recycling and luminescence coupling on III–V single and dual junction photovoltaic devices. *J. Photonics Energy* 5: 053087.
22 Richter, A., Glunz, S.W., Werner, F. et al. (2012). Improved quantitative description of auger recombination in crystalline silicon. *Phys. Rev. B* 86 (16): 165202.
23 Martí, A. and Araújo, G.L. (1996). Limiting efficiencies for photovoltaic energy conversion in multigap systems. *Sol. Energy Mater. Sol. Cells* 43: 203–222.
24 De Wolf, S., Holovsky, J., Moon, S.-J. et al. (2014). Organometallic halide perovskites: sharp optical absorption edge and its relation to photovoltaic performance. *J. Phys. Chem. Lett.* 5: 1035–1039.

25 Swartwout, R., Hoerantner, M.T., and Bulović, V. (2019). Scalable deposition methods for large-area production of perovskite thin films. *Energy Environ. Mater.* 2: 119–145.

26 Albrecht, S. and Rech, B. (2017). Perovskite solar cells: on top of commercial photovoltaics. *Nat. Energy* 2: 16196.

27 Sofia, S.E., Wang, H., Bruno, A. et al. (2020). Roadmap for cost-effective, commercially-viable perovskite silicon tandems for the current and future PV market. *Sustain. Energy Fuels* 2: 852–862.

28 Unger, E.L., Kegelmann, L., Suchan, K. et al. (2017). Roadmap and roadblocks for the band gap tunability of metal halide perovskites. *J. Mater. Chem. A* 5: 11401–11409.

29 Gharibzadeh, S., Nejand, B.A., Jakoby, M. et al. (2019). Record open-circuit voltage wide-bandgap perovskite solar cells utilizing 2D/3D perovskite heterostructure. *Adv. Energy Mater.* 9: 1803699.

30 Rajagopal, A., Stoddard, R.J., Jo, S.B. et al. (2018). Overcoming the photovoltage plateau in large bandgap perovskite photovoltaics. *Nano Lett.* 18: 3985–3993.

31 Jošt, M., Kegelmann, L., Korte, L., and Albrecht, S. (2020). Monolithic perovskite tandem solar cells: a review of the present status and advanced characterization methods toward 30% efficiency. *Adv. Energy Mater.* 10: 1904102.

32 Zheng, X., Hou, Y., Bao, C. et al. (2020). Managing grains and interfaces via ligand anchoring enables 22.3%-efficiency inverted perovskite solar cells. *Nat. Energy* 5: 131–140.

33 Zhu, H., Liu, Y., Eickemeyer, F.T. et al. (2020). Tailored amphiphilic molecular mitigators for stable perovskite solar cells with 23.5% efficiency. *Adv. Mater.* 32: 1907757.

34 Hoke, E.T., Slotcavage, D.J., Dohner, E.R. et al. (2015). Reversible photo-induced trap formation in mixed-halide hybrid perovskites for photovoltaics. *Chem. Sci.* 6: 613–617.

35 Duong, T., Wu, Y., Shen, H. et al. (2017). Rubidium multication perovskite with optimized bandgap for perovskite-silicon tandem with over 26% efficiency. *Adv. Energy Mater.* 7: 1700228.

36 Bush, K.A., Frohna, K., Prasanna, R. et al. (2018). Compositional engineering for efficient wide band gap perovskites with improved stability to photoinduced phase segregation. *ACS Energy Lett.* 3: 428–435.

37 Löper, P., Moon, S.-J., de Nicolas, S.M. et al. (2015). Organic–inorganic halide perovskite/crystalline silicon four-terminal tandem solar cells. *Phys. Chem. Chem. Phys.* 17: 1619–1629.

38 Bailie, C.D., Christoforo, M.G., Mailoa, J.P. et al. (2015). Semi-transparent perovskite solar cells for tandems with silicon and CIGS. *Energy Environ. Sci.* 8: 956–963.

39 Mailoa, J.P., Bailie, C.D., Johlin, E.C. et al. (2015). A 2-terminal perovskite/silicon multijunction solar cell enabled by a silicon tunnel junction. *Appl. Phys. Lett.* 106: 121105.

40 Todorov, T., Gershon, T., Gunawan, O. et al. (2015). Monolithic perovskite-CIGS tandem solar cells via in situ band gap engineering. *Adv. Energy Mater.* 5: 1500799.

41 Bush, K.A., Palmstrom, A.F., Yu, Z.J. et al. (2017). 23.6%-efficient monolithic perovskite/silicon tandem solar cells with improved stability. *Nat. Energy* 2: 17009.

42 Sahli, F., Werner, J., Kamino, B.A. et al. (2018). Fully textured monolithic perovskite/silicon tandem solar cells with 25.2% power conversion efficiency. *Nat. Mater.* 17: 820–826.

43 Bush, K.A., Manzoor, S., Frohna, K. et al. (2018). Minimizing current and voltage losses to reach 25% efficient monolithic two-terminal perovskite–silicon tandem solar cells. *ACS Energy Lett.* 3: 2173–2180.

44 Chen, B., Yu, Z., Liu, K. et al. (2019). Grain engineering for perovskite/silicon monolithic tandem solar cells with efficiency of 25.4%. *Joule* 3: 177–190.

45 Jošt, M., Köhnen, E., Morales-Vilches, A.B. et al. (2018). Textured interfaces in monolithic perovskite/silicon tandem solar cells: advanced light management for improved efficiency and energy yield. *Energy Environ. Sci.* 11: 3511–3523.

46 Mazzarella, L., Lin, Y.-H., Kirner, S. et al. (2019). Infrared light management using a nanocrystalline silicon oxide interlayer in monolithic perovskite/silicon heterojunction tandem solar cells with efficiency above 25%. *Adv. Energy Mater.* 9: 1803241.

47 Köhnen, E., Jošt, M., Morales-Vilches, A.B. et al. (2019). Highly efficient monolithic perovskite silicon tandem solar cells: analyzing the influence of current mismatch on device performance. *Sustain. Energy Fuels* 3: 1995–2005.

48 P.V. Oxford (2018). Oxford PV sets world record for perovskite solar cell. https://www.oxfordpv.com/news/oxford-pv-sets-world-record-perovskite-solar-cell (accessed 28 April 2021).

49 Xu, J., Boyd, C.C., Yu, Z.J. et al. (2020). Triple-halide wide–band gap perovskites with suppressed phase segregation for efficient tandems. *Science* 367: 1097–1104.

50 Al-Ashouri, A., Köhnen, E., Li, B. et al. (2020). Monolithic perovskite/silicon tandem solar cell with >29% efficiency by enhanced hole extraction. *Science* 370: 1300–1309.

51 Kim, D., Jung, H.J., Park, I.J. et al. (2020). Efficient, stable silicon tandem cells enabled by anion-engineered wide-bandgap perovskites. *Science* 368: 155–160.

52 Nogay, G., Sahli, F., Werner, J. et al. (2019). 25.1%-efficient monolithic perovskite/silicon tandem solar cell based on a p-type monocrystalline textured silicon wafer and high-temperature passivating contacts. *ACS Energy Lett.* 4: 844–845.

53 Lamanna, E., Matteocci, F., Calabrò, E. et al. (2020). Mechanically stacked, two-terminal graphene-based perovskite/silicon tandem solar cell with efficiency over 26%. *Joule* 4: 865–881.

54 Chen, B., Yu, Z.J., Manzoor, S. et al. (2020). Blade-coated perovskites on textured silicon for 26%-efficient monolithic perovskite/silicon tandem solar cells. *Joule* 4: 850–864.

55 Hou, Y., Aydin, E., Bastiani, M.D. et al. (2020). Efficient tandem solar cells with solution-processed perovskite on textured crystalline silicon. *Science* 367: 1135–1140.

56 Wu, Y., Yan, D., Peng, J. et al. (2017). Monolithic perovskite/ silicon-homojunction tandem solar cell with over 22% efficiency. *Energy Environ. Sci.* 10: 2472–2479.

57 Zheng, J., Mehrvarz, H., Ma, F.-J. et al. (2018). 21.8% efficient monolithic perovskite/homo-junction-silicon tandem solar cell on 16 cm^2. *ACS Energy Lett.* 3: 2299–2300.

58 Shen, H., Omelchenko, S.T., Jacobs, D.A. et al. (2018). In situ recombination junction between p-Si and TiO_2 enables high-efficiency monolithic perovskite/Si tandem cells. *Sci. Adv.* 4: eaau9711.

59 Zheng, J., Mehrvarz, H., Liao, C. et al. (2019). Large-area 23%-efficient monolithic perovskite/homojunction-silicon tandem solar cell with enhanced UV stability using down-shifting material. *ACS Energy Lett.* 4: 2623–2631.

60 Quiroz, C.O.R., Shen, Y., Salvador, M. et al. (2018). Balancing electrical and optical losses for efficient 4-terminal Si–perovskite solar cells with solution processed percolation electrodes. *J. Mater. Chem. A* 6: 3583–3592.

61 Jaysankar, M., Raul, B.A.L., Bastos, J. et al. (2019). Minimizing voltage loss in wide-bandgap perovskites for tandem solar cells. *ACS Energy Lett.* 4: 259–264.

62 Duong, T., Pham, H., Kho, T.C. et al. (2020). High efficiency perovskite-silicon tandem solar cells: effect of surface coating versus bulk incorporation of 2D perovskite. *Adv. Energy Mater.*: 1903553.

63 Chen, B., Baek, S.-W., Hou, Y. et al. (2020). Enhanced optical path and electron diffusion length enable high-efficiency perovskite tandems. *Nat. Commun.* 11: 3.

64 Han, Q., Hsieh, Y.-T., Meng, L. et al. (2018). High-performance perovskite/Cu(In,Ga)Se_2 monolithic tandem solar cells. *Science* 361: 904–908.

65 Jošt, M., Bertram, T., Koushik, D. et al. (2019). 21.6%-efficient monolithic perovskite/Cu(In,Ga)Se_2 tandem solar cells with thin conformal hole transport layers for integration on rough bottom cell surfaces. *ACS Energy Lett.* 4: 583–590.

66 Al-Ashouri, A., Magomedov, A., Roß, M. et al. (2019). Conformal monolayer contacts with lossless interfaces for perovskite single junction and monolithic tandem solar cells. *Energy Environ. Sci.* 12: 3356–3369.

67 C. Kaufmann (2020. Highly efficient monolithic 2-terminal cigse/perovskite tandem solar cells, in *Virtual Chalcogenide PV Conference*. Berlin, Germany (25–28 May 2020).

68 Shen, H., Duong, T., Peng, J. et al. (2018). Mechanically-stacked perovskite/CIGS tandem solar cells with efficiency of 23.9% and reduced oxygen sensitivity. *Energy Environ. Sci.* 11: 394–406.

69 H. Degans, Perovskite/CIGS tandem cell with record efficiency of 24.6 percent paves the way for flexible solar cells and high-efficiency building-integrated PV. https://www.imec-int.com/en/articles/perovskite-cigs-tandem-cell-with-record-efficiency-of-24-6-percent (accessed 28 April 2021).

70 Kim, D.H., Muzzillo, C.P., Tong, J. et al. (2019). Bimolecular additives improve wide-band-gap perovskites for efficient tandem solar cells with CIGS. *Joule* 3: 1734–1745.

71 Sprafke, A.N. and Wehrspohn, R.B. (2013). Light trapping concepts for photon management in solar cells. *GREEN* 2: 177–187.

72 Lal, N.N., White, T.P., and Catchpole, K.R. (2014). Optics and light trapping for tandem solar cells on silicon. *IEEE J. Photovoltaics* 4: 1380–1386.

73 Chen, R., Tong, H., Zhu, H. et al. (2020). 23.83% efficient mono-PERC incorporating advanced hydrogenation. *Prog. Photovoltaics Res. Appl.* 28: 1239–1247.

74 Green, M.A., Hishikawa, Y., Dunlop, E.D. et al. (2019). Solar cell efficiency tables (version 53). *Prog. Photovoltaics Res. Appl.* 27: 3–12.

75 Haase, F., Hollemann, C., Schäfer, S. et al. (2018). Laser contact openings for local poly-Si-metal contacts enabling 26.1%-efficient POLO-IBC solar cells. *Sol. Energy Mater. Sol. Cells* 186: 184–193.

76 A. Richter, J. Benick, F. Feldmann, et al. (2019). Both sides contacted silicon solar cells: options for approaching 26% efficiency, in 36th European Photovoltaic Solar Energy Conference and Exhibition. Marseille, France, 90–95 (9–13 September 2019).

77 C. X. Xu (2019). Micro-crystalline silicon oxide front contact layers for silicon heterojunction solar cells, in 2nd International Workshop on Silicon Heterojunction Solar Cell Technology. Chengdu, China (11–12 November 2019).

78 Quiroz, C.O.R., Spyropoulos, G.D., Salvador, M. et al. (2019). Interface molecular engineering for laminated monolithic perovskite/silicon tandem solar cells with 80.4% fill factor. *Adv. Funct. Mater.* 29: 1901476.

79 Battaglia, C., Cuevas, A., and Wolf, S.D. (2016). High-efficiency crystalline silicon solar cells: status and perspectives. *Energy Environ. Sci.* 9: 1552–1576.

80 Hermle, M., Feldmann, F., Bivour, M. et al. (2020). Passivating contacts and tandem concepts: approaches for the highest silicon-based solar cell efficiencies. *Appl. Phys. Rev.* 7: 021305.

81 Werner, J., Weng, C.-H., Walter, A. et al. (2016). Efficient monolithic perovskite/silicon tandem solar cell with cell area >1 cm^2. *J. Phys. Chem. Lett.* 7: 161–166.

82 Tai, Q., Tang, K.-C., and Yan, F. (2019). Recent progress of inorganic perovskite solar cells. *Energy Environ. Sci.* 12: 2375–2405.

83 Torabi, N., Behjat, A., Zhou, Y. et al. (2019). Progress and challenges in perovskite photovoltaics from single- to multi-junction cells. *Mater. Today Energy* 12: 70–94.

84 Song, Z., Chen, C., Li, C. et al. (2019). Wide-bandgap, low-bandgap, and tandem perovskite solar cells. *Semicond. Sci. Technol.* 34: 093001.

85 Hu, Y., Song, L., Chen, Y., and Huang, W. (2019). Two-terminal perovskites tandem solar cells: recent advances and perspectives. *Solar RRL* 3: 1900080.

86 Shen, H., Walter, D., Wu, Y. et al. (2019). Monolithic perovskite/Si tandem solar cells: pathways to over 30% efficiency. *Adv. Energy Mater.* 10: 1902840.

87 Taguchi, M., Yano, A., Tohoda, S. et al. (2014). 24.7% record efficiency HIT solar cell on thin silicon wafer. *IEEE J. Photovoltaics* 4: 96–99.

88 Albrecht, S., Saliba, M., Correa Baena, J.P. et al. (2016). Monolithic perovskite/silicon-heterojunction tandem solar cells processed at low temperature. *Energy Environ. Sci.* 9 (1): 81–88.

89 Werner, J., Niesen, B., and Ballif, C. (2017). Perovskite/silicon tandem solar cells: marriage of convenience or true love story? – an overview. *Adv. Mater. Interfaces* 5: 1700731.

90 Jošt, M., Albrecht, S., Lipovšek, B. et al. (2016). Back- and front-side texturing for light-management in perovskite/silicon-heterojunction tandem solar cells. *Energy Procedia* 102: 43–48.

91 Santbergen, R., Mishima, R., Meguro, T. et al. (2016). Minimizing optical losses in monolithic perovskite/c-Si tandem solar cells with a flat top cell. *Opt. Express* 24: A1288–A1299.

92 Werner, J., Barraud, L., Walter, A. et al. (2016). Efficient near-infrared-transparent perovskite solar cells enabling direct comparison of 4-terminal and monolithic perovskite/silicon tandem cells. *ACS Energy Lett.* 1: 474–480.

93 Werner, J., Walter, A., Rucavado, E. et al. (2016). Zinc tin oxide as high-temperature stable recombination layer for mesoscopic perovskite/silicon monolithic tandem solar cells. *Appl. Phys. Lett.* 109: 233902.

94 Zheng, J., Lau, C.F.J., Mehrvarz, H. et al. (2018). Large area efficient interface layer free monolithic perovskite/homo-junction-silicon tandem solar cell with over 20% efficiency. *Energy Environ. Sci.* 11: 2432–2443.

95 Wang, Z., Lin, Q., Chmiel, F.P. et al. (2017). Efficient ambient-air-stable solar cells with 2D–3D heterostructured butylammonium-caesium-formamidinium lead halide perovskites. *Nat. Energy* 2: 8.

96 Lee, J.-W., Dai, Z., Han, T.-H. et al. (2018). 2D perovskite stabilized phase-pure formamidinium perovskite solar cells. *Nat. Commun.* 9: 8.

97 Kamino, B.A., Paviet-Salomon, B., Moon, S.-J. et al. (2019). Low-temperature screen-printed metallization for the scale-up of two-terminal perovskite–silicon tandems. *ACS Appl. Energy Mater.* 2: 3815–3821.

98 Li, J., Wang, H., Chin, X.Y. et al. (2020). Highly efficient thermally co-evaporated perovskite solar cells and mini-modules. *Joule* 4: 1035–1053.

99 B. A. Kamino, A. Paracchino, S.-J. Moon, et al. (2019). Scale-up technologies towards large area 2-terminal perovskite-silicon tandems, in *36th European Photovoltaic Solar Energy Conference and Exhibition*. Marseille, France (9–13 September 2019).

100 Chen, Y., Zhang, L., Zhang, Y. et al. (2018). Large-area perovskite solar cells – a review of recent progress and issues. *RSC Adv.* 8: 10489–10508.

101 Burger, B., Kiefer, K., Kost, C. et al. (2020). *Photovoltaics Report*. Freiburg, Germany: Fraunhofer Institute for Solar Energy Systems ISE.

102 Lang, F., Jošt, M., Frohna, K. et al. (2020). Proton radiation hardness of perovskite tandem photovoltaics. *Joule* 4: 1054–1069.

103 Solar Frontier Achieves World Record Thin-Film Solar Cell Efficiency of 23.35%. https://www.solar-frontier.com/eng/news/2019/0117_press.html (accessed 28 April 2021).

104 Eisenlohr, J., Lee, B.G., Benick, J. et al. (2015). Rear side sphere gratings for improved light trapping in crystalline silicon single junction and silicon-based tandem solar cells. *Sol. Energy Mater. Sol. Cells* 142: 60–65.

105 Meng, L., You, J., Guo, T.-F., and Yang, Y. (2015). Recent advances in the inverted planar structure of perovskite solar cells. *Acc. Chem. Res.* 49: 155–165.

106 Jošt, M., Al-Ashouri, A., Lipovšek, B. et al. (2020). Perovskite/CIGS tandem solar cells – can they catch up with perovskite/c-Si tandems? In: *47th IEEE Photovoltaic Specialists Conference (PVSC)*, Calgary, AB, Canada (15 June-21 Aug. 2020).

107 Jaysankar, M., Paetel, S., Ahlswede, E. et al. (2019). Toward scalable perovskite-based multijunction solar modules. *Prog. Photovoltaics Res. Appl.* 6 https://doi.org/10.1002/pip.3153.

108 Domanski, K., Roose, B., Matsui, T. et al. (2017). Migration of cations induces reversible performance losses over day/night cycling in perovskite solar cells. *Energy Environ. Sci.* 10: 604–613.

109 Futscher, M.H. and Ehrler, B. (2016). Efficiency limit of perovskite/Si tandem solar cells. *ACS Energy Lett.* 1: 863–868.

110 Schmager, R., Langenhorst, M., Lehr, J. et al. (2019). Methodology of energy yield modelling of perovskite-based multi-junction photovoltaics. *Opt. Express* 27: A507.

111 Calcabrini, A., Ziar, H., Isabella, O., and Zeman, M. (2019). A simplified skyline-based method for estimating the annual solar energy potential in urban environments. *Nat. Energy* 4: 206–215.

112 Tucher, N., Eisenlohr, J., Gebrewold, H. et al. (2016). Optical simulation of photovoltaic modules with multiple textured interfaces using the matrix-based formalism OPTOS. *Opt. Express* 24: A1083–A1093.

113 Chen, D., Manley, P., Tockhorn, P. et al. (2018). Nanophotonic light management for perovskite–silicon tandem solar cells. *J. Photonics Energy* 8: 022601.

114 Jin, J.-M. (2014). *The Finite Element Method in Electromagnetics*. Wiley.

115 Tiedje, T., Yablonovitch, E., Cody, G.D., and Brooks, B.G. (1984). Limiting efficiency of silicon solar cells. *IEEE Trans. Electron Devices* 31: 711–716.

116 Green Martin, A. (2002). Lambertian light trapping in textured solar cells and light-emitting diodes: analytical solutions. *Prog. Photovoltaics Res. Appl.* 10: 235–241.

117 Burgelman, M., Nollet, P., and Degrave, S. (2000). Modelling polycrystalline semiconductor solar cells. *Thin Solid Films* 361–362: 527–532.

118 Varache, R., Leendertz, C., Gueunier-Farret, M.E. et al. (2015). Investigation of selective junctions using a newly developed tunnel current model for solar cell applications. *Sol. Energy Mater. Sol. Cells* 141: 14–23.

119 Hörantner, M.T. and Snaith, H.J. (2017). Predicting and optimising the energy yield of perovskite-on-silicon tandem solar cells under real world conditions. *Energy Environ. Sci.* 10 (9): 1983–1993.

120 Jäger, K., Korte, L., Rech, B., and Albrecht, S. (2017). Numerical optical optimization of monolithic planar perovskite-silicon tandem solar cells with regular and inverted device architectures. *Opt. Express* 25: A473–A482.

121 Jošt, M., Lipovšek, B., Glažar, B. et al. (2020). Perovskite solar cells go outdoors: field testing and temperature effects on energy yield. *Adv. Energy Mater.* 10: 2000454.

122 Bouhlel, M.A., Hwang, J.T., Bartoli, N. et al. (2019). A python surrogate modeling framework with derivatives. *Adv. Eng. Software* 135: 102662.

123 Rozza, G., Huynh, D.B.P., and Patera, A.T. (2008). Reduced basis approximation and a posteriori error estimation for affinely parametrized elliptic coercive partial differential equations. *Arch. Comput. Meth. Eng.* 15: 229–275.

124 Jabbar, M.A. and Azeman, A.B. (2004). Fast optimization of electromagnetic-problems: the reduced-basis finite element approach. *IEEE Trans. Magn.* 40: 2161–2163.

125 Hammerschmidt, M. (2016). Optical simulation of complex nanostructured solar cells with a reduced basis method. PhD Thesis, Freie Universität Berlin.

126 Futscher, M.H. and Ehrler, B. (2017). Modeling the performance limitations and prospects of perovskite/Si tandem solar cells under realistic operating conditions. *ACS Energy Lett.* 2: 2089–2095.

18

All Perovskite Tandem Solar Cells

Zhaoning Song and Yanfa Yan

The University of Toledo, Wright Center for Photovoltaics Innovation and Commercialization, Department of Physics and Astronomy, 2801 W. Bancroft St., Toledo, OH 43606, USA

18.1 Introduction

In recent years, organic–inorganic metal halide perovskite solar cells (PSCs) have emerged as a promising photovoltaic (PV) technology, extending the frontiers of research and development in renewable energy generation [1]. The unprecedented rapid development of PSCs has attracted tremendous academic and industrial interest and extensive research efforts [2]. Within a decade, the power conversion efficiency (PCE) of single-junction solar cells has climbed from 3.8% [3] to a certified 25.2% [4]. As the PCE record of PSCs is quickly approaching the thermodynamical efficiency limits (~33%) of single-junction cells set by the Shockley–Queisser (S–Q) detailed balance theory [5], the strategies to develop the next-generation PSCs that have a potential to surpass such an efficiency ceiling have been proposed and studied. So far, the only PV technology that has demonstrated such potential is multijunction tandem solar cells [6]. Notably, the metal halide perovskite materials are an ideal candidate for fabricating tandem solar cells [7–10], due to their bandgap tunability in a wide range from 1.1 to ~3 eV [11], low-temperature (less than 150 °C) solution-based processing methods [12–14], and low estimated manufacturing costs of ~US$ 0.4 per peak watt [15, 16].

Tandem solar cells combining a wide-bandgap (1.6–1.9 eV) perovskite top cell and a low-bandgap (1.1–1.3 eV) bottom cell, such as mixed tin (Sn)-lead (Pb) perovskite, crystalline silicon (Si), or copper indium gallium selenide (CIGS), offer an extraordinary opportunity to achieve PCEs higher than *S–Q* limit for single-junction solar cells [7–10]. Among various perovskite tandem configurations, all-perovskite tandem solar cells (APTSCs) consisting of two perovskite subcells with complementary bandgaps possess unique advantages [10], including low-temperature processes for all subcells [17], compatibility with flexible and lightweight applications [18], low life-cycle environmental impacts and embodied energy [19–21], and potentially low fabrication costs [22, 23]. To date, APTSCs with two-terminal (2-T) and four-terminal (4-T) tandem configurations have demonstrated PCEs of up to 24.8% [24] and 25%

Hybrid Perovskite Solar Cells: Characteristics and Operation,
First Edition. Edited by Hiroyuki Fujiwara.

[25], close to the record PCE of single-junction PSCs. Such tandem combinations hold a promise to achieve theoretical maximum PCEs of more than 40% [7, 26].

One of the distinct advantages of metal halide perovskite materials that enable the fabrication of APTSCs is bandgap tunability via adjusting the stoichiometry of perovskites. A common metal halide perovskite can be described by a formula ABX_3, where A, B, and X represent a monovalent cation (e.g. methylammonium $(MA)^+$, formamidinium $(FA)^+$, Cs^+, and Rb^+), a divalent cation (e.g. Pb^{2+}, Sn^{2+}, Ge^{2+}), and a halogen anion (e.g. I^-, Br^-, and Cl^-), respectively. The crystal structure is constructed by a framework of corner-sharing metal halide (BX_3^-) octahedra with the voids filling by monovalent cations A^+, as depicted in Figure 18.1a. Methylammonium lead iodide ($MAPbI_3$) with a bandgap of ~1.55 eV is the prototypical and most widely used compound [27]. Substituting A, B, and X sites with different elements can tune the optical absorption band edge in the range of ~400 to 1100 nm, corresponding to a bandgap range of 3.1 to ~1.1 eV, as shown in Figure 18.1b. The wide-bandgap (>1.7 eV) perovskites are typically synthesized with halide substitution using smaller-sized anions. For example, substituting I^- with Br^- and Cl^- shifts the bandgap of $MAPbX_3$ from 1.6 ($MAPbI_3$) to 2.3 ($MAPbBr_3$) and 3.1 eV ($MAPbCl_3$), respectively [28, 29]. The low-bandgap (1.1–1.3 eV) perovskites are achieved by intermixing of Sn- and Pb-based iodide perovskites, which allows narrowing of bandgap to the near-infrared (NIR) range of ~1.1 eV with approximately 50% Sn substitution [30–32]. This wide range of bandgap tunability, covering from NIR to ultraviolet (UV), allows the combination of two complementary bandgaps of perovskites, which utilizes the solar spectrum more efficiently than a single bandgap absorber, enabling the applications of APTSCs.

In this chapter, we will review the theory and major advancements of APTSCs in recent years. We first discuss the working principles of tandem solar cells. After

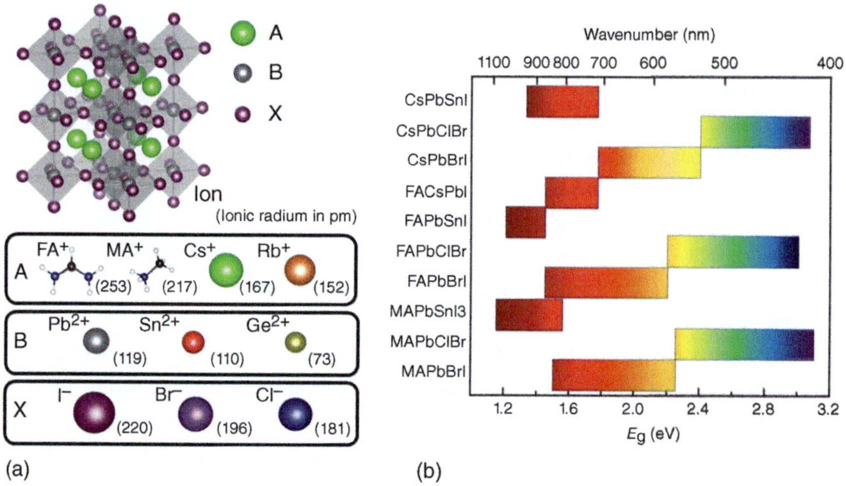

Figure 18.1 (a) Crystal structure of an ABX_3 perovskite and representative elements. (b) Band gap ranges for ABX_3 perovskites with different combinations of elements.

that, recent progress and issues in wide- and low-bandgap single-junction PSCs are summarized. We then provide a survey of current state-of-the-art APTSCs and discuss the limitations and challenges for APTSCs. Lastly, we conclude with an outlook for the future development of APTSCs.

18.2 Working Principles of Tandem Solar Cells

18.2.1 Why to Use Tandem Solar Cells

Figure 18.2a plots the maximum achievable PCE for a single-junction solar cell governed by the bandgap (E_g) of the absorber material, which is known as the *S–Q* limit [5]. Limited by the intrinsic thermodynamic energy loss, a bandgap value in the range of 1.1–1.4 eV is optimal for achieving the maximum PCE values of ~33%. Figure 18.2b illustrates a simple graphic analysis of intrinsic energy losses in a single-junction solar cell [33], including optical (no absorption) and thermal (thermalization) losses. In brief, photons with energies lower than the bandgap of a semiconductor absorber ($h\upsilon < E_g$) cannot be absorbed, whereas photons with energies higher than the absorber bandgap ($h\upsilon > E_g$) excite electron–hole pairs to higher energy states and then lose the excess energy above the bandgap by releasing heat (Figure 18.2c). As a result, an absorber material with a narrower bandgap is desired to harvest more photons but suffers more severely from the thermalization loss of high-energy photons. This trade-off between the optical and thermal losses determines the theoretical limit of a solar cell with an absorber layer.

Multijunction tandem solar cells have been proved to be an effective approach to break the thermodynamic efficiency limit of single-junction cells. The purpose of utilizing multiple absorber layers with different bandgaps is to mitigate intrinsic energy losses from the thermalization of photoexcited carriers [10]. For example, a double-junction tandem solar cell consisting of two absorber materials with complementary bandgaps can utilize the whole solar spectrum effectively (Figure 18.2d). A top cell with a wider bandgap is used to harvest high-energy photons, while a bottom cell with a lower bandgap is used to absorb low-energy photons that penetrate deeper into the bottom cell. This tandem combination ensures the maximum harvest of the full solar spectrum with lower intrinsic thermalization losses than a single-junction solar cell and therefore, holds a promise to achieve higher PCEs than the *S–Q* limit of single-junction cells. Similar to the single-junction device, there exist thermodynamic limits for the double-junction tandem solar cells [26], which are determined by the combination of bandgap values of the top and bottom cells and the constraints on subcell integration. Using multijunction (more than two junctions) tandem configurations can further raise the efficiency ceiling from ~46% for double junction to more than 60% for septuple and octuple junction (Figure 18.2e) [34]. So far, the highest PCE exceeding 50% has been demonstrated by sextuple-junction III–V tandem cells [35]. Despite the promising for ultrahigh PCE, it is challenging to fabricate multijunction tandem solar cell devices due to processing and structural complexity. Moreover, in practice, the potential efficiency gains typically are insufficient to compensate for the additional processing time and costs. Therefore, for the low-cost and

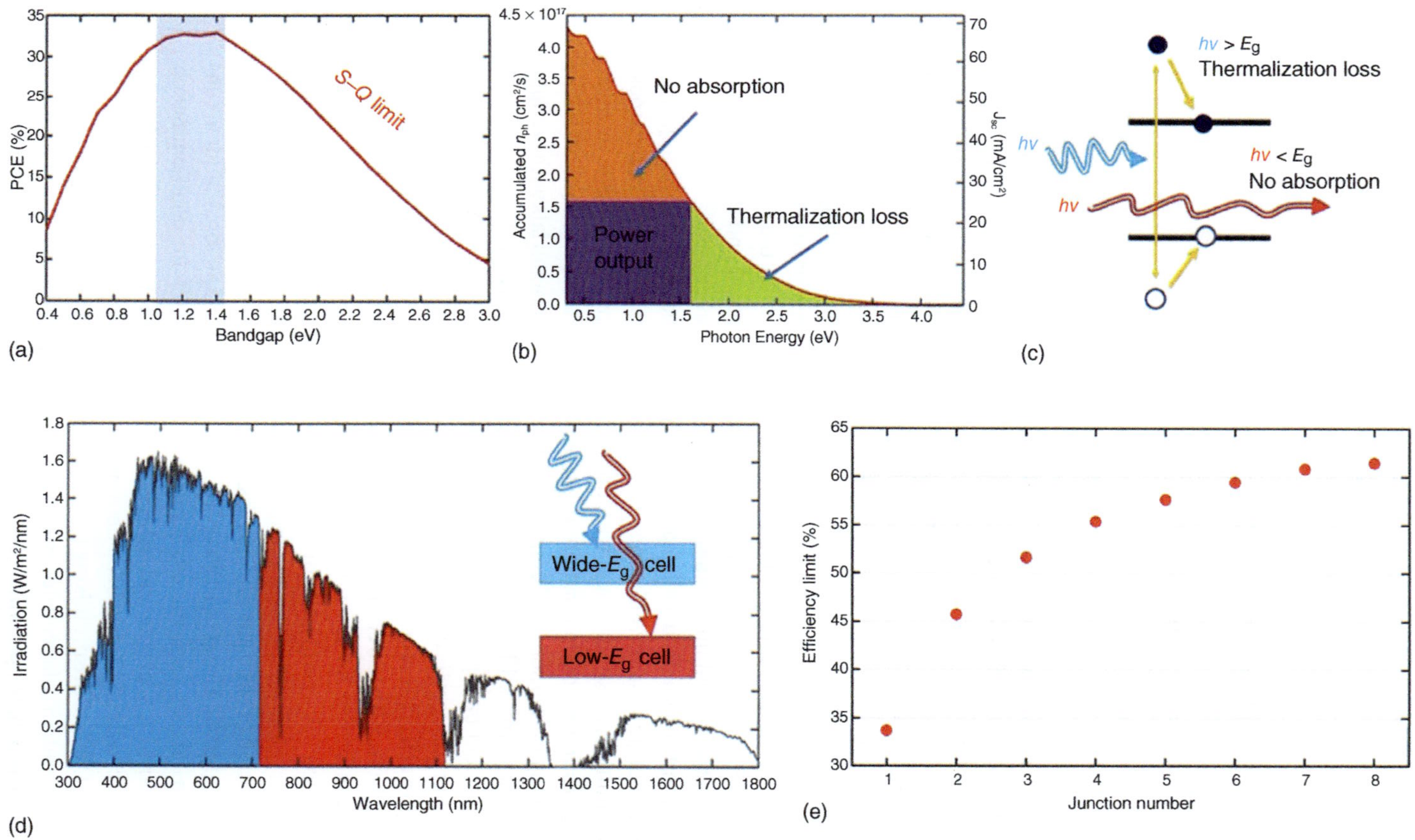

Figure 18.2 (a) Thermodynamic *S–Q* efficiency limit for single-junction solar cells. (b) Graphical analysis and (c) schematic illustration of primary sources of energy losses in a single-junction solar cell. (d) AM1.5 solar spectrum and spectrum utilization of a double-junction tandem solar cell. (e) Efficiency limit for multijunction tandem solar cells connected in series. *Source*: (c) and (d) Song et al. [10].

high-efficiency terrestrial PV applications, double-junction tandem configurations are commonly considered for perovskite-based tandem solar cells.

18.2.2 Tandem Device Architectures

Double-junction tandem solar cells are typically categorized based on the number of external terminals in a tandem device. A 4-T tandem device comprises two separate solar cells, each with two independent external electrodes. The two cells can be typically assembled into a tandem system by mechanically stacking (Figure 18.3a). The back electrode of the top cell is usually transparent, allowing long-wavelength photons to penetrate the top cell and enter the bottom one. Alternatively, an optical splitter can be used to divide the solar spectrum into two parts and distribute the light to individual solar cells (Figure 18.3b). For practical use, the mechanically stacked 4-T configuration is more widely adopted than the optical-splitting one due to its simplicity. Figure 18.3c shows a schematic structure of a 2-T tandem solar cell, consisting of two subcells fabricated on the same substrate and monolithically integrated. A 2-T tandem device has two external electrodes: a transparent front electrode and an opaque back electrode. The two subcells are connected internally by an interconnecting layer (ICL). Different tandem configurations have their pros and cons. For instance, 4-T tandem cells possess design and fabrication simplicity but require more materials (extra substrate and transparent electrodes) and more balance of system components (frames, wires, invertors, etc.) to support the operation of solar panels. In contrast, 2-T tandem cells are fabricated monolithically in a concise device architecture but have unique challenges in the design and device fabrication.

The subcells of a tandem solar cell can be regarded as individual single-junction solar cells. High-efficiency single-junction PSCs are fabricated in the so-called n-i-p and p-i-n device structures, as shown in Figure 18.4. The naming convention of PSCs represents the sequence of device preparation. Since most PSCs are in the superstrate configuration, i.e. light incident from the glass substrate, the fabrication sequence is

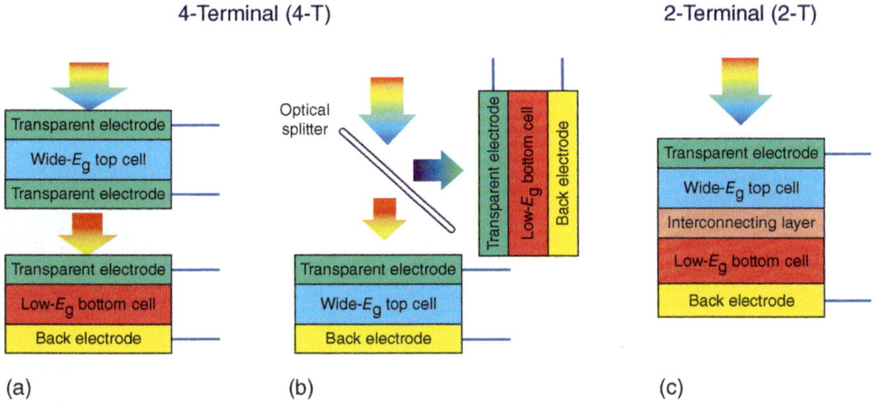

Figure 18.3 Schematic illustration of (a) 4-T mechanically stacked, (b) 4-T optically splitting, and (c) 2-T monolithically integrated tandem solar cells. *Source:* Song et al. [10].

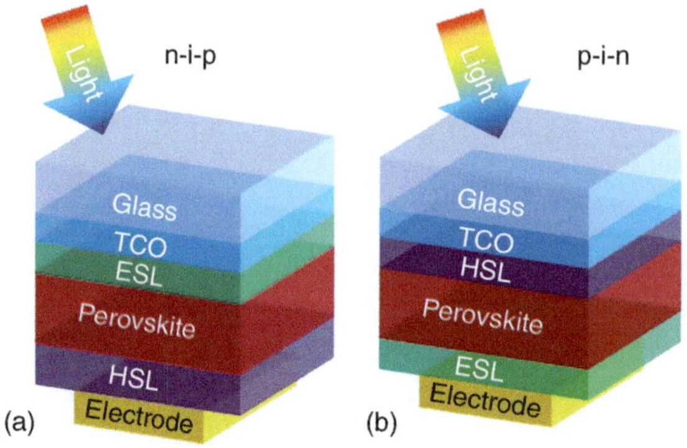

Figure 18.4 General device structures of perovskite solar cells with (a) n-i-p and (b) p-i-n device configurations. *Source:* Song et al. [10].

the same as the light propagation direction. For the n-i-p structure (Figure 18.4a), a film stack consisting of a transparent conductive oxide (TCO) front electrode, an n-type electron selective layer (ESL), a perovskite absorber layer, a p-type hole selective layer (HSL), and a back electrode is sequentially deposited on a glass substrate. For the p-i-n structure, the sequences of ESL and HSL are exchanged (Figure 18.4b). High-efficiency wide-bandgap PSCs are made in both configurations. Low-bandgap PSCs, however, are mostly made in the p-i-n configurations.

18.2.3 PCE of Tandem Solar Cells

The operation principles of 4-T and 2-T tandem solar cells are different. Figure 18.5 plots the J–V curves of 4-T and 2-T tandem solar cells and their corresponding subcells. For a 4-T tandem solar cell, the two subcells are optically coupled but not electronically connected. A 4-T tandem solar cell is not a single working device but two individual cells operating separately at the same time (Figure 18.5a). Typically, the total PCE of a 4-T tandem device can be calculated by summing up the PCEs of individual subcells, as to express in Eqs. (18.1)–(18.3):

$$\eta_{\text{tandem}} = \eta_{\text{top}} + \eta_{\text{bottom}} \tag{18.1}$$

$$\eta_{\text{top}} = V_{\text{OC, top}} \times J_{\text{SC, top}} \times \text{FF}_{\text{top}} \tag{18.2}$$

$$\eta_{\text{bottom}} = V_{\text{OC, bottom}} \times J_{\text{SC, bottom}} \times \text{FF}_{\text{bottom}} \tag{18.3}$$

where η_{tandem}, η_{top}, and η_{bottom} are PCEs for tandem, top, and bottom cells, respectively; $V_{\text{OC, top}}$, $V_{\text{OC, bottom}}$, $J_{\text{SC, top}}$, $J_{\text{SC, bottom}}$, FF_{top}, and $\text{FF}_{\text{bottom}}$ are open-circuit voltage (V_{OC}), short-circuit current density (J_{SC}), and fill factor (FF) of the top and bottom subcells, respectively. Because the total tandem PCE is just a mathematical sum of two subcells, 4-T tandem solar cells are also referred as unconstraint tandem solar cells.

For a 2-T tandem solar cell, the two subcells are electrically connected in series. Therefore, current and voltage behaviors of the system follow the Kirchhoff's law, i.e. the voltage of tandem is the sum of that of two subcells, whereas the current is the same for the whole device (Figure 18.5b). Note that a 2-T tandem solar cell generally

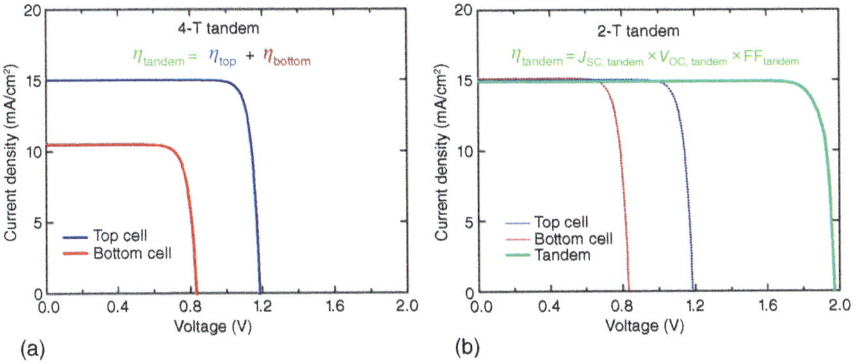

Figure 18.5 Simulated J–V curves of (a) two subcells in a 4-T tandem cell and (b) a 2-T tandem cell and its subcells.

outputs only one J–V curve that represents the performance of the whole device. The J–V curves of corresponding subcells are only used for device analysis and cannot be directly measured. The total PCE of a 2-T tandem is governed by Eqs. (18.4)–(18.6):

$$V_{tandem} = V_{top} + V_{bottom} \tag{18.4}$$

$$J_{tandem} = J_{top} = J_{bottom} \tag{18.5}$$

$$\eta_{tandem} = V_{OC,\,tandem} \times J_{SC,\,tandem} \times FF_{tandem} \tag{18.6}$$

where V_{tandem}, V_{top}, V_{bottom}, J_{tandem}, J_{top}, and J_{bottom}, are voltage and current density of tandem, top, and bottom cells, respectively. Because of the current match condition, 2-T tandem solar cells are also named as constraint tandem solar cells.

Figure 18.6 plots the theoretical limits of PCEs of 2-T and 4-T tandem solar cells with respect to the bandgap energies of top and bottom cells. For 4-T tandem cells, because the subcells operate individually, there is no constraint on the selection of bandgaps. As a result, the PCE limit of 4-T tandem cells is not very sensitive to the variation in bandgap (Figure 18.6a). PCEs of more than 40% are achievable for a wide range of bandgap combinations. For monolithic 2-T tandem cells, the

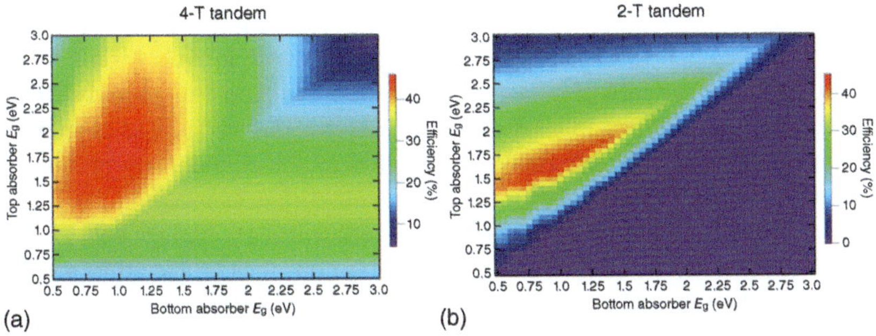

Figure 18.6 Theoretical efficiency limits of (a) 4-T and (b) 2-T tandem solar cells.

series connection requires both subcells to generate identical current. This current match condition presents constraints on the design of absorber bandgap and thickness, limiting the optimal bandgap combination to a narrower attainable range (Figure 18.6b). For APTSC applications with a bottom cell bandgap of ~1.2 eV, both 4-T and 2-T configurations have the potential to reach PCEs over 40%.

In real-world applications, the design of tandem solar cells also needs to consider the spectrum variation from the standard measurement conditions. Geographic factors, such as attitude, irradiance, temperature, and humidity, may influence the annual energy yield of tandem solar cells [36–39]. A recent study compared the energy yield at different locations and concluded that matching bandgap pairs exactly at the maximum PCE points are not essential for the fabrication of high-performance perovskite tandem solar cells [36]. This result shows some tolerances on the compositional control during the preparation of perovskite absorber layers for tandem solar cells. Therefore, a variety of bandgap combinations developed in the laboratory can be used for the fabrication of perovskite tandem solar cells.

18.3 Wide-Bandgap Perovskite Solar Cells

18.3.1 Wide-Bandgap Mixed I-Br Perovskites

APTSCs use wide-bandgap perovskite subcells to minimize the thermalization loss of high-energy photons. A high-efficiency wide-bandgap PSC is essential to maximize the voltage and performance of a tandem device. To achieve high-performance APTSCs, it is critical to judiciously select the bandgap of the top cell. According to the theoretical PCE calculation shown in Figure 18.6, a top cell bandgap in the range of 1.65–1.85 eV is desired to pair with low-bandgap mixed Sn-Pb perovskite (E_g = ~1.2 eV). This bandgap range is readily accessible for mixed halide perovskites by partially substituting Br for I in iodide-based perovskites (Figure 18.7). Increasing the Br proportion typically increases the bandgap of perovskite. Noh et al. [28] and Eperon et al. [40] investigated the bandgap tunability of single-monovalent-cation perovskites, $MAPb(I_{1-x}Br_x)_3$ and $FAPb(I_{1-x}Br_x)_3$, respectively. They demonstrated that the bandgap of mixed halide perovskites could be tuned in a wide range from 1.5 to 2.3 eV. Beal et al. investigated inorganic $CsPb(I_{1-x}Br_x)_3$ perovskites, showing a tunable bandgap range of 1.8–2.3 eV [41]. Although these mixed I-Br perovskites show continuously tunable bandgap range, for perovskites with a relatively high Br proportion (>30 mol%), it is challenging to maintain their bandgap energy and deliver the expected high V_{OC} values in corresponding devices. This V_{OC} loss is mainly attributed to photoinduced halide phase segregation and a high density of defects in the mixed halide perovskites [42].

The commonly used absorber compositions for efficient wide-bandgap PSCs can be described as $APb(Br_xI_{1-x})_3$, where A is a combination of MA, FA, and Cs. The bandgap is typically adjusted by tuning the halide ratio. Br molar ratios of 20–40% are typically used for wide-bandgap PSCs. The ternary combination space of monovalent cations (Cs, FA, MA) and the corresponding PCEs for wide-bandgap PSCs

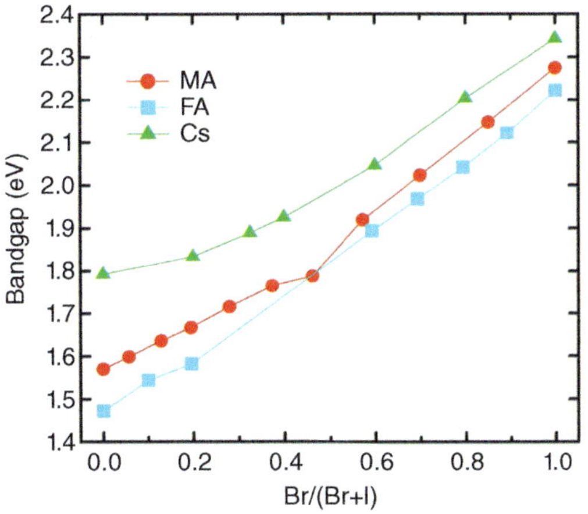

Figure 18.7 Bandgap variations of mixed I-Br perovskites. Bandgap values of MA, FA, Cs-based mixed I-Br perovskites are adapted from Refs. [28, 40, 41], respectively. Sources: Adapted from Noh et al. [28]; Eperon et al. [40]; and Beal et al. [41].

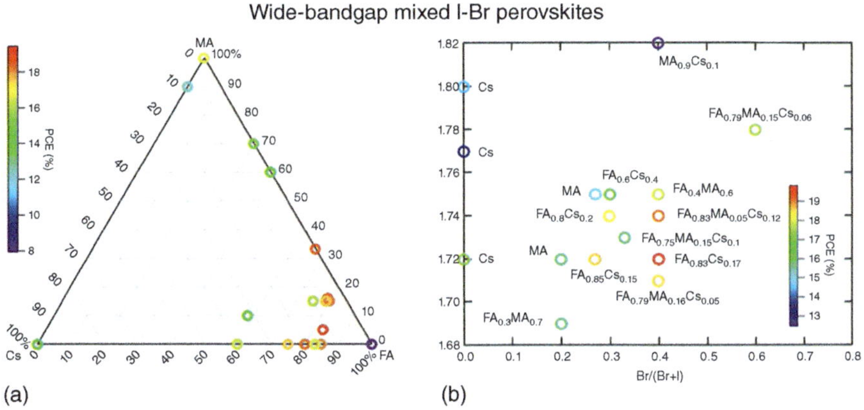

Figure 18.8 (a) Ternary monovalent cation composition of wide-bandgap mixed I-Br perovskites and the PCEs for corresponding PV devices. (b) Bandgap as a function of Sn proportion for commonly used wide-bandgap mixed I-Br perovskites and the PCEs for corresponding PV devices.

reported in the literature are summarized in Figure 18.8a. The bandgap values of the mostly used wide-bandgap perovskites as a function of the Br proportion are depicted in Figure 18.8b. It is worth noting that various bandgap values have been reported for mixed I-Br perovskites with the same Br proportions because the methods to determine the bandgap are different. For a straightforward comparison, it is proposed that a unique and easily applicable approach should be used as a general reference [43]. It can be seen in Figure 18.8 that the most commonly used absorbers for wide-bandgap PSCs are FA-based (FA > 75 mol%) perovskites with small proportions of Cs and/or MA (MA + Cs < 25%) to stabilize the perovskite phase and with Br proportions of 20–40%. These compositions were selected from a complicated compositional space for fabricating efficient and stable wide-bandgap PSCs.

It is worth noting that wide-bandgap PSCs typically exhibit high V_{OC}-deficits, defined as $E_g/e–V_{OC}$, where e is the unit charge. Hoke et al. found the reversible photoinduced trap formation in mixed halide hybrid perovskites [44]. They showed that the high Br content in $MAPb(I_{1-x}Br_x)_3$ films caused the intense subbandgap photoemission, which is detrimental to the device performance. The photoinstability of mixed halide perovskite is later known as the halide phase segregation [45, 46], which is the primary limitation that causes the V_{OC}-deficits and greatly limited the progress of wide-bandgap PSCs. Many research efforts have been paid to overcoming the phase segregation of the mixed halide perovskites. Among them, McMeekin et al. demonstrated a series of efficient wide-bandgap PSCs with FA/Cs-based mixed cation perovskites [47]. The FA/Cs-based perovskites exhibit better phase stability than the pure MA- or FA-based perovskites under heat and light. Due to the excellent phase stability and device performance, the $Cs_yFA_{1-y}Pb(I_{1-x}Br_x)_3$ perovskite is now one of the most popular choices for the wide-bandgap PSCs.

18.3.2 Current State of Wide-Bandgap Perovskite Solar Cells

High-efficiency wide-bandgap PSCs are fabricated in both n-i-p and p-i-n device configurations. The n-i-p device architecture typically consists of an indium tin oxide (ITO) or fluorine-doped tin oxide (FTO)-coated glass substrate, an SnO_2 or TiO_2 ESL, a wide-bandgap mixed I-Br perovskite absorber layer, a 2,2′,7,7′-Tetrakis[*N*,*N*-di(4-methoxyphenyl)amino]-9,9′-spirobifluorene (Spiro-OMeTAD) HSL, and a gold (Au) back electrode (Figure 18.9a). The p-i-n device architecture typically consists of an ITO-coated glass substrate, a poly[bis(4-phenyl)(2,4,6-trimethylphenyl)amine] (PTAA) HSL, a wide-bandgap mixed I-Br perovskite absorber layer, a fullerene-based ESL (e.g. C_{60} or [6,6]-Phenyl C_{61} butyric acid methyl ester [PCBM]), a bathocuproine (BCP) hole-blocking layer, and a silver (Ag) back electrode (Figure 18.9b). When wide-bandgap PSCs are used for the top cell in a tandem application, the back electrode is replaced by a TCO layer, e.g. ITO [48] or indium-doped zinc oxide (IZO) [49].

Currently, the state-of-the-art wide-bandgap PSCs with absorber bandgaps of more than 1.7 eV are mainly based on Cs/FA perovskites with proper bulk and

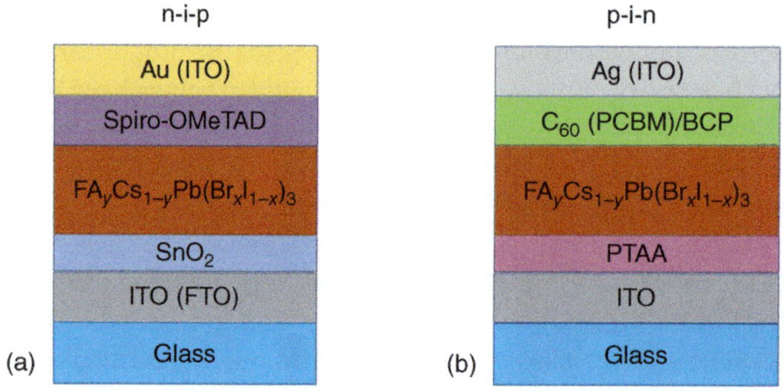

Figure 18.9 Most common material selection for wide-bandgap PSCs in (a) n-i-p and (b) p-i-n device configurations.

surface passivation. Tan et al. modified FA/Cs-based mixed cation perovskites by introducing dipolar cations (MA), which healed the deep traps in wide-bandgap perovskites [50]. They fabricated wide-bandgap PSCs in the n-i-p structure using $Cs_{0.12}MA_{0.05}FA_{0.83}Pb(I_{0.6}Br_{0.4})_3$ absorber layers with a bandgap of 1.74-eV, demonstrating a high PCE of 19.1% with a high V_{OC} of 1.25 V. Gharibzadeh et al. passivated the top surface of $Cs_{0.17}FA_{0.83}Pb(I_{0.6}Br_{0.4})_3$ perovskite (E_g = 1.72 eV) with butylammonium bromide (BABr) [51]. Their champion device with the n-i-p structure exhibited a PCE of 19.4%, with a remarkably high V_{OC} of 1.31 V.

The wide-bandgap PSCs in the p-i-n structure currently exhibit inferior performance than the n-i-p counterparts, likely due to interface recombination/band misalignment at the perovskite/ESL interface. Strategies have been developed to mitigate the charge extraction at the ESL interface. Lin et al. fabricated 1.71-eV p-i-n type wide-bandgap PSCs with $(FA_{0.83}MA_{0.17})_{0.95}Cs_{0.05}Pb(I_{0.6}Br_{0.4})_3$ absorber layers [52]. An isolating isomer of indene-C_{60} bisadduct (ICBA-trans3) ESL was used to tune energy level alignment with wide-bandgap perovskites, resulting in a PCE of 18.5% with a V_{OC} of 1.21 V. Chen et al. used guanidinium bromide (GABr) to from a 2D/3D graded homojunction interface on the mixed halide perovskite films. The best device demonstrated a PCE of 18.1% with a high V_{OC} of 1.24 V [53]. It is worth noting that current 2-T APTSC requires a p-i-n type device configuration due to the limitation of low-bandgap PSCs. Therefore, there is a great need to further improve the performance of p-i-n wide-bandgap PSCs. Palmstrom et al. incorporated dimethylammonium (DMA) into perovskite composition to form $FA_{0.6}Cs_{0.3}DMA_{0.1}PbI_{2.4}Br_{0.6}$ with a bandgap of 1.7 eV, demonstrating a PCE of 19.2% for the champion device [39].

18.3.3 Critical Issues of Wide-Bandgap Perovskite Cells

One of the most significant shortcomings and limitations of current wide-bandgap PSCs is the relatively large V_{OC}-deficits, consequently leading to inferior PCEs. Figure 18.10 shows the V_{OC} and PCE values of representative wide-bandgap PSCs reported in the literature. Despite the rapid progress in recent years, most wide-bandgap PSCs still exhibit V_{OC}-deficits in the order of 0.5 V with only one exception of ~0.4 V [51] (Figure 18.10a). Such V_{OC}-deficits are higher than their lower bandgap counterparts, which typically have V_{OC}-deficits as low as 0.3 V [54]. For the desired bandgap range (1.7–1.9 eV) for APTSCs, a higher V_{OC} can hardly be realized by increasing the perovskite bandgap due to the increasing V_{OC}-deficit. As a result, wide-bandgap PSCs with a higher bandgap typically exhibit a lower PCE, as shown in Figure 18.10b. The V_{OC} limitation presents a critical challenge for further improving the performance of APTSCs.

The large V_{OC}-deficit in wide-bandgap PSCs is mainly attributed to the intrinsic photoinduced phase segregation [55–58]. Under illumination, the compositionally uniform mixed I-Br perovskite alloy tends to segregate into a mixture of I-rich and Br-rich domains. The I-rich domains trap photoexcited charge carriers, limiting the quasi-Fermi energy splitting, and therefore, limiting the photovoltage generated in the perovskite absorber layer. Additionally, phase segregation and halide migration

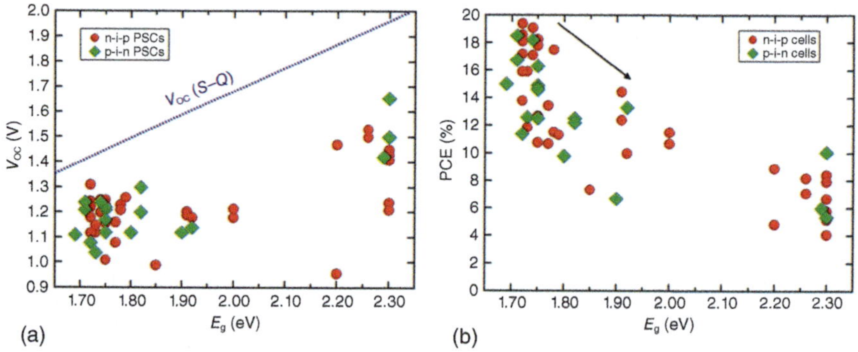

Figure 18.10 Reported (a) V_{OC} and (b) PCE of wide-bandgap PSCs as a function of absorber bandgap. *Source*: Song et al. [10].

under electric field also degrade the device performance of wide-bandgap PSCs [45, 59–63].

In recent years, a variety of strategies have been developed to mitigate the phase segregation in mixed I-Br wide-bandgap perovskites and hence improve the PCE. Some success has been achieved, including the modification of perovskite stoichiometry (e.g. high Cs^+ proportion [64] and Sn^{2+} incorporation [65]), grain boundary passivation using potassium (K^+) [66], perovskite surface passivation [67], improvement in crystallinity and grain sizes of perovskite films [68], synthesis of size-tuned nanocrystals [69], and formation of 2D/3D perovskites [70–72]. Among these strategies, the 2D/3D perovskite bulk and surface passivation are the most effective one. For instance, Xiao et al. incorporated *n*-butylammonium iodide (BAI) into mixed halide perovskites to form a self-assembled capping layer to stabilize bandgaps of perovskites [71]. Stoddard et al. showed enhanced defect tolerance and phase stability of wide-bandgap perovskite films via guanidinium (GA) alloying [70]. Rajagopal et al. incorporated small amounts of phenylethylammonium (PEA) cations in $MAPb(I_{0.6}Br_{0.4})$ perovskites, improving the V_{OC} of wide-bandgap PSCs to 1.35 V [72].

Despite the aforementioned approaches to enhance the phase stability and performance of wide-bandgap mixed I-Br perovskite films and devices, there still needs a way to address the phase instability issue. To achieve this goal, the entire ABX_3 stoichiometric space, including all the possible 2D/3D combinations, needs to be systematically explored. This may need to rely on machine learning [73] and automatic high-throughput robots [74] to synthesize and evaluate the phase stability of newly developed wide-bandgap perovskite films.

18.4 Low-Bandgap Perovskite Solar Cells

18.4.1 Low-Bandgap Mixed Sn-Pb Perovskites

High-performance APTSCs require high-efficiency and stable low-bandgap perovskites. Low-bandgap perovskites with a bandgap of ~1.1 to 1.3 eV are typically synthesized by partially substituting Pb by Sn in the Pb-iodide perovskites. Mixed Sn-Pb perovskites possess some unique optoelectronic properties, such as the bandgap bowing effect [75] and high intrinsic carrier concentrations [30]. Additionally, low-bandgap mixed Sn-Pb perovskites suffer from the instability issue due to the ease of oxidation of Sn^{2+} to Sn^{4+}, resulting in much-shortened device durability and poor operational stability compared with their wider bandgap (>1.4 eV) Pb-based counterparts.

A variety of compositions have been developed for the mixed Sn-Pb perovskites. The commonly used cation combinations (FA, MA, and Cs) and the best reported PCE of the corresponding PSCs are shown in Figure 18.11. The best-performing compositions are based on FA/MA mixed stoichiometry with MA proportion of 30–50% and Sn proportion of 50–60%. This compositional space was rationally selected as a result of extensive research efforts over the past few years.

In 2014, Hayase and coworkers found that $MASn_xPb_{1-x}I_3$ perovskites synthesized by mixing $MASnI_3$ and $MAPbI_3$ exhibited bandgap values as low as 1.17 eV, lower than that of both $MASnI_3$ (~1.35 eV) and $MAPbI_3$ (1.55 eV) [76]. Kanatzidis group then systematically investigated the structural and optical properties of $MASn_xPb_{1-x}I_3$ perovskites, revealing an anomalous bandgap bowing effect in the mixed Sn-Pb perovskites (Figure 18.12) [31, 75]. The origins of the bandgap bowing have been attributed to the spin–orbit coupling (SOC) and the lattice distortion [75] and mismatched band edges formed by Pb-*p* and I-*p*/Sn-*s* [80, 81]. The bandgap bowing effect has also been shown in mixed Sn-Pb perovskites with other monovalent cations (FA and Cs) (Figure 18.12) [77, 78].

Among various mixed Sn-Pb perovskites, mixed organic cation (MA/FA) low-bandgap perovskites exhibit the best device performance. In 2016, Yang et al. explored mixed cation $MA_{1-x}FA_xPb_{0.75}Sn_{0.25}I_3$ perovskites for low-bandgap PSC

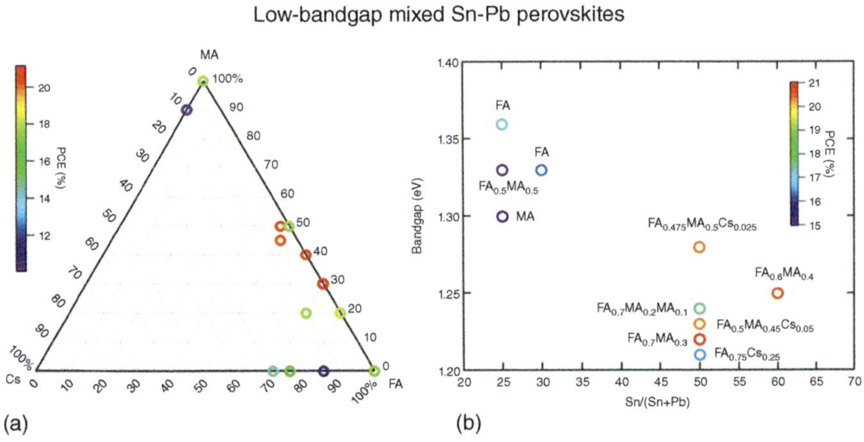

Figure 18.11 (a) Ternary monovalent cation composition of low-bandgap mixed Sn-Pb perovskites and the PCEs for corresponding PV devices. (b) Bandgap as a function of Sn proportion for commonly used low-bandgap mixed Sn-Pb perovskites and the PCEs for corresponding PV devices.

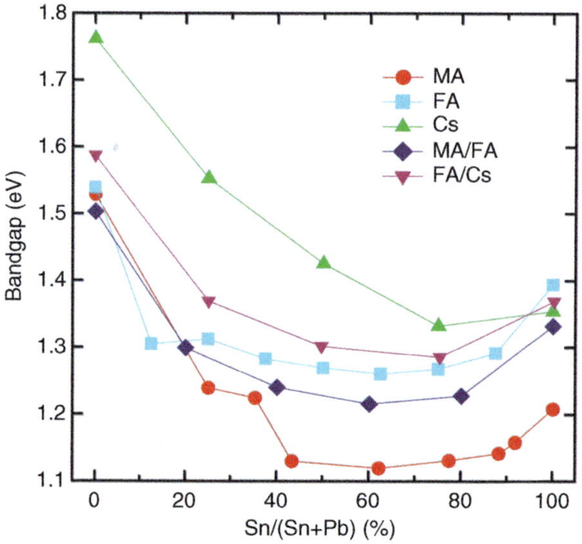

Figure 18.12 Bandgap bowing effect of low-bandgap mixed Sn-Pb perovskites with different monovalent cations, including the MA [75], FA [77], Cs [78], MA/FA [79], and FA/Cs [78]. Sources: Im et al. [75]; Eperon et al. [77]; Rajagopal et al. [78]; Liao et al. [79].

applications. Their champion devices based on 1.33-eV $MA_{0.5}FA_{0.5}Pb_{0.75}Sn_{0.25}I_3$ PSCs exhibited a PCE of 14.2% [82]. Yan and coworkers developed a single-step spin-coating method to synthesize $(FASnI_3)_{1-x}(MAPbI_3)_x$ perovskites by combining $FASnI_3$ and $MAPbI_3$ precursor solutions (Figure 18.13a) [79]. The best-performing devices (PCE = 15%) were fabricated using $(FASnI_3)_{0.6}(MAPbI_3)_{0.4}$ perovskite with a bandgap of 1.25 eV. By increasing the concentration of the mixed precursor solution, perovskite grain sizes were increased to 1 μm, and the thickness of the mixed Sn-Pb absorber was increased to more than 600 nm (Figure 18.13b) [83]. The thicker perovskite films exhibited longer carrier lifetimes of up to ~250 ns (Figure 18.13c), allowing more efficient collection of photoexcited minority charge carriers. The PSCs with thicker absorber layers showed higher absorption of NIR photons, resulting in higher external quantum efficiencies (EQEs) in the long wavelength region (750–1000 nm) (Figure 18.13d). The best-performing device exhibited a certified PCE of ~17% (Figure 18.13e). Yan and coworkers further demonstrated that bulk and grain boundary passivation of $(FASnI_3)_{0.6}(MAPbI_3)_{0.4}$ perovskite via halide (Cl and Br) incorporation led to improved electronic properties and reduced recombination in the mixed Sn-Pb PSCs, enabling high PCEs exceeding 19% [84, 85]. The desirable optoelectronic characteristics of $(FASnI_3)_{0.6}(MAPbI_3)_{0.4}$ were later confirmed by many groups. Using this perovskite composition, Prasanna et al. [86], Xu et al. [87], and Jiang et al. [88] independently demonstrated low-bandgap PSCs with PCEs of 17.8, 18, and 20.7%, respectively.

Although FA/MA-based mixed Sn-Pb perovskites show decent PV performance, their photostability is greatly limited due to containing MA. MA cations are known unstable under illumination [89, 90]. As a thermally stable alternative, low-bandgap mixed Sn-Pb perovskites with FA/Cs-based compositions, i.e. $FA_{1-y}Cs_ySn_xPb_{1-x}I_3$, have been developed. Compared with FA/MA-based mixed Sn-Pb perovskites, FA/Cs-based mixed Sn-Pb perovskites show much better

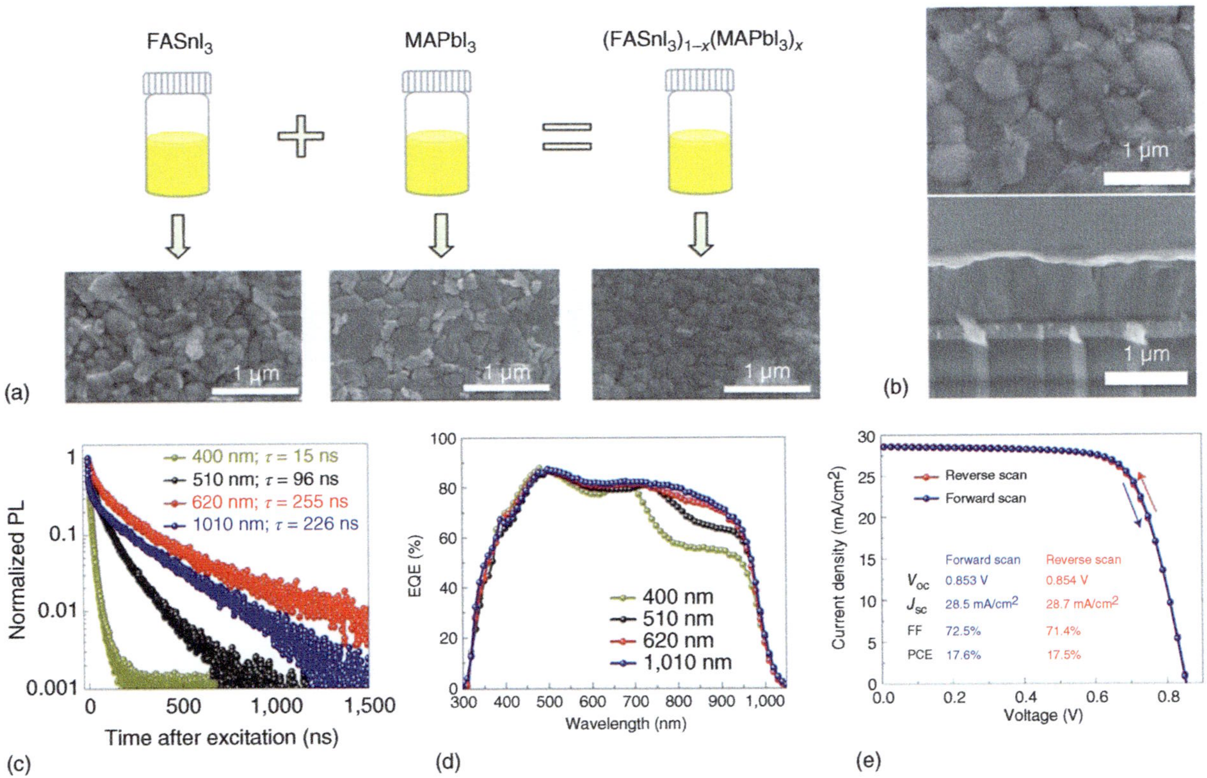

Figure 18.13 (a) Schematic illustration and top-view SEM images of the synthesis of $(FASnI_3)_{1-x}(MAPbI_3)_x$ perovskites by mixing $FASnI_3$ and $MAPbI_3$ precursor solutions. *Source*: Liao et al. [79]. (b) Top-view and cross-sectional SEM images of a 650 nm perovskite film. (c) TRPL lifetimes of perovskite films with different thicknesses and (d) EQE curves of corresponding devices. (e) The $J-V$ curves of the best-performing device. *Source*: (b)–(e) Zhao et al. [83].

thermal- and photostability, resulting in more stable operation under maximum power point (MPP) tracking. In 2016, Eperon et al. fabricated 1.2-eV low-bandgap PSCs based on $FA_{0.75}Cs_{0.25}Sn_{0.5}Pb_{0.5}I_3$ [77]. The best-performing device exhibited a PCE of 14.8%, much higher than the PCE of 10.9% for the control device based on $FASn_{0.5}Pb_{0.5}I_3$. Recently, Leijtens et al. raised the PCE to 15.6% after incorporating MACl and formic acid into the precursor solution [91]. Zong et al. systematically investigated the compositional space of $(FAPbI_3)_{1-x}(CsSnI_3)_x$ [92]. The $(FAPbI_3)_{0.7}(CsSnI_3)_{0.3}$ perovskite with an ideal bandgap of ~1.3 eV for a single-junction PSCs was used as the absorber layer of low-bandgap PSCs, achieving a PCE of up to 14.6%. They recently incorporated Lewis-adduct additive SnF_2–3FACl to the precursor solution, improving the PCE of $(FAPbI_3)_{0.7}(CsSnI_3)_{0.3}$ PSCs to 15.8% [93].

18.4.2 Current State of Low-Bandgap Perovskite Solar Cells

State-of-the-art low-bandgap PSCs are mainly fabricated in the p-i-n device configuration, consisting of an ITO-coated glass substrate, a poly(3,4-ethylenedioxythiophene) polystyrene sulfonate (PEDOT:PSS) HSL, a low-bandgap mixed Sn-Pb perovskite absorber layer, a C_{60}/BCP ESL, and an Ag back electrode (Figure 18.14a). The n-i-p configuration similar to Pb-based PSCs (Figure 18.14b) was explored in the early development of low-bandgap mixed Sn-Pb PSCs. However, the n-i-p device architecture has rarely been used since 2014 [94]. High-efficiency low-bandgap PSCs with high Sn proportion (>50%) in the n-i-p structure are yet to be demonstrated. The challenges of the lack of effective charge selective layers (both ESL and HSL) and intrinsic oxygen sensitivity of mixed Sn-Pb perovskites currently limit the development of low-bandgap n-i-p type PSCs.

The best-performing low-bandgap PSCs are all based on the FA/MA perovskite synthesized by the single-step spin-coating method developed by Yan and coworkers [79, 83]. Several groups demonstrated low-bandgap PSCs with PCEs exceeding 20%. Most recently, Yang et al. incorporated Cd^{2+} additives into $FA_{0.5}MA_{0.45}Cs_{0.05}Pb_{0.5}Sn_{0.5}I_3$ to enhance the electron diffusion length, achieving

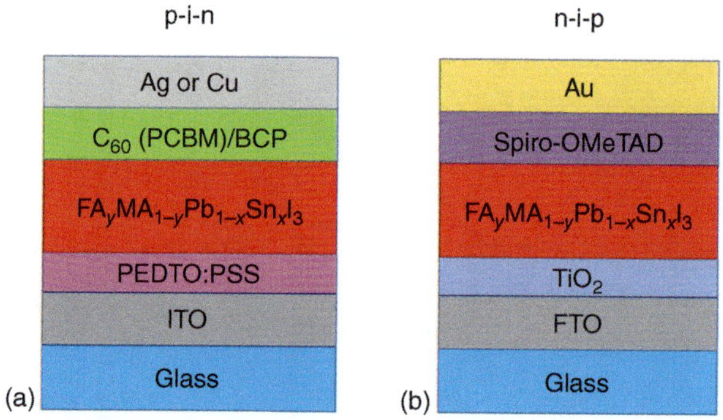

Figure 18.14 Most common material selection for low-bandgap PSCs in (a) p-i-n and (b) n-i-p device configurations.

a PCE of 20.3% for the best-performing device [95]. Tong et al. used guanidinium thiocyanate (GASCN) to passivate the grain boundary of $(FASnI_3)_{0.6}(MAPbI_3)_{0.4}$ perovskite [25]. A long carrier lifetime of more than 1 μs was achieved, which enabled low-bandgap PSCs with a PCE of 20.4%. Lin et al. introduced elemental Sn into the precursor solution to suppress Sn^{2+} oxidation in the precursor ink [24]. They fabricated low-bandgap PSCs using $FA_{0.7}MA_{0.3}Pb_{0.5}Sn_{0.5}I_3$ absorber layer, demonstrating a remarkably high PCE of 21.1%.

18.4.3 Critical Issues of Low-Bandgap Perovskite Cells

Despite remarkable progress in recent years, there remain some critical issues of low-bandgap PSCs and APTSCs. The performance of low-bandgap mixed Sn-Pb PSCs is inferior to their pure Pb perovskite counterparts, due to higher non-radiative recombination losses in bulk and at the interfaces of the mixed Sn-Pb perovskite films. Particularly, due to relatively low absorption coefficients, the thickness of mixed Sn-Pb perovskite layer needs to be on the order of 1 μm. However, the relatively higher carrier and defect densities in mixed Sn-Pb perovskite films than pure-Pb perovskites make the minority carrier lifetime and diffusion length shorter than that in their pure-Pb counterparts.

We surveyed the low-bandgap mixed Sn-Pb PSCs reported in the literature and plot their J_{SC} and PCE values vs. the bandgap value in Figure 18.15. Photocurrent in low-bandgap PSCs still shows a large gap from the expected J_{SC} values from the S–Q limit. The highest reported J_{SC} value of the low-bandgap PSCs that is confirmed by the EQE measurements is 31.4 mA/cm^2 [25], realizing 78.5% of its theoretical potential. In comparison, PSCs with wider bandgap values show high J_{SC} values exceeding 92% of the corresponding S–Q limits [96, 97]. Unlike the obvious J_{SC} loss in the mixed Sn-Pb PSCs, the highest reported V_{OC} value of 0.888 V for a 1.27-eV perovskite accounts for almost 90% of the S–Q limit [85]. The EQE analysis shows that the photon-to-electron conversation efficiency is less than 85% for a most of the spectral range (Figure 18.16). Moreover, most mixed Sn-Pb PSCs show a significant photocurrent collection drops at long wavelengths due to insufficient light

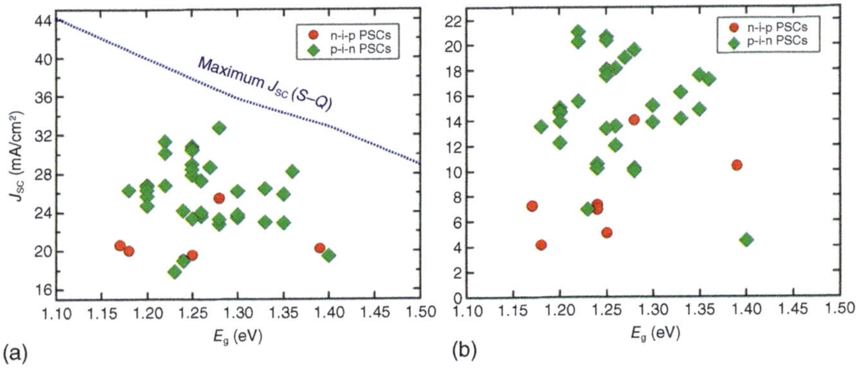

Figure 18.15 Reported (a) J_{SC} and (b) PCE of low-bandgap PSCs as a function of absorber bandgap. *Source:* Song et al. [10].

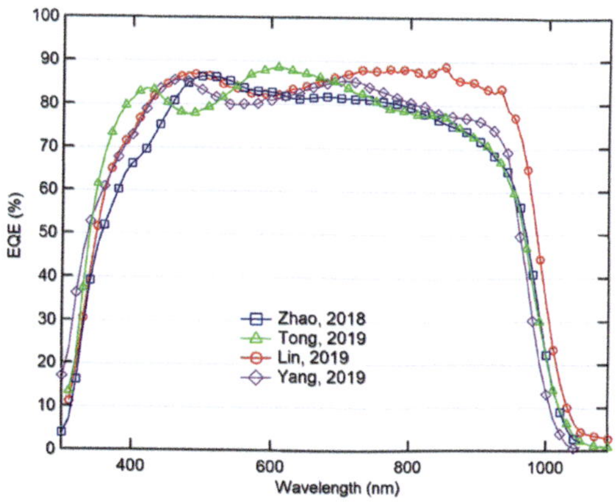

Figure 18.16 EQE spectra of state-of-the-art low-bandgap PSCs. *Source*: Data from Lin et al. [24]; Tong et al. [25]; Zhao et al. [84]; and Yang et al. [95].

absorption. Therefore, achieving an adequate conversion of NIR photons requires the use of thick low-bandgap mixed Sn-Pb perovskite absorber layers. However, when the thickness is increased to certain values, other PV parameters such as V_{OC} and FF start to decrease due to the presence of a significant number of defects in the mixed Sn-Pb perovskites, which lead to short carrier lifetimes and diffusion lengths [10]. Therefore, it is critical to passivate bulk and grain boundaries defects and improve the lifetimes and diffusion lengths in the mixed Sn-Pb perovskites are needed.

Another serious issue for low-bandgap mixed Sn-Pb PSCs is the lack of material and device stability, especially under the working conditions. The material instability is attributed to the ease of oxidation of Sn^{2+} to Sn^{4+}, which rapidly deteriorates the optoelectronic properties of mixed Sn-Pb perovskite films, leading to a high density of holes, a short minority carrier lifetime, and a short minority carrier diffusion length in the degraded mixed Sn-Pb films [94]. Additionally, high-efficiency mixed Sn-Pb PSCs typically contain a high proportion of MA (>30 mol%), which caused significant photoinstability [89, 98]. The PSCs based on MA-contained mixed Sn-Pb perovskites show much more severe photoinduced decay than FA/Cs-based devices under MPP measurements [91]. The ease of material degradation also limits the fabrication route and the design of low-bandgap single-junction PSCs and APTSCs. Currently, most high-efficiency low-bandgap PSCs have been fabricated in the p-i-n device architectures (Figure 18.14a) with PEDOT:PSS and C_{60}/BCP as the HSL and ESL, respectively. For the n-i-p device structure (Figure 18.14b), the commonly used metal oxide ESLs (e.g. SnO_2 and TiO_2) and the doped HSLs (e.g. spiro-OMeTAD) are likely to induce the degradation of the mixed Sn-Pb perovskites, limiting the device performance.

In recent years, many efforts have been made to mitigate the stability of low-bandgap mixed Sn-Pb PSCs. The most successful strategies include the use of more stable HSL/ESL and encapsulation. Chi et al. demonstrated better thermal- and photostability of mixed Sn-Pb PSCs based on $FAPb_{0.75}Sn_{0.25}I_3$ absorber after replacing PEDOT:PSS HSL by NiO_x [99]. Leijtens et al. demonstrated that mixed Sn-Pb PSCs with an ITO capping electrode survived after a thermal stability test at 85 °C in the air for more than 300 hours [91]. More recently, Prasanna et al. eliminated the PEDOT:PSS HSL and demonstrated remarkable device stability for mixed Sn-Pb PSCs [100]. Remarkably, the mixed Sn-Pb PSCs retained 95% of their initial efficiencies after 1000 hours at 85 °C in air with no encapsulation and in a damp heat test (85 °C with 85% relative humidity) with encapsulation. Additionally, the mixed Sn-Pb PSCs showed negligible decay after MPP tracking under 1 sun illumination for over 1000 hours.

Another effective approach to mitigate the instability issue is the incorporation of antioxidant agents. Jen and coworkers used ascorbic acid as an antioxidant agent in the mixed Sn/Pb PSCs, demonstrating enhanced device stability compared with the control devices processed with regular reducing additive SnF_2 [101]. Additionally, they developed another approach of forming stable mixed Sn-Pb perovskite by using a galvanic displacement reaction [102]. By incorporating elemental Sn to the Pb-based perovskite precursor solutions, they demonstrated a PCE of 18.21% for $(FAPb_{0.6}Sn_{0.4}I_3)_{0.85}(MAPb_{0.6}Sn_{0.4}Br_3)_{0.15}$-based PSCs. The encapsulated cells retained more than 90% of their initial efficiencies after the storage in ambient air at room temperature or 80 °C for 1000 or 120 hours, respectively. Lin et al. incorporated metallic Sn powders in the conventional mixed Sn-Pb perovskite precursor solution to suppress Sn^{2+} oxidation and demonstrated better device performance and stability [24]. This method was later confirmed by Jiang et al. in their study [88].

18.5 All-Perovskite Tandem Solar Cells

APTSCs possess unique advantages, including low-temperature processes for all subcells, compatibility with flexible and lightweight applications, and potentially low fabrication and environmental costs [19, 20]. Though potentially promising, it is challenging to fabricate efficient APTSCs due to the lack of highly efficient wide- and low-bandgap perovskite subcells and difficulty of monolithically integrating two perovskite subcells in the 2-T tandem configuration. The recent progress has propelled the PCEs of 4-T and 2-T APTSCs to 25% [25] and 24.8% [24], respectively (Figure 18.17), approaching the record of Pb-based single-junction PSCs. Clearly, the fabrication of high-efficiency and stable high- and low-bandgap PSCs is critical for realizing the full potential of APTSCs.

18.5.1 4-T All-Perovskite Tandem Solar Cells

The potential of using perovskite materials with complementary bandgaps for tandem solar cell applications has been first explored in the 4-T tandem configuration. In 2016, Li et al. demonstrated the proof-of-concept 4-T APTSCs by pairing 2.3-eV $MAPbBr_3$ and 1.6-eV $MAPbI_3$ as the top and bottom cells [103], achieving a PCE of ~9.5%. Yang et al. demonstrated the first 4-T APTSC with a low-bandgap subcell by combining a 1.55-eV $MAPbI_3$ top cell and a 1.33-eV $MA_{0.5}FA_{0.5}Pb_{0.75}Sn_{0.25}I_3$ bottom cell, achieving a high PCE of 19.1% [82]. A thin

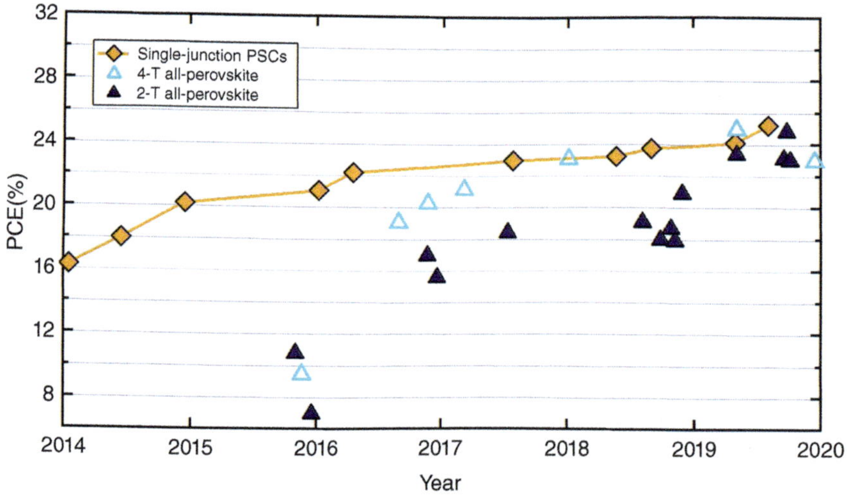

Figure 18.17 Evolution of PCEs for single-junction PSCs and APTSCs. *Source*: Song et al. [10].

absorber layer thickness of the $MAPbI_3$ was used to allow sufficient photons to enter the bottom cell. As a result, PCEs of 13.5% and 5.6% (Figure 18.18a), with J_{SC} values of 16.7 and 9.1 mA/cm^2, were obtained for the top and filtered bottom cells, respectively. Later, Eperon et al. constructed 4-T tandem cells with a better absorber combination consisting of a 1.6-eV $FA_{0.83}Cs_{0.17}Pb(I_{0.83}Br_{0.17})_3$ top cell and a 1.2-eV $FA_{0.75}Cs_{0.25}Sn_{0.5}Pb_{0.5}I_3$ bottom cell [77]. With processing optimization, they achieved PCEs of 15.8% and 4.5% for the top and filtered bottom cells, respectively, resulting in a PCE of 20.3% for the 4-T tandem cells (Figure 18.18b) [91]. In early 2018, Zhao et al. further enhanced the record 4-T APTSC efficiencies to 23.1% by combining a $FA_{0.8}Cs_{0.2}Pb(I_{0.7}Br_{0.3})_3$ top cell with a $(FASnI_3)_{0.6}(MAPbI_3)_{0.4}$ bottom cell (Figure 18.18c) [48]. It is noticeable that this was the first time APTSCs demonstrated a PCE higher than the then-record PCE (22.7%) of single-junction PSCs (Figure 18.17). The incorporation of paraffin oil as an optical coupling layer between the two subcells reduces the optical loss at the air/glass interfaces, enabling PCEs of 15.7% and 7.4% for the top and flittered bottom cells, respectively. Most recently, Tong et al. paired a 1.63-eV $Cs_{0.05}FA_{0.8}MA_{0.15}Pb(I_{0.85}Br_{0.15})_3$ top cell with a 1.25-eV GA-passivated $(FASnI_3)_{0.6}(MAPbI_3)_{0.4}$ bottom cell to achieve a PCE of more than 25% (Figure 18.18d) [25]. This PCE was higher than the best-performing single-junction PSCs at the time, demonstrating a great potential of 4-T APTSCs.

18.5.2 2-T All-Perovskite Tandem Solar Cells

Compared with 4-T tandem cells, 2-T APTSCs are much difficult to design and fabricate. First, the interconnecting (IL) that integrates two subcells is a critical component in 2-T APTSCs. The IL not only provides an optical and electrical interconnection between two subcells but also protects the underlying subcell from any damages introduced by processing the second cell. Second, to achieve the desired high PCE,

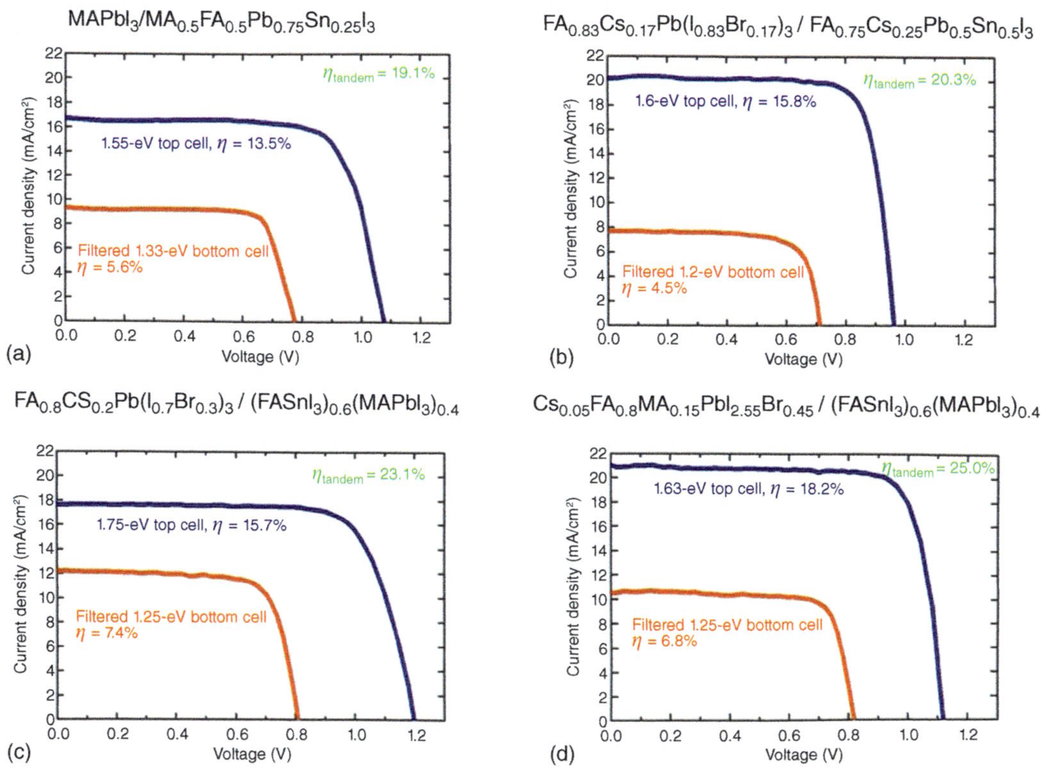

Figure 18.18 Replotted *J*–*V* curves of notable 4-T APTSCs with various perovskite bandgap combination, including (a) 1.55-eV/1.33-eV. *Source*: Based on Yang et al. [82], (b) 1.6-eV/1.2-eV. *Source*: Based on Leijtens et al. [91], (c) 1.75-eV/1.25-eV. *Source*: Based on Zhao et al. [48], and (d) 1.63-eV/1.25-eV. *Source*: Based on Tong et al. [25].

the bandgap and thickness of absorber layers need to be finely adjusted to satisfy the current match condition. Finally, the complicated device architectures consisting of more than a dozen of layers are more prone to the processing flaws in each component layer, which will deteriorate the device performance and stability.

Due to the lagged developed of low-bandgap mixed Sn-Pb PSCs, the early work on 2-T APTSCs started with using 1.55-eV $MAPbI_3$ as the bottom cell. In 2015, Jiang et al. demonstrated the first proof-of-concept monolithic 2-T APTSCs [104]. The tandem device comprises two $MAPbI_3$ cells connected by organic ILs consisting of PEDOT:PSS/polyethylenimine (PEI)/PCBM:PEI (Figure 18.19a). Although the tandem device only exhibited a mediocre PCE of 7%, the high V_{OC} value of 1.89 V was close to the sum of the two subcells, demonstrating the concept of 2-T APTSCs. Meanwhile, Heo and Im fabricated 2-T mechanically-stacked APTSCs by physically laminating two separated subcells ($MAPbBr_3$ and $MAPbI_3$) together (Figure 18.19b) [105]. This stacked 2-T tandem device delivered a PCE of 10.8% with a high V_{OC} of 2.25 V. However, the J_{SC} of this premature tandem device (8.3 mA/cm^2) is limited by the relatively large bandgaps of the $MAPbBr_3$ (2.2 eV) and $MAPbI_3$ (1.55 eV) subcells and the physically contacted interconnecting interface. Bolink and coworkers constructed monolithic 2-T APTSCs using solution-processed wide-bandgap $FA_{0.75}Cs_{0.25}Pb(I_{0.3}Br_{0.7})_3$ perovskite top cells, vacuum-deposited $MAPbI_3$ bottom cells, and doped/undoped organic ILs (Figure 18.19c). The champion tandem cell exhibited a PCE of 18.1%, with a high V_{OC} of 2.14 V, a J_{SC} of 9.7 mA/cm^2, and an FF of 75.6% [106]. More recently, they also demonstrated all-vacuum processed monolithic $MAPbI_3$/$MAPbI_3$ tandem cells, which yielded a PCE exceeding 18% with a remarkably high V_{OC} of 2.30 V [107].

The first 2-T monolithic APTSCs using a low-bandgap mixed Sn-Pb subcell was reported by Eperon et al. in 2016 [77]. The 2-T tandem cells were constructed by combining a 1.8-eV $FA_{0.83}Cs_{0.17}Pb(I_{0.5}Br_{0.5})_3$ top cell with a 1.2-eV $FA_{0.75}Cs_{0.25}Sn_{0.5}Pb_{0.5}I_3$ bottom cell using a zinc-tin oxide (ZTO)/SnO_2/ITO ILs (Figure 18.19d). The metal oxide ILs prepared by atomic layer deposition (ALD) and sputtering show decent optical and electrical properties and provide sufficient protection for processing the low-bandgap mixed Sn-Pb subcell. The rationally selected two subcells with complementary bandgaps (1.8/1.2 eV) allow broad absorption of the solar spectrum. The champion device delivered a PCE of 16.9%, with a V_{OC} of 1.66 V, a J_{SC} of 14.5 mA/cm^2, and an FF of 70%. Leijtens et al. then adjusted the top cell composition to more photostable $FA_{0.83}Cs_{0.17}Pb(I_{0.7}Br_{0.3})_3$ and incorporated MACl and formic acid to improve the quality of the $FA_{0.75}Cs_{0.25}Sn_{0.5}Pb_{0.5}I_3$ bottom cells, demonstrating APTSCs with a PCE of 19.1%, a V_{OC} of 1.81 V, a J_{SC} of 14.8 mA/cm^2, and an FF of 71.3% [91]. Palmstrom et al. then improved the top cell by cation tuning (DMA and Cs), demonstrating PCEs of 23.1% and 21.3% for APTSCs built on rigid and flexible substrates [18] (Figure 18.20). This is the first and only time that flexible APTSCs have been reported.

Rajagopal et al. fabricated 2-T APTSCs consisting of a 1.82-eV $MA_{0.9}Cs_{0.1}Pb(I_{0.6}Br_{0.4})_3$ top cell and a 1.22-eV $MASn_{0.5}Pb_{0.5}I_3$ connected by a sputtered ITO recombination layer (Figure 18.19e), achieving a PCE of 18.4% [108]. Their 2-T APTSCs demonstrated a high V_{OC} of 1.98 V, which is approaching 80% of

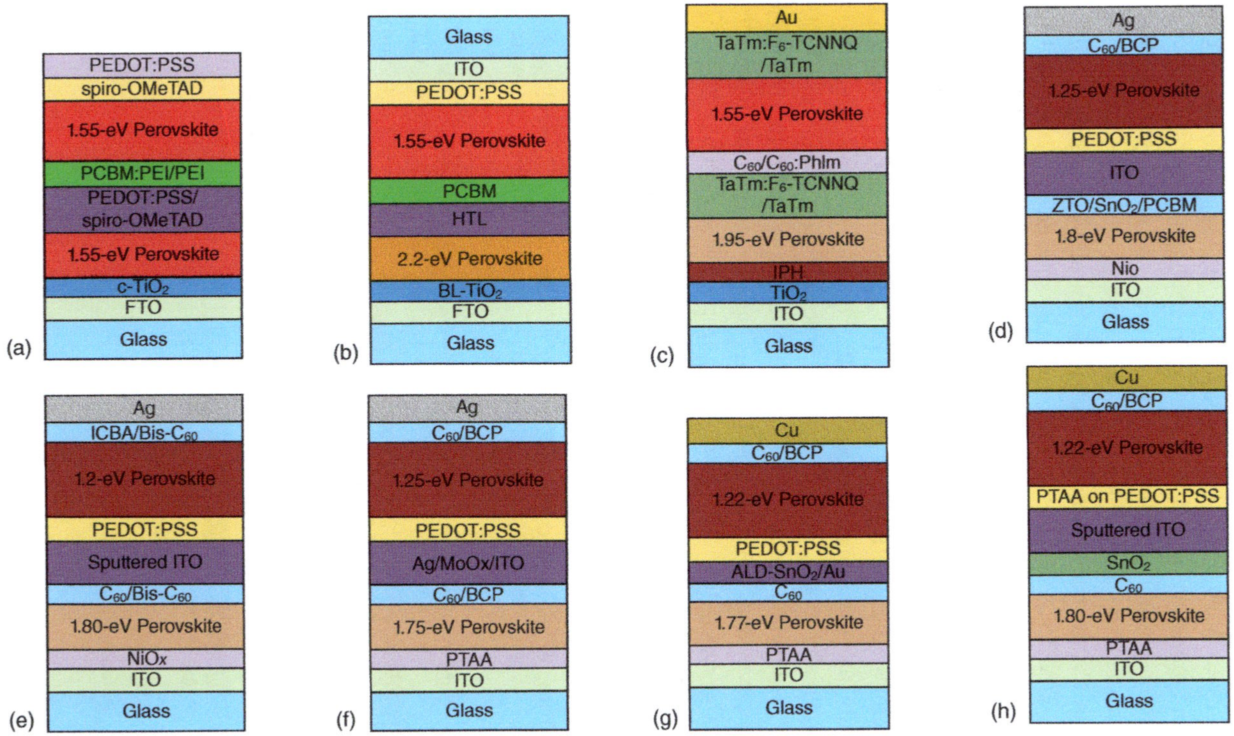

Figure 18.19 (a)–(h) Various device structures developed for 2-T APTSCs.

the theoretical voltage limit for this absorber combination. The tandem cell performance was limited by a relatively low J_{SC} of ~12.5 mA/cm^2. Zhao et al. demonstrated efficient 2-T APTSCs using a 1.75-eV $FA_{0.8}Cs_{0.2}Pb(I_{0.7}Br_{0.3})_3$ top cell and a 1.25-eV $(FASnI_3)_{0.6}(MAPbI_3)_{0.4}$ bottom cell connected by a simple vacuum-processed MoO_3/Ag/ITO ILs (Figure 18.19f) [84]. Chlorine was incorporated in the mixed Sn-Pb perovskite to enlarge the grain size, increase carrier mobility, reduce electronic disorder, and suppress trap-assisted recombination. The improved electronic quality led to the improved performance of low-bandgap PSCs with an absorber layer thickness of ~750 nm and enabled the fabrication of efficient 2-T APTSCs. The best-performing cells showed a V_{OC} of 1.92 V, a J_{SC} of 14.0 mA/cm^2, and an FF of 78.1%, yielding a PCE of 21% [84]. The tandem cells showed decent operational stability, retaining over 85% of initial PCEs after 80 hours of continuous operation under MPP tracking conditions. Beyond the commonly used ITO-based IL, Li et al. recently developed an ultrathin Cu/Au metal thermionic emission-based IL for 2-T APTSCs [109], demonstrating a new route to monolithically integrate two perovskite subcells.

Efficiency breakthrough for 2-T APTSCs was enabled by the fabrication of high-quality 1-μm-thick mixed Sn-Pb perovskite films in 2019. By combining GASCN-passivated $(FASnI_3)_{0.6}(MAPbI_3)_{0.4}$ low-bandgap cells with 1.75-eV $FA_{0.8}Cs_{0.2}Pb(I_{0.7}Br_{0.3})_3$ top cells, Tong et al. reported monolithic 2-T APTSCs with a PCE of 23.4%, a V_{OC} of 1.942 V, a J_{SC} of 15.01 mA/cm^2, and an FF of 80.3% [25] (Figure 18.20). The operational stability of a 2-T APTSC was monitored by MPP tracking under simulated one sun illumination. The tandem cell retained more than 88% of its initial efficiency after 100 hours of continuous operation. Yang et al. fabricated 2-T APTSCs using 0.03% Cd^{2+}-incorporated

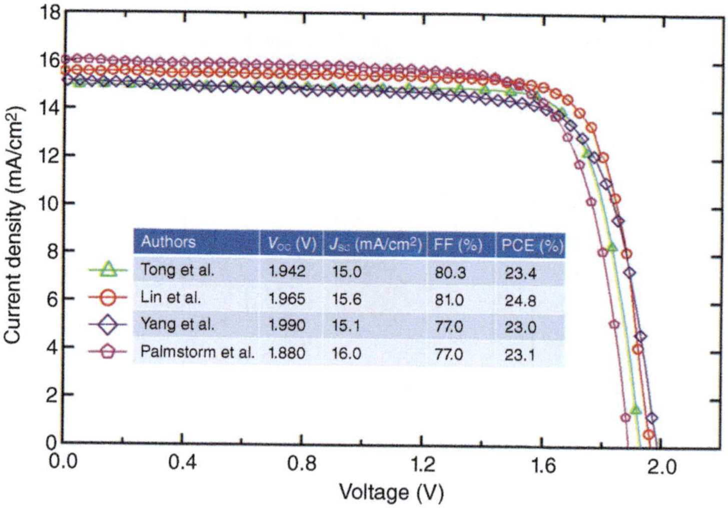

Authors	V_{OC} (V)	J_{SC} (mA/cm^2)	FF (%)	PCE (%)
Tong et al.	1.942	15.0	80.3	23.4
Lin et al.	1.965	15.6	81.0	24.8
Yang et al.	1.990	15.1	77.0	23.0
Palmstorm et al.	1.880	16.0	77.0	23.1

Figure 18.20 *J–V* curves of notable 2-T APTSCs. *Source:* Data from Palmstrom et al. [18]; Lin et al. [24]; Tong et al. [25]; and Yang et al. [95].

$Cs_{0.05}MA_{0.45}FA_{0.5}Pb_{0.5}Sn_{0.5}I_3$ bottom cells and 1.8-eV $FA_{0.6}Cs_{0.4}Pb(I_{0.65}Br_{0.35})_3$ top cells (Figure 18.19g) [95]. The best 2-T tandem cell exhibited a PCE of 23.0%, with a V_{OC} of 1.99 V, a J_{SC} of 15.1 mA/cm^2, and an FF of 77% (Figure 18.20). The device exhibited decent operational stability, retaining more than 90% of its initial PCE after 200 hours of operation. Lin et al. fabricated 2-T APTSCs consisting of a 1.77-eV $FA_{0.8}Cs_{0.2}Pb(I_{0.6}Br_{0.4})_3$ top cell and a 1.22-eV $FA_{0.7}MA_{0.3}Pb_{0.5}Sn_{0.5}I_3$ bottom cell connected by an ultrathin ALD SnO_2/Au ILs (Figure 18.19h) [24]. The record efficiency 2-T APTSC delivered a PCE of 24.8% under reverse scan, with a V_{OC} of 1.965 V, a J_{SC} of 15.6 mA/cm^2, and a high FF of 81% (Figure 18.20). More importantly, the tandem cells retain 90% of their initial PCEs after more than 450 hours of continuous operation at the MPPs under one sun illumination.

18.5.3 Limitations and Challenges of All-Perovskite Tandem Solar Cells

The recent research efforts in APTSCs have significantly advanced this promising PV technology. However, APTSCs are still in the premature stage of the development and far from commercial readiness. First, the scaling-up fabrication presents a critical challenge for APTSCs. So far, most APTSCs reported in the literature have small active areas of less than 0.1 cm^2, with only one exception with active areas of up to 1 cm^2 [24]. These small device areas allow the commonly used spin-coating processes to be applied for the fabrication of APTSCs. However, industrial manufacturing requires scaling-up fabrication techniques for wide- and low-bandgap PSCs and ILs. Moreover, the monolithic integration designs of APTSCs rely on the separation and interconnection of individual cells. The thin-film tandem cell architecture designs and the methods for module integration are yet to be demonstrated. It is expected that more research efforts will be paid to the development in the design and fabrication of large-area tandem cells and modules in the coming years.

Device reliability is another major concern for the applications of APTSCs. While APTSCs hold the promise to deliver high module efficiencies exceeding 30%, the requirement of robust lifetime of up to 25 years for commercial solar panels is a great challenge to overcome in their path toward commercialization. Although some promising stability testing results have been reported in the literature in recent years, the long-term stability of PSCs is yet to be fully demonstrated. Especially for APTSCs, there are much smaller numbers of publications compared with reports of single-junction PSCs. More importantly, standardized operational stability tests for APTSCs have rarely been reported. APTSCs are limited by the relatively poor stability of low-bandgap mixed Sn-Pb PSCs. Due to the vulnerability of mixed Sn-Pb perovskites against oxygen and moisture, low-bandgap mixed Sn-Pb PSCs require complete encapsulation to work in the ambient conditions for a long time. Recently, MPP tracking of APTSCs under one sun illumination for up to hundreds of hours has been demonstrated by few groups [24, 25, 95]. Although such operational lifetime is not comparable with the single-junction PSCs that have shown thousands of hours of MPP stability, these stability testing results are encouraging for further improvements of APTSCs. Nonetheless, stability issues of tandem PSCs need to be fully addressed before translating these promising PV technologies demonstrated in laboratories to real-world applications [110].

Beyond the double-junction APTSCs, multijunction tandem cells with more than two absorber materials will also be considered in the future development due to their potential to achieve higher theoretical limits. Werner et al. have demonstrated a high V_{OC} of 2.69 V in a triple-junction perovskite/perovskite/Si tandem cell [111]. McMeekin et al. fabricated the proof-of-concept triple-junction APTSCs using a combination of 1.94 eV/1.57 eV/1.34 eV perovskites [17]. These attempts have opened up many new avenues for exploration of high-efficiency low-cost multijunction APTSCs. With more technology advances in the future, the full potential of multijunction APTSCs is expected to be realized.

18.6 Conclusion and Outlooks

The past decades have witnessed the rapid progress of metal halide perovskite PV technologies at an unprecedented speed. The success of Pb-iodide single-junction PSCs has propelled research advances in wide-bandgap mixed I-Br and low-bandgap mixed Sn-Pb PSCs, as well as high-efficiency 2-T and 4-T APTSCs. Particularly, the recent breakthrough in high-efficiency low-bandgap mixed Sn-Pb PSCs enabled the fabrication of 2-T and 4-T APTSCs with PCEs approaching the current record of single-junction PSCs. Such a demonstration of high-performance APTSCs is a long-awaited sign showing the path of realizing the PCE potential of thin-film tandem solar cells.

As the PCE of APTSCs is expected to further increase in the near future, the commercialization of this promising PV technology needs to be seriously considered. Future research efforts will be devoted to developing device architectures and deposition processes that are applicable for large-area tandem perovskite PV module fabrication. The processing and environmental sensitivity of perovskite materials presents a major challenge for the design and fabrication of APTSCs. Particularly for the monolithically integrated 2-T tandem cells, depositing the second perovskite subcells on the first-prepared subcells requires special caution and protective remedies to mitigate the possible processing damages. This requires researchers to explore alternative tandem-specific deposition approaches. One of the possible approaches is the vacuum-based deposition [106, 107], which is not limited by the compactness of the top surface and solvent compatibility and can be adapted to the mild-vacuum, low-temperature, high-throughput industrial production. The remaining challenges are how to achieve high-quality perovskite films with different perovskite compositions, which require more research efforts.

Future work also needs to focus on developing more stable perovskite tandem modules. The desired device stability requires not only the advanced engineering on perovskite compositions and cell structures to enhance the stability of PSCs but also the module frame and encapsulation designs that prevent water and oxygen ingress [112–116] as well as the leakage of water-soluble toxic Pb and Sn compounds [19, 117–119]. Once robust encapsulation can be implemented, APTSCs will have a chance to achieve sufficiently long-term stability, which will benefit the adaption of these promising tandem techniques in the market. In the long future, APTSCs hold the potential to become the leader in the perovskite PV field due to their ultrahigh efficiency and low cost [120]. However, the route toward this ultimate promising perovskite PV technology is the most challenging one because of the performance and stability issues of the low-bandgap PSCs. If these critical remaining issues can be resolved, APTSCs will fulfill their promise to deliver more electrical power at lower costs than today's PV technologies.

Acknowledgments

This material is based upon work supported by the U.S. Department of Energy's Office of Energy Efficiency and Renewable Energy (EERE) under the Solar Energy Technologies Office Award Number DE-EE0008753. The U.S. Government is authorized to reproduce and distribute reprints for Governmental purposes notwithstanding any copyright notation thereon.

References

1 Correa-Baena, J.-P., Saliba, M., Buonassisi, T. et al. (2017). *Science* 358: 739–744.
2 Correa-Baena, J.-P., Abate, A., Saliba, M. et al. (2017). *Energy Environ. Sci.* 10: 710–727.
3 Kojima, A., Teshima, K., Shirai, Y., and Miyasaka, T. (2009). *J. Am. Chem. Soc.* 131: 6050–6051.
4 NREL Solar Cell Efficiency Chart. https://www.nrel.gov/pv/cell-efficiency.html (accessed 6 July 2019).
5 Shockley, W. and Queisser, H.J. (1961). *J. Appl. Phys.* 32: 510–519.
6 Green, M.A., Dunlop, E.D., Hohl-Ebinger, J. et al. (2020). *Prog. Photovolt.* 28: 3–15.
7 Leijtens, T., Bush, K.A., Prasanna, R., and McGehee, M.D. (2018). *Nat. Energy* 3: 828–838.
8 Eperon, G.E., Hörantner, M.T., and Snaith, H.J. (2017). *Nat. Rev. Chem.* 1: 0095.
9 Anaya, M., Lozano, G., Calvo, M.E., and Míguez, H. (2017). *Joule* 1: 769–793.
10 Song, Z., Chen, C., Li, C. et al. (2019). *Semicond. Sci. Technol.* 34: 093001.
11 Saliba, M., Correa-Baena, J.-P., Grätzel, M. et al. (2018). *Angew. Chem. Int. Ed.* 57: 2554–2569.
12 Li, Z., Klein, T.R., Kim, D.H. et al. (2018). *Nat. Rev. Mater.* 3: 18017.
13 Song, Z., Watthage, S.C., Phillips, A.B., and Heben, M.J. (2016). *J. Photon. Energy* 6: 022001.
14 Yang, D., Yang, R., Priya, S., and Liu, S. (2019). *Angew. Chem. Int. Ed.* 58: 4466–4483.

15 Song, Z., McElvany, C.L., Phillips, A.B. et al. (2017). *Energy Environ. Sci.* 10: 1297–1305.
16 Chang, N.L., Ho-Baillie, A.W.Y., Vak, D. et al. (2018). *Sol. Energy Mater. Sol. Cells* 174: 314–324.
17 McMeekin, D.P., Mahesh, S., Noel, N.K. et al. (2019). *Joule* 3: 387–401.
18 Palmstrom, A.F., Eperon, G.E., Leijtens, T. et al. (2019). *Joule* 3: 2193–2204.
19 Celik, I., Phillips, A.B., Song, Z. et al. (2017). *Energy Environ. Sci.* 10: 1874–1884.
20 Celik, I., Phillips, A.B., Song, Z. et al. (2017). *IEEE J. Photovol.* 8: 305–309.
21 Celik, I., Song, Z., Cimaroli, A.J. et al. (2016). *Sol. Energy Mater. Sol. Cells* 156: 157–169.
22 Li, Z., Zhao, Y., Wang, X. et al. (2018). *Joule* 2: 1559–1572.
23 Song, Z., Phillips, A. B., Celik, I., Liyanage, G. K., Zhao, D., Apul, D., Yan, Y. and Heben, M. J.(2018). In Manufacturing Cost Analysis of Perovskite Solar Modules in Single-Junction and All-Perovskite Tandem Configurations. In *2018 IEEE 7th World Conference on Photovoltaic Energy Conversion (WCPEC)* (A Joint Conference of 45th IEEE PVSC, 28th PVSEC & 34th EU PVSEC) (10–15 June), pp 1134–1138.
24 Lin, R., Xiao, K., Qin, Z. et al. (2019). *Nat. Energy* 4: 864–873.
25 Tong, J., Song, Z., Kim, D.H. et al. (2019). *Science* 364: 475–479.
26 Vos, A.D. (1980). *J. Phys. D. Appl. Phys.* 13: 839.
27 Song, Z., Watthage, S.C., Phillips, A.B. et al. (2015). *Chem. Mater.* 27: 4612–4619.
28 Noh, J.H., Im, S.H., Heo, J.H. et al. (2013). *Nano Lett.* 13: 1764–1769.
29 Sadhanala, A., Ahmad, S., Zhao, B. et al. (2015). *Nano Lett.* 15: 6095–6101.
30 Stoumpos, C.C., Malliakas, C.D., and Kanatzidis, M.G. (2013). *Inorg. Chem.* 52: 9019–9038.
31 Hao, F., Stoumpos, C.C., Chang, R.P.H., and Kanatzidis, M.G. (2014). *J. Am. Chem. Soc.* 136: 8094–8099.
32 Noel, N.K., Stranks, S.D., Abate, A. et al. (2014). *Energy Environ. Sci.* 7: 3061–3068.
33 Henry, C.H. (1980). *J. Appl. Phys.* 51: 4494–4500.
34 Bremner, S.P., Levy, M.Y., and Honsberg, C.B. (2008). *Prog. Photovolt.* 16: 225–233.
35 Geisz, J.F., Steiner, M.A., Jain, N. et al. (2018). *IEEE J. Photovol.* 8: 626–632.
36 Hosseinian Ahangharnejhad, R., Phillips, A.B., Ghimire, K. et al. (2019). *Sustain. Energy Fuels* 3: 1841–1851.
37 Hörantner, M.T. and Snaith, H.J. (2017). *Energy Environ. Sci.* 10: 1983–1993.
38 Lehr, J., Langenhorst, M., Schmager, R. et al. (2018). *Sustain. Energy Fuels* 2: 2754–2761.
39 Langenhorst, M., Sautter, B., Schmager, R. et al. (2019). *Prog. Photovolt.* 27: 290–298.
40 Eperon, G.E., Stranks, S.D., Menelaou, C. et al. (2014). *Energy Environ. Sci.* 7: 982–988.

41 Beal, R.E., Slotcavage, D.J., Leijtens, T. et al. (2016). *J. Phys. Chem. Lett.* 7: 746–751.

42 Yang, T.C.-J., Fiala, P., Jeangros, Q., and Ballif, C. (2018). *Joule* 2: 1421–1436.

43 Krückemeier, L., Rau, U., Stolterfoht, M., and Kirchartz, T. (2020). *Adv. Energy Mater.* 10: 1902573.

44 Hoke, E.T., Slotcavage, D.J., Dohner, E.R. et al. (2015). *Chem. Sci.* 6: 613–617.

45 Brennan, M.C., Draguta, S., Kamat, P.V., and Kuno, M. (2018). *ACS Energy Lett.* 3: 204–213.

46 Draguta, S., Sharia, O., Yoon, S.J. et al. (2017). *Nat. Commun.* 8: 200.

47 McMeekin, D.P., Sadoughi, G., Rehman, W. et al. (2016). *Science* 351: 151–155.

48 Zhao, D., Wang, C., Song, Z. et al. (2018). *ACS Energy Lett.* 3: 305–306.

49 Werner, J., Dubuis, G., Walter, A. et al. (2015). *Sol. Energy Mater. Sol. Cells* 141: 407–413.

50 Tan, H., Che, F., Wei, M. et al. (2018). *Nat. Commun.* 9: 3100.

51 Gharibzadeh, S., Abdollahi Nejand, B., Jakoby, M. et al. (2019). *Adv. Energy Mater.* 9: 1803699.

52 Lin, Y., Chen, B., Zhao, F. et al. (2017). *Adv. Mater.* 29: 1700607.

53 Chen, C., Song, Z., Xiao, C. et al. (2019). *Nano Energy* 61: 141–147.

54 Liu, Z., Krückemeier, L., Krogmeier, B. et al. (2019). *ACS Energy Lett.* 4: 110–117.

55 Bischak, C.G., Hetherington, C.L., Wu, H. et al. (2017). *Nano Lett.* 17: 1028–1033.

56 Knight, A.J., Wright, A.D., Patel, J.B. et al. (2018). *ACS Energy Lett.* 4: 75–84.

57 Braly, I.L., Stoddard, R.J., Rajagopal, A. et al. (2017). *ACS Energy Lett.* 2: 1841–1847.

58 Lin, Y., Chen, B., Fang, Y. et al. (2018). *Nat. Commun.* 9: 4981.

59 Slotcavage, D.J., Karunadasa, H.I., and McGehee, M.D. (2016). *ACS Energy Lett.* 1: 1199–1205.

60 Ruth, A., Brennan, M.C., Draguta, S. et al. (2018). *ACS Energy Lett.* 3: 2321–2328.

61 Samu, G.F., Janáky, C., and Kamat, P.V. (2017). *ACS Energy Lett.* 2: 1860–1861.

62 Barker, A.J., Sadhanala, A., Deschler, F. et al. (2017). *ACS Energy Lett.* 2: 1416–1424.

63 Gualdrón-Reyes, A.F., Yoon, S.J., and Mora-Seró, I. (2018). *Curr. Opin. Electrochem.* 11: 84–90.

64 Bush, K.A., Frohna, K., Prasanna, R. et al. (2018). *ACS Energy Lett.* 3: 428–435.

65 Yang, Z., Rajagopal, A., Jo, S.B. et al. (2016). *Nano Lett.* 16: 7739–7747.

66 Abdi-Jalebi, M., Andaji-Garmaroudi, Z., Cacovich, S. et al. (2018). *Nature* 555: 497.

67 Belisle, R.A., Bush, K.A., Bertoluzzi, L. et al. (2018). *ACS Energy Lett.* 3: 2694–2700.

68 Hu, M., Bi, C., Yuan, Y. et al. (2016). *Adv. Sci.* 3: 1500301.

69 Gualdrón-Reyes, A.F., Yoon, S.J., Barea, E.M. et al. (2018). *ACS Energy Lett.* 4: 54–62.

70 Stoddard, R.J., Rajagopal, A., Palmer, R.L. et al. (2018). *ACS Energy Lett.* 3: 1261–1268.
71 Xiao, Z., Zhao, L., Tran, N.L. et al. (2017). *Nano Lett.* 17: 6863–6869.
72 Rajagopal, A., Stoddard, R.J., Jo, S.B. et al. (2018). *Nano Lett.* 18: 3985–3993.
73 Howard, J.M., Tennyson, E.M., Neves, B.R.A., and Leite, M.S. (2018). *Joule* 3: 325–337. https://doi.org/10.1016/j.joule.2018.11.010.
74 Chen, S., Hou, Y., Chen, H. et al. (2018). *Adv. Energy Mater.* 8: 1701543.
75 Im, J., Stoumpos, C.C., Jin, H. et al. (2015). *J. Phys. Chem. Lett.* 6: 3503–3509.
76 Ogomi, Y., Morita, A., Tsukamoto, S. et al. (2014). *J. Phys. Chem. Lett.* 5: 1004–1011.
77 Eperon, G.E., Leijtens, T., Bush, K.A. et al. (2016). *Science* 354: 861–865.
78 Rajagopal, A., Stoddard, R.J., Hillhouse, H.W., and Jen, A.K.Y. (2019). *J. Mater. Chem. A* 7: 16285–16293.
79 Liao, W., Zhao, D., Yu, Y. et al. (2016). *J. Am. Chem. Soc.* 138: 12360–12363.
80 Umari, P., Mosconi, E., and De Angelis, F. (2014). *Sci. Rep.* 4: 4467.
81 Goyal, A., McKechnie, S., Pashov, D. et al. (2018). *Chem. Mater.* 30: 3920–3928.
82 Yang, Z., Rajagopal, A., Chueh, C.-C. et al. (2016). *Adv. Mater.* 28: 1602696.
83 Zhao, D., Yu, Y., Wang, C. et al. (2017). *Nat. Energy* 2: 17018.
84 Zhao, D., Chen, C., Wang, C. et al. (2018). *Nat. Energy* 3: 1093–1100.
85 Li, C., Song, Z., Zhao, D. et al. (2019). *Adv. Energy Mater.* 9: 1803135.
86 Prasanna, R., Gold-Parker, A., Leijtens, T. et al. (2017). *J. Am. Chem. Soc.* 139: 11117–11124.
87 Xu, G., Bi, P., Wang, S. et al. (2018). *Adv. Funct. Mater.* 28: 1804427.
88 Jiang, T., Chen, Z., Chen, X. et al. (2019). *Sol. RRL*: 1900467.
89 Song, Z., Wang, C., Phillips, A.B. et al. (2018). *Sustain. Energy Fuels* 2: 2460–2467.
90 Juarez-Perez, E.J., Ono, L.K., Maeda, M. et al. (2018). *J. Mater. Chem. A* 6: 9604–9612.
91 Leijtens, T., Prasanna, R., Bush, K.A. et al. (2018). *Sustain. Energy Fuels* 2: 2450–2459.
92 Zong, Y., Wang, N., Zhang, L. et al. (2017). *Angew. Chem. Int. Ed.* 56: 12658–12662.
93 Zong, Y., Zhou, Z., Chen, M. et al. (2018). *Adv. Energy Mater.* 8: 1800997.
94 Wang, C., Song, Z., Li, C. et al. (2019). *Adv. Funct. Mater.* 29: 1808801.
95 Yang, Z., Yu, Z., Wei, H. et al. (2019). *Nat. Commun.* 10: 4498.
96 Jung, E.H., Jeon, N.J., Park, E.Y. et al. (2019). *Nature* 567: 511–515.
97 Yoo, J.J., Wieghold, S., Sponseller, M.C. et al. (2019). *Energy Environ. Sci.* 12: 2192–2199.
98 Lang, F., Shargaieva, O., Brus, V.V. et al. (2018). *Adv. Mater.* 30: 1702905.
99 Chi, D., Huang, S., Zhang, M. et al. (2018). *Adv. Funct. Mater.* 28: 1804603.
100 Prasanna, R., Leijtens, T., Dunfield, S.P. et al. (2019). *Nat. Energy* 4: 939–947.
101 Xu, X., Chueh, C.-C., Yang, Z. et al. (2017). *Nano Energy* 34: 392–398.
102 Zhu, Z., Li, N., Zhao, D. et al. (2018). *Adv. Energy Mater.* 9: 1802774.
103 Li, Z., Boix, P.P., Xing, G. et al. (2016). *Nanoscale* 8: 6352–6360.
104 Jiang, F., Liu, T., Luo, B. et al. (2016). *J. Mater. Chem. A* 4: 1208–1213.

105 Heo, J.H. and Im, S.H. (2016). *Adv. Mater.* 28: 5121–5125.
106 Forgács, D., Gil-Escrig, L., Pérez-Del-Rey, D. et al. (2017). *Adv. Energy Mater.* 7: 1602121.
107 Ávila, J., Momblona, C., Boix, P. et al. (2018). *Energy Environ. Sci.* 11: 3292–3297.
108 Rajagopal, A., Yang, Z., Jo, S.B. et al. (2017). *Adv. Mater.* 29: 1702140.
109 Li, C., Wang, Z.S., Zhu, H.L. et al. (2018). *Adv. Energy Mater.* 8: 1801954.
110 Meng, L., You, J., and Yang, Y. (2018). *Nat. Commun.* 9: 5265.
111 Werner, J., Sahli, F., Fu, F. et al. (2018). *ACS Energy Lett.* 3: 2052–2058.
112 Song, Z., Abate, A., Watthage, S.C. et al. (2016). *Adv. Energy Mater.* 6: 1600846.
113 Matteocci, F., Cinà, L., Lamanna, E. et al. (2016). *Nano Energy* 30: 162–172.
114 Huang, J., Tan, S., Lund, P.D., and Zhou, H. (2017). *Energy Environ. Sci.* 10: 2284–2311.
115 Song, Z., Werner, J., Watthage, S.C. et al. (2017). *IEEE J. Photovol.* 7: 1563–1568.
116 Cheacharoen, R., Rolston, N., Harwood, D. et al. (2018). *Energy Environ. Sci.* 11: 144–150.
117 Babayigit, A., Ethirajan, A., Muller, M., and Conings, B. (2016). *Nat. Mater.* 15: 247–251.
118 Celik, I., Song, Z., Heben, M.J., and Apul, D. (2017). *J. Clean. Prod.* 186: 632–639.
119 Babayigit, A., Duy Thanh, D., Ethirajan, A. et al. (2016). *Sci. Rep.* 6: 18721.
120 Snaith, H.J. (2018). *Nat. Mater.* 17: 372–376.

A

Optical Constants of Hybrid Perovskite Materials

Yukinori Nishigaki[1], Akio Matsushita[2], Alvaro Tejada[3,4], Taisuke Matsui[2], and Hiroyuki Fujiwara[1]

[1] *Gifu University, Department of Electrical, Electronic and Computer Engineering, 1-1 Yanagido, Gifu, 501-1193, Japan*
[2] *Panasonic Corporation, Technology Innovation Division, 3-1-1 Yagumo-naka-machi, Moriguchi City, Osaka 570-8501, Japan*
[3] *Pontificia Universidad Católica del Perú, Departamento de Ciencias, Sección Física, Av. Universitaria 1801, Lima, 15088, Peru*
[4] *Helmholtz-Zentrum Berlin für Materialien und Energie GmbH, Young Investigator Group for Perovskite Tandem Solar Cells, Kekuléstraße 5 Berlin, 12489, Germany*

The basic performances of solar cell devices are ultimately determined by the light absorption characteristics of semiconducting light absorbers and accurate optical data are crucial in both interpreting and simulating the device performance. In this appendix, we show the tabulated optical-constant data for all hybrid perovskites shown in Figures 1.5, 4.1, and 4.15 (i.e. $MAPbI_3$ [1], $FAPbI_3$ [2], $Cs_{0.17}FA_{0.83}Pb(I_{1-x}Br_x)_3$ (x = 0.0–0.4) [3], $Cs_{0.05}FA_{0.8}MA_{0.15}Pb(I_{1-x}Br_x)_3$ (x = 0.05–0.4), $FA_{0.55}MA_{0.45}(Sn_{0.55}Pb_{0.45})I_3$, $MAPbBr_3$ [4], and $MAPbCl_3$ [4]), which have been determined from careful spectroscopic ellipsometry analyses. For the optical constants, the refractive index n and extinction coefficient k, defined by Eq. (4.1), are indicated for each photon energy E in a range of 1.0–4.6 eV with a step of 0.02 eV. The E can be converted to the wavelength by λ [nm] = 1239.8/(E [eV]). From the corresponding k value, the absorption coefficient (α) in the unit of cm^{-1} can be obtained as $\alpha = 4\pi k \cdot 10^7/(\lambda\ [nm])$ [see also Eq. (4.2)]. Based on (n, k), the dielectric function ($\varepsilon = \varepsilon_1 - i\varepsilon_2$) can also be calculated by applying Eqs. (4.4) and (4.5).

Furthermore, to remove the spectral noises and to better describe the perovskite optical properties, all the experimental optical functions of hybrid perovskites have been parameterized using the Tauc–Lorentz dielectric function model [5]. Through careful modeling, we established that all hybrid perovskite materials can be modeled almost perfectly by combining several Tauc–Lorentz peaks (see also Figure 4.4). Importantly, from this scheme, the optical constants of various perovskite alloys can be calculated systematically, as described further. Our dielectric function modeling allows the accurate calculation of α spectra in a quite wide range extending from 10^2 to 10^6 cm^{-1} [6]. Thus, in the tabulated data, (n, k) values calculated from the

Hybrid Perovskite Solar Cells: Characteristics and Operation,
First Edition. Edited by Hiroyuki Fujiwara.

Tauc–Lorentz model parameters are summarized. Unfortunately, the Tauc–Lorentz calculation is slightly complex and, when one intends to calculate the optical functions from the Tauc–Lorentz model parameters listed in this appendix, a free software shown in Figure 11.13 [7] can be used.

In the Tauc–Lorentz parameterization, the ε_2 spectrum of a semiconductor material is modeled as a sum of the Tauc–Lorentz peaks:

$$\varepsilon_2(E) = \sum_{j=1}^{m} \frac{A_j C_j E_{0,j}(E - E_{g,j})^2}{[(E^2 - E_{0,j}^2)^2 + C_j^2 E^2]E} \tag{A.1}$$

A single Tauc–Lorentz peak is expressed by four parameters: namely, the amplitude parameter A_j, broadening parameter C_j, peak transition energy $E_{0,j}$, and optical gap $E_{g,j}$ of the jth Tauc–Lorentz peak. From the parameters of Eq. (A.1), $\varepsilon_1(E)$ can be calculated from the Kramers–Kronig integration of each Tauc–Lorentz peak:

$$\varepsilon_1(E) = \sum_{j=1}^{m} \left(\varepsilon_{1,j}(\infty) + \frac{2}{\pi} P \int_{E_g}^{\infty} \frac{E' \varepsilon_{2,j}(E')}{E'^2 - E^2} dE' \right) \tag{A.2}$$

where $\varepsilon_{1,j}(\infty)$ represents a constant contribution to $\varepsilon_1(E)$ at high energies. The second term in the parenthesis shows the Kramers–Kronig integration of the jth Tauc–Lorentz peak, which can be calculated using the exact equation described below:

$$\begin{aligned}
\varepsilon_1(E) = & \frac{AC}{\pi \xi^4} \cdot \frac{a_{\ln}}{2\beta E_0} \ln \left(\frac{E_0^2 + E_g^2 + \beta E_g}{E_0^2 + E_g^2 - \beta E_g} \right) \\
& - \frac{A}{\pi \xi^4} \cdot \frac{a_{\tan}}{E_0} \left[\pi - \tan^{-1}\left(\frac{2E_g + \beta}{C} \right) + \tan^{-1}\left(\frac{-2E_g + \beta}{C} \right) \right] \\
& + 2\frac{AE_0}{\pi \xi^4 \beta} E_g (E^2 - \gamma^2) \left[\pi + 2\tan^{-1}\left(2\frac{\gamma^2 - E_g^2}{\beta C} \right) \right] \\
& - \frac{AE_0 C}{\pi \xi^4} \cdot \frac{E^2 + E_g^2}{E} \ln \left(\frac{|E - E_g|}{E + E_g} \right) + \frac{2AE_0 C}{\pi \xi^4} E_g \ln \left[\frac{|E - E_g|(E + E_g)}{\sqrt{(E_0^2 - E_g^2)^2 + E_g^2 C^2}} \right]
\end{aligned} \tag{A.3}$$

where

$$a_{\ln} = (E_g^2 - E_0^2)E^2 + E_g^2 C^2 - E_0^2(E_0^2 + 3E_g^2) \tag{A.4}$$

$$a_{\tan} = (E^2 - E_0^2)(E_0^2 + E_g^2) + E_g^2 C^2 \tag{A.5}$$

$$\xi^4 = (E^2 - \gamma^2)^2 + \beta^2 C^2/4 \tag{A.6}$$

$$\beta = \sqrt{4E_0^2 - C^2} \tag{A.7}$$

$$\gamma = \sqrt{E_0^2 - C^2/2} \tag{A.8}$$

From the calculated (ε_1, ε_2) spectra, (n, k) spectra can then be obtained from well-known expression:

$$n = \{[\varepsilon_1 + (\varepsilon_1^2 + \varepsilon_2^2)^{1/2}]/2\}^{1/2} \tag{A.9}$$

$$k = \{[-\varepsilon_1 + (\varepsilon_1^2 + \varepsilon_2^2)^{1/2}]/2\}^{1/2} \tag{A.10}$$

From the Tauc–Lorentz parameters summarized for each perovskite material in this appendix, the smoother optical functions can be calculated. More importantly, the optical constants of hybrid perovskite alloys with arbitrary composition can further be calculated based on the Tauc–Lorentz parameters (Section 4.5.2). In this case, all the transition peak energies (i.e. $E_{g,j}$ and $E_{0,j}$ in Eq. (A.1)) of a base material are simply increased (or decreased) to model the optical constants of a material with a higher (or lower) E_g. In Tables A.1–A.24, the tabulated optical constants and the Tauc–Lorentz parameters for various hybrid perovskite crystals and their alloys are summarized. For $MAPbBr_3$ and $MAPbCl_3$, the Tauc–Lorentz parameters can be found elsewhere [6].

Table A.1 Optical constants of $MAPbI_3$ reported in Ref. [1].

E (eV)	*n*	*k*	*E* (eV)	*n*	*k*	*E* (eV)	*n*	*k*
1.00	2.275	0	1.96	2.585	0.174	3.32	2.399	1.559
1.02	2.279	0	2.00	2.601	0.188	3.36	2.265	1.566
1.04	2.282	0	2.04	2.618	0.204	3.40	2.157	1.540
1.06	2.286	0	2.08	2.636	0.223	3.44	2.078	1.503
1.08	2.289	0	2.12	2.655	0.243	3.48	2.022	1.469
1.10	2.293	0	2.16	2.673	0.265	3.52	1.976	1.444
1.12	2.297	0	2.20	2.692	0.289	3.56	1.932	1.427
1.14	2.301	0	2.24	2.711	0.317	3.60	1.884	1.410
1.16	2.306	0	2.28	2.730	0.347	3.64	1.836	1.388
1.18	2.310	0	2.32	2.749	0.381	3.68	1.792	1.360
1.20	2.315	0	2.36	2.767	0.420	3.72	1.754	1.329
1.22	2.319	0	2.40	2.783	0.465	3.76	1.723	1.297
1.24	2.324	0	2.44	2.793	0.518	3.80	1.698	1.267
1.26	2.330	0	2.48	2.793	0.576	3.84	1.676	1.238
1.28	2.335	0	2.52	2.779	0.634	3.88	1.658	1.212
1.30	2.340	0	2.56	2.751	0.684	3.92	1.642	1.187
1.32	2.346	0	2.60	2.717	0.718	3.96	1.627	1.163
1.34	2.352	0	2.64	2.687	0.738	4.00	1.614	1.141
1.36	2.359	0	2.68	2.666	0.751	4.04	1.603	1.120

(continued)

Table A.1 (Continued)

E (eV)	*n*	*k*	*E* (eV)	*n*	*k*	*E* (eV)	*n*	*k*
1.38	2.366	0	2.72	2.653	0.763	4.08	1.592	1.100
1.40	2.373	0	2.76	2.646	0.776	4.12	1.583	1.081
1.44	2.388	0	2.80	2.644	0.792	4.16	1.575	1.062
1.48	2.406	0	2.84	2.644	0.810	4.20	1.568	1.045
1.52	2.427	0	2.88	2.647	0.832	4.24	1.562	1.028
1.56	2.453	0	2.92	2.652	0.856	4.28	1.557	1.012
1.60	2.494	0.00696	2.96	2.659	0.884	4.32	1.553	0.998
1.64	2.536	0.0502	3.00	2.670	0.918	4.36	1.549	0.984
1.68	2.535	0.0935	3.04	2.683	0.958	4.40	1.547	0.972
1.72	2.528	0.114	3.08	2.699	1.010	4.44	1.545	0.961
1.76	2.528	0.125	3.12	2.713	1.078	4.48	1.543	0.951
1.80	2.533	0.133	3.16	2.719	1.167	4.52	1.541	0.943
1.84	2.543	0.141	3.20	2.701	1.277	4.56	1.538	0.937
1.88	2.555	0.151	3.24	2.643	1.398	4.60	1.535	0.931
1.92	2.569	0.161	3.28	2.536	1.501			

Refractive index *n* and extinction coefficient *k* for each photon energy *E* are shown. The result is consistent with those shown in Figures 4.1 and 4.4. The optical data calculated based on the Tauc–Lorentz parameterization are indicated.
Source: Based on Shirayama et al. [1].

Table A.2 Tauc–Lorentz parameters of $MAPbI_3$. The model parameters reported in Ref. [1] are shown.

Peak	*A* (eV)	*C* (eV)	E_0 (eV)	E_g (eV)	$\varepsilon_1(\infty)$
$j=1$	1.621	0.024	1.593	1.565	1.486
$j=2$	47.677	0.137	1.607	1.593	0
$j=3$	2.909	0.349	2.553	1.684	0
$j=4$	39.202	2.391	3.046	1.563	0
$j=5$	4.955	0.329	3.278	1.764	0
$j=6$	1.369	0.354	3.581	2.246	0
$j=7$	1.622	0.745	4.726	2.823	0
$j=8$	4.137	0.666	5.648	3.839	0
$j=9$	11.256	2.053	7.408	2.276	0

Source: Based on Shirayama et al. [1].

Table A.3 Optical constants of $FAPbI_3$ reported in Ref. [2].

E (eV)	*n*	*k*	*E* (eV)	*n*	*k*	*E* (eV)	*n*	*k*
1.00	2.125	0	1.96	2.379	0.105	3.32	2.269	1.204
1.02	2.128	0	2.00	2.396	0.117	3.36	2.200	1.216
1.04	2.131	0	2.04	2.414	0.130	3.40	2.134	1.221
1.06	2.133	0	2.08	2.432	0.147	3.44	2.072	1.218
1.08	2.136	0	2.12	2.450	0.165	3.48	2.014	1.210
1.10	2.139	0	2.16	2.468	0.187	3.52	1.961	1.197
1.12	2.142	0	2.20	2.485	0.212	3.56	1.913	1.180
1.14	2.145	0	2.24	2.499	0.238	3.60	1.869	1.161
1.16	2.148	0	2.28	2.513	0.267	3.64	1.830	1.139
1.18	2.152	0	2.32	2.525	0.297	3.68	1.795	1.116
1.20	2.155	0	2.36	2.537	0.331	3.72	1.764	1.092
1.22	2.159	0	2.40	2.548	0.371	3.76	1.736	1.067
1.24	2.163	0	2.44	2.550	0.424	3.80	1.712	1.041
1.26	2.167	0	2.48	2.524	0.480	3.84	1.692	1.015
1.28	2.171	0	2.52	2.480	0.503	3.88	1.674	0.988
1.30	2.175	0	2.56	2.456	0.502	3.92	1.662	0.961
1.32	2.179	0	2.60	2.449	0.504	3.96	1.654	0.937
1.34	2.184	0	2.64	2.447	0.512	4.00	1.649	0.914
1.36	2.189	0	2.68	2.445	0.523	4.04	1.646	0.893
1.38	2.194	0	2.72	2.441	0.533	4.08	1.645	0.875
1.40	2.199	0	2.76	2.441	0.538	4.12	1.645	0.859
1.44	2.211	0	2.80	2.447	0.542	4.16	1.647	0.846
1.48	2.224	0	2.84	2.462	0.545	4.20	1.649	0.835
1.52	2.241	0	2.88	2.487	0.553	4.24	1.651	0.825
1.56	2.266	0.00510	2.92	2.524	0.570	4.28	1.654	0.817
1.60	2.282	0.0249	2.96	2.570	0.606	4.32	1.656	0.811
1.64	2.289	0.0381	3.00	2.613	0.668	4.36	1.658	0.807
1.68	2.297	0.0481	3.04	2.641	0.754	4.40	1.660	0.803
1.72	2.306	0.0569	3.08	2.639	0.854	4.44	1.661	0.801
1.76	2.315	0.0647	3.12	2.608	0.952	4.48	1.662	0.800
1.80	2.325	0.0719	3.16	2.553	1.035	4.52	1.663	0.800
1.84	2.336	0.0790	3.20	2.485	1.100	4.56	1.663	0.800
1.88	2.349	0.0865	3.24	2.413	1.147	4.60	1.662	0.801
1.92	2.364	0.0950	3.28	2.340	1.181			

Refractive index *n* and extinction coefficient *k* for each photon energy *E* are shown. The result is consistent with that shown in Figure 4.1. The optical data calculated based on the Tauc–Lorentz parameterization are indicated.
Source: Based on Kato et al. [2].

Table A.4 Tauc–Lorentz parameters of $FAPbI_3$.

Peak	*A* (eV)	*C* (eV)	E_0 (eV)	E_g (eV)	$\varepsilon_1(\infty)$
$j = 1$	10.018	0.099	1.557	1.543	1.462
$j = 2$	8.849	0.498	1.566	1.521	0
$j = 3$	9.041	0.794	2.344	1.618	0
$j = 4$	1.028	0.171	2.486	1.740	0
$j = 5$	20.435	0.755	2.712	2.093	0
$j = 6$	66.516	0.417	3.083	2.726	0
$j = 7$	98.353	0.657	3.190	2.918	0
$j = 8$	131.316	3.018	3.897	3.882	0

The model parameters reported in Ref. [2] are shown.
Source: Based on Kato et al. [2].

Table A.5 Optical constants of $Cs_{0.17}FA_{0.83}PbI_3$.

E (eV)	*n*	*k*	*E* (eV)	*n*	*k*	*E* (eV)	*n*	*k*
1.00	2.146	0	1.96	2.363	0.227	3.32	2.271	1.043
1.02	2.149	0	2.00	2.367	0.242	3.36	2.211	1.086
1.04	2.153	0	2.04	2.369	0.257	3.40	2.143	1.117
1.06	2.156	0	2.08	2.372	0.270	3.44	2.072	1.136
1.08	2.160	0	2.12	2.375	0.283	3.48	2.001	1.143
1.10	2.163	0	2.16	2.380	0.295	3.52	1.932	1.138
1.12	2.167	0	2.20	2.387	0.308	3.56	1.869	1.124
1.14	2.171	0	2.24	2.396	0.322	3.60	1.813	1.103
1.16	2.175	0	2.28	2.407	0.340	3.64	1.763	1.079
1.18	2.180	0	2.32	2.420	0.363	3.68	1.718	1.053
1.20	2.185	0	2.36	2.431	0.393	3.72	1.674	1.024
1.22	2.189	0	2.40	2.437	0.429	3.76	1.633	0.988
1.24	2.195	0	2.44	2.434	0.471	3.80	1.601	0.946
1.26	2.200	0	2.48	2.418	0.513	3.84	1.579	0.904
1.28	2.206	0	2.52	2.390	0.548	3.88	1.564	0.863
1.30	2.212	0	2.56	2.354	0.570	3.92	1.555	0.825
1.32	2.218	0	2.60	2.318	0.576	3.96	1.551	0.790
1.34	2.225	0	2.64	2.291	0.570	4.00	1.551	0.759
1.36	2.232	0	2.68	2.275	0.558	4.04	1.553	0.731

(continued)

Table A.5 (Continued)

E (eV)	*n*	*k*	*E* (eV)	*n*	*k*	*E* (eV)	*n*	*k*
1.38	2.240	0	2.72	2.270	0.545	4.08	1.557	0.706
1.40	2.249	0	2.76	2.275	0.535	4.12	1.563	0.684
1.44	2.270	0	2.80	2.287	0.531	4.16	1.570	0.665
1.48	2.298	0	2.84	2.304	0.534	4.20	1.577	0.649
1.52	2.340	0.0125	2.88	2.324	0.543	4.24	1.584	0.636
1.56	2.388	0.0537	2.92	2.345	0.560	4.28	1.590	0.625
1.60	2.384	0.140	2.96	2.366	0.585	4.32	1.597	0.616
1.64	2.349	0.157	3.00	2.385	0.617	4.36	1.602	0.609
1.68	2.334	0.163	3.04	2.400	0.657	4.40	1.606	0.603
1.72	2.329	0.166	3.08	2.410	0.703	4.44	1.609	0.598
1.76	2.331	0.170	3.12	2.413	0.756	4.48	1.610	0.594
1.80	2.336	0.176	3.16	2.407	0.813	4.52	1.610	0.591
1.84	2.343	0.186	3.20	2.391	0.873	4.56	1.609	0.587
1.88	2.351	0.198	3.24	2.363	0.934	4.60	1.606	0.582
1.92	2.357	0.212	3.28	2.323	0.992			

Refractive index *n* and extinction coefficient *k* for each photon energy *E* are shown. The optical function has been obtained by performing slight extra roughness correction [8] for the reported result of Ref. [3]. The tabulated data are consistent with those shown in Figures 4.1 and 4.15. The optical data calculated based on the Tauc–Lorentz parameterization are indicated.

Table A.6 Tauc–Lorentz parameters of $Cs_{0.17}FA_{0.83}PbI_3$.

Peak	*A* (eV)	*C* (eV)	E_0 (eV)	E_g (eV)	$\varepsilon_1(\infty)$
$j = 1$	33.711	0.171	1.549	1.511	1.076
$j = 2$	8.315	0.066	1.495	1.481	0
$j = 3$	10.410	0.856	2.005	1.547	0
$j = 4$	6.397	0.449	2.526	1.742	0
$j = 5$	16.002	0.826	3.346	1.824	0
$j = 6$	0.149	0.228	3.717	2.124	0
$j = 7$	3.622	1.401	4.687	1.782	0
$j = 8$	196.258	0.105	8.263	6.351	0
$j = 9$	10.731	0.764	5.924	4.260	0
$j = 10$	11.996	0.040	1.587	1.556	0

Table A.7 Optical constants of $Cs_{0.17}FA_{0.83}Pb(I_{0.9}Br_{0.1})_3$.

E (eV)	*n*	*k*	*E* (eV)	*n*	*k*	*E* (eV)	*n*	*k*
1.00	2.223	0	1.96	2.481	0.224	3.32	2.305	1.106
1.02	2.226	0	2.00	2.493	0.237	3.36	2.253	1.141
1.04	2.229	0	2.04	2.506	0.254	3.40	2.195	1.168
1.06	2.233	0	2.08	2.518	0.274	3.44	2.135	1.187
1.08	2.237	0	2.12	2.528	0.297	3.48	2.074	1.196
1.10	2.241	0	2.16	2.537	0.320	3.52	2.013	1.198
1.12	2.245	0	2.20	2.544	0.345	3.56	1.955	1.191
1.14	2.249	0	2.24	2.551	0.369	3.60	1.900	1.178
1.16	2.253	0	2.28	2.557	0.395	3.64	1.849	1.159
1.18	2.258	0	2.32	2.565	0.423	3.68	1.804	1.135
1.20	2.262	0	2.36	2.572	0.454	3.72	1.763	1.108
1.22	2.267	0	2.40	2.576	0.492	3.76	1.728	1.078
1.24	2.272	0	2.44	2.574	0.534	3.80	1.698	1.047
1.26	2.278	0	2.48	2.563	0.579	3.84	1.674	1.015
1.28	2.283	0	2.52	2.540	0.623	3.88	1.654	0.983
1.30	2.289	0	2.56	2.506	0.657	3.92	1.638	0.951
1.32	2.295	0	2.60	2.467	0.678	3.96	1.626	0.921
1.34	2.302	0	2.64	2.431	0.684	4.00	1.618	0.892
1.36	2.309	0	2.68	2.403	0.680	4.04	1.614	0.865
1.38	2.316	0	2.72	2.386	0.672	4.08	1.611	0.840
1.40	2.324	0	2.76	2.380	0.663	4.12	1.611	0.816
1.44	2.342	0	2.80	2.384	0.659	4.16	1.613	0.796
1.48	2.363	0	2.84	2.393	0.661	4.20	1.616	0.777
1.52	2.390	0	2.88	2.406	0.670	4.24	1.620	0.761
1.56	2.430	0.00184	2.92	2.421	0.686	4.28	1.625	0.747
1.60	2.494	0.0291	2.96	2.435	0.710	4.32	1.630	0.735
1.64	2.523	0.119	3.00	2.446	0.741	4.36	1.635	0.725
1.68	2.474	0.167	3.04	2.454	0.778	4.40	1.639	0.717
1.72	2.453	0.173	3.08	2.456	0.820	4.44	1.643	0.710
1.76	2.450	0.177	3.12	2.452	0.867	4.48	1.645	0.705
1.80	2.453	0.184	3.16	2.440	0.917	4.52	1.647	0.700
1.84	2.457	0.193	3.20	2.419	0.968	4.56	1.648	0.696
1.88	2.463	0.203	3.24	2.389	1.018	4.60	1.648	0.692
1.92	2.471	0.212	3.28	2.351	1.064			

Refractive index *n* and extinction coefficient *k* for each photon energy *E* are shown. The optical function has been obtained by performing slight extra roughness correction [8] for the reported result of Ref. [3]. The tabulated data are consistent with those shown in Figure 4.15. The optical data calculated based on the Tauc–Lorentz parameterization are indicated.

Table A.8 Tauc–Lorentz parameters of $Cs_{0.17}FA_{0.83}Pb(I_{0.9}Br_{0.1})_3$.

Peak	*A* (eV)	*C* (eV)	E_0 (eV)	E_g (eV)	$\varepsilon_1(\infty)$
$j = 1$	15.894	0.191	1.622	1.541	1.237
$j = 2$	6.808	0.693	2.195	1.573	0
$j = 3$	6.165	0.489	2.559	1.616	0
$j = 4$	20.290	0.999	3.338	1.803	0
$j = 5$	3.194	1.421	4.745	1.588	0
$j = 6$	173.991	1.277	7.269	5.916	0
$j = 7$	25.425	0.096	1.620	1.580	0
$j = 8$	24.412	0.013	1.644	1.637	0
$j = 9$	7.188	0.354	1.790	1.643	0
$j = 10$	5.696	1.124	6.038	2.405	0

Table A.9 Optical constants of $Cs_{0.17}FA_{0.83}Pb(I_{0.8}Br_{0.2})_3$.

E (eV)	*n*	*k*	*E* (eV)	*n*	*k*	*E* (eV)	*n*	*k*
1.00	2.184	0	1.96	2.418	0.207	3.32	2.349	0.961
1.02	2.187	0	2.00	2.426	0.215	3.36	2.312	1.001
1.04	2.190	0	2.04	2.437	0.225	3.40	2.270	1.036
1.06	2.193	0	2.08	2.448	0.240	3.44	2.223	1.065
1.08	2.197	0	2.12	2.459	0.259	3.48	2.172	1.088
1.10	2.200	0	2.16	2.466	0.279	3.52	2.119	1.103
1.12	2.204	0	2.20	2.470	0.299	3.56	2.065	1.112
1.14	2.207	0	2.24	2.475	0.317	3.60	2.012	1.113
1.16	2.211	0	2.28	2.481	0.334	3.64	1.961	1.108
1.18	2.215	0	2.32	2.490	0.354	3.68	1.913	1.097
1.20	2.219	0	2.36	2.500	0.378	3.72	1.869	1.081
1.22	2.223	0	2.40	2.508	0.406	3.76	1.829	1.062
1.24	2.227	0	2.44	2.513	0.439	3.80	1.793	1.039
1.26	2.232	0	2.48	2.512	0.475	3.84	1.761	1.014
1.28	2.237	0	2.52	2.505	0.513	3.88	1.734	0.988
1.30	2.242	0	2.56	2.489	0.550	3.92	1.711	0.961
1.32	2.247	0	2.60	2.466	0.581	3.96	1.692	0.934
1.34	2.253	0	2.64	2.437	0.604	4.00	1.677	0.907
1.36	2.258	0	2.68	2.407	0.616	4.04	1.666	0.881

(continued)

Table A.9 (Continued)

E (eV)	*n*	*k*	*E* (eV)	*n*	*k*	*E* (eV)	*n*	*k*
1.38	2.265	0	2.72	2.381	0.619	4.08	1.657	0.856
1.40	2.271	0	2.76	2.364	0.615	4.12	1.651	0.832
1.44	2.285	0	2.80	2.356	0.609	4.16	1.648	0.810
1.48	2.301	0	2.84	2.356	0.605	4.20	1.646	0.790
1.52	2.321	0	2.88	2.363	0.606	4.24	1.647	0.772
1.56	2.345	0	2.92	2.374	0.613	4.28	1.648	0.756
1.60	2.379	0.00114	2.96	2.387	0.626	4.32	1.650	0.741
1.64	2.431	0.0182	3.00	2.400	0.646	4.36	1.653	0.729
1.68	2.481	0.0871	3.04	2.411	0.673	4.40	1.656	0.718
1.72	2.450	0.161	3.08	2.420	0.705	4.44	1.659	0.709
1.76	2.414	0.173	3.12	2.424	0.742	4.48	1.661	0.701
1.80	2.405	0.171	3.16	2.423	0.783	4.52	1.663	0.694
1.84	2.409	0.177	3.20	2.415	0.827	4.56	1.664	0.688
1.88	2.413	0.188	3.24	2.400	0.872	4.60	1.665	0.683
1.92	2.415	0.198	3.28	2.378	0.917			

Refractive index *n* and extinction coefficient *k* for each photon energy *E* are shown. The optical function has been obtained by performing slight extra roughness correction [8] for the reported result of Ref. [3]. The tabulated data are consistent with those shown in Figures 1.5, 4.1, and 4.15. The optical data calculated based on the Tauc–Lorentz parameterization are indicated.

Table A.10 Tauc–Lorentz parameters of $Cs_{0.17}FA_{0.83}Pb(I_{0.8}Br_{0.2})_3$.

Peak	*A* (eV)	*C* (eV)	E_0 (eV)	E_g (eV)	$\varepsilon_1(\infty)$
$j = 1$	4.453	0.075	1.622	1.583	1.238
$j = 2$	1.749	0.506	2.448	1.586	0
$j = 3$	4.895	0.521	2.629	1.579	0
$j = 4$	21.897	1.081	3.403	1.883	0
$j = 5$	2.774	1.417	4.786	1.592	0
$j = 6$	358.128	0.903	7.083	6.345	0
$j = 7$	35.978	0.124	1.683	1.621	0
$j = 8$	2.984	0.445	2.182	1.571	0
$j = 9$	6.114	1.329	6.119	2.333	0
$j = 10$	4.670	0.332	1.883	1.617	0

Table A.11 Optical constants of $Cs_{0.17}FA_{0.83}Pb(I_{0.7}Br_{0.3})_3$.

E (eV)	*n*	*k*	*E* (eV)	*n*	*k*	*E* (eV)	*n*	*k*
1.00	2.163	0	1.96	2.387	0.179	3.32	2.387	0.840
1.02	2.166	0	2.00	2.393	0.188	3.36	2.364	0.880
1.04	2.168	0	2.04	2.399	0.197	3.40	2.334	0.918
1.06	2.171	0	2.08	2.406	0.207	3.44	2.300	0.952
1.08	2.174	0	2.12	2.414	0.219	3.48	2.260	0.981
1.10	2.177	0	2.16	2.423	0.232	3.52	2.218	1.006
1.12	2.180	0	2.20	2.432	0.247	3.56	2.173	1.024
1.14	2.183	0	2.24	2.441	0.264	3.60	2.126	1.037
1.16	2.186	0	2.28	2.447	0.283	3.64	2.080	1.044
1.18	2.190	0	2.32	2.453	0.301	3.68	2.034	1.046
1.20	2.193	0	2.36	2.460	0.320	3.72	1.990	1.042
1.22	2.197	0	2.40	2.467	0.339	3.76	1.949	1.034
1.24	2.200	0	2.44	2.475	0.361	3.80	1.910	1.022
1.26	2.204	0	2.48	2.482	0.388	3.84	1.874	1.007
1.28	2.208	0	2.52	2.486	0.419	3.88	1.842	0.989
1.30	2.213	0	2.56	2.483	0.453	3.92	1.813	0.970
1.32	2.217	0	2.60	2.473	0.486	3.96	1.788	0.949
1.34	2.222	0	2.64	2.457	0.514	4.00	1.766	0.927
1.36	2.226	0	2.68	2.439	0.537	4.04	1.747	0.904
1.38	2.231	0	2.72	2.420	0.555	4.08	1.732	0.882
1.40	2.237	0	2.76	2.399	0.569	4.12	1.719	0.860
1.44	2.248	0	2.80	2.379	0.577	4.16	1.709	0.838
1.48	2.261	0	2.84	2.362	0.577	4.20	1.701	0.818
1.52	2.275	0	2.88	2.354	0.573	4.24	1.696	0.798
1.56	2.292	0	2.92	2.355	0.570	4.28	1.692	0.780
1.60	2.312	0.0000151	2.96	2.364	0.572	4.32	1.690	0.763
1.64	2.338	0.000183	3.00	2.376	0.580	4.36	1.689	0.748
1.68	2.380	0.00417	3.04	2.390	0.596	4.40	1.690	0.734
1.72	2.435	0.0397	3.08	2.402	0.618	4.44	1.690	0.722
1.76	2.445	0.122	3.12	2.412	0.647	4.48	1.692	0.711
1.80	2.407	0.160	3.16	2.418	0.680	4.52	1.693	0.701
1.84	2.386	0.165	3.20	2.419	0.717	4.56	1.695	0.692
1.88	2.381	0.167	3.24	2.415	0.757	4.60	1.697	0.685
1.92	2.382	0.172	3.28	2.404	0.798			

Refractive index n and extinction coefficient k for each photon energy E are shown. The optical function has been obtained by performing slight extra roughness correction [8] for the reported result of Ref. [3]. The tabulated data are consistent with those shown in Figure 4.15. The optical data calculated based on the Tauc–Lorentz parameterization are indicated.

Table A.12 Tauc–Lorentz parameters of $Cs_{0.17}FA_{0.83}Pb(I_{0.7}Br_{0.3})_3$.

Peak	*A* (eV)	*C* (eV)	E_0 (eV)	E_g (eV)	$\varepsilon_1(\infty)$
$j = 1$	8.040	0.048	1.687	1.663	1.130
$j = 2$	2.652	0.435	2.613	1.576	0
$j = 3$	1.370	0.379	2.790	1.581	0
$j = 4$	27.646	1.178	3.448	2.062	0
$j = 5$	2.395	1.389	4.887	1.703	0
$j = 6$	635.240	0.913	6.809	6.457	0
$j = 7$	48.237	0.123	1.733	1.690	0
$j = 8$	2.608	0.526	2.321	1.604	0
$j = 9$	4.247	1.184	6.154	1.791	0
$j = 10$	5.858	0.612	1.977	1.657	0

Table A.13 Optical constants of $Cs_{0.17}FA_{0.83}Pb(I_{0.6}Br_{0.4})_3$.

E (eV)	*n*	*k*	*E* (eV)	*n*	*k*	*E* (eV)	*n*	*k*
1.00	2.091	0	1.96	2.322	0.174	3.32	2.306	0.768
1.02	2.093	0	2.00	2.326	0.180	3.36	2.292	0.803
1.04	2.096	0	2.04	2.331	0.189	3.40	2.274	0.837
1.06	2.098	0	2.08	2.337	0.199	3.44	2.251	0.869
1.08	2.101	0	2.12	2.344	0.210	3.48	2.223	0.899
1.10	2.104	0	2.16	2.352	0.222	3.52	2.191	0.926
1.12	2.107	0	2.20	2.360	0.235	3.56	2.157	0.949
1.14	2.110	0	2.24	2.369	0.251	3.60	2.119	0.968
1.16	2.113	0	2.28	2.376	0.268	3.64	2.081	0.982
1.18	2.116	0	2.32	2.382	0.286	3.68	2.041	0.991
1.20	2.119	0	2.36	2.387	0.304	3.72	2.002	0.997
1.22	2.122	0	2.40	2.393	0.322	3.76	1.964	0.997
1.24	2.126	0	2.44	2.398	0.342	3.80	1.927	0.994
1.26	2.129	0	2.48	2.405	0.362	3.84	1.892	0.988
1.28	2.133	0	2.52	2.411	0.386	3.88	1.859	0.978
1.30	2.137	0	2.56	2.416	0.414	3.92	1.828	0.966
1.32	2.141	0	2.60	2.417	0.446	3.96	1.800	0.952
1.34	2.145	0	2.64	2.411	0.479	4.00	1.775	0.937

(continued)

Table A.13 (Continued)

E (eV)	*n*	*k*	*E* (eV)	*n*	*k*	*E* (eV)	*n*	*k*
1.36	2.149	0	2.68	2.398	0.511	4.04	1.752	0.920
1.38	2.154	0	2.72	2.380	0.538	4.08	1.732	0.902
1.40	2.159	0	2.76	2.359	0.559	4.12	1.715	0.884
1.44	2.169	0	2.80	2.338	0.573	4.16	1.700	0.866
1.48	2.180	0	2.84	2.318	0.583	4.20	1.687	0.848
1.52	2.192	0	2.88	2.301	0.587	4.24	1.676	0.830
1.56	2.206	0	2.92	2.290	0.587	4.28	1.668	0.813
1.60	2.221	0	2.96	2.285	0.587	4.32	1.660	0.796
1.64	2.240	0	3.00	2.287	0.589	4.36	1.655	0.780
1.68	2.262	0	3.04	2.292	0.596	4.40	1.651	0.765
1.72	2.294	0	3.08	2.300	0.607	4.44	1.648	0.751
1.76	2.347	0.0124	3.12	2.307	0.624	4.48	1.646	0.738
1.80	2.401	0.0778	3.16	2.314	0.646	4.52	1.644	0.726
1.84	2.373	0.156	3.20	2.318	0.672	4.56	1.644	0.714
1.88	2.336	0.171	3.24	2.318	0.702	4.60	1.644	0.703
1.92	2.323	0.172	3.28	2.314	0.734			

Refractive index *n* and extinction coefficient *k* for each photon energy *E* are shown. The optical function has been obtained by performing slight extra roughness correction [8] for the reported result of Ref. [3]. The tabulated data are consistent with those shown in Figures 4.1 and 4.15. The optical data calculated based on the Tauc–Lorentz parameterization are indicated.

Table A.14 Tauc–Lorentz parameters of $Cs_{0.17}FA_{0.83}Pb(I_{0.6}Br_{0.4})_3$.

Peak	*A* (eV)	*C* (eV)	E_0 (eV)	E_g (eV)	$\varepsilon_1(\infty)$
$j = 1$	5.388	0.048	1.759	1.723	0.936
$j = 2$	2.652	0.435	2.684	1.746	0
$j = 3$	1.724	0.445	2.840	1.751	0
$j = 4$	27.524	1.261	3.501	2.142	0
$j = 5$	2.724	1.834	5.001	1.773	0
$j = 6$	635.240	0.913	6.879	6.527	0
$j = 7$	50.340	0.117	1.800	1.752	0
$j = 8$	7.484	0.725	2.351	1.774	0
$j = 9$	4.158	1.234	6.198	1.861	0
$j = 10$	6.320	0.541	1.982	1.727	0

Table A.15 Optical constants of $Cs_{0.05}FA_{0.8}MA_{0.15}Pb(I_{0.95}Br_{0.05})_3$.

E (eV)	*n*	*k*	*E* (eV)	*n*	*k*	*E* (eV)	*n*	*k*
1.00	2.240	0	1.96	2.498	0.208	3.32	2.217	1.317
1.02	2.243	0	2.00	2.512	0.214	3.36	2.143	1.324
1.04	2.247	0	2.04	2.531	0.221	3.40	2.073	1.321
1.06	2.251	0	2.08	2.553	0.232	3.44	2.009	1.308
1.08	2.255	0	2.12	2.579	0.247	3.48	1.952	1.290
1.10	2.259	0	2.16	2.607	0.269	3.52	1.902	1.267
1.12	2.263	0	2.20	2.635	0.297	3.56	1.859	1.241
1.14	2.267	0	2.24	2.663	0.334	3.60	1.822	1.214
1.16	2.272	0	2.28	2.688	0.379	3.64	1.791	1.188
1.18	2.277	0	2.32	2.705	0.434	3.68	1.764	1.161
1.20	2.282	0	2.36	2.711	0.495	3.72	1.741	1.136
1.22	2.287	0	2.40	2.703	0.558	3.76	1.721	1.112
1.24	2.293	0	2.44	2.679	0.617	3.80	1.704	1.089
1.26	2.298	0	2.48	2.643	0.666	3.84	1.689	1.068
1.28	2.304	0	2.52	2.600	0.700	3.88	1.676	1.048
1.30	2.311	0	2.56	2.558	0.718	3.92	1.664	1.029
1.32	2.318	0	2.60	2.522	0.724	3.96	1.653	1.011
1.34	2.325	0	2.64	2.497	0.721	4.00	1.643	0.995
1.36	2.332	0	2.68	2.484	0.715	4.04	1.634	0.979
1.38	2.340	0	2.72	2.481	0.711	4.08	1.626	0.964
1.40	2.349	0	2.76	2.487	0.712	4.12	1.619	0.949
1.44	2.369	0	2.80	2.500	0.719	4.16	1.611	0.935
1.48	2.394	0	2.84	2.516	0.735	4.20	1.605	0.922
1.52	2.432	0.00305	2.88	2.534	0.760	4.24	1.599	0.909
1.56	2.478	0.0339	2.92	2.552	0.793	4.28	1.593	0.896
1.60	2.498	0.0853	2.96	2.568	0.838	4.32	1.587	0.884
1.64	2.492	0.128	3.00	2.577	0.893	4.36	1.582	0.872
1.68	2.479	0.150	3.04	2.575	0.956	4.40	1.578	0.860
1.72	2.473	0.163	3.08	2.560	1.024	4.44	1.573	0.848
1.76	2.472	0.172	3.12	2.531	1.093	4.48	1.570	0.836
1.80	2.474	0.181	3.16	2.488	1.158	4.52	1.566	0.825
1.84	2.477	0.190	3.20	2.432	1.216	4.56	1.563	0.813
1.88	2.481	0.197	3.24	2.365	1.263	4.60	1.561	0.801
1.92	2.487	0.203	3.28	2.292	1.297			

Refractive index *n* and extinction coefficient *k* for each photon energy *E* are shown. The result is consistent with those shown in Figures 4.14 and 4.15. The optical data calculated based on the Tauc–Lorentz parameterization are indicated.

Table A.16 Tauc–Lorentz parameters of $Cs_{0.05}FA_{0.8}MA_{0.15}Pb(I_{0.95}Br_{0.05})_3$.

Peak	*A* (eV)	*C* (eV)	E_0 (eV)	E_g (eV)	$\varepsilon_1(\infty)$
$j = 1$	29.582	0.243	1.572	1.502	1.323
$j = 2$	14.424	0.551	2.469	1.715	0
$j = 3$	14.063	0.772	3.206	1.598	0
$j = 4$	124.944	1.365	5.833	5.010	0
$j = 5$	107.184	2.060	2.984	2.903	0
$j = 6$	9.610	0.432	1.777	1.645	0

Table A.17 Optical constants of $Cs_{0.05}FA_{0.8}MA_{0.15}Pb(I_{0.8}Br_{0.2})_3$.

E (eV)	*n*	*k*	*E* (eV)	*n*	*k*	*E* (eV)	*n*	*k*
1.00	2.216	0	1.96	2.458	0.178	3.32	2.368	1.177
1.02	2.219	0	2.00	2.468	0.185	3.36	2.299	1.207
1.04	2.222	0	2.04	2.480	0.192	3.40	2.229	1.224
1.06	2.225	0	2.08	2.493	0.201	3.44	2.161	1.229
1.08	2.228	0	2.12	2.509	0.210	3.48	2.097	1.222
1.10	2.232	0	2.16	2.527	0.221	3.52	2.040	1.208
1.12	2.235	0	2.20	2.547	0.234	3.56	1.990	1.189
1.14	2.239	0	2.24	2.570	0.252	3.60	1.948	1.166
1.16	2.243	0	2.28	2.595	0.275	3.64	1.912	1.141
1.18	2.247	0	2.32	2.621	0.304	3.68	1.882	1.116
1.20	2.251	0	2.36	2.645	0.341	3.72	1.857	1.092
1.22	2.255	0	2.40	2.664	0.386	3.76	1.837	1.068
1.24	2.259	0	2.44	2.676	0.439	3.80	1.820	1.047
1.26	2.264	0	2.48	2.676	0.497	3.84	1.806	1.027
1.28	2.269	0	2.52	2.662	0.553	3.88	1.793	1.009
1.30	2.274	0	2.56	2.634	0.604	3.92	1.783	0.992
1.32	2.279	0	2.60	2.598	0.642	3.96	1.773	0.977
1.34	2.284	0	2.64	2.559	0.665	4.00	1.765	0.963
1.36	2.290	0	2.68	2.524	0.675	4.04	1.757	0.951
1.38	2.296	0	2.72	2.498	0.676	4.08	1.750	0.939
1.40	2.302	0	2.76	2.483	0.672	4.12	1.743	0.929
1.44	2.316	0	2.80	2.478	0.669	4.16	1.736	0.919
1.48	2.331	0	2.84	2.482	0.669	4.20	1.730	0.910
1.52	2.349	0	2.88	2.493	0.676	4.24	1.724	0.901

(continued)

Table A.17 (Continued)

E (eV)	*n*	*k*	*E* (eV)	*n*	*k*	*E* (eV)	*n*	*k*
1.56	2.370	0	2.92	2.508	0.690	4.28	1.718	0.892
1.60	2.399	0	2.96	2.524	0.713	4.32	1.712	0.884
1.64	2.444	0.0118	3.00	2.540	0.745	4.36	1.706	0.876
1.68	2.485	0.0550	3.04	2.553	0.785	4.40	1.700	0.869
1.72	2.493	0.112	3.08	2.562	0.834	4.44	1.695	0.861
1.76	2.475	0.150	3.12	2.561	0.891	4.48	1.690	0.853
1.80	2.455	0.166	3.16	2.549	0.954	4.52	1.685	0.846
1.84	2.445	0.170	3.20	2.523	1.019	4.56	1.680	0.838
1.88	2.444	0.171	3.24	2.483	1.080	4.60	1.675	0.831
1.92	2.449	0.174	3.28	2.430	1.134			

Refractive index *n* and extinction coefficient *k* for each photon energy *E* are shown. The result is consistent with that shown in Figure 4.15. The optical data calculated based on the Tauc–Lorentz parameterization are indicated.

Table A.18 Tauc–Lorentz parameters of $Cs_{0.05}FA_{0.8}MA_{0.15}Pb(I_{0.8}Br_{0.2})_3$.

Peak	*A* (eV)	*C* (eV)	E_0 (eV)	E_g (eV)	$\varepsilon_1(\infty)$
$j = 1$	30.731	0.236	1.690	1.603	1.313
$j = 2$	14.298	0.543	2.567	1.832	0
$j = 3$	14.321	0.756	3.286	1.722	0
$j = 4$	128.241	1.370	5.950	5.136	0
$j = 5$	125.656	2.296	3.103	3.031	0
$j = 6$	9.334	0.616	1.955	1.755	0

Table A.19 Optical constants of $Cs_{0.05}FA_{0.8}MA_{0.15}Pb(I_{0.6}Br_{0.4})_3$.

E (eV)	*n*	*k*	*E* (eV)	*n*	*k*	*E* (eV)	*n*	*k*
1.00	2.125	0	1.96	2.337	0.169	3.32	2.415	0.866
1.02	2.127	0	2.00	2.340	0.170	3.36	2.404	0.910
1.04	2.130	0	2.04	2.345	0.172	3.40	2.384	0.957
1.06	2.133	0	2.08	2.351	0.175	3.44	2.355	1.003

(*continued*)

Table A.19 (Continued)

E (eV)	*n*	*k*	*E* (eV)	*n*	*k*	*E* (eV)	*n*	*k*
1.08	2.135	0	2.12	2.360	0.179	3.48	2.316	1.046
1.10	2.138	0	2.16	2.370	0.183	3.52	2.271	1.082
1.12	2.141	0	2.20	2.382	0.189	3.56	2.219	1.111
1.14	2.144	0	2.24	2.397	0.196	3.60	2.164	1.131
1.16	2.147	0	2.28	2.413	0.204	3.64	2.108	1.142
1.18	2.150	0	2.32	2.431	0.216	3.68	2.054	1.144
1.20	2.153	0	2.36	2.451	0.231	3.72	2.003	1.139
1.22	2.157	0	2.40	2.471	0.249	3.76	1.956	1.128
1.24	2.160	0	2.44	2.490	0.271	3.80	1.914	1.113
1.26	2.164	0	2.48	2.509	0.298	3.84	1.877	1.094
1.28	2.168	0	2.52	2.526	0.328	3.88	1.845	1.074
1.30	2.172	0	2.56	2.540	0.363	3.92	1.819	1.052
1.32	2.176	0	2.60	2.550	0.402	3.96	1.796	1.031
1.34	2.180	0	2.64	2.554	0.442	4.00	1.777	1.011
1.36	2.184	0	2.68	2.552	0.484	4.04	1.761	0.992
1.38	2.189	0	2.72	2.544	0.524	4.08	1.748	0.974
1.40	2.193	0	2.76	2.530	0.561	4.12	1.737	0.957
1.44	2.203	0	2.80	2.512	0.594	4.16	1.727	0.942
1.48	2.214	0	2.84	2.492	0.622	4.20	1.719	0.928
1.52	2.227	0	2.88	2.471	0.644	4.24	1.711	0.916
1.56	2.240	0	2.92	2.451	0.661	4.28	1.704	0.904
1.60	2.256	0	2.96	2.435	0.674	4.32	1.697	0.894
1.64	2.275	0	3.00	2.422	0.684	4.36	1.690	0.884
1.68	2.297	0	3.04	2.414	0.694	4.40	1.684	0.874
1.72	2.329	0.000121	3.08	2.410	0.705	4.44	1.677	0.865
1.76	2.380	0.0124	3.12	2.410	0.719	4.48	1.671	0.856
1.80	2.438	0.0750	3.16	2.413	0.737	4.52	1.665	0.847
1.84	2.413	0.171	3.20	2.417	0.761	4.56	1.659	0.838
1.88	2.355	0.184	3.24	2.420	0.790	4.60	1.654	0.829
1.92	2.337	0.173	3.28	2.420	0.825			

Refractive index *n* and extinction coefficient *k* for each photon energy *E* are shown. The result is consistent with that shown in Figure 4.15. The optical data calculated based on the Tauc–Lorentz parameterization are indicated.

Table A.20 Tauc–Lorentz parameters of $Cs_{0.05}FA_{0.8}MA_{0.15}Pb(I_{0.6}Br_{0.4})_3$.

Peak	*A* (eV)	*C* (eV)	E_0 (eV)	E_g (eV)	$\varepsilon_1(\infty)$
$j = 1$	21.495	0.121	1.823	1.714	0.499
$j = 2$	31.182	0.521	1.796	1.780	0
$j = 3$	25.449	0.872	2.767	2.009	0
$j = 4$	32.412	0.836	3.488	2.493	0
$j = 5$	25.692	1.778	4.275	3.274	0
$j = 6$	5.400	0.995	5.997	2.226	0
$j = 7$	248.363	13.019	6.513	5.748	0

Table A.21 Optical constants of $FA_{0.55}MA_{0.45}(Sn_{0.55}Pb_{0.45})I_3$.

E (eV)	*n*	*k*	*E* (eV)	*n*	*k*	*E* (eV)	*n*	*k*
1.00	2.348	0	1.96	2.688	0.309	3.32	2.100	1.078
1.02	2.354	0	2.00	2.709	0.343	3.36	2.075	1.075
1.04	2.359	0	2.04	2.726	0.385	3.40	2.052	1.070
1.06	2.366	0	2.08	2.736	0.432	3.44	2.031	1.066
1.08	2.372	0	2.12	2.738	0.481	3.48	2.011	1.061
1.10	2.379	0	2.16	2.732	0.530	3.52	1.992	1.056
1.12	2.387	0	2.20	2.719	0.576	3.56	1.974	1.052
1.14	2.395	0	2.24	2.701	0.618	3.60	1.957	1.047
1.16	2.405	0	2.28	2.678	0.655	3.64	1.940	1.042
1.18	2.415	0	2.32	2.652	0.686	3.68	1.923	1.038
1.20	2.427	0	2.36	2.627	0.710	3.72	1.906	1.032
1.22	2.441	0	2.40	2.603	0.729	3.76	1.890	1.027
1.24	2.462	0.00233	2.44	2.584	0.745	3.80	1.874	1.021
1.26	2.485	0.0152	2.48	2.568	0.760	3.84	1.858	1.014
1.28	2.502	0.0369	2.52	2.555	0.776	3.88	1.843	1.006
1.30	2.510	0.0615	2.56	2.543	0.793	3.92	1.829	0.998
1.32	2.509	0.0835	2.60	2.532	0.812	3.96	1.816	0.989
1.34	2.503	0.100	2.64	2.520	0.833	4.00	1.804	0.979

(continued)

Table A.21 (Continued)

E (eV)	*n*	*k*	*E* (eV)	*n*	*k*	*E* (eV)	*n*	*k*
1.36	2.496	0.111	2.68	2.508	0.855	4.04	1.794	0.969
1.38	2.490	0.118	2.72	2.494	0.879	4.08	1.784	0.960
1.40	2.486	0.122	2.76	2.478	0.903	4.12	1.776	0.950
1.44	2.483	0.125	2.80	2.460	0.927	4.16	1.769	0.941
1.48	2.487	0.126	2.84	2.439	0.951	4.20	1.764	0.933
1.52	2.496	0.128	2.88	2.416	0.974	4.24	1.759	0.926
1.56	2.509	0.132	2.92	2.390	0.995	4.28	1.754	0.921
1.60	2.525	0.138	2.96	2.363	1.014	4.32	1.749	0.918
1.64	2.543	0.147	3.00	2.334	1.031	4.36	1.742	0.916
1.68	2.562	0.159	3.04	2.305	1.045	4.40	1.733	0.914
1.72	2.581	0.174	3.08	2.275	1.057	4.44	1.721	0.911
1.76	2.600	0.191	3.12	2.244	1.066	4.48	1.707	0.905
1.80	2.618	0.212	3.16	2.214	1.073	4.52	1.693	0.896
1.84	2.636	0.234	3.20	2.184	1.078	4.56	1.680	0.883
1.88	2.653	0.258	3.24	2.155	1.080	4.60	1.670	0.868
1.92	2.669	0.284	3.28	2.126	1.080			

Refractive index *n* and extinction coefficient *k* for each photon energy *E* are shown. The result is consistent with those shown in Figures 1.5 and 4.1. The optical data calculated based on the Tauc–Lorentz parameterization are indicated.

Table A.22 Tauc–Lorentz parameters of $FA_{0.55}MA_{0.45}(Sn_{0.55}Pb_{0.45})I_3$.

Peak	*A* (eV)	*C* (eV)	E_0 (eV)	E_g (eV)	$\varepsilon_1(\infty)$
$j = 1$	22.267	0.202	1.268	1.227	0.759
$j = 2$	15.807	1.117	2.135	1.349	0
$j = 3$	74.884	0.987	2.034	1.942	0
$j = 4$	71.786	1.275	2.734	2.384	0
$j = 5$	23.386	0.001	8.006	3.077	0
$j = 6$	7.064	1.002	5.820	2.917	0
$j = 7$	16.562	0.975	3.598	3.293	0
$j = 8$	1.997	0.495	4.440	3.241	0

Table A.23 Optical constants of $MAPbBr_3$ reported in Ref. [4].

E (eV)	*n*	*k*	*E* (eV)	*n*	*k*	*E* (eV)	*n*	*k*
1.00	2.045	0	1.96	2.199	0	3.32	2.470	0.472
1.02	2.046	0	2.00	2.212	0	3.36	2.465	0.493
1.04	2.048	0	2.04	2.227	0	3.40	2.459	0.512
1.06	2.050	0	2.08	2.243	0	3.44	2.452	0.527
1.08	2.052	0	2.12	2.262	0	3.48	2.446	0.539
1.10	2.053	0	2.16	2.284	0	3.52	2.443	0.547
1.12	2.055	0	2.20	2.311	0	3.56	2.443	0.554
1.14	2.057	0	2.24	2.349	0	3.60	2.448	0.561
1.16	2.059	0	2.28	2.409	0.0152	3.64	2.457	0.569
1.18	2.061	0	2.32	2.486	0.0733	3.68	2.470	0.581
1.20	2.063	0	2.36	2.491	0.228	3.72	2.487	0.598
1.22	2.065	0	2.40	2.384	0.275	3.76	2.506	0.621
1.24	2.067	0	2.44	2.332	0.257	3.80	2.525	0.652
1.26	2.070	0	2.48	2.314	0.239	3.84	2.540	0.691
1.28	2.072	0	2.52	2.310	0.227	3.88	2.549	0.737
1.30	2.074	0	2.56	2.312	0.219	3.92	2.549	0.787
1.32	2.077	0	2.60	2.319	0.216	3.96	2.539	0.838
1.34	2.079	0	2.64	2.327	0.216	4.00	2.520	0.887
1.36	2.081	0	2.68	2.337	0.219	4.04	2.495	0.930
1.38	2.084	0	2.72	2.348	0.223	4.08	2.467	0.968
1.40	2.087	0	2.76	2.359	0.230	4.12	2.437	1.000
1.44	2.092	0	2.80	2.371	0.238	4.16	2.408	1.028
1.48	2.098	0	2.84	2.382	0.248	4.20	2.380	1.055
1.52	2.104	0	2.88	2.394	0.259	4.24	2.353	1.082
1.56	2.110	0	2.92	2.406	0.272	4.28	2.326	1.111
1.60	2.117	0	2.96	2.418	0.286	4.32	2.295	1.143
1.64	2.124	0	3.00	2.429	0.302	4.36	2.258	1.177
1.68	2.131	0	3.04	2.439	0.319	4.40	2.213	1.210
1.72	2.139	0	3.08	2.449	0.338	4.44	2.159	1.238
1.76	2.147	0	3.12	2.457	0.359	4.48	2.097	1.259
1.80	2.156	0	3.16	2.464	0.381	4.52	2.029	1.268
1.84	2.166	0	3.20	2.469	0.403	4.56	1.961	1.264
1.88	2.176	0	3.24	2.472	0.427	4.60	1.897	1.249
1.92	2.187	0	3.28	2.472	0.450			

Refractive index n and extinction coefficient k for each photon energy E are shown. The result is consistent with those shown in Figures 4.1 and 4.15. The optical data calculated based on the Tauc–Lorentz parameterization are indicated.
Source: Based on Leguy et al. [4].

Table A.24 Optical constants of $MAPbCl_3$ reported in Ref. [4].

E (eV)	*n*	*k*	*E* (eV)	*n*	*k*	*E* (eV)	*n*	*k*
1.00	1.863	0	1.96	1.925	0	3.32	2.047	0.196
1.02	1.863	0	2.00	1.929	0	3.36	2.058	0.189
1.04	1.864	0	2.04	1.933	0	3.40	2.068	0.186
1.06	1.865	0	2.08	1.938	0	3.44	2.077	0.186
1.08	1.866	0	2.12	1.942	0	3.48	2.084	0.187
1.10	1.867	0	2.16	1.947	0	3.52	2.091	0.189
1.12	1.868	0	2.20	1.952	0	3.56	2.097	0.192
1.14	1.868	0	2.24	1.957	0	3.60	2.102	0.195
1.16	1.869	0	2.28	1.963	0	3.64	2.107	0.199
1.18	1.870	0	2.32	1.968	0	3.68	2.112	0.203
1.20	1.871	0	2.36	1.974	0	3.72	2.117	0.206
1.22	1.872	0	2.40	1.981	0	3.76	2.122	0.210
1.24	1.873	0	2.44	1.987	0	3.80	2.127	0.213
1.26	1.874	0	2.48	1.994	0	3.84	2.133	0.217
1.28	1.875	0	2.52	2.002	0	3.88	2.139	0.220
1.30	1.876	0	2.56	2.010	0	3.92	2.145	0.224
1.32	1.877	0	2.60	2.019	0	3.96	2.152	0.228
1.34	1.879	0	2.64	2.028	0	4.00	2.159	0.232
1.36	1.880	0	2.68	2.038	0	4.04	2.167	0.236
1.38	1.881	0	2.72	2.049	0	4.08	2.176	0.241
1.40	1.882	0	2.76	2.062	0	4.12	2.186	0.246
1.44	1.884	0	2.80	2.076	0	4.16	2.197	0.252
1.48	1.887	0	2.84	2.092	0	4.20	2.208	0.258
1.52	1.889	0	2.88	2.110	0	4.24	2.221	0.266
1.56	1.892	0	2.92	2.133	0	4.28	2.235	0.275
1.60	1.895	0	2.96	2.163	0	4.32	2.250	0.286
1.64	1.898	0	3.00	2.205	0	4.36	2.266	0.299
1.68	1.901	0	3.04	2.279	0	4.40	2.284	0.315
1.72	1.904	0	3.08	2.463	0.0856	4.44	2.302	0.335
1.76	1.907	0	3.12	2.402	0.446	4.48	2.321	0.360
1.80	1.910	0	3.16	2.053	0.440	4.52	2.338	0.391
1.84	1.914	0	3.20	2.009	0.299	4.56	2.351	0.430
1.88	1.917	0	3.24	2.020	0.239	4.60	2.357	0.476
1.92	1.921	0	3.28	2.034	0.211			

Refractive index *n* and extinction coefficient *k* for each photon energy *E* are shown. The optical data calculated based on the Tauc–Lorentz parameterization are indicated.
Source: Based on Leguy et al. [4].

References

1 Shirayama, M., Kadowaki, H., Miyadera, T. et al. (2016). *Phys. Rev. Appl.* 5: 014012.

2 Kato, M., Fujiseki, T., Miyadera, T. et al. (2017). *J. Appl. Phys.* 121: 115501.

3 Tejada, A., Braunger, S., Korte, L. et al. (2018). *J. Appl. Phys.* 123: 175302.

4 Leguy, A.M.A., Azarhoosh, P., Alonso, M.I. et al. (2016). *Nanoscale* 8: 6317.

5 Jellison, G.E. Jr., and Modine, F.A. (1996). *Appl. Phys. Lett.* 69: 371; Erratum: (1996) **69**: 2137.

6 Fujiwara, H. and Collins, R.W. (2018). *Spectroscopic Ellipsometry for Photovoltaics, Applications and Optical Data of Solar Cell Materials*, vol. 2. Springer, Cham.

7 The software can be downloaded at the web site of https://unit.aist.go.jp/rpd-envene/PV/en/service/e-ARC_en/index_en.html. This web site can be found using "e-ARC AIST" as key words.

8 Fujiwara, H., Fujimoto, S., Tamakoshi, M. et al. (2017). *Appl. Surf. Sci.* 421: 276.

B

Numerical Values of Shockley–Queisser Limit

Yoshitsune Kato and Hiroyuki Fujiwara

Gifu University, Department of Electrical, Electronic and Computer Engineering, 1-1 Yanagido, Gifu 501-1193, Japan

Shockley–Queisser (SQ) theory described in Section 12.2 allows a simple and straightforward calculation of a solar-cell efficiency limit that depends only on the band gap (E_g) of a semiconducting light absorber. In particular, SQ theory enables us to calculate all solar cell parameters (i.e. open-circuit voltage (V_{oc}), short-circuit current density (J_{sc}), fill factor (FF), and conversion efficiency (η)) for a given E_g. The SQ limit is quite important, as it provides the estimates of the V_{oc}, J_{sc}, and FF deficits in experimental solar cells. Table B.1 summarizes the SQ parameters calculated for each E_g using Eqs. (12.1), (12.12) and (12.13). The reverse saturation current density under the SQ limit ($J_{0,SQ}$) obtained from Eq. (12.12) is also indicated. These numerical values have been shown in Figures 12.1 and 12.8. In Table B.1, even though the SQ parameters are calculated for each E_g with a step of 0.02 eV, the SQ limit with a step of 0.01 eV can easily be estimated by simply interpolating the values in the table. It needs to be emphasized that the SQ limit overestimates the solar-cell efficiency limit due to the simple assumptions (see Section 12.3 for more details).

Table B.1 Solar cell characteristics and $J_{0,SQ}$ calculated for each E_g by Shockley–Queisser theory.

Band gap E_g (eV)	$J_{0,SQ}$ (mA/cm²)	J_{sc} (mA/cm²)	V_{oc} (V)	FF	η (%)
0.80	1.010×10^{-8}	54.46	0.5793	0.8230	25.96
0.82	4.867×10^{-9}	53.21	0.5976	0.8269	26.29
0.84	2.352×10^{-9}	52.47	0.6160	0.8307	26.85
0.86	1.143×10^{-9}	52.16	0.6345	0.8343	27.61
0.88	5.482×10^{-10}	52.07	0.6535	0.8379	28.51
0.90	2.660×10^{-10}	52.07	0.6722	0.8412	29.44

(continued)

Hybrid Perovskite Solar Cells: Characteristics and Operation,
First Edition. Edited by Hiroyuki Fujiwara.

Table B.1 (Continued)

Band gap E_g (eV)	$J_{0,SQ}$ (mA/cm^2)	J_{sc} (mA/cm^2)	V_{oc} (V)	*FF*	η (%)
0.92	1.279×10^{-10}	52.07	0.6911	0.8444	30.38
0.94	6.099×10^{-11}	51.46	0.7099	0.8475	30.96
0.96	2.969×10^{-11}	50.45	0.7280	0.8503	31.23
0.98	1.440×10^{-11}	49.37	0.7462	0.8530	31.42
1.00	6.772×10^{-12}	48.21	0.7651	0.8557	31.56
1.02	3.277×10^{-12}	47.12	0.7832	0.8582	31.67
1.04	1.589×10^{-12}	46.15	0.8014	0.8606	31.83
1.06	7.484×10^{-13}	45.20	0.8203	0.8631	32.00
1.08	3.538×10^{-13}	44.56	0.8393	0.8654	32.36
1.10	1.745×10^{-13}	44.24	0.8574	0.8676	32.91
1.12	8.075×10^{-14}	43.80	0.8771	0.8698	33.41
1.14	3.917×10^{-14}	42.93	0.8953	0.8718	33.50
1.16	1.850×10^{-14}	41.94	0.9141	0.8738	33.50
1.18	8.866×10^{-15}	40.97	0.9325	0.8757	33.45
1.20	4.320×10^{-15}	40.01	0.9504	0.8775	33.37
1.22	2.053×10^{-15}	39.04	0.9690	0.8793	33.26
1.24	9.513×10^{-16}	38.04	0.9883	0.8811	33.12
1.26	4.498×10^{-16}	37.09	1.007	0.8828	32.97
1.28	2.179×10^{-16}	36.33	1.025	0.8844	32.94
1.30	1.031×10^{-16}	35.82	1.044	0.8860	33.14
1.32	5.015×10^{-17}	35.46	1.063	0.8875	33.43
1.34	2.386×10^{-17}	35.05	1.081	0.8890	33.70
1.36	1.109×10^{-17}	34.34	1.101	0.8905	33.66
1.38	5.327×10^{-18}	33.68	1.119	0.8919	33.62
1.40	2.504×10^{-18}	32.88	1.138	0.8933	33.43
1.42	1.222×10^{-18}	32.09	1.156	0.8946	33.18
1.44	5.492×10^{-19}	31.22	1.176	0.8960	32.89
1.46	2.737×10^{-19}	30.49	1.193	0.8972	32.64
1.48	1.253×10^{-19}	29.67	1.213	0.8985	32.33
1.50	6.001×10^{-20}	28.98	1.231	0.8996	32.10
1.52	2.817×10^{-20}	28.33	1.250	0.9008	31.91
1.54	1.391×10^{-20}	27.66	1.268	0.9019	31.63
1.56	6.267×10^{-21}	26.89	1.288	0.9031	31.27
1.58	2.977×10^{-21}	26.18	1.306	0.9042	30.92
1.60	1.386×10^{-21}	25.45	1.325	0.9052	30.53

(continued)

Table B.1 (Continued)

Band gap E_g (eV)	$J_{0,SQ}$ (mA/cm^2)	J_{sc} (mA/cm^2)	V_{oc} (V)	*FF*	η (%)
1.62	6.848×10^{-22}	24.83	1.343	0.9062	30.22
1.64	3.067×10^{-22}	24.37	1.363	0.9073	30.14
1.66	1.461×10^{-22}	23.69	1.382	0.9083	29.73
1.68	6.830×10^{-23}	23.03	1.401	0.9092	29.33
1.70	3.420×10^{-23}	22.48	1.418	0.9101	29.00
1.72	1.542×10^{-23}	21.90	1.438	0.9111	28.69
1.74	7.466×10^{-24}	21.37	1.456	0.9120	28.37
1.76	3.554×10^{-24}	20.77	1.474	0.9128	27.95
1.78	1.663×10^{-24}	20.19	1.493	0.9137	27.54
1.80	7.643×10^{-25}	19.64	1.513	0.9146	27.16
1.82	3.813×10^{-25}	19.14	1.530	0.9153	26.80
1.84	1.692×10^{-25}	18.53	1.550	0.9162	26.31
1.86	8.175×10^{-26}	17.99	1.568	0.9170	25.87
1.88	3.889×10^{-26}	17.47	1.587	0.9177	25.44
1.90	1.820×10^{-26}	16.97	1.605	0.9185	25.03
1.92	8.381×10^{-27}	16.46	1.625	0.9193	24.59
1.94	4.252×10^{-27}	16.01	1.641	0.9199	24.18
1.96	1.894×10^{-27}	15.49	1.662	0.9207	23.70
1.98	9.339×10^{-28}	15.07	1.679	0.9214	23.31
2.00	4.021×10^{-28}	14.56	1.700	0.9221	22.83

Index

Hybrid Perovskite Solar Cells: Characteristics and Operation, First Edition. Edited by Hiroyuki Fujiwara.

d

m

Printed and bound by CPI Group (UK) Ltd, Croydon, CR0 4YY